Mathematical
Surveys
and
Monographs
Volume 233

Sobolev and Viscosity Solutions for Fully Nonlinear Elliptic and Parabolic Equations

N. V. Krylov

AMERICAN
MATHEMATICAL
SOCIETY
Providence, Rhode Island

2010 *Mathematics Subject Classification*. Primary 35-02; Secondary 35J60, 35K55.

For additional information and updates on this book, visit
www.ams.org/bookpages/surv-233

Library of Congress Cataloging-in-Publication Data

Names: Krylov, N. V. (Nikolaĭ Vladimirovich), author.
Title: Sobolev and viscosity solutions for fully nonlinear elliptic and parabolic equations / N.V. Krylov.
Description: Providence, Rhode Island: American Mathematical Society, [2018] | Series: Mathematical surveys and monographs; volume 233 | Includes bibliographical references and index.
Identifiers: LCCN 2018018540 | ISBN 9781470447403 (alk. paper)
Subjects: LCSH: Viscosity solutions. | Elliptic functions. | Differential equations, Parabolic. | AMS: Partial differential equations – Research exposition (monographs, survey articles). msc | Partial differential equations – Elliptic equations and systems – Nonlinear elliptic equations. msc | Partial differential equations – Parabolic equations and systems – Nonlinear parabolic equations. msc
Classification: LCC QA316 .K6945 2018 | DDC 515/.353–dc23
LC record available at https://lccn.loc.gov/2018018540

∞ The paper used in this book is acid-free and falls within the guidelines
established to ensure permanence and durability.
Visit the AMS home page at `https://www.ams.org/`

10 9 8 7 6 5 4 3 2 1 23 22 21 20 19 18

Contents

Preface

The main sources of nonlinear second-order elliptic and parabolic equations are the theory of differential equations and its numerous applications. At present, the class of quasilinear equations, which are linear with respect to the second-order derivatives of the unknown function, is studied almost as thoroughly as the class of linear equations. The results of these studies are summarized in many articles and books; see, for instance, Ladyzhenskaya and Ural'tseva [**86**], Ladyženskaja, Solonnikov, and Ural'ceva [**87**], Ivanov [**49**], Gilbarg and Trudinger [**45**], and Lieberman [**89**].

For the fully nonlinear equations, the main subject of this book, only sparse results were available for a long time, not allowing one to build any general theory. However, the equations on the plane were an exception. There, using some specific features available only for two-dimensional space, one was able to develop the solvability theory in $W^{1,2}_p$ and W^2_p for any p close to 2 (see, for example, [**53**]).

With no general theory at hand, some classes of fully nonlinear equations were investigated in the areas where they arose by using available methods in these areas. The first class of such equations is related to the famous Monge-Ampère equation which arose in geometry and was studied for quite some time by using the theory of convex surfaces. Aleksandrov [**1**] in 1958 introduced the notion of generalized solutions of Monge-Ampère equations. Its smoothness, before 1971, was proved only in dimension two (see Bakel'man [**7**]). Then, Pogorelov [**98, 99**] proved *interior* smoothness for any dimension (see also Cheng and Yau [**19**], who pointed out a vicious circle in some arguments of Pogorelov). In the multidimensional case, the smoothness up to the boundary was proved only after an approach to *general* nonlinear equations was developed in the early 1980s.

Another class of nonlinear elliptic and parabolic equations, so-called Bellman's equations, arose from the probabilistic theory of controlled stochastic diffusion processes. When the diffusion is not controlled, Bellman's equations become quasilinear (see Fleming and Rishel [**43**] and the reference therein). It turned out that by using probabilistic methods, it was possible to prove the solvability of general degenerate nonlinear elliptic Bellman's equations in the whole space in the class of functions with bounded second-order spacial derivatives. Some of these results are presented in [**57**]. Probably the latest results about general possibly degenerate Bellman's equations in smooth domains are presented in [**58**] and [**63**]. In [**63**], in particular, the solvability of possibly degenerate complex Monge-Ampère equations in strictly pseudo-convex domains is established (this result was earlier obtained by PDE methods in [**17**] for *nondegenerate* case). Part of [**63**] is translated into the pure PDE language in [**64**].

These probabilistic results enabled us to deal with a very large class of fully nonlinear equations. The technique was based to a large degree on the fact that the

Monge-Ampère equation is a particular case of Bellman's equations. However, this theory was very hard to comprehend for people from pure PDEs: it was all presented in probabilistic terms. Up until 1979 no other approaches were found. The probabilistic approach was developed by the author, Nisio, Pragarauskas, Safonov, Lions, Menaldi, and many others.

In 1979 Brézis and Evans [**11**] proved the existence of a $C^{2+\alpha}$-solution for Bellman's equation with two elliptic operators. The probabilistic approach only provided solutions with two *bounded* derivatives. The results of Evans and Friedman [**36**], Evans and Lions [**38**], and Lions [**91**] published in about 1979 were obtained by the PDE methods, but the solutions there had only bounded derivatives, and the only novelty in comparison with earlier results was that purely analytical methods were used.

A real breakthrough came in 1982 when Evans [**34**] and the author [**59**] independently proved the solvability in $C^{2+\alpha}$ of a wide class of fully nonlinear equations. In [**34**] elliptic equations are considered like

$$H\big(u(x), Du(x), D^2u(x), x\big) = 0, \tag{1}$$

and in [**59**] both elliptic equations like (1) and parabolic ones like

$$\partial_t u(t,x) + H\big(u(t,x), Du(t,x), D^2u(t,x), t, x\big) = 0 \tag{2}$$

are investigated. The proofs were presented in pure PDE terms. The starting point for these proofs was the fact that the solutions of *linear* equations with measurable coefficients are Hölder continuous. This result was obtained by Safonov and the author in [**83**] and [**84**]. We present the results of [**84**] in this book in almost the same way as it is done in [**61**]; however, we go slightly deeper, which enables us to obtain Fang-Hua Lin's type estimates for elliptic and parabolic equations. These estimates play a crucial role in our approach to the Sobolev space theory.

The results in [**84**] opened the way to build a solvability theory in classes $C^{2+\alpha}$ of equations like (1) and (2) with H which is *convex or concave* with respect to D^2u and is in C^2 with respect to x or in $C^{1,2}$ with respect to (t,x). We refer the reader to [**45**] and [**61**] and the references therein for information on what was achieved in this area.

The next breakthrough came when Safonov ([**102**], [**103**], and [**104**]) came up with a technique allowing one to treat the H's that are only Hölder continuous with respect to the independent variables, which made the whole $C^{2+\alpha}$-theory of fully nonlinear equations as satisfactory as in the case of linear equations. Safonov's approach is also remarkable in the respect that it works equally well for linear and fully nonlinear equations. In our exposition here we also use these techniques.

Then, naturally arose the interest in equations when H is neither convex nor concave with respect to D^2u. These equations often appear as the Isaacs equations in the theory of stochastic differential games. By now we know that generally one can only hope that their solutions are in $C^{1+\alpha}$ for an $\alpha \in (0,1)$ (see [**96**]). In these circumstances the notion of the so-called viscosity solutions introduced by Lions [**92**] turned out to be quite effective. This notion does not require the solutions to have any derivatives. Viscosity solutions later turned out to exist and be unique in a wide variety of cases. At present the theory of viscosity solutions is quite a popular subject. The reader will find three chapters in this book devoted to this theory.

The next dramatic turn of the theory came from groundbreaking results by Caffarelli [**12**] in 1989 (see also [**14**]) about a priori estimates in Sobolev W_p^2 classes. His discovery is comparable with the discoveries in [**34**] and [**59**], which were the starting points of the $C^{2+\alpha}$-theory. He works in the framework of the theory of viscosity solutions and proves that under some conditions any viscosity solution is in $W^2_{p,\text{loc}}$. The reader is referred to [**23**], [**51**], [**52**], and [**25**] for more detailed information about the results obtained by many authors in this theory. We also provide some additional information later in the text while comparing our results with the ones obtained by using the theory of viscosity solutions. The major difference is well seen when the equations are just linear. Then the main coefficients in the theory of viscosity solutions are assumed to be uniformly close to uniformly continuous functions (usual requirement in the classical L_p-theory for linear equations), and in our theory it quite suffices for them to only be in VMO or almost in VMO; that is, to just have sufficiently small integral oscillations over small balls (elliptic case) or cylinders (parabolic case).

The main object of studies in this book is the first boundary-value problem for general fully nonlinear second-order uniformly elliptic and parabolic equations like (1) or (2). Apart from the simplest degenerate Bellman's equations in the whole space with constant coefficients, like equations of Monge-Ampère type, which admit solutions with bounded second-order derivatives, we do not say much about any *specific* class of equation degenerate or not, although we provide quite a few examples of equations to illustrate our assumptions and results. These are given in 1.2.9, 4.1.20, 4.1.21, 6.6.11, 6.6.12, 10.1.1, 10.1.2, 10.1.5, 10.1.18 10.1.24, 10.1.25, 12.1.1, 12.1.14, 12.1.20, 16.2.10, 17.1.6, 18.1.4, C.1.3, C.1.4. Some examples are also given at the end of Section 1.2.

We are looking for solutions in Sobolev classes or for C-viscosity or L_p-viscosity solutions and present *basic solvability results* about such *general* equations. In particular, we do not discuss the remarkable results of Caffarelli [**13**] about Sobolev solutions of the Monge-Ampère equations continued by many researchers (see [**28**] and the references therein).

Most of the auxiliary results are taken from old sources ([**61**], [**103**], and [**104**]), and the core of the main new results was obtained in the last few years. Exposition of these results is based on a generalization of the Fefferman-Stein theorem, Fang-Hua Lin's type estimates, and the so-called "ersatz" existence theorems, saying that one can slightly modify "any" equation and get a "cut-off" equation that will have solutions with bounded derivatives. Here is an example of such "ersatz" existence theorem (cf. Theorem 6.6.3).

1. Assumption. We are given a domain $\Omega \subset \mathbb{R}^d$ which is bounded and which satisfies the exterior ball condition (see Definition 5.5.3). We are also given a function $g \in C^{1,1}(\mathbb{R}^d)$.

Let $\mathbb{S}$ be the set of symmetric $d \times d$ matrices. Suppose that we are given a Borel measurable function $H(\mathsf{u}, x)$,

$$\mathsf{u} = (\mathsf{u}', \mathsf{u}''), \quad \mathsf{u}' = (\mathsf{u}_0', \mathsf{u}_1', ..., \mathsf{u}_d') \in \mathbb{R}^{d+1}, \quad \mathsf{u}'' \in \mathbb{S}, \quad x \in \mathbb{R}^d.$$

2. Assumption. (i) The function $H(\mathsf{u}, x)$ is Lipschitz continuous with respect to u'', and at all points of differentiability of H with respect to u'' we have $D_{\mathsf{u}''} H \in \mathbb{S}_\delta$ (a bounded set of strictly positive symmetric $d \times d$ matrices),

(ii) For a constant $K_0 \in [0, \infty)$, the number

$$\bar{H} := \sup_{\mathsf{u}',x} \big(|H(\mathsf{u}', 0, x)| - K_0 |\mathsf{u}'| \big)$$

is finite,

(iii) The function $H(\mathsf{u}, x)$ is continuous with respect to u for any x,

(iv) The function $H(\mathsf{u}, x)$ is nonincreasing with respect to u'_0.

3. THEOREM. *Under the above assumptions, there exists a function $P(\mathsf{u}'')$ which is positive homogeneous of order one and convex with respect to u'' such that for any $K > 0$ the equation*

$$\max \Big[H\big(u(x), Du(x), D^2u(x), x\big), P\big(D^2u(x)\big) - K \Big] = 0$$

(a.e.) in Ω with boundary condition $u = g$ on $\partial\Omega$ has a solution which is continuous in $\bar{\Omega}$, has bounded gradient, and whose second-order derivatives are locally bounded in Ω.

Theorems of that kind allow us to prove the solvability in Sobolev classes for equations that are quite far from the ones that are convex or concave with respect to the Hessian of the unknown functions. They play the role of starting points, which in the $C^{2+\alpha}$-theory play linear equations allowing one by using the method of continuity to go from linear to fully nonlinear equations. We will see how this works in Chapter 13. The method of continuity consists of moving step-by-step along the parameter connecting a linear equation to the one of interest. It does not work in Sobolev space setting, and the results like Theorem 3 allow one not to use any steps but just send K to infinity and having a priori estimates arrive at solutions of (1).

In the framework of viscosity solutions these approximation theorems also allow us to deal with classical approximating solutions, thus avoiding sometimes very involved constructions from the usual theory of viscosity solutions.

Comments on the book's structure

One of fundamental results of the book is the above Theorem 3 and its parabolic counterpart, Theorem 8.3.3. Their particular cases, when H depends not on the full Hessian D^2u but on finitely many pure second-order derivatives in fixed directions, are given as Theorems 6.1.5 and 8.1.4, respectively. For understanding the proofs of these particular cases the reader need not know *anything* from the theory of linear or nonlinear partial differential equations. Everything is based on rather simple arithmetical manipulations with finite-differences presented in Chapters 5 and 7. Therefore, there was a strong temptation to start the book from scratch with Chapter 5 and in some 35 pages to arrive at a solvability result like Theorem 3 for particular operators H.

Actually, it could even be a good idea for the reader to start with Chapter 5 in order to see why the previous material is needed if one wants to go to more general H.

After we prove the solvability of "cut-off" equations, we send K to infinity, but first we need to obtain some estimates of solutions uniform with respect to K. First, in Chapter 9, come C^α estimates of Krylov-Safonov and Fang-Hua Lin's type estimates of the integrals of small powers of the magnitude of the second-order derivatives of solutions. The Krylov-Safonov estimates and the fact that we know the solvability of "cut-off" equations allow us in Chapter 10 to prove Evans-Krylov $C^{2+\alpha}$-estimates for model equations like $F(D^2u) = 0$ and further, by combining them with Fang-Hua Lin's type estimates from Section 9.4 and a generalization of the Fefferman-Stein theorem from Appendix C for not necessarily even locally summable functions, to establish in Theorem 10.1.14 the solvability in $W^2_{p,\mathrm{loc}}$ for elliptic equations under rather weak conditions without assuming that H is convex or concave with respect to D^2u. This line of arguments:

Evans-Krylov $C^{2+\alpha}$-estimates, Fang-Hua Lin's
type estimates, Fefferman-Stein theorem

is repeated several times like in rondo in different circumstances, each time requiring some new initial information.

For instance, Chapter 11, "Nonlinear elliptic equations in $C^{2+\alpha}_{\mathrm{loc}}(\Omega) \cap C(\bar{\Omega})$", appeared, in particular, because of the need to obtain Evans-Krylov $C^{2+\alpha}$-estimates for model parabolic equations like $\partial_t u + F(D^2u) = 0$. There were two possibilities: either to reprove Evans-Krylov $C^{2+\alpha}$-estimates for parabolic equations (following what was done in [**61**]), or to rewrite the equation as $F(D^2u) = -\partial_t u$, first estimate the C^{α}-norm of $\partial_t u$, and then consider the equation for any fixed t as an elliptic one. We preferred the second choice because it is more elegant and richer in ideas, but then we needed a theory of solvability of equations $F(D^2u) = f$ in $C^{2+\alpha}$-spaces, which is insightful in its own rights.

After that in Chapter 12 we are able to basically repeat the reasoning in Chapter 10 and develop the solvability theory for *parabolic* equations in $W^{1,2}_{p,\mathrm{loc}}$.

Then comes the need for solvability theory in global Sobolev spaces W^2_p for elliptic and $W^{1,2}_p$ for parabolic equations. Everything here depends on a priori estimates near the boundary, so model equations should be considered in Hölder classes of functions in half balls or half cylinders, and this gave rise to a comparatively long Chapter 13, "Elements of the $C^{2+\alpha}$-theory of fully nonlinear elliptic and parabolic equations".

Once the boundary a priori estimates are obtained for model equations, again mimicking the line of arguments in Chapter 10, we can then prove the solvability results in W^2_p spaces in Chapter 14 and in $W^{1,2}_p$ spaces in Chapter 15.

The remaining chapters contain applications of the parabolic counterpart of Theorem 3 to various issues in the theory of viscosity solutions for parabolic equations. It would be interesting to know if all of them, especially Theorem 17.1.5 on the $C^{1+\alpha}$-regularity of the L_p-viscosity solutions for Isaacs equations with almost VMO leading coefficients, can be proved by means of the theory of viscosity solution.

There are no historical notes or comments after each chapter. These are spread throughout the text and only a few times presented in a more condensed form in Remarks 3.0.3, 3.1.7, 3.3.8, 4.2.10, 6.6.22, 9.1.3, 9.1.7, 9.4.8, 9.6.9, 10.1.23, 10.1.26, 10.2.5, 11.2.5, 12.1.12, 12.1.21, 14.1.9, 15.1.6, 16.2.9, 17.1.7, 18.1.3, 18.3.6, and 18.4.3; and Examples 10.1.24, 16.2.10, and 18.1.4.

Comments on the exposition

The theorems, lemmas, remarks, and such which are part of the main body of the text are numbered serially in a single system that proceeds by section. Theorem 1.7.6 is the sixth numbered unit in the seventh section in the first chapter. In the course of Chapter 1, this theorem is referred to as Theorem 7.6, and in the course of the seventh section of Chapter 1, it is referred to as Theorem 6. Similarly and independently of these units formulas are numbered and cross-referenced.

In the proofs of various results in this book we use the symbol N sometimes with indices to denote finite nonnegative constants which may change from one occurrence to another, and we do not always specify on which data these constants depend. In these cases the reader should remember that, if in the statement of a result there are constants called N which are claimed to depend only on certain parameters, then in the proof of the result the constants N also depend only on the same parameters unless specifically stated otherwise. Of course, if we write

$$N = N(...),$$

this means that N depends only on what is inside the parentheses.

A complete reference list of notation can be found in the index at the end of the book.

In the case of elliptic equations we concentrate only on equations in the Euclidean space

$$\mathbb{R}^d = \{x = (x^1, ..., x^d) : x^i \in (-\infty, \infty)\}$$

or in domains $\Omega \subset \mathbb{R}^d$. By domains we mean general open sets. If Γ is a measurable subset of $\mathbb{R}^d$, by $|\Gamma|$ we denote its Lebesgue measure.

The scalar product in $\mathbb{R}^d$ is denoted by

$$\langle \cdot, \cdot \rangle.$$

If a is a matrix of any dimension (for instance, column-vector), we set

$$|a| = [\operatorname{tr} aa^*]^{1/2},$$

where a^* is the transpose of a.

On some occasions, we allow ourselves to use different symbols for the same objects, for example,

$$u_{x^i} = \frac{\partial u}{\partial x^i} = D_i u, \quad u_{x^i x^j} = \frac{\partial^2 u}{\partial x^i \partial x^j} = D_{ij} u,$$

$$u_x = Du = (D_1 u, ..., D_d u), \quad u_{xx} = D^2 u = (D_{ij} u)_{i,j=1}^d .$$

Somewhat ambiguously we set

$$D_i uv := (D_i u)v.$$

In the case of parabolic equations we work with

$$\mathbb{R}^{d+1} = \{(t, x) : t \in \mathbb{R}, x \in \mathbb{R}^d\}.$$

If Γ is a measurable subset of $\mathbb{R}^{d+1}$, by $|\Gamma|$ we denote its Lebesgue measure. If $z_1 = (t_1, x_1), z_2 = (t_2, x_2) \in \mathbb{R}^{d+1}$, we set

$$\rho(z_1, z_2) = |x_1 - x_2| + |t_1 - t_2|^{1/2}$$

and call $\rho(z_1, z_2)$ the parabolic distance between z_1 and z_2.

For functions $u(t, x)$ given on subdomains of $\mathbb{R}^{d+1}$ we use the above notation only for the derivatives in x; sometimes we write $D_x u$, $D_x^2 u$ to emphasize this, and denote

$$\partial_t u = \frac{\partial u}{\partial t} = u_t, \quad \partial_t D_k u = u_{tx^k} = u_{x^k t},$$

and so on.

In various parts of the book we speak about measurable functions and sets: Borel measurable and Lebesgue measurable, and if we say that a particular function or a set is *measurable*, we always mean that it is Lebesgue measurable. On many occasions we consider functions like $F(\mathsf{u}, x)$, where u belongs to a Euclidean space, say $\mathfrak{U}$, and $x \in \mathbb{R}^d$, and require them to be Borel measurable in (u, x). This is done for simplicity of presentation. Actually, what is needed is for them to be measurable with respect to the product of the Borel σ-field in $\mathfrak{U}$ and the Lebesgue σ-field in $\mathbb{R}^d$. Generally, it is useful to keep in mind that for any Lebesgue measurable function there exists a Borel measurable one coinciding with it almost everywhere.

If Ω is a domain in $\mathbb{R}^d$ and $p \in [1, \infty]$, by $L_p(\Omega)$ we mean the set of all (Lebesgue) measurable real-valued functions f for which

$$\|f\|_{L_p(\Omega)} := \Big(\int_\Omega |f(x)|^p \, dx \Big)^{1/p} < \infty$$

with the standard extension of this formula if $p = \infty$. One knows that $L_p(\Omega)$ is a Banach space.

By $C_0^\infty(\Omega)$ we mean the set of all infinitely differentiable functions on Ω with compact support (contained) in Ω. By the *support* of a function we mean the closure of the set where the function is different from zero. We call a subset of $\mathbb{R}^d$ *compact* if it is closed and bounded. Of course, saying "compact support" is the same as saying "bounded support"; we keep "compact" just to remind us that we are talking about *closed* sets. We set

$$C_0^\infty = C_0^\infty(\mathbb{R}^d).$$

Similar notation is introduced for domains in $\mathbb{R}^{d+1}$ and for functions of (t, x) rather than x alone.

We always use the summation convention with respect to repeated indices no matter whether they are at the same level or at different ones, and if there are signs of inf or sup followed by the sums of products of terms with repeated indices, then the summation is performed first, inside the signs of inf or sup, for instance,

$$\inf_\alpha \big[a_i^\alpha u_i + b_j^\alpha v_j\big] := \inf_\alpha \Big[\sum_i a_i^\alpha u_i + \sum_j b_j^\alpha v_j\Big].$$

By $\mathbb{S}$ we mean the set of $d \times d$ symmetric matrices. If $a \in \mathbb{S}$ we usually represent its entries by a^{ij}. However, on many occasions a itself has superscripts, and then we use a_{ij} for the entries of a. In the same vein for numbers u we use the notation

$$u^\pm = u_\pm = (1/2)\big(|u| \pm u\big).$$

We remind the reader that

$$a \wedge b = \min(a, b), \quad a \vee b = \max(a, b).$$

One more general stipulation is that

positive=nonnegative, negative=nonpositive,

decreasing=nonincreasing, increasing=nondecreasing.

Acknowledgments

My deepest gratitude is due to Sungbin Lim and Hyunwoo Kwon, who on their own initiative typeset in LaTeX my book [**61**] and sent me the TeX file, allowing me to use it if I ever go back to the subject of [**61**]. I used their file in a few places. I am also grateful to the NSF for supporting me in the early stages of writing the book through grant DMS-1160569. Finally, the referees made some comments, which certainly helped improve the presentation.

Nicolai Krylov
Minneapolis, MN
August 2017

CHAPTER 1

Bellman's equations with constant "coefficients" in the whole space

In this short chapter we prove the solvability of Bellman's equations with constant "coefficients" in the whole space. On the one hand, this provides a large class of nonlinear equations whose solvability is obtained by elementary means. On the other hand, the results of this chapter are instrumental in proving the elliptic and parabolic Aleksandrov estimates in Chapter 3. The exposition here is close to that in [**61**] but is different in a very important respect that, instead of the quite hard to follow translation of the controlled diffusion theory into the PDEs language in [**61**], we use a quite transparent approach based on finite-difference approximations for nonlinear equations. Because of that significant part Chapter 2 of [**61**] is written differently.

1. Finite-difference equations

Given a set A and finite disjoint sets I and J, let d-dimensional nonzero vectors $l_i(\alpha)$ be defined for $i \in I \cup J$ and $\alpha \in A$. Also let real-valued

$$a_i(\alpha),\ b_j(\alpha),\ c^\alpha,\ f^\alpha(x)$$

be defined for all $i \in I$, $j \in J$, $\alpha \in A$, and $x \in \mathbb{R}^d$.

1. Assumption. There are constants $K, \delta \in (0, \infty)$ such that for all possible values of indices and arguments we have

$$0 \leq a_i,\ b_j,\ c,\ |l_i|,\ |f| \leq K, \quad c \geq \delta.$$

For any $x, l \in \mathbb{R}^d$, $h > 0$, and functions ϕ on $\mathbb{R}^d$ introduce

$$T_{h,l}\phi(x) = \phi(x + hl), \quad \delta_{h,l} = \frac{1}{h}(T_{h,l} - 1), \quad \Delta_{h,l} = \frac{1}{h^2}(T_{h,l} - 2 + T_{h,-l}). \tag{1}$$

Notice that $\delta_{h,l}$ and $\Delta_{h,l}$ are the usual approximations for the first and second-order derivatives along l.

Also set

$$\mathcal{L}_h^\alpha \phi(x) = \sum_{i \in I} a_i(\alpha) \Delta_{h,l_i(\alpha)} \phi(x) + \sum_{i \in J} b_i(\alpha) \delta_{h,l_i(\alpha)} \phi(x).$$

For $h \in (0, 1)$ we are going to consider the equation

$$\sup_{\alpha \in A} \left[\mathcal{L}_h^\alpha \phi(x) - c^\alpha \phi(x) + f^\alpha(x) \right] = 0 \tag{2}$$

in $\mathbb{R}^d$. Observe that the left-hand side of (2) makes sense for any bounded function ϕ on $\mathbb{R}^d$. The reader understands, of course, that the theory of equations like (2) with the inf in place of the sup is quite parallel to that with the sup. It just suffices to change the signs of f and ϕ.

Next introduce

$$M_h^\alpha \phi(x) = \mathcal{L}_h^\alpha \phi(x) + \phi(x)\big[2h^{-2}\sum_{i\in I} a_i(\alpha) + h^{-1}\sum_{i\in J} b_i(\alpha)\big],$$

$$\hat{c}^\alpha = c^\alpha + 2h^{-2}\sum_{i\in I} a_i(\alpha) + h^{-1}\sum_{i\in J} b_i(\alpha),$$

and notice that, owing to the fact that $\hat{c}^\alpha \geq \delta$, equation (2) is equivalent to the following equations

$$\sup_{\alpha\in A} \hat{c}^\alpha \big[(\hat{c}^\alpha)^{-1} M_h^\alpha \phi(x) - \phi(x) + (\hat{c}^\alpha)^{-1} f^\alpha(x)\big] = 0,$$

$$\sup_{\alpha\in A} \big[(\hat{c}^\alpha)^{-1} M_h^\alpha \phi(x) - \phi(x) + (\hat{c}^\alpha)^{-1} f^\alpha(x)\big] = 0,$$

$$\phi(x) = P\phi(x),$$

where

$$P\phi(x) := \sup_{\alpha\in A} \big[(\hat{c}^\alpha)^{-1} M_h^\alpha \phi(x) + (\hat{c}^\alpha)^{-1} f^\alpha(x)\big].$$

2. THEOREM. *The operator P is a contraction in the space $B(\mathbb{R}^d)$ of bounded functions on $\mathbb{R}^d$ with the sup norm and equation* (2) *has a unique solution ϕ in this space.*

Proof. As is easy to see, if $\phi \geq 0$ on $\mathbb{R}^d$, then $M_h^\alpha \phi \geq 0$ on $\mathbb{R}^d$. This means that the operators M_h^α are monotone or order preserving. Furthermore, by denoting by d_1 the number of elements in $I \cup J$ we have

$$M_h^\alpha 1 = 2h^{-2}\sum_{i\in I} a_i(\alpha) + h^{-1}\sum_{i\in J} b_i(\alpha) =: \gamma^\alpha \hat{c}^\alpha \leq 3h^{-2}Kd_1$$

(recall that $h \in (0,1)$), where $\gamma^\alpha \in (0,1)$ defined by the last equation satisfies

$$(1-\gamma^\alpha)M_h^\alpha 1 = \gamma^\alpha c^\alpha, \quad 3h^{-2}Kd_1(1-\gamma^\alpha) \geq \gamma^\alpha \delta,$$

$$\gamma^\alpha \leq 3Kd_1/(\delta h^2 + 3Kd_1) =: \gamma < 1.$$

Thus

$$M_h^\alpha 1 \leq \gamma \hat{c}^\alpha. \tag{3}$$

Next, we use the simple inequality

$$\sup_{\alpha\in A} \xi^\alpha - \sup_{\alpha\in A} \eta^\alpha \leq \sup_{\alpha\in A}(\xi^\alpha - \eta^\alpha) \tag{4}$$

valid for any numbers ξ^α and η^α and conclude that for any functions ϕ and ψ we have

$$P\phi(x) - P\psi(x) \leq \sup_{\alpha\in A}(\hat{c}^\alpha)^{-1} M_h^\alpha(\phi - \psi)(x).$$

This along with the monotonicity of M_h^α and (3) yield that

$$P\phi(x) - P\psi(x) \leq \gamma \sup_{\mathbb{R}^d}(\phi - \psi)_+,$$

which finishes the proof of the theorem. □

Now we prove a global comparison principle.

3. THEOREM. *Assume that we are also given functions $g^\alpha(x)$ satisfying the same conditions as $f^\alpha(x)$ and let ϕ and ψ be bounded functions on $\mathbb{R}^d$ satisfying*

$$\sup_{\alpha\in A}\left[\mathcal{L}_h^\alpha\phi(x)-c^\alpha\phi(x)+f^\alpha(x)\right]\ge\sup_{\alpha\in A}\left[\mathcal{L}_h^\alpha\psi(x)-c^\alpha\psi(x)+g^\alpha(x)\right]$$

on $\mathbb{R}^d$. Then on $\mathbb{R}^d$ we have

$$\phi\le\psi+\delta^{-1}\sup_{A\times\mathbb{R}^d}\left[f^\alpha(x)-g^\alpha(x)\right]_+.$$

Proof. By virtue of (4) the function $u=\phi-\psi$ satisfies

$$\sup_{\alpha\in A}\left[\mathcal{L}_h^\alpha u(x)-c^\alpha u(x)+v^\alpha(x)\right]\ge 0,$$

where $v^\alpha=f^\alpha-g^\alpha$. It follows that

$$w(x):=u(x)-\delta^{-1}\sup_{A\times\mathbb{R}^d}\left[v^\alpha(y)\right]_+$$

satisfies

$$\sup_{\alpha\in A}\left[\mathcal{L}_h^\alpha w(x)-c^\alpha w(x)\right]\ge 0.$$

This relations can be dealt with as before Theorem 2 and transforms into

$$w(x)\le\sup_{\alpha\in A}\left(\hat{c}^\alpha\right)^{-1}M_h^\alpha w(x).$$

According to (3) and the monotonicity of M_h^α we conclude that

$$w\le\gamma\sup_{\mathbb{R}^d}w_+,\quad w_+=0,\quad u\le\delta^{-1}\sup_{A\times\mathbb{R}^d}\left[v^\alpha(y)\right]_+$$

and the theorem is proved. □

In the remaining part of the section by ϕ we mean the solution ϕ of (2) from Theorem 2.

The estimate of u from the above proof can be applied to ϕ and it leads to the second inequality in (5). The first inequality is obtained from the inf version of the above argument.

4. COROLLARY. *We have*

$$-\delta^{-1}\sup_{A\times\mathbb{R}^d}\left[f^\alpha(y)\right]_-\le\phi\le\delta^{-1}\sup_{A\times\mathbb{R}^d}\left[f^\alpha(y)\right]_+. \tag{5}$$

5. COROLLARY. *Assume that, for a vector $l\in\mathbb{R}^d$ and a constant $K_1\in[0,\infty)$ and any $x\in\mathbb{R}^d$ and $\alpha\in A$, the function $f^\alpha(x+tl)+K_1t$ is an increasing function of t. Then $\psi(t,x):=\phi(x+tl)+\delta^{-1}K_1t$ is also an increasing function of t for any x.*

Indeed, $\psi(t,x)$ satisfies

$$\sup_{\alpha\in A}\left[\mathcal{L}_h^\alpha\psi(t,x)-c^\alpha\psi(t,x)+f^\alpha(t,x)\right]=0,$$

where

$$f^\alpha(t,x)=f^\alpha(x+tl)+c^\alpha\delta^{-1}K_1t$$

is an increasing function of t.

6. REMARK. If the functions $f^\alpha(x \pm tl) + K_1 t$ are increasing function of t, that is, if $f^\alpha(x+tl)$ is Lipschitz continuous in t with Lipschitz constant K_1, then $\phi(x+tl)$ is Lipschitz continuous in t with Lipschitz constant $\delta^{-1}K_1$.

7. COROLLARY. *Assume that, for a vector $l \in \mathbb{R}^d$ and a constant $K_2 \in [0,\infty)$ and any $x \in \mathbb{R}^d$ and $\alpha \in A$, the function $f^\alpha(x+tl) + K_2t^2$ is a convex function of t. Then*

$$\psi(t,x) := \phi(x+tl) + \delta^{-1}K_2t^2$$

is also a convex function of t for any x.

To prove this, we use the elementary inequality

$$\sup_{\alpha\in A} \xi^\alpha - \sup_{\alpha\in A} \eta^\alpha \geq 2\sup_{\alpha\in A}\gamma^\alpha + \inf_{\alpha\in A}\left(\xi^\alpha - 2\gamma^\alpha + \eta^\alpha\right) \tag{6}$$

valid for any numbers ξ^α, η^α, and γ^α. Then we see that, for any $\kappa, t \in \mathbb{R}$, the function

$$\psi(\kappa,t,x) = \phi(x+(t+\kappa)l) - 2\phi(x+tl) + \phi(x+(t-\kappa)l)$$

satisfies

$$\inf_{\alpha\in A}\left[\mathcal{L}_h^\alpha\psi(\kappa,t,x) - c^\alpha\psi(\kappa,t,x) + f^\alpha(\kappa,t,x)\right] \leq 0,$$

where

$$f^\alpha(\kappa,t,x) = f^\alpha(x+(t+\kappa)l) - 2f^\alpha(x+tl) + f^\alpha(x+(t-\kappa)l) \geq -2K_2\kappa^2.$$

Hence

$$\inf_{\alpha\in A}\left[\mathcal{L}_h^\alpha\psi(\kappa,t,x) - c^\alpha\psi(\kappa,t,x) - 2K_2\kappa^2\right] \leq 0 \leq \inf_{\alpha\in A}\left[\mathcal{L}_h^\alpha(-N) - c^\alpha(-N) - 2K_2\kappa^2\right],$$

where $N = 2\delta^{-1}K_2\kappa^2$. By Theorem 3 applied with the inf formulation we get that $\psi(\kappa,t,x) \geq N$, which means exactly that $\phi(x+tl) + \delta^{-1}K_2t^2$ is a convex function of t.

Finally, we need a maximum principle. Introduce the so-called stencil:

$$\Lambda = \left\{l_i(\alpha) : \alpha \in A, i \in I \cup J\right\}$$

and for any *closed* set $Q \subset \mathbb{R}^d$ define its interior $Q^o_{h,\Lambda}$ and its boundary $\partial_{h,\Lambda}Q$ relative to Λ as

$$Q^o_{h,\Lambda} = \{x \in Q : x + h\Lambda \subset Q\}, \quad \partial_{h,\Lambda}Q = Q \setminus Q^o_{h,\Lambda}.$$

8. THEOREM. *Let Q be a bounded closed set with $Q^o_{h,\Lambda} \neq \emptyset$ and let u be a continuous function in Q satisfying*

$$\sup_{\alpha\in A}\left[\mathcal{L}_h^\alpha u(x) - c^\alpha u(x) + f^\alpha(x)\right] \geq 0 \tag{7}$$

in $Q^o_{h,\Lambda}$. Then

$$\max_Q u \leq \sup_{\partial_{h,\Lambda}Q} u_+ + \delta^{-1}\sup_{A\times Q^o_{h,\Lambda}} f^\alpha_+(x). \tag{8}$$

In particular (the maximum principle), if $f^\alpha \equiv 0$ (or $f^\alpha \leq 0$), then the maximum of u_+ in Q is attained on its boundary $\partial_{h,\Lambda}Q$.

Furthermore, the same conclusions hold if we replace the sign sup *in* (7) *with* inf.

Proof. Set

$$N=\delta^{-1}\sup_{A\times Q^o_{h,\Lambda}} f^\alpha_+(x),\qquad v=u-N$$

and observe that v satisfies

$$\sup_{\alpha\in A}\big[\mathcal{L}^\alpha_h v(x)-c^\alpha v(x)\big]=\sup_{\alpha\in A}\big[\mathcal{L}^\alpha_h u(x)-c^\alpha u(x)+Nc^\alpha\big]\geq 0.$$

If the maximum principle holds (that is if (8) is true for $f^\alpha\equiv 0$), then in Q

$$v\leq \sup_{\partial_{h,\Lambda}Q} v_+\leq \sup_{\partial_{h,\Lambda}Q} u_+,$$

where the inequality between the extreme terms coincides with (8).

It follows that in the rest of the proof we may assume that $f^\alpha\equiv 0$. In that case, denote by x_0 a point in Q where u attains its maximum over Q. If $x_0\in Q^o_{h,\Lambda}$, then, obviously, $\mathcal{L}^\alpha_h u(x_0)\leq 0$ and (7) yields

$$\sup_{\alpha\in A}\big[-c^\alpha u(x_0)\big]\geq 0,\quad u(x_0)\sup_{\alpha\in A}\big[-c^\alpha\big]\geq 0,\quad u(x_0)\leq 0.$$

In that case (8) holds and the maximum of u_+ in Q is indeed attained on its boundary. However if $x_0\in\partial_{h,\Lambda}Q$, (8) is obvious and the theorem is proved. □

9. COROLLARY. *Let Q be a bounded closed set with $Q^o_{h,\Lambda}\neq\emptyset$ and let u and v be continuous functions in Q, such that $v\leq u$ on $\partial_{h,\Lambda}Q$ and $u<v$ at at least one point in $Q^o_{h,\Lambda}$. Then there is at least one point in $Q^o_{h,\Lambda}$ at which*

$$\sup_{\alpha\in A}\big[\mathcal{L}^\alpha_h u(x)-c^\alpha u(x)+f^\alpha(x)\big]>\sup_{\alpha\in A}\big[\mathcal{L}^\alpha_h v(x)-c^\alpha v(x)+f^\alpha(x)\big].\tag{9}$$

Indeed, it suffices to denote by ξ and η the left-hand side and the right-hand side of (9), respectively, and observe that the function $w=v-u$ satisfies $w\leq 0$ on $\partial_{h,\Lambda}Q$ and in $Q^o_{h,\Lambda}$

$$\sup_{\alpha\in A}\big[\mathcal{L}^\alpha_h w-c^\alpha w\big]\geq\eta-\xi,\quad \sup_{\alpha\in A}\big[\mathcal{L}^\alpha_h w-c^\alpha w+\xi-\eta\big]\geq 0.$$

2. Solvability of Bellman's equations

Let A be a set, and let, for all $\alpha\in A$ and $x\in\mathbb{R}^d$, a nonnegative symmetric $d\times d$ matrix $a(\alpha)=\big(a^{ij}(\alpha)\big)$, a d-dimensional vector $b(\alpha)=\big(b^i(\alpha)\big)$, and real-valued c^α and $f^\alpha(x)$ be defined.

1. ASSUMPTION. There are constants $K,\delta\in(0,\infty)$ such that for all values of indices and arguments we have

$$|a^{ij}|,\ |b^i|,\ c,\ |f|\leq K,\quad c\geq\delta.$$

2. ASSUMPTION. There are constants $K_1,K_2\in(0,\infty)$ such that the functions f^α are Lipschitz continuous with respect to x with Lipschitz constant K_1 and $f^\alpha(x)+K_2|x|^2$ are convex functions of x.

Observe that these assumptions are weaker than in Section 2.1 of [**61**].

For $\mathsf{u} = (\mathsf{u}', \mathsf{u}'')$, where

$$\mathsf{u}' = \left(\mathsf{u}'_0, \mathsf{u}'_1, ..., \mathsf{u}'_d\right) \in \mathbb{R}^{d+1}, \quad \mathsf{u}'' \in \mathbb{S},$$

we set

$$F(\mathsf{u}, x) = \sup_{\alpha \in A} \left[a^{ij}(\alpha)\mathsf{u}''_{ij} + b^i(\alpha)\mathsf{u}'_i - c(\alpha)\mathsf{u}'_0 + f(\alpha, x)\right],$$

where the summations over the repeated indices i and j from 1 to d is performed before the sup is calculated. Also define

$$F[u](x) = F(u, Du, D^2u, x)$$

and consider the problem of finding a sufficiently smooth function $v = v(x)$ such that, in $\mathbb{R}^d$, the following Bellman's equation is satisfied:

$$F[v] = 0. \tag{1}$$

Introduce $I = \{1, ..., d\}$ and for $i \in I$ let $l_i(\alpha)$ be the unit eigenvectors of $a(\alpha)$ and $a_i(\alpha)$ be their corresponding eigenvalues. Also let $J = \{0\}$ and for $i \in J$ let $b_i(\alpha) = |b(\alpha)|$ and $l_i(\alpha) = b(\alpha)/|b(\alpha)|$ $(0/0 := 0)$. With this notation and with $h \in (0, 1)$ we denote by ϕ_h the unique bounded solution of (1.2) (existing thanks to Theorem 1.2).

The rational behind choosing our particular $l_i(\alpha)$, $a_i(\alpha)$, and $b_i(\alpha)$ is that, for any smooth $u(x)$ as $h \downarrow 0$, we have

$$\mathcal{L}_h^\alpha u \to a_i(\alpha)l_i^k(\alpha)l_i^r(\alpha)D_{kr}u + b^i(\alpha)D_iu$$

$$= a^{kr}(\alpha)D_{kr}u + b^i(\alpha)D_iu =: \mathcal{L}^\alpha u, \tag{2}$$

so that (1) is rewritten as

$$\sup_{\alpha \in A} \left[\mathcal{L}^\alpha u - c^\alpha u + f^\alpha\right] = 0.$$

By Corollary 1.4 and Remark 1.6 the family $\{\phi_h, h \in (0, 1)\}$ is uniformly bounded and uniformly Lipschitz continuous with Lipschitz constant $\delta^{-1}K_1$.

Therefore, there exists a sequence $h_n \downarrow 0$ such that the functions

$$v_n = \phi_{h_n}$$

converge uniformly on any ball in $\mathbb{R}^d$ to a function, which we denote by v. Note for the future the following.

3. REMARK. If ψ is a twice continuously differentiable function on $\mathbb{R}^d$ such that $D\psi$ and $D^2\psi$ are bounded and uniformly continuous on $\mathbb{R}^d$ and $F[\psi] \leq 0$ on $\mathbb{R}^d$, then $\psi \geq v$.

Indeed, then for any $\gamma > 0$, any α, and all suficiently small $h > 0$ we have (cf. (2)) $\mathcal{L}_h^\alpha(\psi + \gamma) - c^\alpha(\psi + \gamma) + f^\alpha \leq 0$, which by Theorem 1.3 implies that $v_h \leq \psi + \gamma$, $v \leq \psi + \gamma$, and $v \leq \psi$.

Corollaries 1.4, 1.5 and 1.7 and Remark 1.6 lead to the first assertion in the following lemma, in which

$$D_{(l)}v := l^i D_i v, \quad D_{(l_1)(l_2)}v := D_{(l_1)}D_{(l_2)}v = l_1^i l_2^j D_{ij}v,$$

where $D_i v, D_{ij}v$ are the generalized functions representing the generalized derivatives of v and $l, l_1, l_2 \in \mathbb{R}^d$.

4. LEMMA. *(a) The function v is Lipschitz continuous with Lipschitz constant $\delta^{-1}K_1$ and $v + \delta^{-1}K_2|x|^2$ is a convex function and on $\mathbb{R}^d$*

$$-\delta^{-1} \sup_{A\times\mathbb{R}^d} \left[f^\alpha(y)\right]_- \leq v \leq \delta^{-1} \sup_{A\times\mathbb{R}^d} \left[f^\alpha(y)\right]_+.$$

(b) The function v has the first-order Sobolev derivatives and $|Dv| \leq \delta^{-1}K_1$ (a.e.). Furthermore, all generalized second order derivatives of v are measures, finite on each ball, and

$$D_{(l)(l)}v(dx) \geq -2\delta^{-1}K_2\,dx$$

for any unit $l \in \mathbb{R}^d$.

The assertion of the lemma about Dv follows from the well-known fact that the first-order Sobolev derivatives of a function satisfying the Lipschitz condition exist and are bounded by the Lipschitz constant. Concerning the second-order derivatives, recall that the generalized pure second-order derivatives of convex functions are nonnegative measures finite on each ball. This fact applies to $v + \delta^{-1}K_2|x|^2$. Mixed generalized derivatives $D_{(l_1)(l_2)}v$ are easily expressed through pure ones (see, for instance, Lemma 3.6.1 in [**61**]).

It is also known (see, for instance, Theorem 1.2.1 in [**61**]) that, by the Aleksandrov-Busemann-Feller theorem, the first-order Sobolev derivatives and the second-order generalized derivatives of convex functions allow one to write the second-order Taylor formula at almost any point $x_0 \in \mathbb{R}^d$. The function v is the difference of $v + \delta^{-1}K_2|x|^2$ and $\delta^{-1}K_2|x|^2$, which are convex. Therefore, the above comments about convex functions is valid for v as well. This proves the following theorem.

5. THEOREM. *The Sobolev derivatives $D_i v$ exist and are bounded on $\mathbb{R}^d$. For $i, j = 1, ..., d$, there exist the measures $D_{ij}v(dx)$ which are finite on every bounded Borel set. These measures are the generalized derivatives of v with respect to $x^i x^j$. For almost every $x_0 \in \mathbb{R}^d$ as $x \to x_0$*

$$v(x) = v(x_0) + D_i v(x_0)(x^i - x_0^i)$$

$$+ \frac{1}{2}D_{ij}v^{(0)}(x_0)(x^i - x_0^i)(x^j - x_0^j) + o(|x - x_0|^2), \tag{3}$$

where, for a measure μ on $\mathbb{R}^d$, we denote by $\mu^{(0)}$ the Radon-Nikodym derivative of the absolutely continuous part of μ with respect to Lebesgue measure.

6. THEOREM. *Almost everywhere in $\mathbb{R}^d$, for $D^2v^{(0)} := \left(D_{ij}v^{(0)}\right)$, we have*

$$F\left(v(x), Dv(x), D^2v^{(0)}(x), x\right) = 0. \tag{4}$$

Proof. Take any point x_0 at which (3) holds and represent its right-hand side as a quadratic polynomial u plus $o(|x-x_0|^2)$. Because of (3) for any $\kappa>0$ one can find $r_\kappa \in (0,1)$ such that, for any $r\in(0,r_\kappa]$, for all large enough n we have

$$u_{\kappa,r} := u + \kappa\zeta_r \le v_n(x) \tag{5}$$

in $\bar{B}_{3r}(x_0)\setminus B_{2r}(x_0)$, where

$$\zeta_r(x) = r^2 - |x-x_0|^2, \quad B_R(x_0) = \{x\in\mathbb{R} : |x-x_0|<R\}.$$

Furthermore, for all large enough n we have

$$u_{\kappa,r}(x_0) = v(x_0) + \kappa r^2 > v_n(x_0). \tag{6}$$

For large enough n by Corollary 1.9 (with $Q=\bar{B}_{3r}(x_0)$ and n so large that $\partial_{h_n,\Lambda}Q\subset \bar{B}_{3r}(x_0)\setminus B_{2r}(x_0)$) there are points in $\bar{B}_{3r}(x_0)$ such that

$$\sup_{\alpha\in A}\left[\mathcal{L}^\alpha_{h_n}u(x) - c^\alpha u(x) + \kappa(\mathcal{L}^\alpha_{h_n} - c^\alpha)\zeta_r(x) + f^\alpha(x)\right] \le 0. \tag{7}$$

Observe that, as for any quadratic function, we have

$$\delta_{h,l}u = \langle Du, l\rangle + h\operatorname{tr}\left[D^2 u l l^*\right], \quad \Delta_{h,l}u = \operatorname{tr}\left[D^2 u l l^*\right].$$

Furthermore

$$a_i(\alpha)l_i(\alpha)l_i^*(\alpha) = a(\alpha).$$

It follows that

$$\mathcal{L}^\alpha_h u(x) = \operatorname{tr}\left[a(\alpha)D^2u(x)\right] + b^i(\alpha)D_iu(x) + h\operatorname{tr}\left[b(\alpha)b^*(\alpha)D^2u(x)\right]\left|b(\alpha)\right|^{-1}. \tag{8}$$

Note that here the matrix $D^2u(x)$ is constant, so that $\mathcal{L}^\alpha_h u(x)$ are Lipschitz continuous in x with constant independent of h and α. This combined with the rough estimate $\left|(\mathcal{L}^\alpha_{h_n} - c^\alpha)\zeta_r\right| \le N(\Lambda, K, d)$ and (7) shows, in particular, that

$$\sup_{\alpha\in A}\left[\mathcal{L}^\alpha_{h_n}u(x_0) - c^\alpha u(x_0) + f^\alpha(x_0)\right] \le N(\kappa + r),$$

where and below N is independent of n, κ, and r. Then using (8) we conclude

$$\sup_{\alpha\in A}\left[\mathcal{L}^\alpha u(x_0) - c^\alpha u(x_0) + f^\alpha(x_0)\right] \le N(\kappa + r + h_n).$$

Now, by setting $n\to\infty$, then $r\downarrow 0$, and finally $\kappa\downarrow 0$ we get that the left-hand side of (4) is nonpositive at x_0.

To prove that it is nonnegative, we take the same $u_{\kappa,r}$ but take $\kappa<0$. Then in (5), (6), and (7) the signs of the inequalities will be reversed and this would lead to the desired result. The theorem is proved. □

The following is useful while dealing with measures $D_{ij}v(dx)$.

7. THEOREM. *For any $\alpha\in\mathbb{R}^d$, we have*

$$a^{ij}(\alpha)D_{ij}v(dx) + \left[b^i(\alpha)D_iv + c(\alpha)v + f(\alpha)\right]dx \le 0. \tag{9}$$

Proof. Take a nonnegative symmetric $\zeta \in C_0^\infty(\mathbb{R}^d)$ with unit integral and for $\gamma > 0$ and locally summable functions η define $\eta^{(\gamma)} = \eta * \zeta_\gamma$, where $\zeta_\gamma(x) = \gamma^{-d}\zeta_\gamma(x/\gamma)$.

Observe that for any $\alpha \in A$ and $h, \gamma \in (0,1)$

$$\mathcal{L}_h^\alpha \phi_h - c^\alpha \phi_h + f^\alpha \leq 0, \quad \mathcal{L}_h^\alpha \phi_h^{(\gamma)} - c^\alpha \phi_h^{(\gamma)} + (f^\alpha)^{(\gamma)} \leq 0.$$

We set here $h = h_n$ and let $n \to \infty$. By noting that, for instance, by the dominated convergence theorem

$$\Delta_{h_n,l}\phi_{h_n}^{(\gamma)}(x) = \gamma^{-d}\int_{\mathbb{R}^d} \phi_{h_n}(y)\Delta_{h_n,l}\zeta_\gamma\big((x-y)/\gamma\big)\,dy$$

$$\to \gamma^{-d-2}\int_{\mathbb{R}^d} v(y)(D_{(l)(l)}\zeta)\big((x-y)/\gamma\big)\,dy = l^k l^r D_{kr}v^{(\gamma)}(x),$$

we conclude that

$$a_i(\alpha)l_i^k(\alpha)l_i^r(\alpha)D_{kr}v^{(\gamma)} + b^i(\alpha)D_i v^{(\gamma)} - c^\alpha v^{(\gamma)} + \big(f^\alpha\big)^{(\gamma)} \leq 0.$$

Hence the convolution of the measure standing on the left-hand side of (9) with ζ_γ is negative for any $\gamma > 0$. By passing to the (weak) limit as $\gamma \downarrow 0$ we obtain (9) and the theorem is proved. □

8. REMARK. As a complement to Theorem 7 recall that the set of signed measures on the Borel σ-field is a lattice and any collection μ_α of them bounded from above by a measure has the least upper bound denoted by $\bigvee_\alpha \mu_\alpha$. It turns out see Section 3.7 of [**61**] that in the general case

$$\bigvee_{\alpha\in A}\Big[a^{ij}(\alpha)D_{ij}v(dx) + \big[b^i(\alpha)D_i v + c(\alpha)v + f(\alpha)\big]\,dx\Big] = 0$$

in $\mathbb{R}^d$ (which is simple owing to (4) and (9)) and (which is much harder) v is a unique bounded Lipschitz continuous function such that $v + K|x|^2$ is convex for a constant K and which satisfies the above equation in $\mathbb{R}^d$.

Assertion (b) of Lemma 4 provides estimates from below for the pure second-order derivatives of v. In some cases, in particular, important in Chapter 2, one can also estimate them from above on the basis of Theorem 6. The idea is the following. Imagine that

$$a^{11}D_{11}v + a^{22}D_{22}v \leq 1 \quad \text{and} \quad D_{22}v \geq -1.$$

Then

$$a^{11}D_{11}v \leq 1 - a^{22}D_{22}v \leq 1 + a^{22}, \quad D_{11}v \leq \big(a^{11}\big)^{-1}(1 + a^{22}).$$

In other words, we can estimate the derivative $D_{(l)(l)}v$ from above if it is "actually" present in $\mathcal{L}(\alpha)v$ for at least one α or if it is "actually" present in $F[v]$.

Note that, if A is a singleton, then the assumption that f, Df, D^2f are bounded alone implies that all second-order derivatives of v are bounded on $\mathbb{R}^d$ whether or not any of them enters $\mathcal{L}(\alpha)v$. The situation is completely different if A contains at least two elements. Here is an example.

9. EXAMPLE. Let $d = 2$, $A = \{-1\} \cup \{+1\}$,

$$f(\alpha, x) = (x^1)^2\big(1 + (x^1)^2\big)^{-1},$$

and let $\mathcal{L}(\alpha)u = \alpha D_1 u + D_{22}u - u$. In this case, it can be shown (but cf. Section 3.7 in [**61**]) that

$$v(x^1, x^2) = e^{|x^1|}\int_{|x^1|}^{\infty} \frac{t^2}{1+t^2} e^{-t}\,dt.$$

As is easy to see, the second-order derivative of v with respect to x^1 does not exist on the x^2-axis, and the measure $D_{11}v(dx)$ has a singular component concentrated on this axis and therefore cannot be estimated from above through Lebesgue measure.

Denote

$$F_0(\mathsf{u}'') = \sup_{\alpha} a^{ij}(\alpha)\mathsf{u}''_{ij} \quad \Big(= \lim_{t\to\infty} \frac{1}{t} F(\mathsf{u}', t\mathsf{u}'', u, x)\Big), \tag{10}$$

$$\mu(l) = \inf_{\xi:\langle l,\xi\rangle = 1} F_0(\xi\xi^*), \quad l \in \mathbb{R}^d, \quad l \neq 0, \tag{11}$$

$$\mu = \inf_{\xi:|\xi|=1} F_0(\xi\xi^*) = \inf_{|\xi|=1} \sup_{\alpha\in A} a^{ij}(\alpha)\xi^i\xi^j \quad \big(= \inf_{|l|=1} \mu(l)\big). \tag{12}$$

For $l \neq 0$, $\mu(l)$ is called the *measure of involvement* of $D_{(l)(l)}v$ in $F[v]$. Observe that in Example 9 we have $\mu(l) > 0$ only if $l \parallel e_2 = (0, 1)$. This shows that our definition is quite natural.

10. LEMMA. *Let $\mathsf{u}'' \in \mathbb{S}$, I be a unit $d \times d$ matrix, and let ψ be a number. Suppose that $\langle \mathsf{u}''\xi, \xi\rangle \geq \psi|\xi|^2$ for all $\xi \in \mathbb{R}^d$. Then for all unit vectors l,*

$$\mu(l)\langle \mathsf{u}''l, l\rangle \leq F_0(\mathsf{u}'') + \psi_- F_0(I),$$
$$\mu\langle \mathsf{u}''l, l\rangle \leq F_0(\mathsf{u}'') + \psi_- F_0(I).$$

Proof. First we consider the case that $\psi > 0$. Introduce

$$\eta = \sqrt{\mathsf{u}''}\, l, \quad \zeta = |\eta|^{-2}\eta, \quad \xi = \sqrt{\mathsf{u}''}\,\zeta.$$

Note that

$$\langle \xi, l\rangle = \langle \zeta, \eta\rangle = 1, \quad a^{ij}\mathsf{u}''_{ij} = \operatorname{tr} a\mathsf{u}'' = \operatorname{tr}\sqrt{\mathsf{u}''}\, a\sqrt{\mathsf{u}''},$$

and, for $\tilde{a} := \sqrt{\mathsf{u}''}\, a\sqrt{\mathsf{u}''}$,

$$\operatorname{tr}\tilde{a} \geq |\zeta|^{-2}\langle \tilde{a}\zeta, \zeta\rangle = \langle \mathsf{u}''l, l\rangle\langle a\xi, \xi\rangle.$$

It follows that $\operatorname{tr} a(\alpha)\mathsf{u}'' \geq \langle \mathsf{u}''l, l\rangle\langle a(\alpha)\xi, \xi\rangle$. Taking the supremums on both sides and recalling that $\langle l, \xi\rangle = 1$, we obtain that

$$F_0(\mathsf{u}'') \geq \langle \mathsf{u}''l, l\rangle F_0(\xi\xi^*) \geq \langle \mathsf{u}''l, l\rangle\mu(l) \geq \langle \mathsf{u}''l, l\rangle\mu.$$

This proves the lemma if $\psi > 0$. In case $\psi \geq 0$ it suffices to use continuity.

In the remaining case $\psi < 0$. Observe that by the above, the inequality $\mathsf{u}'' - \psi I \geq 0$ implies that

$$F_0(\mathsf{u}'') \geq F_0(\mathsf{u}'' - \psi I) + \inf_\alpha \psi \operatorname{tr} a(\alpha)$$

$$\geq \mu(l)\big[\langle \mathsf{u}''l, l\rangle - \psi\big] - \psi_- \sup_\alpha \operatorname{tr} a(\alpha)$$

$$\geq \mu\big[\langle \mathsf{u}''l, l\rangle - \psi\big] - \psi_- F_0(I).$$

This shows that the assertions of the lemma are true if $\psi < 0$ as well. The lemma is proved. □

11. THEOREM. *(a) Let $\mu(l) > 0$ for a unit vector l. Then there exists the Sobolev derivative $D_{(l)(l)}v$ which is bounded in $\mathbb{R}^d$. Moreover, given a constant $N \geq 0$ (for instance $N = \delta^{-1}K_2$) such that $v + N|x|^2$ is convex in $\mathbb{R}^d$, the inequalities*

$$\mu(l)D_{(l)(l)}v \leq 2N \sup_\alpha \operatorname{tr} a(\alpha) + |Dv| \sup_\alpha \big|b(\alpha)\big|$$

$$+ |v| \sup_\alpha \big|c(\alpha)\big| + \sup_\alpha \big|f(\alpha)\big|, \tag{13}$$

$$D_{(l)(l)}v \geq -2N$$

hold almost everywhere.

(b) If $\mu > 0$, then all Sobolev derivatives $D_{ij}v$ exist and are bounded on $\mathbb{R}^d$. Furthermore, $D_{ij}v = D_{ij}v^{(0)}$ (a.e.), and almost everywhere in $\mathbb{R}^d$,

$$F\big(v(x), Dv(x), D^2v(x), x\big) = 0.$$

Proof. According to Theorem 5 one can choose a measure ν, with respect to which $D_{ij}v(dx)$ for $i, j = 1, ..., d$ and Lebesgue measure are absolutely continuous. Introduce

$$\mathsf{u}''_{ij} = D_{ij}v(dx)/\nu(dx), \quad \rho = dx/\nu(dx).$$

By Theorem 7, for any $\alpha \in A$,

$$a^{ij}(\alpha)\mathsf{u}''_{ij} \leq \Big(|Dv| \sup_\alpha \big|b(\alpha)\big| + |v| \sup_\alpha \big|c(\alpha)\big| + \sup_\alpha \big|f(\alpha)\big|\Big)\rho =: g\rho \tag{14}$$

ν-almost everywhere.

Next, the set $\big\{a(\alpha) : \alpha \in A\big\}$ is separable as a subset of an appropriate d^2-dimensional Euclidean space. Let $\big\{a(\alpha_n)\big\}$ be a countable everywhere dense subset of $\big\{a(\alpha) : \alpha \in A\big\}$. For any matrix u'' we have

$$F_0(\mathsf{u}'') = \sup_n a^{ij}(\alpha_n)\mathsf{u}''_{ij}.$$

Since the countable intersection of sets of full measure is a set of full measure and (14) hold on a set of full ν-measure for any $\alpha = \alpha_n$, (14) also hold on a set of full ν-measure for all α_n at once. In light of (14) we now obtain $F_0(\mathsf{u}'') \leq g\rho$ (ν-a.e.). On the other hand,

$$D_{(l)(l)}v(dx) + 2N\,dx \geq 0 \quad \text{and} \quad \mathsf{u}''_{ij}l^i l^j \geq -2N\rho$$

(ν-a.e.). It follows by Lemma 10 that (ν-a.e.)

$$-2N\rho \le \mathsf{u}''_{ij}l^i l^j \le \mu^{-1}(l)g\rho + \mu^{-1}(l)2N\rho \sup_\alpha \operatorname{tr} a(\alpha).$$

Coming back from the densities to the original measures, we see that

$$-2N dx \le D_{ij}v(dx)l^i l^j \le \mu^{-1}(l)\big[g + 2N \sup_\alpha \operatorname{tr} a(\alpha)\big]\, dx. \tag{15}$$

This shows that $D_{ij}v(dx)l^i l^j$ is absolutely continuous (with respect to dx) and since the density of this measure is the Sobolev derivative $D_{(l)(l)}v$, the latter exists indeed. Assertion (a) is proved because now (15) and (13) are equivalent.

To prove (b) observe that

$$D_{ij}v = \frac{1}{2}\Big[D_{(l^{ij}_+)(l^{ij}_+)} - D_{(l^{ij}_-)(l^{ij}_-)}\Big]v,$$

where

$$l^{ij}_\pm = 2^{-\frac{1}{2}}(e_i \pm e_j)$$

and e_i is the ith basis vector. Therefore, all Sobolev derivatives $D_{ij}v$ are bounded. They can certainly be used as the densities of measures $D_{ij}v(dx)$. Hence $D_{ij}v = D_{ij}v^{(0)}$ (a.e.) and the last assertion now follows from Theorem 6.

The theorem is proved. □

12. REMARK. It is easy to see that $\mu > 0$ if

$$\sup_\alpha \inf_{|\xi|=1} a^{ij}(\alpha)\xi^i\xi^j > 0$$

that is if there is $\alpha \in A$ for which

$$\inf_{|\xi|=1} a^{ij}(\alpha)\xi^i\xi^j > 0. \tag{16}$$

Condition (16) cannot be expressed in terms of F defining (1). However, at the same time, (16) is "essentially equivalent" to $\mu > 0$.

To show this suppose that $\mu > 0$. Since $a(\alpha)$ is bounded on A, $a^{ij}(\alpha)\xi^i\xi^j$ are Lipschitz continuous with respect to ξ on the unit sphere with Lipschitz constant independent of α, that is, there exists a constant N such that

$$\big|a^{ij}(\alpha)\xi^i\xi^j - a^{ij}(\alpha)\eta^i\eta^j\big| \le N|\xi - \eta|$$

for all $\alpha \in A$ and $|\xi| = |\eta| = 1$. Let $\{\xi_1, ..., \xi_n\}$ be a $4^{-1}\mu N^{-1}$-net on the unit sphere of $\mathbb{R}^d$ and, for every $k = 1, ..., n$, let α_k be such that $a^{ij}(\alpha_k)\xi^i_k\xi^j_k \ge \mu/2$. Introduce

$$\hat{a} = \frac{a(\alpha_1) + ... + a(\alpha_n)}{n}.$$

Observe that, if $\xi \in \mathbb{R}^d$, $|\xi| = 1$, and ξ_k chosen in such a way that $|\xi - \xi_k| \le 4^{-1}\mu N^{-1}$, then

$$\hat{a}^{ij}\xi^i\xi^j \ge \frac{1}{n}\big(a^{ij}(\alpha_k)\xi^i\xi^j - a^{ij}(\alpha_k)\xi^i_k\xi^j_k\big) + \frac{1}{n}a^{ij}(\alpha_k)\xi^i_k\xi^j_k$$

$$\ge -\frac{1}{n}\frac{1}{4}\mu + \frac{1}{n}\frac{1}{2}\mu = \frac{1}{4n}\mu. \tag{17}$$

Finally, add an extra element $\hat{\alpha} \notin A$ to the set A and define $a^{ij}(\hat{\alpha}) = \hat{a}^{ij}$,

$$b(\hat{\alpha}) = \frac{b(\alpha_1) + ... + b(\alpha_n)}{n}, \quad c(\hat{\alpha}) = \frac{c(\alpha_1) + ... + c(\alpha_n)}{n},$$

$$f^{\hat{\alpha}} = \frac{f^{\alpha_1} + ... + f^{\alpha_n}}{n} - 1.$$

Now, if we replace A with $\{\hat{\alpha}\} \cup A$, condition (16) will be satisfied and, in addition, the function F will not be affected. We see that, if $\mu > 0$, it is always possible, to extend the set A so that (16) holds for the extended A and this extension does not affect (1).

13. REMARK. To evaluate $F(\mathsf{u}, x)$, it suffices to know u' and $\mathsf{u}''_{ij} l^i l^j$ only for all $l \in \{l : \mu(l) > 0\}$. In other words,

$$\mathsf{u}''_{ij} l^i l^j = \mathsf{v}''_{ij} l^i l^j \quad \forall l \in \{l : \mu(l) > 0\} \Longrightarrow F(\mathsf{u}', \mathsf{u}'', x) = F(\mathsf{u}', \mathsf{v}'', x). \qquad (18)$$

In other words, the condition $\mu(l) > 0$ describes all second-order derivatives $D_{(l)(l)}v$ which are "actually" present in $F[v]$.

To prove this, it suffices to show that under the condition in (18) we have $a^{ij}(\alpha)\mathsf{u}''_{ij} = a^{ij}(\alpha)\mathsf{v}''_{ij}$ for all $\alpha \in A$. Introduce $\hat{\mathsf{u}} = \mathsf{u} - \mathsf{v}$ and take $\alpha \in A$. Obviously,

$$\mu(l) \geq \inf\{a^{ij}(\alpha)\xi^i\xi^j : \xi \in \mathbb{R}^d, \langle l, \xi\rangle = 1\},$$

and it is easy to see that the right-hand side is strictly positive whenever $l \in \sqrt{a(\alpha)}\,\mathbb{R}^d$, $l \neq 0$. It follows that

$$\hat{\mathsf{u}}''_{ij} l^i l^j = 0$$

on $\sqrt{a(\alpha)}\,\mathbb{R}^d$, specifically, $\hat{\mathsf{u}}''_{ij} l^i_k(\alpha) l^j_k(\alpha) = 0$ for all unit eigenvectors $l_k(\alpha)$ of $a(\alpha)$ corresponding to nonzero eigenvalues $a_k(\alpha)$. Since

$$a^{ij}(\alpha) = a_k(\alpha) l^i_k(\alpha) l^j_k(\alpha),$$

we have $a^{ij}(\alpha)\hat{\mathsf{u}}''_{ij} = 0$ and $a^{ij}(\alpha)\mathsf{u}''_{ij} = a^{ij}(\alpha)\mathsf{v}''_{ij}$, which implies the desired result.

For instance, if $F(\mathsf{u}'') = \mathsf{u}''_{11} + 2\mathsf{u}''_{12} + \mathsf{u}''_{22}$, when $F[u] = D_{11}u + 2D_{12}u + D_{22}u$, the derivatives $D_{11}u$ and $D_{22}u$ are not "actually" present in $F[u]$.

In conclusion, we give some examples of (1) to which the above results are applicable. If A consists of two points, equation (1) is equivalent to the set of three relations:

$$\mathcal{L}^{\alpha_1}v + f^{\alpha_1} \leq 0, \quad \mathcal{L}^{\alpha_2}v + f^{\alpha_2} \leq 0, \quad (\mathcal{L}^{\alpha_1}v + f^{\alpha_1})(\mathcal{L}^{\alpha_2}v + f^{\alpha_2}) = 0.$$

In a particular case when $\mathcal{L}^{\alpha_2} = -\delta$, these reduce to what is typical for free-boundary problems arising in the theory of differential equations and the theory of optimal stopping of stochastic diffusion processes. In even more particular one-dimensional case, if the operator $\mathcal{L}^{\alpha_1}$ is nondegenerate, and $\mathcal{L}^{\alpha_2} = -\delta$, then $\mu > 0$ and one easily sees that the second derivatives of v are generally discontinuous though bounded, in a good agreement with Theorem 11.

The equation $|Dv| - \delta v = f$ is a one more example of equation (1) because

$$|Dv| = \sup\{\xi^i D_i v : |\xi| = 1\}.$$

Lemma 4 guarantees only an estimate from below for the second-order derivatives of v for this equation. The set of three relations

$$\Delta v - \delta v + f^{\alpha_1} \leq 0, \quad |Dv| - \delta v \leq f^{\alpha_2},$$

$$(\Delta v - \delta v + f^{\alpha_1})\big(|Dv| - \delta v - f^{\alpha_2}\big) = 0$$

can be easily combined into a single equation (1). Here $\mu > 0$ and all second-order generalized derivatives of v are bounded.

We give these examples only as an illustration. In Section 2.2, the reader will see the most important applications of the above results.

CHAPTER 2

Estimates in L_p for solutions of the Monge-Ampère type equations

In this chapter we present part of general tools of the theory which will allow us to prove later the Aleksandrov estimates, the maximum principle, and uniqueness theorems. The whole exposition revolves around elliptic and parabolic Monge-Ampère type equations considered in the whole space. The results here are based on the papers [**55, 56**] and closely follow parts of Chapter 3 of [**61**].

1. Some properties of λ-concave functions

If a function $u \in C^{\infty}_{\text{loc}}(\mathbb{R}^d)$ and a number $\lambda > 0$ are such that for all $x \in \mathbb{R}^d$ we have

$$\big(\lambda u(x)\delta_{ij} - D_{ij}u(x)\big) \geq 0,$$

then we call u a λ-concave function. Throughout the section we fix a λ-concave function u.

This notion generalizes the notion of concave function and has the advantage that there exist sufficiently many λ-concave *bounded* functions whereas any bounded concave function on $\mathbb{R}^d$ is just constant. Geometrically, the concavity of a function is expressed by saying that any chord of its graph lies below the graph. It turns out that the λ-concavity can be similarly expressed by using the function

$$c_1 e^{t\sqrt{\lambda}} + c_2 e^{-t\sqrt{\lambda}}$$

($t \in (-\infty, \infty)$) which is a solution to the equation $u_{tt} - \lambda u = 0$.

1. Lemma. (*a*) *Given* $x, y \in \mathbb{R}^d$, $x \neq y$, *and* $r = |x-y|/2$, *the inequality*

$$u\Big(\frac{x+y}{2} + t\frac{y-x}{2r}\Big) \geq u(x)\frac{\sinh\big((r-t)\sqrt{\lambda}\big)}{\sin(2r\sqrt{\lambda})} + u(y)\frac{\sinh\big((r+t)\sqrt{\lambda}\big)}{\sinh(2r\sqrt{\lambda})}, \tag{1}$$

is satisfied for $t \in [-r, r]$. *Specifically,* (*if* $t = 0$)

$$u\Big(\frac{x+y}{2}\Big) \geq \frac{u(x)+u(y)}{2\cosh(r\sqrt{\lambda})};$$

(*b*) $u(y) \geq u(x)e^{-\sqrt{\lambda}|x-y|}$ *for all* $x, y \in \mathbb{R}^d$;
(*c*) $\big|Du(x)\big| \leq \sqrt{\lambda}u(x)$ *for all* $x \in \mathbb{R}^d$;
(*d*) *if* $u(y) = 0$ *for some* y, *then* $u \equiv 0$.

Proof. Let $\phi(t)$ be the left-hand side of (1) and $\psi(t)$ be its right-hand side. Notice that

$$\lambda\varphi - \varphi_{tt} = (\lambda u\delta_{ij} - D_{ij}u)\frac{y^i - x^i}{|y-x|}\frac{y^j - x^j}{|y-x|} \geq 0,$$

$\lambda\psi - \psi_{tt} = 0$, $\varphi(\pm r) = \psi(\pm r)$. By the maximum principle, $\varphi \geq \psi$ on $[-r, r]$. We have proved (a).

Substitute $x + s(y-x)/|y-x|$ in (1) in place of y and set $t = |y-x| - s/2$ with $s \geq |y-x|$. This yields

$$u(y) \geq u(x)\frac{\sinh\left[(s - |y-x|)\sqrt{\lambda}\right]}{\sinh(s\sqrt{\lambda})}$$

$$+ u\Big(x + s\frac{y-x}{|y-x|}\Big)\frac{\sinh\left(|y-x|\sqrt{\lambda}\right)}{\sinh(s\sqrt{\lambda})}.$$

The last term is nonnegative and, by setting $s \to \infty$, we come to (b). Assertion (b) implies that for any $l \in \mathbb{R}^d$ with $|l| = 1$

$$D_{(l)}u(x) := \lim_{t\to 0}\frac{u(x) - u(x-lt)}{t} \leq u(x)\lim_{t\to 0}\frac{1 - e^{-t\sqrt{\lambda}}}{t} = \sqrt{\lambda}u(x).$$

It follows that $\left|Du(x)\right| \leq \sqrt{\lambda}u(x)$. Assertion (b), certainly, implies (d). The lemma is proved. □

2. Remark. One easily proves that, if u is smooth enough and (1) holds for all $x, y \in \mathbb{R}^d$, then $(\lambda u\delta_{ij} - D_{ij}u) \geq 0$.

We need certain information concerning the matrix $(\lambda u\delta_{ij} - D_{ij}u)$. For any $\mathsf{u}'' \in \mathbb{S}$ and $m = 0, ..., d$, we introduce the following, in which I is the unit $d \times d$ matrix:

$$\mathcal{D}(\mathsf{u}'') = \det(\mathsf{u}''), \quad c^m(\mathsf{u}'') = \frac{1}{(d-m)!}\frac{\partial^{d-m}}{\partial\mu^{d-m}}\mathcal{D}(\mu I - \mathsf{u}'')\big|_{\mu=0},$$

$$c^m_{ij}(\mathsf{u}'') = D_{\mathsf{u}''_{ij}}c^m(\mathsf{u}''), \quad c^m[u] = c^m(D^2u), \quad c^m_{ij}[u] = c^m_{ij}(D^2u).$$

3. Lemma. *(a)* $\mathcal{D}(\lambda I - \mathsf{u}'') = \sum_{m=0}^d \lambda^{d-m}c^m(\mathsf{u}'')$,

(b) For any $v \in (-\infty, \infty)$,

$$c^m(\mathsf{u}'') = \sum_{k=0}^m \binom{d-k}{m-k}(-1)^{m-k}v^{m-k}c^k(-vI + \mathsf{u}'').$$

(c) If the matrix u'' *is symmetric and negative, the matrices* $\left(c^m_{ij}(\mathsf{u}'')\right)$ *are also symmetric and negative.*

(d) $D_{\mathsf{u}''_{ir}\mathsf{u}''_{kj}}\mathcal{D} = -D_{\mathsf{u}''_{ij}\mathsf{u}''_{kr}}\mathcal{D}$ *for any* $i, j, k, r = 1, ..., d$, *where* $D_{\mathsf{u}''_{ir}\mathsf{u}''_{kj}}\mathcal{D}$ *are the mixed derivatives of* $\mathcal{D}(\mathsf{u}'')$ *with respect to* $\mathsf{u}''_{ir}, \mathsf{u}''_{kj}$.

Proof. The formula in (a) is just a usual formula for a polynomial of degree d in λ. It follows from (a) that for any μ

$$\mathcal{D}(\mu I - \mathsf{u}'') = \mathcal{D}\big((\mu - v)I + (vI - \mathsf{u}'')\big)$$

$$= \sum_{k=0}^{d} (\mu - v)^{d-k} c^k(-vI + \mathsf{u}'').$$

By differentiating this with respect to μ we get (b).

Coming to (c) observe that, if $m = 0$, obviously, $c_{ij}^m = 0$. If $m \geq 1$, $c^m(\mathsf{u}'')$ is the sum of the m-order principal minors of $-\mathsf{u}''$. Note also that the matrix with entries

$$D_{\mathsf{u}''_{ij}} \det(\mathsf{u}'')$$

is positive if u'' is symmetric and positive (it equals $(\mathsf{u}'')^{-1} \det \mathsf{u}''$ if $\det \mathsf{u}'' > 0$). The same applies to each mth order principal minor of $-\mathsf{u}''$, which (minor) is positive as well. Therefore

$$\big(c_{ij}^m(\mathsf{u}'')\big) = \big(D_{\mathsf{u}''_{ij}} c^m(\mathsf{u}'')\big) \leq 0.$$

By analyzing the definition of the determinant we easily come to (d). This proves the lemma. □

4. LEMMA. *Let a sequence $a_0, ..., a_d$ of numbers be such that*

$$\sum_{i=0}^{n} \binom{n}{i} (-1)^i a_{d-n+i} \geq 0 \tag{2}$$

for $n = 0, ..., d-1$. Then, for every $x \in \mathbb{R}^d$, the symmetric matrix with entries

$$\sum_{m=0}^{d} a_m \lambda^{d-m} u^{d-m} c_{ij}^m[u] = \sum_{m=1}^{d} a_m \lambda^{d-m} u^{d-m} c_{ij}^m[u]$$

is negative definite.

Proof. Introduce

$$b_k = \sum_{m=k}^{d} \binom{d-k}{m-k} (-1)^{m-k} a_m, \quad k = 1, ..., d.$$

By assumption, $b_k \geq 0$. By Lemma 3 (b)

$$\sum_{m=0}^{d} a_m u^{d-m} \lambda^{d-m} c_{ij}^m[u]$$

$$= \sum_{m=1}^{d} \sum_{k=1}^{m} a_m u^{d-m} \lambda^{d-m} \binom{d-k}{m-k} (-1)^{m-k} (\lambda u)^{m-k} c_{ij}^k(-\lambda u I + D^2 u)$$

$$= \sum_{k=1}^{d} b_k (\lambda u)^{d-k} c_{ij}^k(-\lambda u I + D^2 u).$$

Applying Lemma 3 (c) finishes the proof. □

5. REMARK. The sequence of numbers $a_0, a_1, ...,$ such that

$$\sum_{i=0}^{n} \binom{n}{i} (-1)^i a_{k+i} \geq 0$$

for all $n, k \geq 0$, is called *completely monotone* (see Feller [**42**], Chapter 7). Hausdorff's theorem says that, if we are given a completely monotone sequence $a_0, a_1, ...,$ one can find a positive measure F on $[0, 1]$ such that

$$a_m = \int_{[0,1]} t^m \, F(dt)$$

for all $m \geq 0$. For completely monotone sequences $a_0, a_1, ...,$ the proof of the lemma takes only one line:

$$\sum_{m=0}^{d} a_m \lambda^{d-m} u^{d-m} c_{ij}^m[u] = -\int_{[0,1]} t^d D_{\mathrm{u}_{ij}''}(\lambda t^{-1} u I - D^2 u) \, F(dt).$$

6. REMARK. Let $a \in C^{d-1}([1, d])$ be such that

$$(-1)^n a^{(n)}(t) \geq 0 \tag{3}$$

for $n = 0, 1, ..., d-1$, $t \in [d-n, d]$. Define $a_n = a(n)$, $n = 0, ..., d$. Then the sequence a_n satisfies the condition of Lemma 4.

Indeed, for $n = 0$, (2) and (3) coincide. Generally, the left-hand side of (2) equals $(-1)^n$ times the nth difference for $a(t)$ on $[d-n, d]$. This and the well-known formula,

$$\sum_{i=0}^{n} \binom{n}{i} (-1)^i a_{d-n+1}$$
$$= (-1)^n \int_0^1 dt_1 ... \int_0^1 a^{(n)}(d - n + t_1 + ... + t_n) \, dt_n \geq 0,$$

yield (2).

The next lemma follows from Lemma 4 and Remark 6 because $(p-t-1)/(t+1)$ and $t^{-1}(t+1)^{-1}$ satisfy (3) if $p \geq d+1$.

7. LEMMA. *If $p \geq d+1$, $p_1 \geq d$, and $u > 0$, the matrices with entries*

$$\sum_{m=1}^{d} \frac{p-m+1}{m+1} \lambda^{d-m} u^{p-m-2} c_{ij}^m[u],$$

$$\sum_{m=1}^{d} \frac{1}{m(m+1)} \lambda^{d-m} u^{p-m-1} c_{ij}^m[u],$$

$$\sum_{m=1}^{d} \frac{p_1 - m}{m} \lambda^{d-m} u^{p_1-m-1} c_{ij}^m[u]$$

are negative definite.

The following is a particular case of the lemma in Section 12, Chapter 5 of Dunford and Schwartz [**31**].

8. LEMMA. *Let $w(x)$ be a three times continuously differentiable function on $\mathbb{R}^d$. Then for all $m = 0, ..., d$ and $i = 1, ..., d$ we have on $\mathbb{R}^d$ that*

$$\frac{\partial}{\partial x^j} c^m_{ij}[w] = \frac{\partial}{\partial x^j} c^m_{ji}[w] = 0. \tag{4}$$

Proof. Obviously, for $\mu \in (-\infty, \infty)$ and $v = \mu|x|^2/2$ we have

$$\sum_{m=0}^{d} \mu^{d-m} \frac{\partial}{\partial x^j} c^m_{ij}[w] = \frac{\partial}{\partial x^j} D_{\mathsf{u}''_{ij}} \mathcal{D}(\mu I - D^2 w)$$

$$= \frac{\partial}{\partial x^j} D_{\mathsf{u}''_{ij}} \mathcal{D}\big(D^2(v - w)\big)$$

It follows that we need to prove (4) only for $m = d$. Owing to Lemma 3 (d),

$$\frac{\partial}{\partial x^j} D_{\mathsf{u}''_{ij}} \mathcal{D}(D^2 w) = D_{\mathsf{u}''_{ij}\mathsf{u}''_{kr}} \mathcal{D}(D^2 w) D_{krj} w$$

$$= -D_{\mathsf{u}''_{ir}\mathsf{u}''_{kj}} \mathcal{D}(D^2 w) D_{kjr} w = -\frac{\partial}{\partial x^r} D_{\mathsf{u}''_{ir}} \mathcal{D}(D^2 w).$$

This proves the equality between the extreme terms in (4). The remaining equality is proved similarly. The lemma is proved. □

Here is the main result of this section. For $R \in (0, \infty)$, set

$$B_R = \big\{x \in \mathbb{R}^d : |x| < R\big\}.$$

9. THEOREM. *Let constants $T, K \in [0, \infty)$ and let $u(t, x)$ be a function on $\mathbb{R}^{d+1}$ which is infinitely differentiable with respect to (t, x), λ-concave in x (for any t), equal to zero for $t \geq T$ and is such that*

$$\lambda u - \partial_t u \geq 0, \quad |u| + |\partial_t u| \leq K e^{-|x|/K}, \tag{5}$$

$$\sum_{i=1}^{d} |\partial_t D_i u| + \sum_{i,j=1}^{d} \big(|D_{ij} u| + |\partial_t D_{ij} u|\big) \leq K.$$

Let $p \geq d + 1$. Then for any $(t, y) \in \mathbb{R}^{d+1}$,

$$p\lambda^{-d} e^{\lambda pt} \int_t^\infty ds \int_{\mathbb{R}^d} e^{-\lambda ps} u^{p-d-1} (\lambda u - \partial_s u) \mathcal{D}(\lambda u I - D^2 u)\, dx$$

$$\geq \big\|u(t, \cdot)\big\|^p_{L_p(\mathbb{R}^d)} \geq \kappa_d d! \big(p\sqrt{\lambda}\big)^{-d} u^p(t, y), \tag{6}$$

where κ_d is the volume of B_1 and $u^0 := 1$ (for $p = d + 1$).

Proof. Introduce $v(t, x) = u(t, x) e^{-\lambda t}$. Then the first term in (6) is rewritten as $p\lambda^{-d} e^{\lambda pt} A$, where

$$A = -\int_t^\infty ds \int_{\mathbb{R}^d} v^{p-d-1} \mathcal{D}(\lambda v I - D^2 v) \partial_s v\, dx$$

$$= -\sum_{m=0}^{d} \int_{\mathbb{R}^d} \int_t^\infty v^{p-m-1} \lambda^{d-m} c^m[v] \partial_s v\, ds dx =: \sum_{m=0}^{d} A_m. \tag{7}$$

We integrate by parts with respect to s in each term of the sum and use the fact that, owing to (5), we get absolutely convergent integrals with respect to $dsdx$ to which we can apply Fubini's theorem. Then we obtain

$$A_m = \frac{1}{p-m}\int_{\mathbb{R}^d} \lambda^{d-m} v^{p-m}(t,x) c^m[v](t,x)\,dx + \frac{1}{p-m}\int_t^\infty ds \int_{\mathbb{R}^d} \lambda^{d-m} v^{p-m} c^m_{ij}[v]\partial_s D_{ij} v\,dx.$$

Now integrate the latter integral by parts twice with respect to x^i and x^j, bringing D_{ij} from $\partial_s v$ to $v^{p-m}c^m_{ij}[v]$. Observe that by Lemma 8 we need not differentiate $c^m_{ij}[v]$ with respect to x^i and x^j. Also use (5) which guarantees that integrated out terms vanish. Finally, apply the Euler theorem on homogeneous functions by which $c^m_{ij}[v]D_{ij}v = mc^m[v]$. This yields

$$A_m = \frac{1}{p-m}\int_{\mathbb{R}^d} \lambda^{d-m} v^{p-m}(t,x) c^m[v](t,x)\,dx - mA_m + (p-m-1)\int_t^\infty ds \int_{\mathbb{R}^d} \lambda^{d-m} v^{p-m-2}\partial_s v D_i v D_j v\, c^m_{ij}[v] dx. \tag{8}$$

It is worth saying that, if for some s, $v(s,\cdot) \not\equiv 0$, then $v^{p-m-2}|D_i v D_j v| \le \lambda v^{p-m}$ by Lemma 1 (c), and if $v(s,\cdot)\equiv 0$, the last integral over $\mathbb{R}^d$ in (8) is set to be zero.

In (8) we have an equation about A_m from which we find it and substitute the result into (7). We also use Lemma 7 and take into account the fact that $\partial_s v \le 0$. Then we obtain

$$\sum_{m=1}^d \frac{p-m-1}{m+1}\lambda^{d-m} v^{p-m-2}\partial_s v D_i v D_j v\, c^m_{ij}[v] \ge 0.$$

It follows that

$$A \ge \sum_{m=0}^d \frac{1}{(p-m)(m+1)}\int_{\mathbb{R}^d} \lambda^{d-m} v^{p-m}(t,x) c^m[v](t,x)\,dx. \tag{9}$$

It turns out that the sum of terms in (9) corresponding to $m \ge 1$ is nonnegative. To see this replace $c^m[v]$ by $m^{-1}c^m_{ij}[v]D_{ij}v$ and integrate by parts with respect to x^i. Then use the inequality (see Lemma 7)

$$-\sum_{m=1}^d \frac{1}{m(m+1)}\lambda^{d-m} v^{p-m-1} D_i v D_j v\, c^m_{ij}[v] \ge 0.$$

Hence, A is greater than the first term in (9), i.e.

$$A \ge p^{-1}\lambda^d \|v(t,\cdot)\|^p_{L_p(\mathbb{R}^d)}.$$

This proves the first estimate in (6). Lemma 1 (b) yields the second one:

$$\int_{\mathbb{R}^d} u^p(t,x)\,dx \ge u^p(t,y)\int_{\mathbb{R}^d} e^{-p\sqrt{\lambda}|x-y|}\,dx = u^p(t,y)\kappa_d d!\big(p\sqrt{\lambda}\big)^{-d}.$$

The theorem is proved. ☐

10. THEOREM. *Suppose that* $p \ge d$, $K > 0$, *the function* u *is* $\lambda-$*concave and*

$$u(x) \le Ke^{-|x|/K}, \quad \sum_{i,j} \big|D_{ij}u(x)\big| \le K \tag{10}$$

for all $x \in \mathbb{R}^d$. *Then for any* $y \in \mathbb{R}^d$,

$$\int_{\mathbb{R}^d} u^{p-d}\mathcal{D}(\lambda uI - D^2u)\,dx \ge \lambda^d \int_{\mathbb{R}^d} u^p\,dx \ge u^p(y)\lambda^{d/2}\kappa_d d!p^{-d}, \tag{11}$$

where (for $p = d$*)* $u^0 := 1$.

Proof. First let $p > d$. This assumption in combination with (10) is used while integrating by parts and guarantees that the integrated out terms vanish. We have

$$\int_{\mathbb{R}^d} u^{p-d}\mathcal{D}(\lambda uI - D^2u)\,dx = \sum_{m=0}^{d}\int_{\mathbb{R}^d} \lambda^{d-m}u^{p-m}c^m[u]\,dx =: \sum_{m=0}^{d} A_m.$$

For $m \ge 1$, as in the proof of Theorem 9,

$$\begin{aligned} A_m &= \frac{1}{m}\int_{\mathbb{R}^d} \lambda^{d-m}u^{p-m}D_{ij}u\,c^m_{ij}[u]\,dx \\ &= -\frac{p-m}{m}\int_{\mathbb{R}^d} \lambda^{d-m}u^{p-m-1}D_iuD_ju\,c^m_{ij}[u]\,dx. \end{aligned}$$

It follows by Lemma 7 that

$$\begin{aligned} &\int_{\mathbb{R}^d} u^{p-d}\mathcal{D}(\lambda uI - D^2u)\,dx \\ &\quad = \lambda^d\int_{\mathbb{R}^d} u^p\,dx - \int_{\mathbb{R}^d}\sum_{m=1}^{d}\frac{p-m}{m}\lambda^{d-m}u^{p-m-1}D_iuD_ju\,c^m_{ij}\,dx \\ &\quad \ge \lambda^d\int_{\mathbb{R}^d} u^p\,dx. \end{aligned}$$

The last integral dominates $u(x)$ as in the proof of Theorem 9 and to finish the proof we only need to consider the case $p = d$.

Obviously, it suffices to concentrate on the case that

$$\int_{\mathbb{R}^d} \mathcal{D}(\lambda uI - D^2u)\,dx < \infty, \quad u \not\equiv 0.$$

If $u \not\equiv 0$, then $u(x) > 0$ for all x and, by the Lebesgue dominated convergence theorem,

$$\int_{\mathbb{R}^d} \mathcal{D}(\lambda uI - D^2u)\,dx = \lim_{p\downarrow d}\int_{\mathbb{R}^d} u^{p-d}\mathcal{D}(\lambda uI - D^2u)\,dx.$$

Therefore, by passing to the limit in (11) as $p \downarrow d$ we arrive at the desired result. The theorem is proved. □

2. Monge-Ampère type elliptic and parabolic equations in the whole space

Here we derive certain estimates, used in many places in the book, of the solutions to elliptic and parabolic equations (13) and (20) for λ-concave functions, which are analogs of the Monge-Ampère equations. The existence of solutions of (13) and (20) is proved by using the results from Chapter 1.

Throughout the section $\lambda > 0$ is a fixed number. Introduce $\mathbb{S}_+$ as the set of nonnegative symmetric $d \times d$ matrices and let

$$\mathbb{S}_+^{\leq 1} = \{\alpha \in \mathbb{S}_+ : \operatorname{tr} \alpha \leq 1\}, \quad \mathbb{S}_+^1 = \{\alpha \in \mathbb{S}_+ : \operatorname{tr} \alpha = 1\}.$$

For $\alpha \in \mathbb{S}_+^{\leq 1}$, set

$$r(\alpha) := 1 - \operatorname{tr} \alpha.$$

1. LEMMA. *(a) If $\alpha, \beta \in \mathbb{S}_+$, then*

$$\frac{1}{d} \operatorname{tr} \alpha\beta \geq \sqrt[d]{\det \alpha}\, \sqrt[d]{\det \beta}. \tag{1}$$

(b) If $\beta \in \mathbb{S}_+$, then

$$\sqrt[d]{\det \beta} = \frac{1}{d} \inf_{\mathbb{S}_+ \cap \{\det \alpha = 1\}} \operatorname{tr} \alpha\beta, \tag{2}$$

and $\sqrt[d]{\det \beta}$ is a concave function on $\mathbb{S}_+$;

(c) If $\alpha \in \mathbb{S}_+$, $i = 1, ..., d-1$, and $c_i(\alpha)$ is the determinant of the submatrix which results from crossing out the first i rows and i columns of the matrix α, then (Hadamard's inequality)

$$\det \alpha \leq \alpha^{11} \cdot ... \cdot \alpha^{ii} c_i(\alpha), \tag{3}$$

where there is no summation over i.

Proof. (a) As usual, if β is a symmetric nonnegative matrix, $\sqrt{\beta}$ is its symmetric nonnegative square root. Observe that the inequality between the arithmetic and geometric means implies that $\operatorname{tr} \alpha \geq d \sqrt[d]{\det \alpha}$ for every $\alpha \in \mathbb{S}_+$.

It follows that

$$\operatorname{tr} \alpha\beta = \operatorname{tr} \sqrt{\beta}\alpha\sqrt{\beta} \geq d \sqrt[d]{\det(\sqrt{\beta}\alpha\sqrt{\beta})} = d \sqrt[d]{\det \alpha \det \beta}.$$

We have proved (a). Coming to (b) note that (2) represents $\sqrt[d]{\det \alpha}$ as the inf of linear functions, which implies its concavity. Next, owing to (1), (2) holds with $\leq$ in place of $=$. The opposite inequality holds in light of

$$\lim_{\varepsilon \downarrow 0} \operatorname{tr} \left[(\beta + \varepsilon I)^{-1} \sqrt[d]{\det(\beta + \varepsilon I)}\, \beta\right]$$

$$= \sqrt[d]{\det \beta} \lim_{\varepsilon \downarrow 0} \operatorname{tr} \left[(\beta + \varepsilon I)^{-1} \beta\right] \leq d \sqrt[d]{\det \beta}.$$

In the proof of (c) one can obviously use the induction on i and, therefore, it suffices to concentrate on $i = 1$. Let $\mathbb{S}'_+$ be the set of $\beta \in \mathbb{S}_+$ such that $\det \beta = 1$

and $\beta^{1i} = 0$ if $i = 1$. It follows from (2) that

$$d\sqrt[d]{\det \alpha} \le \inf_{\beta \in \mathbb{S}'_+} \operatorname{tr} \alpha\beta$$

$$= \inf_{\beta^{11}>0} \inf_{c_1(\beta)=(\beta^{11})^{-1}} \Big(\beta^{11}\alpha^{11} + \sum_{i,j\ge 2} \beta^{ij}\alpha^{ij}\Big)$$

$$= \inf_{\beta^{11}>0} \Big(\beta^{11}\alpha^{11} + (d-1)(\beta^{11})^{-1/(d-1)} \sqrt[d-1]{c_1(\alpha)}\Big)$$

$$= d\sqrt[d]{\alpha^{11} c_1(\alpha)}.$$

The lemma is proved. □

The following lemma shows that the Monge-Ampère type equations are Bellman's equations and this will allow us to prove their solvability.

2. LEMMA. *Let* $\mathsf{u}'' \in \mathbb{S}$ *and let* f *and* u_0 *be some numbers. Then*

(*a*) *For* $d \ge 2$ *we have*

$$\max_{\alpha \in \mathbb{S}^1_+} \big(\operatorname{tr} \alpha \mathsf{u}'' + \sqrt[d]{\det \alpha}\, f\big) = 0, \tag{4}$$

if and only if $\mathsf{u}'' \le 0$, $(f_+)^d = d^d \det(-\mathsf{u}'')$; *this assertion is also trivially true if* $d = 1$ *and* $f \ge 0$;

(*b*) *The equality*

$$\max_{\alpha \in \mathbb{S}^{\le 1}_+} \big(r(\alpha)\mathsf{u}_0 + \operatorname{tr} \alpha \mathsf{u}'' + \sqrt[d+1]{r(\alpha) \det \alpha\, f}\big) = 0, \tag{5}$$

holds if and only if $\mathsf{u}_0 \le 0$, $\mathsf{u}'' \le 0$, *and*

$$(f_+)^{d+1} = (d+1)^{d+1}(-\mathsf{u}_0)\det(-\mathsf{u}'').$$

Proof. Obviously, we may assume that u'' is diagonal. In this case first assume that $f > 0$. Then $\operatorname{tr} \alpha \mathsf{u}'' = \alpha^{ii} \mathsf{u}''_{ii}$ and, by Lemma 1 (c), the maximum in (4) is less than the maximum over the diagonal matrices $\alpha \in \mathbb{S}^1_0$. On the other hand, obviously the opposite inequality holds. Similar argument is valid for (5). It follows that (if $f > 0$), assertions (a) and (b) will be proved if we can show that, for $n \ge 1$,

$$\max\Big(\sum_{i=1}^n \alpha_i u_i + \Big(\prod_{i=1}^n \alpha_i\Big)^{1/n} f \,:\, \alpha_i \ge 0, \sum_{i=1}^n \alpha_i = 1\Big) = 0 \tag{6}$$

if and only if

$$u_i \le 0, \quad f^n = n^n \prod_{i=1}^n (-u_i). \tag{7}$$

If $n = 1$, (6) is obviously equivalent to (7) even if $f \ge 0$. In case $n \ge 2$ we first show that (6) implies (7). For $\alpha_1 = 1$ and $\alpha_2 = ... = \alpha_n = 0$, (6) shows that $u_1 \le 0$. Likewise, $u_i \le 0$ for all i. The assumption $u_1 = 0$ leads to a contradiction because then the expression on the left in (6) is strictly positive, which one shows by comparing the rates of convergence to zero of

$$\sum_{i=2}^{n} tu_i, \quad \Big[(1-(n-1)t)\prod_{i=2}^{n} t\Big]^{1/n} f,$$

as $t \downarrow 0$.

Thus $u_1 < 0$ and $u_i < 0$ for all i. Let $(\hat{\alpha}_1, ..., \hat{\alpha}_n)$ be a point at which the maximum in (6) is achieved. Since $u_i < 0$ for all i, $\hat{\alpha}_i > 0$ for all i and

$$0 = \sum_{i=1}^{n} \hat{\alpha}_i u_i + \Big(\prod_{i=1}^{n} \hat{\alpha}_i\Big)^{1/n} f \le \Big(\prod_{i=1}^{n} \hat{\alpha}_i\Big)^{1/n} \Big(f - n\Big(\prod_{i=1}^{n}(-u_i)\Big)^{1/n}\Big). \tag{8}$$

This shows that the second relation in (7) holds with $\ge$ in place of $=$. The opposite inequality is obtained from (6) for $\alpha_i = u_i^{-1}\big(\sum u_j^{-1}\big)^{-1}$.

We now show that (7) implies (6). In light of (8) and (7) the left-hand side of (6) is negative. On the other hand, observe that $u_i < 0$ for all i, since $f > 0$. Furthermore, for $\hat{\alpha}_i = u_i^{-1}\big(\sum u_j^{-1}\big)^{-1}$, the inequality in (8) becomes an equality. This shows that the left-hand side of (6) is positive.

It remains to treat the case $f \le 0$. For diagonal α with only one diagonal entry positive (4) shows that $\mathsf{u}''_{ii} \le 0$, and then, obviously, at least one of u''_{ii}'s should vanish, that is $\det(-\mathsf{u}'') = 0$. This proves the necessity in (a). The necessity in (b). is proved similarly. The sufficiency in (a) and (b) for $f \le 0$ is almost obvious. The lemma is proved. □

In the following theorem, for a domain Q in $\mathbb{R}^d$, by $C^2(Q)$ we mean the set of twice continuously differentiable functions g on Q for which

$$\|g\|_{C^2(Q)} := \sup_Q \big(|g| + |Dg| + |D^2 g|\big) < \infty.$$

3. THEOREM. *Let $f \in C^2(\mathbb{R}^d)$, $f \ge 0$, and f be a function with compact support. Then there exists a nonnegative function $v(x)$ defined on $\mathbb{R}^d$ such that*

(a) $v \le Ne^{-|x|/N}$ for all x and a constant N; the generalized derivatives $D_i v$ and $D_{ij} v$, $i, j = 1, ..., d$, are bounded on $\mathbb{R}^d$;

(b) for any $\alpha \in \mathbb{S}_+$, (a.e.)

$$-\lambda v \operatorname{tr} \alpha + \alpha^{ij} D_{ij} v + \sqrt[d]{\det \alpha} f \le 0, \quad |Dv| \le \sqrt{\lambda} v,$$

$$(\lambda v \delta_{ij} - D_{ij} v) \ge 0; \tag{9}$$

(c) for any $x \in \mathbb{R}^d$ and $p \ge d$,

$$\|v\|_{L_p(\mathbb{R}^d)} \le \frac{1}{\lambda d} \|f\|_{L_p(\mathbb{R}^d)},$$

$$v(x) \le N\lambda^{d/(2p)-1} \|f\|_{L_p(\mathbb{R}^d)}, \tag{10}$$

where $N = d^{-1}\big(\kappa_d d! p^{-d}\big)^{-1/p}$.

Proof. We take $\mathbb{S}_+^1$ and λ as A and δ, respectively, in Section 1.2 and, for $\alpha \in \mathbb{S}_+^1$, introduce

$$a(\alpha) = \alpha, \quad b(\alpha) = 0, \quad c^\alpha = -\lambda \operatorname{tr} \alpha = -\lambda, \quad f^\alpha = f \sqrt[d]{\det \alpha}.$$

We take v from Section 1.2 and prove that v possesses the properties (a) to (c). Observe that $v \geq 0$ in light of Lemma 1.2.4.

We want to apply Theorems 1.2.5 and 1.2.11 in our situation. Note that $F_0(\mathsf{u}'')$ (see (1.2.10)) is the largest eigenvalue of u'' which implies (see (1.2.12)) that

$$\mu = \inf_{|\xi|=1} F_0(\xi\xi^*) = \inf_{|\xi|=1} |\xi|^2 = 1.$$

Theorems 1.2.5 and 1.2.11 now guarantee that all Sobolev's derivatives Dv and D^2v exist and are bounded on $\mathbb{R}^d$. Furthermore,

$$\max_{\alpha\in\mathbb{S}^1_+} \left(\alpha^{ij}D_{ij}v - \lambda v + \sqrt[d]{\det\alpha} f\right) = 0 \quad \text{(a.e.)}. \tag{11}$$

This certainly proves the first inequality in (9). Using it with $\xi \in \mathbb{R}^d$ and $\alpha^{ij} = \xi^i\xi^j$ shows that the matrix $(\lambda v\delta_{ij} - D_{ij}v)$ is positive (a.e.).

To prove the rest in (b), take a $\zeta \in C_0^\infty(\mathbb{R}^d)$, $\zeta \geq 0$, which integrates to one, and introduce the notation

$$u^{(\gamma)}(x) = u(x) * \gamma^{-d}\zeta(\gamma^{-1}x) \tag{12}$$

for any locally summable function u and $\gamma > 0$. Obviously

$$0 \leq (\lambda v\delta_{ij} - D_{ij}v)^{(\gamma)} = \left(\lambda v^{(\gamma)}\delta_{ij} - D_{ij}(v^{(\gamma)})\right)$$

and, therefore, $v^{(\gamma)}$ is λ-concave. Therefore, by Lemma 1.1 (c)

$$\left|D(v^{(\gamma)})\right| \leq \sqrt{\lambda} v^{(\gamma)}.$$

By sending $\gamma \downarrow 0$, we get $|Dv| \leq \sqrt{\lambda} v$ (a.e.), which proves (b).

We now deal with (a). It is not hard to find a constant N_0 such that for all $x \in \mathbb{R}^d$ we have

$$f(x) \leq N_0\lambda\cosh^{-3}\left(|x|\sqrt{\lambda}\right),$$

$$f(x) \leq N_0\sqrt{\lambda}\cosh^{-2}\left(|x|\sqrt{\lambda}\right)\frac{\sinh\left(|x|\sqrt{\lambda}\right)}{|x|}$$

Then we introduce $u(x) := N_0\cosh^{-1}\left(|x|\sqrt{\lambda}\right)$, note that u is infinitely differentiable, take $\alpha \in \mathbb{S}^1_+$, and find

$$\mathcal{L}^\alpha u - c^\alpha u + f^\alpha = \frac{\lambda N_0}{\cosh^3\left(|x|\sqrt{\lambda}\right)}\left[\sinh^2\left(|x|\sqrt{\lambda}\right)\frac{\alpha^{ij}x^ix^j}{|x|^2} - \cosh^2\left(|x|\sqrt{\lambda}\right)\right]$$

$$+ f\sqrt[d]{\det\alpha} - \lambda N_0\frac{1}{\cosh^3\left(|x|\sqrt{\lambda}\right)}\frac{\alpha^{ij}x^ix^j}{|x|^2}$$

$$- N_0\sqrt{\lambda}\frac{1}{\cosh^2\left(|x|\sqrt{\lambda}\right)}\frac{\sinh\left(|x|\sqrt{\lambda}\right)}{|x|}\left(1 - \frac{\alpha^{ij}x^ix^j}{|x|^2}\right).$$

Observe that the contents of the brackets is negative because $|\sinh t| \leq \cosh t$ and $\operatorname{tr}\alpha = 1$. The way N_0 is introduced guarantees that the remaining terms on the right are dominated by

$$f\left[\sqrt[d]{\det\alpha}-\frac{\alpha^{ij}x^ix^j}{|x|^2}-\Big(1-\frac{\alpha^{ij}x^ix^j}{|x|^2}\Big)\right]=f(\sqrt[d]{\det\alpha}-1)\le 0,$$

where the inequality is due to $\sqrt[d]{\det\alpha}\le d^{-1}\mathrm{tr}\,\alpha=d^{-1}\le 1$. Hence, $\mathcal{L}^\alpha u-c^\alpha u+f^\alpha\le 0$ and $v\le u$ by Remark 1.2.3. This takes care of (a).

Note that in (11) we have $\alpha^{ij}D_{ij}v-\lambda v=\alpha^{ij}(D_{ij}v-\lambda v\delta_{ij})$ and by Lemma 2

$$\det(\lambda vI-D^2v)=d^{-1}f^d \quad \text{(a.e.)}. \tag{13}$$

Next, for $\gamma>0$, introduce

$$f_\gamma=d\sqrt[d]{\det\big(\lambda v^{(\gamma)}I-D^2v^{(\gamma)}\big)}.$$

The above arguments show that the functions f_γ are uniformly bounded, $f_\gamma\to f$ (a.e.) as $\gamma\downarrow 0$, and, thanks to the above estimate: $v\le u$ and the formulas like

$$D_i(v^{(\gamma)})=\gamma^{-d-1}v*(D_i\zeta)(\gamma^{-1}x),$$

there exists a constant N such that, for all $x\in\mathbb{R}^d$ and $\gamma\in(0,1]$,

$$v^{(\gamma)}(x)\le Ne^{-|x|/N} \tag{14}$$

and, for any $\gamma\in(0,1]$, $v^{(\gamma)}$ satisfies the assumptions of Theorem 1.10.

By Theorem 1.10, for any $y\in\mathbb{R}^d$,

$$d^{-d}\int_{\mathbb{R}^d}(v^{(\gamma)})^{p-d}f_\gamma^d\,dx\ge\lambda^d\int_{\mathbb{R}^d}(v^{(\gamma)})^p\,dx\ge\big(v^{(\gamma)}(y)\big)^p\lambda^{d/2}\kappa_d d!p^{-d}.$$

Let $\gamma\downarrow 0$ when $p>d$. Then applying (14) and the Lebesgue dominated convergence theorem, we conclude that

$$d^{-d}\int_{\mathbb{R}^d}v^{p-d}f^d\,dx\ge\lambda^d\int_{\mathbb{R}^d}v^p\,dx,\quad \int_{\mathbb{R}^d}v^p\,dx\ge v^p(y)\lambda^{-d/2}\kappa_d d!p^{-d}.$$

Now the first estimate in (10) is obtained by simply applying Hölder's inequality to $\int v^{p-d}f^d\,dx$ in the first inequality above. The second inequality above combined with the obtained estimate of $\int v^p\,dx$ leads to the second estimate in (10). This proves (c) for $p>d$. If $p=d$, it suffices to pass to the limit in (10) by using the dominated convergence theorem. The theorem is proved. □

4. THEOREM. *Let $f(t,x)\ge 0$ be a function on $\mathbb{R}^{d+1}$ which is equal to zero for $t\ge T$ and for $|x|>R$ for some constants T and R. Suppose that f and all its derivatives with respect to t and x up to and including the second order are bounded. Then on $\mathbb{R}^{d+1}$, there exists a nonnegative function $v(t,x)$ such that*

(a) all Sobolev derivatives $\partial_t v$, D_iv, $D_{ij}v$ exist and are bounded on $\mathbb{R}^{d+1}$ and $v\le Ne^{-|x|/N}$ for all t, x and a constant N;

(b) for any $\alpha\in\mathbb{S}_+$ and $r\ge 0$,

$$r\partial_t v+\alpha^{ij}D_{ij}v-\lambda(r+\mathrm{tr}\,\alpha)v+\sqrt[d+1]{r\det\alpha}\,f\le 0,$$

$$\partial_t v-\lambda v\le 0,\quad (\lambda v\delta_{ij}-D_{ij}v)\ge 0,\quad |Dv|\le\sqrt{\lambda}v\quad (a.e.); \tag{15}$$

(c) for any $y\in\mathbb{R}^d$, $t\in(-\infty,\infty)$, $p\ge d+1$, we have

$$v(t,y) \le N\Big(\int_0^\infty ds\int_{\mathbb{R}^d} e^{-\lambda(d+1)s} f^p(t+s,x)\,dx\Big)^{1/p}, \tag{16}$$

$$\|v(t,\cdot)\|^{d+1}_{L_p(\mathbb{R}^d)} \le \frac{1}{\lambda^d(d+1)^d}\int_0^\infty e^{-\lambda(d+1)s}\|f(t+s,\cdot)\|^{d+1}_{L_p(\mathbb{R}^d)}\,ds, \tag{17}$$

$$\|v\|_{L_p(\mathbb{R}^{d+1})} \le \frac{1}{\lambda(d+1)}\|f\|_{L_p(\mathbb{R}^{d+1})}, \tag{18}$$

where

$$N = \lambda^{(d+2)/(2p)-1}p^{d/p}\big(\kappa_d d!\big)^{-1/p}(d+1)^{1/p-1}.$$

Proof. The line of argument here is similar to that in the proof of Theorem 3. In Section 1.2, let δ and A be λ and $\mathbb{S}_+^{\le 1}$, respectively, and let t be the zeroth coordinate of (t,x). Also for $\alpha\in\mathbb{S}_+^{\le 1}$ introduce

$$a^{ij}(\alpha)=\alpha^{ij},\quad i,j\ge 1,\quad a^{0i}(\alpha)=a^{i0}(\alpha)=0,\quad i\ge 0,$$

$$b^0=r(\alpha)=1-\operatorname{tr}\alpha,\quad b^i(\alpha)=0,\quad i\ge 1,\quad c^\alpha=-\lambda,$$

$$f^\alpha=\sqrt[d+1]{r(\alpha)\det\alpha}\,f.$$

We take v from Section 1.2 and prove that v possesses properties (a) to (c). As above $v\ge 0$ by Lemma 1.2.4.

Let $\mathsf{u}''=\big(\mathsf{u}''_{ij};i,j=0,...,d\big)$ be a symmetric $(d+1)\times(d+1)$-matrix. Then it is easy to see that $F_0(\mathsf{u}'')$ from (1.2.10) is the largest eigenvalue of $(\mathsf{u}''_{ij};i,j=1,...,d)$. Hence, for $l\in\mathbb{R}^{d+1}$, such that $l^0=0$ and $|l|=1$, $\mu(l)$ in (1.2.12) equals 1 and then by Theorems 1.2.5 and 1.2.11 all Sobolev derivatives $\partial_t v$, $D_i v$, and $D_{ij}v$ $(i,j\ge 1)$ exist, are bounded on $\mathbb{R}^{d+1}$, and

$$\max_{\alpha\in\mathbb{S}_+^{\le 1}}\Big[r(\alpha)(v_t-\lambda v)+\sum_{i,j=1}^d\alpha^{ij}(D_{ij}v-\lambda\delta_{ij}v)+\sqrt[d+1]{r(\alpha)\det\alpha}f\Big]=0 \tag{19}$$

on $\mathbb{R}^{d+1}$ (a.e.). This equation is similar to (11) and assertion (b) is derived from it just as above. The estimate $v\le Ne^{-|x|/N}$ is proved in the same way as in the proof of Theorem 3 (using the same u).

It only remains to prove (c). We start with the case that $p>d+1$. Observe that there exists a constant T' such that $v(t,x)=0$ for $t\ge T'$ and $x\in\mathbb{R}^d$. Indeed, choose $g\in C^\infty(\mathbb{R})$ satisfying the following three conditions: $g(t)\ge f(t,x)$, g vanishes for large t, and

$$u(t):=e^{\lambda t}\int_t^\infty e^{-\lambda s}g(s)ds\ge\frac{1}{\lambda}f(t,x)$$

for all (t,x). Our assumption that $f=0$ for $t\ge T$ guarantees that such g exists indeed. Then for $\alpha\in\mathbb{S}_+^{\le 1}$, we have

$$1=r(\alpha)+\operatorname{tr}\alpha\ge\sqrt[d+1]{r(\alpha)\det\alpha}$$

and

$$\mathcal{L}^\alpha u-c^\alpha u+f^\alpha=-r(\alpha)g-\lambda\big(1-r(\alpha)\big)u+\sqrt[d+1]{r(\alpha)\det\alpha}f\le 0.$$

Remark 1.2.3 now implies that $v \le u$ and, therefore, $v(t,x) = 0$ for $t \ge T'$ if T' is such that $g(t) = 0$ for $t \ge T'$.

Lemma 2 allows us to rewrite (19) as

$$(\lambda v - v_t)\det(\lambda vI - D^2 v) = (d+1)^{-d-1} f^{d+1} \quad \text{(a.e.).} \tag{20}$$

Next, we take a nonnegative $\zeta \in C_0^\infty(\mathbb{R}^{d+1})$ with unit integral and for $\gamma > 0$ we set

$$v^{(\gamma)}(t,x) = v(t,x) * \left[\gamma^{-d-1}\zeta(\gamma^{-1}t,\, \gamma^{-1}x)\right], \tag{21}$$

$$f_\gamma = (d+1)\Big[\big(\lambda v^{(\gamma)} - \partial_t v^{(\gamma)}\big)\det\big(\lambda I v^{(\gamma)} - D^2(v^{(\gamma)})\big)\Big]^{1/(d+1)}.$$

The above allows us to conclude that $f_\gamma \to f$ (a.e.) as $\gamma \downarrow 0$ and f_γ are uniformly bounded. Moreover, formulas like

$$\partial_t D_i(v^{(\gamma)}) = \gamma^{-d-3} v * (\partial_t D_i\zeta)(\gamma^{-1}t, \gamma^{-1}x)$$

combined with $v \le Ne^{-|x|/N}$ imply that $v^{(\gamma)}$ satisfies the assumptions of Theorem 1.9 for each $\gamma > 0$. Therefore,

$$\begin{aligned}(d+1)^{-d-1}p\lambda^{-d}e^{\lambda pt}\int_t^\infty ds\int_{\mathbb{R}^d} e^{-\lambda ps}(v^{(\gamma)})^{p-d-1} f_\gamma^{d+1}\,dx \\ \ge \big\|v^{(\gamma)}(t,\cdot)\big\|^p_{L_p(\mathbb{R}^d)} \ge \kappa_d d!\big(p\sqrt{\lambda}\big)^{-d}\big(v^{(\gamma)}(t,x)\big)^p.\end{aligned} \tag{22}$$

We set $\gamma \downarrow 0$ as in the proof of Theorem 3 and replace $v^{(\gamma)}$ and f_γ by v and f, respectively. After that we introduce

$$w(t) = e^{-\lambda t}\big\|v(t,\cdot)\big\|_{L_p(\mathbb{R}^d)}, \quad h(t) = e^{-\lambda t}\big\|f(t,\cdot)\big\|_{L_p(\mathbb{R}^d)},$$

$$K = p\lambda^{-d}(d+1)^{-d-1}.$$

To estimate the first term in (22) in terms of the above functions, observe that by Hölder's inequality

$$\int_{\mathbb{R}^d} v^{p-d-1} f^{d+1}\,dx \le e^{\lambda ps} w^{p-d-1} h^{d+1}.$$

It follows from (22) that

$$w^p(t) \le K\int_t^\infty w^{p-d-1}(s)h^{d+1}(s)\,ds.$$

To transform this inequality, take $\delta > 0$ and $T_1 \in (-\infty, \infty)$ and define an operator Φ_δ acting on measurable functions $\psi \ge 0$ defined on $[T_1, \infty)$ by the formula

$$\Phi_\delta\psi(t) = \Big(K\int_t^\infty \psi^{p-d-1}(s)h^{d+1}(s)\,ds + \delta\Big)^{1/p}.$$

Obviously, Φ_δ is an order preserving operator: $0 \le \psi^1 \le \psi^2$ on $[T_1,\infty)$ implies that $0 \le \Phi_\delta\psi^1 \le \Phi_\delta\psi^2$ on $[T_1,\infty)$. Furthermore, if we have uniformly bounded $\psi^n \ge 0$ and the $\lim \psi^n$ exists for each t, then $\lim \Phi_\delta\psi^n = \Phi_\delta \lim \psi^n$. This is true because

$h(t)$ is a bounded function with compact support. Finally, for $\psi(t) \equiv N$, we have $\Phi_\delta \psi \leq \psi$ provided that N is sufficiently large.

Then take N_1 sufficiently large so that $\Phi_\delta N_1 \leq N_1$ and $w(t) \leq N_1$ on $[T_1, \infty)$ and write

$$w(t) \leq \Phi_\delta w(t) \leq ... \leq \Phi_\delta^n w(t) \leq N_1.$$

It follows that

$$\lim_{n \to \infty} \Phi_\delta^n w =: w_\delta$$

exists and $w \leq w_\delta$, $w_\delta = \Phi_\delta w_\delta$. We thus obtained an equation with respect to w_δ. By knowing that $w_\delta \geq \delta^{\frac{1}{p}}$ we easily solve the equation and get

$$\big(w_\delta(t)\big)^{d+1} = \frac{K(d+1)}{p} \int_t^\infty h^{d+1}(s)\, ds + \delta^{(d+1)/p}.$$

By recalling that $w \leq w_\delta$ and sending $\delta \to 0$, we conclude

$$w^{d+1}(t) \leq \frac{K(d+1)}{p} \int_t^\infty h^{d+1}(s)\, ds =: N_1 \int_t^\infty h^{d+1}(s)\, ds \tag{23}$$

for $t \in [T_1, \infty)$ and since T_1 is arbitrary, (23) holds for all $t \in (-\infty, \infty)$. Inequality (23) coincides with (17).

Minkowski's inequality applied to (23) yields

$$\begin{aligned}
&\Big(\int_{-\infty}^\infty e^{\lambda pt} w^p\, dt\Big)^{1/p} \\
&\leq N_1^{1/(d+1)} \Big(\int_{-\infty}^\infty e^{\lambda pt} \Big(\int_0^\infty h^{d+1}(t+s)\, ds\Big)^{p/(d+1)} dt\Big)^{1/p} \\
&\leq N_1^{1/(d+1)} \Big(\int_0^\infty ds \Big(\int_{-\infty}^\infty e^{\lambda pt} h^p(t+s)\, dt\Big)^{(d+1)/p}\Big)^{1/(d+1)} \\
&= N_1^{1/(d+1)} \Big(\int_{-\infty}^\infty e^{\lambda pt} h^p(t)\, dt\Big)^{1/p} \Big(\int_0^\infty e^{-\lambda(d+1)s}\, ds\Big)^{1/(d+1)}.
\end{aligned}$$

The inequality between the extreme terms in this sequence is just another form of (18). Finally, Hölder's inequality applied to (23) shows that

$$\begin{aligned}
w^p(t) &\leq N_1^{p/(d+1)} \Big(\int_t^\infty e^{-\lambda(d+1)s} \big(e^{\lambda s} h(s)\big)^{d+1} ds\Big)^{p/(d+1)} \\
&\leq N_1^{p/(d+1)} \Big(\int_t^\infty e^{-\lambda(d+1)s}\, ds\Big)^{(p-d-1)/(d+1)} \int_t^\infty e^{-\lambda(d+1)s} e^{\lambda ps} h^p(s) ds.
\end{aligned}$$

Combining this with the second inequality in (22) leads to (16). We have proved (c) assuming that $p > d+1$, which allowed us to ignore dealing with 0^0. When $p = d+1$, we get (16) - (18) by sending $p \downarrow d+1$. The theorem is proved. □

CHAPTER 3

The Aleksandrov estimates

A.D. Aleksandrov [**2**–**4**] estimated the maximum of solutions to the Dirichlet problem for elliptic equations $\mathcal{L}u = f$ in terms of the L_p-norm of f. Although among the estimates provided in this chapter, only one (see Remark 3.8) was proved by A.D. Aleksandrov himself, nevertheless, his brilliant idea of using convex envelops and the Monge-Ampère equations immensely influenced our approach to proving other estimates in this chapter. This explains the title of the chapter. We have to say that our approach is different from that of Aleksandrov and our analytic manipulations might be hard to grasp at first reading in contrast with clear geometric arguments of Aleksandrov. However, we generalize elliptic Aleksandrov's estimates and provide their parabolic counterparts. Most of the results of this chapter is taken from [**61**] and [**62**].

To give the reader an idea about the results of this chapter, recall that according to the agreement adopted in the Preface, the points in $\mathbb{R}^{d+1}$ are denoted by the pairs (t, x) with $t \in (-\infty, \infty)$ and $x \in \mathbb{R}^d$. Fix $T \in (0, \infty)$ and a bounded domain $\Omega \subset \mathbb{R}^d$ and set

$$\Pi = [0, T) \times \Omega, \quad \partial' \Pi = \bar{\Pi} \setminus (\{0\} \times \bar{\Omega})$$

(the notation Π is partly justified by the fact that, if the t-axis is vertical, then $[0, T) \times \Omega$ looks like a pie).

If Q is an open subset of a Euclidean space by $C^\infty(Q)$ we denote the set of infinitely differentiable functions on Q, for which any derivative of any order is bounded in Q. Then $C^\infty(\bar{Q})$ is the subset of $C^\infty(Q)$ consisting of functions for which any derivative of any order admits a continuous extension to $\bar{Q}$.

Let Q be a domain in $\mathbb{R}^{d+1}$, let $u = u(t, x)$ be a function on Q, and let $p \in [1, \infty)$. Assume that there exist functions $u_n \in C^\infty(\bar{Q})$, $n = 1, 2, \ldots$, such that $u_n \to u$ in $L_p(Q)$ and the sequences Du_n, $D^2 u_n$, $\partial_t u_n$ are Cauchy sequences in $L_p(Q)$. In that case we write $u \in W^{1,2}_p(Q)$.

It is known that, if $u \in W^{1,2}_p(Q)$, its generalized derivatives $\partial_t u, Du, D^2 u$ exist as usual functions and belong to $L_p(Q)$. Furthermore, the space $W^{1,2}_p(Q)$ is a Banach space with norm

$$\|u\|_{W^{1,2}_p(Q)} = \|u\|_{L_p(Q)} + \|Du\|_{L_p(Q)} + \|D^2 u\|_{L_p(Q)} + \|\partial_t u\|_{L_p(Q)}.$$

To be more specific, for any matrix a (in particular, vector) we set

$$|a|^2 = \operatorname{tr} aa^*$$

and

$$\|Du\|_{L_p(Q)} := \big\| \, |Du| \, \big\|_{L_p(Q)}, \quad \|D^2 u\|_{L_p(Q)} := \big\| \, |D^2 u| \, \big\|_{L_p(Q)}.$$

We write $u \in W^{1,2}_{p,\text{loc}}(Q)$ if $u \in W^{1,2}_p(Q')$ for any bounded domain $Q' \subset \bar{Q}' \subset Q$.

Likewise, for domains $\Omega \subset \mathbb{R}^d$, we introduce $W_p^2(\Omega)$ as the closure of the subset of $C^\infty(\bar{\Omega})$ consisting of functions having finite norm

$$\|u\|_{W_p^2(\Omega)} = \|u\|_{L_p(\Omega)} + \|Du\|_{L_p(\Omega)} + \|D^2u\|_{L_p(\Omega)}$$

with respect to this norm. The set $W^2_{p,\text{loc}}(\Omega)$ is introduced in an obvious way.

One knows that, (embedding theorems) if $u \in W^{1,2}_{p,\text{loc}}(Q)$ ($u \in W^2_{p,\text{loc}}(\Omega)$) and p is large enough, then there is a continuous function v on Q (respectively, Ω) such that $v = u$ in Q (a.e.) (in Ω (a.e.)). In such cases, speaking about $u \in W^{1,2}_{p,\text{loc}}(Q)$ ($u \in W^2_{p,\text{loc}}(\Omega)$) we will always mean its continuous modification.

One also knows that, if Q (respectively Ω) is regular enough, $W_p^{1,2}(Q)$ ($W_p^2(\Omega)$) coincides with the set of functions u such that $u, Du, D^2u, \partial_t u \in L_p(Q)$ (respectively, $u, Du, D^2u \in L_p(Q)$), where $Du, D^2u, \partial_t u$ are the generalized derivatives of u.

1. THEOREM. *Let $\Omega \subset B_R$ and let*

$$\mathcal{L}u = \partial_t u + a^{ij}D_{ij}u + b^iD_iu - cu$$

be an operator whose coefficients are measurable functions given in Π. Assume that a^{ij} are bounded and for all $\lambda \in \mathbb{R}^d$, $k = 1, ..., d$ and all values of arguments we have

$$a^{ij}\lambda^i\lambda^j \geq \delta|\lambda|^2, \quad |b^k| \leq K/R, \quad c \geq 0, \tag{1}$$

where $\delta > 0$ and $K \in [0, \infty)$ are some fixed constants. Then there exists a constant N depending only on d, δ, and K, such that, for any $u \in W^{1,2}_{d+1,\text{loc}}(\Pi) \cap C(\bar{\Pi})$, it holds on $\bar{\Pi}$ that

$$u_\pm \leq \sup_{\partial'\Pi} u_\pm + NR^{d/d+1}\|(\mathcal{L}u)_\mp\|_{L_{d+1}(\Pi)}, \tag{2}$$

$$|u| \leq \sup_{\partial'\Pi} |u| + NR^{d/d+1}\|\mathcal{L}u\|_{L_{d+1}(\Pi)}. \tag{3}$$

This theorem is proved in Section 1 before Theorem 1.14. For elliptic operators we have a similar result.

2. THEOREM. *Let*

$$\mathcal{L}u = a^{ij}D_{ij}u + b^iD_iu - cu$$

be an operator whose coefficients are measurable functions given in $\Omega \subset B_R$. Assume that a^{ij} are bounded and for all $\lambda \in \mathbb{R}^d$, $k = 1, ..., d$, and all values of arguments condition (1) is satisfied, where $\delta > 0$ and $K \in [0, \infty)$ are some fixed constants. Then there exists a constant N depending only on d, δ, and K, such that, for any $u \in W^2_{d,\text{loc}}(\Omega) \cap C(\bar{\Omega})$, it holds on $\bar{\Omega}$ that

$$u_\pm \leq \sup_{\partial\Omega} u_\pm + NR\|(\mathcal{L}u)_\mp\|_{L_d(\Omega)}, \tag{4}$$

$$|u| \leq \sup_{\partial\Omega} |u| + NR\|\mathcal{L}u\|_{L_d(\Omega)}. \tag{5}$$

3. REMARK. This theorem is proved in Remark 3.2. Note that estimate (4) was also published by Bakel'man ([**6**], formally a year later than [**2**]) in a particular case of convex Ω and $b^i = c = 0$. In that case (4) may be called the Aleksandrov-Bakel'man estimate. But estimates (3) and (5) in the general case we call the

Aleksandrov estimates in the same vein as the Cauchy-Schwarz inequality in the case of general powers is called Hölder's inequality.

It is also worth noting that in **[22]** and **[41]** it is proved that one can replace the power d in Theorems 1 and 2 with a smaller number depending only on δ, d, and the upper bounds of $|a^{ij}|$.

The results of the chapter are much more subtle than the above two theorems, they also include the maximum principle, the case of degenerate operators, case of b summable to some power, and so on. Sometimes here we encounter expressions like c^{d-p} or $0^{-1/p}(Lu)_-$, which are not defined if $c = 0$ and $d = p$ and $Lu \geq 0$. In this connection we define

$$0^0 = 1, \quad (\mathcal{L}u)_- = -I_{\mathcal{L}u<0}\mathcal{L}u, \tag{6}$$

where the latter definition allows us to integrate $(\mathcal{L}u)_-$ with various factors only over the set $\{\mathcal{L}u < 0\}$, outside of which these factors need not be even defined (we remind the reader the standard notation $I_\Gamma(x)$ which equals one if $x \in \Gamma$ and zero otherwise).

1. Parabolic uniformly nondegenerate case

In this section $Q \subset \mathbb{R}^{d+1}$ is a bounded domain, $R \in (0, \infty)$, and we suppose that

$$Q \subset \mathbb{R} \times B_R = \{(t, x) : |x| < R\}. \tag{1}$$

Assume that, for any $(t, x) \in Q$, the following objects are defined: nonnegative definite $a(t,x) = (a^{ij}(t,x)) \in \mathbb{S}$, $b(t,x) = (b^i(t,x)) \in \mathbb{R}^d$ and numbers $r(t,x) \geq 0$ and $c(t,x)$. Suppose that these functions are measurable and introduce

$$\mathcal{L}u = r\partial_t u + a^{ij}D_{ij}u + b^i D_i u - cu.$$

According to the way the operator $\mathcal{L}$ is defined we introduce the notion of parabolic boundary as follows.

1. Definition. Given a domain $G \subset \mathbb{R}^{d+1}$, the (right) *parabolic boundary* of G is the set of all points (t_0, x_0) belonging to the boundary ∂G of G (as a set in $\mathbb{R}^{d+1}$) for each of which one can find a continuous function $x_t \in \mathbb{R}^d$ and a number $\delta > 0$ such that $x_{t_0} = x_0$ and $(t, x_t) \in G$ for $t \in [t_0 - \delta, t_0)$. The parabolic boundary of G is denoted by $\partial' G$.

2. Definition. Given a domain $G \subset \mathbb{R}^{d+1}$ and a point $(t_0, x_0) \in \bar{G}$, *the set dominating* (t_0, x_0) in G is the set of all points $(t_1, x_1) \in G$ such that $t_1 > t_0$ and on $[t_0, t_1]$, there exists a continuous function $x_t \in \mathbb{R}^d$ such that $x_{t_0} = x_0$, $x_{t_1} = x_1$, $(t, x_t) \in G$ for $t \in (t_0, t_1]$. The set dominating (t_0, x_0) in the domain G is denoted by $G(t_0, x_0)$.

It is easy to see that $G(t_0, x_0)$ is open and $\partial' G(t_0, x_0) \subset \partial' G$.

3. Assumption. The functions a, b, c, r are bounded and, for a constant $\delta > 0$, we have $a \geq \delta(\delta^{ij})$, $c \geq 0$, $r \geq \delta$.

Fix some constants

$$K_1 \in [0, \infty), \quad \varepsilon \in [0, R^{-1}).$$

4. Assumption. There exists a nonnegative $h \in L_{d+1}(Q)$ such that in Q

$$|b| \le \varepsilon \operatorname{tr} a + K_1\Big[a^{ij}\frac{x^i x^j}{|x|^2} + c\Big] + h\sqrt[d+1]{r \det a} \quad (0/0 := 0). \tag{2}$$

Here is the main result of this section, where and throughout the section

$$p \ge d. \tag{3}$$

5. Theorem. *Let $u \in W^{1,2}_{p+1,\mathrm{loc}}(Q) \cap C(\bar{Q})$. Then for any $(t_0, x_0) \in \bar{Q}$, such that $Q(t_0, x_0) \neq \emptyset$,*

$$u(t_0, x_0) \le \sup_{\partial' Q(t_0,x_0)} u_+ + N\big\|c^{(d-p)/(p+1)}(r \det a)^{-1/(p+1)}(\mathcal{L}u)_-\big\|_{L_{p+1}(Q(t_0,x_0))}, \tag{4}$$

where the constant N depends only on R, K_1, ε, d, and the norm of h in $L_{d+1}(Q)$.

Just in case, recall that in (4), the norm is taken only over $\{\mathcal{L}u > 0\} \cap Q(t_0, x_0)$ even if the factor of $(\mathcal{L}u)_-$ is 0^{-1} outside this set. Also recall that, if $p = d$, $c^0 \equiv 1$.

In a special case we have a simpler statement.

6. Theorem. *Suppose that Assumption 4 is satisfied with $K_1 = \varepsilon = 0$, that is*

$$|b| \le h\sqrt[d+1]{r \det a}, \tag{5}$$

where $h \in L_{d+1}(Q)$ and let $u \in W^{1,2}_{p+1,\mathrm{loc}}(Q) \cap C(\bar{Q})$. Then for any $(t_0, x_0) \in \bar{Q}$, such that $Q(t_0, x_0) \neq \emptyset$, estimate (4) holds true with

$$N = N(d)\Big(R + \|h\|^{d+1}_{L_{d+1}(Q)}\Big)^{d/(p+1)}.$$

These theorems will be proved after appropriate preparations.

7. Remark. Theorem 6 has much in common with the results in [**95**]. It is proved in [**62**] that if $|b| \le h\sqrt[d]{\det a}$ and h depends *only* on x and belong to L_d, then again the assertion of Theorem 6 holds true however with a different N. This is generalized in [**94**] to the case of $|b|$ summable to different powers with respect to t and x.

In the following lemma Assumptions 3 and 4 are not used.

8. Lemma. *Assume that $c \ge 0$ and $a = (a^{ij}) \ge 0$. For $x \neq 0$, denote*

$$\lambda = \frac{x}{|x|}, \quad \rho = |x|, \quad f = \varepsilon \operatorname{tr} a + K_1(a^{ij}\lambda^i\lambda^j + c).$$

Then there exists a function $\psi \in C^\infty_{\mathrm{loc}}(\mathbb{R}^d)$, which depends only on R, K_1, ε, and x, such that $\psi \ge 0$ in B_R and

$$a^{ij}D_{ij}\psi + f|D\psi| - c\psi + f + \operatorname{tr} a \le 0$$

on $\mathbb{R} \times B_R \setminus \{(t, x) : x = 0\}$.

Proof. Since for $\phi = R^2 - |x|^2$ we have

$$a^{ij}D_{ij}\phi + f|D\phi| - c\phi + \operatorname{tr} a \le -\operatorname{tr} a + 2Rf \le 2Rf,$$

we need only find $\psi \in C^\infty_{\mathrm{loc}}(\mathbb{R}^d)$ such that $\psi \ge 0$ in B_R and

$$J := a^{ij}D_{ij}\psi + f|D\psi| - c\psi + f \le 0$$

on $Q := \mathbb{R} \times B_R \setminus \{(t,x) : x = 0\}$.

For $\alpha > 0$ and $\beta > \cosh(\alpha R)$ to be specified later, define $\psi = \beta - \cosh\big(\alpha|x|\big)$. Then for $x \ne 0$

$$\begin{aligned}
I :&= (1 + \alpha \sinh \alpha\rho)^{-1}(-J)\\
&= (1 + \alpha \sin \alpha\rho)^{-1}\Big\{c(\beta - \cosh \alpha\rho) - \alpha f \sinh \alpha\rho\\
&\quad + \alpha^2(\alpha\lambda, \lambda)\cosh \alpha\rho + \frac{\alpha}{\rho}\big[\operatorname{tr} a - (a\lambda, \lambda)\big]\sinh \alpha\rho - f\Big\}\\
&= c\frac{\beta - \cosh \alpha\rho}{1 + \alpha \sinh \alpha\rho} + (\alpha\lambda, \lambda)\frac{\alpha^2 \cosh \alpha\rho}{1 + \alpha \sinh \alpha\rho}\\
&\quad + \frac{\alpha \sinh \alpha\rho}{1 + \sinh \alpha\rho}\,\frac{1}{\rho}\,\big[\operatorname{tr} a - (a\lambda, \lambda)\big] - f. \qquad (6)
\end{aligned}$$

Observe that $\cosh \alpha\rho \ge 1$, $\cosh \alpha\rho \ge \sinh \alpha\rho$, $\alpha \sinh \alpha\rho \ge \alpha^2\rho$, and for $(t,x) \in Q$, we have $\rho \le R$. Therefore,

$$\frac{\beta - \cosh \alpha\rho}{1 + \alpha \sinh \alpha\rho} \ge \frac{\beta - \cosh \alpha R}{1 + \alpha \sinh \alpha R},$$

$$\frac{\alpha^2 \cosh \alpha\rho}{1 + \alpha \sinh \alpha\rho} \ge \frac{\alpha^2 \cosh \alpha\rho}{\cosh \alpha\rho + \alpha \cosh \alpha\rho} = \frac{\alpha^2}{\alpha + 1},$$

$$\frac{1}{\rho}\,\frac{\alpha \sinh \alpha\rho}{1 + \alpha \sinh \alpha\rho} \ge \frac{1}{\rho}\,\frac{\alpha^2\rho}{1 + \alpha^2\rho} \ge \frac{\alpha^2}{1 + \alpha^2 R}.$$

Now, it follows from (6) that

$$I \ge c\frac{\beta - \cosh \alpha R}{1 + \alpha \sinh \alpha R} + (a\lambda, \lambda)\frac{\alpha^2}{1 + \alpha} + \frac{\alpha^2}{1 + \alpha^2 R}\big[\operatorname{tr} a - (a\lambda, \lambda)\big] + f.$$

Next, we choose α large enough so that

$$\frac{\alpha^2}{1 + \alpha^2 R} \ge \varepsilon, \qquad \frac{\alpha^2}{1 + \alpha} \ge K_1 + \varepsilon.$$

Then we choose β large enough so that

$$\frac{\beta - \cosh \alpha R}{1 + \alpha \sinh \alpha R} \ge K_1.$$

After that we see that

$$I \ge cK_1 + (\alpha\lambda, \lambda)(K_1 + \varepsilon) + \varepsilon\big[\operatorname{tr} a - (a\lambda, \lambda)\big] - f = 0,$$

and the lemma is proved. □

The following lemma provides the first step on our way of proving Theorems 5 and 6. Introduce

$$C_{T,R} = [0,T) \times B_R.$$

9. LEMMA. *Let $h \equiv 0$, $T \in (0,\infty)$ and $Q = C_{T,R}$ and let $u \in C^\infty(\bar{Q})$ be such that $u(t,x) \le 0$ on $\partial' Q$. Then on $\bar{Q}$, we have*

$$u \le N \|(r \det a)^{-1/(d+1)}(\mathcal{L}u)_-\|_{L_{d+1}(Q)}, \tag{7}$$

where N depends only on R, ε, K_1, and d.

Proof. The form of (7) allows passing to various limits. Therefore, we may assume that $b(t,x) = 0$ for sufficiently small $|x|$. This additional assumption allows us to replace a, b, c, and r by their mollified versions and preserve inequalities (2) modulo replacing K_1 with $2K_1$. Hence, we may assume that $a, b, c, r \in C^\infty(\mathbb{R}^{d+1})$. Also, we can replace $\mathcal{L}u$ by $r^{-1}\mathcal{L}u$, which shows that we may also assume that

$$r \equiv 1.$$

Set

$$f = -(\det a)^{-1/(d+1)}\mathcal{L}u$$

and find a sequence of nonnegative functions $f^n \in C_0^\infty(\mathbb{R}^{d+1})$ such that

$$|f^n - f_+| \le 1/n \tag{8}$$

in Q and $f^n \to 0$ in $L_{d+1}(Q^c)$. Observe that since f^n are smooth, there are constants N_n such that $f^n + N_n|x|^2$ and

$$f^n_+ + N_n|x|^2 = \max\big(f^n + N_n|x|^2, N_n|x|^2\big)$$

are convex. This allows us to introduce v_n as the functions which are associated with $\lambda = 1$ and $e^t f^n_+(t,x)$ by Theorem 2.2.4. Next, we want to compare u and v_n by using the classical maximum principle. One of obstacles is that we only know that the second-order derivatives of v_n with respect to x are bounded rather than continuous. That is the reason why we introduce $v_n^{(\gamma)}$ according to notation (2.2.21) and observe that owing to Theorem 2.2.4 in $\mathbb{R}^{d+1}$

$$\partial_t v_n^{(\gamma)} + a^{ij}D_{ij}v_n^{(\gamma)} - (1+\operatorname{tr} a)v_n^{(\gamma)} + \sqrt[d+1]{\det a}\,\big(e^t f^n_+\big)^{(\gamma)} \le 0,$$

$$e^{-t}v_n(t,x) \le N(d)\Big(\int_0^\infty ds \int_{\mathbb{R}^d} \big(f^n_+\big)^{d+1}(t+s,y)\,dy\Big)^{1/(d+1)}. \tag{9}$$

By recalling that f^n are compactly supported and recalling (8) we obtain that the functions

$$u^{n\gamma}(t,x) = e^{-t}v_n^{(\gamma)}(t,x)$$

in Q satisfy

$$\partial_t u^{n\gamma} + a^{ij}D_{ij}u^{n\gamma} - \operatorname{tr} a u^{n\gamma} + \sqrt[d+1]{\det a}\, f_+ \le \omega(n,\gamma) + 1/n, \tag{10}$$

where $\omega(n,\gamma)$ is independent of (t,x) and tends to zero as $\gamma \downarrow 0$ for any n.

Next set

$$\mathcal{K}^{n\gamma} = \max_{\bar{Q}} u^{n\gamma},$$

take the function ψ from Lemma 8, and observe that, due to (10) and the fact that $|Du^{n\gamma}| \le u^{n\gamma}$, we have in Q that

$$\begin{aligned}\mathcal{L}\big(u^{n\gamma}+\mathcal{K}^{n\gamma}\psi\big) + &\sqrt[d+1]{\det a}\, f_+ \\ =&\big[\partial_t u^{n\gamma} + a^{ij}D_{ij}u^{n\gamma} - \operatorname{tr} a u^{n\gamma} + \sqrt[d+1]{\det a}\, f_+\big] \\ &+ \operatorname{tr} a u^{n\gamma} + b^i D_i u^{n\gamma} - c u^{n\gamma} + \mathcal{K}^{n\gamma}\mathcal{L}\psi \\ \le& u^{n\gamma}\big(\operatorname{tr} a + |b|\big) - \mathcal{K}^{n\gamma}\big(\operatorname{tr} a + |b|\big) + \omega(n,\gamma) + 1/n \\ \le& \omega(n,\gamma) + 1/n\end{aligned}$$

if $x \ne 0$ and, since the the above functions are continuous, this is also true for $x = 0$.

Now since $u^{n\gamma} \ge 0$ and $u \le 0$ on $\partial' Q$, by the maximum principle

$$u \le u^{n\gamma} + \mathcal{K}^{n\gamma}\psi + T\big[\omega(n,\gamma) + 1/n\big] \le N\mathcal{K}^{n\gamma} + T\big[\omega(n,\gamma) + 1/n\big]$$

in $\bar{Q}$, where N depends only on R, ε, K_1, and d. By sending $\gamma \downarrow 0$ and using (9) we get

$$u \le N\Big(\int_0^\infty ds \int_{\mathbb{R}^d} \big(f_+^n\big)^{d+1}(s,y)\,dy\Big)^{1/(d+1)} + T/n$$

in $\bar{Q}$. Letting $n \to \infty$ proves (7) and the lemma. □

10. REMARK. We stated Lemma 9 for $u \in C^\infty(\bar{Q})$. By approximation one easily proves that it is also true if $u \in W^{1,2}_{d+1}(Q)$ (keep in mind that $W^{1,2}_{d+1}(C_{T,R}) \subset C(\bar{C}_{T,R})$).

This remark and (7) immediately lead to the following.

11. COROLLARY. *Let $h \equiv 0$, $T \in (0,\infty)$, and $Q = C_{T,R}$ and let $u \in W^{1,2}_{d+1}(Q)$ be such that $u \le 0$ on $\partial' Q$. Finally, assume that $u(0) > 0$. Then the measure of the set $\{\mathcal{L}u < 0\} \cap Q$ is strictly positive.*

Now comes the case of a general bounded domain

$$Q \subset \mathbb{R} \times B_R.$$

12. LEMMA. *Let $h \equiv 0$ and let $u \in C^\infty(\bar{Q})$ be such that $u \le 0$ on $\partial' Q$. Then*

$$u \le N\big\|(r\det a)^{-1/(d+1)}(\mathcal{L}u)_-\big\|_{L_{d+1}(Q)} \tag{11}$$

in $\bar{Q}$, where N depends only on R, K_1, ε, and d.

Proof. As before we may assume that $a^{ij}, b^i, c \in C^\infty(\mathbb{R}^{d+1})$ and $r \equiv 1$. Furthermore, by having in mind an obvious passage to the limit, we may assume that $c \ge \kappa$, where the constant $\kappa > 0$.

Introduce

$$f = (\mathcal{L}u)_- I_Q$$

and introduce v as a solution of the problem

$$\mathcal{L}v = -f \quad \text{in} \quad (-\infty, T) \times B_R \text{ (a.e.)}, \quad v = 0 \quad \text{on} \quad \partial'(-\infty, T) \times B_R,$$

where T is any number such that $f(t,x) = 0$ for $t \geq T$. By linear theory such v exists and belongs to $W^{1,2}_{p+1}((t_0, T) \times B_R)$ for any $t_0 < T$. By applying Lemma 9 (and Remark 10) to $(-v)$, we find that $v \geq 0$ in $(-\infty, T] \times \bar{B}_R$.

Now we introduce

$$w = v - u$$

with the goal to show that $w \geq 0$ on $\bar{Q}$. Assume the contrary. Then, for a $\xi > 0$, we have

$$\Gamma := \{(t,x) \in \bar{Q} : w(t,x) \leq -\xi\} \neq \emptyset.$$

Since Γ is closed and Q is bounded, there exists $(t_0, x_0) \in \Gamma$ with the largest possible t_0 for which $w(t_0, x_0) \leq -\xi$ and $w(t,x) > -\xi$ as long as $t > t_0$ and $(t,x) \in \bar{Q}$. There are three cases:

(a) $(t_0, x_0) \in \overline{\partial' Q}$;
(b) $(t_0, x_0) \in \partial Q \setminus \overline{\partial' Q}$;
(c) $(t_0, x_0) \in Q$.

The inequalities $w \geq v \geq 0$ on $\partial' Q$ and on $\overline{\partial' Q}$ exclude (a).

In case (b) we claim that

$$G_\gamma := \{(t,x) : t_0 < t < t_0 + \gamma \text{ and } |x - x_0| < \gamma\} \subset Q$$

for a sufficiently small $\gamma > 0$. Indeed, if we assume that for any $\gamma > 0$, there exists $(t_\gamma, x_\gamma) \in G_\gamma \setminus Q$, then, since $(t_0, x_0) \in \bar{Q}$, one can find $(\tilde{t}_\gamma, \tilde{x}_\gamma) \in Q$ such that $(\tilde{t}_\gamma, \tilde{x}_\gamma) \to (t_0, x_0)$ and, in addition, $\tilde{t}_\gamma < t_\gamma$. Then on the straight segment with endpoints $(\tilde{t}_\gamma, \tilde{x}_\gamma)$ and (t_γ, x_γ), one can find at least one points from ∂Q which automatically is on $\partial' Q$ and goes to (t_0, x_0) as $\gamma \downarrow 0$. However, $(t_0, x_0) \notin \overline{\partial' Q}$, and this proves the claim.

Take any γ_0 for which $G_{\gamma_0} \subset Q$. In G_{γ_0} (a.e.) it holds that

$$\mathcal{L}w = \mathcal{L}v - \mathcal{L}u = -(\mathcal{L}u)_- - \mathcal{L}u = -(\mathcal{L}u)_+.$$

Furthermore, since $w \geq -\xi$ in G_{γ_0} and $w(t_0 + \gamma_0, x) > -\xi$ for $|x - x_0| \leq \gamma_0$, for all small enough $\varepsilon > 0$ we have that the function

$$\phi = -\xi - w + \varepsilon\psi,$$

where $\psi(x) = \gamma^2 - |x - x_0|^2$, is negative on $\partial' G_{\gamma_0}$. In addition $\phi(t_0, x_0) \geq \varepsilon\gamma^2 > 0$. By Corollary 11 the measure of the set $\{\mathcal{L}\phi < 0\} \cap G_{\gamma_0}$ is strictly positive. However,

$$\mathcal{L}\phi = c\xi + (\mathcal{L}u)_+ + \varepsilon\mathcal{L}\psi$$

and this expression is strictly positive if ε is small enough because $\xi > 0$ and $c \geq \kappa$. This excludes (b).

Case (c): $(t_0, x_0) \in Q$, is also impossible since then, the cylinder G_γ as above obviously exists and this, as above, leads to a contradiction.

Hence, $w \geq 0$ on $\bar{Q}$ and Lemma 9 (and Remark 10) imply that, for $(t_0, x_0) \in \bar{Q}$,

$$u(t_0, x_0) \leq v(t_0, x_0) \leq N\big\|(r \det a)^{-1/(d+1)}(\mathcal{L}v)_-\big\|_{L_{d+1}((t_0,T)\times B_R)}.$$

It only remains to recall that $\mathcal{L}v = -(\mathcal{L}u)_- I_Q$. The lemma is proved. □

We now weaken our condition on u.

13. LEMMA. *Let $h \equiv 0$ and assume that $u \in W^{1,2}_{d+1,\mathrm{loc}}(Q) \cap C(\bar{Q})$ is such that $u \leq 0$ on $\partial' Q$. Then the assertion of Lemma 12 holds true again.*

Proof. We may assume that, for a constant $\gamma > 0$, we have $u \leq -\gamma$ on $\partial' Q$. Indeed, generally we have that $u - \gamma \leq -\gamma$ on $\partial' Q$ and, if estimate (11) holds for $u - \gamma$, then using the fact that

$$\mathcal{L}(u-\gamma) = \mathcal{L}u + \gamma c \geq \mathcal{L}u, \quad \big(\mathcal{L}(u-\gamma)\big)_- \leq (\mathcal{L}u)_-$$

and letting $\gamma \to 0$, we obtain (11) as is.

Thus, we assume that $u \leq -\gamma$ on $\partial' Q$ and, for $\rho > 0$, we denote by $Q^{(\rho)}$ the set of $(t,x) \in Q$ for which $\mathrm{dist}\,((t,x), \partial Q) > \rho$. It turns out that $\mathrm{dist}\,(\partial' Q, \partial' Q^{(\rho)}) = \rho$ if $Q^{(\rho)} \neq \emptyset$. Indeed, if $(t_1, x_1) \in \partial' Q^{(\rho)}$ and (t, x_t), $t \in [t_1 - \kappa, t_1]$, is a continuous trajectory such that $(t, x_t) \in Q^{(\rho)}$ for $t \in [t_1 - \kappa, t_1)$ and $x_{t_1} = x_1$, then the point (t_1, x_1) is in $\partial Q^{(\rho)}$ and its distance to ∂Q equals ρ. It follows that one can shift the trajectory (t, x_t) as a rigid body by distance ρ so that the shifted trajectory lies in Q apart from its end point which is on ∂Q and hence on $\partial' Q$. Therefore, $\mathrm{dist}\,\big(\partial' Q, (t_1, x_1)\big) = \rho$ as claimed.

Since u is continuous in $\bar{Q}$ and $u \leq -\gamma$ on $\partial' Q$, it follows that, if ρ is small enough, then $u \leq -\gamma/2$ on $\partial' Q^{(\rho)}$. Then take the mollifications $u^{(\kappa)}$ and observe that, for small enough κ, the functions $u^{(\kappa)}$ are infinitely differentiable in $\bar{Q}^{(\rho)}$ and $u^{(\kappa)} \leq 0$ on $\partial' Q^{(\rho)}$. By Lemma 12 in $\bar{Q}^{(\rho)}$

$$u^{(\kappa)} \leq N\big\|(r \det a)^{-1/(d+1)}\big(\mathcal{L}u^{(\kappa)}\big)_-\big\|_{L_{d+1}(Q^{(\rho)})}$$

for all small κ. By letting $\kappa \downarrow 0$ we get that (11) holds in $\bar{Q}^{(\rho)}$ in place of $\bar{Q}$, and it only remains to use again the fact that u is continuous in $\bar{Q}$.

The lemma is proved. □

Proof of Theorem 0.1. First observe that the self-similarity transformation $(t,x) \to (\alpha^2 t, \alpha x)$ allows us to assume that $R = 1$. Then notice that (0.2) trivially implies (0.3). While proving (0.2) it suffices to concentrate on upper signs and show that in Π

$$u \leq M_+ + N\|(\mathcal{L}u)_-\|_{L_{d+1}(Q)}, \tag{12}$$

where N is as in the statement of the theorem and

$$M_+ = \sup_{\partial' \Pi} u_+, \quad Q = \big((0,T) \times \Omega\big) \cap \{u > M_+\}.$$

Of course, it suffices to prove (12) only on Q. Also note that the norm in (12) will only decrease if we set $c = 0$, because the integration is performed over the set where $u \geq 0$ and $c \geq 0$. With $c = 0$ estimate (12) will be proved if we can prove that $v := u - M_+$ is less than the second term on the right in (12) in which u is replaced with v. This shows that while proving (12) we may assume that $c = 0$ and $u \leq 0$ on $\partial' Q$. However, in that case (12) is a direct consequence of Lemma 13 which proves the theorem. □

In the following result Assumption 4 is not used. Recall that $p \geq d$.

14. THEOREM. *Assume that there exist functions* $\mathsf{A}, \mathsf{B} \in W^{1,2}_{d+1,\mathrm{loc}}(Q) \cap C(\bar{Q})$ *such that* $\mathsf{A}, \mathsf{B} \geq 0$ *and in* Q *(a.e.)*

$$\mathcal{L}\mathsf{A} + \operatorname{tr} a \leq 0, \quad \mathcal{L}\mathsf{B} + |b| \leq 0.$$

Let $u \in W^{1,2}_{p+1,\mathrm{loc}}(Q) \cap C(\bar{Q})$ *and* $u \leq 0$ *on* $\partial' Q$. *Then for any* $(t_0, x_0) \in \bar{Q}$, *such that* $Q(t_0, x_0) \neq \emptyset$, *estimate* (4) *holds if we replace the constant* N *with*

$$N(d)\big(\mathsf{B}^2 + \mathsf{A}\big)^{d/(2p+2)}.$$

Proof. First observe that it suffices to prove that (4) holds for any $(t_0, x_0) \in \bar{Q}$ with Q in place of $Q(t_0, x_0)$. This is seen from the fact that, for $(t_0, x_0) \in \bar{Q}$, such that $Q(t_0, x_0) \neq \emptyset$, we have $(t_0, x_0) \in \bar{Q}(t_0, x_0)$ and $u \leq 0$ on $\partial' Q(t_0, x_0)$ since $\partial' Q(t_0, x_0) \subset \partial' Q$.

Next, if the norm on the right in (4) is infinite, we have nothing to prove. Therefore, we assume that it is finite and first concentrate on the case that

$$p > d.$$

Then on the set $\{c = 0\}$ we have $(\mathcal{L}u)_- = 0$ (a.e.). It is seen that one can find nonnegative $f_n \in C_0^\infty(\mathbb{R}^{d+1})$ such that

$$\lim_{n\to\infty} \big\| f_n^{(d+1)/(p+1)} - c^{(d-p)/(p+1)} (r \det a)^{-1/(p+1)} (\mathcal{L}u)_- I_Q \big\|_{L_{p+1}(\mathbb{R}^{d+1})} = 0. \quad (13)$$

Then as in the proof of Lemma 9, by Theorem 2.2.4, for any $\lambda > 0$ and $n \geq 1$, there exists a nonnegative function v_n on $\mathbb{R}^{d+1}$ having bounded Sobolev derivatives $\partial_t v_n, Dv_n, D^2 v_n$ such that

$$r\partial_t v_n + a^{ij} D_{ij} v_n - \lambda(r + \operatorname{tr} a) v_n + \sqrt[d+1]{r \det a}\, e^{\lambda t} f_n \leq 0, \quad |Dv_n| \leq \sqrt{\lambda}\, v_n \quad \text{(a.e.)},$$

$$v_n(t,x) \leq N(d) e^{\lambda t} \lambda^{-d/(2d+2)} \Big(\int_0^\infty ds \int_{\mathbb{R}^d} f_n^{d+1}(t+s, y)\, dy \Big)^{1/(d+1)}$$

in $\mathbb{R}^{d+1}$. Introduce $u_n = v_n e^{-\lambda t}$. Then

$$r\partial_t u_n + a^{ij} D_{ij} u_n - \lambda \operatorname{tr} a u_n + \sqrt[d+1]{r \det a}\, f_n \leq 0, \quad \text{(a.e.)}, \quad (14)$$

$$|Du_n| \leq \sqrt{\lambda}\, u_n \quad \text{(a.e.)}, \quad (15)$$

$$u_n \leq N(d) \lambda^{-d/(2d+2)} \|f_n\|_{L_{d+1}(\mathbb{R}^{d+1})}. \quad (16)$$

Next, fix $\gamma > 0$ and denote

$$\xi = u - \big[\|u_n\|_{C(\mathbb{R}^{d+1})} (\lambda \mathsf{A} + \sqrt{\lambda} \mathsf{B}) + u_n \big] \gamma \frac{d+1}{p+1} - \frac{p-d}{p+1} \gamma^{-(d+1)/(p-d)}.$$

Owing to the assumptions about A and B and owing to (14) and (15), in Q (a.e.)

$$\mathcal{L}\xi \geq -(\mathcal{L}u)_- + \Big[\|u_n\|_{C(\mathbb{R}^{d+1})}\big(\lambda \operatorname{tr} a + \sqrt{\lambda}\,|b|\big) - \lambda \operatorname{tr} a u_n$$

$$+ (r\det a)^{1/(d+1)} f_n - \sqrt{\lambda}\, u_n |b|\Big]\gamma\frac{d+1}{p+1} + c\frac{p-d}{p+1}\gamma^{-(d+1)/(p-d)}$$

$$\geq \Big[(r\det a)^{1/(d+1)} f_n \gamma\frac{d+1}{p+1} + c\frac{p-d}{p+1}\gamma^{-(d+1)/(p-d)}\Big] - (\mathcal{L}u)_-.$$

Here the infimum with respect to $\gamma > 0$ of the expression in the last brackets equals

$$\big(c^{p-d} f_n^{d+1} r\det a\big)^{1/(p+1)}.$$

Hence, in Q (a.e.)

$$\mathcal{L}\xi \geq c^{(p-d)/(p+1)} f_n^{(d+1)/(p+1)} (r\det a)^{1/(p+1)} - (\mathcal{L}u)_-. \tag{17}$$

Observe that $\xi \leq 0$ on $\partial' Q$ and that Assumption 4 is satisfied with $h \equiv 0$, $\varepsilon = 0$, and some K_1 owing to Assumption 3. Therefore, by Lemma 13 (recall that $r \geq \delta$ and $a \geq \delta(\delta^{ij})$), ξ in $\bar{Q}$ is dominated by a constant independent of n times the $L_{d+1}(Q)$-norm of the right-hand side of (17). By Hölder's inequality this norm can be estimated through the $L_{p+1}(Q)$-norm of the right-hand side of (17). In light of (13) and the fact that $(\mathcal{L}u)_- = 0$ on the set $\{c = 0\}$, this last norm tends to zero as $n \to \infty$, which yields that in $\bar{Q}$

$$u \leq (\lambda\mathsf{A} + \sqrt{\lambda}\mathsf{B} + 1)N(d)\lambda^{-d/(2d+2)} C\gamma + \gamma^{-(d+1)/(p-d)}, \tag{18}$$

where

$$C = \lim_{n\to\infty} \|f_n\|_{L_{d+1}(\mathbb{R}^{d+1})} = \big\|c^{(d-p)/(p+1)}(r\det a)^{-1/(p+1)}(\mathcal{L}u)_-\big\|_{L_{p+1}(Q)}^{(p+1)/(d+1)}.$$

After getting (18) at any point in $\bar{Q}$ we minimize the right-hand side with respect to $\lambda, \gamma > 0$. Observe that, if $\mathsf{A} < \mathsf{B}^2$ at a point $(t, x) \in \bar{Q}$, then for $\lambda = \mathsf{B}^{-2}$ at this point we have

$$(\lambda\mathsf{A} + \sqrt{\lambda}\mathsf{B} + 1)\lambda^{-d/(2d+2)} \leq 3\mathsf{B}^{d/(d+1)} \leq 3(\mathsf{B}^2 + \mathsf{A})^{d/(2d+2)}. \tag{19}$$

If $\mathsf{A} \geq \mathsf{B}^2$ and $\mathsf{A} > 0$, then the inequality between the extreme terms in (19) holds for $\lambda = \mathsf{A}^{-1}$. Finally, if $\mathsf{A} = \mathsf{B} = 0$, then the limit as $\lambda \to \infty$ of the first expression in (19) is less than the last expression.

Hence,

$$u \leq N(d)\big(\mathsf{B}^2 + \mathsf{A}\big)^{d/(2d+2)} C\gamma + \gamma^{-(d+1)/(p-d)}. \tag{20}$$

At those points in $\bar{Q}$ at which $C(\mathsf{B}^2 + \mathsf{A}) \neq 0$, by finding γ from the equation

$$N(d)\big(\mathsf{B}^2 + \mathsf{A}\big)^{d/(2d+2)} C\gamma = \gamma^{-(d+1)/(p-d)},$$

we obtain the assertion of the theorem with $2N^\alpha(d)$ in place of $N(d)$, where $\alpha = (d+1)/(p+1) < 1$. Since $N^\alpha \leq N + 1$, the assertion holds true as stated.

At those points in $\bar{Q}$ at which $C(\mathsf{B}^2 + \mathsf{A}) = 0$, it suffices to let $\gamma \to \infty$ in (20) and this proves the theorem if $p > d$. In case $p = d$ one need only set $\gamma = 1$ in the above argument. The theorem is proved. □

Observe that in Theorem 14 the size of Q plays no role and, if $\operatorname{tr} a \leq K^2 c$ and $|b| \leq Kc$, where K is a constant, then for any Q one can take $\mathsf{A} = K^2, \mathsf{B} = K$. In that case $\mathsf{B}^2 + \mathsf{A} = 2K^2$.

15. REMARK. In our case that $Q \subset \mathbb{R} \times B_R$, if only function B exists satisfying the assumption of Theorem 14, then its assertion still holds, one need only replace $\mathsf{B}^2 + \mathsf{A}$ with $(\mathsf{B} + R)^2$.

Indeed, introduce $\mathsf{A} = (1/2)\big(R^2 - |x|^2\big) + R\mathsf{B}$. Then

$$\mathcal{L}\mathsf{A} \leq -\operatorname{tr} a - b^i x^i + R\mathcal{L}\mathsf{B} \leq -\operatorname{tr} a.$$

Hence, this function can be used in Theorem 14 and it only remains to notice that $\mathsf{A} \leq R^2 + R\mathsf{B}$ and $\mathsf{B}^2 + \mathsf{A} \leq (\mathsf{B} + R)^2$.

For a domain $Q \subset \mathbb{R}^{d+1}$, by $C^{1,2}(Q)$ we mean the set of continuous functions g on Q, for which $\partial_t g$, Dg, $D^2 g$ are continuous and

$$\|g\|_{C^{1,2}(Q)} = \sup_Q \big(|g| + |Dg| + |D^2 g| + |\partial_t g|\big) < \infty.$$

Proof of Theorem 5. By replacing u with

$$v = u - \sup_{\partial' Q(t_0,x_0)} u_+$$

and noting that $(\mathcal{L}v)_- \leq (\mathcal{L}u)_-$ we reduce the general case to the one in which $u \leq 0$ on $\partial' Q(t_0, x_0)$. Also as in the proof of Theorem 14 we can replace $Q(t_0, x_0)$ with Q. After these reductions we are going to rely on Remark 15. As in the proof of Lemma 9 we may assume that r, a, b, c, h are smooth functions and $b, h = 0$ in Q^c. Introduce $\mathsf{B}, \mathsf{B}_1 \in C^{1,2}(\mathbb{R} \times B_R)$ as solutions of

$$\mathcal{L}\mathsf{B} + |b| = 0, \quad \mathcal{L}\mathsf{B}_1 + h \sqrt[d+1]{r \det a} = 0$$

in $\mathbb{R} \times B_R$ with zero boundary condition for $|x| = R$.

Set

$$f = \varepsilon \operatorname{tr} a + K_1 \Big[a^{ij} \frac{x^i x^j}{|x|^2} + c\Big].$$

Take $\psi = \beta - \cosh\big(\alpha|x|\big)$ from Lemma 8. Then in $\mathbb{R} \times B_R$, $\psi \geq 0$ and (a.e.)

$$a^{ij} D_{ij}\psi + f|D\psi| - c\psi + f \leq 0.$$

Next, introduce

$$N_1 = 1 + \|\psi\|_{C(B_R)}, \quad \mathsf{B}_2 = \psi + N_1 \mathsf{B}_1.$$

Then in $\mathbb{R} \times B_R$ (a.e.)

$$\mathcal{L}\mathsf{B}_2 = a^{ij} D_{ij}\psi + b^i D_i \psi - c\psi - N_1 h \sqrt[d+1]{r \det a}$$

$$\leq -f + \big(\|D\psi\|_{C(B_R)} - N_1\big) h \sqrt[d+1]{r \det a} \leq -|b|.$$

By the maximum principle $\mathsf{B} \leq \mathsf{B}_2$ in $\mathbb{R} \times B_R$. Furthermore, since h has compact support, $\mathsf{B}_1(T, x) = 0$ for large enough T on B_R. By Remark 15 and Theorem 14 applied to B_1 we find that in $\mathbb{R} \times B_R$

$$\mathsf{B} \leq \psi + N_1 \mathsf{B}_1 \leq \psi + N_1 N(d) (\mathsf{B} + R)^{d/(d+1)} \|h\|_{L_{d+1}(Q)}. \tag{21}$$

This inequality certainly implies that $\mathsf{B} \le N$, where N depends only on R, K_1, ε, d, and the norm of h in $L_{d+1}(Q)$.

It only remains to refer to Remark 15 and Theorem 14. The theorem is proved. □

Proof of Theorem 6. Under the assumptions of the theorem in the above proof one can take $\psi = 0$, $N_1 = 1$ and then it follows from (21) that either $\mathsf{B} \le R$ or, if $\mathsf{B} \ge R$, then $\mathsf{B} \le N(d)\|h\|^{d+1}_{L_{d+1}(Q)}$. After that again it only remains to refer to Remark 15 and Theorem 14. The theorem is proved. □

2. General parabolic case

We borrow the setting (including (1.1) and (1.3)) from Section 1 and replace Assumptions 1.3 and 1.4 with the following.

1. Assumption. There exist constants $K_1 \in [0, \infty)$, $K \in \mathbb{R}$, and $\varepsilon \in [0, R^{-1})$ and a nonnegative $h \in L_{d+1}(Q)$ such that in Q

$$c_K := c + Kr \ge 0, \ |b| \le \varepsilon \operatorname{tr} a + K_1 \Big[a^{ij} \frac{x^i x^j}{|x|^2} + c_K\Big] + h \sqrt[d+1]{r \det a} \quad (0/0 := 0). \quad (1)$$

2. Assumption. In Q (a.e.)

$$r + \operatorname{tr} a + c_K > 0. \quad (2)$$

The above assumptions are supposed to be satisfied below in this section apart from Theorem 8. Recall that the definitions (0.6) are used throughout the chapter.

3. Theorem. *If $u \le 0$ on $\partial' Q$ and $K = 0$, Theorems 1.5 and 1.6 continue to hold with the same constants N.*

To prove the theorem we need two lemmas.

4. Lemma. *If $u \le 0$ on $\partial' Q$ and $K = 0$, Theorems 1.5 and 1.6 continue to hold with the same constants N if we additionally assume that $u \in W^{1,2}_{p+1}(Q) \cap C(\bar{Q})$ and h is bounded.*

Proof. We first deal with what concerns Theorem 1.5 and note that the norm in (1.4) is not affected if we multiply all coefficients of $\mathcal{L}$ by the same strictly positive function. Therefore, since by assumption $r + \operatorname{tr} a + c > 0$, we may suppose that

$$r + \operatorname{tr} a + c = 1 \quad \text{in } Q \text{ (a.e.)} \quad (3)$$

without loss of generality. Now since h is bounded, all r, a, b, and c are bounded.

For $\delta > 0$, we introduce

$$\mathcal{L}^\delta = \mathcal{L} + \delta(\partial_t + \Delta - 1)$$

and denote by $a(\delta)$, $b(\delta)$, $c(\delta)$, and $r(\delta)$ the coefficients of $\mathcal{L}^\delta$. Clearly,

$$|b(\delta)| = |b| \le \varepsilon \operatorname{tr} a(\delta) + K_1 \Big[a^{ij}(\delta) \frac{x^i x^j}{|x|^2} + c(\delta)\Big] + h \sqrt[d+1]{r \det a(\delta)}$$

if $x \ne 0$. It follows by Theorem 1.5 that, for any $w \in W^{1,2}_{p+1}(Q) \cap C(\bar{Q})$ such that $w \le 0$ on $\partial' Q$,

$$w \le N \big\| g^{-1/(p+1)}(\delta) \big[\mathcal{L} w + \delta(\partial_t w + \Delta w - w)\big]_- \big\|_{L_{p+1}(Q)} \quad (4)$$

in $\bar{Q}$, where

$$g(\delta) = (c+\delta)^{p-d}(r+\delta)\det(a+\delta I).$$

Furthermore, since h is bounded, by the inequality between geometric an arithmetic means, for any $\kappa > 0$ there exists an N such that

$$h\sqrt[d+1]{r\det a} \leq r + \kappa \operatorname{tr} a + N a^{ij}\frac{x^i x^j}{|x|^2}.$$

Hence owing to Assumption 1, there exist constants $\varepsilon_1 \in [0, R^{-1})$ and $N \in [0, \infty)$ such that

$$|b| \leq \varepsilon_1 \operatorname{tr} a + N\Big[a^{ij}\frac{x^i x^j}{|x|^2} + c + r\Big] =: f_1$$

and $f_1 > 0$. By using Lemma 1.8 we find a function ψ such that $\psi > 0$ and

$$a^{ij}D_{ij}\psi + f_1|D\psi| - (c+r)\psi + f_1 + \operatorname{tr} a \leq 0 \tag{5}$$

in $\mathbb{R} \times B_R$.

After that we take $\gamma > 0$ and set

$$w = u^\gamma := u + \gamma\psi_1$$

in (4), where

$$\psi_1 = -e^{-t}(1+\psi).$$

In light of (5), one easily sees that in Q

$$\mathcal{L}\psi_1 = e^{-t}\big[r(1+\psi) - \mathcal{L}(1+\psi)\big] \geq re^{-t} + ce^{-t} + \operatorname{tr} ae^{-t} > 0.$$

Next, introduce

$$g = \lim_{\delta \downarrow 0} g(\delta), \quad \Gamma = \big\{(t,x) \in Q \,:\, g(t,x) = 0\big\},$$

$$f^\gamma = \partial_t u^\gamma + \Delta u^\gamma - u^\gamma.$$

Then (4) yields

$$u^\gamma \leq N\big\|g^{-1/(p+1)}(\delta)[\mathcal{L}u^\gamma + \delta f^\gamma]_-\big\|_{L_{p+1}(\Gamma)}$$

$$- N\big\|g^{-1/(p+1)}(\delta)[\mathcal{L}u^\gamma + \delta f^\gamma]_-\big\|_{L_{p+1}(Q\setminus\Gamma)}. \tag{6}$$

If $|\{\mathcal{L}u < 0\} \cap \Gamma| > 0$, then the right-hand side of (1.4) is infinite and the estimate holds. In the remaining case $\mathcal{L}u \geq 0$ on Γ (a.e.) and $\mathcal{L}u^\gamma > 0$ on Γ (a.e.). Observe that $g(\delta) \geq \delta^{p+1}$ so that the first norm on the right in (6) is less than

$$\big\|\,[\delta^{-1}\mathcal{L}u^\gamma + f^\gamma]_-\big\|_{L_{p+1}(\Gamma)}.$$

By the Lebesgue dominated convergence theorem this term tends to zero as $\delta \downarrow 0$ since

$$\mathcal{L}u^\gamma > 0 \quad \text{on } \Gamma \text{ (a.e.)}, \quad [\delta^{-1}\mathcal{L}u^\gamma + f^\gamma]_- \leq |f^\gamma| \in L_{p+1}(Q).$$

In the second term in (6), we have $g > 0$ on $Q \setminus \Gamma$ implying that $\det a > 0$ and

$$1 \geq \frac{\delta^{p+1}}{g(\delta)} = \frac{\delta^{p-d}}{(c+\delta)^{p-d}}\frac{\delta}{r+\delta}\frac{\delta^d}{\det(a+\delta I)} \to 0$$

on $Q \setminus \Gamma$ as $\delta \downarrow 0$. Also observe that $g(\delta) \geq g$ and

$$[\mathcal{L}u^\gamma + \delta f^\gamma]_- \le (\mathcal{L}u^\gamma)_- + \delta|f^\gamma| \le (\mathcal{L}u)_- + \delta|f^\gamma|.$$

This and the monotone convergence theorem and the Lebesgue dominated convergence theorem convince us that, as $\delta \downarrow 0$, the limit of the second norm in (6) is less than

$$\lim_{\delta\to 0}\left\{\|g^{-1/(p+1)}(\delta)(\mathcal{L}u)_-\|_{L_{p+1}(Q\setminus\Gamma)} + \|\delta g^{-1/(p+1)}(\delta)f^\gamma\|_{L_{p+1}(Q\setminus\Gamma)}\right\}$$

$$= \|g^{-1/(p+1)}(\mathcal{L}u)_-\|_{L_{p+1}(Q\setminus\Gamma)} \le \|g^{-1/(p+1)}(\mathcal{L}u)_-\|_{L_{p+1}(Q)}.$$

We emphasize that precisely these arguments lead to our definitions surrounding (0.6). Finally, we get from (6) that for any $(t_0, x_0) \in \bar{Q}$

$$u^\gamma(t_0, x_0) \le N\|g^{-1/(p+1)}(\mathcal{L}u)_-\|_{L_{p+1}(Q)}.$$

If $Q(t_0, x_0) \neq \emptyset$, one can replace here Q with $Q(t_0, x_0)$ since, as is mentioned before, $\partial' Q(t_0, x_0) \subset \partial' Q$. After that by letting $\gamma \to 0$, we arrive at (1.4). We argued in the framework of Theorem 1.5. One obtains the assertion of Theorem 1.6 just by setting $\varepsilon = K_1 = 0$ in the above. The lemma is proved. □

5. Lemma. *Lemma 4 holds true without the assumption that h is bounded (but still $K = 0$ and $u \le 0$ on $\partial' Q$).*

Proof. Introduce

$$\mathcal{L}_n = I_{h\le n}\mathcal{L} + I_{h>n}(\partial_t + \Delta - 1\ )$$

and let r_n, a_n, b_n, c_n be the coefficients of $\mathcal{L}_n$. Obviously, r_n, a_n, b_n, c_n satisfy Assumption 1.4 with the same ε and K_1 and $hI_{h\le n}$ in place of h. They also satisfy (2). Furthermore,

$$\|c_n^{(d-p)/(p+1)}(r_n \det a_n)^{-1/(p+1)}(\mathcal{L}_n u)_-\|_{L_{p+1}(Q(t_0,x_0))}$$

$$\le \|(\partial_t u + \Delta u - u)I_{h>n}\|_{L_{p+1}(Q(t_0,x_0))}$$

$$+ \|c^{(d-p)/(p+1)}(r \det a)^{-1/(p+1)}(\mathcal{L}u)_- I_{h\le n}\|_{L_{p+1}(Q(t_0,x_0))}$$

and the latter tends to the norm in the right-hand side of (1.4) owing to the dominated and monotone convergence theorems. Here the fact that $\partial_t u + \Delta u - u \in L_{p+1}(Q)$ is crucial. The combination of this argument and Lemma 4 proves the current lemma. □

Proof of Theorem 3. As in the proof of Lemma 1.13 we may assume that $u \le -\gamma$ on $\partial' Q$, where $\gamma > 0$, and, by choosing $\rho > 0$ sufficiently small, that $u \in W^{1,2}_{d+1}(Q^{(\rho)})$ and $u \le 0$ on $\partial' Q^{(\rho)}$. Then by applying Lemma 5 we get that u in $\bar{Q}^{(\rho)}$ is dominated by the right-hand side of (1.4) with Q in place of $Q(t_0, x_0)$. Setting $\rho \downarrow 0$ proves the same inequality in Q and, by the continuity of u, in $\bar{Q}$ as well. After that we recall that in the beginning of the proof of Theorem 1.14 we saw that we could assume that $Q(t_0, x_0) = Q$ from the start. The theorem is proved. □

Now comes the most general result of this section.

6. THEOREM. *Under the assumptions stated at the beginning of this section, fix a point* $(t_0, x_0) \in \bar{Q}$, *such that* $G := Q(t_0, x_0) \neq \emptyset$ *and let* $u \in W^{1,2}_{p+1,\mathrm{loc}}(G) \cap C(\bar{G})$. *Introduce a constant* m *such that* $mr + c \geq 0$ *in* G *and set*

$$M = \sup \{ e^{m(t-t_0)} u_+, \partial' G \}.$$

Finally, take an $l \geq M$ *and set*

$$G_+ = \{ (t, x) \in G : e^{m(t-t_0)} u(t, x) > l \}.$$

Then

$$u(t_0, x_0) \leq l + N \|g\|_{L_{p+1}(G_+)}, \tag{7}$$

where

$$g = e^{K(t-t_0)} c_K^{(d-p)/(p+1)} (r \det a)^{-1/(p+1)} (\mathcal{L}u)_-,$$

N *is the same constant as in* (1.4) *and we recall* (0.6) *for an explanation of formally undefined expressions.*

Proof. Without loss of generality assume that $t_0 = 0$. Introduce

$$\mathcal{L}_K v = e^{Kt} \mathcal{L}(e^{-Kt} v), \quad w(t, x) = e^{Kt} u(t, x) - e^{(K-m)t} l \tag{8}$$

and assume that we can prove that

$$w(t_0, x_0) \leq N \| c_K^{(d-p)/(p+1)} (r \det a)^{-1/(p+1)} (\mathcal{L}_K w)_- \|_{L_{p+1}(G_+)}. \tag{9}$$

Then we can show how to prove (7). Observe that in G_+

$$\mathcal{L}_K e^{(K-m)t} = -e^{(K-m)t} (c + mr) \leq 0.$$

Therefore ($l \geq 0$ since $l \geq M$),

$$\mathcal{L}_K w \geq e^{Kt} \mathcal{L}u, \quad (\mathcal{L}_K w)_- \leq e^{Kt} (\mathcal{L}u)_-.$$

We replace $(\mathcal{L}_K w)_-$ in (9) by $e^{Kt} (\mathcal{L}u)_-$ and come to (7) (with $t_0 = 0$).

Thus, we need only prove (9), where $\mathcal{L}_K 1 = -c_K$.

If $w(t_0, x_0) \leq 0$, (9) is trivial. Then let $w(t_0, x_0) > 0$ and $(t_0, x_0) \in \bar{G}_+$. In light of $l \geq M$ we have $w \leq 0$ on $\partial' G$. Therefore,

$$(t_1, x_1) \in \partial' G_+ \cap \partial' G \Longrightarrow w(t_1, x_1) \leq 0.$$

Also for $(t_1, x_1) \in \partial' G_+ \setminus \partial' G$ we have

$$(t_1, x_1) \notin \partial G, \ (t_1, x_1) \in G \cap \partial \{ w > 0 \} \Longrightarrow w(t_1, x_1) = 0.$$

We have proved that $w \leq 0$ on $\partial' G_+$. By adding that $w \in W^{1,2}_{p+1,\mathrm{loc}}(G_+) \cap C(\bar{G}_+)$ we see that (9) follows from Theorem 3. The theorem is proved. □

7. COROLLARY (maximum principle). *Assume that* $c \geq 0$. *Also assume that* $u \in W^{1,2}_{d+1,\mathrm{loc}}(Q) \cap C(\bar{Q})$ *and* $\mathcal{L}u \geq 0$ *in* Q *(a.e.). Then in* $\bar{Q}$ *we have*

$$u \leq \sup\{u_+, \partial' Q\}. \tag{10}$$

Indeed, it suffices to prove (10) in Q, and for $(t_0, x_0) \in Q$ it suffices to apply Theorem 6 with $p = d$ and $m = 0$ and observe that $\partial' G = \partial' Q(t_0, x_0) \subset \partial' Q$.

Here is the maximum principle of Bony's type for parabolic degenerate equations. For its validity Assumptions 1 and 2 are not needed. It is stated in the framework of Section 1 described before Definition 1.1.

8. THEOREM. *Let* $c \geq 0$ *and let the coefficients of* $\mathcal{L}$ *be bounded in* Q. *Let* $u \in W^{1,2}_{d+1,\mathrm{loc}}(Q) \cap C(\bar{Q})$ *and* $(t_0, x_0) \in \bar{Q}$ *be such that* $Q(t_0, x_0) \neq \emptyset$, $(t_0, x_0) \notin \overline{\partial' Q}(t_0, x_0)$, *and*

$$u(t_0, x_0) \geq \sup\big(u_+, \partial' Q(t_0, x_0)\big).$$

Then

$$\text{ess inf}\big(\mathcal{L}u, Q(t_0, x_0)\big) \leq 0.$$

Proof. Obviously,

$$\text{ess inf}\big(\mathcal{L}u - \delta u, Q(t_0, x_0)\big) \to \text{ess inf}\big(\mathcal{L}u, Q(t_0, x_0)\big)$$

as $\delta \downarrow 0$. Therefore, we assume that $c \geq \delta > 0$. This guarantees that Assumptions 1 and 2 are satisfied with a K_1 and $K = \varepsilon = h = 0$. Furthermore, the possibility to replace u with $u + \gamma$ and then send $\gamma \downarrow 0$ allows us to assume that $u(t_0, x_0) > 0$. Then we may assume that

$$u(t_0, x_0) > \sup\big(u_+, \partial' Q(t_0, x_0)\big),$$

since one can first consider $u - \sigma\big(|x - x_0|^2 + |t - t_0|^2\big)$, where the constant $\sigma > 0$, and then send $\sigma \downarrow 0$. This works owing to $(t_0, x_0) \notin \overline{\partial' Q(t_0, x_0)}$.

In that case for $m = 0$, $p = d$, and $l = M$ in Theorem 6 we have that $u(t_0, x_0) > M = l$, hence the norm on the right in (7) is strictly positive and our assertion follows. The theorem is proved. □

9. REMARK. There are examples in which the coefficients of $\mathcal{L}$ are not bounded and the assertion of Theorem 8 is false.

3. Elliptic operators

Recall that definitions (0.6) are used throughout the chapter. Fix a number $R \in (0, \infty)$, a domain $\Omega \subset \mathbb{R}^d$, and suppose that

$$\Omega \subset B_R = \big\{x : |x| < R\big\}.$$

Suppose that, for any $x \in \Omega$, the following objects are defined: nonnegative definite $a(x) = \big(a^{ij}(x)\big) \in \mathbb{S}$, $b(x) = \big(b^i(x)\big) \in \mathbb{R}^d$ and a number $c(x) \geq 0$. Suppose that these functions are measurable and denote

$$\mathcal{L}u = a^{ij} D_{ij} u + b^i D_i u - cu.$$

1. THEOREM. *Assume that there exist* $\varepsilon \in (0, R^{-1})$ *and* $K_1 \in [0, \infty)$ *such that on* $\Omega \setminus \{0\}$.

$$c + \operatorname{tr} a > 0, \quad |b| \leq \varepsilon \operatorname{tr} a + K_1 \Big[a^{ij} \frac{x^i x^j}{|x|^2} + c \Big]. \tag{1}$$

Let $p \geq d$ *and* $v \in W^2_{p,\mathrm{loc}}(\Omega) \cap C(\bar{\Omega})$. *Then*

$$v \leq \|v_+\|_{C(\partial\Omega)} + N \big\| c^{(d-p)/p} (\det a)^{-1/p} (\mathcal{L}v)_- \big\|_{L_p(\Omega_+)}, \tag{2}$$

in $\bar{\Omega}$, *where*

$$\Omega_+ = \big\{ x \in \Omega : v(x) > \|v_+\|_{C(\partial\Omega)} \big\},$$

and N *depends only on* R, K_1, ε, *and* d.

In particular (the maximum principle), if $\mathcal{L}v \geq 0$ *in* Ω *(a.e.), then* $v \leq \|v_+\|_{C(\partial\Omega)}$ *in* Ω.

Proof. Clearly, we need only prove (2) on Ω_+. If we introduce

$$w_0 = v - \|v_+\|_{C(\partial\Omega)}$$

and observe that $\mathcal{L}w_0 \geq \mathcal{L}v$ $(c \geq 0)$ and $(\mathcal{L}w_0)_- \leq (\mathcal{L}v)_-$ on Ω_+, then we see that it suffices to prove that, for any $w \in W^2_{p,\mathrm{loc}}(\Omega_+) \cap C(\bar{\Omega}_+)$ such that $w \leq 0$ on $\partial\Omega_+$, we have

$$w \leq N \big\| c^{(d-p)/p} (\det a)^{-1/p} (\mathcal{L}w)_- \big\|_{L_p(\Omega_+)} \tag{3}$$

on Ω_+.

For such w if, for any $\gamma > 0$,

$$w - \gamma \leq N \big\| c^{(d-p)/p} (\det a)^{-1/p} \big(\mathcal{L}(w - \gamma) \big)_- \big\|_{L_p(\Omega_+ \cap \{w > \gamma\})}$$

on $\Omega_+ \cap \{w > \gamma\}$, then we come to (3) sending $\gamma \downarrow 0$ and observing that $\big(\mathcal{L}(w - \gamma)\big)_- \leq (\mathcal{L}w)_-$ and $\Omega_+ \cap \{w > \gamma\} \subset \Omega_+$. Since $\overline{\Omega_+ \cap \{w > \gamma\}} \subset \Omega_+$, we have $w - \gamma \in W^2_p(\Omega_+ \cap \{w > \gamma\})$, and our task is reduced to showing that

$$w \leq N \big\| c^{(d-p)/p} (\det a)^{-1/p} (\mathcal{L}w)_- \big\|_{L_p(D)} \tag{4}$$

in D if D is a subdomain of Ω, $w \in W^2_p(D) \cap C(\bar{D})$, $w > 0$ in D, and $w = 0$ on ∂D.

First observe that (4) is not affected if we multiply $\mathcal{L}$ by a strictly positive function. Therefore we may assume that $c + \operatorname{tr} a \equiv 1$, so that all coefficients of $\mathcal{L}$ are bounded.

Then we prove (4) under the additional assumption that

$$c \geq \delta, \quad a \geq \delta(\delta^{ij}),$$

where the constant $\delta > 0$. In this situation one could prove "elliptic" version of Lemma 1.9 and then use arguments similar to those from the proofs of Lemma 1.12 applying Theorem 2.2.3 instead of Theorem 2.2.4. Interestingly enough it turns out to be possible to get (4) from Theorem 2.3 imposing one more assumption that $w \in C^\infty(\bar{D})$, which is harmless since one can use approximations.

Take $T \in (0, \infty)$ to be sent to infinity. In Theorem 2.3 set

$$r = w^{-1}(\mathcal{L}w)_-, \quad M_0 = \sup_D w, \quad u(t,x) = e^{-t}w(x) - e^{-T}M_0,$$

$$\mathcal{P}u = r\partial_t u + \mathcal{L}u, \quad Q = (0,T) \times D, \quad K = 0, \quad h = 0.$$

Take $t_0 = 0$, $x_0 \in D$. Observe that now in Theorem 2.3 we have $Q(t_0, x_0) = Q$, and on $Q(t_0, x_0)$

$$\mathcal{P}u = e^{-t}(-rw + \mathcal{L}w) + ce^{-T}M_0 \geq e^{-t}\big(-(\mathcal{L}w)_- + \mathcal{L}w\big) \geq -2e^{-t}(\mathcal{L}w)_-,$$

$$(r \det a)^{-1/(p+1)}(\mathcal{P}u)_- \leq 2e^{-t}(r \det a)^{-1/(p+1)}(\mathcal{L}w)_-$$

$$= 2e^{-t}(\det a)^{-1/(p+1)}(\mathcal{L}w)_-^{p/(p+1)}w^{1/(p+1)}. \tag{5}$$

By Theorem 2.3,

$$w(x_0) \leq e^{-T}M_0 + N_1\|c^{(d-p)/(p+1)}(r \det a)^{-1/(p+1)}(\mathcal{P}u)_-\|_{L_{p+1}(Q)}.$$

By letting $T \to \infty$ we eliminate the first term on the right. According to (5), the second term is dominated by

$$2(p+1)^{-1/(p+1)}N_1 M_0^{1/(p+1)}\|c^{(d-p)/p}(\det a)^{-1/p}(\mathcal{L}w)_-\|_{L_p(D)}^{p/(p+1)}.$$

Hence on D

$$w \leq 2(p+1)^{-1/(p+1)}N_1 M_0^{1/(p+1)}\|c^{(d-p)/p}(\det a)^{-1/p}(\mathcal{L}w)_-\|_{L_p(D)}^{p/(p+1)}.$$

Here we can replace w with its sup over D yielding

$$M_0 \leq 2^{(p+1)/p}(p+1)^{-1/p}N_1^{(p+1)/p}\|c^{(d-p)/p}(\det a)^{-1/p}(\mathcal{L}w)_-\|_{L_p(D)}.$$

It only remains to observe that the factor of the norm on the right admits an estimate independent of p.

In order to drop the additional assumptions we take $\gamma, \delta > 0$, take ψ from Lemma 1.8, introduce $\mathcal{L}^\delta = \mathcal{L} + \delta(\Delta - 1)$, and apply the above result to $w^\gamma = w - \gamma(\psi + 1)$ and $\mathcal{L}^\delta$. Then following almost word for word the appropriate parts of the proof of Lemma 2.4, we arrive at (4). The theorem is proved. □

2. REMARK. By applying (2) with $R = 1$ to $v = \pm u$ we, obviously, prove Theorem 0.2. The case of general R is taken care of by using scaling.

Sometimes the Bony maximum principle (see Bony [**9**]) can be useful.

3. THEOREM. *Let a, b, c be bounded and let $u \in W_d^2(\Omega)$, $x_0 \in \Omega$, and*

$$u(x_0) \geq \sup_{\partial\Omega} u_+.$$

Then

$$\operatorname{ess\,inf}(\mathcal{L}u, \Omega) \leq 0.$$

This theorem is similar to Theorem 2.8 and its proof is obtained by an obvious modification of the proof of the latter.

Here is the classical Aleksandrov estimate, in which we however, unlike Aleksandrov, do not explicitly provide the values of constants depending only on d.

4. THEOREM. *Let $v \in W^2_{d,\mathrm{loc}}(\Omega) \cap C(\bar{\Omega})$. Assume that in B_R*

$$\operatorname{tr} a > 0, \quad |b| \le h\sqrt[d]{\det a},$$

where $h \in L_d(B_R)$. Then in Ω

$$v \le \|v_+\|_{C(\partial\Omega)} + N_1 \big\|(\det a)^{-1/d}(\mathcal{L}v)_-\big\|_{L_d(\Omega_+)}, \tag{6}$$

where $\Omega_+ = \{x \in \Omega : v(x) > \|v_+\|_{C(\partial\Omega)}\}$ and

$$N_1 = N(d)R\exp\big(N(d)\|h\|_{L_d(B_R)}\big).$$

In particular (the maximum principle), if $\mathcal{L}v \ge 0$ in Ω (a.e.), then $v \le \|v_+\|_{C(\partial\Omega)}$ in Ω.

The argument in the beginning of the proof of Theorem 1 shows that to prove Theorem 4 it suffices to show that if a nonnegative $v \in W^2_d(\Omega) \cap C(\bar{\Omega})$ and $v = 0$ on $\partial\Omega$, then

$$v \le N_1 \big\|(\det a)^{-1/d}(\mathcal{L}v)_-\big\|_{L_d(\Omega)}. \tag{7}$$

Since $\mathcal{L}v + cv \ge \mathcal{L}v$ and $(\mathcal{L}v + cv)_- \le (\mathcal{L}v)_-$ in Ω it suffices to prove (7) under the additional assumption that $c \equiv 0$. Then due to $\operatorname{tr} a > 0$ and the invariance of (7) under the multiplication of the coefficients by functions > 0, we may assume that

$$\operatorname{tr} a \equiv 1.$$

Next, as in the short proof of Lemma 2.5, we see that while proving (7) we may assume that h is bounded, so that b is bounded as well. Thus, to prove Theorem 4, it suffices to prove the following.

5. THEOREM. *Let $v \in W^2_d(\Omega) \cap C(\bar{\Omega})$ and $v \le 0$ on $\partial\Omega$. Assume that in B_R*

$$\operatorname{tr} a \equiv 1, \quad c \equiv 0, \quad |b| \le h\sqrt[d]{\det a}, \tag{8}$$

where h is a bounded function. Then in Ω

$$v \le N_1 \big\|(\det a)^{-1/d}(\mathcal{L}v)_-\big\|_{L_d(\Omega)}. \tag{9}$$

We prove Theorem 5 after a few auxiliary results. The starting assumptions in the following lemma are the basic ones from the beginning of the section. For a domain Q in $\mathbb{R}^d$ and integer $n \ge 1$, by $C^n(Q)$ we mean the set of continuous functions g on Q, which have n continuous derivatives and

$$\|g\|_{C^n(Q)} = \sup_Q \big(|g| + |Dg| + ... + |D^n g|\big) < \infty,$$

where $D^k g$ is the collection of the kth order derivatives and the norm $|D^k g(x)|$ is defined arbitrarily as long as $|D^k g(x)| = 0$ implies that all kth order derivatives vanish at x. By $C^n(\bar{Q})$ we mean a subset of $C^n(Q)$ consisting of functions which admit continuous extension to $\bar{Q}$ along with their derivatives up to the nth order.

6. LEMMA. *Let $a, b, c, f \in C^1(\bar{B}_R)$ and let $a \geq \delta(\delta^{ij})$ and $f \geq 0$, where the constant $\delta > 0$. Then there exists $u \in C^2(\bar{B}_R)$ such that $u = 0$ on ∂B_R and $\mathcal{L}u + f = 0$ in B_R. Furthermore,*

$$0 \leq u \leq N_1 \|(\det a)^{-1/d} f\|_{L_d(B_R)}, \tag{10}$$

where

$$N_1 = N(d) R \exp\big(N(d) \|(\det a)^{-1/d} |b| \,\|_{L_d(B_R)}\big).$$

Proof. The existence (and uniqueness) of u with the stated properties is a classical result of the Hölder space theory of linear elliptic equations. Therefore we concentrate on proving (10), where the first inequality follows by the classical maximum principle. If $u \equiv 0$, we have nothing to prove. Therefore, we assume that $u > 0$ at some points in B_R.

Next, it is convenient to arbitrarily extend u outside B_R so that $u \in C^2(\mathbb{R}^d)$. Then by Theorem 2 in Section 3.4.2 of [**37**] about the change of variables for any Borel set $\Gamma \subset \mathbb{R}^d$

$$\int_\Gamma \big(1 + |Du|\big)^{-d} |\det D^2 u| \, dx = \int_{\mathbb{R}^d} \big(1 + |p|\big)^{-d} \#\{x \in \Gamma : Du(x) = p\} \, dp, \tag{11}$$

where by $\#G$ we mean the number of points in G. Here the right-hand side is clearly bigger than

$$\int_{Du(\Gamma)} \big(1 + |p|\big)^{-d} \, dp.$$

where $Du(\Gamma) = \big\{Du(x) : x \in \Gamma\big\}$ is the normal image of Γ introduced by Aleksandrov. We indeed follow very closely the original arguments of Aleksandrov.

We are going to take Γ as the subset of B_R such that for any $x_0 \in \Gamma$ we have on B_R that

$$u(x) \leq u(x_0) + \big(x^i - x_0^i\big) D_i u(x_0), \tag{12}$$

which geometrically means that the tangent plane at point $\big(x_0, u(x_0)\big)$ to the graph of u lies above the graph of u over B_R. Obviously, the intersection of Γ with any closed subset of B_R is closed and, therefore, Γ is a Borel set.

Then, observe that owing to (12) we have $D^2 u(x_0) \leq 0$ at any point $x_0 \in \Gamma$. Furthermore, on Γ

$$\begin{aligned} d \sqrt[d]{\det a} \, \sqrt[d]{\det(-D^2 u)} &\leq -a^{ij} D_{ij} u \\ &= -\mathcal{L}u + b^i D_i u - cu \\ &= f + b^i D_i u - cu \\ &\leq f + b^i D_i u \\ &\leq \big(f + |b|\big)\big(1 + |Du|\big), \end{aligned}$$

$$\big(1 + |Du|\big)^{-d} |\det D^2 u| \leq d^{-d} (\det a)^{-1} \big(f + |b|\big)^d.$$

Thus (11) yields

$$\int_{Du(\Gamma)} \big(1 + |p|\big)^{-d} \, dp \leq d^{-d} \int_{B_R} (\det a)^{-1} \big(f + |b|\big)^d \, dx. \tag{13}$$

To estimate from below the left-hand side we find a point $x_0 \in B_R$ at which u attains its maximum over B_R (which is strictly positive by assumption). Then for any $p \in \mathbb{R}^d$ such that

$$|p| \leq u(x_0)/(2R) =: r,$$

the graph of the plane $y = u(x_0) + \langle p, x - x_0 \rangle$ lies above the graph of u over ∂B_R. There are two possibilities: either $p = 0$, or $p \neq 0$. In the first case $Du(x_0) = 0 = p$ and $p \in Du(\Gamma)$. In the second case in the neighborhood of x_0 there are points such that

$$u(x_0) + \langle p, x - x_0 \rangle < u(x)$$

since $y = u(x_0) + \langle p, x - x_0 \rangle$ is not a tangent plane at $\big(x_0, u(x_0)\big)$. Therefore, we can move the plane $y = u(x_0) + \langle p, x - x_0 \rangle$ up in a parallel way until it touches the graph of u over B_R for the last time. If this happens at a point x_1, then obviously $p = Du(x_1)$ and $x_1 \in \Gamma$, $p \in Du(\Gamma)$. Thus,

$$Du(\Gamma) \supset B_r,$$

and the left-hand side of (13) is greater than

$$\int_{B_r} \frac{dp}{\big(1 + |p|\big)^d} = N \int_0^r \frac{s^{d-1}}{(1+s)^d}\, ds =: N\phi(r).$$

Observe that by the l'Hospital rule

$$\lim_{r \to \infty} \frac{\phi(r)}{\ln r} = \lim_{r \to \infty} \frac{r^d}{(1+r)^d} = 1.$$

From (13) we conclude that

$$\phi\big(\sup_{B_R} u/(2R)\big) \leq N\Big(\int_{B_R} (\det a)^{-1} |b|^d\, dx + F^d\Big),$$

where

$$F^d = \int_{B_R} (\det a)^{-1} f^d\, dx.$$

If $F = 0$, estimate (10) is trivial. For $F > 0$ by considering u/F, f/F in place of u, f, we conclude that

$$\phi\big(\sup_{B_R} u/(2RF)\big) \leq N\Big(\int_{B_R} (\det a)^{-1} |b|^d\, dx + 1\Big).$$

Finally, if $\sup_{B_R} u/(2RF) \leq 2$, then (10) obviously holds. In the remaining case

$$\ln\big(\sup_{B_R} u/(2RF)\big) \leq N\Big(\int_{B_R} (\det a)^{-1} |b|^d\, dx + 1\Big)$$

and (10) holds again. The lemma is proved. □

7. LEMMA. *Theorem* 5 *holds under the additional assumption that* $a \geq \delta(\delta^{ij})$ *with a constant* $\delta > 0$.

Proof. The additional assumption, the boundedness of a and b (recall that $c = 0$) and standard approximations allows us to assume that $v \in C^\infty(\bar{\Omega})$. Then by using mollifiers we approximate a and b by smooth functions replacing h in the condition (8) with the function

$$\big(h\sqrt[d]{\det a}\big)^{(\varepsilon)} / \sqrt[d]{\det a^{(\varepsilon)}}$$

which is bounded uniformly with respect to ε and tends to h (a.e. B_R) as $\varepsilon \downarrow 0$ by the Lebesgue theorem. Therefore, its $L_d(B_R)$-norm will become less than twice that of h for all sufficiently small ε. It follows that we may assume that a and b are infinitely differentiable in $\bar{B}_R$. One more reduction is achieved by observing that $\big|\{x \in \Omega : v(x) = \gamma\}\big| \neq 0$ only for countably many γ and the only nontrivial case is when $v > 0$ at some points in Ω. Therefore, replacing v and Ω with $v - \gamma$ and $\{x \in \Omega : v(x) > \gamma\}$, if necessary, where $\gamma > 0$ and then sending $\gamma \downarrow 0$, we may assume that $|\partial\Omega| = 0$

After that we find a sequence $f^n \in C^\infty(\bar{B}_R)$, $n = 1, 2, ...$, such that

$$f^n \downarrow (\mathcal{L}v)_- I_{\bar{\Omega}} \quad \big(= (\mathcal{L}v)_- I_\Omega \quad \text{(a.e.)}\big)$$

in B_R. Let u^n be the corresponding functions from Lemma 6. By the classical maximum principle applied for Ω and functions u^n and v we have that $v \le u^n$ in Ω. Now we apply estimate (10) with u^n and f^n in place of u and f and let $n \to \infty$, which brings the proof of the lemma to an end. □

Proof of Theorem 5. We basically repeat the proof of Lemma 2.4 with some obvious simplifications. For $\delta > 0$, we set

$$\mathcal{L}^\delta = \mathcal{L} + \delta\Delta.$$

By Lemma 7 in Ω

$$w \le N_1 \big\|(\det(a + \delta I)^{-1/d}(\mathcal{L}w + \delta\Delta v)_-\big\|_{L_d(\Omega)} \tag{14}$$

if $w \in W^2_d(\Omega) \cap C(\bar{\Omega})$ and $w \le 0$ on $\partial\Omega$.

Furthermore, since h is bounded, there exist constants $\varepsilon \in [0, R^{-1})$ and $N \in [0, \infty)$ such that in $B_R \setminus \{0\}$

$$|b| \le \varepsilon \operatorname{tr} a + N a^{ij}\frac{x^i x^j}{|x|^2} =: f$$

and $f > 0$. By using Lemma 1.8 we find a function ψ such that $\psi > 0$ and

$$a^{ij}D_{ij}\psi + f|D\psi| + \operatorname{tr} a \le 0 \tag{15}$$

in $B_R \setminus \{0\}$. Next, we plug

$$w = v^\gamma := v - \gamma\psi,$$

where $\gamma > 0$, into (14). Observe that, in light of (15), in B_R

$$\mathcal{L}\psi < 0.$$

Introduce

$$\Gamma = \{x \in \Omega : \det a(x) = 0\},$$

and note that (14) implies that in Ω

$$v^\gamma \le N_1 \big\| (\det(a+\delta I)^{-1/d}(\mathcal{L}v^\gamma + \delta\Delta v^\gamma)_- \big\|_{L_d(\Gamma)}$$

$$+ N_1 \big\| (\det(a+\delta I)^{-1/d}(\mathcal{L}v^\gamma + \delta\Delta v^\gamma)_- \big\|_{L_d(\Omega\setminus\Gamma)}. \tag{16}$$

If $|\{\mathcal{L}v < 0\} \cap \Gamma| > 0$, then the right-hand side of (9) is infinite and (9) holds. In the remaining case $\mathcal{L}v \ge 0$ on Γ (a.e.) and $\mathcal{L}v^\gamma > 0$ on Γ (a.e.). Since $\det(a+\delta I) \ge \delta^d$, the first norm in (16) is less than

$$\big\| [\delta^{-1}\mathcal{L}v^\gamma + \Delta v^\gamma]_- \big\|_{L_d(\Gamma)}.$$

This term goes to zero as $\delta \to 0$ by the Lebesgue dominated convergence theorem and the fact that

$$\mathcal{L}v^\gamma > 0 \quad \text{on } \Gamma \text{ (a.e.)}, \quad [\delta^{-1}\mathcal{L}v^\gamma + \Delta v^\gamma]_- \le |\Delta v^\gamma| \in L_d(\Omega).$$

In the second term in (16) we have $\det a > 0$ on $\Omega \setminus \Gamma$ and

$$\frac{\delta^d}{\det(a+\delta I)} \to 0$$

as $\delta \to 0$. We also note that $\det(a+\delta) \ge \det a$ and

$$[\mathcal{L}v^\gamma + \delta\Delta v^\gamma]_- \le (\mathcal{L}v^\gamma)_- + \delta|\Delta v^\gamma| \le (\mathcal{L}v)_- + \delta|\Delta v^\gamma|.$$

These arguments allow us to apply the monotone convergence theorem and the Lebesgue dominated convergence theorem, and convince us that the limit of the second norm in (16), as $\delta \downarrow 0$, is dominated by

$$\lim_{\delta\to 0} \Big\{ \big\| \big(\det(a+\delta I)\big)^{-1/d}(\mathcal{L}v)_- \big\|_{L_d(\Omega\setminus\Gamma)} + \big\| \delta\big(\det(a+\delta I)\big)^{-1/d}\Delta v^\gamma \big\|_{L_d(\Omega\setminus\Gamma)} \Big\}$$

$$= \big\| (\det a)^{-1/d}(\mathcal{L}v)_- \big\|_{L_d(\Omega\setminus\Gamma)} \le \big\| (\det a)^{-1/d}(\mathcal{L}v)_- \big\|_{L_d(\Omega)}.$$

Finally, (16) implies that in Ω

$$v^\gamma \le N_1 \big\| (\det a)^{-1/d}(\mathcal{L}v)_- \big\|_{L_d(\Omega)}.$$

Sending $\gamma \downarrow 0$ yields (9). The theorem is proved. □

8. REMARK. A.D. Aleksandrov [**2**–**4**] proved Theorem 5 for a broader class of functions than $W^2_{d,\text{loc}}(\Omega) \cap C(\bar\Omega)$ and used the assumption $\det a > 0$ in place of $\operatorname{tr} a > 0$. If $\det a \equiv 0$, then $b \equiv 0$, one can take $h \equiv 0$, and the right-hand side of (6) is finite iff $a^{ij}D_{ij}v - cv \ge 0$ in Ω_+ (a.e.). The result of Theorem 5 in this case is contained also in Theorem 1.

The case that $\operatorname{tr} a \equiv 0$ (first-order operators) also is somewhat covered by Theorem 1 on the account of c. In that case again the right-hand side of (2) is finite iff $b^iD_iv - cv \ge 0$ in Ω_+ (a.e.). The result in that case is rather sharp in the sense that there exists $v \in W^2_p(B_1)$, $p \in [1,d)$ (but not $p = d$), $\mathbb{R}^d$-valued $b(x)$ and $c(x)$ on B_1 such that $|b| \equiv 1 \le c$,

$$b^iD_iv - cv = 0$$

in B_1 (a.e.), $v = 0$ on ∂B_1, and $v \ne 0$ in B_1. An example of such situation is given by $v = 1 - |x|$, $b = -x/|x|$, and $c = 1/v$.

CHAPTER 4

First results for fully nonlinear equations

1. The maximum principle and uniqueness of solutions in Sobolev classes

For linear elliptic operators $\mathcal{L}$ the maximum principle in its usual form is a statement that if $\mathcal{L}u \geq 0$ and $\mathcal{L}1 \leq 0$ in a domain, then u_+ in the domain is less than its maximum on the boundary. Because of linearity of $\mathcal{L}$, the above form of the maximum principle can be restated as the fact that $u_1 - u_2$ in the domain is less than the maximum of $(u_1 - u_2)_+$ on the boundary provided that $\mathcal{L}u_1 \geq \mathcal{L}u_2$, and $\mathcal{L}1 \leq 0$. The second version is stronger for nonlinear operators and we concentrate on it in this section.

We start with some well-known facts allowing us to reduce proving the maximum principle for nonlinear operators to that for linear ones. Recall that Lipschitz continuous functions on an open set are differentiable almost everywhere and their first-order partial derivatives are bounded by the Lipschitz constant (Rademacher's theorem).

1. LEMMA. *Let $F(u)$ be a real-valued Lipschitz continuous function defined in the closure* Dom_F *of an open convex set* $\mathrm{Dom}_F^o \subset \mathbb{R}^n$. *Assume that there is a convex closed bounded set $A \subset \mathbb{R}^n$ and a set of full measure $D'_F \subset \mathrm{Dom}_F^o$ such that at all points $u \in D'_F$ the function F is differentiable and $DF(u) \in A$. Then for any $u, v \in \mathrm{Dom}_F$ there exists an $a \in A$ such that*

$$F(u) - F(v) = a^i(u^i - v^i). \tag{1}$$

In particular,

$$\min_{a \in A} \left[a^i(u^i - v^i)\right] \leq F(u) - F(v) \leq \max_{a \in A} \left[a^i(u^i - v^i)\right]. \tag{2}$$

Proof. Fix $v \in \mathrm{Dom}_F^o$. By using the Fubini theorem in polar coordinates we obtain that, for almost all points $\omega \in \partial B_1$, $v + t\omega \in D'_F$ for almost all t belonging to the interval $\{s : v + s\omega \in \mathrm{Dom}_F^o\}$. In addition, $F(v + t\omega)$ is a Lipschitz and absolutely continuous function of t on that interval. It follows that, for almost all points $\omega \in \partial B_1$, for almost all t in $\{s : v + s\omega \in \mathrm{Dom}_F^o\}$ we have

$$\partial_t F(v + t\omega) = \omega^i [D_{u^i} F](v + t\omega)$$

and for all $t > 0$ in $\{s : v + s\omega \in \mathrm{Dom}_F^o\}$

$$F(v + t\omega) = F(v) + ta^i\omega^i,$$

where

$$a^i = \frac{1}{t} \int_0^t [D_{u^i} F](v + s\omega)\, ds.$$

Since A is closed and convex, $a \in A$, and we obtain (1) for almost all $u \in \mathrm{Dom}^o_F$. Then by compactness of A and the continuity of F we extend (1) to all $u, v \in \mathrm{Dom}_F$. The lemma is proved. □

2. LEMMA. *Under the assumptions of Lemma 1 suppose that $0 \in \mathrm{Dom}_F$ and F also depends on parameters $p \in \mathbb{R}^d$ and $q \in \mathbb{R}$: $F = F(q, p, u)$. Assume that $F(q, p, 0)$ is a nonincreasing function of q and*

$$\big|F(0, p, 0)\big| \leq K_0|p| + \bar{F}$$

for all $p \in \mathbb{R}^d$, where $K_0 \geq 0$ and $\bar{F} \geq 0$ are some constants. Then for any $u \in \mathrm{Dom}_F$, $p \in \mathbb{R}^d$, and $q \in \mathbb{R}$ there exist $a \in A$, $b \in \mathbb{R}^d$, $c, f \in \mathbb{R}$ such that

$$|b| \leq K_0, \quad c \geq 0, \quad |f| \leq \bar{F},$$

$$F(q, p, u) = a^i u^i + b^j p^j - cq + f.$$

Proof. By Lemma 1 we have $F(q, p, u) = a^i u^i + F(q, p, 0)$, where

$$F(q, p, 0) = F(0, p, 0) - cq$$

with

$$c = \big(F(q, p, 0) - F(0, p, 0)\big)/q \geq 0 \quad (0/0 := 0).$$

To prove the existence of appropriate b and f such that $F(0, p, 0) = b^j p^j + f$ observe that

$$-\bar{F} - K_0 b^i p^i \leq F(0, p, 0) \leq \bar{F} + K_0 b^i p^i,$$

where $b^i = p^i/|p|$ $(0/0 := 0)$ and

$$F(0, p, 0) = t\big(\bar{F} + K_0 b^i p^i\big),$$

where $t = F(0, p, 0)/\big(\bar{F} + K_0 b^i p^i\big)$ and $|t| \leq 1$. The lemma is proved. □

The following is in a sense a converse statement to Lemma 1.

3. LEMMA. *Let A and Dom_F be as in Lemma 1 and let $F(u)$ be a real-valued function such that for all $u, v \in \mathrm{Dom}_F$*

$$F(u) - F(v) \leq \max_{a \in A} \big[a^i(u^i - v^i)\big] =: U(u - v). \tag{3}$$

Then F is Lipschitz continuous on Dom_F and at all points $u \in \mathrm{Dom}^o_F$, at which F is differentiable, we have $DF(u) \in A$.

Proof. Obviously $|U(w)| \leq N|w|$, where N is a constant independent of w. By interchanging u and v in (3) we see that $F(u) - F(v) \geq -U(v - u)$, which along with (3) yield the Lipschitz continuity of F. In addition, if F is differentiable at $v \in \mathrm{Dom}^o_F$, then

$$F(u) - F(v) = D_{u^i} F(v)(u^i - v^i) + o\big(|u - v|\big)$$

and using (3) divided by $|u - v|$ and setting $u - v \to 0$ we get

$$D_{u^i} F(v)\omega^i \leq \max_{a \in A} [a^i \omega^i]$$

for any unit ω. This is only possible if $DF(v) \in A$ owing to the fact that A is closed, bounded, and convex. The lemma is proved. □

4. COROLLARY. *The set of functions $F(u)$ satisfying the assumptions of Lemma 1 is closed under taking sups and infs, that is, if F_α, $\alpha \in \mathfrak{A}$, where $\mathfrak{A}$ is an arbitrary set, is a family of functions each of which satisfies the assumptions of Lemma 1 with fixed A and Dom_F (and, perhaps, different D'_F), then so do the functions*

$$\bar{F} := \sup_{\alpha \in \mathfrak{A}} F_\alpha, \quad \underline{F} := \inf_{\alpha \in \mathfrak{A}} F_\alpha,$$

provided that they are finite. Furthermore, at each point $u \in D_F$, at which $\bar{F}$ ($\underline{F}$) is differentiable, we have $D\bar{F}(u) \in A$ (respectively, $D\underline{F}(u) \in A$).

This assertion immediately follows from Lemmas 1 and 3 and the fact that

$$\sup_{\alpha \in \mathfrak{A}} F_\alpha(u) - \sup_{\alpha \in \mathfrak{A}} F_\alpha(v) \le \sup_{\alpha \in \mathfrak{A}} \big[F_\alpha(u) - F_\alpha(v)\big],$$

$$\inf_{\alpha \in \mathfrak{A}} F_\alpha(u) - \inf_{\alpha \in \mathfrak{A}} F_\alpha(v) \le \sup_{\alpha \in \mathfrak{A}} \big[F_\alpha(u) - F_\alpha(v)\big].$$

In the following lemmas by measurable functions, as always, we mean Lebesgue measurable ones.

5. LEMMA. *Assume that, for each $x \in \mathbb{R}^m$, we are given a nonempty convex closed bounded set $A(x) \subset \mathbb{R}^n$ and $A(x)$ is a measurable function of x, that is, for any $u \in \mathbb{R}^n$, the function*

$$H(u,x) := \max\big[\langle a,u\rangle : a \in A(x)\big]$$

is measurable.

Then for any $\mathbb{R}^m$-valued measurable function $w(x)$ defined on $\mathbb{R}^m$ there exists a measurable function $a(x)$ such that $a(x) \in A(x)$ and (a.e.)

$$a^i(x)w^i(x) = H\big(w(x),x\big).$$

Proof. Observe that the function

$$1 + \max_{A(x)} |a| = 1 + \max\big\{U(u,x) : |u| = 1\big\}$$

is measurable since $U(x,u)$ is measurable in x and (Lipschitz) continuous in u. By dividing all elements of $A(x)$ by this function we see that we may assume that $|a| \le 1$ for any $a \in A(x)$ and any x. By dividing w by $|w|+1$ we reduce the problem to the one with $|w| \le 1$.

Then, for $n = 1, 2, ...$ consider the function

$$H_n(w,x) = \max\big[\langle a,w\rangle - n^{-1}|a|^2 : a \in A(x)\big]. \tag{4}$$

By the minimax theorem

$$H_n(w,x) = \min_{b \in \mathbb{R}^d} \big[H(w - 2n^{-1}b, x) + n^{-1}|b|^2\big],$$

which implies that $H_n(w,x)$ is a measurable function of (w,x). Next, since $\langle a,w\rangle - n^{-1}|a|^2$ is a strictly concave function and A is compact and convex, for each x and w, the maximum in (4) is attained at only one point

$$a = a_n(w,x).$$

Because of the uniqueness of $a_n(w,x)$ and compactness of A, $a_n(w,x)$ is a continuous function of w for each x. We want to prove that it is also measurable in x for any w. To do that fix $h, w \in \mathbb{R}^n$ and observe that, if x is such that $\langle a_n(w,x), h\rangle \geq 1$, then

$$H_n(w,x) = H_n(w,x,h), \tag{5}$$

where

$$H_n(w,x,h) = \max\big[\langle a, w\rangle - n^{-1}|a|^2 : a \in A(x) \cap \big\{b : \langle b, h\rangle \geq 1\big\}\big]$$

$(\max\emptyset =: -\infty)$. On the other hand, if (5) holds, then $A(x) \cap \big\{b : \langle b,h\rangle \geq 1\big\} \neq \emptyset$ and, owing to the uniqueness of $a_n(w,x)$, we have $a_n(w,x) \in A(x)\cap\big\{b : \langle b,h\rangle \geq 1\big\}$, so that $\langle a_n(w,x), h\rangle \geq 1$. Thus,

$$\big\{x : \langle a_n(w,x), h\rangle \geq 1\big\} = \big\{x : H_n(w,x) = H_n(w,x,h)\big\}. \tag{6}$$

It is not hard to see that the function $H_n(w,x,h)$ admits a different representation

$$\begin{aligned} H_n(w,x,h) &= \lim_{\lambda\to\infty} \max\Big[\langle a, w\rangle - n^{-1}|a|^2 - \lambda\big[\langle a,h\rangle - 1\big]_- : a \in A(x)\Big] \\ &= \lim_{\lambda\to\infty} \min_{t\in[0,1]} \big[H_n(w + t\lambda h, x) - t\lambda\big], \end{aligned}$$

which shows, in particular, that $H_n(w,x,h)$ is a measurable function of x and along with (6), which holds for any h, yield that $a_n(w,x)$ is a measurable function of x for any w. Since it is continuous in w, we have that $a_n(x) := a_n(w(x),x)$ is a measurable function. By definition, for all x,

$$H_n\big(w(x),x\big) = a_n^i(x)w^i(x) - n^{-1}\big|a_n(x)\big|^2,$$

so that

$$\big|H\big(w(x),x\big) - a_n^i(x)w^i(x)\big| \leq n^{-1}.$$

Moreover, since $a_n(x)$ are uniformly bounded, there exists a subsequence which converges weakly in L_2 on any of the balls $x + B_1$ to, say a_0, where $x \in \mathbb{Z}^m$ and B_1 is the unit ball in $\mathbb{R}^m$. Without loss of generality we assume that this is true for the original a_n's. Then the Banach-Sacs theorem guarantees that there exists a subsequence of a_n, again denoted by a_n for simplicity of notation, for which

$$\frac{1}{n}\sum_{k=1}^n a_k =: a^{(n)}$$

converge to a_0 in L_2 on any of the balls. In particular, $a^{(n)} \to a_0$ in measure on any of the balls and one of their subsequences converges almost everywhere in $\mathbb{R}^m$.

Since $A(x)$ are convex, we have $a^{(n)}(x) \in A(x)$ for any x, which implies that $a_0(x) \in A(x)$ for almost all x. We modify a_0 arbitrarily on the set of measure zero where this inclusion might not hold to have $a_0(x) \in A(x)$ for all x. Finally, for any of the balls, say B we have

$$\begin{aligned} \big\|H\big(w(\cdot),\cdot\big) - \langle a_0. w\rangle\big\|_{L_2(B)} &= \lim_{n\to\infty}\Big\|H\big(w(\cdot),\cdot\big) - \frac{1}{n}\sum_{i=k}^n \langle a_k, w\rangle\Big\|_{L_2(B)} \\ &\leq \lim_{n\to\infty}\frac{1}{n}\sum_{k=1}^n \big\|H\big(w(\cdot),\cdot\big) - \langle a_k, w\rangle\big\|_{L_2(B)} = 0. \end{aligned}$$

The lemma is proved. □

6. LEMMA. *Let $A(x)$ be as in Lemma 5, let Dom_F^o be an open convex subset of $\mathbb{R}^n$ containing the origin, and let $F(q,p,u,x)$ be a real-valued Borel measurable function defined on $\mathbb{R}\times\mathbb{R}^d\times\mathrm{Dom}_F\times\mathbb{R}^m$, where Dom_F is the closure of Dom_F^o. Assume that, for any p,q,x, the function $F(q,p,u,x)$ as a function of u is Lipschitz continuous on Dom_F and at all points $u\in\mathrm{Dom}_F^o$ of its differentiability with respect to u we have*

$$D_uF(q,p,u,x)\in A(x).$$

Also assume that $F(q,p,0,x)$ is a nonincreasing function of q. Finally, suppose that for a constant $K_0\geq 0$ and a function $\bar{F}(x)\geq 0$ defined on $\mathbb{R}^m$ we have

$$\big|F(0,p,0,x)\big|\leq K_0|p|+\bar{F}(x)$$

for all $p\in\mathbb{R}^d$ and $x\in\mathbb{R}^m$.

Then for any measurable Dom_F-valued functions $u(x)$ and $v(x)$, $\mathbb{R}^d$-valued $p(x)$, and real-valued $q(x)$ defined on $\mathbb{R}^m$ there exist measurable $\mathbb{R}^n$-valued functions $a(x)$ and $\bar{a}(x)$, $\mathbb{R}^d$-valued $b(x)$, and real-valued $c(x)$ and $f(x)$ defined on $\mathbb{R}^m$ such that

$$a(x),\bar{a}(x)\in A(x),\quad |b(x)|\leq K_0,\quad c(x)\geq 0,\quad |f(x)|\leq\bar{F}(x)$$

for all x and (a.e.)

$$F\big(q(x),p(x),u(x),x\big)-F\big(q(x),p(x),v(x),x\big)=a^i(x)\big[u^i(x)-v^i(x)\big],\tag{7}$$

$$F\big(q(x),p(x),u(x),x\big)=a^i(x)u^i(x)+b^i(x)p^i(x)-c(x)q(x)+f(x).\tag{8}$$

Proof. To prove (7) set

$$F\big(u(x),x\big)=F\big(q(x),p(x),u(x),x\big).$$

Observe that by Lemma 1

$$m(x):=\min_{a\in A(x)}a^iw^i(x)\leq F\big(u(x),x\big)-F\big(v(x),x\big)\leq\max_{a\in A(x)}a^iw^i(x)=M(x),$$

where $w=u-v$. By Lemma 5 applied to $\pm w$, there exist two measurable functions $\bar{a}(x),\underline{a}(x)$ with values in $A(x)$ for each x such that (a.e.)

$$\max_{a\in A(x)}a^iw^i(x)=\bar{a}^i(x)w^i(x),\qquad\min_{a\in A(x)}a^iw^i(x)=\underline{a}^i(x)w^i(x).$$

To obtain (7), it only remains to set $a(x)=t(x)\underline{a}(x)+\big(1-t(x)\big)\bar{a}(x)$, where

$$t(x)=\frac{1}{M(x)-m(x)}\Big(M(x)-\big[F\big(u(x),x\big)-F\big(v(x),x\big)\big]\Big)\quad(0/0:=0)$$

and observe that $t(x)$ is measurable and takes values in $[0,1]$.

To prove (8) write that by (7)

$$F\big(q(x),p(x),u(x),x\big)=a^i(x)u^i(x)+F\big(q(x),p(x),0,x\big),$$

and just repeat the proof of Lemma 2, using along the way that the function $F\big(q(x),p(x),0,x\big)$ is measurable. The lemma is proved. □

7. REMARK. The result of Lemma 5 have an obvious version if $A(x)$ are defined for x in a measurable set Γ rather than in the whole $\mathbb{R}^m$. To obtain this version it suffices just to define $A(x) = A(x_0)$ and $w(x) = w(x_0)$ for $x \notin \Gamma$, where x_0 is a fixed element of Γ.

Similar observation can be made regarding Lemma 6. One more almost irrelevant comment is that our representations of $H(w(x), x)$ or $F(q(x), p(x), u(x), x)$ hold everywhere rather than only almost everywhere because on the zero sets where a chosen representation does not hold we can always adjust it in any way preserving (Lebesgue) measurability.

We have already proved the maximum principle for linear operators. The case of nonlinear operators we split into two subcases. Throughout the rest of this section (unless stated otherwise)

$$R > 0, \quad K_1 \geq 0, \quad K \geq 0, \quad \varepsilon \in (0, R^{-1})$$

are fixed constants, a domain $\Omega \subset \mathbb{R}^d$ is such that $\Omega \subset B_R$, a bounded subdomain Q of

$$\mathbb{R}^{d+1} = \{(t,x) : t \in (-\infty, \infty), \, x \in \mathbb{R}^d\}$$

is such that

$$Q \subset \mathbb{R} \times B_R.$$

1:1. Parabolic operators. Recall that $\mathbb{S}$ is the set of symmetric $d \times d$ matrices. Any point of

$$\mathfrak{U}^t := \mathbb{R}^{d+1} \times \mathbb{S} \times \mathbb{R}$$

we represent as

$$\mathsf{u} = (\mathsf{u}', \mathsf{u}'', \mathsf{u}_0), \quad \mathsf{u}' = (\mathsf{u}'_0, \mathsf{u}'_1, ..., \mathsf{u}'_d) \in \mathbb{R}^{d+1}, \quad \mathsf{u}'' \in \mathbb{S}, \quad \mathsf{u}_0 \in \mathbb{R}.$$

8. ASSUMPTION. We are given a real-valued function $F(\mathsf{u}, t, x)$ defined for $(t,x) \in \mathbb{R}^{d+1}$ and for u in the closure $\mathrm{Dom}_F(t,x)$ of an an open convex set

$$\mathrm{Dom}_F^o(t,x) \subset \mathfrak{U}^t.$$

It is locally Lipschitz continuous with respect to $\mathsf{u} \in \mathrm{Dom}_F(t,x)$ for any (t,x).

9. NOTATION. For any (t,x), at all points of differentiability of F with respect to $\mathsf{u} \in \mathrm{Dom}_F^o(t,x)$ define

$$a^{ij} = D_{\mathsf{u}''_{ij}} F, \quad b^i = D_{\mathsf{u}'_i} F, \quad i = 1, ..., d, \quad r = D_{\mathsf{u}_0} F,$$

$$c = -D_{\mathsf{u}'_0} F, \quad a = (a^{ij})_{i,j=1}^d, \quad b = (b^i)_{i=1}^d.$$

At those points at which F is not differentiable with respect to u and which compose a set of measure zero we set $a = (\delta^{ij})$, $r = 1$, $b = 0$, and $c = 0$. In this way we obtain the functions

$$(r, a, b, c) = (r, a, b, c)(\mathsf{u}, t, x).$$

Next, for regular enough functions $w(t,x)$, such that

$$\big(w(t,x), Dw(t,x), D^2w(t,x), \partial_t w(t,x)\big) \in \mathrm{Dom}_F(t,x) \tag{9}$$

for almost all $(t,x) \in Q$, introduce

$$F[w](t,x) = F\big(w(t,x), Dw(t,x), D^2w(t,x), \partial_t w(t,x), t, x\big).$$

10. REMARK. We want to explain what we mean by $D_{\mathsf{u}''_{12}}F(\mathsf{u}'')$ if we are given a differentiable function $F(\mathsf{u}'')$ on $\mathbb{S}$. For instance, if $F = \mathsf{u}''_{12}$, then $D_{\mathsf{u}''_{12}}F$ turns out to be $1/2$ and not 1, as one might expect. The point is that by $D_{\mathsf{u}''_{ij}}F(\mathsf{u}'')$ we mean a *symmetric* $d \times d$ matrix such that, for any symmetric $d \times d$ matrix v'' we have

$$\frac{d}{dt}F(\mathsf{u}'' + t\mathsf{v}'')\big|_{t=0} = \mathsf{v}''_{ij}D_{\mathsf{u}''_{ij}}F(\mathsf{u}'').$$

In case $F = \mathsf{u}''_{12}$ the left-hand side is $\mathsf{v}''_{12} = (1/2)\big(\mathsf{v}''_{12} + \mathsf{v}''_{21}\big)$ and hence $D_{\mathsf{u}''_{12}}F = D_{\mathsf{u}''_{21}}F = 1/2$.

The following theorem allows one to use the maximum principle, in particular, for degenerate parabolic operators, one of examples of which the reader will find in the proof of Lemma 17.3.3 when we investigate $C^{1+\alpha}$-regularity of viscosity solutions of the Isaacs equations with almost VMO coefficients.

11. THEOREM. *Assume that for all $(t,x) \in Q$ and $\mathsf{u} \in \mathrm{Dom}^o_F(t,x)$, the above introduced functions satisfy*

$$r \geq 0, \quad a = a^* \geq 0, \quad Kr + c \geq 0, \tag{10}$$

$$|b| \leq \varepsilon \operatorname{tr} a + K_1\Big[a^{ij}\frac{x^i x^j}{|x|^2} + Kr + c\Big] + h\sqrt[d+1]{r \det a}, \quad (0/0 := 0) \tag{11}$$

where $h \in L_{d+1}(Q)$, $h \geq 0$. Also assume that for any $N \in (0,\infty)$ there exists a measurable function $\delta_N = \delta_N(t,x) > 0$ on Q such that for any $(t,x) \in Q$ and $\mathsf{u} \in \mathrm{Dom}^o_F(t,x)$, satisfying $|\mathsf{u}| \leq N$, we have

$$\operatorname{tr} a + (K+1)r + c \geq \delta_N. \tag{12}$$

Let $u, v \in W^{1,2}_{d+1,\mathrm{loc}}(Q) \cap C(\bar{Q})$ and assume that condition (9) is satisfied with $w = u, v$ in Q (a.e.) and $F[u] \geq F[v]$ in Q (a.e.). Then

(a) *if $(t_0,x_0) \in Q$ and $u \leq v$ on $\partial' Q(t_0,x_0)$, then $u(t_0,x_0) \leq v(t_0,x_0)$;*
(b) *if $(t_0,x_0) \in Q$ and $c \geq 0$ in $Q(t_0,x_0)$ (a.e.), then*

$$u(t_0,x_0) - v(t_0,x_0) \leq \sup_{\partial' Q(t_0,x_0)} [u-v]_+.$$

Proof. For $(t,x) \in Q$ denote by $A(t,x)$ the set of all points $\alpha = (r,a,b,c)$, where r, c are real numbers, $a \in \mathbb{S}$, and $b \in \mathbb{R}^d$, such that they satisfy conditions (10) and (11). By $A_1(t,x)$ we denote the subset of $A(t,x)$ consisting of points for which

$$\operatorname{tr} a + (K+1)r + c = 1. \tag{13}$$

The sets $A(t,x)$ and $A_1(t,x)$ are obviously closed. They are also convex since $\sqrt[d+1]{r \det a}$ is a concave function of (r,a) on the domain $\{r \geq 0, a = a^* \geq 0\}$. In addition the sets $A_1(t,x)$ are bounded. Let A_0 be the set of all points (r,a,c) (note, b is excluded) with the same meaning of r,a,c, such that (10) and (13) are satisfied.

Observe that

$$H_0(\mathsf{u}) := \max\big[r\mathsf{u}_0 + a^{ij}\mathsf{u}''_{ij} - c\mathsf{u}'_0 : (r,a,c) \in A_0\big]$$

is a Lipschitz continuous function of $\mathsf{u} \in \mathfrak{U}^t$. Therefore,

$$H_1(\mathsf{u},t,x) := \max\big[r\mathsf{u}_0 + a^{ij}\mathsf{u}''_{ij} + b^i\mathsf{u}'_i - c\mathsf{u}'_0 : (r,a,b,c) \in A_1(t,x)\big]$$

$$= H_0(\mathsf{u}) + \bar{b}(t,x)\big|[\mathsf{u}']\big| \tag{14}$$

is a measurable function of (t,x), where

$$[\mathsf{u}'] := (\mathsf{u}'_1, ..., \mathsf{u}'_d), \tag{15}$$

$$\bar{b}(t,x) = \max\Big\{\varepsilon \operatorname{tr} a + K_1\Big[a^{ij}\frac{x^ix^j}{|x|^2} + Kr + c\Big] + h(t,x)\sqrt[d+1]{r \det a} : (r,a,c) \in A_0\Big\}.$$

For $u \in W^{1,2}_{d+1,\mathrm{loc}}(Q)$, we set

$$H_1[u](t,x) = H_1(u, Du, D^2u, \partial_t u, t, x).$$

Next, we introduce $w = u - v$ and claim that in Q (a.e.)

$$H_1[w] \geq 0. \tag{16}$$

Assume the contrary and let $\Gamma = \{H_1[w] < 0\} \cap Q$. Then $|\Gamma| > 0$. For $M \in (0,\infty)$ introduce the following measurable subsets of Q:

$$\Gamma_M = \big\{(t,x) \in \Gamma : u^2 + |Du|^2 + |D^2u|^2 + (\partial_t u)^2,$$

$$v^2 + |Dv|^2 + |D^2v|^2 + (\partial_t v)^2 < M^2\big\}$$

and we will concentrate on M only so large that the Lebesgue measure of Γ_M is strictly positive.

Fix $(t,x) \in \Gamma_M$, such that (9) holds at (t,x) for $w = u, v$, and observe that, since $F(\mathsf{u},t,x)$ is locally Lipschitz continuous with respect to $\mathsf{u} \in \mathrm{Dom}_F(t,x)$, there exists a constant N such that at all points of differentiability of $F(\mathsf{u},t,x)$ with respect to u

$$\big|D_{\mathsf{u}}F(\mathsf{u},t,x)\big| \leq N$$

as long as $\mathsf{u} \in \mathrm{Dom}^o_F(t.x) \cap \big\{|\mathsf{u}| < M\big\}$. Then by Lemma 1, in which we take

$$A(t,x) \cap \big\{|\alpha| \leq N\big\} \cap \big\{(r,a,b,c) : \operatorname{tr} a + (K+1)r + c \geq \delta_M(t,x)\big\}$$

and $\mathrm{Dom}^o_F(t,x) \cap \big\{|\mathsf{u}| < M\big\}$ as A and Dom^o_F, respectively, there exists $\alpha = (r,a,b,c)$ with $|\alpha| \leq N$, satisfying conditions (10), (11), and (12) and such that

$$F[u](t,x) - F[v](t,x) = r\partial_t w(t,x) + a^{ij}D_{ij}w(t,x) + b^iD_iw(t,x) - cw(t,x),$$

where $w = u - v$, and $\delta := \operatorname{tr} a + (K+1)r + c \geq \delta_M(t,x) > 0$. Note that we have $\delta^{-1}\alpha \in A_1(t,x)$. Hence

$$F[u](t,x) - F[v](t,x) \leq \delta H_1[w](t,x) < 0.$$

This holds at any point $(t,x) \in \Gamma_M$, so that $F[u] < F[v]$ on a set of positive measure. We got a contradiction with one of our hypothesis and thus proved (16).

Now Lemma 5 guarantees the existence of $\mathcal{L}$, satisfying the assumptions of Theorem 3.2.6, for which $\mathcal{L}w \geq 0$ in Q (a.e.). Finally, Theorem 3.2.6 applied to $\mathcal{L}$ and w leads to both assertions (a) and (b). The theorem is proved. □

12. COROLLARY. *Let the conditions of the theorem be satisfied and let $F[u] = F[v]$ in Q (a.e.) and $u = v$ on $\partial' Q$. Then $u = v$ in $\bar{Q}$.*

This corollary follows from assertion (a) by interchanging u and v.

13. REMARK. Condition (12) turns out to be necessary even in the case of linear operators. Let $Q = (0,T) \times B_R$, and let $\Gamma \subset Q$ satisfy $|\Gamma| > 0$. From the linear theory we know that there exists $u \in W^{1,2}_{d+1}(Q)$ satisfying

$$\partial_t u + \Delta u - u + I_\Gamma = 0$$

in Q and $u = 0$. Also $u \geq 0$ in Q (see e.g., Theorem 11), and $u \not\equiv 0$. Then introduce

$$\mathcal{L} = I_{Q \setminus \Gamma} (\partial_t + \Delta - 1).$$

We have $\mathcal{L}u = \mathcal{L}0 = 0$, but the assertions (a) and (b) are false if $v = 0$.

Theorem 11 allows us to prove uniqueness for equations like $F[u] = f$ (cf. Corollary 12), and it turns out that it can also be used while treating problems with two free boundaries.

14. THEOREM. *Let F satisfy the conditions stated in and before Theorem 11. Let $u, v \in W^{1,2}_{d+1,\mathrm{loc}}(Q) \cap C(\bar{Q})$ and assume that condition (9) is satisfied with $w = u, v$ in Q (a.e.) and $u = v$ on $\partial' Q$. Also let two real-valued functions $\psi_1 \geq \psi_2$ be given on Q.*

Finally, suppose that $\psi_1 \geq u \geq \psi_2$ and $\psi_1 \geq v \geq \psi_2$ in Q and

$$(-1)^i F[u] \geq 0 \quad (a.e.) \text{ on the set } \big\{(t,x) \in Q : (-1)^i (u - \psi_i) \geq 0\big\}, \quad i = 1,2,$$

$$(-1)^i F[v] \geq 0 \quad (a.e.) \text{ on the set } \big\{(t,x) \in Q : (-1)^i (v - \psi_i) \geq 0\big\}, \quad i = 1,2.$$

Then $u = v$ in $\bar{Q}$.

Proof. For $\gamma \geq 0$, $\delta \geq 0$, and $\mathsf{u} \in \mathfrak{U}^t$ introduce

$$F^{\gamma\delta}(\mathsf{u}, t, x) = \frac{1}{\gamma + \delta + 1}\big[F(\mathsf{u}, t, x) - \gamma(\mathsf{u}_0' - \psi_1) - \delta(\mathsf{u}_0' - \psi_2)\big],$$

$$G(\mathsf{u}, t, x) = \inf_{\gamma \geq 0} \sup_{\delta \geq 0} F^{\gamma\delta}(\mathsf{u}, t, x).$$

One easily checks that $F^{\gamma\delta}$ satisfies the same conditions with the same constants as F does (perhaps in (12) one should replace δ_N with $\delta_N \wedge 1$). The same is true for G, which follows from Corollary 4 applicable in $\mathrm{Dom}^o_F(t,x) \cap \{|\mathsf{u}| < M\}$, since, as in the proof of Theorem 11 for any $M \in (0,\infty)$ there exists a constant $N \in (0,\infty)$ such that at all points of differentiability of $F(\mathsf{u}, t, x)$ with respect to u

$$|D_{\mathsf{u}} F(\mathsf{u}, t, x)| \leq N$$

as long as $\mathsf{u} \in \mathrm{Dom}^o_F(t,x) \cap \{|\mathsf{u}| < M\}$.

To prove the theorem, it now suffices to show that $G[u] = G[v] = 0$ in Q (a.e.). Observe that

$$\sup_{\delta \geq 0} \frac{1}{\gamma + \delta + 1} \big[F[u] - \gamma(u - \psi_1) - \delta(u - \psi_2) \big]$$

$$= \max \Big[\psi_2 - u, \ \frac{1}{\gamma + 1} \big(F[u] - \gamma(u - \psi_1) \big) \Big],$$

$$G[u] = \max \Big(\psi_2 - u, \ \min \big(F[u], \ \psi_1 - u \big) \Big).$$

A simple analysis of this representation of $G[u]$ and the assumptions of the theorem enables the reader to prove that $G[u] = 0$ in Q (a.e.). Similarly, $G[v] = 0$ in Q (a.e.). The theorem is proved. □

Here comes the maximum principle for parabolic equations in a strip. Recall that

$$C_{T,R} = [0, T) \times B_R.$$

15. THEOREM. *Take $T \in (0, \infty)$. For every $R > 0$, let there exist constants $\varepsilon \in (0, R^{-1})$, $K_1 \geq 0$, and $K \geq 0$ and functions h and $\delta_N > 0$ such that the conditions associated with* (10), (11), *and* (12) *hold with $C_{T,R}$ in place of Q.*

Let $u, v \in W^{1,2}_{d+1,\text{loc}}\big((0, T) \times \mathbb{R}^d\big)$ be continuous in $[0, T] \times \mathbb{R}^d$. Suppose that (9) *holds in $(0, T) \times \mathbb{R}^d$ (a.e.) for $w = u, v$, $F[u] \geq F[v]$ in $(0, T) \times \mathbb{R}^d$ (a.e.), and $u(T, x) \leq v(T, x)$ on $\mathbb{R}^d$.*

Finally, let there exist a continuous in $[0, T] \times \mathbb{R}^d$ function $\psi \geq 0$ such that $\psi \in W^{1,2}_{d+1,\text{loc}}\big((0, T) \times \mathbb{R}^d\big)$, let $H_1[\psi] \leq 0$ (see (14)*) in $(0, T) \times \mathbb{R}^d$ (a.e.), and for any R_1, $\delta > 0$, let there exist $R > R_1$ such that $u \leq v + \delta\psi$ whenever $|x| = R$. Then $u \leq v$ in $[0, T] \times \mathbb{R}^d$.*

Proof. Take a point $(t_0, x_0) \in (0, T) \times \mathbb{R}^d$, a constant $\delta > 0$, and find $R > |x_0|$ for which on $\partial' C_{T,R}$ we have

$$u - (v + \delta\psi) \leq 0.$$

In light of $H_1[\psi] \leq 0$, one proves that

$$H_1\big[u - (v + \delta\psi)\big] \geq 0$$

in $(0, T) \times \mathbb{R}^d$ (a.e.) as in the proof of Theorem 11. By that theorem, $u - (v + \delta\psi) \leq 0$ in $C_{T,R}$, and of course, $u(t_0, x_0) \leq v(t_0, x_0) + \delta\psi(t_0, x_0)$. This proves the theorem since δ and (t_0, x_0) are arbitrary. □

1:2. Elliptic operators. Any point of

$$\mathfrak{U} = \mathbb{R}^{d+1} \times \mathbb{S}$$

we represent as

$$\mathsf{u} = (\mathsf{u}', \mathsf{u}''), \quad \mathsf{u}' = \big(\mathsf{u}'_0, \mathsf{u}'_1, ..., \mathsf{u}'_d\big) \in \mathbb{R}^{d+1}, \quad \mathsf{u}'' \in \mathbb{S}.$$

Recall that in this section Ω is a domain in $\mathbb{R}^d$ such that

$$\Omega \subset B_R.$$

16. ASSUMPTION. We are given a real-valued function $F(\mathsf{u},x)$ defined for $x\in\mathbb{R}^d$ and for u in the closure Dom_F of an open convex set $\mathrm{Dom}^o_F(x)\subset\mathfrak{U}$. This function is locally Lipschitz continuous with respect to u in $\mathrm{Dom}_F(x)$ for any x.

17. NOTATION. For any x, at all points of differentiability of F with respect to $\mathsf{u}=(\mathsf{u}',\mathsf{u}'')\in\mathrm{Dom}^o_F(x)$ define

$$a^{ij}=D_{\mathsf{u}''_{ij}}F,\quad b^i=D_{\mathsf{u}'_i}F,\quad c=-D_{\mathsf{u}'_0}F,$$

$$a=\big(a^{ij}\big)_{i,j=1}^d,\quad b=\big(b^i\big)_{i=1}^d.$$

At those points at which F is not differentiable with respect to u and which compose a set of measure zero we set $a=(\delta^{ij})$, $b=0$, and $c=0$. In this way we obtain the functions

$$(a,b,c)=(a,b,c)(\mathsf{u},x).$$

Next, for regular enough functions $w(x)$, such that

$$\big(w(x),Dw(x),D^2w(x)\big)\in\mathrm{Dom}_F(x)\tag{17}$$

for almost all $x\in D$, introduce

$$F[u](x)=F\big(u(x),Du(x),D^2u(x),x\big).$$

We state the following two theorems on the maximum principle for elliptic operators without proofs, that are easily obtained by mimicking the proof of Theorem 11 and taking into account that, in the proofs of Theorem 18 and 19, Theorems 3.3.1 and 3.3.4, respectively, should be used instead of Theorem 3.2.3.

18. THEOREM. *Assume that for all $x\in\Omega$ and $\mathsf{u}\in\mathrm{Dom}^o_F(x)$, the above introduced functions satisfy*

$$a=a^*\ge 0,\quad c\ge 0,\tag{18}$$

$$|b|\le\varepsilon\,\mathrm{tr}\,a+K_1\Big[a^{ij}\frac{x^ix^j}{|x|^2}+c\Big]\quad(0/0:=0).\tag{19}$$

Also assume that for any $N\in(0,\infty)$ there exists a measurable function $\delta_N=\delta_N(x)>0$ on Ω such that for all $x\in\Omega$ and $\mathsf{u}\in\mathrm{Dom}^o_F(x)$, with $|\mathsf{u}|\le N$, we have

$$\mathrm{tr}\,a+c\ge\delta_N.\tag{20}$$

Let $u,v\in W^2_{d,\mathrm{loc}}(\Omega)\cap C(\bar\Omega)$, and assume that (17) *holds in Ω (a.e.) with $w=u,v$ and $F[u]\ge F[v]$ in Ω (a.e.). Then in Ω we have*

$$u-v\le\max_{\partial\Omega}(u-v)_+.$$

19. THEOREM. *Theorem* 18 *holds true if conditions* (19) *and* (20) *are replaced with*

$$|b|\le h\sqrt[d]{\det a},\quad \mathrm{tr}\,a\ge\delta_N,$$

where $h\in L_d(\Omega)$.

Results analogous to Theorem 14 and 15 can also be stated for elliptic operators. We do not do that in order not to clutter the exposition and instead present two examples in which the reader will see the need for the objects like Dom^o_F.

20. EXAMPLE. Take $p > d$ and a sufficiently regular Ω for which $W^2_p(\Omega) \subset C^1(\bar{\Omega})$. Also let $f > 0$ on Ω.

We claim that, in $W^2_p(\Omega)$, the equation

$$\det(D_{ij}u - D_iuD_ju - \delta_{ij}u) = f \quad \text{in } \Omega \text{ (a.e.)} \tag{21}$$

can have only one solution for which the matrix

$$\big(D_{ij}u - D_iuD_ju - \delta_{ij}u\big)_{i,j=1}^d \geq 0 \quad \text{in } \Omega \text{ (a.e.)}$$

and u takes prescribed boundary values.

To prove this, suppose that we have two solutions u, v with the described properties and with the same boundary data. Define

$$K = 1 + \sup_\Omega \big\{|Du| + |Dv|\big\}.$$

On $\mathfrak{U}$ introduce

$$P(\mathsf{u}) := \big(\mathsf{u}''_{ij} - \mathsf{u}'_i\mathsf{u}'_j - \delta_{ij}\mathsf{u}'_0\big)_{i,j=1}^d, \qquad F(\mathsf{u}) := \det P(\mathsf{u}),$$

and define $\mathrm{Dom}^o_F(x)$ as the set of $\mathsf{u} \in \mathfrak{U}$ such that the matrix $P(\mathsf{u})$ is strictly positive definite and

$$\det P(\mathsf{u}) > f(x), \quad \big|[\mathsf{u}']\big| < K$$

($[\mathsf{u}']$ is defined in (15)). As is easy to see

$$P\big((\mathsf{u}+\mathsf{v})/2\big) \geq (1/2)\big(P(\mathsf{u}) + P(\mathsf{v})\big)$$

and since $\sqrt[d]{\det a}$ is a concave functions on $\{a = a^* \geq 0\}$, the sets $\mathrm{Dom}^o_F(x)$ are convex.

For $\mathsf{u} \in \mathrm{Dom}^o_F(x)$ set $M(\mathsf{u}) = P^{-1}(\mathsf{u})\det P(\mathsf{u})$. Then in the notation introduced before Theorem 18, for $\mathsf{u} \in \mathrm{Dom}^o_F(x)$ we have

$$a(\mathsf{u}, x) = M(\mathsf{u}) > 0, \quad b^i(\mathsf{u}, x) = -2\sum_{j=1}^d M_{ij}(\mathsf{u})\mathsf{u}'_j, \quad c(\mathsf{u}, x) = \mathrm{tr}\, M(\mathsf{u}),$$

$$\big|b(\mathsf{u}, x)\big| \leq 2K \mathrm{tr}\, M(\mathsf{u}) = 2Kc(\mathsf{u}, x).$$

Furthermore, by the inequality between the geometric and the arithmetic means

$$(1/d)c(\mathsf{u}, x) \geq \sqrt[d]{M(\mathsf{u})} = \big(\det P(\mathsf{u})\big)^{(d-1)/d} > f^{(d-1)/d}(x).$$

Finally, by the choice of K, condition (17) is satisfied. By Theorem 18 we conclude that $u = v$ in Ω, as claimed.

In the above discussion, the assumption $f > 0$ is needed only to ensure that (20) is satisfied.

In case $f \geq 0$ a different argument works based on the fact that, by Lemma 2.2.2, in our class of function (21) is equivalent to

$$\inf\big(a^{ij}D_{ij}u - a^{ij}D_iuD_ju - u - d\sqrt[d]{f\det a}\big) = 0 \quad \text{in } \Omega \text{ (a.e.)}, \tag{22}$$

where the infimum is taken over nonnegative definite $a \in \mathbb{S}$ with $\mathrm{tr}\, a = 1$. Now one can apply Theorem 18 and Corollary 4 to

$$F(\mathsf{u}, x) = \inf\big(a^{ij}\mathsf{u}''_{ij} - a^{ij}\mathsf{u}'_i\mathsf{u}'_j - \mathsf{u}'_0 - d\sqrt[d]{f\det a}\big)$$

with

$$\operatorname{Dom}^o_F(x) = \{\mathsf{u} \in \mathfrak{U} : |[\mathsf{u}']| < K\}.$$

21. EXAMPLE. Take the same p and Ω as in Example 20 and concentrate on the equation

$$\det(D_{ij}u + D_iuD_ju - \delta_{ij}u) = f \quad \text{in } \Omega \text{ (a.e.)} \tag{23}$$

for $u \in W^2_p(\Omega)$ satisfying

$$\big(D_{ij}u + D_iuD_ju - \delta_{ij}u\big)^d_{i,j=1} \geq 0$$

in Ω (a.e.). Here using $F(\mathsf{u}) = \det P(\mathsf{u})$, where

$$P(\mathsf{u}) = \big(\mathsf{u}''_{ij} + \mathsf{u}'_i\mathsf{u}'_j - \delta_{ij}\mathsf{u}'_0\big)^d_{i,j=1},$$

does not work because even if $P(\mathsf{u}) \geq 0$, $P(\mathsf{v}) \geq 0$ the inequality $P\big((1/2)(\mathsf{u}+\mathsf{v})\big) \geq 0$ can be false.

Therefore, we rewrite (23) in our class of functions as

$$\inf\big(a^{ij}D_{ij}u + a^{ij}D_iuD_ju - u - d\sqrt[d]{f\det a}\big) = 0, \tag{24}$$

and can easily apply Theorem 18.

It may look surprising that Theorem 18 does not work for (23) but can be applied to an *equivalent* equation (24) (when $P(\mathsf{u}) \geq 0$). However, in the case of equations like $F[u] = 0$, say, the function $a(\mathsf{u}, x)$ is determined by the values of F at points (u, x) lying also outside of the set $\{F(\mathsf{u}, x) = 0\}$ and there are plenty of F's, for which this set is the same and, hence, the equations are equivalent. Therefore, it may happen that the condition $a(\mathsf{u}, x) \geq 0$ is satisfied only for some of those F's.

2. Passage to the limit for nonlinear operators in the Sobolev classes

While investigating functions $F(x)$ it is often needed to know that $F(x_n) \to F(x)$ as $x_n \to x$. This requires some kind of continuity. However, in applications when x is an element of a Banach space the convergence $x_n \to x$ may be quite weak and one has plenty of examples of continuous F's for which weak convergence $x_n \to x$ does not imply that $F(x_n) \to F(x)$. Remarkably, fully nonlinear elliptic and parabolic second-order operators present an exception which we discuss in this section.

First we show the key idea on a model case.

1. DEFINITION. Let X be a Banach space and let $F[x]$ be an operator defined on $D_F \subset X$ with values in X. The *operator* F is said to be *antipositive*, if there exists a constant $K \geq 0$ such that

$$\lambda\|x - y\| \leq K\big\|\lambda(x - y) - \big(F[x] - F[y]\big)\big\|$$

for all $\lambda \geq K$ and $x, y \in D_F$.

2. THEOREM. *Let F be an antipositive operator in X, let D_F be a linear subset dense in X, and let $F[x+ty]$ be continuous with respect to $t \in \mathbb{R}$ for any $x, y \in D_F$. Let $x_0 \in D_F$ and $f_0 \in X$. Then $F[x_0] = f_0$ if and only if*

$$\lambda\|x_0 - y\| \leq K\big\|\lambda(x_0 - y) - \big(f_0 - F[y]\big)\big\| \tag{1}$$

for any $y \in D_F$ *and* $\lambda \geq K$. *Furthermore, if* $x_n \in D_F$, $f_n \in X$, $n = 0, 1, ...,$ *are such that* $F[x_n] = f_n$ *for* $n \geq 1$ *and* $x_n \to x_0$, $f_n \to f_0$, *then* $F[x_0] = f_0$.

Proof. "If" part. Take $z \in D_F$, $K^{-1} \geq t > 0$, $\lambda = t^{-1}$, and plug $y = x_0 + tz$ into (1). Then we find

$$\|z\| \leq K\big\|z + f_0 - F[x_0 + tz]\big\|.$$

As $t \downarrow 0$, we see that $\|z\| \leq K\big\|z + f_0 - F[x_0]\big\|$ for every $z \in D_F$. By assumption D_F is dense in X and therefore this inequality holds for any $z \in X$. By taking $z = -\big(f_0 - F[x_0]\big)$, we get $\|z\| = 0$ and $F[x_0] = f_0$. This proves the "if" part in the first assertion. The "only if" part follows by definition.

To prove the second assertion it suffices to observe that the inequality

$$\lambda\|x_n - y\| \leq K\big\|\lambda(x_n - y) - \big(F[x_n] - F[y]\big)\big\|$$

as $n \to \infty$, implies (1). The theorem is proved □

To show how this idea is implemented for differential operators fix constants $K, T, R > 0$ and a bounded domain $Q \subset \mathbb{R}^{d+1}$.

3. DEFINITION. Let $\delta \geq 0$ and let an operator $F : W^{1,2}_{d+1}(Q) \to L_{d+1}(Q)$ be given. The operator F is called a δ-*nondegenerate* $\mathcal{L}$-*type operator* in Q if for any $u, v \in W^{1,2}_{d+1}(Q)$ there exists a linear operator

$$\mathcal{L} = r\partial_t + a^{ij}D_{ij} + b^i D_i - c \tag{2}$$

with measurable coefficients in Q such that in Q (a.e.)

$$F[u] - F[v] = \mathcal{L}(u - v),$$

$$a = a^*, \quad r + \operatorname{tr} a + |b| + |c| \leq K, \tag{3}$$

$$r \geq \delta, \quad \langle a\lambda, \lambda\rangle \geq \delta|\lambda|^2 \quad \forall\lambda \in \mathbb{R}^d, \tag{4}$$

where $a = (a^{ij})$, $b = (b^i)$. When $\delta = 0$, the operator F is merely called an $\mathcal{L}$-*type operator* in Q.

Obviously, δ-nondegenerate $\mathcal{L}$-type operators compose a convex set. The following lemma can be used to find examples of $\mathcal{L}$-type operators.

4. LEMMA. (*a*) *Let* F *be a* δ-*nondegenerate* $\mathcal{L}$-*type operator in* Q. *Then* $G[u] := \big(-F[-u]\big)$ *is also a* δ-*nondegenerate* $\mathcal{L}$-*type operator in* Q *and there exists a constant* N *which depends only on* K *and* d *such that*

$$\big\|F[u] - F[v]\,\big\|_{L_{d+1}(Q)} \leq N\|u - v\|_{W^{1,2}_{d+1}(Q)}.$$

(*b*) *If* $\mathcal{L}$ *is an operator of type* (2), *whose coefficients are measurable in* Q *and satisfy* (3) *and* (4) *in* Q (*a.e.*), *and* $f \in L_{d+1}(Q)$, *then* $F : u \to \mathcal{L}u + f$ *is a* δ-*nondegenerate* $\mathcal{L}$-*type operator.*

(*c*) *Let* $F(\mathsf{u}, t, x)$ *be Borel measurable and satisfy Assumption* 1.8 *with* $\operatorname{Dom}^o_F(t, x) = \mathfrak{U}^t$. *In Notation* 1.9, *let it satisfy* $F(0, \cdot, \cdot) \in L_{d+1}(Q)$, (3), *and* (4) *for all* $(t, x) \in Q$ *and* $\mathsf{u} \in \mathfrak{U}^t$. *Then* $F[u]$ *is a* δ-*nondegenerate* $\mathcal{L}$-*type operator in* Q.

(*d*) *Given a collection* F^n, $n = 1, 2, ...,$ *of* δ-*nondegenerate* $\mathcal{L}$-*type operators in* Q, *the operators*

$$\underline{F}[u] := \inf_n F^n[u], \quad \bar{F}[u] := \sup_n F^n[u]$$

are δ-nondegenerate $\mathcal{L}$-type operators in Q, provided that $\underline{F}[0] \in L_{d+1}(Q)$ *and* $\bar{F}[0] \in L_{d+1}(Q)$, *respectively.*

Proof. Assertions (a) and (b) are obvious. In what concerns (c), observe that Lemma 1.6 guarantees that $F[u] - F[v]$ admits a representation required in Definition 3. This and the fact that $F[0] \in L_{d+1}(Q)$ implies that the operator F maps $W^{1,2}_{d+1}(Q)$ into $L_{d+1}(Q)$.

To prove (d) introduce A as the set of $(r, a, b, c) \in \mathbb{R} \times \mathbb{S} \times \mathbb{R}^d \times \mathbb{R}$ satisfying (3) and (4) and introduce

$$H(\mathsf{u}) = \sup_A \left[r\mathsf{u}_0 + a^{ij}\mathsf{u}''_{ij} + b^i\mathsf{u}'_i - c\mathsf{u}'_0 \right].$$

Clearly

$$\bar{F}[u] - \bar{F}[v] \leq H\big(u - v, D(u - v), D^2(u - v), \partial_t(u - v)\big).$$

This, Lemma 1.5, and the proof of (1.7) easily convince us that $\bar{F}[u] - \bar{F}[v]$ admits a representation required in Definition 3. Then the fact that $\bar{F}[0] \in L_{d+1}(Q)$ imply that the operator $\bar{F}$ maps $W^{1,2}_{d+1}(Q)$ into $L_{d+1}(Q)$.

Similar argument works for $\underline{F}[u]$. The lemma is proved. □

It turns out that δ-nondegenerate $\mathcal{L}$-type operators are antipositive.

5. LEMMA. *Let* $\overset{0}{W}{}^{1,2}_{d+1}(C_{T,R})$ *be the subset of* $W^{1,2}_{d+1}(C_{T,R})$ *consisting of functions which are equal to zero on* $\partial' C_{T,R}$. *Let* $\delta > 0$ *and* F *be a δ-nondegenerate* $\mathcal{L}$*-type operator in* $C_{T,R}$. *Then* F *defined on* $\overset{0}{W}{}^{1,2}_{d+1}(C_{T,R})$ *is an antipositive operator in* $L_{d+1}(C_{T,R})$. *Moreover, there exists a constant* N *which depends only on* K, d, *and* δ, *such that*

$$\lambda \|(u - v)_\pm\|_{L_{d+1}(C_{T,R})} \leq N \big\| \big(\lambda(u - v) - \big(F[u] - F[v]\big)\big)_\pm \big\|_{L_{d+1}(C_{T,R})} \tag{5}$$

for all $\lambda \geq N$, $u, v \in \overset{0}{W}{}^{1,2}_{d+1}(C_{T,R})$.

Proof. Definition 3 shows that we only need to prove (5) with $\mathcal{L}(u-v)$ in place of $F[u] - F[v]$. Introducing $w = \pm(u - v)$ reduces our task to proving that for a constant $N = N(K, d, \delta)$ we have

$$\lambda \|w_+\|_{L_{d+1}(C_{T,R})} \leq N \big\| (\lambda w - \mathcal{L}w)_+ \big\|_{L_{d+1}(C_{T,R})} \tag{6}$$

as long as $\lambda \geq N$, $w \in \overset{0}{W}{}^{1,2}_{d+1}(C_{T,R})$, and $\mathcal{L}$ is like in (2) with the coefficients satisfying (3) and (4).

Bearing in mind obvious approximations allows us to suppose that r, a^{ij}, b^i, c are in $C^\infty(\mathbb{R}^{d+1})$. Furthermore, from linear theory we know that any $w \in \overset{0}{W}{}^{1,2}_{d+1}(C_{T,R})$ can be written as $w = w_{(+)} - w_{(-)}$ with $w_{(\pm)} \in \overset{0}{W}{}^{1,2}_{d+1}(C_{T,R})$ such that in $C_{T,R}$ (a.e.)

$$\lambda w_{(\pm)} - \mathcal{L}w_{(\pm)} = (\lambda w - \mathcal{L}w)_\pm.$$

Applying the maximum principle (see Theorem 3.0.1) yields that $w_{(\pm)} \geq 0$ in $C_{T,R}$ and, hence, to prove (6) it suffices to show that

$$\lambda \|w_+\|_{L_{d+1}(C_{T,R})} \leq N \|\lambda w - \mathcal{L}w\|_{L_{d+1}(C_{T,R})} \tag{7}$$

additionally assuming that $\lambda w - \mathcal{L}w \geq 0$ in $C_{T,R}$ (a.e.). Finally, one can approximate $\lambda w - \mathcal{L}w$ by smooth functions f which shows that while proving (7) we may suppose that

$$\lambda w - \mathcal{L}w = f \text{ (a.e.) in } C_{T,R}, \quad f \geq 0, \quad f \in C_0^\infty(\mathbb{R}^{d+1}).$$

Then replace λ in Theorem 2.2.4 by μ defined from

$$K\mu + K\sqrt{\mu} + K = \lambda,$$

and take v from that theorem. Then in $C_{T,R}$ (a.e.)

$$\mathcal{L}v - \lambda v = r\partial_t v + a^{ij}D_{ij}v + b^i D_i v - cv - \lambda v$$

$$\leq \big[\mu(r + \operatorname{tr} a) + |b|\sqrt{\mu} + K - \lambda\big]v - \sqrt[d+1]{r \det a}\, f$$

$$\leq -\delta f.$$

By the maximum principle, $v \geq \delta w$ in $C_{T,R}$ and

$$\delta \|w_+\|_{L_{d+1}(C_{T,R})} \leq \|v\|_{L_{d+1}(C_{T,R})} \leq \frac{1}{\mu(d+1)}\|f\|_{L_{d+1}(\mathbb{R}^{d+1})}.$$

Observe that the inequality between the extreme terms holds for any smooth extension of f outside $C_{T,R}$. It follows that

$$\delta \|w_+\|_{L_{d+1}(C_{T,R})} \leq \frac{1}{\mu(d+1)}\|\lambda w - \mathcal{L}w\|_{L_{d+1}(C_{T,R})}.$$

It only remains to observe that $\mu \geq (3K)^{-1}\lambda$ if $\lambda \geq 3K$. □

Here is a starting result about passing to the limit in fully nonlinear parabolic operators.

6. THEOREM. *Let $\delta > 0$ and let F be a δ-nondegenerate $\mathcal{L}$-type operator in $C_{T,R}$, $u^n \in \overset{0}{W}{}^{1,2}_{d+1}(C_{T,R})$, $n = 0, 1, \ldots$, and $u^n \to u^0$ in $L_{d+1}(C_{T,R})$ as $n \to \infty$. Then*

(a) *if $F[u^n] \to f$ in $L_{d+1}(C_{T,R})$ as $n \to \infty$, then in $C_{T,R}$ (a.e.)*

$$F[u^0] = f;$$

(b) *if $\sup_n F[u^n] \in L_{d+1}(C_{T,R})$, then in $C_{T,R}$ (a.e.)*

$$\varlimsup_{n\to\infty} F[u^n] \geq F[u^0];$$

(c) *if $\inf_n F[u^n] \in L_{d+1}(C_{T,R})$, then in $C_{T,R}$ (a.e.)*

$$F[u^0] \geq \varliminf_{n\to\infty} F[u^n].$$

Proof. Assertion (a) is a simple corollary of Lemmas 4 and 5 and Theorem 2, assertions (c) and (b) are equivalent, which is seen after replacing $F[u]$ with $(-F[-u])$ (see Lemma 4). To prove (b), introduce

$$f^n = F[u^n], \quad f = \varlimsup_{n\to\infty} f^n$$

and follow the proof of Theorem 2. Dropping the obvious subscripts we get by Lemma 5 that

$$\lambda\|(u-u^n)_+\| \le N\|\big(\lambda(u-u^n)-F[u]+f^n\big)_+\|$$

whenever $\lambda \ge N$ and $u \in \overset{0}{W}{}^{1,2}_{d+1}(C_{T,R})$.

Observe that $f^n \le \sup_n f^n \in L_{d+1}(C_{T,R})$, which allows us to use Fatou's lemma on the right. It follows that

$$\lambda\|(u-u^0)_+\| \le N\|\big(\lambda(u-u^0)-F[u]+f\big)_+\|.$$

We take $w \in \overset{0}{W}{}^{1,2}_{d+1}(C_{T,R})$, substitute $u = u^0 + \lambda^{-1}w$, and let $\lambda \to \infty$, which yields

$$\|w_+\| \le N\|\big(w-F[u^0]+f\big)_+\|.$$

This holds for any $w \in \overset{0}{W}{}^{1,2}_{d+1}(C_{T,R})$ and, by continuity, for any $w \in L_{d+1}(C_{T,R})$. By taking $w = F[u^0] - f \vee k$ we see that $w_+ = 0$, $w \le 0$. Thus, $F[u^0] \le f \vee k$ in $C_{T,R}$ (a.e.) for any k. The theorem is proved. □

The assumption in Theorem 6 that the u^n's equal zero on $\partial' C_{T,R}$ is certainly unnatural. We remove it by using the following lemma.

7. LEMMA. *Let an operator $\mathcal{L}$ be as in (2) and let in $C_{T,R}$,*

$$c>0, \quad r+\operatorname{tr} a+|b| \le K_1 c, \quad r \ge 0, \quad a = a^* \ge 0,$$

where K_1 is a constant. Take a function $u \in W^{1,2}_{d+1}(C_{T,R})$ such that $\mathcal{L}u \ge 0$ in $C_{T,R}$ (a.e.) and define a number μ as the positive root of the equation

$$1 = K_1(\mu+\mu^2).$$

Then in $C_{T,R}$,

$$u(t,x) \le \frac{\cosh \mu|x|}{\cosh \mu R} \sup_{[0,T]\times\partial B_R} u_+ + e^{(t-T)/K_1} \sup_{\{T\}\times B_R} u_+ =: w. \tag{8}$$

Proof. By the maximum principle, it suffices to check that $\mathcal{L}w \le 0$ in $C_{T,R}$. By using the fact that $\sinh\mu|x| \le \cosh\mu|x|$ and $\sinh\mu|x| \le \mu|x|\cosh\mu|x|$, one gets

$$\begin{aligned}\mathcal{L}\cosh\mu|x| &= -\,c\cosh\mu|x| + \mu b^i\frac{x^i}{|x|}\sinh\mu|x| \\ &\quad + \mu^2 a^{ij}\frac{x^i x^j}{|x|^2}\cosh\mu|x| + \mu^2\frac{1}{\mu|x|}\Big[\operatorname{tr} a - a^{ij}\frac{x^i x^j}{|x|^2}\Big]\sinh\mu|x| \\ &\le \big(-c+\mu|b|+\mu^2\operatorname{tr} a\big)\cosh\mu|x| \le 0.\end{aligned}$$

Furthermore,

$$\mathcal{L}e^{(t-T)/K_1} = e^{(t-T)/K_1}(K_1^{-1}r - c) \le 0.$$

The lemma is proved. □

8. LEMMA. *Let $\delta > 0$ and let F be a δ-nondegenerate $\mathcal{L}$-type operator in $C_{T,R}$. Take $\tau \in [0,T]$ and $\rho \in [0,R]$. Then there exists a constant N, which depends only on K, d, and δ, such that, for any $\lambda \ge N$ and $u, v \in W^{1,2}_{d+1}(C_{T,R})$,*

$$\begin{aligned}\lambda\|(u-v)_\pm\|_{L_{d+1}(C_{\tau,\rho})} &\le \lambda\|\varphi_\lambda\|_{L_{d+1}(C_{\tau,\rho})}\|(u-v)_\pm\|_{C(\partial' C_{T,R})} \\ &\quad + N\|\big(\lambda(u-v)-(F[u]-F[v])\big)_\pm\|_{L_{d+1}(C_{T,R})},\end{aligned}$$

where

$$\varphi_\lambda(t,x) = \frac{\cosh \mu|x|}{\sinh \mu R} + \exp\big((\lambda K^{-1} - 1)(t-T)\big)$$

and μ is the positive root of the equation $\lambda - K = K(\mu + \mu^2)$.

Proof. As in Lemma 5, we may replace F with a linear operator $\mathcal{L}$ of the form (2) whose coefficients are smooth. In short, it suffices to prove that

$$\lambda\|w_+ |_{L_{d+1}(C_{\tau,\rho})} \leq \lambda\|\varphi_\lambda\|_{L_{d+1}(C_{\tau,\rho})}\|w_+\|_{C(\partial' C_{T,R})}$$

$$+ N\big\|(\lambda w - \mathcal{L}w)_+\big\|_{L_{d+1}(C_{T,R})}$$

provided that $\lambda \geq N$ and $w \in W^{1,2}_{d+1}(C_{T,R})$.

We write $w = w_0 + w_1$, where $w_1 \in W^{1,2}_{d+1}(C_{T,R})$, $w_0 = 0$ and $w_1 = w$ on $\partial' C_{T,R}$,

$$\mathcal{L}w_0 - \lambda w_0 = \mathcal{L}w - \lambda w, \quad \mathcal{L}w_1 - \lambda w_1 = 0$$

in $C_{T,R}$ (a.e.). Then Lemma 5 implies that

$$\lambda\big\|(w_0)_+\big\|_{L_{d+1}(C_{T,R})} \leq N\big\|(\lambda w - \mathcal{L}w)_+\big\|_{L_{d+1}(C_{T,R})} \tag{9}$$

for $\lambda \geq N$. Observe that, if $\lambda > K$ and $K_1 := K(\lambda - K)^{-1}$, then $c + \lambda > 0$ and

$$r + \operatorname{tr} a + |b| \leq K_1(c + \lambda).$$

This allows us to use Lemma 7 and yields

$$w_1 \leq \varphi_\lambda \|w_+\|_{C(\partial' C_{T,R})}.$$

It only remains to recall (9) and use

$$w_+ = (w_0 + w_1)_+ \leq (w_0)_+ + (w_1)_+.$$

The lemma is proved. □

Here is one of the main results of this section.

9. THEOREM. *Let $\delta > 0$ and let F be a δ-nondegenerate $\mathcal{L}$-type operator in Q. Let F_0 be an $\mathcal{L}$-type operator in Q, $u^n \in W^{1,2}_{d+1,\mathrm{loc}}(Q) \cap C(\bar{Q})$, $n = 0, 1, \ldots$, and assume that u^n converges to u^0 in Lebesgue measure on Q as $n \to \infty$. Then*

(a) if $\sup_n \big\|u^n_-\big\|_{C(\partial' Q)} < \infty$ and $\sup_n F[u^n] \in L_{d+1}(Q)$, then in Q (a.e.)

$$\overline{\lim_{n\to\infty}}\, F_0[u^n] \geq F_0[u^0]; \tag{10}$$

(b) if $\inf_n \big\|u^n_+\big\|_{C(\partial' Q)} < \infty$ and $\inf_n F[u^n] \in L_{d+1}(Q)$, then in Q (a.e.)

$$F_0[u^0] \geq \underline{\lim_{n\to\infty}}\, F_0[u^n].$$

Proof. The assertions (a) and (b) are equivalent (see Lemma 4). We prove (a) first in case that $F_0 = F$. By using Definition 3, for each n, we find $\mathcal{L}^n$ for which

$$\mathcal{L}^n(-u^n) = -\mathcal{L}^n u^0 - F[u^n] + F(u^0),$$

in Q (a.e.). Then observe that by Theorem 3.2.6 the functions $(-u^n)$ are uniformly bounded from above on Q because $\sup_n F[u^n] \in L_{d+1}(Q)$, i.e.,

$$\sup_n \|u^n_-\|_{C(Q)} < \infty.$$

Note also that (10) holds in Q if it holds in every cylinder contained in Q. Take $\bar{C}_{T,R} \subset Q$, $\tau \in [0,T)$, and $\rho \in [0,R)$. Introduce

$$f^n = F[u^n], \quad f = \overline{\lim} f^n$$

and mimick the proof of Theorem 6. Lemma 8 shows that

$$\lambda\|(u-u^n)_+\|_{L_{d+1}(C_{T,\rho})} \le \lambda\|\varphi_\lambda\|_{L_{d+1}(C_{\tau,\rho})}\|(u-u^n)_+\|_{C(Q)}$$
$$+ N\|\big(\lambda(u-u^n) - F[u] + f^n\big)_+\|_{L_{d+1}(C_{T,\rho})} \tag{11}$$

if $\lambda \ge N$ and $u \in W^{1,2}_{d+1}(C_{T,R})$. Here the uniformly bounded functions $(u-u^n)_+$ converge to $(u-u^0)_+$ in measure. This allows us to apply the Lebesgue dominated convergence theorem to the left-hand side of (11). In the first term in the right-hand side replace f^n with $\sup\{f^n : n \ge n_0\}$, then let $n \to \infty$ and $n_0 \to \infty$ in that order both times using the dominated convergence theorem. This leads us to

$$\lambda\|(u-u^0)_+\|_{L_{d+1}(C_{T,\rho})} \le \lambda\|\varphi_\lambda\|_{L_{d+1}(C_{\tau,\rho})}\big(\|u\|_{C(Q)} + N\big)$$
$$+ N\|\big(\lambda(u-u^0) - F[u] + f\big)_+\|_{L_{d+1}(C_{T,\rho})}. \tag{12}$$

Take $w \in W^{1,2}_{d+1}(Q)$, plug $u = u^0 + \lambda^{-1}w$ into (12), and let first $\lambda \to \infty$ and then $\tau \uparrow T$ and $\rho \uparrow R$. The explicit form of φ_λ (see Lemma 8) shows that

$$\lim_{\lambda\to\infty} \lambda\|\varphi_\lambda\|_{L_{d+1}(C_{\tau,\rho})} = 0$$

for any $\tau < T$ and $\rho < R$. Therefore, the first term on the right in (12) drops out and we finish the proof of this theorem for $F_0 = F$ as that of Theorem 6.

To consider the general case take constants $m > 0$ and $\varepsilon \in (0,1]$ and notice that

$$m : u \to m, \quad u \to \min\big(F_0[u], m\big), \quad u \to \varepsilon F[u] + (1-\varepsilon)\min\big(F_0[u], m\big)$$

are $\mathcal{L}$-type operators in Q in light of Lemma 4. Also, the third one is clearly $\varepsilon\delta$-nondegenerate and, in addition,

$$\sup_n \Big(\varepsilon F[u^n] + (1-\varepsilon)\big(m \wedge F_0[u^n]\big)\Big) \le \varepsilon \sup_n F[u^n] + m \in L_{d+1}(Q).$$

Hence, by the above result established in a particular case (a.e.)

$$\begin{aligned} &\varepsilon \sup_n F[u^n] + (1-\varepsilon)\big(m \wedge \varlimsup_{n\to\infty} F_0[u^n]\big) \\ &\geq \varlimsup_{n\to\infty} \Big(\varepsilon F[u^n] + (1-\varepsilon)\big(m \wedge F_0[u^n]\big)\Big) \\ &\geq \varepsilon F[u^0] + (1-\varepsilon)\big(m \wedge F_0[u^0]\big). \end{aligned}$$

Upon letting $\varepsilon \downarrow 0$ and $m \to \infty$, we get (10). The theorem is proved. □

10. REMARK. There are examples showing that under the assumptions under which (a) is proved it may happen that $\sup \|u^n\|_{C(Q)} = \infty$ and that $u^n \not\to u^0$ in $L_{d+1}(Q)$. Generally, the results of this section are taken from Section 3.5 of [**61**]. They generalize earlier results for elliptic equations by the author [**54**] (1971) and Evans [**33**] (1978). The methods and assumptions in [**54**] and [**33**] are different. In [**54**] only the Isaacs equations are considered and in [**33**], in contrast with [**54**], u^n is required to weakly converge to u^0 in W^2_p with $p > d$, which by embedding theorems implies that $u^n \to u$ in C^1. Oddly enough, the authors of [**8**] claim that the same weak convergence is required in [**54**]. The methods in [**54**] are quite transparent and are based on expressing the solution of the equation $F[u] = -f$ in the form $u = R_\lambda(\lambda u + f)$, $\lambda > 0$, where R_λ is a nonlinear integral operator continuous in L_p. It is easy to pass to the limit under the sign of R_λ. In addition, it turns out that if $u \in W^2_p$, then $\lambda\big[R_\lambda(\lambda u + f) - u\big] \to F[u] + f$ as $\lambda \to \infty$. Later on it became clear that the above integral representations are equivalent to having (1) that possesses the same features as the integral representations.

11. COROLLARY. *Let $\delta > 0$ and introduce $A(\delta)$ as the set of $(r, a, b, c) \in \mathbb{R} \times \mathbb{S} \times \mathbb{R}^d \times \mathbb{R}$ satisfying* (3) *and* (4). *Let $u^n \in W^{1,2}_{d+1,\mathrm{loc}}(Q) \cap C(\bar{Q})$, $n \geq 0$. Assume that u^n converges to u^0 in measure in Q and u^n are uniformly bounded on $\partial' Q$. Also, for any $n \geq 1$, let there exist an operator $\mathcal{L}^n$, whose measurable coefficients belong to $A(\delta)$ for any (t, x), such that $\mathcal{L}^n u^n = 0$ in Q (a.e.). Then there exists an operator $\mathcal{L}^0$, with measurable coefficients belonging to $A(\delta)$ for any (t, x) such that $\mathcal{L}^0 u^0 = 0$ in Q (a.e.).*

Proof. Indeed, introduce

$$H(\mathsf{u}) = \sup_{A(\delta)} \big[r\mathsf{u}_0 + a^{ij}\mathsf{u}''_{ij} + b^i\mathsf{u}'_i - c\mathsf{u}'_0\big],$$

$$H[u] = H(u, Du, D^2u, \partial_t u),$$

and observe that for any n,

$$-H[-u^n] \leq \mathcal{L}^n u^n = 0, \quad H[u^n] \geq \mathcal{L}^n u^n = 0$$

on Q. In the definition of H on can replace $A(\delta)$ with any countable dense subset and, as $n \to \infty$, it follows from Lemma 4 and Theorem 9 that on Q (a.e.)

$$-H[-u^0] \leq 0 \leq H[u^0].$$

Next, find a measurable $s \in [0, 1]$ for which $F[u^0] = 0$ in Q (a.e.), where

$$F[u] := -sH[-u] + (1-s)H[u].$$

We know that F is a δ-nondegenerate $\mathcal{L}$-type operator. Therefore by definition there exists an operator $\mathcal{L}^0$ with the coefficients from $A(\delta)$ such that $0 = F[u^0] - F[0] = \mathcal{L}^0 u^0$ in Q (a.e.). □

This corollary has some relations to the G-convergence of linear operators (see Zhikov and Sirazhudinov [**117**], Zhikov, Kozlov, and Oleinik [**118**] and Krylov [**60**]).

12. COROLLARY. *Let $\delta > 0$ and let $\mathcal{L}^n$, $n = 1, 2, \ldots$, be a sequence of operators with measurable coefficients belonging to $A(\delta)$ for any (t, x) ($A(\delta)$ are introduced in Corollary 11). Let $u^n \in W^{1,2}_{d+1,\mathrm{loc}}(Q) \cap C(\bar{Q})$ and let $f \geq 0$ be a function on Q such that*

$$\mathcal{L}^n u^n \leq -f$$

(a.e.) for all n. Suppose that $u^n \to 0$ in measure in Q and u^n are uniformly bounded on $\partial' Q$. Then $f = 0$ (a.e.).

Indeed, for $F[u] = -H[-u]$, where $H[u]$ is taken from the proof of Corollary 11, we have $F[u^n] \leq -(f \wedge m)$ for any m. Theorem 9 (a) implies that $F[0] \leq -(f \wedge m)$, $0 \leq -(f \wedge m)$, and $f = 0$ (a.e.).

This corollary is a rather weak version of Corollary 9.1.5.

13. THEOREM. *Let $\delta > 0$ and let F be a δ-nondegenerate operator of $\mathcal{L}$-type in Q. Let $u^n \in W^{1,2}_{d+1,\mathrm{loc}}(Q) \cap C(\bar{Q})$, $n = 0, 1, \ldots$, be such that u^n converge to u^0 in measure in Q and $F[u^n] \to f$ in $L_{d+1}(Q)$. Finally, assume that u^n are uniformly bounded on $\partial' Q$. Then $F[u^0] = f$ in Q (a.e.).*

Proof. Let a subsequence $n_k \to \infty$ be such that

$$\big\| f - F[u^{n_k}] \big\|_{L_{d+1}(Q)} \leq 2^{-k}.$$

From the sequence $F[u^{n_k}]$ we can extract a further subsequence converging to f almost everywhere. This brings us to a sequence $v^n \in W^{1,2}_{d+1,\mathrm{loc}}(Q) \cap C(\bar{Q})$ for which $F[v^n] \to f$ in Q (a.e.), $v^n \to u^0$ in measure, and

$$\big\| f - F[v^n] \big\|_{L_{d+1}(Q)} \leq 2^{-n}.$$

Then

$$\big\| \sup_n |F[v^n]| \, \big\|_{L_{d+1}(Q)} \leq \|f\|_{L_{d+1}(Q)} + \big\| \sup_n |f - F[v^n]| \, \big\|_{L_{d+1}(Q)}$$

$$\leq |f\|_{L_{d+1}(Q)} + \sum_n \big\| f - F[v^n] \big\|_{L_{d+1}(Q)} < \infty.$$

Hence,

$$\sup_n |F[v^n]| \in L_{d+1}(Q).$$

Now, by Theorem 9,

$$f = \varlimsup_{n \to \infty} F[v^n] \geq F[u^0] \geq \varliminf_{n \to \infty} F[v^n] = f \quad \text{in } Q \text{ (a.e.)}.$$

The theorem is proved. □

14. COROLLARY. *Let $\delta > 0$ and let F be a δ-nondegenerate operator of $\mathcal{L}$ type in Q. Let Γ be a subset of $W^{1,2}_{d+1,\mathrm{loc}}(Q) \cap C(\bar{Q})$ which is closed and compact as a subset of $L_{d+1}(Q)$. Finally, let the restriction of functions in Γ to $\partial' Q$ make up a bounded set in $C(\partial' Q)$. Then the set $H := \{F[u] : u \in \Gamma\}$ is closed in $L_{d+1}(Q)$.*

Proof. If $f^n \in H$ and $f^n \to f$ in $L_{d+1}(Q)$, there exist $u^n \in \Gamma$ such that $F[u^n] = f^n$. By assumption there is a $u^0 \in \Gamma$ for which $u^n \to u^0$ in $L_{d+1}(Q)$ possibly along a subsequence. Theorem 13 says that $F[u^0] = f$ and $f \in H$ as asserted. □

We now extend the above results to nonlinear elliptic operators.

15. THEOREM. *Let Ω be a bounded domain in $\mathbb{R}^d$. In Definition 3, we replace r, $W^{1,2}_{d+1}(Q)$, $L_{d+1}(Q)$, and Q by 0, $W^2_d(\Omega)$, $L_d(\Omega)$, and Ω, respectively, drop the condition $r \geq \delta$, and regard the resulting definition as the definition of a δ-nondegenerate $\mathcal{L}$-type operator in Ω. Then after replacing*

$$d+1, \quad C_{T,R}, \quad \partial' C_{T,R}, \quad \partial' Q, \quad Q, \quad W^{1,2}_{d+1}, \quad L_{d+1}$$

by

$$d, \quad B_R, \quad \partial B_R, \quad \partial\Omega, \quad \Omega, \quad W^2_d, \quad L_d,$$

respectively, Lemmas 5, 7, *and* 8, *Theorems* 6, 9, 13, *and Corollaries* 11, 12, *and* 14 *remain true.*

Proof. To obtain this result, it suffices to repeat almost word for word the proofs of the above results numbered as 5, 6, 9, and 11-14 applying Theorem 2.2.3 in places where Theorem 2.2.4 is applied, to restate Lemma 7 in an appropriate obvious way, and to set $\varphi_\lambda = (\cosh \mu|x|)(\cosh \mu R)^{-1}$ in Lemma 8. □

CHAPTER 5

Finite-difference equations of elliptic type

The results of this chapter could have been obtained from those of Chapter 7 devoted to parabolic equations (as we do such thing in Section 6.6). However, the arguments in Chapter 7 are much more involved and just to better explain the underlying simple ideas the author decided to put the elliptic case first.

The goal of this chapter is to prepare some results which are necessary for proving existence theorems for fully nonlinear elliptic *differential* equations without assuming any structure of the equations but instead slightly modifying them by using special cut-off operators.

For the moment there is no other approaches to these results apart from going through finite-difference approximations. Therefore we deal with the necessary issues first at the level of finite-difference equations. This chapter presents an extended exposition of part of [**79**].

1. Some elementary technical tools

Recall that the operators $T_{h,l}$, $\delta_{h,l}$, and $\Delta_{h,l}$ are introduced in (1.1.1) for $l \in \mathbb{R}^d$ and $h > 0$.

1. Lemma. *For any $\xi, \eta \in \mathbb{R}^d$ and functions $a(x), u(x)$*

$$\delta_{h,\xi}(au) = (\delta_{h,\xi}a)u + (T_{h,\xi}a)\delta_{h,\xi}u = a\delta_{h,\xi}u + u\delta_{h,\xi}a + h(\delta_{h,\xi}a)\delta_{h,\xi}u,$$

$$\begin{aligned}\delta_{h,\xi}\delta_{h,\eta}(au) &= a\delta_{h,\xi}\delta_{h,\eta}u + (\delta_{h,\xi}a)\delta_{h,\eta}u + (\delta_{h,\eta}a)\delta_{h,\xi}u \\ &\quad + h[\delta_{h,\eta}a + \delta_{h,\xi}a]\delta_{h,\xi}\delta_{h,\eta}u + (\delta_{h,\xi}\delta_{h,\eta}a)T_{h,\xi+\eta}u,\end{aligned} \tag{1}$$

$$\Delta_{h,\xi}(au) = a\Delta_{h,\xi}u + u\Delta_{h,\xi}a + (\delta_{h,\xi}a)\delta_{h,\xi}u + (\delta_{h,-\xi}a)\delta_{h,-\xi}u.$$

In particular,

$$\Delta_{h,\xi}(u^2) = 2u\Delta_{h,\xi}u + (\delta_{h,\xi}u)^2 + (\delta_{h,-\xi}u)^2, \quad \delta_{h,\xi}(u^2) \geq 2u\delta_{h,\xi}u.$$

This lemma is proved by straightforward computations. Before stating one more lemma we introduce the following.

2. Definition. Let B be a finite subset of $\mathbb{R}^d$, $0 \in B$, and let $p(x,y)$ be a real-valued function on $\mathbb{R}^d \times \mathbb{R}^d$. For an $x_0 \in \mathbb{R}^d$ we say that the operator

$$S : u \to Su, \quad Su(x) = \sum_{y \in B} p(x, x+y)u(x+y), \tag{2}$$

respects the maximum principle at x_0 relative to B if, for any function $\phi(x)$ such that $\phi(x_0 + y) \geq \phi(x_0)$ for all $y \in B$, we have $Su(x_0) \geq 0$.

3. LEMMA. *If an operator*

$$Su(x) = \sum_{y \in B} p(x, x+y)u(x+y)$$

respects the maximum principle at a point $x_0 \in \mathbb{R}^d$ *relative to* B *and* u *is a function such that* $u(x_0) \le 0$, *then*

$$-Su \le S(u_-)$$

at x_0. *In particular,*

$$(Su)_- \le \big(S(u_-)\big)_+, \quad \phi_- S\phi \ge -\phi_- S(\phi_-), \quad \phi_-(S\phi)_- \le \phi_- \big|S(\phi_-)\big|$$

at x_0 *for any function* ϕ.

This follows from the definition and the fact that $u + u_- \ge 0$ on $x_0 + B$ and $u + u_- = 0$ at x_0.

For $\rho > 0$ and open $\Omega \subset \mathbb{R}^d$ introduce

$$\Omega^\rho = \{x : \rho_\Omega(x) > \rho\}, \quad \rho_\Omega(x) = \text{dist}\,(x, \mathbb{R}^d \setminus \Omega), \quad \partial_\rho \Omega = \bar{\Omega} \setminus \Omega^\rho$$

and denote by $\rho_{\text{int}}(\Omega)$ the internal radius of Ω, that is the supremum of ρ for which $\Omega^\rho \ne \emptyset$.

4. LEMMA. *Let* Ω *be a domain in* $\mathbb{R}^d$. *Then there exists a constant* $N \in (0, \infty)$ *depending only on* d *and for any* $\rho \in \big(0, \rho_{\text{int}}(\Omega)/2\big)$ *there exists a nonnegative* $\eta_\rho \in C_0^\infty(\Omega)$ *satisfying*

$$\eta_\rho = 1 \quad \textit{on} \quad \Omega^{2\rho}, \quad \eta_\rho = 0 \quad \textit{outside} \quad \Omega^\rho,$$

$$\eta_\rho \le 1, \quad |D\eta_\rho| \le N/\rho, \quad |D^2\eta_\rho| \le N/\rho^2.$$

Proof. Take $\rho \in \big(0, \rho_{\text{int}}(\Omega)/2\big)$ and take a nonnegative $\zeta \in C_0^\infty(\mathbb{R}^d)$ which integrates to one and have support in the unit ball centered at the origin. Then set $\zeta_\rho(x) = (2/\rho)^d \zeta(2x/\rho)$ and

$$\eta_\rho = \zeta_\rho * I_{\Omega^{3\rho/2}}.$$

Then the stated properties of η_ρ are easily obtained. □

2. Solvability of elliptic finite-difference equations

Fix some constants $m, d \in \{1, 2, ...\}$ and $\delta \in (0, 1]$. Also fix a bounded domain $\Omega \subset \mathbb{R}^d$.

1. ASSUMPTION. We are given vectors $l_i \in \bar{B}_1 \subset \mathbb{R}^d$, $i = \pm 1, ..., \pm m$, such that $l_{-i} = -l_i$ and $\text{Span}\,\Lambda = \mathbb{R}^d$, where

$$\Lambda = \{l_i, |i| = 1, ..., m\}.$$

2. REMARK. The reason to use $-l_i$ along with l_i comes from the last formula in (1.1) showing that in our calculations $-l_i$ would appear anyway.

3. ASSUMPTION. (i) We are given a function $\mathcal{H}(z,x)$, $x\in\mathbb{R}^d$,

$$z=(z',z''),\quad z'=(z'_0,z'_{\pm1},...,z'_{\pm m})\in\mathbb{R}^{2m+1},\quad z''=(z''_{\pm1},...,z''_{\pm m})\in\mathbb{R}^{2m},$$

which is Lipschitz continuous with respect to z for any x. At all point of its differentiability with respect to z introduce

$$a_j=a_j(z,x)=D_{z''_j}\mathcal{H}(z,x),\quad b_j=b_j(z,x)=D_{z'_j}\mathcal{H}(z,x),\quad j=\pm1,...,\pm m,$$

$$c=c(z,x)=-D_{z'_0}\mathcal{H}(z,x),$$

and at points of non-differentiability set $a_j=\delta$, $b_j=0$, $c=1$.

(ii) The above introduced functions and $f(x):=\mathcal{H}(0,x)$ are bounded, $|b_j|\le N'$, $|j|=1,...,m$, where N' is a constant, and

$$\delta^{-1}\ge a_j\ge\delta \tag{1}$$

for all indices and values of the arguments.

(iii) We have $c\ge0$.

For any function v on $\mathbb{R}^d$ and $h>0$ define

$$\begin{aligned}\mathcal{H}_h[v](x)&=\mathcal{H}(v,\delta_h v,\delta_h^2 v,x)\\&=\mathcal{H}(v,\delta_{h,l_{\pm1}}v,...,\delta_{h,l_{\pm m}}v,\Delta_{h,l_{\pm1}}v,...,\Delta_{h,l_{\pm m}}v,x),\end{aligned} \tag{2}$$

where

$$\delta_h v=(\delta_{h,l_{\pm1}}u,...,\delta_{h,l_{\pm m}}u), \tag{3}$$

$$\delta_h^2 v=(\Delta_{h,l_{\pm1}}u,...,\Delta_{h,l_{\pm m}}u). \tag{4}$$

Fix a bounded function g on $\bar\Omega$ and consider the equation

$$\mathcal{H}_h[v]=0\quad\text{in}\quad\Omega^h \tag{5}$$

with boundary condition

$$v=g\quad\text{on}\quad\partial_h\Omega. \tag{6}$$

4. REMARK. We do not assume that $\mathcal{H}$ is measurable with respect to x and do not expect the solutions of (5) to be measurable either.

Of course, we will only consider h such that $\Omega^h\ne\emptyset$. The following simple result can be found in [**85**] or in [**71**] or else in [**76**].

5. THEOREM. *Under the above assumptions, there exists a constant* $h_0=h_0(\Omega,\delta,N',m)>0$ *such that for any* $h\in(0,h_0]$ *the problem* (5)-(6) *has a unique bounded solution* v_h.

We prove the theorem after we prove the following, in which

$$D_l=l^iD_i$$

(previously also denoted by $D_{(l)}$).

6. LEMMA. *There exists a function $\Psi_0 \in C^\infty_{\text{loc}}(\mathbb{R}^d)$ such that, in Ω, $\Psi_0 \ge 1$ and*

$$a_j D^2_{l_j}\Psi_0 + \delta^{-1}\sum_j |D_{l_j}\Psi_0| \le -2 \tag{7}$$

and for $0 < h \le h_0 = h_0(\Omega, \delta, N', m)$ in Ω^h

$$a_j \Delta_{h,l_j}\Psi_0 + \delta^{-1}\sum_j |\delta_{h,l_j}\Psi_0| \le -1 \tag{8}$$

whenever $\delta^{-1} \ge a_j \ge \delta$.

Proof. The global barrier Ψ_0 we construct as in the proof of Lemma 3.1.8:

$$\Psi_0(x) = 1 + \cosh(\mu R) - \cosh\big(\mu|x|\big),$$

where R is twice the radius of the smallest ball centered at the origin containing Ω. Then in Ω we have $\Psi_0 \ge 1$ and

$$J := a_j D^2_{l_j}\Psi_0 + \delta^{-1}\sum_j |D_{l_j}\Psi_0| = -\mu|x|^{-1}a_j\big[|l_j|^2\sinh\big(\mu|x|\big) - |x|^{-2}\langle l_j, x\rangle^2\big]$$

$$-\mu^2|x|^{-2}a_j\langle l_j, x\rangle^2\cosh\big(\mu|x|\big) + \mu\delta^{-1}|x|^{-1}\sum_j\big|\langle l_j, x\rangle\big|\sinh\big(\mu|x|\big).$$

Since $\big|l_j\big|^2 \ge |x|^{-2}\langle l_j, x\rangle^2$,

$$J \le -\mu^2|x|^{-2}a_j\langle l_j, x\rangle^2\cosh\big(\mu|x|\big) + \mu\delta^{-1}|x|^{-1}\sum_j\big|\langle l_j, x\rangle\big|\sinh\big(\mu|x|\big).$$

Here $a_j\big|(l_j, x)\big|^2 > 0$ if $x \ne 0$ since $a_j \ge \delta$ and $\text{Span}\,\{l_j\} = \mathbb{R}^d$. Hence, for an $\varepsilon > 0$, $a_j\big|(l_j, x)\big|^2 \ge \varepsilon|x|^2$. Adding to that that $\big|(l_j, x)\big| \le |x|$ and $\sinh\big(\mu|x|\big) \le \cosh\big(\mu|x|\big)$ we conclude that

$$J \le -\mu^2[\varepsilon - \mu^{-1}\delta^{-1}2m]\cosh\big(\mu|x|\big) \le -2,$$

in Ω if μ is large enough. This yields (7).

Finally, we observe that (7) is almost preserved if we replace D_{l_j} and $D^2_{l_j}$ with Δ_{h,l_j} and δ_{h,l_j}, respectively, and choose $h > 0$ sufficiently small. This is true owing to the continuity of the derivatives of and Ψ_0 in $\mathbb{R}^d$. The lemma is proved. □

Proof of Theorem 5. First assume that $c(z, x) \ge \varepsilon$ for $x \in \Omega$ and any z, where the constant $\varepsilon > 0$. In that case, we proceed almost like in Section 1.1 and, as is usual for monotone finite-difference equations, we rely on the method called Jacobi's iteration method in [**85**] and the Banach fixed point theorem in the space $B(\Omega)$ of bounded functions on Ω provided with the sup norm.

By Lemma 4.1.1, applied for each $x \in \Omega$, for any functions u and v on Ω

$$\big(\mathcal{H}_h[u] - \mathcal{H}_h[v]\big)(x) = \frac{1}{h^2}a_j(x)\big[w(x + hl_j) + w(x - hl_j)\big] + \frac{1}{h}b_j(x)w(x + hl_j)$$

$$-\frac{1}{h^2}\Big[h^2c(x) + h\sum_j b_j(x) + 2\sum_j a_j(x)\Big]w(x),$$

where $w = u - v$ and the functions a_j, b_j, c, $j = 1, ..., m$, are such that, in light of Assumption 3 (ii), $\delta^{-1} \geq a_j \geq \delta$, $|b_j| \leq N'$, $\varepsilon \leq c$ and c is bounded by a constant independent of u, w, and x. Observe that there exists a constant N_0 such that for $h \in (0, 1]$

$$N_0 - h^2 c - h \sum_j b_j - 2 \sum_j a_j \geq 0.$$

Furthermore, $a_j \geq \delta$, $|b_j| \leq N'$,

$$a_j(x)\big[w(x + hl_j) + w(x - hl_j)\big] + hb_j(x)w(x + hl_j)$$

$$= \big[a_j(x) + hb_j(x)\big]w(x + hl_j) + a_j(x)w(x - hl_j)$$

and all the coefficients on the right are nonnegative if $h \in \big(0, h(\delta, N')\big)$, where $h(\delta, N')$ is sufficiently small. It follows that in Ω^h

$$N_0(u - v) + h^2\big(\mathcal{H}_h[u] - \mathcal{H}_h[v]\big)$$

$$\leq \big(N_0 - h^2 c - h \sum_j b_j - 2 \sum_j a_j\big) \sup_\Omega w_+$$

$$+ \sum_j (a_j + hb_j) \sup_\Omega w_+ + \sum_j a_j \sup_\Omega w_+$$

$$= (N_0 - ch^2) \sup_\Omega w_+ \leq (N_0 - \varepsilon h^2) \sup_\Omega (u - v)_+. \tag{9}$$

Next, note that, as is easy to see, (5) with boundary condition (6) is equivalent to the following single equation:

$$v(x) = \big[N_0^{-1} h^2 \mathcal{H}_h[v](x) + v(x)\big] I_{\Omega^h} + g I_{\partial_h \Omega}. \tag{10}$$

Finally, introduce an operator $S_h : v \to S_h v$ given by

$$S_h v(x) := \big[N_0^{-1} h^2 \mathcal{H}_h[v](x) + v(x)\big] I_{\Omega^h} + g I_{\partial_h \Omega}. \tag{11}$$

By the above, for sufficiently small $h > 0$, S_h is a contraction operator in the space $B(\Omega)$ with contraction constant less than $1 - N_0^{-1} \varepsilon h^2 < 1$. By the Banach fixed point theorem, equation (10) has a unique bounded solution and the theorem is proved in our particular case that $c \geq \varepsilon$.

In the general case, take Ψ_0 from Lemma 6, in which we replace δ with a smaller one δ_1 in order to have $|b_j| \leq \delta_1^{-1}$, and for $h > 0$ set

$$\bar{\mathcal{H}}^h(z, x) = \mathcal{H}\big(Z(z, x), x\big),$$

where

$$Z(z, x) = \big(Z'_0(z, x), Z'_{\pm 1}(z, x), ..., Z'_{\pm m}(z, x), Z''_{\pm 1}(z, x), ..., Z''_{\pm m}(z, x)\big),$$

$Z'_0(z, x) = z'_0 \Psi_0(x)$ and, for $j = \pm 1, ..., \pm m$,

$$Z'_j(z, x) = z'_j T_{h,l_j} \Psi_0(x) + z'_0 \delta_{h,l_j} \Psi_0(x),$$

$$Z''_j(z, x) = z''_j \Psi_0(x) + z'_j \delta_{h,l_j} \Psi_0(x) + z'_{-j} \delta_{h,l_{-j}} \Psi_0(x) + z'_0 \Delta_{h,l_j} \Psi_0(x).$$

Observe the fundamental property of $\bar{\mathcal{H}}^h$ which follows from (1.1): For any function u,

$$\bar{\mathcal{H}}^h[u] = \mathcal{H}_h[u\Psi_0]. \tag{12}$$

Furthermore, for each x, the mapping $z \to Z(z,x)$ is linear and, obviously, is one-to-one, therefore its inverse is sending into the sets of measure zero only the sets of measure zero. It follows that, if we denote by $\big(\bar{a}_j^h, \bar{b}_j^h, \bar{c}^h\big)(z,x)$ the functions corresponding to $\bar{\mathcal{H}}^h$ as in Assumption 3, then this functions can be found (a.e.) by means of usual formulas of differentiating compositions and hence obviously the boundedness and nondegeneracy conditions in Assumption 3 will be satisfied, and, due to Lemma 6,

$$\bar{c}^h = c\Psi_0 - b_j\delta_{h,l_j}\Psi_0 - a_j\Delta_{h,l_j}\Psi_0 \geq 1,$$

where we dropped obvious values of the arguments for simplicity.

By the above, for sufficiently small $h > 0$, equation $\bar{\mathcal{H}}^h[u] = 0$ in Ω^h with boundary condition $u = g/\Psi_0$ has a unique bounded solution. It only remains to set $v = u\Psi_0$ and use (12). The theorem is proved. □

7. Corollary (comparison principle). *If $h \in (0, h_0]$, then for any bounded functions u, v on $\bar{\Omega}$ such that $u \leq v$ on $\partial_h\Omega$ and $\mathcal{H}_h[u] \geq \mathcal{H}_h[v]$ in Ω^h we have $u \leq v$ in $\bar{\Omega}$.*

Indeed, if $c(z,x) \geq \varepsilon > 0$, it follows from the assumptions and (9) that

$$N_0 \sup_{\Omega^h}(u-v)_+ \leq (N_0 - \delta h^2)\sup_{\Omega}(u-v)_+,$$

$$N_0 \sup_{\Omega}(u-v)_+ \leq (N_0 - \delta h^2)\sup_{\Omega}(u-v)_+,$$

and $(u-v)_+ = 0$ in Ω. In the general case it suffices to use the argument in the end of the proof of Theorem 5. □

In the future we will need an estimate of v_h.

8. Theorem. *Fix a constant $K_0 \in [0, \infty)$ and in addition to Assumptions 1 and 3 suppose that the number*

$$\hat{\mathcal{H}} = \sup_{x\in\Omega, z':z'_0=0}\big(\big|\mathcal{H}(z',0,x)\big| - K_0|z'|\big)$$

is finite. Then there is a constant N, depending only on K_0, δ, Λ, and $\operatorname{diam}(\Omega)$, *such that for $h \in (0, h_0]$*

$$\sup_{\bar{\Omega}}|v_h| \leq N\hat{\mathcal{H}} + \sup_{\partial_h\Omega}|g|.$$

9. Remark. Under Assumption 3, owing to the Lipschitz continuity of $\mathcal{H}$ with respect to z (uniform with respect to x in light of Assumption 3 (ii)), there always exists a K_0 such that $\hat{\mathcal{H}} < \infty$. In the future, however, it will be important not to tighten up K_0 to the Lipschitz continuity.

The proof of Theorem 8 is based on the following adaptation of Lemma 4.1.2 in which the boundedness of c is guaranteed by the Lipschitz continuity of $\mathcal{H}$ with respect to z.

10. LEMMA. *Under the assumptions of Theorem* 8 *let* ϕ *be a function on* Ω. *Then for any* $h > 0$, *such that* $\Omega^h \neq \emptyset$, *there exist bounded functions* a_j, b_j, c, f, $|j| = 1, ..., m$, *on* Ω^h *(generally different from the ones in Assumption* 3) *such that on* Ω^h

$$\delta \le a_j \le \delta^{-1}, \quad |b_j| \le K_0, \quad c \ge 0, \quad |f| \le \hat{\mathcal{H}},$$

$$\mathcal{H}_h[\phi] = a_k \Delta_{h,l_k}\phi + b_k \delta_{h,l_k}\phi - c\phi + f. \tag{13}$$

Proof of Theorem 8. In Lemma 6 reduce δ, if necessary, in such a way that the new one, δ_1, will satisfy $K_0 \le \delta_1^{-1}$, and then take Ψ_0 from that lemma and set

$$\phi = \Psi_0 \hat{\mathcal{H}} + \sup_{\partial_h \Omega} |g|.$$

Then by using Lemmas 10 and 6 we see that for $h \in (0, h_0]$ on Ω^h

$$\mathcal{H}_h[\phi] = a_k \Delta_{h,l_k}\phi + b_k \delta_{h,l_k}\phi - c\phi + f \le 0.$$

Furthermore, $v^h = g \le \phi$ on $\partial_h \Omega$. It follows by Corollary 7 that $v^h \le \phi$ on $\bar{\Omega}$. By replacing ϕ with $-\phi$ we get that $v^h \ge -\phi$ on $\bar{\Omega}$. The theorem is proved. □

3. The way we proceed further

Assume that $\mathcal{H}$ is independent of z'. Having the solvability of the problem (2.5)-(2.6) for any small $h > 0$, it is natural to send $h \downarrow 0$ and try to show that the solutions of (2.5)-(2.6) converge to the solutions of the corresponding partial differential boundary-value problem. In order to do that we need to show that at least the first and the second finite differences of the solutions of (2.5)-(2.6) are under control independent of h. We only know from Chapter 1 how to achieve this if $\mathcal{H}$ is a convex positive homogeneous or concave function of z'' preferably independent of x. Then the idea came to cut-off $\mathcal{H}$ for large $|z|$ and replace (2.5) with

$$\max\big(\mathcal{H}_h[v], \mathcal{P}_h[v] - K\big) = 0, \tag{1}$$

where $\mathcal{P}_h[v] = \mathcal{P}(\delta_h^2 v)$ and $\mathcal{P}(z) = \mathcal{P}(z'')$ is a convex positive homogeneous function of z satisfying $\hat{\delta} \le D_{z_i''}\mathcal{P} \le \hat{\delta}^{-1}$ for a constant $\hat{\delta}$, such that, roughly speaking, one the set where

$$\mathcal{H}(z, x) \ge \mathcal{P}(z) - K \tag{2}$$

we have

$$|z''| \le N_1, \tag{3}$$

where N_1 is a constant perhaps depending on K. In that situation we only need to estimate the second finite differences of v on the set, say

$$G = \big\{|\delta_h^2 v| > N_1\big\}. \tag{4}$$

Naturally, on the boundary of this set, as everywhere in Ω outside G we have $|\delta_h^2 v| \le N_1$. On the other hand on G equation (1) becomes

$$\mathcal{P}_h[v] - K = 0 \tag{5}$$

and as we said before, we know how to deal with such equations and estimate the second finite differences in G through their estimates on the boundary of G, where they are (or seem to be) already under control. These estimates will be obtained in this chapter. Then we will be able to pass to the limit as $h \downarrow 0$ in (1) and get the solvability of the PDE counterpart of the cut-off equation (1) in Chapter 6. After that, much later, by developing some a priori W^2_p-estimates we will be able to set $K \to \infty$ and pass from the cut-off equations like (1) to the original ones.

One of substantial difficulties in estimating the second-order differences of solutions of (5) lies in the fact that part of the boundary of G may, actually, lie on the boundary of Ω, and there is absolutely no hope to estimate second-order differences of solutions on this part. Therefore, we need to obtain local estimates which may blow up near the boundary. Here we use Bernstein's method who suggested in the case of linear elliptic equations to consider the maximum points of a function $w = |D^2v|^2\zeta^2 + \nu|Dv|^2$, where ζ is a smooth function with compact support in Ω and ν is a sufficiently large constant. In the case of linear equations one differentiates the equation twice and relatively easily obtains that w cannot attain its maximum inside the domain (cf. the proof of Lemma 13.3.3). It was realized in the theory of fully nonlinear equations like (5) (with convex $\mathcal{P}$) that a good substitute for w is

$$\sup_{|l|=1} \big[(D_{(l)(l)}v)_-\big]^2\zeta^2 + \nu|Dv|^2$$

and, if the equation does not explicitly involve x (as (5)), then, for each particular unit l, considering

$$\big[(D_{(l)(l)}v)_-\big]^2\zeta^2 + \nu|D_lv|^2$$

allows one to obtain local estimates from below of $D_{(l)(l)}v$ (see [**61**]).

Naturally, we are going to deal with $w = \big[(\Delta_{h,l_r}v)_-\big]^2\zeta^2 + \nu|\delta_{h,l_r}v|^2$. Now the main difficulty is how to obtain a finite-difference inequality for w similar to differential inequalities which appear in the case of partial differential equations, when one just differentiates the given equation twice and gets a differential inequality by combining the results of the differentiations. Here applying separately Δ_{h,l_r} and δ_{h,l_r} to (5) seems not to yield anything tractable, and the idea came to introduce the operator

$$L\phi = \zeta^2(\Delta_{h,l_r}v)_-\Delta_{h,l_r}\phi - \nu\zeta\delta_{h,l_r}v\delta_{h,l_r}\phi$$

(no summation in r). It turns out that in the main case of interest L respects the maximum principle and, since for any point $x_0 \in G$ there are (have in mind a tangent plane to the graph of $\mathcal{P}$) a_k, $|k| = 1, ..., m$, such that $\hat{\delta} \le a_k \le \hat{\delta}^{-1}$ and for x close to x_0

$$K - a_k\Delta_{h,l_k}v(x) = \mathcal{P}_h[v](x) - a_k\Delta_{h,l_k}v(x) \ge 0$$

with the equality for $x = x_0$, we have

$$L[a_k\Delta_{h,l_k}v](x_0) \le 0. \tag{6}$$

This is a perfect substitute of what one gets in the case of differential equations and we extract what is needed from it.

Observe that technically the exposition could be made shorter if we assumed that the continuous counterpart of (1) has a solution which is four times continuously differentiable. Then we could treat the continuous counterpart of (5) just as in, for instance, [**61**]. However, unlike the situation in [**61**] there is no known way to show that solutions of the continuous counterpart of (1) are four times continuously differentiable, no matter how smooth H is with respect to x and so no approximation argument works. This is the reason why we deal with finite-difference equations.

4. Some estimates for finite-difference elliptic equations with constant coefficients

Take an $h \in (0,1]$, let $m \geq 1$ be an integer and let $l_{-m}, ..., l_{-1}, l_1, ..., l_m$ be some fixed vectors in $\mathbb{R}^d$ such that, for all k,

$$l_{-k} = -l_k.$$

Next denote $\Lambda = \{l_k : k = \pm 1, ..., \pm m\}$,

$$\Lambda_1 = \Lambda, \quad \Lambda_{n+1} = \Lambda_n + \Lambda, \quad n \geq 1, \quad \Lambda_\infty = \bigcup_n \Lambda_n, \quad \Lambda^h_\infty = h\Lambda_\infty .$$

Let Q^o be a nonempty *finite* subset of Λ^h_∞. Introduce

$$Q = Q^o \cup \{x + h\Lambda : x \in Q^o\}. \tag{1}$$

Let $A = \{\alpha = (a, c)\}$ be a closed bounded set in $\mathbb{R}^{2m} \times \mathbb{R}$ such that

$$a = (a_{-m}, a_{-m+1}, ..., a_{-1}, a_1, ..., a_m) \in \mathbb{R}^{2m}, \quad c \in \mathbb{R}.$$

Also let $f(\alpha, x)$ be a real-valued function defined for $\alpha \in A$ and $x \in \mathbb{R}^d$. It is worth mentioning that in applications of the results of this section we take f independent of x and α.

Recall (1.1.1) and for $k = \pm 1, ..., \pm m$ for brevity of notation set

$$\delta_{h,k} = \delta_{h,l_k} = \delta_k, \quad \Delta_{h,k} = \Delta_{h,l_k} = \Delta_k.$$

Fix an $r \in \{1, ..., m\}$.

1. ASSUMPTION. There are constants $\delta \in (0,1)$ and $K_1, K_2 \in [0, \infty)$ such that
(i) For any $(a, c) \in A$ and all k we have

$$a_k = a_{-k}, \quad \delta \leq a_k \leq \delta^{-1}, \quad c \geq 0;$$

(ii) The function f is continuous in α for any x and on $\mathbb{R}^d$

$$|\delta_r f| \leq K_1, \quad \Delta_r f \geq -K_2.$$

For $z = (z_0', z'')$ with

$$z_0' \in \mathbb{R}, \quad z'' = (z''_{-m}, ..., z''_{-1}, z''_1, ..., z''_m) \in \mathbb{R}^{2m},$$

introduce

$$\mathcal{P}(z, x) = \max_{\alpha=(a,c)\in A} \Big(\sum_{|k|=1}^{m} a_k z''_k - c z_0' + f(\alpha, x) \Big).$$

For any function u on $\mathbb{R}^d$ define

$$\mathcal{P}[u](x) = \mathcal{P}_h[u](x) = \mathcal{P}\big(u(x), \delta_h^2 u(x), x\big), \tag{2}$$

where δ_h^2 is introduced in (2.4).

In connection with this notation a natural question arises as to why use l_k along with $l_{-k} = -l_k$ since $\Delta_k = \Delta_{-k}$ and

$$a_k\Delta_k = 2\sum_{k\geq 1} a_k\Delta_k$$

owing to the assumption that $a_k = a_{-k}$. This is done for the sake of convenience of computations. For instance,

$$\Delta_k(uv) = u\Delta_k v + v\Delta_k u + (\delta_k u)(\delta_k v) + (\delta_{-k}u)(\delta_{-k}v)$$

(no summation in k). At the same time

$$a_k\Delta_k(uv) = ua_k\Delta_k v + va_k\Delta_k u + 2a_k(\delta_k u)(\delta_k v)$$

as if we were dealing with usual partial derivatives. Another reason was mentioned in Remark 2.2.

Next, take a function $\eta \in C_0^\infty(\mathbb{R}^d)$, such that $|\eta| \leq 1$ and set $\zeta = \eta^2$,

$$\big|\eta'(x)\big| = \big|\eta'(x)\big|_h = \sup_k \big|\delta_k\eta(x)\big|, \quad \big|\eta''(x)\big| = \big|\eta''(x)\big|_h = \sup_k \big|\Delta_k\eta(x)\big|,$$

$$\|\eta'\| = \|\eta'\|_h = \sup_{\Lambda^h_\infty} |\eta'|_h, \quad \|\eta''\| = \|\eta''\|_h = \sup_{\Lambda^h_\infty} |\eta''|_h.$$

Finally, let u be a function on $\mathbb{R}^d$ which satisfies

$$\mathcal{P}[u] = 0 \quad \text{in } Q^o \tag{3}$$

and

$$\mathcal{P}[u] \leq 0 \quad \text{on } Q \setminus Q^o. \tag{4}$$

Here is a one-sided interior estimate of pure second-order differences of u. This result is a particular case of Theorem 5.1 of [**72**].

2. THEOREM. *There exists a constant $N = N(m, \delta) \geq 1$ such that for any constant ν satisfying*

$$\nu \geq N\big(\|\eta''\| + \|\eta'\|^2\big),$$

we have in Q that (recall that $a^\pm = (1/2)\big(|a| \pm a\big)$)

$$\zeta^2\big[(\Delta_r u)^-\big]^2 \leq \sup_{Q\setminus Q^o} \zeta^2\big[(\Delta_r u)^-\big]^2 + N\nu\bar{W}_r + N\nu^{-2}K_2^2 + N\nu^{-1}K_1^2, \tag{5}$$

where

$$\bar{W}_r = \sup_Q \big(|\delta_r u|^2 + |\delta_{-r}u|^2\big). \tag{6}$$

To prove the theorem we need a few auxiliary results.

3. LEMMA. *For any $(a,c) \in A$ and $n \in \{\pm 1, ..., \pm m\}$*

$$a_k \Delta_k \zeta = 2\eta a_k \Delta_k \eta + 2a_k(\delta_k \eta)^2, \quad |\Delta_n \zeta| \le 2\big(|\eta''| + |\eta'|^2\big). \tag{7}$$

Furthermore, there exists an absolute constant N such that for any $k \in \{\pm 1, ..., \pm m\}$

$$\delta_k \zeta = 2\eta \delta_k \eta + h(\delta_k \eta)^2, \quad (\delta_k \zeta)^2 \le N|\eta'|^2 \zeta + Nh^2|\eta'|^4 \le N|\eta'|^2, \tag{8}$$

and for any $\varepsilon \in (0,1]$, function ψ, and $i \in \{\pm 1, ..., \pm m\}$

$$h(\delta_k \zeta)^2 |\psi \delta_i \psi| \le \varepsilon \big|\delta_i(\zeta\psi)\big|^2 + \varepsilon|\eta'|^2 \zeta \psi^2 + N\varepsilon^{-1}|\eta'|^4\Big[(h\psi)^2 + \big(h^2 \delta_i \psi\big)^2\Big]. \tag{9}$$

Proof. Equations (7) and (8) obviously follow from Lemma 1.1 (in the second inequality in (8) use that $h|\eta'| \le 2$).

By (8) the left-hand side of (9) is less than

$$Nh|\eta'|^2 \zeta |\psi \delta_i \psi| + Nh^3|\eta'|^4 |\psi \delta_i \psi| =: I + Nh^3 |\eta'|^4 |\psi \delta_i \psi|,$$

where

$$h^3|\psi\delta_i\psi| = \big(h|\psi|\big)\big(h^2|\delta_i\psi|\big) \le (h\psi)^2 + \big(h^2|\delta_i\psi|\big)^2.$$

By taking into account that

$$\zeta\delta_i\psi = \delta_i(\zeta\psi) - \psi\delta_i\zeta - h(\delta_i\zeta)\delta_i\psi, \tag{10}$$

we find that

$$I \le Nh|\eta'|^2\big|\psi\delta_i(\zeta\psi)\big| + Nh|\eta'|^2\psi^2|\delta_i\zeta| + Nh^2|\eta'|^2\big|\psi(\delta_i\zeta)\delta_i\psi\big| =: I_1 + I_2 + I_3.$$

Obviously,

$$I_1 \le \varepsilon\big|\delta_i(\zeta\psi)\big|^2 + \varepsilon^{-1}N|\eta'|^4(h\psi)^2,$$

$$I_2 \le Nh|\eta'|^2\psi\big(|\eta'|\zeta^{1/2}\psi\big) + Nh^2|\eta'|^4\psi^2 \le \varepsilon|\eta'|^2\zeta\psi^2 + N\varepsilon^{-1}|\eta'|^4(h\psi)^2.$$

Finally, by (8)

$$I_3 \le N\big(|\eta'|\zeta^{1/2}|\psi|\big)\big(|\eta'|^2h^2\delta_i\psi|\big) + N|\eta'|^4(h\psi)\big(|h^2\delta_i\psi|\big),$$

which is clearly dominated by the right-hand side of (9). Upon combining the above estimates we come to (9). The lemma is proved. □

In the remaining part of this section, concerned with the proof of Theorem 2, no summation with respect to r is performed. Recall that the number r is fixed at the beginning of the section. For simplicity of notation set

$$u_{rr} = \Delta_r u, \quad u_r = \delta_r u, \quad u_{kr} = -\delta_{-k}\delta_r u.$$

Notice that in the above line the last notation when $k = r$ is consistent with the first one.

Define

$$u_{rr}^- = (u_{rr})^-.$$

In the following lemma the fact that u is a solution of (3) is not used.

4. LEMMA. *There exists $N = N(m,\delta)$ such that on Q^o for any $\alpha = (a,c) \in A$ we have*

$$-2\zeta^2 u_{\bar{r}r} a_k \Delta_k u_{\bar{r}r} \geq -a_k \Delta_k\big(\zeta^2 (u_{\bar{r}r})^2\big) - N\big(|\eta''| + |\eta'|^2\big)\zeta\big(u_{\bar{r}r}\big)^2 - N|\eta'|^4 \bar{W}_r, \quad (11)$$

$$-2\zeta u_r a_k \Delta_k u_r \geq -a_k \Delta_k\big(\zeta u_r^2\big) + \zeta a_k u_{kr}^2 - N\big(|\eta''| + |\eta'|^2\big)\bar{W}_r. \quad (12)$$

Proof. Observe that for any $\alpha = (a,c) \in A$ by Lemma 3 we have

$$a_k \Delta_k\big(\zeta^2 (u_{\bar{r}r})^2\big) = 2\zeta u_{\bar{r}r} a_k \Delta_k\big(\zeta u_{\bar{r}r}\big) + 2a_k\big|\delta_k\big(\zeta u_{\bar{r}r}\big)\big|^2$$

$$= 2\zeta^2 u_{\bar{r}r} a_k \Delta_k u_{\bar{r}r} + 2\zeta\big(u_{\bar{r}r}\big)^2 a_k \Delta_k \zeta + 4\zeta u_{\bar{r}r} a_k (\delta_k \zeta)\delta_k u_{\bar{r}r} + 2a_k\big|\delta_k(\zeta u_{\bar{r}r})\big|^2$$

and for each k

$$\zeta u_{\bar{r}r}(\delta_k\zeta)\delta_k u_{\bar{r}r} = u_{\bar{r}r}(\delta_k\zeta)\delta_k(\zeta u_{\bar{r}r}) - \big(u_{\bar{r}r}\big)^2(\delta_k\zeta)^2 - h u_{\bar{r}r}(\delta_k\zeta)^2\delta_k u_{\bar{r}r},$$

so that

$$a_k \Delta_k\big(\zeta^2 (u_{\bar{r}r})^2\big) = 2\zeta^2 u_{\bar{r}r} a_k \Delta_k u_{\bar{r}r} + 2a_k\big[\delta_k(\zeta u_{\bar{r}r})\big]^2 - 4h u_{\bar{r}r} a_k (\delta_k\zeta)^2 \delta_k u_{\bar{r}r}$$

$$+ 4a_k u_{\bar{r}r}(\delta_k\zeta)\delta_k\big(\zeta u_{\bar{r}r}\big) + 2a_k\big(u_{\bar{r}r}\big)^2\big[\zeta\Delta_k\zeta - 2(\delta_k\zeta)^2\big]. \quad (13)$$

By using also (8) we see that for any $\varepsilon > 0$,

$$\big|4a_k u_{\bar{r}r}(\delta_k\zeta)\delta_k\big(\zeta u_{\bar{r}r}\big)\big| \leq \varepsilon a_k\big[\delta_k\big(\zeta u_{\bar{r}r}\big)\big]^2 + N\varepsilon^{-1}\big(u_{\bar{r}r}\big)^2\big(|\eta'|^2\zeta + h^2|\eta'|^4\big).$$

To estimate the last term note that

$$\big(u_{\bar{r}r}\big)^2 h^2 \leq |h u_{rr}|^2 = |u_r + u_{-r}|^2, \quad (14)$$

so that $|\eta'|^4 (u_{\bar{r}r})^2 h^2 \leq 4|\eta'|^4 \bar{W}_r$ in Q^o. Therefore, in Q^o

$$\big|4a_k u_{\bar{r}r}(\delta_k\zeta)\delta_k\big(\zeta u_{\bar{r}r}\big)\big| \leq \varepsilon a_k\big[\delta_k\big(\zeta u_{\bar{r}r}\big)\big]^2 + N\varepsilon^{-1}|\eta'|^2\zeta\big(u_{\bar{r}r}\big)^2 + N\varepsilon^{-1}\bar{W}_r|\eta'|^4.$$

Similarly,

$$\big|2a_k\big(u_{\bar{r}r}\big)^2\big[\zeta\Delta_k\zeta - 2(\delta_k\zeta)^2\big]\big| \leq N\big(|\eta''| + |\eta'|^2\big)\zeta\big(u_{\bar{r}r}\big)^2 + N\bar{W}_r|\eta'|^4.$$

According to (9) for any $\varepsilon \in (0,1]$

$$J_{ik} := h(\delta_k\zeta)^2\big|u_{\bar{r}r}\delta_i u_{\bar{r}r}\big| \leq \varepsilon\big|\delta_i\big(\zeta u_{\bar{r}r}\big)\big|^2 + \varepsilon|\eta'|^2\zeta\big(u_{\bar{r}r}\big)^2$$

$$+ N\varepsilon^{-1}|\eta'|^4\big[\big(h u_{\bar{r}r}\big)^2 + \big(h^2\delta_i u_{\bar{r}r}\big)^2\big].$$

Estimate (14) leads to

$$J_{ik} \leq \varepsilon\big|\delta_i\big(\zeta u_{\bar{r}r}\big)\big|^2 + |\eta'|^2\zeta\big(u_{\bar{r}r}\big)^2$$

$$+ N\varepsilon^{-1}\bar{W}_r|\eta'|^4 + N\varepsilon^{-1}|\eta'|^4\big(h^2\delta_i u_{\bar{r}r}\big)^2$$

on Q^o, where the last term is estimated by using the fact that $|\delta_i\psi^-| \leq |\delta_i\psi|$ for any function ψ implying that

$$|\eta'|^4\big(h^2\delta_i u_{rr}^-\big)^2 \le |\eta'|^4\big|h\delta_i(u_r+u_{-r})\big|^2 = |\eta'|^4\big|(T_{h,l_i}-1)(u_r+u_{-r})\big|^2 \le N|\eta'|^4\bar W_r.$$

Hence on Q^o

$$J_{ik} \le \varepsilon\big|\delta_i\big(\zeta u_{rr}^-\big)\big|^2 + |\eta'|^2\zeta\big(u_{rr}^-\big)^2 + N\varepsilon^{-1}\bar W_r|\eta'|^4.$$

Upon combining these estimates, choosing ε appropriately, and coming back to (13), we find on Q^o that

$$-2\zeta^2 u_{rr}^- a_k\Delta_k u_{rr}^- \ge -\,a_k\Delta_k\big(\zeta^2\big(u_{rr}^-\big)^2\big) + a_k\big[\delta_k\big(\zeta u_{rr}^-\big)\big]^2$$
$$-\,N\big(|\eta''|+|\eta'|^2\big)\zeta\big(u_{rr}^-\big)^2 - N\bar W_r|\eta'|^4,$$

which yields (11).

To prove (12) observe that (recall that $\delta_k u_r = -u_{-kr}$ and $a_{-k}=a_k$)

$$a_k\Delta_k\big(\zeta u_r^2\big) = a_k\zeta\big[2u_r\Delta_k u_r + 2u_{kr}^2\big] + a_k u_r^2\Delta_k\zeta + 2a_k(\delta_k\zeta)\big(hu_{-kr}^2 - 2u_r u_{-kr}\big),$$

where

$$u_r^2|\Delta_k\zeta| \le N\big(|\eta''|+|\eta'|^2\big)\bar W_r,$$

and owing to (14)

$$\big|2a_k(\delta_k\zeta)\big(hu_{-kr}^2 - 2u_r u_{-kr}\big)\big| \le Na_k\big(|\eta'|\zeta^{1/2} + h|\eta'|^2\big)\big(|u_r u_{kr}| + hu_{kr}^2\big)$$
$$\le Na_k|u_{kr}|\big(|\eta'|\zeta^{1/2}+h|\eta'|^2\big)\bar W_r^{1/2} \le Na_k|u_{kr}||\eta'|\zeta^{1/2}\bar W_r^{1/2} + N|\eta'|^2\bar W_r$$
$$\le \zeta a_k u_{kr}^2 + N|\eta'|^2\bar W_r.$$

It follows that in Q^o

$$-2\zeta a_k u_r\Delta_k u_r \ge -a_k\Delta_k\big(\zeta u_r^2\big) + \zeta a_k u_{kr}^2 - N\big(|\eta''|+|\eta'|^2\big)\bar W_r,$$

which yields (12) and the lemma is proved. □

For a constant $\nu\ge 0$ introduce an operator (recall that r is fixed)

$$L_\nu\phi = \zeta^2 u_{rr}^-\Delta_r\phi - \nu\zeta u_r\delta_r\phi.$$

Observe that

$$L_\nu u = -\zeta^2(u_{rr}^-)^2 - \nu\zeta u_r^2 =: -V_\nu. \tag{15}$$

In the next lemma the fact that u is a solution of (3) is not used either. In order not to lose the orientation, the reader is adviced to come back to our explanation surrounding (3.6).

5. LEMMA. *There exists $N=N(m,\delta)\ge 1$ such that, if*

$$\nu \ge N\big(\|\eta''\| + \|\eta'\|^2\big), \tag{16}$$

then on Q^o for any $\alpha=(a,c)\in A$ we have

$$2L_\nu[a_k\Delta_k u] \ge -a_k\Delta_k V_\nu - N\nu^2\bar W_r + (\nu/2)\zeta a_k u_{kr}^2. \tag{17}$$

Proof. Since, for each k, the operators $a_k\Delta_k$ respect the maximum principle, it follows by Lemma 1.3 that

$$u_{rr}^- a_k \Delta_k u_{rr} \geq -u_{rr}^- a_k \Delta_k u_{rr}^-.$$

Hence,

$$I := \zeta^2 u_{rr}^- \Delta_r [a_k \Delta_k u] = \zeta^2 u_{rr}^- a_k \Delta_k u_{rr} \geq -\zeta^2 u_{rr}^- a_k \Delta_k u_{rr}^-,$$

which by Lemma 4 and the observation that

$$\zeta \big(u_{rr}^-\big)^2 \leq N \zeta a_k u_{kr}^2,$$

yields

$$2I \geq -a_k \Delta_k \big(\zeta^2 (u_{rr}^-)^2\big) - N\big(|\eta''| + |\eta'|^2\big)\zeta a_k u_{kr}^2 - N|\eta'|^4 \bar{W}_r. \tag{18}$$

Furthermore, by Lemma 4

$$\begin{aligned} -2\nu\zeta u_r \delta_r [a_k \Delta_k u] &= -\, 2\nu\zeta u_r a_k \Delta_k u_r \\ &\geq -\, a_k \Delta_k \big(\nu\zeta u_r^2\big) + \nu\zeta a_k u_{kr}^2 - N\nu\big(|\eta''| + |\eta'|^2\big)\bar{W}_r. \end{aligned}$$

By combining this with (18) and recalling (15) we find

$$\begin{aligned} 2L_\nu [a_k \Delta_k u] \geq &-\, a_k \Delta_k V_\nu + \big(\nu - N_1 |\eta''| - N_1 |\eta'|^2\big)\zeta a_k u_{kr}^2 \\ &-\, N\big(|\eta'|^4 + \nu|\eta''| + \nu|\eta'|^2\big)\bar{W}_r. \end{aligned} \tag{19}$$

We may assume that $N_1 \geq 1$ and then, if

$$\nu \geq 2N_1 \big(|\eta''| + |\eta'|^2\big),$$

we have that $|\eta'|^2 \leq \nu$, $|\eta''| \leq \nu$, and

$$|\eta'|^4 + \nu|\eta''| + \nu|\eta'|^2 \leq 3\nu^2.$$

After that (19) clearly yields (17) and the lemma is proved. □

Proof of Theorem 2. Denote by N_0 the constant N in Lemma 5 and take and fix a ν satisfying (16) (with N_0 in place of N). Obviously we need only prove (5) in Q^o.

Notice that in Q^o

$$|u_{rr}| = h^{-1}|u_r + u_{-r}| \leq 2h^{-1}\bar{W}_r^{1/2},$$

which shows that (5) holds if $h \geq \nu^{-1/2}$. Therefore below we assume that

$$h \leq \nu^{-1/2}. \tag{20}$$

Let $x_0 \in Q$ be a point such that

$$V_\nu(x_0) = \max_Q V_\nu.$$

If $x_0 \notin Q^o$, then

$$\max_Q V_\nu = \max_{Q \setminus Q^o} V_\nu$$

and (5) follows owing to definition (15). Therefore, in the rest of the proof we may assume that

$$x_0 \in Q^o.$$

We may also assume that

$$\zeta(x_0)u_{rr}^-(x_0) > \nu h u_r(x_0). \tag{21}$$

Indeed, if the opposite inequality holds, then in light of (20) in Q^o

$$\zeta^2\big[u_{rr}^-\big]^2 \le V_\nu(x_0) \le \nu^2 h^2 u_r^2(x_0) + \nu\bar{W}_r \le 2\nu\bar{W}_r.$$

To extract some consequences of (21), we notice that if a function $\phi(x)$ is such that $\phi(x) \le \phi(x_0)$ for $x \in x_0 + h\Lambda$, then at x_0 we have

$$\begin{aligned} h^2 L_\nu\phi(x_0) &= \zeta\big[\phi(x_0 + hl_r)\big(\zeta u_{rr}^- - \nu h u_r\big) + \phi(x_0 - hl_r)\zeta u_{rr}^-\big] \\ &\quad - \zeta\big[2\zeta u_{rr}^- - \nu h u_r\big]\phi \\ &\le \zeta\big[\big(\zeta u_{rr}^- - \nu h u_r\big)\phi + \zeta u_{rr}^-\phi\big] - \zeta\big[2\zeta u_{rr}^- - \nu h u_r\big]\phi, \end{aligned}$$

where the last expression is zero. Thus

$$L_\nu\phi(x_0) \le 0,$$

which means that L_ν respects the maximum principle.

Furthermore, we can find an $\hat{\alpha} = (\hat{a}, \hat{c}) \in A$ such that

$$\hat{a}_k\Delta_k u(x_0) - \hat{c}u(x_0) + f(\hat{\alpha}, x_0) = \mathcal{P}[u](x_0) = 0.$$

Since $P[u] \le 0$ in Q, we have that

$$\phi(x) := \hat{a}_k\Delta_k u(x) - \hat{c}u(x) + f(\hat{\alpha}, x) \le 0$$

for $x \in x_0 + h\Lambda$. Hence, $0 \ge 2L_\nu\phi(x_0)$, which owing to (15) and (17), yields

$$0 \le [\hat{a}_k\Delta_k - 2\hat{c}]V_\nu(x_0) - (\nu/2)\zeta\hat{a}_k u_{kr}^2(x_0) + N\nu^2\bar{W}_r - 2L_\nu f(\hat{\alpha}, \cdot)(x_0).$$

Here the last term is dominated by

$$\begin{aligned} &2K_2\zeta^2 u_{rr}^-(x_0) + 2\nu|u_r(x_0)|K_1 \\ &\le N\nu^{-1}K_2^2 + (\nu/4)\zeta\hat{a}_k u_{kr}^2(x_0) + K_1^2 + \nu^2\bar{W}_r. \end{aligned}$$

Furthermore, since $V_\nu(x) \ge 0$ attains its maximum at x_0,

$$[\hat{a}_k\Delta_k - 2\hat{c}]V_\nu(x_0) \le 0.$$

We now conclude that

$$(\nu/4)\zeta\hat{a}_k u_{kr}^2(x_0) \le N\nu^2\bar{W}_r + N\nu^{-1}K_2^2 + K_1^2,$$

which implies that in Q^o

$$\begin{aligned} \zeta^2\big(u_{rr}^-\big)^2 &\le V_\nu(x_0) \le N\zeta\bar{a}_k u_{kr}^2(x_0) + \nu\bar{W}_r \\ &\le N\nu\bar{W}_r + N\nu^{-2}K_2^2 + N\nu^{-1}K_1^2. \end{aligned}$$

Thus, estimate (5) holds on Q^o in all cases and this proves the theorem. □

5. Estimates for finite-difference elliptic equations of cut-off type with variable coefficients

The most important results of this section are summarized in Theorem 10. Fix some constants $K_0, K \in [0,\infty)$. We use the notation and assumptions introduced in the beginning of Section 2, however, we append Assumptions 2.1 and 2.3 with the following.

1. ASSUMPTION. The coordinates of $l_j, |j| = 1, ..., m$, are rational numbers.

2. ASSUMPTION. The number

$$\bar{\mathcal{H}} = \sup_{x,z'} \big(|\mathcal{H}(z', 0, x)| - K_0|z'|\big)$$

is finite.

To state one more assumption we need the following definition before which we note that if $\rho \in (0,\infty)$ and $\varepsilon \in (0,1]$, then

$$B_{\rho(1+\varepsilon)}(\rho e_1) \setminus B_{2\rho} \subset B_{\mu}(2\rho e_1),$$

where e_1 is the first basis vector in $\mathbb{R}^d$ and

$$\mu^2 = \rho^2(4\varepsilon + 2\varepsilon^2) \le 6\rho^2\varepsilon.$$

This fact is proved by elementary computations.

3. DEFINITION. Let Ω be an open subset of $\mathbb{R}^d$. We say that it satisfies the exterior ball condition with radius $\rho_{\text{ext}}(\Omega)$ (> 0) if any point $y_0 \in \partial\Omega$ is the single point of the intersection of $\bar{\Omega}$ and $\bar{B}_{\rho_{\text{ext}}(\Omega)}(z)$, where $z = z(y_0) \notin \Omega$. Without loosing generality we suppose that

$$\rho_{\text{ext}}(\Omega) \le 1.$$

Furthermore, what is said before the definition shows that, by taking one half of the original $\rho_{\text{ext}}(\Omega)$ as the new $\rho_{\text{ext}}(\Omega)$, if necessary, we may assume that, for any $\varepsilon \in (0,1]$ and $y_0 \in \partial\Omega$ we have

$$\Omega \cap B_{\rho_{\text{ext}}(\Omega)(1+\varepsilon)}\big(z(y_0)\big) \subset B_{3\rho_{\text{ext}}(\Omega)\sqrt{\varepsilon}}(y_0).$$

We also impose the following, in which, for a domain G in $\mathbb{R}^d$ by $C^{0,1}(G)$ we mean the set of bounded Lipschitz continuous functions on G or, equivalently, of bounded continuous functions u on G for which $D_i u$ (the generalized derivatives) are bounded. By $C^{1,1}(G)$ we mean the subset of $C^{0,1}(G)$ consisting of functions u for which $Du \in C^{0,1}(G)$. We set

$$\|g\|_{C^{1,1}(G)} = \sup_G \big(|g| + |Dg|\big) + \|D^2 g\|_{L_\infty(G)} < \infty.$$

4. ASSUMPTION. The domain Ω is bounded and satisfies the exterior ball condition (see Definition 3). We are given a function $g \in C^{1,1}(\mathbb{R}^d)$.

Define

$$\mathcal{P}(z'') = \mathcal{P}_{m,\delta/2}(z'') = \max_{\substack{\delta/2 \le a_k \le 2\delta^{-1} \\ |k|=1,\dots,m}} \sum_{|k|=1}^{m} a_k z''_k. \tag{1}$$

Observe that

$$\mathcal{H}(z,x) \le \mathcal{P}(z'') - (\delta/2)\sum_{|k|=1}^{m} |z''_k| + K_0|z'| + \bar{\mathcal{H}}. \tag{2}$$

Indeed, from Assumption 2.3 and Lemma 4.1.1 it follows that

$$\mathcal{H}(z,x)-\mathcal{H}(z',0,x)=\sum_{|k|=1}^{m}a_k z''_k,$$

where $\delta\le a_k\le\delta^{-1}$ for all k. Hence,

$$\mathcal{H}(z,x)-\mathcal{H}(z',0,x)+(\delta/2)\sum_{|k|=1}^{m}|z''_k|=\sum_{|k|=1}^{m}b_k z''_k,$$

where $b_k:=a_k+(\delta/2)\operatorname{sign}z''_k$ lie in $(\delta/2,\delta^{-1}+\delta/2)\subset(\delta/2,2/\delta)$.

Next, introduce one more function

$$\mathcal{H}_K(z,x)=\max\big(\mathcal{H}(z,x),\mathcal{P}(z'')-K\big). \tag{3}$$

Recall that $\delta_h^2 v$ and $\delta_h v$ are defined in (2.3) and (2.4) and for any function v on Ω set

$$\mathcal{H}_{K,h}[v](x)=\mathcal{H}_K\big(v(x),\delta_h v(x),\delta_h^2 v(x),x\big),$$

and similarly introduce $\mathcal{H}_h[v]$ and $\mathcal{P}_h[v]$.

We will concentrate on sufficiently small h such that $\Omega^h\ne\emptyset$ and will consider the equation

$$\mathcal{H}_{K,h}[v]=0\quad\text{in}\quad\Omega^h \tag{4}$$

with boundary condition

$$v=g\quad\text{on}\quad\partial_h\Omega. \tag{5}$$

By Theorem 2.5 and Corollary 4.1.4 for any sufficiently small $h>0$ there exists a unique bounded solution $v=v_h$ of (4)–(5). By the way, we do not include K in the notation v_h since K is a fixed number. The solution of the PDE version of (4)–(5) will be obtained as the limit of a subsequence of v_h as $h\downarrow 0$. Therefore, we need to have appropriate bounds on v_h and the first- and second-order differences in x of v_h.

Observe that owing to Assumption 2

$$-K_0|z'|-\bar{\mathcal{H}}\le\mathcal{H}(0,z'_{\pm1},...,z'_{\pm m},0,x)\le\mathcal{H}_K(0,z'_{\pm1},...,z'_{\pm m},0,x)$$
$$=\max\big(\mathcal{H}(0,z'_{\pm1},...,z'_{\pm m},0,x),-K\big)\le K_0|z'|+\bar{\mathcal{H}},$$

so that $\hat{\mathcal{H}}_K\le\bar{\mathcal{H}}$, where $\hat{\mathcal{H}}_K$ is taken from Theorem 2.8 with $\mathcal{H}_K$ in place of $\mathcal{H}$. Therefore, the following is a direct consequence of Theorem 2.8 and Corollary 4.1.4.

5. LEMMA. *There is a constant N, depending only on Λ, K_0, δ, and* $\operatorname{diam}(\Omega)$, *and there is $h_0>0$ such that for $h\in(0,h_0]$*

$$\sup_{\Omega}|v_h|\le N\bar{\mathcal{H}}+\|g\|_{C(\partial_h\Omega)}. \tag{6}$$

Lemma 2.10 and Corollary 4.1.4 also yield the following useful result.

6. LEMMA. *Let ϕ be a function on Ω. Then for any $h\in(0,\rho_{\text{int}}(\Omega))$ there exist bounded functions a_j,b_j,c,f, $|j|=1,...,m$, on Ω^h such that on Ω^h*

$$\delta/2\le a_j\le 2\delta^{-1},\quad |b_j|\le K_0,\quad c\ge 0,\quad |f|\le\bar{\mathcal{H}}, \tag{7}$$

$$\mathcal{H}_{K,h}[\phi]=a_k\Delta_{h,l_k}\phi+b_k\delta_{h,l_k}\phi-c\phi+f.$$

Below by h_0 and N with occasional indices we denote various (finite) constants depending only on $\mathrm{diam}(\Omega)$, $\rho_{\text{ext}}(\Omega)$, $\rho_{\text{int}}(\Omega)$, $\Lambda = \{l_i\}$, K_0, d, and δ, unless explicitly stated otherwise.

We need a barrier function. Recall that by definition $\rho_{\text{ext}}(\Omega) \le 1$.

7. LEMMA. *There exists a constant $\beta > 0$, depending only on Λ, δ, $\rho_{\text{ext}}(\Omega)$, and K_0, such that for $\psi(x) = |x|^{-\beta} - \rho_{\text{ext}}^{-\beta}(\Omega)$ we have*

$$a_k D_{l_k}^2 \psi(x) + b_k D_{l_k}\psi(x) - c\psi(x) \ge 1, \tag{8}$$

whenever $2\delta^{-1} \ge a_k \ge \delta/2$, $|b_k| \le K_0$, $|k| = 1, ..., m$, $|c| \le K_0$, and $0 < |x| \le 3$.

Proof. Observe that

$$a_k D_{l_k}^2 \psi = a^{ij} D_{ij}\psi,$$

where

$$a^{ij} = a_k l_k^i l_k^j.$$

For any $\lambda \in \mathbb{R}^d$ it holds that

$$a^{ij}\lambda^i\lambda^j = a_k \langle l_k, \lambda\rangle^2 \ge \delta \sum_{k=1}^m \langle l_k, \lambda\rangle^2 \ge \delta_1 |\lambda|^2,$$

$$a^{ij}\lambda^i\lambda^j \le 2m\delta^{-1}|\lambda|^2 =: N_1|\lambda|^2,$$

where $\delta_1 > 0$ is a constant whose existence is guaranteed by the assumption that $\mathrm{Span}\,\{l_j\} = \mathbb{R}^d$.

For α such that $(\beta + 2)\delta_1 \ge N_1 d + \beta\delta_1/2$ we obtain that

$$a^{ij}D_{ij}\psi = \beta|x|^{-\beta-2}\big((\beta+2)|x|^{-2}a^{ij}x^ix^j - \mathrm{tr}\,(a^{ij})\big) \ge (\delta_1/2)\beta^2|x|^{-\beta-2}.$$

One easily finishes the proof after noticing that

$$\big|b_k D_{l_k}\psi\big| = \beta|x|^{-\beta-2}\big|b_k\langle l_k, x\rangle\big|, \quad |c\psi| \le K_0\big(|x|^{-\beta} + \rho_{\text{ext}}^{-\beta}(\Omega)\big).$$

The lemma is proved. □

8. LEMMA. *There are constants $h_0 > 0$ and N such that for all $h \in (0, h_0]$ on $\bar{\Omega}$ we have*

$$|v_h - g| \le N\big(\bar{\mathcal{H}} + \|g\|_{C^{1,1}(\Omega)}\big)\rho_\Omega. \tag{9}$$

Proof. Note that (5) implies that (9) holds in $\partial_h\Omega$. Therefore, we may concentrate on proving (9) in Ω^h.

Take an $x_0 \in \Omega^h$ (so that $h < \rho_\Omega(x_0)$). If $\rho_\Omega(x_0) \ge 1$, then (9) holds at x_0 in light of Lemma 5. Hence we may assume that $\rho_\Omega(x_0) \le 1$. Let $\bar{x}_0$ be a closest point to x_0 on $\partial\Omega$. Then the closures of the two balls has only one common point $\bar{x}_0$: one is the ball centered at x_0 with radius $\rho_\Omega(x_0)$ and the other is a ball of radius $\rho_{\text{ext}}(\Omega)$, which lies outside Ω and is centered, say at y_0. Then all three points x_0, $\bar{x}_0$, and y_0 lie on the same line. Therefore, without loss of generality, on the account of moving the origin, we may assume that $y_0 = 0$, the straight line passing through the origin and x_0 crosses $\partial\Omega$ at $\bar{x}_0$, and $|\bar{x}_0| = \rho_{\text{ext}}(\Omega)$ (≤ 1 by assumption)

Then set

$$\phi = g - 2N_1\big(\bar{\mathcal{H}} + \|g\|_{C^{1,1}(\Omega)}\big)\psi,$$

where ψ is taken from Lemma 7 and the constant N_1 is such that in Ω^h

$$a_k\Delta_{h,l_k}g + b_k\delta_{h,l_k}g - cg + f \le N_1\big(\bar{\mathcal{H}} + \|g\|_{C^{1,1}(\Omega)}\big)$$

as long as conditions (7) are satisfied.

We increase N_1 if necessary in order to have $v_h \le \phi$ in $\Omega \setminus B_2$, the latter being possible owing to Lemma 5 and the fact that in $\Omega \setminus B_2$ we have

$$-\psi \ge \rho_{\text{ext}}^{-\beta}(\Omega) - 2^{-\beta} \ge 1 - 2^{-\beta}.$$

We also choose $h_0 > 0$ in such a way that finite-difference approximations of $D^2_{l_k}\psi$ and $D_{l_k}\psi$ be sufficiently close to these quantities in $B_3 \setminus B_{\rho_{\text{ext}}(\Omega)}$, so that owing to Lemma 7

$$a_k\Delta_{h,l_k}\psi + b_k\delta_{h,l_k}\psi - c\psi \ge 1/2$$

in $B_3 \setminus B_{\rho_{\text{ext}}(\Omega)}$ for $h \in (0, h_0]$.

Then for $\Omega_1 = \Omega \cap B_3$ (where $\psi < 0$), by using Lemma 6, we find that in Ω_1^h

$$\mathcal{H}_{K,h}[\phi] \le 0.$$

Also by the choice of N_1 (if $h_0 \le 1$), $v_h \le \phi$ on $\partial_h\Omega_1$. Hence by the comparison principle, $v_h \le \phi$ in $\Omega_1 = \Omega \cap B_3$. By the choice of N_1, this inequality, actually, holds in Ω, which yields the desired estimate of $v_h - g$ from above. Similarly one obtains it from below as well. The lemma is proved. □

The proof of the lemma allows us to get control on the boundary behavior of v_h uniformly with respect to h in a relatively easy way. The way to treat the finite-differences of v_h is much more complicated and mostly related to the fact that on the set (3.2) we only have a bleak shadow of (3.3) (see (12) below) and the fact that the boundary of G from (3.4) may contain part of the boundary of Ω^h, so that what is said right after (3.4) may not be quite true.

Define

$$M_h(x) = \sum_{|k|=1}^{m} \big|\delta_{h,l_k}v_h(x)\big|, \quad \bar{M}_h = \sup_{\Omega^h} M_h,$$

$$G := \Big\{x \in \Omega^h : (\delta/2)\sum_{k=1}^{m} \big|\Delta_{h,l_k}v_h(x)\big| > \bar{\mathcal{H}} + K + K_0\big(\big|v_h(x)\big| + M_h(x)\big)\Big\}. \quad (10)$$

and observe a fundamental fact that, thanks to (2),

$$\mathcal{P}_h[v_h] - K = 0 \quad \text{in} \quad G. \quad (11)$$

Also observe that on $\Omega^h \setminus G$ we have

$$(\delta/2)\sum_{k=1}^{m} |\Delta_{h,l_k}v_h| \le \bar{\mathcal{H}} + K + K_0\big(|v_h| + M_h\big). \quad (12)$$

9. LEMMA. *There are constants $h_0 > 0$ and N such that for all $h \in (0, h_0]$ and $r = 1, ..., m$ on Ω^h we have*

$$\big[1 \wedge (\rho_\Omega - 9h)\big]|\Delta_{h,l_r}v_h| \le N\big(\bar{M}_h + \bar{\mathcal{H}} + K + \|g\|_{C^{1,1}(\Omega)}\big). \quad (13)$$

Proof. Having in mind translations, we see that it suffices to prove (13) in $\Omega^h \cap \Lambda^h_\infty$. Then fix r and define

$$Q^o := \Omega^{3h} \cap \Lambda^h_\infty \cap G.$$

In light of Assumption 1 (used for the first time), the set Q^o is finite, since the number of points in Λ^h_∞ lying in any ball is finite, because for an appropriate integer I we have $I\Lambda^h_\infty \subset h\mathbb{Z}^d$.

For Q from (4.1), obviously, $Q \subset \Omega^h$. Next, if $x \in \Omega^h \cap \Lambda^h_\infty$ is such that $x \notin Q^o$, then either $x \notin \Omega^{3h}$, so that $\rho_\Omega(x) \le 3h$ and (13) holds, or else $x \in \Omega^{3h}$ but (12) is valid, in which case (13) holds again owing to (6).

Thus we need only prove (13) on Q^o assuming, of course, that $Q^o \neq \emptyset$. We know that (11) holds and the left-hand side of (11) is nonpositive in $Q \setminus Q^o$ (as everywhere else in Ω^h).

To proceed further use Lemma 1.4 according to which there exists a constant $N \in (0,\infty)$ depending only on d such that for any $\rho \in (0, \rho_{\text{int}}(\Omega)/2)$ there exists an $\eta_\rho \in C_0^\infty(\Omega)$ satisfying

$$\eta_\rho = 1 \quad \text{on} \quad \Omega^{2\rho}, \qquad \eta_\rho = 0 \quad \text{outside} \quad \Omega^\rho,$$

$$|\eta_\rho| \le 1, \quad |D\eta_\rho| \le N/\rho, \quad |D^2\eta_\rho| \le N/\rho^2.$$

By Theorem 4.2 on $Q^o \cap \Omega^{2\rho}$, for $h \in (0, h_0]$,

$$\big[(\Delta_{h,l_r} v_h)^-\big]^2 \le \sup_{Q\setminus Q^o} \eta_\rho \big[(\Delta_{h,l_r} v_h)^-\big]^2 + N\rho^{-2}\bar{M}_h^2.$$

While estimating the last supremum we will only concentrate on

$$h_0 < \rho_{\text{int}}(\Omega)/6 \quad \text{and} \quad \rho \in \big[3h, \rho_{\text{int}}(\Omega)/2\big),$$

when $\eta_\rho = 0$ outside Ω^{3h}. In that case, for any $y \in Q \setminus Q^o$, either $y \notin \Omega^{3h}$ implying that

$$\eta_\rho \big[(\Delta_{h,l_r} v_h)^-\big]^2(y) = 0,$$

or $y \in \Omega^{3h} \cap \Lambda^h_\infty$ but (12) holds at y.

It follows that as long as $h \in (0, h_0]$ and $x \in Q^o \cap \Omega^{2\rho}$ and $\rho \in \big[3h, \rho_{\text{int}}(\Omega)/2\big)$, we have

$$\rho(\Delta_{h,l_r} v_h)^-(x) \le N\big(\bar{\mathcal{H}} + K + \|g\|_{C^{1,1}(\Omega)} + \bar{M}_h\big). \tag{14}$$

If $x \in Q^o$ and x is such that $\rho_\Omega(x) \ge 9h$, take

$$\rho = \big(\rho_{\text{int}}(\Omega)/2\big) \wedge \big(\rho_\Omega(x)/3\big),$$

which is bigger than $3h$ for $h \in (0, h_0]$ since $h_0 \le \rho_{\text{int}}(\Omega)/6$. In that case also $\rho_\Omega(x) \ge 3\rho$, so that $x \in \Omega^{2\rho}$ (but may not belong to $\Omega^{3\rho}$) and we conclude from (14) that

$$\rho(\Delta_{h,l_r} v_h)^-(x) \le N\big(\bar{\mathcal{H}} + K + \|g\|_{C^{1,1}(\Omega)} + \bar{M}_h\big). \tag{15}$$

Obviously, one can replace ρ in (15) with $\rho_\Omega \wedge 1$ or with $(\rho_\Omega - 9h)\wedge 1$. Therefore,

$$\big[1 \wedge (\rho_\Omega(x) - 9h)\big](\Delta_{h,l_r} v_h)^-(x) \le N\big(\bar{\mathcal{H}} + K + \|g\|_{C^{1,1}(\Omega)} + \bar{M}_h\big)$$

for $x \in Q^o$ such that $\rho_\Omega(x) \ge 9h$. However, this relation is obvious for $\rho_\Omega(x) \le 9h$.

As a result of all the above arguments we see that

$$\big[1 \wedge (\rho_\Omega - 9h)\big](\Delta_{h,l_r} v_h)^- \le N\big(\bar{\mathcal{H}} + K + \|g\|_{C^{1,1}(\Omega)} + \bar{M}_h\big) \tag{16}$$

holds in $\Omega^h \cap \Lambda^h_\infty$ for any r whenever $h \in (0, h_0]$.

Finally, since $P_h[v_h] \le K$ in Ω^h, we have that

$$4\delta^{-1} \sum_{r=1}^m (\Delta_r v_h)^+ \le \delta \sum_{r=1}^m (\Delta_r v_h)^- + K,$$

which after being multiplied by $(\rho_\Omega - 9h) \wedge 1$ along with (16) leads to (13) on $\Omega^h \cap \Lambda^h_\infty$. Thus, as is explained at the beginning of the proof, the lemma is proved. □

Now we exclude $\bar{M}_h$ from (13). Here is the summary of our estimates.

10. THEOREM. *(i) There is a constant N, depending only on Λ, K_0, δ, and* $\operatorname{diam}(\Omega)$, *such that for all sufficiently small $h > 0$*

$$\sup_{\bar{\Omega}} |v_h| \le N\big(\bar{\mathcal{H}} + \|g\|_{C(\Omega)}\big). \tag{17}$$

(ii) There is a constant N, depending only on $\operatorname{diam}(\Omega)$, $\rho_{\text{ext}}(\Omega)$, $\rho_{\text{int}}(\Omega)$, Λ, K_0, *d, and δ, such that for all sufficiently small $h > 0$ the estimates*

$$|v_h - g| \le N\big(\bar{\mathcal{H}} + \|g\|_{C^{1,1}(\Omega)}\big)\rho_\Omega, \tag{18}$$

$$|\delta_{h,l_k} v_h|,\ \big[1 \wedge (\rho_\Omega - 9h)\big]|\Delta_{h,l_k} v_h| \le N\big(\bar{\mathcal{H}} + K + \|g\|_{C^{1,1}(\Omega)}\big) \tag{19}$$

hold in Ω^h for all k.

Proof. Estimate (17) is proved in Lemma 5. Estimate (18) coincides with (9). Owing to Lemma 9, (19) would follow if we can prove that

$$|\delta_{h,l_k} v_h| \le N\big(\bar{\mathcal{H}} + K + \|g\|_{C^{1,1}(\Omega)}\big) \tag{20}$$

in Ω^h for all k.

We are going to use interpolation inequalities. Note that if we have a function $u(i)$ on a set $-r+1, ..., 0, 1, ..., r$, where $r \ge 2$ is an integer, which satisfies

$$u(i+1) - 2u(i) + u(i-1) \ge -N_1 \tag{21}$$

for $i = -r+2, ..., r-1$, where N_1 is a constant, then

$$u(i+1) - u(i) \ge u(i) - u(i-1) - N_1.$$

It follows that $w(i) := u(i+1) - u(i) + N_1 i$ is an increasing function of $i = -r+1, ..., r-1$. In particular,

$$u(1) - u(0) = w(0) \le \frac{1}{r-1} \sum_{i=1}^{r-1} w(i)$$

$$= \frac{1}{r-1} \sum_{i=1}^{r-1} \big(u(i+1) - u(i) + N_1 i\big) = \frac{1}{r-1}\big(u(r) - u(1)\big) + \frac{1}{2} N_1 r.$$

On the other hand,

$$u(1)-u(0)\geq\frac{1}{r-1}\sum_{i=-r+1}^{-1}\big(u(i+1)-u(i)+N_1 i\big)$$
$$=\frac{1}{r-1}\big(u(0)-u(-r+1)\big)-\frac{1}{2}N_1 r.$$

Hence,

$$\big|u(1)-u(0)\big|\leq\frac{1}{2}N_1 r+\frac{2}{r-1}\max\big\{|u(i)|:i=-r+1,...,r\big\},$$

and for any function w (use that $(r-1)^{-1}\leq 2r^{-1}$ for $r\geq 2$)

$$\big|w(1)-w(0)\big|\leq\frac{r}{2}\max_{|i|\leq r}\big|w(i+1)-2w(i)+w(i-1)\big|+\frac{4}{r}\max_{|i|\leq r}\big|w(i)\big|. \tag{22}$$

We want to apply (22) to v_h-g. Fix an $\varepsilon\in(0,1]$ and set

$$n(\varepsilon)=13/\varepsilon.$$

Observe that if $x\in\Omega^{n(\varepsilon)h}$ and we take

$$r=\big\lfloor(\varepsilon\rho_\Omega(x)-9h)(2h)^{-1}\big\rfloor$$

($\lfloor a\rfloor$ is the integer part of a), then $r\geq 2$. Furthermore, $\varepsilon\rho_\Omega(x)-2rh\geq 9h$ and for any i we have $\rho_\Omega(x+ihl_k)\geq\rho_\Omega(x)-ih$. Therefore, for $|i|\leq r$ we have

$$\varepsilon\rho_\Omega(x+ihl_k)-rh\geq\varepsilon\rho_\Omega(x)-2rh\geq 9h,$$
$$\varepsilon\big[\rho_\Omega(x+ihl_k)-9h\big]\geq rh. \tag{23}$$

In particular, $x+ihl_k\in\Omega^h$ (even $\in\Omega^{9h}$) for $|i|\leq r$ and it makes sense applying (22) to $w(i)=v_h(x+ihl_k)-g(x+ihl_k)$, which yields

$$\big|\delta_{h,l_k}(v_h-g)(x)\big|\leq\frac{1}{2}rh\max_{|i|\leq r}\big|\Delta_{h,l_k}(v_h-g)(x+ihl_k)\big|$$
$$+\frac{4}{rh}\max_{|i|\leq r}\big|(v_h-g)(x+ihl_k)\big|. \tag{24}$$

Also notice that for $x\in\Omega^{n(\varepsilon)h}$ we have $13h<\varepsilon\rho_\Omega(x)$ and by definition $(r+1)2h\geq\varepsilon\rho_\Omega(x)-9h$. Hence ($|i|\leq r$)

$$2rh\geq\varepsilon\rho_\Omega(x)-11h,\quad 13rh\geq\varepsilon\rho_\Omega(x),$$

$$\rho_\Omega(x+ihl_k)\leq\rho_\Omega(x)+rh\leq rh(13\varepsilon^{-1}+1)\leq 14rh\varepsilon^{-1}. \tag{25}$$

Estimates (23) and (25) allow us to derive from (24) that

$$\big|\delta_{h,l_k}(v_h-g)(t,x)\big|\leq N\varepsilon\max_{|i|\leq r}\big[\rho_\Omega(x+ihl_k)-9h\big]\big|\Delta_{h,l_k}(v_h-g)(t,x+ihl_k)\big|$$
$$+N\varepsilon^{-1}\max_{|i|\leq r}\rho_\Omega^{-1}(x+ihl_k)\big|(v_h-g)(t,x+ihl_k)\big|,$$

which along with Lemmas 8 and 9 show that for all sufficiently small h, $\varepsilon\in(0,1]$, and $x\in\Omega^{n(\varepsilon)h}$

$$\big|\delta_{h,l_k}(v_h-g)(t,x)\big|\leq N\varepsilon\big(\bar{M}_h+\bar{\mathcal{H}}+K+\|g\|_{C^{1,1}(\Omega)}\big)$$
$$+N\varepsilon^{-1}\big(\bar{\mathcal{H}}+K+\|g\|_{C^{1,1}(\Omega)}\big).$$

Hence, for all sufficiently small h we have

$$\bar{M}_h = \sup_{\Omega^h} \sum_{|k|=1}^{m} |\delta_{h,l_k} v_h| \le N_1 \varepsilon \big(\bar{M}_h + \bar{\mathcal{H}} + K + \|g\|_{C^{1,1}(\Omega)} \big)$$

$$+ N\varepsilon^{-1} \big(\bar{\mathcal{H}} + K + \|g\|_{C^2(\Omega)} \big) + \sup_{\Omega^h \setminus \Omega^{n(\varepsilon)h)}} \sum_{|k|=1}^{m} \big| \delta_{h,l_k} (v_h - g) \big|,$$

where the last term is dominated by

$$Nn(\varepsilon) \big(\bar{\mathcal{H}} + \|g\|_{C^{1,1}(\Omega)} \big)$$

in light of (18). To finish proving (20) it now remains only to pick and fix $\varepsilon \in (0,1]$ so that $N_1 \varepsilon \le 1/2$. The theorem is proved. □

11. Remark. Much of what is done above goes through for domains satisfying the exterior cone condition instead of the exterior ball condition. However, then it would be impossible to get the global discrete gradient estimate and to get estimates of $\Delta_{h,l_k} v_h$ in a closed form.

CHAPTER 6

Elliptic differential equations of cut-off type

The goal of this chapter is to give sufficient conditions on the second-order elliptic operator $H[u]$ and an auxiliary operator $P[u]$ guaranteeing that for any given $g \in C^{1,1}(\mathbb{R}^d)$ and a bounded regular domain $\Omega \subset \mathbb{R}^d$ the equation

$$\max\big(H[v], P[v] - K\big) = 0 \quad \text{in} \quad \Omega \quad \text{(a.e.)} \tag{1}$$

with boundary condition $v = g$ on $\partial\Omega$ has a solution $v \in C(\bar{\Omega}) \cap C^{1,1}_{\text{loc}}(\Omega)$.

Recall that if G is a domain in $\mathbb{R}^d$ by $C^{0,1}(G)$ we mean the set of bounded Lipschitz continuous functions on G or, equivalently, of bounded continuous functions u on G for which $D_i u$ (the generalized derivatives) are bounded. By $C^{1,1}(G)$ we mean the subset of $C^{0,1}(G)$ consisting of functions u for which $Du \in C^{0,1}(G)$. The "local" versions of these spaces are defined in an obvious way.

We will also be interested in obtaining the estimates

$$|v| \leq N\bar{H} + \|g\|_{C(\partial\Omega)} \quad \text{in} \quad \Omega, \tag{2}$$

$$|v - g| \leq N\big(\bar{H} + \|g\|_{C^{1,1}(\mathbb{R}^d)}\big)\rho_\Omega \quad \text{in} \quad \Omega, \tag{3}$$

$$|D_i v|, (\rho_\Omega \wedge 1)|D_{ij} v| \leq N\big(\bar{H} + K + \|g\|_{C^{1,1}(\mathbb{R}^d)}\big) \quad \text{in} \quad \Omega \quad \text{(a.e.)}, \tag{4}$$

where $i, j = 1, ..., d$, $D_{ij}v$ are the generalized derivatives of v, and the constants $\bar{H}$ and N are chosen appropriately (recall that $\rho_\Omega(x) = \text{dist}\,(x, \Omega^c)$).

The most significant results of this chapter are given in Section 6; they are further development of the results in [**72**] as presented in [**79**].

1. A particular case of equations in pure derivatives

Fix some constants $m, d \in \{1, 2, ...\}$, $\delta \in (0, 1]$, and $K_0, K, N' \in [0, \infty)$.

1. Assumption. We are given vectors $l_i \in \bar{B}_1 \subset \mathbb{R}^d$, $i = \pm 1, ..., \pm m$, such that $l_{-i} = -l_i$, the coordinates of l_i's are rational numbers, and

$$e_i,\ (1/2)(e_i \pm e_j) \in \Lambda := \{l_k : k = \pm 1, ..., \pm m\}$$

for $i, j = 1, .., d$, $i \neq j$, where the e_i's form the standard orthonormal basis in $\mathbb{R}^d$.

2. Assumption. We are given a domain $\Omega \subset \mathbb{R}^d$ which is bounded and satisfies the exterior ball condition. We are also given a function $g \in C^{1,1}(\mathbb{R}^d)$.

3. Assumption. (i) We are given a continuous function $\mathcal{H}(z, x)$, $x \in \mathbb{R}^d$,

$$z = (z', z''), \quad z' = \big(z'_0, z'_{\pm 1}, ..., z'_{\pm m}\big) \in \mathbb{R}^{2m+1}, \quad z'' = \big(z''_{\pm 1}, ..., z''_{\pm m}\big) \in \mathbb{R}^{2m},$$

which is Lipschitz continuous with respect to $z \in \mathbb{R}^{4m+1}$ with Lipschitz constant N' for any $x \in \mathbb{R}^d$. At all point of its differentiability with respect to z introduce

$$a_j = a_j(z, x) = D_{z_j''}\mathcal{H}, \quad c = c(z, x) = -D_{z_0'}\mathcal{H}$$

and at points of non-differentiability set $a_j = \delta$ and $c = 0$.

(ii) The above introduced functions a_j and c satisfy

$$\delta^{-1} \geq a_j \geq \delta, \quad c \geq 0 \tag{1}$$

for all values of j and the arguments.

(iii) The number

$$\bar{\mathcal{H}} := \sup_{x,z'} \big(|\mathcal{H}(z', 0, x)| - K_0|z'|\big)$$

is finite.

(iv) The function $\mathcal{H}(z, x)$ is Lipschitz continuous with respect to $x \in \mathbb{R}^d$ for any $z \in \mathbb{R}^{4m+1}$ and at any point x of its differentiability with respect to x

$$\big|D_x\mathcal{H}(z, x)\big| \leq N'\big(1 + |z|\big). \tag{2}$$

(v) There is a constant $\rho_0 > 0$ and a function $\mathcal{H}(z)$ such that

$$\mathcal{H}(z, x) = \mathcal{H}(z) \tag{3}$$

for all z if $\rho_\Omega(x) \leq \rho_0$ (recall that $\rho_\Omega(x) = \text{dist}\,(x, \mathbb{R}^d \setminus \Omega)$).

4. Remark. We will see that assuming that $\mathcal{H}(z)$ in (3) depends on the whole of z presents more difficulties than as if it was simply $\mathcal{H}(z'')$. However, in our reduction of the general case to the case of pure second-order derivatives we automatically obtain $\mathcal{H}(z)$ in (3) depending on all coordinates of z even if in the corresponding condition in the general case $\mathcal{H}$ depended only on the second-order derivatives.

For sufficiently smooth functions $u = u(x)$ introduce

$$\begin{aligned}\mathcal{H}[u](x) &= \mathcal{H}\big(u(x), D_{l_j}u(x), D^2_{l_k}u(x), x\big)\\ &= \mathcal{H}\big(u(x), D_{l_{\pm 1}}u(x), ..., D_{l_{\pm m}}u(x), D^2_{l_{\pm 1}}u(x), ..., D^2_{l_{\pm m}}u(x), x\big).\end{aligned}$$

Similarly we introduce $\mathcal{P}[u]$, where $\mathcal{P} = \mathcal{P}_{m,\delta/2}$ is taken from (5.5.1).

Here is the pilot result of this chapter, which will be gradually generalized to a very large extent in the subsequent sections. The first generalization is given in Theorem 10 in which the global Lipschitz condition of $\mathcal{H}(z, x)$ with respect to z is replaced with the local one.

In the following theorem the above assumptions are supposed to be satisfied.

5. Theorem. *Equation* (0.1) *with boundary condition* $v = g$ *on* $\partial\Omega$ *has a unique solution* $v \in C(\bar{\Omega}) \cap C^{1,1}_{\text{loc}}(\Omega)$. *Furthermore, estimate* (0.2) *holds with* N, *depending only on* d, K_0, δ, *and* $\text{diam}(\Omega)$, *and the estimates* (0.3) *and* (0.4) *hold with* N *depending only on* $\text{diam}(\Omega)$, $\rho_{\text{ext}}(\Omega)$, $\rho_{\text{int}}(\Omega)$, $\Lambda = \{l_i\}$, K_0, d, *and* δ *(in particular,* N *is independent of* N' *and* ρ_0*).*

6. Remark. The constant N in the most important estimates (0.3) and (0.4) is independent of N' and ρ_0 which enter Assumptions 3 (i), (iv), and (v). In the future Theorem 6.20 Assumption 3 (i) is substantially relaxed and Assumptions 3 (iv), (v) are dropped. Therefore, it is worth explaining how we use them here. Assumption 3 (i) will allow us to use the results about finite-difference equations while replacing pure second-order derivatives with second-order differences and Assumptions 3 (iv),

(v) will allow us to show that the collection of solutions of finite-difference equations is almost equicontinuous in Ω in the sense specified in Lemma 7.

The proof of Theorem 5 consists of three lemmas. But first of all we observe that by Corollary 4.1.4 and Lemma 4.1.6, if we have two solutions u and v, then in Ω (a.e.)

$$0 = \max\big(\mathcal{H}[u], \mathcal{P}[u] - K\big) - \max\big(\mathcal{H}[v], \mathcal{P}[v] - K\big)$$
$$= a_k D_{l_k}^2(u - v) + b^i D_i(u - v) - c(u - v),$$

where $\delta/2 \leq a_k \leq 2/\delta$, $b = (b^i)$, and c are certain measurable bounded functions, $c \geq 0$. Since Λ contains all the e_i's, the matrix $(a_k l_k^i l_k^j)$ is greater than $(\delta/2)(\delta^{ij})$. Now by Theorem 3.0.2, $u - v = 0$ and hence uniqueness.

As is explained in Section 5.3, we are going to use finite differences in the proof of the existence and the estimates. Observe that Assumptions 5.2.1, 5.5.1, 5.5.2, and 5.5.4 are satisfied due to Assumptions 1, 2, and 3. Assumption 5.2.3 (i) is also satisfied in light of Assumption 3 (i). In Assumption 5.2.3 (iii) the condition $c \geq 0$ is required in (1), (5.2.1) follows from (1), the boundedness of a, b, c follows from Assumption 3 (i), and the fact that $\mathcal{H}(0, x)$ is a bounded function follows from Assumption 3 (iii).

Thus, all the assumptions of Section 5.5 are satisfied and we can use the information obtained in this section about functions v_h defined in $\bar{\Omega}$ for sufficiently small $h > 0$ as bounded solutions of (5.5.4)-(5.5.5).

Estimates (5.5.19) are estimates of v_h on the translates of a multiple of $\mathbb{Z}^d$. They do not tell us anything about the difference $v_h(x) - v_h(y)$ if x and y cannot be connected by a broken line consisting of translates of hl_j, $|j| = 1, ..., m$. However, this information is necessary if we want to let $h \downarrow 0$ and extract a subsequence of v_h converging everywhere in Ω. In the proof of the following result Assumptions 3 (iv) and (v) are crucial.

7. Lemma. *There exists a constant N such that for all sufficiently small $h > 0$ we have*

$$\big|v_h(x) - v_h(y)\big| \leq N\big(|x - y| + h\big) \tag{4}$$

whenever $x, y \in \bar{\Omega}$.

Proof. First we want to estimate $v_h(x) - v_h(y)$ when $|x - y| \leq h$. Take $x_0 \in \Omega$ and $h > 0$ such that $11h < \rho_\Omega(x_0)$. Denote by N_1 the right-hand side of (5.5.19) to which we add the right-hand side of (5.5.17) and with the idea of applying Lemma 4.1.1 introduce

$$D_{\mathcal{H}} = \Big\{(z, x) : |z_k'|, |z_j''| \leq N_1\big[1 \wedge (\rho_\Omega(x_0) - 10h)\big]^{-1},$$
$$|k| = 0, ..., m, |j| = 1, ..., m\Big\} \times \bar{B}_h(x_0).$$

We claim that, on the account of our assumptions, there is a constant N_2, independent of h and x_0, such that, at all points of differentiability of $\mathcal{H}$ with respect to (z, x) belonging to the interior of $D_{\mathcal{H}}$, we have

$$\delta/2 \leq D_{z_j''}\mathcal{H}(z, x) \leq 2\delta^{-1}, \quad |j| = 1, ..., m, \quad 0 \leq D_{z_0'}\mathcal{H}(z, x) \leq N_2,$$
$$\big|D_{z,x}\mathcal{H}(z, x)\big| \leq N_2. \tag{5}$$

Indeed, the first set of inequalities and the second one are given by assumption and the Lipschitz continuity of $\mathcal{H}(z,x)$ in z uniform with respect to x. If $\rho_\Omega(x_0) \le \rho_0/2$, where ρ_0 is taken from Assumption 3 (v), then $B_h(x_0) \subset \partial_{\rho_0}\Omega$ and $D_{z,x}\mathcal{H}(z,x)$ (on $D_{\mathcal{H}}$) coincides with the gradient of the right-hand side of (3), which is bounded due to Assumption 3 (i), implying the last inequality in (5). However, if $\rho_\Omega(x_0) \ge \rho_0/2$, then

$$\rho_\Omega(x_0) - 10h \ge (1/11)\rho_\Omega(x_0) \ge (1/22)\rho_0,$$

and the last inequality in (5) follows from the local Lipschitz continuity of $\mathcal{H}(z,x)$, which in turn is due to the uniform Lipschitz continuity with respect to z and (2).

The same claim holds true, obviously, for $\mathcal{P}(z'') - K$ and, by Corollary 4.1.4, for $\mathcal{H}(z,x) \vee \big(\mathcal{P}(z'') - K\big)$ as well.

Next, it is convenient to continue v_h outside Ω by setting it equal to g and for a fixed $l \in \mathbb{R}^d$, such that $|l| \le h$, introduce the function $w_h(x) = v_h(x+l)$. Observe that, owing to Theorem 5.5.10, both points

$$\big(v_h(x_0), \delta_{h,l_{\pm 1}}v_h(x_0), ..., \delta_{h,l_{\pm m}}v_h(x_0), \Delta_{h,l_{\pm 1}}v_h(x_0), ..., \Delta_{h,l_{\pm m}}v_h(x_0), x_0\big)$$

and

$$\big(w_h(x_0), \delta_{h,l_{\pm 1}}w_h(x_0), ..., \delta_{h,l_{\pm m}}w_h(x_0), \Delta_{h,l_{\pm 1}}w_h(x_0), ..., \Delta_{h,l_{\pm m}}w_h(x_0), x_0 + l\big)$$

are in $D_{\mathcal{H}}$ for all sufficiently small $h > 0$. Furthermore, $x_0 \in \Omega^{11h}$ and

$$\mathcal{H}\big(w_h(x_0), \delta_h w_h(x_0), \delta_h^2 w_h(x_0), x_0 + l\big) \vee \big(\mathcal{P}_h[w_h](x_0) - K\big) = 0.$$

Also

$$\mathcal{H}\big(v_h(x_0), \delta_h v_h(x_0), \delta_h^2 v_h(x_0), x_0\big) \vee \big(\mathcal{P}_h[v_h](x_0) - K\big) = 0.$$

We subtract these two equations and, by using the arbitrariness of x_0, Corollary 4.1.4, and what was said above in the proof, conclude that the function $u_h = w_h - v_h$ satisfies in Ω^{11h}

$$a_k\Delta_{h,l_k}u_h + b_k\delta_{h,l_k}u_h - cu_h + f = 0,$$

where $\delta/2 \le a_k \le 2\delta^{-1}$, $|b_k|$, $c \le N_2$, $c \ge 0$, $|f| \le N_2 h$.

Next, in Lemma 5.2.6 reduce δ if needed, in such a way that the new one, δ_1, will satisfy $N_2 \le \delta_1^{-1}$ and $\delta_1 \le \delta/2$, and then take Ψ_0 from that lemma. Then by the comparison principle we have in $\mathbb{R}^d$ that

$$|w_h - v_h| \le N_2 h \sup_\Omega \Psi_0 + \sup_{\mathbb{R}^d \setminus \Omega^{11h}} |w_h - v_h|.$$

The last supremum is easily estimated by using (5.5.9) and the fact that

$$|w_h(x) - v_h(x)| \le \big|v_h(x+l) - g(x+l)\big| + \big|g(x+l) - g(x)\big| + \big|g(x) - v_h(x)\big|.$$

This yields (4) for $x, y \in \mathbb{R}^d$, such that $|x - y| \le h$.

It only remains to observe that, if $|x-y| \ge h$, one can split the straight segment between x and y into adjacent pieces of length h combined with a remaining one of length less than h and then apply the above result to each piece. The lemma is proved. □

8. LEMMA. *There is a sequence $h_n \downarrow 0$ such that v^{h_n} converge uniformly on $\bar{\Omega}$ to a function $v \in C(\bar{\Omega}) \cap C^{1,1}_{\text{loc}}(\Omega)$ for which the second assertion of Theorem 5 holds true.*

Proof. The functions $v^h(x)$ are uniformly bounded in $\bar\Omega$ if h is small enough. Therefore there exists a sequence $h_n \downarrow 0$ such that v^{h_n} converge at any point of a countable everywhere dense in $\bar\Omega$ set S. This defines v on S. It follows from Lemma 7 that, for $x, y \in S$, we have

$$\big|v(x) - v(y)\big| \le N|x - y|,$$

where N is taken from (4). This estimate allows us to extend v to all of $\bar\Omega$ as a continuous function and allows us to get that $v^{h_n} \to v$ on $\bar\Omega$.

Next, for any $\varepsilon > 0$ and any $x_0 \in \bar\Omega$ there is n_0 such that $|v^{h_n} - v| \le (5N+1)h$ in $B_h(x_0) \cap \bar\Omega$ for $n \ge n_0$ since there is a point $y \in B_h(x_0) \cap S$ and n_0 such that $\big|v^{h_n}(y) - v(y)\big| \le h$ for $n \ge n_0$. One can cover $\bar\Omega$ with finitely many balls $B_h(x_0)$ and then one sees that $v^{h_n} \to v$ uniformly on $\bar\Omega$. Of course, $|v|$ and $|v - g|$ admit estimates (0.2) and (0.3) owing to (5.5.17) and (5.5.18).

Next, for any $\zeta \in C_0^\infty(\Omega)$ and $|k| = 1, ..., m$ we have

$$\Big|\int_\Omega v^h \delta_{h,-l_k}\zeta\,dx\Big| = \Big|\int_\Omega \zeta\delta_{h,l_k}v^h\,dx\Big| \le N_0\big(\bar{\mathcal{H}} + K + \|g\|_{C^{1,1}(\Omega)}\big)\int_\Omega |\zeta|\,dx$$

if h is small enough, where N_0 is the constant from (5.5.19). By passing to the limit along the sequence h_n we conclude that

$$\Big|\int_\Omega vD_{-l_k}\zeta\,dx\Big| \le N_0\big(\bar{\mathcal{H}} + K + \|g\|_{C^{1,1}(\Omega)}\big)\int_\Omega |\zeta|\,dx,$$

which is the same as to say that the generalized derivative $D_{l_k}v$ in Ω is a bounded function satisfying (a.e.)

$$|D_{l_k}v| \le N_0\big(\bar{\mathcal{H}} + K + \|g\|_{C^{1,1}(\Omega)}\big).$$

Since $e_1, ..., e_d \in \Lambda$, v is in $C^{0,1}(\bar\Omega)$ and $D_i v$ admits the desired estimate.

Passing to the second order derivatives, for any $\zeta \in C_0^\infty(\Omega)$, $|k| = 1, ..., m$, and sufficiently small h such that $\zeta = 0$ outside Ω^{11h}, we use (5.5.19) to write

$$\Big|\int_\Omega v^h \Delta_{h,l_k}\zeta\,dx\Big| = \Big|\int_\Omega \zeta\Delta_{h,l_k}v^h\,dx\Big|$$

$$\le N_0(\bar{\mathcal{H}} + K + \|g\|_{C^{1,1}(\Omega)})\int_\Omega |\zeta|\,\big[1 \wedge (\rho_\Omega - 9h)\big]^{-1}\,dx.$$

This yields

$$\Big|\int_\Omega vD_{l_k}^2\zeta\,dx\Big| \le N_0\big(\bar{\mathcal{H}} + K + \|g\|_{C^{1,1}(\Omega)}\big)\int_\Omega |\zeta|\,[1 \wedge \rho_\Omega]^{-1}\,dx,$$

which implies that the generalized derivatives $D_{l_k}^2 v$ are locally bounded functions satisfying

$$\Big|\int_\Omega \zeta D_{l_k}^2 v\,dx\Big| \le N_0\big(\bar{\mathcal{H}} + K + \|g\|_{C^{1,1}(\Omega)}\big)\int_\Omega |\zeta|\,[1 \wedge \rho_\Omega]^{-1}\,dx$$

and hence (a.e.)

$$|D_{l_k}^2 v| \le N_0\big(\bar{\mathcal{H}} + K + \|g\|_{C^{1,1}(\Omega)}\big)[1 \wedge \rho_\Omega]^{-1}.$$

In particular, $D_i^2 v$ admits the desired estimate, since $e_i \in \Lambda$ by assumption. Also for $i \neq j$

$$D_{ij}v = D^2_{(1/2)(e_i+e_j)}v - D^2_{(1/2)(e_i-e_j)}v$$

and, since $(1/2)(e_i \pm e_j) \in \Lambda$ by assumption, $D_{ij}v$ admits the desired estimate. By observing that $D_{ij}v = D_i D_j v$ we conclude that the bounded functions $D_j v$ admit locally bounded first generalized derivatives. By using mollifications one easily concludes that the $D_j v$'s have continuous modifications, whose generalized derivatives $D_{ij}v$ are bounded by

$$2N_0\big(\bar{\mathcal{H}} + K + \|g\|_{C^{1,1}(\Omega)}\big)[1 \wedge \rho_\Omega]^{-1}.$$

It follows that $D_j v \in C^{0,1}_{\text{loc}}(\Omega)$. The lemma is proved. □

Now observe that, since, obviously, $v = g$ on $\partial\Omega$, to finish the proof of Theorem 5 we only need to establish the following.

9. LEMMA. *The function v satisfies* (0.1).

Proof. Since $v \in C^{1,1}_{\text{loc}}(\Omega)$, according to the Aleksandrov-Busemann-Feller theorem (see, for instance, Theorem 1.2.1 in [**61**]), for any point $x_0 \in \Omega$ which is the Lebesgue point of all $D_{ij}v$ (the set of such points has full measure) we have

$$v(x_0 + y) = P^{x_0}(y) + o\big(|y|^2\big)$$

as $y \to 0$, where

$$P^{x_0}(y) = v(x_0) + y^i D_i v(x_0) + (1/2)y^i y^j D_{ij}v(x_0).$$

Fix such an $x_0 \in \Omega$, take an $\varepsilon > 0$, and observe that for all small $r > 0$ we have $\bar{B}_{2r}(x_0) \subset \Omega$ and $\big|o\big((2r)^2\big)\big| \leq \varepsilon r^2$. Then, for

$$u(x) := P^{x_0}(x - x_0) + \varepsilon\big(|x - x_0|^2 - r^2\big),$$

it holds on $\partial B_{2r}(x_0)$ that

$$v - u = o\big((2r)^2\big) - 3\varepsilon r^2 \leq -2\varepsilon r^2.$$

On the other hand

$$v(x_0) = u(x_0) + \varepsilon r^2.$$

Next, we take n_0 so large that for $n \geq n_0$

$$v^{h_n} \leq u$$

on $\partial_{h_n} B_{2r}(x_0)$ and $v^{h_n}(x_0) > u(x_0)$. By Corollary 5.2.7 there is at least one point $x_{n,r} \in B^{h_n}_{2r}(x_0)$ such that

$$0 = \mathcal{H}_{K,h_n}[v^{h_n}](x_{n,r}) < \mathcal{H}_{K,h_n}[u](x_{n,r})$$

(recall that $\mathcal{H}_K$ is introduced in (5.5.3)). Thus, for any sufficiently small $r > 0$ and large enough n there exists $x_{n,r} \in B_{2r}(x_0)$ such that

$$\mathcal{H}\big(u(x_{n,r}), \delta_{h_n}u(x_{n,r}), \delta^2_{h_n}u(x_{n,r}), x_{n,r}\big) \vee \big[\mathcal{P}\big(\delta^2_{h_n}u(x_{n,r})\big) - K\big] > 0.$$

Since u is just a polynomial and $\mathcal{P}(z'')$ and $\mathcal{H}(z,x)$ are locally Lipschitz continuous, by letting $n\to\infty$ and choosing a converging subsequence of $x_{n,r}$, we get that for any sufficiently small $r>0$ there exists $x_r\in\bar{B}_{2r}(x_0)$, such that

$$\mathcal{H}\big(u(x_r),D_{l_j}u(x_r),D^2_{l_k}u(x_r),x_r\big)\vee\big[\mathcal{P}\big(D^2_{l_k}u(x_r)\big)-K\big]\geq 0. \tag{6}$$

Notice that as $r\downarrow 0$

$$u(x_r)\to v(x_0),\ \ D_{l_j}u(x_r)\to D_{l_j}v(x_0),\ \ D^2_{l_k}u(x_r)\to D^2_{l_k}v(x_0)+2\varepsilon|l_k|^2.$$

Hence (6) yields

$$\mathcal{H}\big(v(x_0),D_{l_j}v(x_0),D^2_{l_k}v(x_0)+2\varepsilon|l_k|^2,x_0\big)\vee\big[\mathcal{P}\big(D^2_{l_k}v(x_0)+2\varepsilon|l_k|^2\big)-K\big]\geq 0,$$

which after letting $\varepsilon\downarrow 0$ shows that at x_0 the left-hand side of (0.1) is greater than zero. One proves that it also less than zero by considering

$$u(x)=P^{x_0}(x-x_0)-\varepsilon\big(|x-x_0|^2-r^2\big).$$

The lemma and Theorem 5 are thus proved. □

The following theorem will be used when, after rewriting in terms of pure derivatives general fully nonlinear elliptic operators, even depending in Lipschitz continuous way on the unknown function and its derivatives, we obtain an operator that is only locally Lipschitz as a function of the unknown function and its first-order derivatives.

10. THEOREM. *In Assumption* 3 (*i*) *replace the condition that* $\mathcal{H}(z,x)$

"is Lipschitz continuous with respect to $z\in\mathbb{R}^{4m+1}$ *with Lipschitz constant* N' *for any* $x\in\mathbb{R}^d$*"*

by $\mathcal{H}(z,x)$

"is Lipschitz continuous with respect to z *in any ball lying in* $\mathbb{R}^{4m+1}$ *with Lipschitz constant independent of* x*"*

and suppose that so modified Assumption 3 *along with Assumptions* 1 *and* 2 *are satisfied. Then the assertions of Theorem* 5 *are still true.*

Proof. Take the constants N, that exist by Theorem 5, and denote by N^0 the sum of the right-hand side of (0.2) and the product of $1+\rho_{\text{int}}(\Omega)$ and the right-hand side of (0.4). Note that, if we substitute this product in place of the right-hand side of (0.4), then the resulting inequalities will hold with ρ_Ω in place of $\rho_\Omega\wedge 1$. Then take a smooth increasing function $\chi(t), t\in\mathbb{R}$, such that $\chi'\leq 1$, $\chi(t)=t$ for $|t|\leq N^0$ and $\chi(t)=2N^0\operatorname{sign}t$ for $|t|\geq 3N^0$ and, for $\rho\in(0,1]$, introduce

$$\mathcal{H}^\rho(z,x)=\delta\sum_{|k|=1}^{m}\big[z''_k-\rho^{-1}\chi(\rho z''_k)\big]+\mathcal{H}\big(\chi(z'),\rho^{-1}\chi(\rho z''),x\big),$$

where we use the notation

$$\chi(z')=\big(\chi(z'_0),\chi(z'_{\pm1}),...,\chi(z'_{\pm m})\big),$$
$$\chi(\rho z'')=\big(\chi\big(\rho z''_{\pm1}\big),...,\chi\big(\rho z''_{\pm m}\big)\big).$$

Obviously, $\mathcal{H}^\rho$ is Lipschitz continuous on $\mathbb{R}^{4m+1}\times\mathbb{R}^d$. Also, due to $0\leq\chi'\leq 1$, it is easy to check that Assumption 3 (ii) is satisfied for $\mathcal{H}^\rho$ with the same δ.

Assumption 3 (iii) is obviously satisfied for $\mathcal{H}^\rho$ with the same K_0 and perhaps smaller $\bar{\mathcal{H}}$. Assumption 3 (v) is satisfied for $\mathcal{H}^\rho$ with the same ρ_0 and

$$\delta \sum_{|k|=1}^{m} \left[z_k'' - \rho^{-1}\chi\big(\rho z_k''\big)\right] + \mathcal{H}\big(\chi(z'), \rho^{-1}\chi(\rho z'')\big)$$

in place of $\mathcal{H}(z)$. Finally, Assumption 3 (iv) is satisfied for $\mathcal{H}^\rho$ with $2N'$ in place of N' since $|\chi(z')| \le |z'|$ and $\big|\rho^{-1}\chi\big(\rho z_k''\big)\big| \le \big|z_k''\big|$.

By Theorem 5, there is a solution v^ρ of the equation

$$\mathcal{H}^\rho[v] \vee \big(\mathcal{P}[v] - K\big) = 0$$

in Ω (a.e.) with boundary data $v = g$ on $\partial\Omega$ enjoying the properties listed in Theorem 5. By the choice of N^0 and (0.2) and (0.4) we have $\mathcal{H}^\rho[v^\rho] = \mathcal{H}[v^\rho]$ in Ω^ρ so that v^ρ also satisfies (0.1) in Ω^ρ. Now we find a sequence $\rho_n \downarrow 0$ such that v^{ρ_n} converge uniformly in $\bar{\Omega}$ to a function v. Obviously, $v = g$ on $\partial\Omega$, $v \in C(\bar{\Omega}) \cap C^{1,1}_{\text{loc}}(\Omega)$ and the estimates claimed in Theorem 5 hold. By fixing m and concentrating on Ω^{ρ_m} by Theorems 4.2.15 and 4.2.13 we get that v satisfies (0.1) in Ω^{ρ_m} for any m and hence in Ω. This takes care of the existence and the estimates.

To prove uniqueness it suffices to apply the argument after the statement of Theorem 5 to Ω^ε for small $\varepsilon > 0$ and conclude that in Ω

$$|u - v| \le \max_{\partial\Omega^\varepsilon} |u - v|,$$

which after sending $\varepsilon \searrow 0$ and taking into account that u, v are continuous in $\bar{\Omega}$ and $u = v = g$ on $\partial\Omega$ yields $u = v$ in $\bar{\Omega}$. The theorem is proved. □

2. General elliptic equations with Lipschitz continuous H

Fix some constants $\delta \in (0, 1]$ and $K_0, K_1, K, N' \in [0, \infty)$. Recall that $\mathbb{S}$ is the set of $d \times d$ symmetric matrices and introduce

$$\mathbb{S}_\delta = \big\{a \in \mathbb{S} : \delta|\lambda|^2 \le a^{ij}\lambda^i\lambda^j \le \delta^{-1}|\lambda|^2 \quad \forall \lambda \in \mathbb{R}^d\big\}.$$

Here is a corollary of Theorem 3.1 of [**71**] proved in Appendix A.

1. THEOREM. *There exist an integer $m = m(\delta, d) \ge d$, a set*

$$\Lambda = \{l_{\pm 1}, ..., l_{\pm m}\} \subset \bar{B}_1 \subset \mathbb{R}^d, \tag{1}$$

and there exists a constant

$$\hat{\delta} = \hat{\delta}(\delta, d) \in (0, \delta/4]$$

such that the coordinates of l_k are rational numbers and

(a) We have $l_{-k} = -l_k$, $k = 1, ..., m$, e_i, $(1/2)(e_i + e_j) \in \Lambda$, $i \ne j$, $i, j = 1, ..., d$, where $e_1, ..., e_d$ is the standard orthonormal basis of $\mathbb{R}^d$;

(b) There exist real-analytic functions $\lambda_{\pm 1}(a), ..., \lambda_{\pm m}(a)$ on $\mathbb{S}_{\delta/4}$ such that for any $a \in \mathbb{S}_{\delta/4}$

$$a = \sum_{|k|=1}^{m} \lambda_k(a) l_k l_k^*, \quad \hat{\delta}^{-1} \ge \lambda_k(a) \ge \hat{\delta}, \quad \forall k. \tag{2}$$

For $z'' = (z''_{\pm 1}, ..., z''_{\pm m})$ recall that

$$\mathcal{P}_{m,\hat{\delta}/2}(z'') = \max_{\substack{\hat{\delta}/2 \le a_k \le 2\hat{\delta}^{-1} \\ |k|=1,...,m}} \sum_{|k|=1}^{m} a_k z''_k, \tag{3}$$

and for $\mathsf{u}'' \in \mathbb{S}$ define (recall that $\langle \cdot, \cdot \rangle$ is the scalar product in $\mathbb{R}^d$)

$$P(\mathsf{u}'') = P_{m,\hat{\delta}/2,\Lambda}(\mathsf{u}'') = \mathcal{P}_{m,\hat{\delta}/2}\big(\langle \mathsf{u}'' l_{\pm 1}, l_{\pm 1}\rangle, ..., \langle \mathsf{u}'' l_{\pm m}, l_{\pm m}\rangle\big). \tag{4}$$

2. REMARK. Observe that, by Corollary 4.1.4, $P(\mathsf{u}'')$ is Lipschitz continuous and at all points of its differentiability, for any $\xi \in \mathbb{R}^d$, we have

$$\min_{\substack{\hat{\delta}/2 \le a_k \le 2\hat{\delta}^{-1} \\ |k|=1,...,m}} \sum_{|k|=1}^{m} a_k \langle l_k, \xi \rangle^2 \le \langle \xi, D_{\mathsf{u}''} P(\mathsf{u}'')\xi \rangle \le \max_{\substack{\hat{\delta}/2 \le a_k \le 2\hat{\delta}^{-1} \\ |k|=1,...,m}} \sum_{|k|=1}^{m} a_k \langle l_k, \xi \rangle^2.$$

Since, $e_i \in \Lambda \subset \bar{B}_1$, we see that there is a

$$\bar{\delta} = \bar{\delta}(d, \delta) \in (0, \delta/4]$$

such that

$$D_{\mathsf{u}''} P(\mathsf{u}'') \in \mathbb{S}_{\bar{\delta}}. \tag{5}$$

3. ASSUMPTION. We are given a domain $\Omega \subset \mathbb{R}^d$ which is bounded and satisfies the exterior ball condition. We are also given a function $g \in C^{1,1}(\mathbb{R}^d)$.

Suppose that we are also given a function $H(\mathsf{u}, x)$,

$$\mathsf{u} = (\mathsf{u}', \mathsf{u}''), \quad \mathsf{u}' = \big(\mathsf{u}'_0, \mathsf{u}'_1, ..., \mathsf{u}'_d\big) \in \mathbb{R}^{d+1}, \quad \mathsf{u}'' \in \mathbb{S}, \quad x \in \mathbb{R}^d.$$

4. ASSUMPTION. (i) The function $H(\mathsf{u}, x)$ is Lipschitz continuous with respect to u'', is nonincreasing with respect to u'_0, and $D_{\mathsf{u}''} H \in \mathbb{S}_{\delta/2}$ at all points of differentiability of H with respect to u''.

(ii) The number

$$\bar{H} := \sup_{\mathsf{u}', x} \big(|H(\mathsf{u}', 0, x)| - K_0 |\mathsf{u}'|\big) \quad (\ge 0) \tag{6}$$

is finite.

(iii) The function $H(\mathsf{u}, x)$ is Lipschitz continuous with respect to u' and at all point of differentiability of $H(\mathsf{u}, x)$ with respect to u' we have

$$\big|D_{\mathsf{u}'} H(\mathsf{u}, x)\big| \le N'. \tag{7}$$

(iv) The function $H(\mathsf{u}, x)$ is Lipschitz continuous with respect to x and at all point of differentiability of $H(\mathsf{u}, x)$ with respect to x we have

$$\big|D_x H(\mathsf{u}, x)\big| \le N'\big(1 + |\mathsf{u}|\big). \tag{8}$$

(v) For every u', x there exists a subset of $\mathbb{S}$ of full measure at every point of which $H(\mathsf{u}', \mathsf{u}'', x)$ is differentiable with respect to u'' and

$$\big|H(\mathsf{u}, x) - \langle \mathsf{u}'', D_{\mathsf{u}''} H(\mathsf{u}, x)\rangle\big| \le K_1\big(\bar{H} + K + |\mathsf{u}'|\big). \tag{9}$$

(vi) There is a constant $\rho_0 > 0$ and a function $H(\mathsf{u})$ such that

$$H(\mathsf{u}, x) = H(\mathsf{u}) \tag{10}$$

for all u if $\rho_\Omega(x) \le \rho_0$.

5. Remark. The condition $D_{\mathsf{u}''}H \in \mathbb{S}_{\delta/2}$ may look strange. Why do not use $D_{\mathsf{u}''}H \in \mathbb{S}_{\delta}$, which will indeed be imposed in Theorem 6.20? The point is that we are going to use the same $P(\mathsf{u}'')$ in Theorems 6 and 6.20, but in the proof of the latter we will replace H, satisfying $D_{\mathsf{u}''}H \in \mathbb{S}_{\delta}$ with the one satisfying $D_{\mathsf{u}''}H \in \mathbb{S}_{\delta/2}$.

For smooth enough functions $u(x)$ introduce

$$H[u](x) = H\big(u(x), Du(x), D^2u(x), x\big), \quad P[u](x) = P\big(D^2u(x)\big).$$

Here is the result we are after, in which the above assumptions are supposed to be satisfied.

6. Theorem. *Equation* (0.1) *with boundary condition $v = g$ on $\partial\Omega$ has a unique solution $v \in C(\bar{\Omega}) \cap C^{1,1}_{\text{loc}}(\Omega)$. Furthermore, estimate* (0.2) *holds with N, depending only on d, K_0, δ, and* $\operatorname{diam}(\Omega)$, *and the estimates* (0.4) *hold with N depending only on* $\operatorname{diam}(\Omega)$, $\rho_{\text{ext}}(\Omega)$, $\rho_{\text{int}}(\Omega)$, Λ, K_0, d, δ, *and K_1 (in particular, N is independent of N' and ρ_0 but does depend on K_1).*

7. Remark. If we already know that equation (0.1) in Ω with boundary condition $v = g$ has a solution $v \in C(\bar{\Omega}) \cap C^{1,1}_{\text{loc}}(\Omega)$, then estimate (0.2):

$$\sup_{\Omega} |v| \leq N\bar{H} + \|g\|_{C(\partial\Omega)}, \tag{11}$$

where N depends only on d, K_0, δ, and $\operatorname{diam}(\Omega)$, is obtained on the basis of the maximum principle. Indeed, introduce $H_K(\mathsf{u}, x) = H(\mathsf{u}, x) \vee (P(\mathsf{u}'') - K)$, $H_K[v] = H_K\big(v, Dv, D^2v, x\big)$ and observe that on the set $\Omega_+ = \{x : v(x) > 0\}$ we have $H_K(v, Dv, 0, x) \leq H_K(0, Dv, 0, x) \leq K_0|Dv| + \bar{H}$ and (a.e.)

$$0 = H_K[v] = H_K\big(v, Dv, D^2v, x\big) - H_K(v, Dv, 0, x) + H_K(v, Dv, 0, x)$$

$$\leq a^{ij}D_{ij}v + K_0 b^i D_i v + \bar{H},$$

where (a^{ij}) is an appropriate $\mathbb{S}_{\bar{\delta}}$-valued measurable function and $b = Dv/|Dv|$. After that (11) with v_+ in place of v follows from the Aleksandrov estimates (Theorem 3.0.2) applied to the domain Ω_+. One treats v_- similarly.

8. Remark. The function $H(\mathsf{u}, x)$ is locally Lipschitz continuous with respect to (u, x), since, due to Lemma 4.1.1 and Assumption 4, for any $x, y \in \mathbb{R}^d$, $\mathsf{u}, \mathsf{v} \in \mathbb{R}^{d+1} \times \mathbb{S}$,

$$\big|H(\mathsf{u}, x) - H(\mathsf{v}, y)\big| \leq \big|H(\mathsf{u}, x) - H(\mathsf{v}, x)\big| + N'\big(1 + |\mathsf{v}|\big)|x - y|$$

$$\leq \big|H(\mathsf{u}, x) - H(\mathsf{u}', \mathsf{v}'', x)\big| + N'|\mathsf{u}' - \mathsf{v}'| + N'\big(1 + |\mathsf{v}|\big)|x - y|$$

$$\leq N(\delta, d)|\mathsf{u}'' - \mathsf{v}''| + N'|\mathsf{u}' - \mathsf{v}'| + N'\big(1 + |\mathsf{v}|\big)|x - y|.$$

Theorem 6 will be proved later in Section 5 after we use Assumption 4 (v) to rewrite $H[u]$ in terms of pure second-order derivatives allowing us to use Theorem 1.5. Assumption 4 (iii), (iv), and (vi) are needed to guarantee that the new equation satisfy Assumptions 1.3 (i) (as modified in Theorem 1.10), (iv), and (v).

The reader will see that, in case $K > 0$, apart from the proof of uniqueness, in the proof of other assertions of Theorem 6 none of the results of Chapters 1-4 is used, so that we could have started the book with Chapter 5. However, later generalizations of Theorem 6 rely heavily on the results of Chapter 4.

9. REMARK. Uniqueness in Theorem 6 is proved in the same way as in Theorem 1.5 by using Corollary 4.1.4, Lemma 4.1.6, and Theorem 3.0.2.

Also observe that the (generally irrelevant) case of $K = 0$ is easily treated by using Theorems 4.2.15 and 4.2.9 and passing to the limit as $K \downarrow 0$.

3. Minimax representation of certain functions

The material of this section, originated in [**53**], is taken from [**71**] apart from Theorem 10, that appeared in [**77**]. Fix a closed bounded subset E of $\mathbb{R}^{n+1}$. Recall that, if $H(u)$ is a real-valued Lipschitz continuous function given on $\mathbb{R}^n$, then it is differentiable on a set of full measure.

1. DEFINITION. We denote by $\mathfrak{H}_E$ the set of real-valued Lipschitz continuous functions H given on $\mathbb{R}^n$ for each of which there exists a set $D'_H \subset \mathbb{R}^n$ of full measure such that, at any point $u \in D'_H$, H is differentiable and

$$\Big(H(u) - \big\langle u, DH(u)\big\rangle, DH(u)\Big) \in E.$$

The next remark will not be used in the future.

2. REMARK. If $H \in \mathfrak{H}_E$, introduce a positive homogeneous of degree one function $tH(u/t)$ on $(0,\infty) \times \mathbb{R}^n$. Notice that for $u \in \mathbb{R}^n$ and $t > 0$ such that $u/t \in D'_H$ we have

$$\frac{\partial}{\partial t}\big[tH(u/t)\big] = H(u/t) - \big\langle u/t, (Du)H(u/t)\big\rangle, \quad D_u\big[tH(u/t)\big] = (D_uH)(u/t),$$

so that, owing to the boundedness of E, the gradient of the function $tH(u/t)$ as function of $(t,u) \in (0,\infty) \times \mathbb{R}^n$ exists and is bounded by a constant on a set of full measure. Since $tH(u/t)$ is locally Lipschitz continuous, by Lemma 4.1.1 applied to each set $[1/n, n] \times \bar{B}_n$ as $n \to \infty$, it is globally Lipschitz continuous in $(0,\infty) \times \mathbb{R}^n$ and, by Lemma 4.1.3, at all points in $(0,\infty) \times \mathbb{R}^n$ where the gradient of $tH(u/t)$ exists, it belongs to the convex hull of E. In particular, $tH(u/t)$ has a continuous extension to $[0,\infty) \times \mathbb{R}^n$.

Also note that for positive homogeneous of degree one functions H the quantity $H(u) - \big\langle u, DH(u)\big\rangle$ vanishes. In our case this quantity is assumed to be bounded and that is why H is *boundedly* inhomogeneous if $H \in \mathfrak{H}_E$.

For $u \in \mathbb{R}^n$ and $H \in \mathfrak{H}_E$ also define

$$E(u) = E_H(u) = \big\{(f,l) \in E : f \in \mathbb{R}, l \in \mathbb{R}^n, f + \langle l, u\rangle \le H(u)\big\}.$$

Note that, if $u \in D'_H$ and we take

$$(f,l) = \big(H(u) - \big\langle u, DH(u)\big\rangle, DH(u)\big),$$

then

$$f + \langle l, u\rangle = H(u), \tag{1}$$

so that, in particular, $E(u) \neq \emptyset$. Since D'_H is everywhere dense in $\mathbb{R}^n$ and E is compact, we have that, for any $u \in \mathbb{R}^n$, there exist $(f,l) \in E$ such that (1) holds. Hence, $E(u) \neq \emptyset$, obviously, $E(u)$ are closed, and

$$\max_{(f,l)\in E(u)} \big[f + \langle l, u\rangle\big] = H(u). \tag{2}$$

3. THEOREM. *Let $H \in \mathfrak{H}_E$, then for all $u \in \mathbb{R}^n$ we have*

$$H(u) = \min_{v\in\mathbb{R}^n} \max_{(f,l)\in E(v)} \big[\beta f + \langle l, u\rangle\big]. \tag{3}$$

Proof. In light of (2) we only need to prove that for any $u, v \in \mathbb{R}^n$

$$\max_{(f,l)\in E(v)} \big[f + \langle l, u\rangle\big] \geq H(u).$$

Without losing generality we may assume that $v = 0$ and prove that

$$\max_{(f,l)\in E(0)} \big[f + \langle l, u\rangle\big] \geq H(u) \tag{4}$$

for any $u \in \mathbb{R}^n$. Since both parts of (4) are continuous with respect to u, it suffices to prove (4) on any dense subset of $\mathbb{R}^n$.

Observe that the function $H(tu)$ is at least locally Lipschitz continuous in (t, u), $t \in [0, 1]$, $u \in \mathbb{R}^n$, differentiable at each point (t_0, u_0) such that $H(u)$ is differentiable at t_0u_0 and its partial derivatives can be computed by usual formulas. Since the set of points $(t, u) \in [0, 1] \times \mathbb{R}^n$ such that tu belongs to a subset of $\mathbb{R}^n$ of full measure also is of full measure, we conclude that there is a set $\Lambda \subset \mathbb{R}^n$ of full measure such that, for any $u_0 \in \Lambda$, for almost all $t \in [0, 1]$, we have that $tu_0 \in D'_H$, the function $H(su_0)$ is differentiable at $s = t$, and

$$\partial_t\big[H(tu_0)\big] = \big\langle u_0, DH(tu_0)\big\rangle.$$

Furthermore, Lipschitz continuous functions of one variable are absolutely continuous so that for all $s, t \in [0, 1]$ and $u_0 \in \Lambda$

$$H(tu_0) - H(su_0) = \int_s^t \big\langle u_0, DH(ru_0)\big\rangle\, dr.$$

Now we prove (4) for $u = u_0 \in \Lambda$. Consider the function $H(tu_0)$. There are two possibilities:

(i) For all $t \in [0, 1]$, we have $H(tu_0) \leq tH(u_0) + (1 - t)H(0)$ or

(ii) There exists a $t_0 \in [0, 1)$ such that $H(t_0u_0) = t_0H(u_0) + (1 - t_0)H(0)$ and, for sufficiently small $\delta > 0$, in $(t_0, t_0 + \delta)$ we have $H(tu_0) > tH(u_0) + (1 - t)H(0)$.

In case (ii) for $t \in (t_0, t_0 + \delta)$ we have

$$\int_{t_0}^t \big\langle u_0, DH(su_0)\big\rangle\, ds = H(tu_0) - H(t_0u_0)$$

$$> tH(u_0) + (1 - t)H(0) - H(t_0u_0) = (t - t_0)\big(H(u_0) - H(0)\big)$$

implying that there is a point $s \in (t_0, t)$ such that $su_0 \in D'_H$ and

$$\big\langle u_0, DH(su_0)\big\rangle \geq H(u_0) - H(0).$$

Denote

$$l_s = DH(su_0), \quad f_s = H(su_0) - \big\langle su_0, DH(su_0)\big\rangle.$$

Since $f_s + \langle su_0, l_s\rangle = H(su_0)$ and E is closed, by setting $t \downarrow t_0$, which also forces $s \downarrow t_0$, we find an element $(f, l) \in E$ such that

$$f + \langle t_0u_0, l\rangle = H(t_0u_0), \quad \langle u_0, l\rangle \geq H(u_0) - H(0).$$

Now for two linear functions $\alpha(t) = f + \langle tu_0, l\rangle$ and $\beta(t) = tH(u_0) + (1-t)H(0)$ we have that

$$\alpha(t_0) = H(t_0u_0) = \beta(t_0)$$

and

$$\alpha'(t_0) = \langle u_0, l\rangle \geq \beta'(t_0).$$

It follows that $\alpha(0) \leq \beta(0)$ and $\alpha(1) \geq \beta(1)$ that is

$$f \leq H(0), \quad f + \langle u_0, l\rangle \geq H(u_0).$$

Here the first relations means that $(f, l) \in E(0)$, which combined with the second one yields (4) for $u = u_0$ and finishes our argument in case (ii).

In case (i), for $t \in [0, 1]$, we have

$$\int_t^1 \langle u_0, DH(su_0)\rangle\, ds = H(u_0) - H(tu_0) \geq (1-t)\big(H(u_0) - H(0)\big)$$

and as above we find that there exists $(f, l) \in E$ such that $f + \langle u_0, l\rangle = H(u_0)$ and $\langle u_0, l\rangle \geq H(u_0) - H(0)$ implying that $(f, l) \in E(u_0)$ and

$$\begin{aligned} t\langle u_0, l\rangle + f &= t\langle u_0, l\rangle + H(u_0) - \langle u_0, l\rangle \\ &= (t-1)\langle u_0, l\rangle + H(u_0) \\ &\leq (t-1)\big[H(u_0) - H(0)\big] + H(u_0) \end{aligned}$$

for $t \in [0, 1]$. Substituting $t = 0$ yields that $f \leq H(0)$, $(f, l) \in E(0)$, and again we get (4) at $u = u_0$. The theorem is proved. □

4. REMARK. By mimicking the proof of Theorem 3 one obtains the following maximin representation of H:

$$H(u) = \max_{v \in \mathbb{R}^d} \min_{\substack{(f,l)\in E,\\ f+\langle l,y\rangle \geq H(v)}} \big[f + \langle l, u\rangle\big].$$

5. REMARK. Assume that H is convex and the set

$$\Big\{\Big(H(u) - \langle u, DH(u)\rangle, DH(u)\Big) : DH(u) \text{ exists}\Big\}$$

is bounded. Call its closure E. Then for any $(f, l) \in E$ we have that $f + (l, u) \leq H(u)$ for any $u \in \mathbb{R}^n$. In this case $E(v) = E$ and the min in (3) can be dropped providing the usual representation of convex functions in terms of its supporting planes.

6. REMARK. One can restrict our arguments to H defined on a closed convex set $\Gamma \subset \mathbb{R}^n$ rather than on $\mathbb{R}^n$. Then the right-hand side of (3) (with Γ in place of $\mathbb{R}^n$) provides an extension of H initially defined on Γ to $\mathbb{R}^n$ keeping intact E.

Much later in Chapter 16 it will turn out to be convenient to replace the representation (3), in which $E(v) = E_H(v)$ with a different one, so that the sets over which we take the min and the max are independent of H, and then of course, f and l will depend on H.

One of alternative representation is described below.

7. ASSUMPTION. We are given a set E_0, which is a convex closed bounded subset of $\mathbb{R}^{n+1}$, such that $E_0 \supset E$ and the distance between the relative boundaries of E and E_0 is strictly positive.

For $\alpha \in \mathbb{R}^n$ define

$$G(\alpha) = \inf_{(f,l)\in E_0} \big(f + \langle l, \alpha\rangle\big).$$

Obviously $G \le H$.

8. REMARK. Let $\mathcal{P}$ be the smallest hyperplane containing E_0, and let (f_0, l_0) be a fixed point in E. By using the assumption about the boundaries of E and E_0, define Γ as the intersection of a closed ball in $\mathbb{R}^{n+1}$, with nonzero radius centered at the origin, with $\mathcal{P} - (f_0, l_0)$ such that $(f,l) + \Gamma \subset E_0$ for any $(f,l) \in E$. Define

$$\gamma(\alpha) = \sup_{(f,l)\in\Gamma} \big[f + \langle l, \alpha\rangle\big]$$

and observe that, for a constant $\mu > 0$, we have $\pm\mu\big(E_0 - (f_0, l_0)\big) \subset \Gamma$. Hence, for any $(f', l'), (f'', l'') \in E_0$, it holds that $\mu(f' - f'', l' - l'') \in 2\Gamma$ and for any $\alpha \in \mathbb{R}^n$

$$\mu\big|f' + \langle l', \alpha\rangle - \big(f'' + \langle l'', \alpha\rangle\big)\big| \le 2\gamma(\alpha).$$

In particular,

$$G \le H \le G + (2/\mu)\gamma.$$

Furthermore, for any $(f,l) \in E$, since $(f,l) - \Gamma \subset E_0$, we have that

$$f + \langle l, \alpha\rangle \ge \gamma(\alpha) + \inf_{(f',l')\in E_0} \big(f' + \langle l', \alpha\rangle\big), \quad f + \langle l, \alpha\rangle - G(\alpha) \ge \gamma(\alpha), \tag{5}$$

which shows that

$$\lambda_H^{\alpha\beta} := 1 \wedge \frac{H(\alpha) - G(\alpha)}{f + \langle l, \alpha\rangle - G(\alpha)} \qquad (0/0 := 1)$$

is well defined for $\alpha \in \mathbb{R}^n$ and $\beta = (f,l) \in E$ and, of course, $\lambda_H^{\alpha\beta} \in [0,1]$. We draw the reader's attention to the fact that above we set $0/0 = 1$.

9. REMARK. Since G is concave, at any point $(\alpha_0, G(\alpha_0))$, its graph has a supporting plane $y = G(\alpha_0) + \langle \alpha - \alpha_0, l\rangle$, and since E_0 is closed, bounded, and convex, we have

$$\big(G(\alpha_0) - \langle \alpha_0, l\rangle, l\big) \in E_0.$$

For each α we fix an appropriate l and call it $\partial G(\alpha)$. Of course, $\partial G(\alpha) = DG(\alpha)$ at any point where G is differentiable and thus almost everywhere.

Thereby, for $\alpha \in \mathbb{R}^n$ and $\beta = (f,l) \in E$, the definitions

$$f_H^{\alpha\beta} = \lambda_H^{\alpha\beta} f + \big(1 - \lambda_H^{\alpha\beta}\big)\big[G(\alpha) - \big\langle \alpha, \partial G(\alpha)\big\rangle\big],$$

$$l_H^{\alpha\beta} = \lambda_H^{\alpha\beta} l + \big(1 - \lambda_H^{\alpha\beta}\big)\partial G(\alpha)$$

make sense.

Obviously $\big(f_H^{\alpha\beta}, l_H^{\alpha\beta}\big) \in E_0$. In this way, for each α and β, on the set $\mathfrak{H}_E$ we constructed a mapping sending each $H \in \mathfrak{H}_E$ into the function $\big(f_H^{\alpha\beta}, l_H^{\alpha\beta}\big)$ defined on $\mathbb{R}^n \times E$.

10. THEOREM. *Under Assumption 7 for any $H \in \mathfrak{H}_E$ and any $u \in \mathbb{R}^n$ we have*

$$H(u) = \inf_{\alpha \in \mathbb{R}^n} \sup_{\beta \in E} \left[f_H^{\alpha\beta} + \langle l_H^{\alpha\beta}, u \rangle \right].$$

Furthermore, if $H, F \in \mathfrak{H}_E$, then for any $\alpha \in \mathbb{R}^n$ and $\beta = (f, l) \in E$ we have

$$|f_H^{\alpha\beta} - f_F^{\alpha\beta}| \leq \frac{|H(\alpha) - F(\alpha)|}{\gamma(\alpha)} |f + \langle \alpha, \partial G(\alpha) \rangle - G(\alpha)| \qquad (0/0 := 0),$$

$$|l_H^{\alpha\beta} - l_F^{\alpha\beta}| \leq \frac{|H(\alpha) - F(\alpha)|}{\gamma(\alpha)} |l - \partial G(\alpha)| \qquad (0/0 := 0).$$

Proof. Observe that, for $\beta = (f, l) \in E$, if $f + \langle l, \alpha \rangle \leq H(\alpha)$, then $H(\alpha) - G(\alpha) \geq f + \langle l, \alpha \rangle - G(\alpha)$ and $\lambda_H^{\alpha\beta} = 1$ (no matter $\gamma(\alpha) = 0$ or $\gamma(\alpha) > 0$) and $f = f_H^{\alpha\beta}$ and $l = l_H^{\alpha\beta}$. It follows that

$$\max_{\substack{\beta = (f,l) \in E, \\ f + \langle l, \alpha \rangle \leq H(\alpha)}} [f + \langle l, u \rangle] \leq \sup_{\substack{\beta \in E, \\ \lambda_H^{\alpha\beta} = 1}} \left[f_H^{\alpha\beta} + \langle l_H^{\alpha\beta}, u \rangle \right] \leq \sup_{\beta \in E} \left[f_H^{\alpha\beta} + \langle l_H^{\alpha\beta}, u \rangle \right].$$

Furthermore, for $\beta = (f, l) \in E$, if $\lambda_H^{\alpha\beta} = 1$, then $H(\alpha) - G(\alpha) \geq f + \langle l, \alpha \rangle - G(\alpha)$, $(f_H^{\alpha\beta}, l_H^{\alpha\beta}) = (f, l) \in E_0$, and

$$f_H^{\alpha\beta} + \langle l_H^{\alpha\beta}, \alpha \rangle \leq H(\alpha).$$

In addition, for $\beta = (f, l) \in E$, if $\lambda_H^{\alpha\beta} < 1$, then as always $(f_H^{\alpha\beta}, l_H^{\alpha\beta}) \in E_0$ and

$$f_H^{\alpha\beta} + \langle l_H^{\alpha\beta}, \alpha \rangle = \lambda_H^{\alpha\beta} [f + \langle l, \alpha \rangle] + (1 - \lambda_H^{\alpha\beta}) G(\alpha) = H(\alpha).$$

Hence,

$$\max_{\substack{\beta = (f,l) \in E, \\ f + \langle l, \alpha \rangle \leq H(\alpha)}} [f + \langle l, u \rangle] \leq \sup_{\beta \in E} \left[f_H^{\alpha\beta} + \langle l_H^{\alpha\beta}, u \rangle \right] \leq \max_{\substack{(f,l) \in E_0, \\ f + \langle l, \alpha \rangle \leq H(\alpha)}} [f + \langle l, u \rangle],$$

and the first assertion of the theorem follows from Theorem 3 applied as it is stated to E_0 in place of E.

To prove the second assertion it suffices to note that, for instance,

$$f_H^{\alpha\beta} - f_F^{\alpha\beta} = (\lambda_H^{\alpha\beta} - \lambda_F^{\alpha\beta}) \left[f + \langle \alpha, \partial G(\alpha) \rangle - G(\alpha) \right],$$

where the right-hand side is zero if $\gamma(\alpha) = 0$, and after that use (5). The theorem is proved. □

4. One step closer to nonlinear partial-differential operators

We use the setting and notation from Section 3. In the future, Theorem 3.3 will be applied when $u \in \mathbb{S}$ and $l \in \mathbb{S}_{\delta/4}$ with $\langle l, u\rangle = \operatorname{tr} lu$ and u playing the role of the Hessian $D^2 v$ of a function v. We will represent l as a linear combination with positive coefficients of a fixed collection of rank-one matrices. Then $\langle l, u\rangle$ will be the linear combination with positive coefficients of pure second-order derivatives of v. This would allow us to use Theorem 1.5 right away if no first-order derivatives and space variables x were present in H. Hence, we have to consider H also depending on parameters and figure out what to require from the original H in order to have the transformed equation satisfy the assumptions of Theorem 1.10.

Recall that in Section 3, E is a closed bounded subset of $\mathbb{R}^{n+1}$.

1. ASSUMPTION. There are $E' \subset \mathbb{R}$ and $E'' \subset \mathbb{R}^n$ such that $E = E' \times E''$ and there are points $l_i \in \mathbb{R}^n$, $i = 1, ..., m$, and Lipschitz continuous functions $p_1(l)$,..., $p_m(l)$ on E'' such that for all $l \in E''$

$$l = \sum_{i=1}^{m} p_i(l) l_i.$$

2. REMARK. In our applications when $E'' = \mathbb{S}_{\delta/4}$, Theorem 2.1 guarantees that Assumption 1 is satisfied with l_i being rank-one matrices and strictly positive p_i's.

3. DEFINITION. Let $\beta \in (0, \infty)$. We denote by $\mathfrak{H}_{E,\beta}$ the set of real-valued Lipschitz continuous functions H given on $\mathbb{R}^n$ for each of which there exists a set $D'_H \subset \mathbb{R}^n$ of full measure such that, at any point $u \in D'_H$, H is differentiable and

$$\Big(\beta^{-1} H(u) - \beta^{-1}\langle u, DH(u)\rangle, DH(u)\Big) \in E.$$

For $v \in \mathbb{R}^n$ and $H \in \mathfrak{H}_{E,\beta}$ we denote

$$E_{H,\beta}(v) = \big\{(f, l) \in E : \beta f + \langle l, v\rangle \le H(v)\big\}.$$

Let

$$p(l) = \big(p_1(l), ..., p_m(l)\big)$$

and let $\mathcal{E}$ be the (closed) convex envelope of the set

$$\Big\{\big(f, p(l)\big) : (f, l) \in E\Big\}. \tag{1}$$

4. THEOREM. *Suppose that Assumption* 1 *is satisfied. Then there exists a constant* N_0 *and, for any* $\beta \in (0, \infty)$, *on the set* $\mathfrak{H}_{E,\beta}$ *there is a mapping* $H \to \mathcal{H}$, *where, for* $z \in \mathbb{R}^m$, *the function* $\mathcal{H}(z) = \mathcal{H}(\beta, z)$ *is defined by*

$$\mathcal{H}(z) = \mathcal{H}(\beta, z) := \inf_{v \in \mathbb{R}^n} \max_{(f,l) \in E_{H,\beta}(v)} \big[\beta f + \langle p(l), z\rangle\big].$$

Furthermore, $\mathcal{H}(\beta, z)$ *is a Lipschitz continuous function on* $\mathbb{R}^m$ *with Lipschitz constant* N_0 *and is such that*

(*i*) *If* $u \in \mathbb{R}^n$ *and* $z \in \mathbb{R}^m$ *satisfy*

$$z_j = \langle l_j, u\rangle \quad \forall j,$$

then $H(\nu u) = \mathcal{H}(\nu z)$ *for any* $\nu \in \mathbb{R}$;

(*ii*) *At each point* $z \in \mathbb{R}^m$ *where* $\mathcal{H}$ *is differentiable we have*

$$\Big(\beta^{-1}\mathcal{H}(z) - \beta^{-1}\big\langle z, D\mathcal{H}(z)\big\rangle, D\mathcal{H}(z)\Big) \in \mathcal{E}. \tag{2}$$

Proof. By replacing $H(u)$ with $\beta H(\beta^{-1}u)$, we see that we may assume that $\beta = 1$. Then observe that $\mathcal{H}$ is well defined because E is bounded. The fact that $\mathcal{H}$ is Lipschitz continuous is also due to the boundedness of E.

If $z_j = \langle l_j, u\rangle$, then, for any $l \in E''$,

$$\big\langle p(l), \nu z\big\rangle = \nu \sum_{j=1}^{n} p_j(l) z_j = \nu \Big\langle \sum_{j=1}^{n} p_j(l) l_j, u\Big\rangle = \langle l, \nu u\rangle$$

and (i) follows from Theorem 3.3.

To prove (ii), observe that the gradient, where it exists, of the function

$$\mathcal{H}(t, z) := \inf_{v \in \mathbb{R}^d} \max_{(f,l) \in E_H(v)} \big[tf + \big\langle p(l), z\big\rangle\big]$$

as a function of (t, z) belongs to $\mathcal{E}$ due to Corollary 4.1.4. Since, for $t > 0$, we have

$$\mathcal{H}(t, zt) = t\mathcal{H}(z),$$

for any point z of differentiability of $\mathcal{H}(z)$ the function $\mathcal{H}(t, zt)$ is differentiable with respect to (t, z) at $(1, z)$ and

$$D_z\mathcal{H}(1, z) = D_z\mathcal{H}(z),$$

$$\big\langle z, D_z\mathcal{H}(1, z)\big\rangle + \partial_t\mathcal{H}(1, z) = \mathcal{H}(z).$$

This shows that the inclusion $\big(\partial_t\mathcal{H}(1, z), D_z\mathcal{H}(1, z)\big) \in \mathcal{E}$ is equivalent to (2). The theorem is proved. □

Next we study how dependence of $H(u)$ on parameters reflects on the dependence of $\mathcal{H}$ on the same parameters. We come to the most important result of this and the previous sections.

5. ASSUMPTION. Assumption 1 is satisfied. The closed bounded set E is *convex* and we are given a closed set $E_1 = E_1' \times E_1''$ such that $E_1' \subset E'$, $E_1'' \subset E''$ and E_1 lies in the relative interior of E.

6. ASSUMPTION. We are given a Lipschitz continuous function $\beta(v) \geq \beta_0$ on $\mathbb{R}^r$, where the constant $\beta_0 > 0$. For each $v \in \mathbb{R}^r$ and $x \in \mathbb{R}^d$ a function $H(v, u, x)$, $u \in \mathbb{R}^n$, is given such that

$$H(v, \cdot, x) \in \mathfrak{H}_{E_1, \beta(v)}.$$

The function $H(v, u, x)$ is locally Lipschitz continuous and there exists a constant N' such that at all points of differentiability of $H(v, u, x)$ we have

$$\big|D_x H(v, u, x)\big| \leq N'\big(1 + \beta(v) + |u|\big), \tag{3}$$

$$\big|D_v H(v, u, x)\big| \leq N'. \tag{4}$$

It is worth saying for the reader's orientation that in our applications v will stand for the unknown function and its first-order derivatives and u will be replaced with the Hessian of the unknown function.

7. THEOREM. *Under Assumptions* 5 *and* 6, *for* $v \in \mathbb{R}^r$, $z \in \mathbb{R}^m$, *and* $x \in \mathbb{R}^d$, *define*

$$\mathcal{H}(v, z, x) = \inf_{w \in \mathbb{R}^n} H(v, w, z, x),$$

where

$$H(v, w, z, x) = \max\left\{ \beta(v)f + \langle p(l), z\rangle :\right.$$
$$\left.(f, l) \in E,\ \beta(v)f + \langle l, w\rangle \leq H(v, w, x)\right\} \tag{5}$$

(*note that we use* E *and not* E_1 *in* (5)).

We assert the following.

(*i*) *The function* $\mathcal{H}(v, z, x)$ *is locally Lipschitz continuous and at all points of its differentiability in* z, x, *and* v *we have, respectively,*

$$\Big(\beta^{-1}(v)\mathcal{H}(v, z, x) - \beta^{-1}(v)\big\langle z, D_z\mathcal{H}(v, z, x)\big\rangle, D_z\mathcal{H}(v, z, x)\Big) \in \mathcal{E}, \tag{6}$$

$$\big|D_x\mathcal{H}(v, z, x)\big| \leq N\big(1 + \beta(v) + |z|\big), \quad \big|D_v\mathcal{H}(v, z, x)\big| \leq N\big(1 + \beta(v) + |z|\big), \tag{7}$$

where the constants N *are independent of* v, z, x.

(*ii*) *If* $u \in \mathbb{R}^n$ *and* $z \in \mathbb{R}^m$ *are such that*

$$z_j = \langle l_j, u\rangle \quad \forall j,$$

then $H(v, u, x) = \mathcal{H}(v, z, x)$ *for all* v *and* x.

(*iii*) *If* $H(v, u, x)$ *is a decreasing function of* v^1 *and* $\beta(v)$ *is independent of* v^1, *then* $\mathcal{H}(v, z, x)$ *is also a decreasing function of* v^1.

Proof. Inclusions (6) follow from (2) and assertion (ii) follows from Theorem 4 (i). Assertion (iii) is obvious since, as v^1 increases, the set, over which the maximum in (5) is taken, shrinks.

To prove (7), it suffices to prove that there exist constants $N, \varepsilon_0 > 0$ such that for any $v_0, v \in \mathbb{R}^r$, $x_0, x \in \mathbb{R}^d$, $w_0 \in \mathbb{R}^n$, and $z_0 \in \mathbb{R}^m$ satisfying $|v| + |x| \leq \varepsilon_0$, we have

$$H(v_0 + v, w_0, z_0, x_0 - x) \leq H(v_0, w_0, z_0, x_0) + N\big(|v| + |x|\big)\big(1 + \beta(v_0) + |z_0|\big). \tag{8}$$

Since $H(v_0, \cdot, x_0) \in \mathfrak{H}_{E_1, \beta(v_0)}$, in light of (3.1) (or (3.2)), there exists $(f_0, l_0) \in E_1$ such that

$$\beta(v_0)f_0 + \langle l_0, w_0\rangle = H(v_0, w_0, x_0).$$

For $t \in [0, 1]$ and $(f, l) \in E$ define

$$f_t(f) = (1 - t)f_0 + tf, \quad l_t(l) = (1 - t)l_0 + tl$$

and observe that, since the closed E_1 lies in the relative interior of E and E is convex, there is a constant $\kappa > 0$ (depending only on E_1 and E) such that, for any $t \in [0, 1]$, the distances from $f_t(f)$ and $l_t(l)$ to the boundaries of E' and E'', respectively, are greater than $\kappa(1 - t)$. This is easily seen if, while discussing this issue, one shifts (f_0, l_0) to the origin.

Hence, for $l' = w_0/|w_0|$ $(0/0 := 0)$ we have

$$\big(\hat{f}_t(f), \hat{l}_t(l)\big) := \big(f_t(f) - \kappa(1-t), l_t(l) - \kappa(1-t)l'\big) \in E.$$

Now let

$$(f, l) \in E \quad \text{and} \quad \beta(v_0 + v)f + \langle l, w_0\rangle \le H(v_0 + v, w_0, x_0 + x). \tag{9}$$

Our first goal is to give conditions on t, x, v such that we have

$$I := \beta(v_0)\hat{f}_t(t) + \big\langle \hat{l}_t(l), w_0\big\rangle \le H(v_0, w_0, x_0). \tag{10}$$

To this end observe that

$$I = t\big[\beta(v_0 + v)f + \langle l, w_0\rangle\big] + t\big[\beta(v_0) - \beta(v_0 + v)\big]f$$

$$+ (1-t)H(v_0, w_0, x_0) - \kappa(1-t)\beta(v_0) - \kappa(1-t)|w_0|.$$

Here the first term on the right is less than

$$tH(v_0 + v, w_0, x_0 + x) \le tH(v_0, w_0, x_0) + N'|v|$$

$$+ \big|H(v_0, w_0, x_0 + x) - H(v_0, w_0, x_0)\big|$$

$$\le tH(v_0, w_0, x_0) + N'|v| + N'\big(1 + \beta(v_0) + |w_0|\big)|x|.$$

To estimate the second term we use the Lipschitz continuity of β and conclude that

$$I \le H(v_0, w_0, x_0) + N_1|v| + N_1\big(1 + \beta(v_0) + |w_0|\big)|x| - \kappa(1-t)\big(\beta(v_0) + |w_0|\big).$$

If

$$2N_1\big(|v| + |x|\big) \le \kappa(1-t)\beta_0 \quad \text{and} \quad 2N_1|x| \le \kappa(1-t), \tag{11}$$

then (recall that $\beta \ge \beta_0$) we come to (10) and therefore conditions (11) imply that

$$J(v, x, t) := \max\Big\{\ \beta(v_0)\big[f_t(f) - \kappa(1-t)\big] + \Big\langle p\big(l_t(l) - \kappa(1-t)l'\big), z_0\Big\rangle :$$

$$(f, l) \in E,\ \beta(v_0 + v)f + \langle l, w_0\rangle \le H(v_0 + v, w_0, x_0 + x)\Big\}$$

$$\le \max\Big\{\beta(v_0)f + \langle p(l), z_0\rangle : (f, l) \in E,\ \beta(v_0)f + \langle l, w_0\rangle \le H(v_0, w_0, x_0)\Big\}$$

$$= H(v_0, w_0, z_0, x_0). \tag{12}$$

Furthermore, the p_k's are Lipschitz continuous and

$$\big|p_k\big(l_t(l) - \kappa(1-t)l'\big) - p_k(l)\big| \le N(1-t),$$

Also $\big|f_t(f) - f\big| \le N(1-t)$. It follows that

$$J(v, x, t) \ge -N(1-t)\big(\beta(v_0) + |z_0|\big) + \max\Big\{\beta(v_0)f + \langle p(l), z_0\rangle :$$

$$(f, l) \in E, \beta(v_0 + v)f + \langle l, w_0\rangle \le H(v_0 + v, w_0, x_0 + x)\Big\}.$$

By using the Lipschitz continuity of β we replace $\beta(v_0)$ with $\beta(v_0+v)$ and conclude that

$$J(v,x,t) \geq -N(1-t)\big(\beta(v_0)+|z_0|\big) - N|v| + H(v_0+v,w_0,z_0,x_0+x).$$

We combine this with (12) and obtain that under (11) we have

$$H(v_0,w_0,z_0,x_0) \geq -N(1-t)\big(\beta(v_0)+|z_0|\big) - N|v| + H(v_0+v,w_0,z_0,x_0+x)$$

We now choose

$$\varepsilon_0 = \kappa 2^{-1} N_1^{-1} \min(1,\beta_0),$$

take $|v|+|x| \leq \varepsilon_0$, set

$$1-t = \big(|v|+|x|\big)/\varepsilon_0,$$

and then come to (8). The theorem is proved. □

8. REMARK. If we only needed an estimate of the derivative of $\mathcal{H}$ with respect to v, the condition that only E_1' lies in the relative interior of the interval E' would suffice. Indeed, in that case the term $\kappa(1-t)l'$ above can be dropped.

9. REMARK. Inclusion (6) and the definition of $\mathcal{E}$ implies, in particular, that at all points of differentiability of $\mathcal{H}(v,z,x)$ in z we have

$$\sup_{E''} p_i(l) \geq D_{z_i}\mathcal{H}(v,z,x) \geq \inf_{E''} p_i(l) \quad \forall i.$$

5. Proof of Theorem 2.6

Recall that the l_k's and $\hat{\delta}$ are introduced in Theorem 2.1. According to Remark 2.9 we may assume that $K>0$. Let N_0 be the right-hand side of (2.11) and for $\mathsf{u}' = \big(\mathsf{u}_0', \mathsf{u}_1', ..., \mathsf{u}_d'\big)$ define

$$\beta(\mathsf{u}') = K_1\big(\bar{H} + K + N_0 + \big|[\mathsf{u}']\big|\big),$$

where according to (4.1.15)

$$[\mathsf{u}'] = \big(\mathsf{u}_1', ..., \mathsf{u}_d'\big).$$

Without loss of generality we may assume that $K_1 > 0$ (cf. (2.9)). Then $\beta \geq K_1 K > 0$. Also cut-off the range of the variable u_0' by setting

$$H^0(\mathsf{u},x) = H\big(-N_0 \vee \mathsf{u}_0' \wedge N_0, \mathsf{u}_1', ..., \mathsf{u}_d', \mathsf{u}'', x\big).$$

This is done in order to have that, owing to (2.9),

$$\big|H^0(\mathsf{u},x) - \langle \mathsf{u}'', D_{\mathsf{u}''}H^0(\mathsf{u},x)\rangle\big| \leq \beta(\mathsf{u}')$$

with $\beta(\mathsf{u}')$ independent of u_0', which allows us to use assertion (iii) of Theorem 4.7.

Next, introduce

$$E' = [-2,2], \quad E_1' = [-1,1], \quad E'' = \mathbb{S}_{\delta/4}, \quad E_1'' = \mathbb{S}_{\delta/2}.$$

Then for any $\mathsf{u}' \in \mathbb{R}^{d+1}$ and $x \in \mathbb{R}^d$ we have

$$H^0(\mathsf{u}',\cdot,x) \in \mathfrak{H}_{E_1,\beta(\mathsf{u}')}$$

with a natural identification of v,u in Section 4 with $\mathsf{u}', \mathsf{u}''$ in the current section. By Remark 2.8 the function H^0 is locally Lipschitz continuous.

Furthermore, in light of (2.8), at all points of differentiability of H^0 with respect to x we have

$$\big|D_x H^0(\mathsf{u},x)\big| \le N'\big(1+\beta(\mathsf{u}')+|\mathsf{u}''|\big), \tag{1}$$

with a constant N' perhaps different from the one in (2.8). Finally, condition (2.7) implies that (4.4) holds for H^0 in place of H. Thus, H^0 satisfies Assumption 4.6.

By Theorem 2.1, for $a \in E''$ we have a representation

$$a = \sum_{|k|=1}^{m} \lambda_k(a) l_k l_k^*,$$

with Lipschitz continuous (real-analytic) $\lambda_k(a)$. Hence Assumption 4.1 is satisfied. Also observe that in our situation $\hat\delta \le \lambda_k(a) \le \hat\delta^{-1}$, so that (see (4.1))

$$\mathcal{E} \subset [-2,2] \times [\hat\delta, \hat\delta^{-1}]^{2m}.$$

Now, according to Theorem 4.7, for $\mathsf{u}' \in \mathbb{R}^{d+1}$, $x \in \mathbb{R}^d$, and

$$z'' = \big(z''_{\pm 1}, ..., z''_{\pm m}\big) \in \mathbb{R}^{2m},$$

define the function $\mathcal{H}^0(\mathsf{u}', z'', x)$ by the formula

$$\mathcal{H}^0(\mathsf{u}', z'', x) = \inf_{w \in \mathbb{S}} \max\big\{\beta(\mathsf{u}')f + \lambda_k(a) z''_k :$$

$$(f,a) \in E,\ \beta(\mathsf{u}')f + \operatorname{tr} aw \le H^0(\mathsf{u}', w, x)\big\}. \tag{2}$$

Then $\mathcal{H}^0(\mathsf{u}', z'', x)$ possesses the following properties:

(i) The function $\mathcal{H}^0(\mathsf{u}', z'', x)$ is locally Lipschitz continuous and at all points of its differentiability

$$\hat\delta^{-1} \ge D_{z''_k}\mathcal{H}^0 \ge \hat\delta \quad \forall k, \tag{3}$$

$$\big|\mathcal{H}^0(\mathsf{u}', z'', x) - \big\langle z'', D_{z''}\mathcal{H}^0(\mathsf{u}', z'', x)\big\rangle\big| \le 2\beta(\mathsf{u}'), \tag{4}$$

$$\big|D_x \mathcal{H}^0(\mathsf{u}', z'', x)\big| + \big|D_{\mathsf{u}'}\mathcal{H}^0(\mathsf{u}', z'', x)\big| \le N\big(1+\beta(\mathsf{u}')+|z''|\big), \tag{5}$$

where the constant N is independent of u', z'', x.

(ii) If $\mathsf{u}'' \in \mathbb{S}$ and $z''_j = \big\langle l_j, \mathsf{u}'' l_j\big\rangle$ for $|j| = 1, ..., m$, then

$$H^0(\mathsf{u}', \mathsf{u}'', x) = \mathcal{H}^0(\mathsf{u}', z'', x) \tag{6}$$

for all u', x.

(iii) $\mathcal{H}^0(\mathsf{u}', z'', x)$ is a decreasing function of u'_0.

Now we want to apply Theorem 1.10. Since by assumption $e_i \in \Lambda$, $i = 1, ..., d$, there is an identification of some e_i with some l_k. Let $e_i = l_{k(i)}$, for $i = 1, ..., d$, and then for

$$z = (z', z''), \quad z' = \big(z'_0, z'_{\pm 1}, ..., z'_{\pm m}\big) \in \mathbb{R}^{2m+1}, \quad z'' = \big(z''_{\pm 1}, ..., z''_{\pm m}\big) \in \mathbb{R}^{2m}$$

and $x \in \mathbb{R}^d$, set

$$\check{\mathcal{H}}^0(z, x) = \mathcal{H}^0\big(z'_0, z'_{k(1)}, ..., z'_{k(d)}, z'', x\big).$$

Observe that if $\rho_\Omega(x) \le \rho_0$, $\check{\mathcal{H}}^0(z,x)$ depends only on z, since then $H(\mathsf{u},x)$ depends only on u by Assumption 2.4 (vi). Assumption 1.3 (iv) is satisfied for $\check{\mathcal{H}}^0(z,x)$ due to (5). If we take $z''=0$ in (4), then we see that

$$\big|\check{\mathcal{H}}^0(z',0,x)\big| \le 2K_1\big(\bar{H}+K+N_0+|z'|\big).$$

Hence, Assumption 1.3 (iii) is satisfied for $\check{\mathcal{H}}^0(z,x)$ with a number smaller than

$$2K_1(\bar{H}+K+N_0) \quad \text{and} \quad 2K_1$$

in place of $\bar{H}$ and K_0. Assumption 1.3 (ii) is satisfied for $\check{\mathcal{H}}^0(z,x)$ with $\hat{\delta}$ in place of δ. Finally, the modification of Assumption 1.3 (i) stated in Theorem 1.10 is satisfied for $\check{\mathcal{H}}^0(z,x)$ in light of (3) and (5). Hence, all the assumptions of Theorem 1.10 are satisfied. We draw the reader's attention to the fact that the assumptions of Theorem 1.10 are satisfied with $\hat{\delta}$ in place of δ (among a few other substitutions) and that $P[u]$ in Theorem 2.6 is $P_{m,\hat{\delta}/2,\Lambda}(D^2u)$, where $\hat{\delta}/2$ is in the perfect accordance with the setting of Theorem 1.10.

By Theorem 1.10 the equation

$$\check{\mathcal{H}}^0\big(v(x), D_{l_j}v(x), D^2_{l_k}v(x), x\big) \vee \big(P[v](x)-K\big)=0 \tag{7}$$

in Ω (a.e.) with boundary condition $v=g$ on $\partial\Omega$ has a unique solution $v \in C(\bar{\Omega})\cap C^{1,1}_{\text{loc}}(\Omega)$ and the estimates (0.4) hold true if its right-hand side is replaced with

$$N\big[K_1(\bar{H}+K+N_0)+K+\|g\|_{C^{1,1}(\mathbb{R}^d)}\big]. \tag{8}$$

We recall what N_0 is and estimate (8) from above by

$$N\big(\bar{H}+K+\|g\|_{C^{1,1}(\mathbb{R}^d)}\big)$$

always with N as in the statement of the theorem.

Next, notice that because of (6) equation (7) coincides with

$$H^0\big(v(x), Dv(x), D^2v(x), x\big) \vee \big(P[v](x)-K\big)=0 \tag{9}$$

and since, as pointed out in Remark 2.7, any sufficiently regular solution of (9) in Ω with boundary condition $v=g$ on $\partial\Omega$, satisfies estimate (2.11) (that is (0.2)), we have that $|v|\le N_0$ and then (9) becomes (0.1) by the construction of H^0. Thus, the existence and the claimed estimates have been established. Uniqueness is already taken care of in Remark 2.9. The theorem is proved. □

6. The case of H measurable in space variables

Here we present the main results of this chapter, interesting in their own right and serving as a starting point in many issues of the theory of fully nonlinear elliptic equations treated in this book. Fix some constants $K, K_0, K_1 \in [0,\infty)$, $\delta \in (0,1]$ (K_1 is needed to use Theorem 2.6).

1. ASSUMPTION. We are given a domain $\Omega \subset \mathbb{R}^d$ which is bounded and satisfies the exterior ball condition. We are also given a function $g \in C^{1,1}(\mathbb{R}^d)$.

Suppose that we are also given a function $H(\mathsf{u},x)$,

$$\mathsf{u}=(\mathsf{u}',\mathsf{u}''), \quad \mathsf{u}'=\big(\mathsf{u}'_0,\mathsf{u}'_1,...,\mathsf{u}'_d\big)\in\mathbb{R}^{d+1}, \quad \mathsf{u}''\in\mathbb{S}, \quad x\in\mathbb{R}^d.$$

2. ASSUMPTION. (i) The function $H(\mathsf{u}, x)$ is Borel measurable with respect to (u, x), is Lipschitz continuous with respect to u'', and $D_{\mathsf{u}''}H \in \mathbb{S}_\delta$ at all points of differentiability of H with respect to u''.

(ii) The number

$$\bar{H} := \sup_{\mathsf{u}', x} \big(|H(\mathsf{u}', 0, x)| - K_0|\mathsf{u}'|\big) \quad (\geq 0)$$

is finite.

(iii) The function $H(\mathsf{u}, x)$ is continuous with respect to u for any x.

(iv) The function $H(\mathsf{u}, x)$ is nonincreasing with respect to u'_0.

For smooth enough functions u introduce

$$H[u](x) = H\big(u(x), Du(x), D^2u(x), x\big), \quad P[u](x) = P\big(D^2u(x)\big),$$

where P is defined in the beginning of Section 2.

The main result (Theorem 3) is given with full proof in this chapter and improved in Subsection 6:1 (of this section) on the basis of the results of Section 8.3. Such an organization of the chapter may look strange. Indeed, why not first prove the results of Section 8.3 bearing on parabolic equations and then apply them to the elliptic ones? The point is the same as in Chapter 5. The arguments in parabolic case are much more technically involved than in the elliptic case and we first wanted the reader to get acquainted with the main ideas in a simpler situation. Anyhow, the results of Subsection 6:1 are not used to obtain the results neither in Section 8.3 nor elsewhere in this chapter, so that there is no vicious circle. It is also to be noted that further extensions of Theorem 3 with relaxed Assumption 2 (ii) are presented in Theorems 10.1.16 (when $g \in C(\partial\Omega)$ and $v \in W^2_{p,\text{loc}}(\Omega)$) and 14.1.6 (when $g \in W^2_p(\Omega)$ and $v \in W^2_p(\Omega)$).

In the following theorem the above assumptions are supposed to be satisfied.

3. THEOREM. *Equation* (0.1) *with boundary condition* $v = g$ *on* $\partial\Omega$ *has a solution* $v \in C(\bar{\Omega}) \cap C^{1,1}_{\text{loc}}(\Omega)$ *such that*

(*i*) *in* Ω *we have*

$$|v| \leq N\big(\delta, d, K_0, \text{diam}(\Omega)\big)\bar{H} + \|g\|_{C(\partial\Omega)}, \tag{1}$$

$$|v - g| \leq N\rho_\Omega\big(\bar{H} + \|g\|_{C^{1,1}(\Omega)} + \sup_\Omega |v|\big), \tag{2}$$

where the constant N *depends only on* δ, d, K_0, *and* $\rho_{\text{ext}}(\Omega)$;

(*ii*) *there exists* $\alpha = \alpha(\delta, d) \in (0, 1)$ *such that*

$$|v|_{C^\alpha(\Omega)} \leq N\big(\bar{H} + \|g\|_{C^1(\Omega)} + \sup_\Omega |v|\big),$$

with N *of the same kind as in* (2);

(*iii*) *there is a constant* N, *depending only on* $\text{diam}(\Omega)$, $\rho_{\text{ext}}(\Omega)$, $\rho_{\text{int}}(\Omega)$, Λ, K_0, d, *and* δ, *such that*

$$|Dv|, (\rho_\Omega \wedge 1)|D^2v| \leq N\big(\bar{H} + K + \|g\|_{C^{1,1}(\mathbb{R}^d)} + \sup_\Omega |v|\big) \quad \textit{in} \quad \Omega \quad (\textit{a.e.}); \tag{3}$$

(*iv*) *if* $\Omega \in C^{1,1}$ (*see Definition* 14.1.10), *then* $v \in W^2_p(\Omega)$ *for any* $p \in [1, \infty)$.

The inclusion of the $\sup_\Omega |v|$ into the right-hand sides of (2) and (3) may look unusual in view of (1). We did it to make the statements in what follows look lighter.

6:1. Some generalizations. The results of this subsection are based on the results of Section 8.3, they are more general than Theorem 3, and they are not used in the proof of Theorem 3. We reiterate that the results of this subsection are not used to obtain the results in Section 8.3 either.

4. ASSUMPTION. There exist constants $M_0 \geq 0, \varepsilon > 0$ such that

$$\mp H(\pm M_0 + \varepsilon_0, \varepsilon_1, ..., \varepsilon_d, 0, x) \geq 0 \tag{4}$$

in Ω as long as the length of the vector $(\varepsilon_0, \varepsilon_1, ..., \varepsilon_d)$ in $\mathbb{R}^{d+1}$ is less than ε.

5. THEOREM. *Let Assumptions* 1, 2 *(i)-(iii), and* 4 *be satisfied. Also suppose that* $|g| \leq M_0$ *in* $\bar{\Omega}$. *Then all the assertions of Theorem* 3 *hold true with one exception that* $|v| \leq M_0$ *should replace* (1).

6. REMARK. Assumption 4 does not restrict the behavior of $H(\mathsf{u}, x)$ as $|\mathsf{u}_0'| \to \infty$. For instance, if $d = 1$ and $H(\mathsf{u}, x) = \mathsf{u}'' + (\mathsf{u}_0')^3 - 2\mathsf{u}_0'$, Assumption 4 is satisfied with $M_0 = 1$ and small $\varepsilon > 0$. Theorem 5 provides the solvabiity of the equation with this H as long as $|g| \leq 1$.

The proof of Theorem 5 is deferred until Section 8.3 devoted to parabolic equations. Theorem 3 is proved in the next subsection without using neither Theorem 5 nor the following theorem before which we formulate an assumption.

7. ASSUMPTION. There exist $\bar{G} \in L_d(\Omega)$, $\bar{G} \geq 0$, and a constant $N_0 \in [0, \infty)$ such that, for all values of the arguments,

$$H(\mathsf{u}', 0, x) \operatorname{sign} \mathsf{u}_0' \leq N_0 \big|[\mathsf{u}']\big| + \bar{G}(x) \quad (\operatorname{sign} 0 := \pm 1) \tag{5}$$

(regarding the notation $[\mathsf{u}']$ see (4.1.15)).

8. REMARK. If Assumption 2 (ii) is satisfied and $H(\mathsf{u}', 0, x)$ is a nonincreasing function of u_0', then Assumption 7 is satisfied with $N_0 = K_0$ and $\bar{G} = \bar{H}$. This shows that Theorem 9 is a generalization of Theorem 3.

9. THEOREM. *Let Assumptions* 1, 2 *(i)-(iii), and* 7 *be satisfied. Then all the assertions of Theorem* 3 *hold true with one exception that* (1) *should be replaced with*

$$\sup_\Omega |u| \leq N \|\bar{G}\|_{L_d(\Omega)} + \sup_{\partial\Omega} |g|, \tag{6}$$

where N depends only on d, δ, N_0, and $\operatorname{diam}(\Omega)$.

To derive this theorem from Theorem 5 we need a lemma. As usual for $\mathsf{u} = (\mathsf{u}', \mathsf{u}'')$ set

$$H_K(\mathsf{u}, x) = \max\big(H(\mathsf{u}, x), P(\mathsf{u}'') - K\big).$$

From Remark 2.2 and Corollary 4.1.4 we know that at all points of differentiability of H_K with respect to u'' we have

$$D_{\mathsf{u}''} H_K \in \mathbb{S}_{\bar{\delta}}. \tag{7}$$

We state the following for domains more general than Ω.

10. LEMMA. *Suppose that* Θ *is bounded domain in* $\mathbb{R}^d$ *and Assumptions* 2 (*i*) *and* 7 *are satisfied with* Θ *in place of* Ω. *Then for any* $u \in W^2_{d,\text{loc}}(\Theta) \cap C(\bar{\Theta})$ *that satisfies* (0.1) *in* Θ (*a.e.*) *we have*

$$\sup_{\Theta} u_{\pm} \leq N\|\bar{G}\|_{L_d(\Theta)} + \sup_{\partial\Theta} u_{\pm}, \tag{8}$$

where N *depends only on* d, δ, N_0, *and* $\text{diam}(\Theta)$.

One obtains the proof of the lemma by repeating Remark 2.7 and taking into account that on the set $\Theta_+ = \{x : u(x) > 0\}$ we have $H_K(u, Du, 0, x) \leq N_0|Du| + \bar{G}$.

Proof of Theorem 9. Take an odd increasing continuous function $\xi(t)$ such that $\xi(t) = 0$ for $|t| \leq 1/2$ and $\xi(t) = t$ for $|t| \geq 1$, $|\xi(t)| \leq |t|$. For (large) $\kappa > 0$ define $\xi_\kappa(t) = \kappa(K_0 + 1)\xi(2t/\kappa)$ and

$$H^\kappa(\mathsf{u}, x) = H(\mathsf{u}, x) - \xi_\kappa(\mathsf{u}'_0).$$

Observe that for

$$\varepsilon = (\varepsilon_0, \varepsilon_1, ..., \varepsilon_d), \quad M_0 = (\kappa, 0, ..., 0) \in \mathbb{R}^{d+1}$$

with $|\varepsilon| \leq 1$ and $\kappa > 2$ we have

$$H^\kappa(M_0 + \varepsilon, 0, x) \leq K_0(\kappa + 1) + \bar{H} - \kappa(K_0 + 1),$$

which is negative if $\kappa > K_0 + \bar{H}$. Similarly,

$$H_\kappa(-M_0 - \varepsilon, 0, x) > 0$$

if $\kappa > K_0 + \bar{H}$.

Furthermore,

$$\big|H^\kappa(\mathsf{u}', 0, x)\big| \leq K_0|\mathsf{u}'| + \bar{H} + 2(K_0 + 1)\big|\mathsf{u}'_0\big|,$$

which shows that Assumption 2 (ii) is satisfied for H^κ with $3K_0 + 2$ in place of K_0.

By Theorem 5 for

$$\kappa \geq K_0 + \bar{H} + 1 + \sup_{\Omega} |g| =: \kappa_0$$

there exists a solution $v^\kappa \in C(\bar{\Omega}) \cap C^{1,1}_{\text{loc}}(\Omega)$ of the equation

$$\max\big(H^\kappa[v^\kappa], P[v^\kappa] - K\big) = 0 \tag{9}$$

in Ω (a.e.) satisfying $v^\kappa = g$ on $\partial\Omega$ and for which all assertions (i)-(iv) hold true with only one exception that (1) should be replaced with $|v^\kappa| \leq \kappa$.

However, set

$$H^\kappa_K(\mathsf{u}, x) = \max\big(H^\kappa(\mathsf{u}, x), P(\mathsf{u}'') - K\big)$$

and notice that due to (5), if $\mathsf{u}'_0 \geq 0$, then

$$H^\kappa_K(\mathsf{u}', 0, x) = \max\big(H(\mathsf{u}', 0, x) - \xi_\kappa(\mathsf{u}'_0), -K\big) \leq N_0\big|[\mathsf{u}']\big| + \bar{G}(x),$$

and, if $\mathsf{u}'_0 \leq 0$, then

$$-H^\kappa_K(\mathsf{u}', 0, x) = \min\big(-H(\mathsf{u}', 0, x) + \xi_\kappa(\mathsf{u}'_0), K\big) \leq N_0\big|[\mathsf{u}']\big| + \bar{G}(x).$$

It follows by Lemma 10 that estimate (6) is valid for v^κ, which in turn implies that if κ is larger than twice the right-hand side in (6), then $H^\kappa[v^\kappa] = H[v^\kappa]$, equation (9) shows that v^κ is a solution of (0.1), and the theorem is proved. □

11. EXAMPLE. Let A and B be countable sets and assume that for $\alpha \in A$ and $\beta \in B$ we are given $\mathbb{S}_\delta$-valued measurable functions $a^{\alpha\beta}(\mathsf{u}'_1, ..., \mathsf{u}'_d, x)$ on $\mathbb{R}^{2d}$ and real-valued measurable functions $b^{\alpha\beta}(\mathsf{u}', x)$ on $\mathbb{R}^{2d+1}$, which are continuous with respect to u' uniformly with respect to α, β, and x. Suppose that $b^{\alpha\beta}$ is a nonincreasing function of u'_0 and

$$\left|b^{\alpha\beta}(\mathsf{u}', x)\right| \leq K_0|\mathsf{u}'| + \bar{b},$$

where $\bar{b} \geq 0$ is a constant. Then the equation

$$\inf_{\alpha\in A} \sup_{\beta\in B} \left[a_{ij}^{\alpha\beta}\left(Dv(x), x\right) D_{ij}v(x) + b^{\alpha\beta}\left(v(x), Dv(x), x\right)\right] \vee \left(P[v](x) - K\right) = 0$$

presents a particular case of equation (0.1) and falls into the framework of Theorem 3.

12. EXAMPLE. Let A and B be countable sets and assume that for $\alpha \in A$ and $\beta \in B$ we are given $\mathbb{S}_\delta$-valued measurable functions $a^{\alpha\beta}(\mathsf{u}', x)$ and real-valued measurable functions $b^{\alpha\beta}(\mathsf{u}', x)$ on $\mathbb{R}^{2d+1}$, which are continuous with respect to u' uniformly with respect to α, β, and x. Suppose that

$$b^{\alpha\beta}(\mathsf{u}', x) \operatorname{sign} u'_0 \leq K_0\left|[\mathsf{u}']\right| + \bar{b},$$

where $\bar{b} \geq 0$ is a constant. Then the equation

$$\left(\inf_{\alpha\in A} \sup_{\beta\in B} \mathcal{L}^{\alpha\beta}v(x)\right) \vee \left(P[v] - K\right) = 0,$$

where

$$\mathcal{L}^{\alpha\beta}v(x) = a_{ij}^{\alpha\beta}\left(v(x), Dv(x), x\right) D_{ij}v(x) + b^{\alpha\beta}\left(v(x), Dv(x), x\right),$$

presents another particular case of equation (0.1) and falls into the framework of Theorem 9.

6:2. Proof of Theorem 3.

13. REMARK. Here we derive assertion (iv) of Theorem 3 from (1), (3), and Theorem 14.4.1. The reader will see that there is no vicious circle in our arguments.

We set $u = v - g$ and rewrite (0.1) as

$$P[u] = f, \tag{10}$$

where

$$f = P[v - g] - P[v] - \max\left(H[v] - P[v], -K\right).$$

Observe that $\left|P[v - g] - P[v]\right| \leq N|D^2 g|$, which is bounded in Ω. Moreover,

$$\left|H\left(v(x), Dv(x), 0, x\right)\right| \leq \bar{H} + K_0\left(|v| + |Dv|\right)(x),$$

which is bounded in Ω, and

$$H[v](x) - H\left(v(x), Dv(x), 0, x\right) = a^{ij}(x) D_{ij}v(x),$$

where $(a^{ij}(x)) \in \mathbb{S}_\delta$. By construction $a^{ij}D_{ij}v \leq P[v]$, so that $\max\left(H[v] - P[v], -K\right)$ is bounded from above. It is also trivially bounded from below by $-K$. Hence f is bounded.

By Theorem 14.4.1, for any point $x_0 \in \partial\Omega$, there exists $r > 0$, such that $u \in W_p^2\left(\Omega \cap B_r(x_0)\right)$ and, as it follows from the proof of Theorem 14.4.1, r can be chosen to be independent of x_0 and the $W^2_p(\Omega \cap B_r(x_0))$-norm of u is a bounded function of x_0. This obviously proves the last assertion of Theorem 3.

Interestingly enough in the above argument we did not use Assumptions 2 (iii) and (iv).

14. REMARK. Observe that there is no uniqueness statement in Theorem 3 in contrast with Lemmas 7.1, 7.2, and 7.4 coming later.

The point is that, generally, uniqueness may fail. For instance, in the one-dimensional case the equation

$$\max\big(D^2u + \sqrt{12|Du|}, D^2u - 7\big) = 0$$

for $x \in (-1, 1)$ with zero boundary data has (more than) two solutions: one is identically equal to zero and the other one is $1 - |x|^3$.

The following two remarks will be often used.

15. REMARK. Under Assumption 2, $D_{\mathsf{u}''}H_K \in \mathbb{S}_{\bar\delta}$ (see Remark 2.2 and Corollary 4.1.4). Also by Lemma 4.1.6 for any measurable $\mathsf{u}(x)$, there exist measurable $\mathbb{S}_{\bar\delta}$-valued $\big(a^{ij}(x)\big)$, measurable $\mathbb{R}^d$-valued $b(x)$, and measurable real-valued $\tau(x)$ and $c(x) \geq 0$ such that

$$|b| \leq K_0, \quad |\tau| \leq 1,$$

$$H_K\big(\mathsf{u}(x), x\big) = a^{ij}(x)\mathsf{u}''_{ij}(x) + b^i(x)\mathsf{u}'_i(x) - c(x)\mathsf{u}'_0(x) + \tau(x)\bar H \quad \text{(a.e.)}.$$

16. REMARK. In Remark 15 Assumption 2 (iv) ensured that $c \geq 0$. If we only use Assumption 2 (i), (ii), then for any measurable $\mathsf{u}(x)$, there exist measurable $\mathbb{S}_{\bar\delta}$-valued $\big(a^{ij}(x)\big)$, measurable $\mathbb{R}^d$-valued $b(x)$, and measurable real-valued $\tau(x)$ and $c(x)$ such that

$$|b| \leq K_0, \quad |\tau| \leq 1, \quad |c| \leq K_0,$$

$$H_K\big(\mathsf{u}(x), x\big) = a^{ij}(x)\mathsf{u}''_{ij}(x) + b^i(x)\mathsf{u}'_i(x) + c(x)\mathsf{u}'_0(x) + \tau(x)\bar H \quad \text{(a.e.)}.$$

This follows again from Lemma 4.1.6 in which one drops the dependence on q.

To discuss estimates (1) and (2) we first state the following.

17. LEMMA. *There exists a constant $\beta > 0$, depending only on d, δ, $\rho_{\text{ext}}(\Omega)$, and K_0, such that for $\psi(x) = |x|^{-\beta} - \rho^{-\beta}_{\text{ext}}(\Omega)$ we have*

$$a^{ij}D_{ij}\psi(x) + b^iD_i\psi(x) - c\psi(x) \geq 1, \tag{11}$$

whenever $(a^{ij}) \in \mathbb{S}_\delta$, $|b_i| \leq K_0$, $i = 1, ..., d$, $|c| \leq K_0$, and $0 < |x| \leq 3$.

The proof of this lemma is almost identical to that of Lemma 5.5.7.

18. REMARK. If we already know that there is a $v \in C(\bar\Omega) \cap C^{1,1}_{\text{loc}}(\Omega)$ satisfying equation (0.1) and such that $v = g$ on $\partial\Omega$, then (1) follows directly from Remark 15 and the maximum principle.

Also, according to Remark 16, for appropriate a, b, c, τ, we have

$$\big|a^{ij}(x)D_{ij}(v-g) + b^iD_i(v-g)\big| \leq \big|cv + \tau(x)\bar H\big| + \big|a^{ij}(x)D_{ij}g + b^iD_ig\big|$$

$$\leq K_0 \sup_\Omega |v| + \bar H + N|Dg| + N|D^2g| \tag{12}$$

in Ω (a.e.) with N depending only on δ and d. After that an obvious use of (12), Lemma 17, and Theorem 3.0.2 yields (2) (cf. the proof of Lemma 5.5.8).

19. REMARK. If we already know that there is a $v \in C(\bar{\Omega}) \cap C^{1,1}_{\text{loc}}(\Omega)$ satisfying equation (0.1) and such that $v = g$ on $\partial\Omega$, then assertion (ii) of Theorem 3 is obtained in the following way. Here again we use some results, that are proved later, giving the reader the assurance that no vicious circle arrises.

Similarly to (12) for appropriate a, b, c, τ, we have in Ω (a.e.)

$$\big|a^{ij}(x)D_{ij}v + b^i D_i v\big| \leq \big|cv + \tau(x)\bar{H}\big| \leq K_0 \sup_\Omega |v| + \bar{H}.$$

Hence, by Theorem 9.6.12, there exists $\beta = \beta(d, \delta) \in (0, 1)$ such that if $x_1, x_2 \in \Omega$, $\rho := |x_1 - x_2| \leq 1$, and $B_{2\sqrt{\rho}}(x_1) \subset \Omega$, then (with $r = \sqrt{\rho}$ in (9.6.21))

$$\big|v(x_1) - v(x_2)\big| \leq N\big(\bar{H} + \sup_\Omega |v|\big)|x_1 - x_2|^{\beta/2}. \tag{13}$$

However, if $\rho \leq 1$ and $B_{2\sqrt{\rho}}(x_1) \not\subset \Omega$, then $\rho_\Omega(x_1) \leq 2\sqrt{\rho}$, $\rho_\Omega(x_2) \leq \rho_\Omega(x_1) + \rho \leq 3\sqrt{\rho}$, and in light of (2)

$$\begin{aligned}\big|v(x_1) - v(x_2)\big| &\leq \big|v(x_1) - g(x_1)\big| + \big|g(x_1) - g(x_2)\big| + \big|v(x_2) - g(x_2)\big| \\ &\leq N\sqrt{\rho} + \big|g(x_1) - g(x_2)\big| \\ &\leq N\sqrt{\rho} = N|x_1 - x_2|^{1/2},\end{aligned}$$

where the constants N have obvious meanings.

We thus estimated $\big|v(x_1) - v(x_2)\big|$ when $|x_1 - x_2| \leq 1$. If $|x_1 - x_2| \geq 1$, (13) holds with $N = 2$. It only remains to add that $|v|$ is estimated as well according to Remark 18.

We derive the remaining assertions of Theorem 3 (concerning the existence of solutions and the estimates (3)) in this section from the following slightly weaker result, that is proved in the next section.

20. THEOREM. *Under Assumptions 1 and 2 also suppose that $H(\mathsf{u}, x)$ is Lipschitz continuous with respect to u with a Lipschitz constant independent of x. Then equation (0.1) with boundary condition $v = g$ on $\partial\Omega$ has a unique solution $v \in C(\bar{\Omega}) \cap C^{1,1}_{\text{loc}}(\Omega)$ such that assertion (iii) of Theorem 3 holds true.*

We also need one more intermediate result.

21. LEMMA. *Suppose that Assumptions 1 and 2 are satisfied and $F(\mathsf{u}, x) := H(\mathsf{u}, x) - \delta \operatorname{tr} \mathsf{u}''$ is a bounded function. Then equation (0.1) with boundary condition $v = g$ on $\partial\Omega$ has a solution $v \in C(\bar{\Omega}) \cap C^{1,1}_{\text{loc}}(\Omega)$ such that assertion (iii) of Theorem 3 holds true.*

Proof. Take a nonnegative C_0^∞-function $\zeta(\mathsf{u}')$ with support in $\{|\mathsf{u}'| < 1\}$, which integrates to one over $\mathbb{R}^{d+1}$ and introduce $H^n(\mathsf{u}, x)$ as the convolution of $H(\mathsf{u}', \mathsf{u}'', x)$ and $n^{d+1}\zeta(n\mathsf{u}')$ performed with respect to u'. Note that, as is easy to see, $H^n(\mathsf{u}, x)$ satisfies Assumption 2 (i).

Observe that, owing to Assumption 2 (iv), $H^n(\mathsf{u}, x)$ is a nonincreasing function of u'_0 and for any $k = 0, ..., d$

$$D_{\mathsf{u}'_k} H^n(\mathsf{u}',\mathsf{u}'',x) = n \int_{\mathbb{R}^{d+1}} H(\mathsf{u}' - \mathsf{v}'/n, \mathsf{u}'', x) D_{\mathsf{u}'_k} \zeta(\mathsf{v}')\, d\mathsf{v}',$$

which shows that H^n are Lipschitz continuous with respect to u' with constant independent of u'', x. Since, $D_{\mathsf{u}''} H^n \in \mathbb{S}_\delta$, H^n are also Lipschitz continuous with respect to u'' with constant independent of u', x (and n). Hence, H^n are Lipschitz continuous with respect to u with constant independent of x.

Finally, as is easy to see,

$$\bar{H}^n = \sup_{\mathsf{u}',x} \big(|H^n(\mathsf{u}',0,x)| - K_0|\mathsf{u}'|\big) \le \bar{H} + K_0 n^{-1}.$$

By Theorem 20, for any n, there exists a unique $C(\bar{\Omega}) \cap C^{1,1}_{\text{loc}}(\Omega)$-solution of

$$\max\big(H^n[v^n], P[v^n] - K\big) = 0 \tag{14}$$

in Ω (a.e.) with boundary condition $v^n = g$ on $\partial\Omega$ for which the estimates (3) hold. Combining this with Remark 18, we see that v^n and their gradients are uniformly bounded. Hence by renumbering v^n, if necessary, we may assume that there is a function v, such that $v^n \to v$ uniformly in $\bar{\Omega}$. Then, of course, $v \in C(\bar{\Omega}) \cap C^{1,1}_{\text{loc}}(\Omega)$, $v = g$ on $\partial\Omega$, and the estimates (1) and (3) hold true. Furthermore, since $D^2 v_n$ are locally uniformly bounded and $v_n \to v$, we have that $Dv_n \to Dv$ in Ω (even locally uniformly). This and the continuity of $H(\mathsf{u}, x)$ with respect to u' imply that

$$\check{H}^m(\mathsf{u}'',x) := \sup_{n\ge m} H^n\big(v_n(x), Dv_n(x), \mathsf{u}'', x\big) \to H\big(v(x), Dv(x), \mathsf{u}'', x\big) \tag{15}$$

for any $x \in \Omega$ and $\mathsf{u}'' \in \mathbb{S}$.

After that we observe that for $n \ge m$, obviously, v^n satisfy in Ω (a.e.)

$$\max\big(\check{H}^m[v^n], P[v^n] - K\big) \ge 0. \tag{16}$$

By an elliptic counterpart of Lemma 4.2.4, the operators we are dealing with are $\bar{\delta}$-nondegenerate $\mathcal{L}$-type operators in Ω. Therefore, due to (16), an elliptic version of Theorem 4.2.6 (see Theorem 4.2.15), applied to subdomains of Ω in which all $C^{1,1}$-norms of v^n are bounded, yields that for any m in Ω (a.e.)

$$\max\big(\check{H}^m[v], P[v] - K\big) \ge 0. \tag{17}$$

By setting $m \to \infty$ and using (15) we conclude that

$$\max\big(H[v], P[v] - K\big) \ge 0$$

in Ω (a.e.). The reader understands that the inequality $\max\big(H[v], P[v] - K\big) \le 0$ is obtained similarly by using

$$\inf_{n\ge m} H^n\big(v_n(x), Dv_n(x), \mathsf{u}'', x\big) \to H\big(v(x), Dv(x), \mathsf{u}'', x\big).$$

The lemma is proved. □

Proof of Theorem 3. As we have pointed out before Theorem 20, we only need to deal with the existence of solutions and the estimates (3).

For $r \in \mathbb{R}$ and $\mathsf{u}'' \in \mathbb{S}$ introduce $\chi(r)$ and $\kappa(\mathsf{u}'')$ as the projections of r and u'', respectively, on the unit balls in $\mathbb{R}$ and $\mathbb{S}$, respectively, for $\mathsf{u}' \in \mathbb{R}^{d+1}$ set

$$\chi(\mathsf{u}') = \big(\chi(\mathsf{u}'_0), \chi(\mathsf{u}'_1), ..., \chi(\mathsf{u}'_d)\big),$$

and for $n = 1, 2, ...$ define

$$H^n(\mathsf{u}, x) = \delta \mathrm{tr}\, \big[\mathsf{u}'' - n\kappa(\mathsf{u}''/n)\big] + H\big(n\chi(\mathsf{u}'/n), n\kappa(\mathsf{u}''/n), x\big).$$

Almost obviously, H^n satisfies Assumption 2 with the same δ and, perhaps, smaller $\bar{H}$ than the original one. Furthermore, $H^n(\mathsf{u}, x) - \delta \mathrm{tr}\, \mathsf{u}''$ is bounded.

By Lemma 21 and Remarks 18, for any n, there exists a unique solution of (14) in Ω with boundary condition $v^n = g$ on $\partial\Omega$ for which (1) and (3) hold true. After that we just repeat the rest of the proof of Lemma 21 noting that $H^n(\mathsf{u}, x) = H(\mathsf{u}, x)$ in any ball in the space of u's if n is large enough, so that (15) holds true owing to the continuity of H with respect to u'. The theorem is proved. □

22. REMARK. Cut-off elliptic equations first appeared in [**72**] with P depending on all coordinates of u and H Lipschitz with respect to u. Then in [**74**] in the case of parabolic equations the author found a way to use P only depending on u'' and this allowed to treat H just continuous with respect to u'. Further development for the elliptic equation is presented in [**79**], where also the existence theorems in global Sobolev spaces are given for the original equations and not only for their cut-off versions. Similar cut-off equations are introduced in Theorem 8.1 of [**97**], where the authors prove local W^2_p-estimates in the case that $H = H(\mathsf{u}'', x)$.

7. Proof of Theorem 6.20

The proof of Theorem 6.20 is based on Theorem 2.6 and will be achieved in several steps in the first three of which we drop Assumptions 2.4 (iv), (v), (vi), one by one.

1. LEMMA. *Let the assumptions of Theorem* 2.6, *apart from Assumption* 2.4 (*vi*), *be satisfied. Then the assertions of Theorem* 2.6 *are still true.*

Proof. Define $H(\mathsf{u}'') = H(0, \mathsf{u}'', 0)$ and for $\rho_0 > 0$ find a function $\zeta_{\rho_0} \in C_0^\infty(\Omega)$ such that $\zeta_{\rho_0} = 0$ on $\Omega \setminus \Omega^{\rho_0}$ and $\zeta_{\rho_0} = 1$ on $\Omega^{2\rho_0}$ and $0 \le \zeta_{\rho_0} \le 1$ in Ω. Then introduce

$$H^{\rho_0}(\mathsf{u}, x) = \zeta_{\rho_0} H(\mathsf{u}, x) + (1 - \zeta_{\rho_0}) H(\mathsf{u}'').$$

Obviously, $H^{\rho_0}(\mathsf{u}, x) = H(\mathsf{u}'')$ if $\rho(x) \le \rho_0$ and condition (2.7) and (2.9) are satisfied for H^{ρ_0} with the same N' and K_1. Furthermore,

$$\big|H(\mathsf{u}, x)\big| \le \big|H(\mathsf{u}, x) - H(0, \mathsf{u}'', x)\big| + \big|H(0, \mathsf{u}'', x) - H(0, x)\big|$$

$$+\big|H(0, x)\big| \le N'|\mathsf{u}'| + N|\mathsf{u}''| + \bar{H},$$

where N' is taken from (2.7) and N accounts for the Lipschitz continuity of H in u''. It follows that condition (2.8) is satisfied for H^{ρ_0} with a different constant N' (by the way, depending on ρ_0 but this is irrelevant). Condition (2.6) is satisfied for H^{ρ_0} with the number $\bar{H}^{\rho_0}$ which is at most twice $\bar{H}$ from (2.6).

By Theorem 2.6 the equation

$$\max\big(H^{\rho_0}[v^{\rho_0}], P[v^{\rho_0}] - K\big) = 0 \tag{1}$$

in Ω (a.e.) with boundary condition $v^{\rho_0} = g$ on $\partial\Omega$ has a unique solution v^{ρ_0} with the properties described in that theorem (with the constants N from Theorem 2.6 multiplied by 2, because $\bar{H}^{\rho_0} \le 2\bar{H}$). One sends $\rho_0 \downarrow 0$ and finishes the proof of existence of a solution with desired properties as in the end of the proof of Theorem 1.10. Uniqueness is also shown as there. The lemma is proved. □

In the following lemma Assumption 2.4 (i) is replaced with a stronger Assumption 6.2 (i).

2. LEMMA. *Let Assumption* 6.2 (*i*) *and the assumptions of Theorem* 2.6, *apart from Assumptions* 2.4 (*v*), (*vi*), *be satisfied. Then the assertions of Theorem* 2.6 *are still true and, this time,* N *in the estimates* (0.4) (*implying* (6.3)) *depends only on* $\mathrm{diam}(\Omega)$, $\rho_{\mathrm{ext}}(\Omega)$, $\rho_{\mathrm{int}}(\Omega)$, Λ, K_0, d, *and* δ (*thus independent of* K_1, *which does not even enter the assumptions of the present lemma*).

Proof. Introduce

$$P_0(\mathsf{u}'') = \max_{a \in \mathbb{S}_{\delta/2}} a^{ij} \mathsf{u}''_{ij}$$

(as usual, the summation is performed before the maximum is taken). It is easy to see that $P_0[u]$ is Pucci's operator:

$$P_0(\mathsf{u}'') = -(\delta/2) \sum_{k=1}^{d} \lambda_k^-(\mathsf{u}'') + 2\delta^{-1} \sum_{k=1}^{d} \lambda_k^+(\mathsf{u}''),$$

where $\lambda_1(\mathsf{u}''), ..., \lambda_d(\mathsf{u}'')$ are the eigenvalues of u'' and $a^{\pm} = (1/2)(|a| \pm a)$.

Moreover, owing to Theorem 2.1 (b), the collection of matrices

$$\sum_{|k|=1}^{m} a_k l_k l_k^*$$

such that $\hat{\delta} \leq a_k \leq \hat{\delta}^{-1}$, $|k| = 1, ..., m$, covers $\mathbb{S}_{\delta/4}$. Hence,

$$P(\mathsf{u}'') = \max_{\substack{\hat{\delta}/2 \leq a_r \leq 2\hat{\delta}^{-1} \\ |r|=1,...,m}} \operatorname{tr} \mathsf{u}'' a_k l_k l_k^* \geq P_0(\mathsf{u}''),$$

and therefore,

$$\max(H, P - K) = \max(H_K, P - K),$$

where

$$H_K = \max(H, P_0 - K).$$

It is easy to see that the function H_K satisfies Assumption 2.4 (i) (here Remark 2.5 is relevant). The function H_K also satisfies Assumptions 2.4 (iii), (iv) with the same constant N'. Since

$$\big|H_K(\mathsf{u}', 0, x)\big| = \big|\max\big(H(\mathsf{u}', 0, x), -K\big)\big| \leq \big|H(\mathsf{u}', 0, x)\big|,$$

Assumption 2.4 (ii) is also satisfied for H_K.

Next we claim that Assumption 2.4 (v) is also satisfied for H_K with K_1 depending only on K_0, δ, and d (and $\bar{H}_K$ in place of $\bar{H}$). To prove the claim, first of all observe that there is a constant $\kappa > 0$ depending only on δ and d such that

$$H \leq P_0 - \kappa|\mathsf{u}''| + K_0|\mathsf{u}'| + \bar{H}_K. \tag{2}$$

Indeed, by Lemma 4.1.1 (owing to Assumption 6.2 (i)) there exists $a \in \mathbb{S}_\delta$ (depending on (u, x)) such that

$$H(\mathsf{u}, x) = H(\mathsf{u}', 0, x) + a^{ij} \mathsf{u}''_{ij}.$$

Also $H \leq H_K \leq |H_K|$ and it follows that

$$H(\mathsf{u},x) \leq \big(|H_K(\mathsf{u}',0,x)| - K_0|\mathsf{u}'|\big) + K_0|\mathsf{u}'| + \sup_{a\in\mathbb{S}_\delta} a^{ij}\mathsf{u}''_{ij}.$$

Here the first term on the right is less than $\bar{H}_K$ by definition and the last term equals

$$-\delta\sum_{k=1}^d \lambda_k^-(\mathsf{u}'') + \delta^{-1}\sum_{k=1}^d \lambda_k^+(\mathsf{u}'') = P_0(\mathsf{u}) - (\delta/2)\sum_{k=1}^d \lambda_k^-(\mathsf{u}'') - \delta^{-1}\sum_{k=1}^d \lambda_k^+(\mathsf{u}'')$$

$$\leq P_0(\mathsf{u}) - (\delta/2)\sum_{k=1}^d \big|\lambda_k(\mathsf{u}'')\big|.$$

This certainly implies (2).

Next, fix x and u' and let $\bar{\mathsf{u}}''$ be a point at which $H_K(\mathsf{u}',\mathsf{u}'',x)$, $P_0(\mathsf{u}'')$, and $H(\mathsf{u}',\mathsf{u}'',x)$ are differentiable. The set of such $\bar{\mathsf{u}}''$ has full measure in $\mathbb{S}$ because all three functions are Lipschitz continuous in $\mathbb{S}$.

Note that if

$$H(\mathsf{u}',\bar{\mathsf{u}}'',x) \leq P_0(\bar{\mathsf{u}}'') - K,$$

then, owing to the differentiability of $H_K(\mathsf{u}',\mathsf{u}'',x)$ ($\geq P_0(\mathsf{u}'') - K$) and $P_0(\mathsf{u}'')$ at $\bar{\mathsf{u}}''$, we have $D_{\mathsf{u}''}H_K(\mathsf{u}',\bar{\mathsf{u}}'',x) = D_{\mathsf{u}''}P_0(\bar{\mathsf{u}}'')$. In that case the left-hand side of (2.9) with H_K in place of H and $\bar{\mathsf{u}}''$ in place of u'' is just K owing to the fact that P_0 is positive homogeneous of degree one.

In case

$$H(\mathsf{u}',\bar{\mathsf{u}}'',x) > P_0(\bar{\mathsf{u}}'') - K,$$

due to (2) we have

$$\kappa|\bar{\mathsf{u}}''| < K_0|\mathsf{u}'| + \bar{H}_K + K, \tag{3}$$

and also $H_K(\mathsf{u}',\mathsf{u}'',x) = H(\mathsf{u}',\mathsf{u}'',x)$ in a neighborhood of $\bar{\mathsf{u}}''$, so that the left-hand side of (2.9) with H_K in place of H and $\bar{\mathsf{u}}''$ in place of u'' becomes

$$\big|H(\mathsf{u}',\bar{\mathsf{u}}'',x) - \bar{\mathsf{u}}''_{ij}D_{\mathsf{u}''_{ij}}H(\mathsf{u}',\bar{\mathsf{u}}'',x)\big| = \big|H_K(\mathsf{u}',\bar{\mathsf{u}}'',x) - \bar{\mathsf{u}}''_{ij}D_{\mathsf{u}''_{ij}}H(\mathsf{u}',\bar{\mathsf{u}}'',x)\big| =: I$$

Because Assumptions 2.4 (i), (ii) are satisfied for H_K, we have

$$I \leq \big|H_K(\mathsf{u}',\bar{\mathsf{u}}'',x)\big| + N|\bar{\mathsf{u}}''_{ij}|$$

$$\leq \big|H_K(\mathsf{u}',\bar{\mathsf{u}}'',x) - H_K(\mathsf{u}',0,x)\big| + K_0|\mathsf{u}'| + \bar{H}_K + N\big|\bar{\mathsf{u}}''_{ij}\big|$$

$$\leq K_0|\mathsf{u}'| + \bar{H}_K + N\big|\bar{\mathsf{u}}''_{ij}\big|,$$

where the constants N depend only on δ and d. It follows in light of (3) that in this case

$$I \leq N(K_0+1)\big(|\mathsf{u}'| + K + \bar{H}_K\big),$$

where the constant N depends only on δ and d. This proves our claim.

Now to finish the proof of the lemma it only remains to refer to Lemma 1. The lemma is proved. □

3. REMARK. One might think that in the above proof we could take

$$H_K(\mathsf{u},x) = \max\big(H(\mathsf{u},x), P(\mathsf{u}'') - K\big)$$

replacing P_0 with P. However, then H_K would satisfy Assumption 2.4 (i) with $\bar{\delta}$ in place of $\delta/2$ and this would make Lemma 1 inapplicable.

The following lemma concludes the proof of Theorem 6.20.

4. LEMMA. *Under the assumptions of Theorem* 6.20 *equation* (0.1) *with boundary condition $v = g$ on $\partial\Omega$ has a unique solution $v \in C(\bar{\Omega}) \cap C^{1,1}_{\text{loc}}(\Omega)$ such that assertion (iii) of Theorem* 6.3 *holds true.*

Proof. We only need to show that Assumption 2.4 (iv) about the x-derivative of H can be dropped in Lemma 2. Note again that

$$\big|H(\mathsf{u},x)\big| \leq \big|H(\mathsf{u},x) - H(\mathsf{u}',0,x)\big| + K_0|\mathsf{u}'| + \bar{H}$$

$$\leq \bar{H} + N(K_0,d,\delta)|\mathsf{u}|. \tag{4}$$

Take a nonnegative $\zeta \in C_0^\infty(B_1)$, which integrates to one and introduce $H^n(\mathsf{u},x)$ as the convolution of $H(\mathsf{u},x)$ and $n^d\zeta(nx)$ performed with respect to x. Observe that H^n satisfies Assumptions 6.2 (i) and 2.4 (ii) with the same constants δ and K_0, whereas

$$\big|H^n(\mathsf{u},x) - H^n(\mathsf{u},y)\big| \leq n|B_1|\,|x-y| \sup_z \big|H(\mathsf{u},z)\big| \sup |D\zeta|,$$

where $|B_1|$ is the volume of B_1, and (2.8) is satisfied due to (4) (with N' of course depending on n).

By Lemma 2, for any n, there exists a unique $C(\bar{\Omega}) \cap C^{1,1}_{\text{loc}}(\Omega)$-solution of (0.1) in Ω (a.e.) with boundary condition $v^n = g$ on $\partial\Omega$ for which assertion (iii) of Theorem 6.3 holds true.

By Remark 6.18 the functions v^n are bounded by the same constant. Hence, by renumbering v^n, if necessary, we may assume that there is a function v, such that $v^n \to v$ uniformly in $\bar{\Omega}$. Then, of course, $v \in C(\bar{\Omega}) \cap C^{1,1}_{\text{loc}}(\Omega)$ and the estimates (6.3) hold true.

It still remains to prove that v satisfies (0.1), and that it is the only function in $C(\bar{\Omega}) \cap C^{1,1}_{\text{loc}}(\Omega)$ satisfying (0.1) with the given boundary data.

Observe that for any $n \geq m \geq 1$ the function v^n satisfies (6.16), with

$$\check{H}^m(\mathsf{u},x) = \sup_{n \geq m} H^n(\mathsf{u},x). \tag{5}$$

Several properties of $\check{H}^m(\mathsf{u},x)$ are to be noted. For any $\mathsf{u} \in \mathbb{R}^{d+1} \times \mathbb{S}$, by the Lebesgue differentiation theorem, there exists a set $\Gamma_{\mathsf{u}} \subset \mathbb{R}^d$ of full measure such that

$$\lim_{m\to\infty} \check{H}^m(\mathsf{u},x) = H(\mathsf{u},x) \quad \forall x \in \Gamma_{\mathsf{u}}. \tag{6}$$

Furthermore, $H(\mathsf{u},x)$ and $H^m(\mathsf{u},x)$ are Lipschitz continuous with respect to u with a constant independent of x and m in light of the assumptions of Theorem 6.20. Therefore, choosing a countable everywhere dense set $\{\mathsf{u}(1),\mathsf{u}(2),...\} \subset \mathbb{R}^{d+1} \times \mathbb{S}$ and setting

$$\Gamma = \bigcap_k \Gamma_{\mathsf{u}(k)}$$

we have that Γ has full measure and (6) holds for any $x \in \Gamma$ and $\mathsf{u} \in \mathbb{R}^{d+1} \times \mathbb{S}$.

By an elliptic counterpart of Lemma 4.2.4, the operators we are dealing with are $\bar{\delta}$-nondegenerate operators of $\mathcal{L}$-type in Ω. Therefore, due to (6.16), an elliptic version of Theorem 4.2.6, applied to subdomains of Ω in which all $C^{1,1}$-norms of v^n are bounded, yields that for any m

$$\max\big(\check{H}^m[v], P[v] - K\big) \geq 0 \tag{7}$$

in Ω (a.e.). We denote by Γ_0 the subset of Ω on which (7) holds for any m and then by sending $m \to \infty$ and taking into account what was said about (6) we conclude that

$$\max\big(H[v], P[v] - K\big) \geq 0 \tag{8}$$

on $\Gamma \cap \Gamma_0$, and, in particular, almost everywhere in Ω.

The inequality $\max\big(H[v], P[v] - K\big) \leq 0$ is obtained similarly starting with

$$\underline{H}^m(\mathsf{u}, x) = \inf_{n \geq m} H^n(\mathsf{u}, x).$$

We thus have proved the existence part of the theorem and the claimed estimates.

To prove uniqueness let $u, v \in C(\bar{\Omega}) \cap C^{1,1}_{\text{loc}}(\Omega)$ satisfy (0.1) in Ω and $u = v = g$ on $\partial\Omega$. If $u \not\equiv v$, then one of the sets $\Omega' = \Omega \cap \{u > v\}$ or $\Omega \cap \{v > u\}$ is nonempty. Due to symmetry assume that $\Omega' \neq \emptyset$. On the boundary of Ω', obviously, $u = v$. Furthermore, since $H(\mathsf{u}, x)$ is a nonincreasing function of u_0', we have of Ω' that

$$H[u](x) \leq H\big(v(x), Du(x), D^2u(x), x\big) =: H^{(v)}[u](x)$$

and in Ω' (a.e.)

$$\max\big(H^{(v)}[v], P[v] - K\big) = 0 = \max\big(H[u], P[u] - K\big)$$

$$\leq \max\big(H^{(v)}[u], P[u] - K\big).$$

By the maximum principle, $u \leq v$ in Ω'. This is a contradiction and the lemma is proved. $\square$

CHAPTER 7

Finite-difference equations of parabolic type

The exposition in this chapter follows [**29**].

1. Solvability of parabolic finite-difference equations

Fix some constants $m, d \in \{1, 2, ...\}$, $\delta \in (0, 1]$, and $T \in (0, \infty)$. Also fix a bounded domain $\Omega \subset \mathbb{R}^d$, recall that, for sufficiently small $h > 0$,

$$\Omega^h = \{x : \rho_\Omega(x) > h\}, \quad \rho_\Omega(x) = \operatorname{dist}(x, \mathbb{R}^d \setminus \Omega), \quad \partial_h \Omega = \bar{\Omega} \setminus \Omega^h$$

and introduce

$$\Pi = [0, T) \times \Omega, \quad \Pi_h = [0, T) \times \Omega^h, \quad \partial'_h \Pi = \bar{\Pi} \setminus \Pi_h.$$

1. Assumption. We are given vectors $l_i \in \bar{B}_1 \subset \mathbb{R}^d$, $i = \pm 1, ..., \pm m$, such that $l_{-i} = -l_i$.

2. Assumption. (i) We are given a function $\mathcal{H}(z, t, x)$, $(t, x) \in [0, \infty) \times \mathbb{R}^d$,

$$z = (z', z''), \quad z' = (z'_0, z'_{\pm 1}, ..., z'_{\pm m}) \in \mathbb{R}^{2m+1}, \quad z'' = (z''_{\pm 1}, ..., z''_{\pm m}) \in \mathbb{R}^{2m},$$

which is Lipschitz continuous with respect to z for any (t, x) and measurable with respect to t for any z, x. At all point of its differentiability with respect to z for $j = \pm 1, ..., \pm m$ introduce

$$a_j = a_j(z, t, x) = D_{z''_j} \mathcal{H}(z, t, x), \quad b_j = b_j(z, t, x) = D_{z'_j} \mathcal{H}(z, t, x),$$

$$c = c(z, x) = -D_{z'_0} \mathcal{H}(z, t, x),$$

and at points of non-differentiability set $a_j = \delta$, $b_j = 0$, $c = 1$.

(ii) The above introduced functions and $f(t, x) := \mathcal{H}(0, t, x)$ are bounded, $|b_j| \leq N'$, $|j| = 1, ..., m$, where N' is a constant, $|c| \leq N'$, and

$$\delta^{-1} \geq a_j \geq \delta, \tag{1}$$

for all indices and values of the arguments.

Note that in the parabolic case we have no restrictions on the sign of c.

For any function v on $[0, \infty) \times \mathbb{R}^d$ and $h > 0$ define

$$\mathcal{H}_h[v](t, x) = \mathcal{H}\big(v, \delta_{h, l_{\pm 1}} v, ..., \delta_{h, l_{\pm m}} v, \Delta_{h, l_{\pm 1}} v, ..., \Delta_{h, l_{\pm m}} v, t, x\big). \tag{2}$$

For a function g given on $\bar{\Pi}$ we say that its derivative $\partial_t g$ is bounded in $\bar{\Pi}$ if g is Lipschitz continuous with respect to t with constant independent of x. In this case, indeed, for any $x \in \bar{\Omega}$ the function $g(t, x)$ is absolutely continuous in $[0, T]$ with respect to t and its derivative $\partial_t g(t, x)$, which exists for almost any $t \in (0, T)$, is bounded by the Lipschitz constant mentioned above.

Fix a bounded measurable function g on $\bar{\Pi}$, whose derivative $\partial_t g$ is also bounded in $\bar{\Pi}$, and consider the equation

$$\partial_t u + \mathcal{H}_h[u] = 0 \quad \text{in} \quad \Pi_h \tag{3}$$

with boundary condition

$$u = g \quad \text{on} \quad \partial'_h \Pi. \tag{4}$$

We understand (3) in the sense of the following definition.

3. DEFINITION. For two functions u and f given on Π_h we write

$$\partial_t u \geq f \quad \text{in} \quad \Pi_h$$

if for each $x \in \Omega^h$, $u(t,x)$ is absolutely continuous in $[0,T]$ and its derivative with respect to t is greater than or equal to $f(t,x)$ in $(0,T)$ (a.e.). Similarly we understand the relations $\partial_t u \leq f$ and $\partial_t u = f$ in Π_h.

According to Definition 3 equation (3) with condition $u(T,x) = g(T,x)$ means that for any $(t,x) \in [0,T] \times \Omega^h$

$$u(t,x) = g(T,x) + \int_t^T \mathcal{H}_h[u](s,x)\, ds.$$

4. THEOREM. *Under the above assumptions, for any $h > 0$ such that $\Omega^h \neq \emptyset$, problem (3)-(4) has a unique bounded solution v_h, whose derivative $\partial_t v_h$ is also bounded in $\bar{\Pi}$.*

Proof. Observe that problem (3)-(4) is equivalent to the integral equation

$$u(t,x) = g(T,x) + \int_t^T \left[I_{\Pi_h} \mathcal{H}_h[u] - I_{\partial'_h \Pi} \partial_s g \right](s,x)\, ds \tag{5}$$

considered now in $\bar{\Pi}$. In view of the Lipschitz continuity of $\mathcal{H}$ with respect to z and Picard's method of successive iterations, this integral equation has a unique bounded solution $v = v_h$. Obviously, $\partial_t v_h$ is also bounded. The theorem is proved. □

5. THEOREM (comparison principle). *There is a constant $h_0 > 0$, depending only on Ω, δ, and N', such that, if $h \in (0, h_0]$ and we are given two bounded functions u and v on $\bar{\Pi}$ with bounded derivatives $\partial_t u$ and $\partial_t v$ such that $u \leq v$ on $\partial'_h \Pi$ and*

$$\partial_t u(t,x) + \mathcal{H}_h[u](t,x) \geq \partial_t v(t,x) + \mathcal{H}_h[v](t,x)$$

in Π_h, then $u \leq v$ in $\bar{\Pi}$.

Proof. Set $w = u - v$. By Lemma 4.1.6, for any $x \in \Omega^h$, there exist measurable in t functions $a_k = a_k(t,x)$, $b_k = b_k(t,x)$ and $c = c(t,x)$ such that $\delta^{-1} \geq a_k \geq \delta$, $|b_k|, |c| \leq N'$ and in Π_h

$$\partial_t w + a_k \Delta_{h,l_k} w + b_k \delta_{h,l_k} w - cw =: f \geq 0.$$

Therefore, if N is a constant, $\hat{w}(t,x) = w(t,x)e^{-Nt}$ satisfies

$$\partial_t \hat{w} + \mathcal{L}\hat{w} = \hat{f}$$

where $\hat{w} = we^{-Nt}$, $\hat{f} = fe^{-Nt}$, and

$$\mathcal{L}\phi := a_k \Delta_{h,l_k}\phi + b_k \delta_{h,l_k}\phi + (N-c)\phi.$$

Observe that for $h \le \delta/N'$ and $N \ge 2m\delta^{-1}h^{-2} + 2mN'h^{-1} + N'$ all the coefficients of $\phi(t, x+hl_k)$ and $\phi(t,x)$ in the expression $\mathcal{L}\phi(t,x)$ are nonnegative. Furthermore, similarly to (5) the function $\hat{w}$ satisfies

$$\hat{w}(t,x) = \hat{w}(T,x) + \int_t^T \big(I_{\Pi_h}[\mathcal{L}\hat{w} - \hat{f}] - I_{\partial_h'\Pi}\partial_s \hat{w}\big)(s,x)\,ds.$$

We solve this equation by the method of iterations setting $\hat{w}^0 = 0$ and for $n = 0, 1, \ldots,$ defining

$$\hat{w}^{n+1}(t,x) = \hat{w}(T,x) + \int_t^T \big(I_{\Pi_h}[\mathcal{L}\hat{w}^n - \hat{f}] - I_{\partial_h'\Pi}\partial_s \hat{w}\big)(s,x)\,ds.$$

Notice that if $\hat{w}^n \le 0$, then, by the above mentioned property of $\mathcal{L}$ and the assumption that $\hat{w} \le 0$ on $\partial_h'\Pi$, we have

$$\hat{w}^{n+1}(t,x) \le \hat{w}(T,x) - I_{\partial_h \Omega}(x)\int_t^T \partial_s \hat{w}(s,x)\,ds$$

$$= \hat{w}(T,x) - I_{\partial_h\Omega}(x)\big[\hat{w}(T,x) - \hat{w}(t,x)\big] = I_{\Omega^h}(x)\hat{w}(T,x) + I_{\partial_h\Omega}(x)\hat{w}(t,x) \le 0.$$

Thus $\hat{w}^{n+1} \le 0$ if $\hat{w}^n \le 0$. Since $\hat{w}^0 \le 0$, all $\hat{w}^n \le 0$ and since they converge to $\hat{w}$, it is also negative, which proves the theorem. □

The following result is obtained from Lemma 4.1.2 by deleting there the parameter q.

6. LEMMA. *Fix a constant $K_0 \in [0,\infty)$ and in addition to Assumptions* 1 *and* 2 *suppose that the number*

$$\bar{\mathcal{H}} = \sup_{(t,x)\in\Pi, z'} \big(|\mathcal{H}(z',0,t,x)| - K_0|z'|\big)$$

is finite. Let ϕ be a function on Π. Then for any $h > 0$, such that $\Pi_h \ne \emptyset$, there exist functions a_j, b_j, c, f, $|j| = 1,\ldots,m$, on Π_h such that on Π_h

$$\delta \le a_j \le \delta^{-1}, \quad |b_j| \le K_0, \quad |c| \le K_0, \quad |f| \le \bar{\mathcal{H}},$$

$$\mathcal{H}_h[\phi] = a_k \Delta_{h,l_k}\phi + b_k\delta_{h,l_k}\phi - c\phi + f.$$

Here is a counterpart of Theorem 5.2.8.

7. THEOREM. *Under the assumption of Lemma* 6 *for $h \in (0, h_0]$ we have*

$$\sup_{\bar{\Pi}} |v_h| \le e^{K_0 T}\big(T\bar{\mathcal{H}} + \sup_{\partial_h'\Pi}|g|\big).$$

Proof. Introduce

$$\psi(t,x) = e^{K_0(T-t)}\big((T-t)\bar{\mathcal{H}} + \sup_{\partial_h'\Pi}|g|\big).$$

By using Lemma 6 we see that in Π_h

$$\partial_t\psi + \mathcal{H}_h[\psi] \le \partial_t\psi + K_0\psi + \bar{\mathcal{H}} \le 0.$$

Furthermore, $\psi \geq v_h = g$ on $\partial_h' \Pi$. By Theorem 5 we find that $v_h \leq \psi$ in $\bar{\Pi}$. Similarly one gets $v_h \geq -\psi$ in $\bar{\Pi}$. The theorem is proved. □

To understand better the motivation of the subsequent sections of this chapter we advise the reader to go back to Section 5.3.

2. Some estimates for second-order finite-differences of solutions to parabolic equations with constant coefficients

Take an $h \in (0, 1]$, let $m \geq 1$ be an integer and let $l_{-m}, ..., l_{-1}, l_1, ..., l_m$ be some fixed vectors in $\mathbb{R}^d$ such that

$$l_{-k} = -l_k.$$

Next denote

$$\Lambda = \{l_k : k = \pm 1, ..., \pm m\}, \tag{1}$$

$$\Lambda_1 = \Lambda, \quad \Lambda_{n+1} = \Lambda_n + \Lambda, \quad n \geq 1, \quad \Lambda_\infty = \bigcup_n \Lambda_n, \quad \Lambda_\infty^h = h\Lambda_\infty. \tag{2}$$

Let Q^o be a nonempty bounded subset of $\mathbb{R} \times \Lambda_\infty^h$, which is open in the relative topology of $\mathbb{R} \times \Lambda_\infty^h$ and is such that its projection on Λ_∞^h is a finite set. Introduce $\hat{Q}^o$ as the set of points $(t_0, x_0) \in \mathbb{R} \times \Lambda_\infty^h$ for each of which there exists a sequence $t_n \uparrow t_0$ such that $(t_n, x_0) \in Q^o$. Observe that $Q^o \subset \hat{Q}^o$. Also define

$$Q = \hat{Q}^o \cup \big\{(t, x + h\Lambda) : (t, x) \in Q^o\big\}. \tag{3}$$

Let $A = \big\{\alpha = (a, c)\big\}$ be a closed bounded set in $\mathbb{R}^{2m} \times \mathbb{R}$ such that

$$a = (a_{-m}, a_{-m+1}, ..., a_{-1}, a_1, ..., a_m) \in \mathbb{R}^{2m}, \quad c \in \mathbb{R}.$$

Also let $f(\alpha, t, x)$ be a real-valued function defined for $\alpha \in A$ and $(t, x) \in \mathbb{R}^{d+1}$. In the main case of applications of the results of this section f is just a constant.

Fix an $r \in \{1, ..., m\}$, recall (1.1.1), and for brevity of notation, for $k = \pm 1, ..., \pm m$, set

$$\delta_{h,k} = \delta_{h,l_k} = \delta_k, \quad \Delta_{h,k} = \Delta_{h,l_k} = \Delta_k.$$

1. ASSUMPTION. There are constants $\delta \in (0, 1]$ and $K_1, K_2 \in [0, \infty)$ such that

(i) For any $(a, c) \in A$ and all k we have

$$a_k = a_{-k}, \quad \delta \leq a_k \leq \delta^{-1}, \quad c \geq 0;$$

(ii) The function f is continuous in α for any (t, x), continuous in t uniformly with respect to α for any x, and on $\mathbb{R}^{d+1}$

$$|\delta_r f| \leq K_1, \quad \Delta_r f \geq -K_2.$$

For $z = \big(z_0', z''\big)$ with

$$z_0' \in \mathbb{R}, \quad z'' = \big(z_{-m}'', ..., z_{-1}'', z_1'', ..., z_m''\big) \in \mathbb{R}^{2m},$$

introduce

$$\mathcal{P}(z, t, x) = \max_{\alpha=(a,c)\in A} \Big(\sum_{|k|=1}^{m} a_k z_k'' - c z_0' + f(\alpha, t, x) \Big).$$

For any function u on $\mathbb{R}^{d+1}$ define

$$\mathcal{P}[u](t,x) = \mathcal{P}_h[u](t,x) = \mathcal{P}\big(u(t,x), \delta_h^2 u(t,x), t, x\big), \tag{4}$$

where δ_h^2 is introduced in (5.2.4).

Next, take a function $\eta \in C_0^\infty(\mathbb{R}^d)$, such that $|\eta| \le 1$ and set $\zeta = \eta^2$,

$$|\eta'(x)| = |\eta'(x)|_h = \sup_k |\delta_k \eta(x)|, \quad |\eta''(x)| = |\eta''(x)|_h = \sup_k |\Delta_k \eta(x)|,$$

$$\|\eta'\| = \|\eta'\|_h = \sup_{\Lambda_\infty^h} |\eta'|_h, \quad \|\eta''\| = \|\eta''\|_h = \sup_{\Lambda_\infty^h} |\eta''|_h.$$

Finally, let u be a function on $\mathbb{R}^{d+1}$, which is continuously differentiable with respect to t for any $x \in \mathbb{R}^d$ and satisfies

$$\partial_t u + \mathcal{P}[u] = 0 \quad \text{in } Q^o \tag{5}$$

and

$$\partial_t u + \mathcal{P}[u] \le 0 \quad \text{on } Q \setminus Q^o. \tag{6}$$

2. REMARK. Owing to the continuity assumptions, (5) also holds in $\bar{Q}^o$ and (6) holds in Q^o, $Q \setminus Q^o$, Q and $\bar{Q}$.

Here is the main result of the section providing a one-sided interior estimate of pure second-order differences of u.

3. THEOREM. *There exists a constant $N = N(m, \delta) \ge 1$ such that for any constant ν satisfying*

$$\nu \ge N\big(\|\eta''\| + \|\eta'\|^2\big),$$

we have in Q that

$$\zeta^2\big[(\Delta_r u)^-\big]^2 \le \sup_{Q\setminus Q^o} \zeta^2\big[(\Delta_r u)^-\big]^2 + N\nu \bar{W}_r + N\nu^{-2}K_2^2 + N\nu^{-1}K_1^2, \tag{7}$$

where

$$\bar{W}_r = \sup_Q \big(|\delta_r u|^2 + |\delta_{-r} u|^2\big).$$

The proof of the theorem is based on a lemma before which we postulate that below no summation with respect to r is performed. The number r is fixed at the beginning of the section. Also for simplicity of notation set

$$u_{rr} = \Delta_r u, \quad u_r = \delta_r u, \quad u_{kr} = -\delta_{-k}\delta_r u,$$

define

$$u_{rr}^- = (u_{rr})^-,$$

and for a constant $\nu \geq 0$ introduce an operator (recall that r is fixed)

$$L_\nu \phi = \zeta^2 u^-_{rr} \Delta_r \phi - \nu \zeta u_r \delta_r \phi.$$

Observe that

$$L_\nu u = -\zeta^2 \left(u^-_{rr}\right)^2 - \nu \zeta u_r^2 =: -V_\nu. \tag{8}$$

4. LEMMA. *There exists $N = N(m, \delta) \geq 1$ such that if*

$$\nu \geq N\left(\|\eta''\| + \|\eta'\|^2\right), \tag{9}$$

then on Q^o for any $\alpha = (a, c) \in A$ we have

$$2L_\nu[\partial_t + a_k \Delta_k]u \geq -[\partial_t + a_k \Delta_k]V_\nu - N\nu^2 \bar{W}_r + (\nu/2)\zeta a_k u^2_{kr}\,. \tag{10}$$

Up to the obvious formula $2L_\nu \partial_t u = -\partial_t V_\nu$ this lemma is identical to Lemma 5.4.5.

Proof of Theorem 3. Denote by N_0 the constants N in Lemma 4 and take and fix a ν satisfying (9) (with N_0 in place of N). As in the proof of Theorem 5.4.2 we convince ourselves that it suffices to prove the theorem under the additional assumption that

$$h \leq \nu^{-1/2}. \tag{11}$$

Let $(t_0, x_0) \in \bar{Q}$ be a point such that

$$V_\nu(t_0, x_0) = \sup_Q V_\nu.$$

If $(t_0, x_0) \notin \bar{Q}^o$, then, as is easy to see, this point can be approximated by points lying in $Q \setminus Q^o$, in which case

$$\sup_Q V_\nu = \sup_{Q \setminus Q^o} V_\nu$$

and (7) follows. Therefore, in the rest of the proof we may assume that

$$(t_0, x_0) \in \bar{Q}^o.$$

Next, consider two cases: 1) $\partial_t V_\nu(t_0, x_0) > 0$, 2) $\partial_t V_\nu(t_0, x_0) \leq 0$. In the first case $(t_0, x_0) \notin Q^o$ and there is no sequence $t_n \downarrow t_0$ such that $(t_n, x_0) \in Q^o$. Hence $(t_0, x_0) \in \hat{Q}^o \setminus Q^o$ and in Q

$$\zeta^2 \left[u^-_{rr}\right]^2 \leq V_\nu(t_0, x_0) \leq \sup_{\hat{Q}^o \setminus Q^o} \zeta^2 \left[(\Delta_r u)^-\right]^2 + \nu \bar{W}_r,$$

so that (7) holds. In the remaining case

$$\partial_t V_\nu(t_0, x_0) \leq 0. \tag{12}$$

Now observe that we may assume that

$$\zeta(x_0) u^-_{rr}(t_0, x_0) > \nu h u_r(t_0, x_0). \tag{13}$$

Indeed, if the opposite inequality holds, then in light of (11) in Q

$$\zeta^2 \left[u^-_{rr}\right]^2 \leq V_\nu(t_0, x_0) \leq \nu^2 h^2 u_r^2(t_0, x_0) + \nu \bar{W}_r \leq 2\nu \bar{W}_r.$$

It follows, as in the proof of Theorem 5.4.2, that, owing to (13), the operator $L_\nu : \phi \to L\phi(t_0, \cdot)$, respects the maximum principle at x_0 relative to $x_0 + h\Lambda$. We are going to apply this fact to a special function. Notice that we can find an $\hat{\alpha} = (\hat{a}, \hat{c}) \in A$ such that, owing to Remark 2,

$$\partial_t u(t_0, x_0) + \hat{a}_k \Delta_k u(t_0, x_0) - \hat{c} u(t_0, x_0) + f(\hat{\alpha}, t_0, x_0)$$

$$= \partial_t u(t_0, x_0) + \mathcal{P}[u](t_0, x_0) = 0.$$

Since $\partial_t u + \mathcal{P}[u] \leq 0$ in $\bar{Q}$, we have that

$$\phi(t_0, x) := \partial_t u(t_0, x) + \hat{a}_k \Delta_k u(t_0, x) - \hat{c} u(t_0, x) + f(\hat{\alpha}, t_0, x) \leq 0$$

for $x \in x_0 + h\Lambda$. Hence, $0 \geq 2L_\nu \phi(t_0, x_0)$, which owing to (8), (10), and (12) yields

$$0 \leq [\partial_t + \hat{a}_k \Delta_k - 2\hat{c}] V_\nu(t_0, x_0) - (\nu/2) \zeta \hat{a}_k u_{kr}^2(t_0, x_0)$$

$$+ N\nu^2 \bar{W}_r - 2L_\nu f(\hat{\alpha}, \cdot)(t_0, x_0)$$

$$\leq [\hat{a}_k \Delta_k - 2\hat{c}] V_\nu(t_0, x_0) - (\nu/2) \zeta \hat{a}_k u_{kr}^2(t_0, x_0)$$

$$+ N\nu^2 \bar{W}_r - 2L_\nu f(\hat{\alpha}, \cdot)(t_0, x_0).$$

Here the last term is dominated by

$$2K_2 \zeta^2 u_{rr}^-(t_0, x_0) + 2\nu \big| u_r(t_0, x_0) \big| K_1$$

$$\leq N\nu^{-1} K_2^2 + (\nu/4) \zeta \hat{a}_k u_{kr}^2(t_0, x_0) + K_1^2 + \nu^2 \bar{W}_r.$$

Furthermore, since $V_\nu(t_0, x) \geq 0$ attains its maximum at $(t_0, x_0) \in \bar{Q}^o$, we have $(t_0, x_0 + hl_k) \in \bar{Q}$, $V_\nu(t_0, x_0) \geq V_\nu(t_0, x_0 + hl_k)$, and

$$[\hat{a}_k \Delta_k - 2\hat{c}] V_\nu(t_0, x_0) \leq 0.$$

We now conclude that

$$(\nu/4) \zeta \hat{a}_k u_{kr}^2(t_0, x_0) \leq N\nu^2 \bar{W}_r + N\nu^{-1} K_2^2 + K_1^2,$$

which implies that in Q

$$\zeta^2 \left(u_{rr}^-\right)^2 \leq V_\nu(t_0, x_0) \leq N \zeta \hat{a}_k u_{kr}^2(t_0, x_0) + \nu \bar{W}_r$$

$$\leq N\nu \bar{W}_r + N\nu^{-2} K_2^2 + N\nu^{-1} K_1^2.$$

The theorem is proved. □

3. Maximum principle for semi-discrete parabolic equations

We will work in the framework of Section 2 borrowing from there Λ, Λ^h_∞, Q^o, $\hat{Q}^o$, and Q. For $x \in \Lambda^h_\infty$ we denote by $Q^o_{|x}$ the x-section of Q^o:

$$Q^o_{|x} := \{t : (t,x) \in Q^o\}.$$

1. THEOREM. *Let $(a,b,c)(t,x)$ be a bounded $\mathbb{R}^{2m} \times \mathbb{R}^{2m} \times \mathbb{R}$-valued function on $\mathbb{R}^{d+1}$ satisfying*

$$a_k \geq 0, \quad hb^-_k \leq a_k, \quad |k| = 1, ..., m,$$

and let $v(t,x)$ be a bounded function in Q which is absolutely continuous with respect to t on each open interval belonging to $Q^o_{|x}$ for any $x \in \Lambda^h_\infty$ (if $Q^o_{|x} \neq \emptyset$) and satisfies there

$$\partial_t v + \mathcal{L}v := \partial_t v + a_k \Delta_{h,l_k} v + b_k \delta_{h,l_k} v - cv = -\eta$$

(a.e.), where $\eta = \eta(t,x)$ is a bounded function. Redefine v if necessary for $(t,x) \in \hat{Q}^o \setminus Q^o$ so that

$$v(t,x) = \overline{\lim_{s\uparrow t,(s,x)\in Q^o}} v(s,x).$$

Finally, let S be the width of Q^o in the t-direction. Then in Q^o we have

$$v \leq Se^{\bar{c}S} \sup_{Q^o} \eta^+ + e^{\bar{c}S} \sup_{Q\setminus Q^o} v^+,$$

where

$$\bar{c} := \sup c^-.$$

Proof. Without losing generality we assume that $Q^o \subset (0,S) \times \Lambda^h_\infty$. Observe that the function $u(t,x) = v(t,x)e^{\bar{c}t}$ satisfies

$$\partial_t u + a_k \Delta_{h,l_k} u + b_k \delta_{h,l_k} u - (c + \bar{c})u = -\eta e^{\bar{c}t},$$

where $c + \bar{c} \geq 0$. It is seen that if we can prove the lemma under the additional assumption that $c \geq 0$, then it is also true in the general case. Below in the proof we assume that $c \geq 0$, so that $\bar{c} = 0$.

Then by considering

$$v(t,x) - (T-t)\big[2\varepsilon + \sup_{Q^o} \eta^+\big],$$

where $\varepsilon > 0$, and then sending $\varepsilon \downarrow 0$, we reduce the general case to the one with $\eta \leq -2\varepsilon$. Finally, we make one more harmless assumption that

$$\bar{v} := \sup_Q v > 0.$$

After that take a sequence $(t_n, x_n) \in Q$ such that

$$v(t_n, x_n) \to \bar{v}.$$

If infinitely many points $(t_n, x_n) \notin Q^o$, then we have nothing to prove.

In the opposite case we may assume that $x_n = x_0$, $(t_n, x_n) \in Q^o$ for all n, and the sequence t_n converges, say to t_0. Denote by I_n the connected component (open interval) of $Q^o_{|x_0}$ containing t_n. By using subsequences if needed and taking into account the continuity of v in Q^o, we come to three possibilities: either $(t_0, x_0) \in$

$\hat{Q}^o \setminus Q^o$ and we have nothing to prove, or $(t_0, x_0) \in Q^o$, or else $t_n \downarrow t_0$. Note that the second case can be reduced to the third one by redefining the t_n's.

If the third possibility realizes, we claim that

$$\lim_{n\to\infty} |I_n| = 0, \tag{1}$$

where $|I_n|$ is the length of I_n.

Indeed if (1) fails, then for all large n the intervals I_n coincide. Also in that case there is an open interval $I \in \mathbb{R}$ such that

$$I \times \{x_0\} \subset Q^o, \quad I \times \{x_0 + h\Lambda\} \subset Q.$$

Furthermore, $\partial_t v(t, x_0)$ is bounded on I, so that the limit of $v(t, x_0)$ as $t \downarrow t_0$ exists and

$$\lim_{t\downarrow t_0} v(t, x_0) = \lim_{n\to\infty} v(t_n, x_0) = \bar{v}.$$

In addition, $\bar{v} \geq v(t, x_0)$ for $t > t_0$ and, since

$$\partial_t v(t, x_0) = -\mathcal{L}v(t, x_0) - \eta(t, x_0)$$

for almost all $t \in I$, there exists a sequence of points $s_n \in I$ such that $s_n \downarrow t_0$ and

$$\mathcal{L}v(s_n, x_0) + \eta(s_n, x_0) \geq -\varepsilon$$

implying that (recall that $\eta \leq -2\varepsilon$)

$$\mathcal{L}v(s_n, x_0) \geq \varepsilon. \tag{2}$$

Next, consider the functions $v_n(x) = v(s_n, x)$, $x \neq x_0$, $v_n(x_0) = \bar{v}$, for which $v_n(x_0) \geq v_n(x)$ for all $x \in x_0 + h\Lambda$. On the one hand, by the maximum principle we have $\mathcal{L}v_n(s_n, x_0) \leq 0$ and, on the other hand,

$$0 \geq \mathcal{L}v_n(s_n, x_0) = \mathcal{L}v(s_n, x_0) + \xi_n,$$

where

$$\xi_n = 2h^{-2} \sum_{|k|=1}^{m} a_k(s_n, x_0)\big[v(s_n, x_0) - \bar{v}\big]$$

$$+h^{-1} \sum_{|k|=1}^{m} b_k(s_n, x_0)\big[v(s_n, x_0) - \bar{v}\big] + c(s_n, x_0)\big[v(s_n, x_0) - \bar{v}\big]$$

and $\xi_n \to 0$ as $n \to \infty$. This leads to a contradiction with (2) and proves (1).

It follows that for infinitely many n, as n increases, the value of v at (t_n, x_0) will become closer and closer to its value at the right end points of I_n's since the time derivative of v is bounded and this proves the theorem. □

2. REMARK. It was very tempting to replace ∂_t with its finite-difference approximation in the previous sections, thus considering fully discrete approximations instead of semi-discrete ones. In that situation one would have results very similar to those in Section 1 and 2 with even shorter proofs and a much simpler statement replacing Theorem 1, that (Theorem 1) will be used in a very substantial way while estimating the time derivative of solutions to certain equations. Unfortunately, fully discrete approximations seem to allow only one-sided estimate of the finite-difference approximation of the time derivative.

4. Estimates for finite-difference parabolic equations with variable coefficients

Fix some constants $K_0, K \in [0, \infty)$ and $\delta \in (0, 1]$. We use the notation and assumptions introduced in the beginning of Section 1, however, we append Assumptions 1.1 and 1.2 with the following.

1. ASSUMPTION. The coordinates of $l_j, |j| = 1, ..., m$, are rational numbers and $\text{Span}\,\Lambda = \text{Span}\,\{l_i, |i| = 1, ..., m\} = \mathbb{R}^d$.

2. ASSUMPTION. (i) The function $\mathcal{H}(z, t, x)$ is continuous with respect to t for any z, x;

(ii) The number

$$\bar{\mathcal{H}} = \sup_{(t,x),z'} \left(|\mathcal{H}(z', 0, t, x)| - K_0|z'|\right)$$

is finite.

We also impose the following, in which we write $g \in W^{1,2}_\infty(\mathbb{R}^{d+1})$ to mean that g is a bounded continuous function whose generalized derivatives $\partial_t g$, Dg, and $D^2 g$ are bounded on $\mathbb{R}^{d+1}$. The norm in $W^{1,2}_\infty(\mathbb{R}^{d+1})$ is introduced in an obvious way.

3. ASSUMPTION. The domain Ω is bounded and satisfies the exterior ball condition. We are given a function $g \in W^{1,2}_\infty(\mathbb{R}^{d+1})$.

For $z'' \in \mathbb{R}^{2m}$ introduce $\mathcal{P}(z'') = \mathcal{P}_{m,\delta/2}(z'')$ by (5.5.1). Observe that similarly to (5.5.2)

$$\mathcal{H}(z, t, x) \leq \mathcal{P}(z'') - (\delta/2) \sum_{|k|=1}^{m} |z''_k| + K_0|z'| + \bar{\mathcal{H}}. \tag{1}$$

We need one more function. Set

$$\mathcal{H}_K(z, t, x) = \max\left(\mathcal{H}(z, t, x), \mathcal{P}(z'') - K\right).$$

Recall that $\delta_h^2 v$ and $\delta_h v$ are defined in (5.2.3) and (5.2.4) and for any function v on Π define

$$\mathcal{H}_{K,h}[v](t, x) = \mathcal{H}_K\left(v(t, x), \delta_h v(t, x), \delta_h^2 v(t, x), t, x\right).$$

We will concentrate on sufficiently small h such that $\Pi_h \neq \emptyset$ and will consider the equation

$$\partial_t v + \mathcal{H}_{K,h}[v] = 0 \quad \text{in} \quad \Pi_h \tag{2}$$

with boundary condition

$$v = g \quad \text{on} \quad \partial'_h \Pi. \tag{3}$$

Of course, we understand (2) in the sense of Definition 1.3.

By Theorem 1.4 and Corollary 4.1.4, for any sufficiently small $h > 0$, there exists a unique bounded solution $v = v_h$ of (2)–(3). We do not include K in the notation v_h since K is a fixed number and below only consider sufficiently small $h > 0$.

4. REMARK. One of consequences of Assumption 2 (i) is that $\mathcal{H}_{K,h}[v_h](t, x)$ is continuous with respect to t for any $x \in \Omega^h$ and, owing to (2), the function $\partial_t v(t, x)$ is continuous in $[0, T]$ for any $x \in \Omega^h$.

The solution of the PDE version of (2)–(3) will be obtained as the limit of a subsequence of v_h as $h \downarrow 0$. Therefore, we need to have appropriate bounds on v_h, on its first- and second-order differences in x, and on $\partial_t v_h$.

Observe that owing to Assumption 2

$$-K_0|z'| - \bar{\mathcal{H}} \le \mathcal{H}_K(z', 0, t, x) = \max\big(\mathcal{H}(z', 0, t, x), -K\big) \le K_0|z'| + \bar{\mathcal{H}},$$

so that $\bar{\mathcal{H}}_K \le \bar{\mathcal{H}}$. Therefore, the following is a direct consequence of Theorem 1.7 and Corollary 4.1.4.

5. Lemma. *There is $h_0 > 0$ such that for $h \in (0, h_0]$*

$$\sup_{\bar{\Pi}} |v_h| \le e^{K_0 T}\big(T\bar{\mathcal{H}} + \|g\|_{C(\partial'_h \Pi)}\big). \tag{4}$$

Lemma 1.6 and Corollary 4.1.4 also yield the following useful result (cf. Remark 6.6.16).

6. Lemma. *Let ϕ be a function on Π. Then for any $h \in \big(0, \rho_{\text{int}}(\Omega)\big)$ there exist functions a_j, b_j, c, f, $|j| = 1, ..., m$, on Π_h such that on Π_h*

$$\delta/2 \le a_j \le 2\delta^{-1}, \quad |b_j| \le K_0, \quad |c| \le K_0, \quad |f| \le \bar{\mathcal{H}}, \tag{5}$$

$$\mathcal{H}_{K,h}[\phi] = a_k \Delta_{h,l_k}\phi + b_k \delta_{h,l_k}\phi - c\phi + f.$$

Below by h_0 and N with occasional indices we denote various (finite) constants depending only on $\text{diam}(\Omega)$, $\rho_{\text{ext}}(\Omega)$, $\rho_{\text{int}}(\Omega)$, $\Lambda = \{l_i\}$, K_0, d, δ, unless stated otherwise. Thus, the constants N are independent of T, and this fact plays a big role in obtaining some results for elliptic equations from what we can prove for parabolic ones (see the proof of Theorem 6.6.5 on page 162).

Now comes a counterpart of Lemma 5.5.8.

7. Lemma. *There are constants $h_0 > 0$ and N such that for all $h \in (0, h_0]$*

$$|v_h - g| \le N\big(\bar{\mathcal{H}} + \|g\|_{W^{1,2}_\infty(\mathbb{R}^{d+1})} + \sup_{\bar{\Pi}} |v_h|\big)\rho_\Omega \tag{6}$$

on $\bar{\Pi}$.

Proof. As in the proof of Lemma 5.5.8, it suffices to prove that (6) holds at points $(t_0, x_0) \in \Pi_h$ such that $\rho_\Omega(x_0) \le 1$. As there, without loss of generality, on the account of moving the origin, we may assume that the origin is at the distance $\rho_{\text{ext}}(\Omega)$ from Ω and the straight line passing through the origin and x_0 crosses $\partial\Omega$ at the closest to x_0 point of $\partial\Omega$.

After that we set

$$M = \bar{\mathcal{H}} + \|g\|_{W^{1,2}_\infty(\mathbb{R}^{d+1})} + \sup_{\bar{\Pi}} |v_h|, \quad \phi = g - 2N_1 M\psi,$$

where ψ is taken from Lemma 5.5.7 and the constant N_1 is such that in Π_h

$$\partial_t g + a_k \Delta_{h,l_k} g + b_k \delta_{h,l_k} g - cg + f \le N_1 M$$

as long as conditions (5) are satisfied.

We increase N_1 if necessary in order to have $v_h \le \phi$ in $[0,T] \times (\bar{\Omega} \setminus B_2)$, the latter being possible owing to the fact that in $[0,T] \times (\bar{\Omega} \setminus B_2)$ we have

$$-\psi \ge \rho_{\text{ext}}^{-\beta}(\Omega) - 2^{-\beta} \ge 1 - 2^{-\beta}.$$

We also choose $h_0 > 0$ in such a way that finite-difference approximations of $D_{l_k}^2 \psi$ and $D_{l_k}\psi$ are sufficiently close to these quantities in $B_3 \setminus B_{\rho_{\text{ext}}(\Omega)}$, so that owing to Lemma 5.5.7

$$a_k \Delta_{h,l_k} \psi + b_k \delta_{h,l_k} \psi - c\psi \ge 1/2$$

in $B_3 \setminus B_{\rho_{\text{ext}}(\Omega)}$ for $h \in (0, h_0]$.

Then for $\Sigma = \Omega \cap B_3$ (where $\psi < 0$) we find that in $[0,T) \times \Sigma$

$$\partial_t \phi + a_k \Delta_{h,l_k} \phi + b_k \delta_{h,l_k} \phi - c\phi + f \le 0 \tag{7}$$

as long as conditions (5) are satisfied. It follows by Lemma 6 that

$$\partial_t \phi + \mathcal{H}_{K,h}[\phi] \le 0.$$

Also by the choice of N_1 (if $h_0 \le 1$), $v_h \le \phi$ on $\partial_h'\big([0,T) \times \Sigma\big)$. Hence by the comparison principle, $v_h \le \phi$ in $[0,T] \times \bar{\Sigma}$. By the choice of N_1, this inequality, actually, holds in $\bar{\Pi}$, which yields the desired estimate of $v_h - g$ from above. Similarly one obtains it from below as well. The lemma is proved. □

We also need to know how fast v_h approaches g at the top lid of Π.

8. LEMMA. *There are constants $h_0 > 0$ and N such that for all $h \in (0, h_0]$*

$$|v_h - g| \le N\big(\bar{\mathcal{H}} + \|g\|_{W_\infty^{1,2}(\mathbb{R}^{d+1})} + \sup_{\bar{\Pi}} |v_h|\big)\rho_T$$

on $\bar{\Pi}$, where

$$\rho_T(t,x) = (T-t)_+.$$

Proof. Take the constants M and N_1 from the proof of Lemma 7 and set

$$\phi = g + 2N_1 M \rho_T.$$

Observe that for this ϕ, as is easy to see in light of Lemma 6 and the choice of N_1, the left-hand side of (7) is dominated by

$$N_1(2K_0\rho_T - 1)M + \bar{\mathcal{H}}.$$

The latter is negative for $4K_0\rho_T \le 1$ if $N_1 \ge 2$, that we may harmlessly assume. It follows that in

$$\Pi_h \cap \big\{(t,x) : (4K_0 + 1)\rho_T(t,x) \le 1\big\}$$

inequality (7) holds. By the comparison principle

$$v_h \le \phi = g + 2N_1 M \rho_T$$

in

$$\bar{\Pi} \cap \big\{(t,x) : (4K_0 + 1)\rho_T(t,x) \le 1\big\}.$$

On the set where $(4K_0 + 1)\rho_T > 1$ we have $\rho_T > (4K_0 + 1)^{-1}$ and the estimate from above of $v_h - g$ holds because M dominates $\sup_{\bar{\Pi}} |v_h|$.

The estimate of $v_h - g$ from below is similar and the lemma is proved. □

Now we proceed with estimating $\partial_t v_h$ and the finite-differences of v_h in x. Define

$$M_h(t,x) = \sum_{|k|=1}^{m} \big|\delta_{h,l_k} v_h(t,x)\big|, \quad \bar{M}_h = \sup_{\Pi_h} M_h,$$

$$G := \Big\{(t,x) \in \Pi_h : (\delta/2)\sum_{k=1}^{m} \big|\Delta_{h,l_k} v_h(t,x)\big| > \bar{\mathcal{H}} + K + K_0\big(|v_h| + M_h\big)(t,x)\Big\}, \quad (8)$$

and observe a fundamental fact that, thanks to (1),

$$\partial_t v_h + \mathcal{P}_h[v_h] - K = 0 \quad \text{in} \quad G. \quad (9)$$

Also note that on $\Pi_h \setminus G$ we have

$$(\delta/2)\sum_{k=1}^{m} \big|\Delta_{h,l_k} v_h\big| \le \bar{\mathcal{H}} + K + K_0\big(|v_h| + M_h\big). \quad (10)$$

While estimating $\partial_t v_h$ and the finite differences of v_h we rely on the results of Sections 2 and 3. Therefore we draw the reader's attention to the fact that Λ and Λ^h_∞ are introduced in (2.1) and (2.2), and, if we are given a set Q^o with the properties described in the beginning of Section 2, then the sets $\hat{Q}^o$ and Q are constructed as described before (2.3) and in (2.3).

9. LEMMA. *There are constants $h_0 > 0$ and N such that for all $h \in (0, h_0]$ on $\bar{\Pi}$ we have*

$$|\partial_t v_h| \le N\big(\bar{M}_h + \bar{\mathcal{H}} + K + \|g\|_{W^{1,2}_\infty(\mathbb{R}^{d+1})} + \sup_{\bar{\Pi}} |v_h|\big). \quad (11)$$

Proof. Since v_h satisfies (3), estimate (11) obviously holds on $[0,T] \times \partial_h \Omega$. It also holds on $\{T\} \times \Omega^h$ because there $\partial_t v_h = \mathcal{H}_{K,h}[g]$ and

$$|\mathcal{H}_{K,h}[g]| \le N\big(\bar{\mathcal{H}} + \|g\|_{W^{1,2}_\infty(\mathbb{R}^{d+1})}\big)$$

due to Lemma 6. Therefore, we need only prove (11) on $[0,T) \times \Omega^h$ or due to the continuity of $\partial_t v_h$ (Remark 4) on $(0,T) \times \Omega^h$. Having in mind the possibility of translating the origin in $\mathbb{R}^d$ we see that it suffices to prove (11) on $(0,T) \times \big[\Omega^h \cap \Lambda^h_\infty\big]$. By the way, as in the elliptic case, $\Omega^h \cap \Lambda^h_\infty$ is a finite set in light of Assumption 1.

Recall that G is defined in (8) and introduce

$$Q^o = \big\{(0,T) \times \big[\Omega^h \cap \Lambda^h_\infty\big]\big\} \cap G.$$

Observe that on $(0,T) \times \big[\Omega^h \cap \Lambda^h_\infty\big] \setminus Q^o$, we have (10) which together with (2) and Lemma 6 implies (11). Therefore, it only remains to establish (11) on Q^o assuming that $Q^o \ne \emptyset$.

Recall that (9) holds in G and hence in Q^o. Furthermore, every nonempty x-section of Q^o is the union of open intervals on which $\partial_t v_h$ is Lipschitz continuous by virtue of (9). By subtracting the left-hand sides of (9) evaluated at points t and $t + \varepsilon$, then transforming the difference by using Corollary 4.1.4, and finally dividing by ε and letting $\varepsilon \to 0$, we get that there exist functions a_k such that

$\delta/2 \le a_k \le 2\delta^{-1}$ and on every x-section of Q^o (a.e.) we have

$$\partial_t(\partial_t v_h) + a_k \Delta_{h,l_k}(\partial_t v_h) = 0.$$

By Theorem 3.1 this yields

$$\sup_{Q^o} |\partial_t v_h| \le \sup_{(0,T]\times\left[\Omega\cap\Lambda^h_\infty\right]\setminus Q^o} |\partial_t v_h|,$$

which along with what was said above implies that (11) holds on Q^o. The lemma is proved. □

10. LEMMA. *There are constants $h_0 > 0$ and N such that for all $h \in (0, h_0]$ and $r = 1, ..., m$ on Π_h we have*

$$\big[1 \wedge (\rho_\Omega - 9h)\big]|\Delta_{h,l_r} v_h| \le NM, \tag{12}$$

where

$$M = \bar{M}_h + \bar{\mathcal{H}} + K + \|g\|_{W^{1,2}_\infty(\mathbb{R}^{d+1})} + \sup_{\bar{\Pi}} |v_h|.$$

Proof. As in the proof of Lemma 9 we may focus on proving (12) in $(0, T) \times [\Omega^h \cap \Lambda^h_\infty]$. Then fix r and define

$$Q^o := \big\{(0,T) \times \big[\Omega^{3h} \cap \Lambda^h_\infty\big]\big\} \cap G.$$

For Q defined in (2.3), obviously, $Q \subset (0,T] \times \Omega^h$. Next, if $t \in (0,T)$, and $x \in \Omega^h \cap \Lambda^h_\infty$ is such that $(t,x) \notin Q^o$, then either $x \notin \Omega^{3h}$, so that $\rho_\Omega(x) \le 3h$ and (12) holds, or else $x \in \Omega^{3h}$ but (10) is valid, in which case (12) holds automatically.

Thus we need only prove (12) on Q^o assuming, of course, that $Q^o \neq \emptyset$. We know that (9) holds in Q^o and the left-hand side of (9) is nonpositive in $Q \setminus Q^o$.

To proceed further use Lemma 5.1.4 according to which there exists a constant $N \in (0,\infty)$ depending only on d such that for any $\rho \in (0, \rho_{\text{int}}(\Omega)/2)$ there exists an $\eta_\rho \in C^\infty_0(\Omega)$ satisfying

$$\eta_\rho = 1 \quad \text{on} \quad \Omega^{2\rho}, \quad \eta_\rho = 0 \quad \text{outside} \quad \Omega^\rho,$$

$$|\eta_\rho| \le 1, \quad |D\eta_\rho| \le N/\rho, \quad \big|D^2\eta_\rho\big| \le N/\rho^2.$$

By Theorem 2.3 on $Q^o \cap \Pi_{2\rho}$

$$\big[(\Delta_{h,l_r} v_h)^-\big]^2 \le \sup_{Q\setminus Q^o} \eta_\rho\big[(\Delta_{h,l_r} v_h)^-\big]^2 + N\rho^{-2}\bar{M}_h^2.$$

While estimating the last supremum we will only concentrate on

$$h_0 < \rho_{\text{int}}(\Omega)/6 \quad \text{and} \quad \rho \in \big[3h, \rho_{\text{int}}(\Omega)/2\big),$$

when $\eta_\rho = 0$ outside Ω^{3h}. In that case, for any $(s,y) \in Q \setminus Q^o$, either $s = T$ and

$$\eta_\rho\big[(\Delta_{h,l_r} v_h)^-\big]^2(s,y) = \eta_\rho\big[(\Delta_{h,l_r} g)^-\big]^2(s,y),$$

or $s \in (0,T)$, but $y \notin \Omega^{3h}$ implying that

$$\eta_\rho\big[(\Delta_{h,l_r} v_h)^-\big]^2(y) = 0,$$

or else $(s,y) \in \Omega^{3h} \cap \Lambda^h_\infty$ but $(s,y) \notin G$ and (10) holds at (s,y).

It follows that as long as $h \in (0,h_0]$ and $(t,x) \in Q^o \cap \Pi_{2\rho}$ and $\rho \in [3h, \rho_{\text{int}}(\Omega)/2)$, we have

$$\rho(\Delta_{h,l_r} v_h)^-(t,x) \le NM. \tag{13}$$

If $(t,x) \in Q^o$ and x is such that $\rho_\Omega(x) \ge 9h$, take

$$\rho = \big(\rho_{\text{int}}(\Omega)/2\big) \wedge \big(\rho_\Omega(x)/3\big),$$

which is bigger than $3h$ for $h \in (0,h_0]$ since $h_0 \le \rho_{\text{int}}(\Omega)/6$. In that case also $\rho_\Omega(x) \ge 3\rho$, so that $x \in \Omega^{2\rho}$ and we conclude from (13) that

$$\rho(\Delta_{h,l_r} v_h)^-(t,x) \le NM. \tag{14}$$

Obviously, one can replace ρ in (13) with $\rho_\Omega \wedge 1$ or with $(\rho_\Omega - 9h) \wedge 1$. Therefore,

$$\big[1 \wedge (\rho_\Omega(x) - 9h)\big](\Delta_{h,l_r} v_h)^-(t,x) \le NM$$

for $(t,x) \in Q^o$ such that $\rho_\Omega(x) \ge 9h$. However, this relation is obvious if $\rho_\Omega(x) \le 9h$.

As a result of all the above arguments we see that

$$(\rho_\Omega - 9h)(\Delta_{h,l_r} v_h)^- \le NM \tag{15}$$

holds in $(0,T) \times \big[\Omega^h \cap \Lambda^h_\infty\big]$ for any r whenever $h \in (0,h_0]$.

Finally, since $\mathcal{P}_h[v_h] \le K - \partial_t v_h$ in Π_h, we have that

$$4\delta^{-1}\sum_{r=1}^m (\Delta_r v_h)^+ \le -\partial_t v_h + \delta \sum_{r=1}^m (\Delta_r v_h)^- + K + |\partial_t v_h|,$$

which after being multiplied by $(\rho_\Omega - 9h) \wedge 1$ along with (11) and (15) lead to (12) on $(0,T) \times \big[\Omega^h \cap \Lambda^h_\infty\big]$. Thus, as is explained at the beginning of the proof, the lemma is proved. □

By repeating the proof of Theorem 5.5.10 we arrive at the following.

11. THEOREM. *(i) For all sufficiently small $h > 0$*

$$\sup_{\bar\Pi} |v_h| \le e^{K_0 T}\big(T\bar{\mathcal{H}} + \|g\|_{C(\Pi)}\big).$$

(ii) There is a constant N, depending only on $\operatorname{diam}(\Omega)$, $\rho_{\text{int}}(\Omega)$, $\rho_{\text{ext}}(\Omega)$, Λ, K_0, *d, and δ (thus, independent of T), such that for all sufficiently small $h > 0$ the estimates*

$$|v_h - g| \le N\big(\bar{\mathcal{H}} + \|g\|_{W^{1,2}_\infty(\mathbb{R}^{d+1})} + \sup_{\bar\Pi}|v_h|\big)\rho_T \wedge \rho_\Omega,$$

$$|\delta_{h,l_k} v_h|, |\partial_t v_h|, \big[1 \wedge (\rho_\Omega - 9h)\big]|\Delta_{h,l_k} v_h|$$

$$\le N\big(\bar{\mathcal{H}} + K + \|g\|_{W^{1,2}_\infty(\mathbb{R}^{d+1})} + \sup_{\bar\Pi}|v_h|\big) \tag{16}$$

hold in Π_h for all k.

CHAPTER 8

Parabolic differential equations of cut-off type

Here we present further development of some results originated in [**29**] and [**74**].

1. A particular case of equations in pure derivatives

Fix some constants $m, d \in \{1, 2, ...\}$, $\delta \in (0, 1]$, $T \in (0, \infty)$, and $K_0, K, N' \in [0, \infty)$.

1. ASSUMPTION. We are given vectors $l_i \in \bar{B}_1 \subset \mathbb{R}^d$, $i = \pm 1, ..., \pm m$, such that $l_{-i} = -l_i$, the coordinates of l_i's are rational numbers, and

$$e_i,\ (1/2)(e_i \pm e_j) \in \Lambda := \{l_k : k = \pm 1, ..., \pm m\}$$

for $i, j = 1, .., d$, $i \neq j$, where the e_i's form the standard orthonormal basis in $\mathbb{R}^d$.

2. ASSUMPTION. We are given a domain $\Omega \subset \mathbb{R}^d$ that is bounded and satisfies the exterior ball condition. We are also given a function $g \in W^{1,2}_{\infty}(\mathbb{R}^{d+1})$.

3. ASSUMPTION. (i) We are given a continuous function $\mathcal{H}(z, t, x)$, $t \in \mathbb{R}$, $x \in \mathbb{R}^d$,

$$z = (z', z''), \quad z' = \big(z'_0, z'_{\pm 1}, ..., z'_{\pm m}\big) \in \mathbb{R}^{2m+1}, \quad z'' = \big(z''_{\pm 1}, ..., z''_{\pm m}\big) \in \mathbb{R}^{2m},$$

which is Lipschitz continuous with respect to $z \in \mathbb{R}^{4m+1}$ with Lipschitz constant N' for any $(t, x) \in \mathbb{R}^{d+1}$. At all point of its differentiability with respect to z introduce

$$a_j = a_j(z, t, x) = D_{z''_j}\mathcal{H},$$

and at points of non-differentiability set $a_j = \delta$.

(ii) The above introduced functions a_j satisfy

$$\delta^{-1} \geq a_j \geq \delta \tag{1}$$

for all values of j and the arguments.

(iii) The number

$$\bar{\mathcal{H}} = \sup_{t,x,z'} \big(|\mathcal{H}(z', 0, t, x)| - K_0|z'|\big)$$

is finite.

(iv) The function $\mathcal{H}(z, t, x)$ is Lipschitz continuous with respect to x for any $(z, t) \in \mathbb{R}^{4m+2}$ and at any point x of its differentiability with respect to x

$$\big|D_x\mathcal{H}(z, t, x)\big| \leq N'\big(1 + |z|\big). \tag{2}$$

(v) There is a constant $\rho_0 > 0$ and a function $\mathcal{H}(z)$ such that

$$\mathcal{H}(z, t, x) = \mathcal{H}(z) \tag{3}$$

for all z, t if $\rho_\Omega(x) \leq \rho_0$.

For sufficiently smooth functions $u = u(t,x)$ introduce

$$\mathcal{H}[u](t,x) = \mathcal{H}\big(u(t,x), D_{l_j}u(t,x), D^2_{l_k}u(t,x), t, x\big)$$

$$= \mathcal{H}\big(u(t,x), D_{l_{\pm 1}}u(t,x), ..., D_{l_{\pm m}}u(t,x), D^2_{l_{\pm 1}}u(t,x), ..., D^2_{l_{\pm m}}u(t,x), t, x\big). \quad (4)$$

Similarly we introduce $\mathcal{P}[u]$, where $\mathcal{P}(z'') = \mathcal{P}_{m,\delta/2}(z'')$ is defined by (5.5.1). We will be interested in the solvability of the equation

$$\partial_t v + \max\big(\mathcal{H}[v], \mathcal{P}[v] - K\big) = 0. \quad (5)$$

Here is the pilot result of this chapter, which will be gradually generalized to a very large extent in the subsequent sections. The first generalization is given in Theorem 7 in which the global Lipschitz condition of $\mathcal{H}(z,x)$ with respect to z is replaced with the local one.

4. THEOREM. *Under the above assumptions equation* (5) *in*

$$\Pi = [0,T) \times \Omega$$

(*a.e.*) *with boundary condition* $v = g$ *on* $\partial'\Pi$ *has a unique solution* $v \in C(\bar{\Pi}) \cap W^{1,2}_{\infty,loc}(\Pi)$. *In addition,*

$$\sup_{\Pi} |v| \leq e^{K_0 T}\big(T\bar{\mathcal{H}} + \|g\|_{C(\partial'\Pi)}\big),$$

and there is a constant N, *depending only on* $\operatorname{diam}(\Omega)$, $\rho_{\text{int}}(\Omega)$, $\rho_{\text{ext}}(\Omega)$, $\Lambda = \{l_i\}$, K_0, d, *and* δ (*thus, independent of* T, N', *and* ρ_0), *such that*

$$|v - g| \leq N\big(\bar{\mathcal{H}} + \|g\|_{W^{1,2}_\infty(\mathbb{R}^{d+1})} + \sup_{\Pi}|v|\big)(\rho_\Omega \wedge \rho_T)$$

on $\bar{\Pi}$ *and for all* $i,j = 1,...,d$,

$$|D_i v|, \rho_\Omega|D_{ij}v|, |\partial_t v| \leq N\big(\bar{\mathcal{H}} + K + \|g\|_{W^{1,2}_\infty(\mathbb{R}^{d+1})} + \sup_{\Pi}|v|\big) \quad \textit{in} \quad \Pi \quad (\textit{a.e.}),$$

where $D_{ij}v$, $\partial_t v$ *are the generalized derivatives of* v.

As in the elliptic case, we are going to use finite differences in the proof of the existence and the estimates. Again as in the elliptic case, Assumptions 7.4.1 and 7.4.2 are satisfied due to Assumptions 1 and 3 (i), (iii). Assumptions 7.1.1 and 7.1.2 are also satisfied owing to Assumptions 1 and 3. Thus, all the assumptions of Section 7.4 are satisfied and we can use the information stated in Theorem 7.4.11 about the functions v_h defined in $\bar{\Pi}$ for sufficiently small $h > 0$ as bounded solutions of (7.4.2)-(7.4.3). In addition to the estimates from Theorem 7.4.11, as in the elliptic case, we need an estimate of a kind of modulus of continuity of v_h in x.

5. LEMMA. *There exists a constant* N *such that for all sufficiently small* $h > 0$ *we have*

$$\big|v_h(t,x) - v_h(t,y)\big| \leq N\big(|x-y| + h\big) \quad (6)$$

whenever $(t,x), (t,y) \in \bar{\Pi}$.

Proof. By repeating almost word for word the proof of Lemma 6.1.7, where we use Theorem 7.4.11 in place of Theorem 5.5.10 and use a version of Remark 6.6.16 obviously applicable in the parabolic case, we find that for sufficiently small $h > 0$ and $|l| \le h$, the function $u_h(t,x) := v_h(t,x+l) - v_h(t,x)$ satisfies

$$\partial_t u_h + \mathcal{L} u_h := \partial_t u_h + a_k \Delta_{h,l_k} u_h + b_k \delta_{h,l_k} u_h - c u_h + f = 0 \tag{7}$$

in Π_{11h}, where the measurable functions $a_k = a_k(t,x)$, $b_k = b_k(t,x)$, $c = c(t,x)$, and $f = f(t,x)$ satisfy $\delta/2 \le a_k \le 2\delta^{-1}$, $|b_k|$, $|c| \le N_2$, $|f| \le N_2 h$, with a constant N_2 independent of h. Then, we introduce

$$\eta(t) = N_2 h e^{N_2(T-t)}(T-t) + e^{N_2(T-t)} \sup_{\mathbb{R}^d} \big|g(T,x+l) - g(T,x)\big|$$
$$+ e^{N_2(T-t)} \sup_{[0,T]\times(\mathbb{R}^d \setminus \Omega^{11h})} \big|v_h(t,x+l) - v_h(t,x)\big|,$$

where we set $v_h = g$ outside Π. Obviously $\partial_t \eta + \mathcal{L}\eta \le 0$ in Π_{11h}. By the comparison principle (Theorem 7.1.5) we obtain $u_h \le \eta$ in $\bar{\Pi}_{11h}$.

In a similar fashion we get an estimate of u_h from below and conclude that in $\bar{\Pi}_{11h}$

$$\big|v_h(t,x+l) - v_h(t,x)\big| \le N\Big(h + \sup_{\mathbb{R}^d} \big|g(T,x+l) - g(T,x)\big|$$
$$+ \sup_{[0,T]\times(\mathbb{R}^d \setminus \Omega^{11h})} \big|v_h(t,x+l) - v_h(t,x)\big|\Big), \tag{8}$$

where $N = N(N_2, T)$. Furthermore, $|v_h(t,x+l) - v_h(t,x)|$ for $x \in \mathbb{R}^d \setminus \Omega^{11h}$ is dominated by a constant independent of h times h in light of Lemma 7.4.7. The rest is the same as in the proof of Lemma 6.1.7. The lemma is proved. □

6. LEMMA. *There is a sequence $h_n \downarrow 0$ such that v^{h_n} converge uniformly on $\bar{\Pi}$ to a function $v \in C(\bar{\Pi}) \cap W^{1,2}_{\infty,loc}(\Pi)$, for which the estimates in Theorem 4 hold true.*

Proof. Theorem 7.4.11 and Lemma 5 imply that for any $(t,x), (s,y) \in \bar{\Pi}$ and sufficiently small $h > 0$ we have

$$\big|v_h(t,x) - v_h(s,y)\big| \le N\big(|t-s| + |x-y| + h\big),$$

where N is independent of $(t,x), (s,y)$, and h. Then the argument in the proof of Lemma 6.1.8 shows that indeed there is a sequence $h_n \downarrow 0$ such that v^{h_n} converge uniformly on $\bar{\Pi}$ to a continuous function v. One shows that $v, D_i v, D_{ij} v$ admit the claimed estimates just by repeating the corresponding parts of the proof of Lemma 6.1.8. That $\partial_t v$ is also estimated as claimed is proved similarly. The lemma is proved. □

Proof of Theorem 4. In light of Lemma 6 we only need to prove that v satisfies (5) and that v is the only function in $C(\bar{\Pi}) \cap W^{1,2}_{\infty,\text{loc}}(\Pi)$ satisfying (5) which equals g on $\partial' \Pi$. The latter follows directly from Theorem 4.1.11 and we will concentrate on the former.

Since $v \in W^{1,2}_{\infty,\text{loc}}(\Pi)$, at almost any point $(t_0, x_0) \in \Pi$ we have (see, for instance, Appendix 2 in [**61**])

$$v(t_0 + s, x_0 + y) = P^{t_0,x_0}(s,y) + o\big(|s| + |y|^2\big),$$

where

$$P^{t_0,x_0}(s,y) = v(t_0,x_0) + y^i D_i v(t_0,x_0) + \frac{1}{2} y^i y^j D_{ij} v(t_0,x_0) + s\partial_t v(t_0,x_0).$$

For small $r > 0$ introduce

$$Q_r = \{(t,y) : t_0 \le t < t_0 + 4r^2, |y - x_0| < 2r\}.$$

Then take $\varepsilon > 0$ and observe that for all small $r > 0$, $|o(8r^2)| \le 2\varepsilon r^2$, which implies that for

$$u_{\varepsilon,r}(t,x) := P^{t_0,x_0}(t - t_0, x - x_0) + \varepsilon\big(t - t_0 + |x - x_0|^2 - r^2\big)$$

we have on $\partial' Q_r$ that

$$\begin{aligned} v(t,x) - u_{\varepsilon,r}(t,x) &= o\big(|t - t_0| + |x - x_0|^2\big) - \varepsilon\big(t - t_0 + |x - x_0|^2 - r^2\big) \\ &\le o(8r^2) - 3\varepsilon r^2 \le -\varepsilon r^2. \end{aligned}$$

On the other, hand $v(t_0,x_0) = u_{\varepsilon,r}(t_0,x_0) + \varepsilon r^2$. For all large n we will also have $v^{h_n} \le u_{\varepsilon,r}$ on $\partial' Q_r$ and $v^{h_n}(t_0,x_0) > u_{\varepsilon,r}(t_0,x_0)$, which by Theorem 7.1.5 implies that there is at least one point in

$$\{(t,y) : t_0 \le t < t_0 + 4r^2, |y - x_0| < 2r - h_n\}$$

at which $\partial_t u_{\varepsilon,r} + \mathcal{H}_{K,h_n}[u_{\varepsilon,r}] > 0$.

By observing that the $u_{\varepsilon,r}$'s are just simple polynomials, letting first $n \to \infty$, then $r \downarrow 0$ and finally $\varepsilon \downarrow 0$, as in the proof of Lemma 6.1.9 we arrive at the conclusion that

$$\partial_t v + \mathcal{H}_K[v] \ge 0$$

at (t_0,x_0). One obtains the reverse inequality just by changing the sign of ε in the formula for $u_{\varepsilon,r}$. Since (t_0,x_0) was almost any point in Π, the theorem is proved. □

The following result is proved in the same way as Theorem 6.1.10 with obvious modifications.

7. Theorem. *In Assumption* 3 (*i*) *replace the condition that* $\mathcal{H}(z,t,x)$
"is Lipschitz continuous with respect to $z \in \mathbb{R}^{4m+1}$ *with Lipschitz constant* N' *for any* $(t,x) \in \mathbb{R}^{d+1}$*"*
by $\mathcal{H}(z,t,x)$
"is Lipschitz continuous with respect to z *in any ball lying in* $\mathbb{R}^{4m+1}$ *with Lipschitz constant independent of* (t,x)*"*
and suppose that so modified Assumption 3 *along with Assumptions* 1 *and* 2 *are satisfied. Then the assertions of Theorem* 4 *still hold true.*

2. General parabolic equations with Lipschitz continuous H

Fix some constants $\delta \in (0,1)$, $T \in (0,\infty)$, and $K_0, K_1, K, N' \in [0,\infty)$. Recall that $\mathbb{S}$ is the set of $d \times d$ symmetric matrices, and also take the objects $\mathbb{S}_\delta$, m, l_k, Λ, $\hat{\delta}$, $\bar{\delta}$, $\lambda_k(a)$, $\mathcal{P}(z'')$, and $P(\mathsf{u}'')$ from the beginning of Section 6.2.

1. Assumption. We are given a domain $\Omega \subset \mathbb{R}^d$ which is bounded and satisfies the exterior ball condition. We are also given a function $g \in W^{1,2}_\infty(\mathbb{R}^{d+1})$.

Suppose that we are also given a function $H(\mathsf{u}, t, x)$,

$$\mathsf{u} = (\mathsf{u}', \mathsf{u}''), \quad \mathsf{u}' = (\mathsf{u}'_0, \mathsf{u}'_1, ..., \mathsf{u}'_d) \in \mathbb{R}^{d+1}, \quad \mathsf{u}'' \in \mathbb{S}, \quad t \in \mathbb{R}, \quad x \in \mathbb{R}^d.$$

2. Assumption. (i) The function $H(\mathsf{u}, t, x)$ is Lipschitz continuous with respect to u'' and $D_{\mathsf{u}''}H \in \mathbb{S}_{\delta/2}$ at all points of differentiability of H with respect to u'' (cf. Remark 6.2.5).

(ii) The number

$$\bar{H} := \sup_{\mathsf{u}', t, x} \big(|H(\mathsf{u}', 0, t, x)| - K_0 |\mathsf{u}'|\big) \tag{1}$$

is finite.

(iii) The function $H(\mathsf{u}, t, x)$ is Lipschitz continuous with respect to u' and at all point of differentiability of $H(\mathsf{u}, t, x)$ with respect to u' we have

$$\big|D_{\mathsf{u}'} H(\mathsf{u}, t, x)\big| \le N'. \tag{2}$$

(iv) The function $H(\mathsf{u}, t, x)$ is Lipschitz continuous with respect to (t, x) and at all point of differentiability of $H(\mathsf{u}, t, x)$ with respect to (t, x) we have

$$\big|D_{t,x} H(\mathsf{u}, t, x)\big| \le N'\big(1 + |\mathsf{u}|\big). \tag{3}$$

(v) For every u', t, x there exists a subset of $\mathbb{S}$ of full measure at every point of which $H(\mathsf{u}', \mathsf{u}'', t, x)$ is differentiable with respect to u'' and

$$\big|H(\mathsf{u}, t, x) - \langle \mathsf{u}'', D_{\mathsf{u}''} H(\mathsf{u}, t, x)\rangle\big| \le K_1\big(\bar{H} + K + |\mathsf{u}'|\big). \tag{4}$$

(vi) There is a constant $\rho_0 > 0$ and a function $H(\mathsf{u})$ such that

$$H(\mathsf{u}, t, x) = H(\mathsf{u}) \tag{5}$$

for all u, t if $\rho(x) \le \rho_0$.

Set

$$H_K(\mathsf{u}, t, x) = \max\big(H(\mathsf{u}, t, x), P(\mathsf{u}'') - K\big). \tag{6}$$

For smooth enough functions $u(t, x)$ introduce

$$H[u](t,x) = H\big(u(t,x), Du(t,x), D^2u(t,x), t, x\big),$$

$$P[u](t,x) = P\big(D^2u(t,x)\big),$$

$$H_K[u](t,x) = H_K\big(u(t,x), Du(t,x), D^2u(t,x), t, x\big). \tag{7}$$

We will be interested in the solvability of the equation

$$\partial_t v + H_K[v] = 0. \tag{8}$$

3. THEOREM. *Under the above assumptions equation* (8) *in* $\Pi = [0,T) \times \Omega$ *(a.e.) with boundary condition* $v = g$ *on* $\partial'\Pi$ *has a unique solution* $v \in C(\bar{\Pi}) \cap W^{1,2}_{\infty,loc}(\Pi)$. *In addition,*

$$\sup_{\bar{\Pi}} |v| \leq e^{K_0 T}\big(T\bar{H} + \|g\|_{C(\Pi)}\big), \tag{9}$$

and there is a constant N_0, *depending only on* $\rho_{\text{ext}}(\Omega)$, Λ, K_0, d, *and* δ *(thus, independent of* T, N', *and* ρ_0*), such that*

$$|v - g| \leq N_0\big(\bar{H} + \|g\|_{W^{1,2}_\infty(\mathbb{R}^{d+1})} + \sup_{\bar{\Pi}} |v|\big)(\rho_\Omega \wedge \rho_T). \tag{10}$$

on $\bar{\Pi}$, *and there is a constant* N, *depending only on* $\text{diam}(\Omega)$, $\rho_{\text{int}}(\Omega)$, $\rho_{\text{ext}}(\Omega)$, Λ, K_0, K_1, d, *and* δ, *such that for all* $i, j = 1, ..., d$,

$$|D_i v|, \rho_\Omega |D_{ij} v|, |\partial_t v| \leq N\big(\bar{H} + K + \|g\|_{W^{1,2}_\infty(\mathbb{R}^{d+1})} + \sup_{\bar{\Pi}} |v|\big) \quad \textit{in} \quad \Pi \quad \textit{(a.e.)}. \tag{11}$$

We first prove part of the assertions of Theorem 3.

4. LEMMA. *Suppose that only Assumptions* 1 *and* 2 *(i), (ii) are satisfied and equation* (8) *in* Π *(a.e.) with boundary condition* $v = g$ *on* $\partial'\Pi$ *has a solution* $v \in C(\bar{\Pi}) \cap W^{1,2}_{\infty,loc}(\Pi)$. *Then the assertions of Theorem* 3 *about the estimates* (9) *and* (10) *hold true.*

Proof. By Remark 6.6.16, obviously applicable in the parabolic case,

$$\partial_t v + a^{ij} D_{ij} v + b^i D_i v + cv + \tau \bar{H} = 0 \tag{12}$$

(a.e.) in Π for certain measurable $\mathbb{S}_{\bar{\delta}}$-valued (a^{ij}), measurable $\mathbb{R}^d$-valued b, and measurable real-valued τ and c, such that $|b|, |c| \leq K_0$, and $|\tau| \leq 1$.

Then the function

$$\psi(t,x) = e^{K_0(T-t)}\big((T-t)\bar{H} + \sup_{\partial'\Pi} |g|\big).$$

satisfies

$$\partial_t \psi + a^{ij} D_{ij} \psi + b^i D_i \psi + c\psi + \tau \bar{H} \leq 0$$

in Π. In addition, $\psi \geq v = g$ on $\partial'\Pi$. By Theorem 3.0.1, applied to $v - \psi$, we find that $v \leq \psi$ in $\bar{\Pi}$. Similarly one gets $v \geq -\psi$ in $\bar{\Pi}$, and this proves (9).

Relying on (12), the estimate

$$|v - g| \leq N_0\big(\bar{H} + \|g\|_{W^{1,2}_\infty(\mathbb{R}^{d+1})} + \sup_{\bar{\Pi}} |v|\big)\rho_\Omega$$

is proved similarly to the arguments in Remark 6.6.18 by using the same function ψ from Lemma 6.6.17 as there (cf. the proof of Lemma 7.4.7). That one can replace here ρ_Ω with ρ_T is proved by the arguments quite similar to the proof of Lemma 7.4.8. The lemma is proved. □

Proof of Theorem 3. Uniqueness is proved in the same way as in Theorem 6.1.5 by using Corollary 4.1.4, Lemma 4.1.6, and Theorem 3.0.1. The result in case $K = 0$ is a direct consequence of case $K > 0$ and Theorem 4.2.9. Part of the assertions of the theorem is proved in Lemma 4. Therefore we assume that $K > 0$ and only deal with the existence and the estimates (11). We may also obviously assume that $K_1 > 0$, which guarantees that

$$\beta(\mathsf{u}') := K_1\big(\bar{H} + K + |\mathsf{u}'|\big) \geq KK_1 > 0.$$

We interpret u, v, and x in Section 6.4 as u'', u', and (t, x), respectively, and define

$$E' = [-2, 2], \quad E'_1 = [-1, 1], \quad E'' = \mathbb{S}_{\delta/4}, \quad E''_1 = \mathbb{S}_{\delta/2},$$
$$E = E' \times E'', \quad E_1 = E'_1 \times E''_1.$$

Then for any $\mathsf{u}' \in \mathbb{R}^{d+1}$ and $(t, x) \in \mathbb{R}^{d+1}$ we have

$$H(\mathsf{u}', \cdot, t, x) \in \mathfrak{H}_{E_1, \beta(\mathsf{u}')}.$$

As in Remark 6.2.8 the function H is locally Lipschitz continuous.

Furthermore, in light of (3), at all points of differentiability of H with respect to (t, x) we have

$$\big|D_{t,x}H(\mathsf{u}, t, x)\big| \leq N'\big(1 + \beta(\mathsf{u}') + |\mathsf{u}''|\big), \tag{13}$$

with a constant N' perhaps different from the one in (3). Finally, condition (2) implies that (6.4.4) holds true. Thus, H satisfies Assumption 6.4.6.

By Theorem 6.2.1, for $a \in E''$ we have a representation

$$a = \sum_{|k|=1}^{m} \lambda_k(a) l_k l_k^*,$$

with Lipschitz continuous (real-analytic) $\lambda_k(a)$. Hence Assumption 6.4.1 is satisfied. Also observe that in our situation $\hat{\delta} \leq \lambda_k(a) \leq \hat{\delta}^{-1}$, so that (see (6.4.1))

$$\mathcal{E} \subset [-2, 2] \times [\hat{\delta}, \hat{\delta}^{-1}]^{2m}.$$

Now, according to Theorem 6.4.7, for $\mathsf{u}' \in \mathbb{R}^{d+1}$, $(t, x) \in \mathbb{R}^{d+1}$, and

$$z'' = \big(z''_{\pm 1}, ..., z''_{\pm m}\big) \in \mathbb{R}^{2m},$$

define the function $\mathcal{H}(\mathsf{u}', z'', t, x)$ by the formula

$$\mathcal{H}(\mathsf{u}', z'', t, x) = \inf_{w \in \mathbb{S}} \max\big\{\beta(\mathsf{u}')f + \lambda_k(a) z''_k :$$
$$(f, a) \in E, \beta(\mathsf{u}')f + \operatorname{tr} aw \leq H(\mathsf{u}', w, t, x)\big\}. \tag{14}$$

Then the function $\mathcal{H}(\mathsf{u}', z'', t, x)$ possesses the following properties:

(i) The function $\mathcal{H}(\mathsf{u}', z'', t, x)$ is locally Lipschitz continuous and at all points of its differentiability

$$\hat{\delta}^{-1} \geq D_{z''_k}\mathcal{H} \geq \hat{\delta} \quad \forall k, \tag{15}$$

$$\big|\mathcal{H}(\mathsf{u}', z'', t, x) - \big\langle z'', D_{z''}\mathcal{H}(\mathsf{u}', z'', t, x)\big\rangle\big| \leq 2\beta(\mathsf{u}'), \tag{16}$$

$$\big|D_{t,x}\mathcal{H}(\mathsf{u}', z'', t, x)\big| + \big|D_{\mathsf{u}'}\mathcal{H}(\mathsf{u}', z'', t, x)\big| \leq N\big(1 + \beta(\mathsf{u}') + |z''|\big), \tag{17}$$

where the constant N is independent of u', z'', t, x.

(ii) If $\mathsf{u}'' \in \mathbb{S}$ and $z''_j = \langle l_j, \mathsf{u}'' l_j\rangle$ for $|j| = 1, ..., m$, then

$$H(\mathsf{u}', \mathsf{u}'', t, x) = \mathcal{H}(\mathsf{u}', z'', t, x) \tag{18}$$

for all u', t, x.

Now we want to apply Theorem 1.7. Since by assumption $e_i \in \Lambda$, $i = 1, ..., d$, there is an identification of some e_i with some l_k. Let $e_i = l_{k(i)}$, for $i = 1, ..., d$, and then for

$$z = (z', z''), \quad z' = \left(z'_0, z'_{\pm 1}, ..., z'_{\pm m}\right) \in \mathbb{R}^{2m+1}, \quad z'' = \left(z''_{\pm 1}, ..., z''_{\pm m}\right) \in \mathbb{R}^{2m}$$

and $(t, x) \in \mathbb{R}^d$, set

$$\check{\mathcal{H}}(z, t, x) = \mathcal{H}\left(z'_0, z'_{k(1)}, ..., z'_{k(d)}, z'', t, x\right).$$

Observe that, if $\rho_\Omega(x) \le \rho_0$, $\check{\mathcal{H}}(z, t, x)$ depends only on z, since then $H(\mathsf{u}, t, x)$ depends only on u by Assumption 2 (vi). Hence, Assumption 1.3 (v) is satisfied. Assumption 1.3 (iv) is satisfied for $\check{\mathcal{H}}(z, t, x)$ due to (17). If we take $z'' = 0$ in (16), then we see that

$$\left|\check{\mathcal{H}}(z', 0, t, x)\right| \le 2K_1\left(\bar{H} + K + |z'|\right).$$

Hence, Assumption 1.3 (iii) is satisfied for $\check{\mathcal{H}}(z, t, x)$ with a number smaller than

$$2K_1(\bar{H} + K) \quad \text{and} \quad 2K_1$$

in place of $\bar{H}$ and K_0. Assumption 1.3 (ii) is satisfied for $\check{\mathcal{H}}(z, x)$ with $\hat{\delta}$ in place of δ. Also, the modification of Assumption 1.3 (i) stated in Theorem 1.7 is satisfied for $\check{\mathcal{H}}(z, x)$ in light of (15), and (17). Finally, Assumption 1.1 is satisfied due to what is said in the beginning of the section, and Assumption 1.2 coincides with Assumption 1. Hence, all the assumptions of Theorem 1.7 are satisfied. We draw the reader's attention to the fact that the assumptions of Theorem 1.7 are satisfied with $\hat{\delta}$ in place of δ (among a few other substitutions) and that $P[u]$ in the current section (borrowed from section 6.2) is constructed as $P[u]$ in Theorem 1.7 with $\hat{\delta}$ in place of δ.

Introduce $\check{\mathcal{H}}_K[v]$ following the example (1.4). By Theorem 1.7 the equation

$$\partial_t v + \check{\mathcal{H}}_K[v] = 0 \tag{19}$$

in Π (a.e.) with boundary condition $v = g$ on $\partial'\Pi$ has a unique solution $v \in C(\bar{\Pi}) \cap W^{1,2}_{\infty,\text{loc}}(\Pi)$ and the estimates (11) hold true if the right-hand side is replaced with

$$N\left[K_1(\bar{H} + K) + K + \|g\|_{W^{1,2}_\infty(\mathbb{R}^{d+1})} + \sup_{\Pi} |v|\right], \tag{20}$$

where N is a constant depending only on $\operatorname{diam}(\Omega)$, $\rho_{\text{int}}(\Omega)$, $\rho_{\text{ext}}(\Omega)$, Λ, K_0, K_1, d, and δ.

Finally, as is explained in the beginning of the proof, to finish the proof of the theorem, it only remains to note that, due to (18), equations (19) and (8) coincide. The theorem is proved. □

3. The case of H measurable in time and space variables

Here we present the main results of this chapter.

Fix some constants $T \in (0, \infty)$, $\delta \in (0, 1)$, $K, K_0, K_1 \in [0, \infty)$, ($K_1$ is needed to use Theorem 2.3).

1. ASSUMPTION. We are given a domain $\Omega \subset \mathbb{R}^d$ which is bounded and satisfies the exterior ball condition. We are also given a function $g \in W^{1,2}_\infty(\mathbb{R}^{d+1})$.

Suppose that we are also given a function $H(\mathsf{u}, t, x)$,

$$\mathsf{u} = (\mathsf{u}', \mathsf{u}''), \quad \mathsf{u}' = (\mathsf{u}'_0, \mathsf{u}'_1, ..., \mathsf{u}'_d) \in \mathbb{R}^{d+1}, \quad \mathsf{u}'' \in \mathbb{S}, \quad t \in \mathbb{R}, \quad x \in \mathbb{R}^d.$$

2. ASSUMPTION. (i) The function $H(\mathsf{u}, t, x)$ is Borel measurable with respect to (u, t, x), is Lipschitz continuous with respect to u'', and $D_{\mathsf{u}''} H \in \mathbb{S}_\delta$ at all points of differentiability of H with respect to u''.

(ii) The number

$$\bar{H} := \sup_{\mathsf{u}', t, x} \big(|H(\mathsf{u}', 0, t, x)| - K_0 |\mathsf{u}'| \big) \quad (\geq 0)$$

is finite.

(iii) The function $H(\mathsf{u}, t, x)$ is continuous with respect to u for any (t, x).

For smooth enough functions u introduce

$$H[u](t, x) = H\big(u(t, x), Du(t, x), D^2 u(t, x), t, x\big),$$

$$P[u](t, x) = P\big(D^2 u(t, x)\big),$$

where P is defined in (6.2.4).

Here is the major result of this chapter, which admits further extensions with relaxed Assumption 2 (ii) presented in Theorem 12.1.13 (when $g \in C(\overline{\partial' \Pi})$ and $v \in W^{1,2}_{p,\text{loc}}(\Pi)$) and in Theorem 15.1.7 (when $g \in W^{1,2}_p(\Pi)$ and $v \in W^{1,2}_p(\Pi)$).

In the following theorem the above assumptions are supposed to be satisfied.

3. THEOREM. *The equation*

$$\partial_t v + \max\big(H[v], P[v] - K\big) = 0 \tag{1}$$

in $\Pi = [0, T) \times \Omega$ *(a.e.) with boundary condition* $v = g$ *on* $\partial' \Pi$ *has a solution* $v \in C(\bar{\Pi}) \cap W^{1,2}_{\infty,\text{loc}}(\Pi)$ *such that*

(i) in Π *we have*

$$\sup_{\Pi} |v| \leq e^{K_0 T} \big(T \bar{H} + \sup_{\partial' \Pi} |g|\big), \tag{2}$$

$$|v - g| \leq N \big(\bar{H} + \|g\|_{W^{1,2}_\infty(\mathbb{R}^{d+1})} + \sup_{\Pi} |v|\big)(\rho_\Omega \wedge \rho_T), \tag{3}$$

where the constant N*, depends only on* $\rho_{\text{ext}}(\Omega)$, Λ, K_0, d*, and* δ *(thus, independent of* T*),*

(ii) there exists $\alpha = \alpha(\delta, d) \in (0, 1)$ *such that*

$$|v|_{C^\alpha(\Pi)} \leq N \big(\bar{H} + \|g\|_{W^{1,2}_\infty(\mathbb{R}^{d+1})} + \sup_{\Pi} |v|\big), \tag{4}$$

where the constant N *is of the same kind as in* (3)*,*

(iii) there is a constant N*, depending only on* $\text{diam}(\Omega)$, $\rho_{\text{int}}(\Omega)$, $\rho_{\text{ext}}(\Omega)$, Λ, K_0, d*, and* δ*, such that*

$$|Dv|, \rho_\Omega |D^2 v|, |\partial_t v| \leq N \big(\bar{H} + K + \|g\|_{W^{1,2}_\infty(\mathbb{R}^{d+1})} + \sup_{\Pi} |v|\big) \quad \text{in} \quad \Pi \quad (a.e.), \tag{5}$$

(iv) if $\Omega \in C^{1,1}$ *(see Definition* 14.1.10*), then* $v \in W^{1,2}_p(\Pi)$ *for any* $p \in [1, \infty)$*.*

4. REMARK. Here we derive the last assertion (iv) of Theorem 3, which we are going to use only in Section 15.4, from the previous ones (excluding (3) and (4)) and Theorem 14.4.1. The reader will see that there is no vicious circle in our arguments.

We set $u = v - g$ and rewrite (1) as

$$P[u] = f, \tag{6}$$

where

$$f = P[v-g] - P[v] - \max\big(H[v] - P[v], -K\big) - \partial_t v.$$

As in the case of Remark 6.6.13 using also the boundedness of $\partial_t v$ we conclude that f is bounded in Π.

Now we look at equation (6) as an elliptic equation with bounded free term and $t \in (0,T)$ as a parameter. By Theorem 14.4.1, for any point $x_0 \in \partial\Omega$, there exists $r > 0$, such that $u(t,\cdot) \in W^2_p\big(\Omega \cap B_r(x_0)\big)$ and, as it follows from the proof of Theorem 14.4.1, r can be chosen to be independent of t and x_0 and the $W^2_p\big(\Omega \cap B_r(x_0)\big)$-norm of $u(t,\cdot)$ is a bounded function of t and x_0. This obviously proves assertion (iv).

Interestingly enough in the above argument we did not use Assumption 2 (iii).

5. REMARK. Observe that there is no uniqueness statement in Theorem 3 in contrast with Lemmas 4.1, 4.2, and 4.3 below. The point is that, generally, uniqueness may fail. For instance, in one space dimension the (quasilinear) equation

$$\partial_t u + \max\Big(D^2u - (1-t)\big(12|Du|\big)^{1/2} + 2\big(1-|x|^3\big)^{1/2}|u|^{1/2}, D^2u - 7\Big) = 0$$

in $(0,1) \times (-1,1)$ with zero boundary data has two solutions: one is identically equal to zero and the other one is $(1-t)^2\big(1-|x|^3\big)$.

6. REMARK. Despite there could be many solutions, estimates (2) and (3) hold for any such solution. This is shown in Lemma 2.4.

7. REMARK. Here we derive (4) from other assertions of Theorem 3, excluding (iv), and some results, that are proved later, assuring the reader that no vicious circle arrises.

As in Remark 6.6.19

$$\big|\partial_t v + a^{ij}D_{ij}v + b^iD_iv\big| \le K_0 \sup_{\bar\Pi}|v| + \bar H$$

in Π (a.e.) for certain measurable $\mathbb{S}_{\bar\delta}$-valued (a^{ij}) and measurable $\mathbb{R}^d$-valued b such that $|b| \le K_0$.

Next observe that if $z_1 = (t_1,x_1), z_1 = (t_2,x_2) \in \Pi$, $\rho := \rho(z_1,z_2) \le 1$, $t_1 \le t_2$, and $C_{2\sqrt{\rho}}(z_1) \subset \Pi$, then $z_2 \in \bar C_{\sqrt{\rho}}(z_1)$ and by Theorem 9.6.6 (with $r = \sqrt{\rho}$ in (9.6.11)) there exists $\beta = \beta(\delta,d) \in (0,1)$ such that

$$|v(z_1) - v(z_2)| \le N\big(\bar H + \sup_{\bar\Pi}|v|\big)\rho^{\beta/2}(z_1,z_2). \tag{7}$$

However, if $\rho \le 1$ and $C_{2\sqrt{\rho}}(z_1) \not\subset \Pi$, then either $t_1 \ge T - 4\rho$, in which case $t_2 \ge T - 4\rho$ as well, so that

$$\rho_\Omega(x_1) \wedge (T-t_1),\ \rho_\Omega(x_2) \wedge (T-t_2) \le 4\sqrt{\rho}, \tag{8}$$

or $\rho_\Omega(x_1) \le 2\sqrt{\rho}$, in which case, as in Remark 6.6.19, $\rho_\Omega(x_2) \le 3\sqrt{\rho}$ and (8) holds again. This and (3) allow us to finish the argument as in Remark 6.6.19.

8. REMARK. We did not replace the last term in (5) and in other estimates with its estimate from (2) in order to keep N in (5) independent of T and having in mind that there are situations in which $|v|$ admits estimates also independent of T.

To that effect we have the following.

9. THEOREM. *Suppose that Assumptions* 1 *and* 2 *are satisfied and there exist constants* $M_0 \geq 0, \varepsilon > 0$ *such that*

$$\mp H(\pm M_0 + \varepsilon_0, \varepsilon_1, ..., \varepsilon_d, 0, t, x) \geq 0 \tag{9}$$

in Π *as long as the length of the vector* $(\varepsilon_0, \varepsilon_1, ..., \varepsilon_d)$ *in* $\mathbb{R}^{d+1}$ *is less than* ε. *Assume that* $|g| \leq M_0$ *in* $\bar{\Pi}$. *Then there exists a function* v *with the properties described in Theorem* 3 *and such that* $|v| \leq M_0$ *in* $\bar{\Pi}$.

We derive Theorems 3 and 9 in this section from the following result, that is proved in the next section.

10. THEOREM. *Under Assumptions* 1 *and* 2 *also suppose that* $H(\mathsf{u}, t, x)$ *is Lipschitz continuous with respect to* u *with a constant independent of* (t, x). *Then the assertions of Theorem* 3 *hold true and the solution with the described properties is unique. Furthermore, if there exists a constant* $M_0 \geq 0$ *such that* (9) *holds in* Π *with* $\varepsilon_0 = \varepsilon_1 = ... = \varepsilon_d = 0$ *and* $|g| \leq M_0$ *in* $\bar{\Pi}$, *then* $|v| \leq M_0$ *in* $\bar{\Pi}$.

As in the elliptic case we also use one more intermediate result.

11. LEMMA. *Suppose that Assumptions* 1 *and* 2 *are satisfied and* $H(\mathsf{u}, t, x) - \delta \operatorname{tr} \mathsf{u}''$ *is a bounded function. Then the assertions of Theorem* 3 *hold true.*

Furthermore, if the additional assumptions stated in Theorem 9 *are satisfied, then* $|v| \leq M_0$ *in* $\bar{\Pi}$.

Proof. The proof of the first assertion of the lemma starts as that of Lemma 6.6.21 by introducing H^n as the same mollifications with respect to u' and then applying Theorem 10 to the equation

$$\partial_t v^n + \max\big(H^n[v^n], P[v^n] - K\big) = 0. \tag{10}$$

The rest of the proof of the first assertion of the lemma runs along the same lines as in Section 6.6 relying on Theorem 10 although ignoring any kind of monotonicity with respect to u_0' and is left to the reader as a simple exercise.

To prove the second assertion, recall that v is obtained as the limit of a subsequence of v^n. Furthermore, due to our assumptions for the functions $u_{(\pm)} \equiv \pm M_0$ and $n > 1/\varepsilon$ we have $\mp H^n[u_{(\pm)}] \geq 0$ and the second assertion follows again from Theorem 10. The lemma is proved. □

Proof of Theorems 3 and 9. As in the proof of Theorem 6.6.3 we introduce the functions $\chi(r)$ and $\kappa(\mathsf{u}'')$ as projections of r and u'', respectively, on the unit balls in $\mathbb{R}$ and $\mathbb{S}$, respectively, for $\mathsf{u}' \in \mathbb{R}^{d+1}$ set

$$\chi(\mathsf{u}') = \big(\chi(\mathsf{u}_0'), \chi(\mathsf{u}_1'), ..., \chi(\mathsf{u}_d')\big),$$

and for $n = 1, 2, ...$ define

$$H^n(\mathsf{u}, t, x) = \delta \operatorname{tr}\big[\mathsf{u}'' - n\kappa(\mathsf{u}''/n)\big] + H\big(n\chi(\mathsf{u}'/n), n\kappa(\mathsf{u}''/n), t, x\big).$$

Then we note that H^n satisfies Assumption 2 with the same δ and, perhaps, smaller $\bar{H}$ than the original one. Furthermore, $H^n(\mathsf{u}, x) - \delta \operatorname{tr} \mathsf{u}''$ is bounded and if condition

(9) is satisfied for the original H, then it is also satisfied for H^n if $n \geq M_0 + 1$, because the left-hand side does not change if we replace H with H^n.

Hence, Lemma 11 is applicable and the rest is practically the same as in the proof of Theorem 6.6.3. The theorem are proved. □

4. Proof of Theorems 3.10 and 6.6.5

Proof of Theorems 3.10. To prove the last assertion of the theorem observe that for the function $u_{(+)} \equiv M_0$ we have

$$\partial_t u_{(+)} + \max\big(H[u_{(+)}], P[u_{(+)}] - K\big) = \max\big(H[u_{(+)}], -K\big) \leq 0.$$

By subtracting this inequality from (3.1) we obtain that $w := v - u_{(+)}$ satisfies a uniformly parabolic linear inequality in Π (a.e.) with bounded coefficients and the free term ≥ 0. Since $v - u_{(+)} \leq 0$ on $\partial'\Pi$, by the maximum principle $v - u_{(+)} \leq 0$ in $\bar{\Pi}$. The inequality $M_0 + v \geq 0$ is obtained in a similar way, and the proof of uniqueness goes the same way.

The rest of the proof of Theorem 3.10 is based on Theorem 2.3 and is done in several steps in the first three of which we drop Assumptions 2.2 (iv), (v), (vi), one by one. These steps are quite similar to the steps in the proof of Theorem 6.6.20 and therefore we will give only short comments.

1. LEMMA. *Let the assumptions of Theorem* 2.3*, apart from Assumption* 2.2 *(vi), be satisfied. Then the assertions of Theorem* 2.3 *are still true.*

The proof of this lemma is an easy adaptation of that of Lemma 6.7.1.

2. LEMMA. *Let Assumption* 3.2 *(i) and the assumptions of Theorem* 2.3*, apart from Assumptions* 2.2 *(v), (vi), be satisfied. Then the assertions of Theorem* 2.3 *are still true with N in* (2.11) *(coinciding with* (3.5)*), this time, depending only on* $\operatorname{diam}(\Omega)$*,* $\rho_{\text{int}}(\Omega)$*,* $\rho_{\text{ext}}(\Omega)$*,* Λ*,* K_0*,* d*, and* δ *(and independent of K_1, which does not even appear in the statement of the current lemma).*

As in the case of Lemma 6.7.2 the proof of Lemma 2 is achieved by replacing H with $H_K = \max(H, P_0 - K)$, where

$$P_0(\mathsf{u}'') = \max_{a \in \mathbb{S}_{\delta/2}} a^{ij} \mathsf{u}''_{ij}.$$

In light of Remarks 3.4, 3.6, and 3.7, to conclude the proof of Theorem 3.10, it suffices to prove the following.

3. LEMMA. *Under the the assumptions of Theorem* 3.10 *equation* (3.1) *in* Π *(a.e.) with boundary condition $v = g$ on $\partial'\Pi$ has a solution $v \in C(\bar{\Pi}) \cap W^{1,2}_{\infty,\text{loc}}(\Pi)$ such that assertion (iii) of Theorem* 3.3 *holds true.*

This lemma is proved similarly to Lemma 6.7.4 by using the mollifications of H with respect to (t, x) and adding the derivative with respect to t to all equations. This allows us to drop Assumption 2.2 (iv). □

Proof of Theorem 6.6.5. Remarks 6.6.13, 6.6.18, and 6.6.19 show that we need only prove that (under the condition of Theorem 6.6.5) there exists a $C(\bar{\Omega}) \cap C^{1,1}_{\text{loc}}(\Omega)$-solution of (6.0.1) with boundary condition $v = g$ on $\partial\Omega$ such that $|v| \leq M_0$ and the estimates (6.6.3) hold.

First suppose that $H(\mathsf{u}, x)$ is Lipschitz continuous with respect to u with Lipschitz constant independent of x. Then take a smooth function $\zeta(\tau)$ of one variable

τ, such that $0 \le \zeta \le 1$, $\zeta' \ge 0$, $\zeta(\tau) = 0$ for $\tau \le -1$, and $\zeta(\tau) = 1$ for $\tau \ge 0$ and for $T \ge 1$ introduce

$$g_T(t,x) = \big(1 - \zeta(t-T)\big)g(x) + \zeta(t-T)M_0.$$

After that consider the parabolic equation (3.1) in $\Pi_T = [0,T) \times \Omega$ with boundary condition g_T on $\partial'\Pi_T$. Let $u_T(t,x)$ be the unique solution of this problem obtained by using Theorem 3.10. In particular, we have $|u_T| \le M_0$ in $\bar{\Pi}_T$.

Take $h \in (0,1)$ and apply the operator

$$\delta_{h,t} : u(t,x) \to u(t+h,x) - u(t,x)$$

to (3.1). Then by using the assumption that H is Lipschitz continuous in u, we obtain that $\delta_{h,t}u_T(t,x)$ satisfies a linear uniformly nondegenerate homogeneous equation with bounded coefficients in $(0,T-h)\times\Omega$. On the lateral boundary of this cylinder $\delta_{h,t}u_T = \delta_{h,t}g_T \ge 0$, since $g \le M_0$ and $\zeta' \ge 0$. Finally, $\delta_{h,t}u_T(T-h,x) = M_0 - u_T(T-h,x) \ge 0$ for $x \in \bar{\Omega}$. Hence by the maximum principle $\delta_{h,t}u_T \ge 0$ in $(0,T-h) \times \Omega$ implying that $\partial_t u_T \ge 0$ in Π_T.

Next, uniqueness in Theorem 3.10 and the time-homogeneity of (3.1) imply that $u_T(t,x) = u_{T+s}(t+s,x)$, $s \ge 0$, and

$$u_T(t,x) = u_{T-t}(0,x) \tag{1}$$

for $x \in \bar{\Omega}$ and $0 \le t \le T$. By adding to this that $\partial_t u_T \ge 0$ we conclude that $u_T(x) := u_T(0,x)$ is a decreasing function of T, which along with (1) shows that the limit

$$\lim_{T\to\infty} u_T(t,x) =: v(x)$$

exists and is independent of t. Obviously the estimates (6.6.3) are valid for v. Furthermore, by Theorem 4.2.6 the function v satisfies (3.1) and since it is independent of t, it satisfies (6.0.1) in Ω. By noting that $v = g$ on $\partial\Omega$ we conclude the first step of the proof.

Observe that in this step we only used that (6.6.4) hold only for $\varepsilon_0 = \varepsilon_1 = \ldots = \varepsilon_d = 0$.

After that we assume that $H(\mathsf{u},x) - \delta \operatorname{tr} \mathsf{u}''$ is a bounded function and proceed as in the proof of Lemma 3.11 with quite minor changes, in particular, ignoring the time derivative and using the first part of the proof in place of Theorem 3.10.

Then we finish the proof of Theorem 6.6.5 by basically repeating what is in the end of Section 3. The theorem is proved. □

CHAPTER 9

A priori estimates in C^α for solutions of linear and nonlinear equations

In this chapter we prepare some indispensable tools needed for the derivation of estimates in W^2_p and $W^{1,2}_p$ for solutions of nonlinear elliptic and parabolic equations, respectively. Of course, talking about Sobolev's solutions of an equation in a domain we always mean that the equation is satisfied (only) almost everywhere in the domain.

For an inexperience reader the results of this chapter may look rather week. Indeed, we only estimate the Hölder exponent of solutions of *second-order* equations, or estimate integrals of $|D^2u|^\gamma$, where γ is (much) less than one.

Still the reader will see that these are indeed indispensable tools in the Hölder- or Sobolev-space theory of fully nonlinear equations.

1. Estimating the integrals of Green's functions over small sets

For fixed $K \in [0,\infty)$ and $\delta \in (0,1]$ we denote by $\mathbb{L}^t_{\delta,K}$ the set of parabolic operators

$$\mathcal{L} = \partial_t + a^{ij}(t,x)D_{ij} + b^i(t,x)D_i - c(t,x), \tag{1}$$

whose measurable coefficients are defined on $\mathbb{R}^{d+1}$ and satisfy

$$a^{ij} = a^{ji}, \quad \delta|\lambda|^2 \le a^{rs}\lambda^r\lambda^s \le \delta^{-1}|\lambda|^2, \quad |b^i| \le K, \quad 0 \le c \le K \tag{2}$$

for all $(t,x) \in \mathbb{R}^{d+1}$, $\lambda \in \mathbb{R}^d$, $i,j = 1,...,d$.

Recall that, for $T,R > 0$,

$$B_R = \{x \in \mathbb{R}^d : |x| < R\}, \quad C_{T,R} = [0,T) \times B_R$$

and also introduce

$$C_R = [0,R^2) \times B_R, \quad C_{T,R}(t,x) = (t,x) + C_{T,R}, \quad C_R(t,x) = C_{R^2,R}(t,x).$$

One knows that for an operator $\mathcal{L} \in \mathbb{L}^t_{\delta,K}$ with sufficiently regular coefficients there exists a Green's function $G(t,x,s,y) \ge 0$ in C_1 with zero boundary condition on $\partial' C_1$. To be more precise for sufficiently regular f given on C_1 there exists $u \in C^{1,2}_{\text{loc}}(C_1) \cap C(\bar{C}_1)$ such that $\mathcal{L}u = -f$ in C_1 and $u = 0$ on $\partial' C_1$. Furthermore, we have

$$u(t,x) = \int_{C_1} f(s,y)G(t,x,s,y)\,dsdy.$$

The goal of this section is to present some estimates from below for

$$u_\Gamma(t,x) = \int_\Gamma G(t,x,s,y)\,dsdy = \int_{C_1} I_\Gamma(s,y)G(t,x,s,y)\,dsdy,$$

where Γ is a subset of C_1. We want this estimates to be expressed only in terms of the measure of Γ. Of course, because of the zero boundary condition we have to restrict (t,x) to a compact subset of C_1.

The function u_Γ in most cases satisfies the equation $\mathcal{L}u_\Gamma = -I_\Gamma$ and vanishes on $\partial' C_1$. In these cases

$$u_\Gamma(t,x) \geq m(t,x,q),$$

with $q = |\Gamma|/|C_1|$, where $|\Gamma|$ is the Lebesgue measure of Γ and the function $m(t,x,q)$ is introduced as follows.

Let $\mathbb{L}^{t,\infty}_{\delta,K}$ be the subset of $\mathbb{L}^{t}_{\delta,K}$ consisting of operators with infinitely differentiable coefficients. Recall that functions of class $C^{1,2}(C_1)$ have bounded derivatives in (t,x) in C_1 and therefore admit unique continuations onto $\bar{C}_1$. Speaking about functions of class $C^{1,2}(C_1)$ we will always mean these continuations and for $q \in [0,1]$, we denote by $\mathbb{U}^{t,m}_q$ the set of functions $u \in C^{1,2}(C_1)$ vanishing on $\partial' C_1$ for each of which there exists an operator $\mathcal{L} \in \mathbb{L}^{t,\infty}_{\delta,K}$ such that $\mathcal{L}u \leq 0$ on $C_{1,1}$ and

$$\big|C_1 \cap \{\mathcal{L}u \leq -1\}\big| \geq q|C_1|. \tag{3}$$

Then set

$$m(t,x,q) = \inf\big\{u(t,x) : u \in \mathbb{U}^{t,m}_q\big\}, \quad (t,x) \in C_1, \quad q \in [0,1]. \tag{4}$$

To define an even more important function, for $q \in [0,1]$, denote by $\mathbb{U}^{t,M}_q$ the set of functions $u \in C^{1,2}(C_{2,1})$ vanishing on $\partial' C_{2,1}$ for each of which there exists an operator $\mathcal{L} \in \mathbb{L}^{t,\infty}_{\delta,K}$ such that $\mathcal{L}u \leq 0$ on $C_{2,1}$ and

$$\big|C_1(1,0) \cap \{\mathcal{L}u \leq -1\}\big| \geq q|C_1|. \tag{5}$$

Then set

$$M(t,x,q) := \inf\big\{u(t,x) : u \in \mathbb{U}^{t,M}_q\big\}, \quad (t,x) \in C_{2,1}, \quad q \in [0,1]. \tag{6}$$

Observe that $u \geq 0$ if $u \in \mathbb{U}^{t,m}_q$ by the maximum principle. Hence, $m \geq 0$. Similarly, $M \geq 0$. Main properties of the functions M and m are stated in the following theorems.

1. THEOREM. *For any $R \in [0,1)$ there exist constants $N \in (0,\infty)$ and $\gamma \in (0,1)$, depending only on K, δ, d, and R, such that for all $q \in [0,1]$ and $x \in B_R$ we have*

$$NM(0,x,q) \geq q^{1/\gamma}. \tag{7}$$

2. THEOREM. *Let $q \in (0,1]$ and $R \in [0,1)$. Then there exists a constant $m_0 > 0$, which depends only on d, K, δ, q, and R, such that $m(t,x,q) \geq m_0$ if $x \in B_R$ and $0 \leq t \leq qR$.*

3. REMARK. Most of the results and proofs of this and the following two sections is taken from [**84**] and [**61**]. Theorem 1 appeared in [**70**] thanks to a simple iteration argument based on the results in [**84**] and [**61**]. Estimate (7) is proved earlier in [**39**] for M which is defined as the inf over the set of u satisfying (5) with $C_{1,1/2}(1,0)$ in place of $C_1(1,0)$. This is slightly less than what we need in the future. Also in [**39**] there are no lower order terms in $\mathcal{L}$.

The proof of Theorems 1 and 2 is given in Section 3 after we prepare some auxiliary facts in Section 2. We finish this section by showing that the inf's in (4) and (6) can be extended to a wider set without affecting the result.

4. LEMMA. *(i) Let $q \in [0,1)$, $u \in W^{1,2}_{d+1}(C_1)$, and $\mathcal{L} \in \mathbb{L}^t_{\delta,K}$. Assume that $u \geq 0$ on $\partial' C_1$, $\mathcal{L}u \leq 0$ on C_1 (a.e.), and (3) holds. Then $u(t,x) \geq m(t,x,q)$ in C_1.*

(ii) Let $q \in [0,1)$, $u \in W^{1,2}_{d+1}(C_{2,1})$, and $\mathcal{L} \in \mathbb{L}^t_{\delta,K}$. Assume that $u \geq 0$ on $\partial' C_{2,1}$, $\mathcal{L}u \leq 0$ on $C_{2,1}$ (a.e.), and (5) holds. Then $u(t,x) \geq M(t,x,q)$ in $C_{2,1}$.

Proof. We will only prove (i) because the proof of (ii) is similar. First assume that $\mathcal{L} \in \mathbb{L}^{t,\infty}_{\delta,K}$ and introduce

$$\Gamma = C_1 \cap \{\mathcal{L}u \leq -1\}.$$

We claim that there exists a sequence of closed $\Gamma_n \subset C_1$ for which $|\Gamma_n| \geq q|C_1|$ and $I_{\Gamma_n} \to I_\Gamma$ on C_1 (a.e.). Indeed, if $|\Gamma| > q|C_1|$ it suffices to use the regularity of Lebesgue measure and take closed $\Delta_n \subset \Gamma$ such that $I_{\Delta_n} \to I_\Gamma$ (a.e.). In case $|\Gamma| = q|C_1|$ it holds that

$$|C_1 \setminus \Gamma| = (1-q)|C_1| > 0,$$

and there exist closed $\delta_n \subset C_1 \setminus \Gamma$ for which $I_{\delta_n} \to 0$ (a.e.) and $|\delta_n| \geq |\Gamma \setminus \Delta_n|$ for large n. Then for the closed sets $\Gamma_n = \Delta_n \cup \delta_n$ and sufficiently large n it holds that

$$|\Gamma_n| = |\Gamma| - |\Gamma \setminus \Delta_n| + |\delta_n| \geq q|C_1|.$$

Observe that the indicator of any closed set is the decreasing limit of smooth functions. Therefore, there exist uniformly bounded $\zeta_n \in C_0^\infty$ such that $\big|C_1 \cap \{\zeta_n \geq 1\}\big| \geq q|C_1|$ and $\zeta_n \to I_\Gamma$ (a.e.) in C_1.

Next, from the linear theory we know that there exist $v, v_n \in \overset{0}{W}{}^{1,2}_{d+1}(C_1)$ satisfying

$$\mathcal{L}v = -I_\Gamma, \quad \mathcal{L}v_n = -\zeta_n,$$

Clearly $v_n \in C^\infty(\bar{C}_1)$ and $v_n \in \mathbb{U}^m_q$. By the maximum principle, $u \geq v$ and by Theorem 3.0.1, $v - v_n \to 0$. It follows that

$$u(t,x) \geq v_n(t,x) + \big[v(t,x) - v_n(t,x)\big] \geq m(t,x,q) + \big[v(t,x) - v_n(t,x)\big].$$

Sending $n \to \infty$, we get the desired result in our particular case.

For general $\mathcal{L}$ we find a sequence $\mathcal{L}_n \in \mathbb{L}^{t,\infty}_{\delta,K}$ with the coefficients of $\mathcal{L}_n$ converging (a.e.) to the corresponding coefficients of $\mathcal{L}$. Set $f = \mathcal{L}u$, $f_n = \mathcal{L}_n u$, and introduce u_n and w_n as the $\overset{0}{W}{}^{1,2}_{d+1}(C_1)$-solutions of $\mathcal{L}_n u_n = f_n$ and $\mathcal{L}_n w_n = f - f_n$. We have

$$\mathcal{L}_n(u + w_n) = \mathcal{L}_n(u_n + w_n) = f.$$

It follows by the above result combined with the maximum principle that $u + w_n \geq u_n + w_n \geq m(q)$. It only remains to add that $w_n \to 0$ by Theorem 3.0.1 because $f_n \to f$ in $L_{d+1}(C_1)$. The lemma is proved. □

Assuming that Theorem 2 is proved we have the following.

5. COROLLARY. *Let $\mathcal{L}_n \in \mathbb{L}^t_{\delta,K}$ and $u_n \in W^{1,2}_{d+1}(C_1)$. Assume that $u_n \geq 0$ in C_1 and $\mathcal{L}_n u_n = f_n \leq 0$ in C_1 (a.e.), $n = 1, 2, \ldots$. Take x_0 such that $|x_0| < 1$ and assume that $u_n(0, x_0) \to 0$ as $n \to \infty$. Then $f_n \to 0$ in measure as $n \to \infty$.*

Indeed, by Lemma 4 for each $\kappa > 0$

$$\kappa^{-1} u_n(0, x_0) \geq m\big(0, x_0, \big|\{f_n \leq -\kappa\}\big|/|C_1|\big).$$

Theorem 2 implies that the expression on the right goes to zero only if $\big|\{f_n \leq -\kappa\}\big| \to 0$. □

This corollary is a very deep generalization of Corollary 4.2.12.

For elliptic equations we have the following results. Let $\mathbb{L}_{\delta,K}$ be the set of elliptic operators

$$\mathcal{L} = a^{ij}(x) D_{ij} + b^i(x) D_i - c(x),$$

with measurable coefficients defined in $\mathbb{R}^d$ and satisfying (2) for all λ, $x \in \mathbb{R}^d$, $i, j = 1, ..., d$. Let $\mathbb{L}^\infty_{\delta,K}$ be the subset of $\mathbb{L}_{\delta,K}$ consisting of operators with C^∞-coefficients. We introduce sets $\mathbb{U}^m_q$ by replacing C_1, $C^{1,2}(C_1)$, and $\mathbb{L}^{t,\infty}_{\delta,K}$ in the definition of $\mathbb{U}^{t,m}_q$ with B_1, $C^2(B_1)$, and $\mathbb{L}^\infty_{\delta,K}$, respectively. Define

$$m(x, q) = \inf\big\{u(x) : u \in \mathbb{U}^m_q\big\} \quad x \in B_1, \quad q \in [0, 1].$$

6. THEOREM. *(a) We have $m(x, q) \geq M(0, x, q)$.*

(b) If $q \in [0, 1)$, $u \in W^2_d(B_1)$, and $\mathcal{L} \in \mathbb{L}_{\delta,K}$ are such that $u \geq 0$ on ∂B_1, $\mathcal{L}u \leq 0$ in B_1 (a.e.), and $\big|B_1 \cap \{\mathcal{L}u \leq -1\}\big| \geq q|B_1|$, then $u(x) \geq m(x, q)$ on B_1.

Proof. Assertion (a) trivially follows from Lemma 4, since $C^2(B_1)$ is the subset of $C^{1,2}(C_{2,1})$ composed of functions independent of t. The proof of (b) is almost identical to that of Lemma 4. One only needs to use Theorem 3.0.2 in place of Theorem 3.0.1. □

7. REMARK. Theorems 6 (a) and 1 imply that for $|x| \leq R < 1$ we have $Nm(x, q) \geq q^{1/\gamma}$. This fact was first noted in [**35**].

8. COROLLARY. *Let $\mathcal{L}_n \in \mathbb{L}_{\delta,K}$ and $u_n \in W^2_d(B_1)$. Assume that $u_n \geq 0$ in B_1 and $\mathcal{L}_n u_n = f_n \leq 0$ on B_1 (a.e.), $n = 1, 2, ...$ Take $x_0 \in B_1$ and assume that $u_n(x_0) \to 0$ as $n \to \infty$. Then $f_n \to 0$ in measure as $n \to \infty$.*

Thanks to Theorem 6 (a), the proof of this corollary is similar to that of Corollary 5.

2. Auxiliary results

The following is a "lemma on an oblique cylinder".

1. LEMMA. *Let $\kappa \in (0, 1]$, and let Q lying in $C_{8,2}$ be a cylinder whose bases are balls in $\mathbb{R}^d$ of the same radius R such that one base lies in the hyperplane $\big\{(t, x) : t = t_1\big\}$ and the other in the hyperplane $\big\{(t, x) : t = t_2\big\}$ and $t_1 < t_2$. Suppose that the tangent of the angle between the axis of Q and the t-axis does not exceed κ^{-1} and*

$$\kappa(t_2 - t_1) \leq R^2 \leq \kappa^{-1}(t_2 - t_1). \tag{1}$$

Let $\mathcal{L} \in \mathbb{L}^t_{\delta,K}$, let $u \in W^{1,2}_{d+1}(Q)$ satisfy $u \geq 0$ and $\mathcal{L}u \leq 0$ in Q (a.e.), and for a $\gamma \in [0, 1]$, let $u(t_2, x) \geq 1$ for x belonging to the ball of radius γR whose center coincides with that of the base of Q lying in $\{t = t_2\}$. Then there exist constants $\varepsilon > 0$ and $n \geq 0$, which depend only on d, K, δ, and κ, such that

$$u(t_1, x) \geq \varepsilon \gamma^n$$

for x belonging to the ball of radius $(1-\kappa)R$ whose center coincides with that of the base of Q lying in the plane $\{t=t_1\}$.

Proof. Let $\mathcal{L}$ be given in Q as

$$\mathcal{L}u = \partial_t u + a^{ij}D_{ij}u + b^i D_i u - cu,$$

Take $\tilde{b} \in \mathbb{R}^d$ and observe that $\tilde{u}(t,x) := u(t, x+\tilde{b}t)$ satisfies

$$\tilde{\mathcal{L}}\tilde{u}(t,x) = (\mathcal{L}u)(t,x+\tilde{b}t) \quad \text{in} \quad \tilde{Q} := \big\{(t,x) : (t,x+\tilde{b}t) \in Q\big\},$$

where $\tilde{\mathcal{L}}$ has form (1.1) with the coefficients $\tilde{a}$, $\tilde{b}$, and $\tilde{c}$ defined by

$$\tilde{a}^{ij}(t,x) = a^{ij}(t,x+\tilde{b}t),$$

$$\tilde{b}^i(t,x) = b^i(t,x+\tilde{b}t) - \tilde{b}^i, \quad \tilde{c}(t,x) = c(t,x+\tilde{b}t).$$

If $\tilde{b}$ is chosen appropriately, $\tilde{Q}$ is a cylinder with axis parallel to the t-axis. Observe that for such a $\tilde{b}$ the above coefficients satisfy (1.2) with $K+\kappa^{-1}$ in place of K. Hence, we may assume from the start that the cylinder Q is right.

Next, in light of (1) and the fact that $R \le 2$ ($Q \subset C_{8,2}$), one can easily find an $x_0 \in \mathbb{R}^d$ such that for

$$\tilde{u}(t,x) := u\big(t_1 + t(t_2-t_1), x_0 + Rx\big)$$

considered in C_1 there is an $\tilde{\mathcal{L}}$ of type (1.1) such that in C_1

$$\tilde{\mathcal{L}}\tilde{u}(t,x) = (t_2-t_1)(\mathcal{L}u)\big(t_1+t(t_2-t_1), x_0+Rx\big)$$

and its coefficients satisfy (1.2) with $\tilde{\delta} > 0$ and $\tilde{K}$ depending only on δ, K, and κ. This shows that we need only concentrate on the case $Q = C_1$.

In this case, our task is to show that

$$u(0,x_0) \ge \varepsilon\gamma^n$$

if $|x_0| \le 1-\kappa$, with $\varepsilon > 0$ and $n \ge 0$ depending only on d, K, δ, and κ. Consider the oblique cylinder $\tilde{Q}$ with bases $B_\kappa(x_0)$ and B_κ lying in the planes $\{t=0\}$ and $\{t=1\}$, respectively. Clearly, $u \ge 0$, $\mathcal{L}u \le 0$ in $\tilde{Q}$ (a.e.), $u(1,x) \ge 1$ for $x \in B_{\kappa\wedge\gamma}$, and we need to show that $u \ge \varepsilon\gamma^n$ at the center of the lower base.

As in the beginning of the proof we can "straighten" $\tilde{Q}$ and reduce the problem to showing that

$$u(0,0) \ge \varepsilon\gamma^n$$

if $u \in W^{1,2}_{d+1}(C_{1,\kappa})$, $u \ge 0$, $\mathcal{L}u \le 0$ on $C_{1,\kappa}$ (a.e.), where $\mathcal{L} \in \mathbb{L}^t_{\delta,K}$, $\gamma \in [0,1]$, $u(1,x) \ge 1$ for $|x| < \kappa \wedge \gamma$, with $\varepsilon > 0$ and $n \ge 0$ depending only on d, K, δ, and κ. Obviously, we may also assume that

$$\gamma \le \kappa/2.$$

Introduce

$$\xi = \kappa^2 - \gamma^2, \quad \varphi = \varphi(t,x) = \xi(1-t) - |x|^2 + \gamma^2,$$

$$\psi = \psi(t,x) = \varphi^2(t,x)\big(\xi(1-t)+\gamma^2\big)^{-n},$$

$$P = \big\{(t,x) : 0 < t < 1, \varphi(t,x) > 0\big\},$$

where n will be specified later. We have $P \subset C_{1,\kappa}$, $\psi(t,x) = 0$ for $(t,x) \in \partial' P$, $t < 1$ and $\psi(1,x) \leq \gamma^{4-2n}$ for $(1,x) \in \partial' P$. Consequently,

$$\gamma^{2n-4}\psi \leq u \quad \text{on } \partial' P.$$

Now we compute and find that in P

$$\begin{aligned}\big(\xi(1-t)+\gamma^2\big)^n \mathcal{L}\psi = n\xi\big(\xi(1-t)+\gamma^2\big)^{-1}\varphi^2 \\ + 8a^{ij}x^ix^j - [2\xi + 4\operatorname{tr} a + 4b^ix^i + c\varphi]\varphi.\end{aligned} \tag{2}$$

It is not hard to see that $0 \leq \varphi \leq \xi + \gamma^2 = \kappa^2 \leq 1$ and

$$|2\xi + 4\operatorname{tr} a + 4b^ix^i + c\varphi| \leq N_0 := 2 + 4\delta^{-1} + (4d+1)K$$

in P. Therefore, (2) implies that $\mathcal{L}\psi \geq 0$ on the part of P on which $8\delta|x|^2 \geq N_0\varphi(t,x)$. Everywhere else in P we have

$$8\delta\big(\xi(1-t)+\gamma^2-\varphi\big) < N_0\varphi,$$

$$8\delta\big(\xi(1-t)+\gamma^2\big) < (N_0 + 8\delta)\varphi,$$

and

$$\big(\xi(1-t)+\gamma^2\big)^{-1}\varphi^2 > \varphi 8\delta(N_0+8\delta)^{-1}.$$

By the choice of γ, $\xi \geq 3\kappa^2/4$, and on this part of P we get that $\mathcal{L}\phi \geq 0$ by virtue of (2) with

$$n = \frac{4}{3}\kappa^{-2}N_0(N_0+8\delta)(8\delta)^{-1}.$$

We have proved that $\mathcal{L}\psi \geq 0$ in P. The maximum principle implies that $\gamma^{2n-4}\psi \leq u$ in $\bar{P}$ and, in particular,

$$u(0,0) \geq \gamma^{2n-4}\varphi(0,0) = \gamma^{2n-4}\frac{\kappa^4}{\kappa^{2n}} = \kappa^{4-2n}\gamma^{2n-4}.$$

The lemma is proved. □

Here is the first crucial step.

2. LEMMA. *Let $R \in [0,1)$. Then there exist $q_0 \in (0,1)$ and $m_0 > 0$, which depend only on K, δ, d, and R, such that $m(0,x,q) \geq m_0$ for $|x| \leq R$ and $q \in [q_0, 1]$.*

Proof. Take $q \in [0,1]$, $u \in \mathbb{U}_q^{t,m}$, and let $\mathcal{L} \in \mathbb{L}_{\delta,K}^{t,\infty}$ be the operator corresponding to q and u. Introduce

$$\Gamma = C_1 \cap \{\mathcal{L}u \geq -1\}.$$

Also introduce $v, w \in \overset{0}{W}{}^{1,2}_{d+1}(C_1)$ as the solutions of

$$\mathcal{L}v = -I_\Gamma, \quad \mathcal{L}w = -1.$$

We have $\mathcal{L}(w - v) \geq \mathcal{L}u$, which implies by the maximum principle that

$$u \geq w - v.$$

By Theorem 3.0.1,

$$|v| \leq N_1|\Gamma|^{1/(d+1)} \leq N_2(1-q)^{1/(d+1)},$$

where N_1 and N_2 depend only on d, K, and δ. We have estimated v from above and next we are going to estimate w from below.

Introduce $\varphi = 1-|x|^2$ and introduce $\psi \in \overset{0}{W}{}^{1,2}_{d+1}(C_1)$ as the solution of $\mathcal{L}\psi = \mathcal{L}\varphi$. We have $\mathcal{L}(\varphi - \psi) = 0$ in C_1 (a.e.) and $0 \leq \varphi - \psi \leq 1$ on $\partial' C_1$, which implies, by the maximum principle, that $0 \leq \varphi - \psi \leq 1$ in $\bar{C}_1$. Therefore,

$$\mathcal{L}\big[(1-t)(\varphi - \psi)\big] = -(\varphi - \psi) + (1-t)\mathcal{L}(\varphi - \psi) \geq -1$$

in C_1 (a.e.) and, once more by the maximum principle,

$$w \geq (1-t)(\varphi - \psi).$$

Also use again $\mathcal{L}(\varphi - \psi) = 0$ and observe that $\varphi - \psi = \varphi \geq 1/2$ for $t = 1$ and $|x|^2 \leq 1/2$. By Lemma 1, there is an $\varepsilon > 0$, depending only on d, K, δ, and R, for which $(\varphi - \psi)(0, x) \geq \varepsilon$ if $|x| \leq R$. Therefore,

$$u(0,x) \geq \varepsilon - N(1-q)^{1/(d+1)} \tag{3}$$

if $|x| \leq R$. Obviously, there exists a q_0 such that the right-hand side of (3) is greater than $\varepsilon/2$ for any $q \geq q_0$. After that it only remains to to use the arbitrariness of $u \in \mathbb{U}^{t,m}_q$ in (3). The lemma is proved. □

3. Corollary. *For any $R \in (0,1)$ there exist $q_0 \in (0,1)$ (close to 1) and $M_0 > 0$, depending only on δ, d, K, and R, such that*

$$M(t,x,q) \geq M_0$$

for $q \in [q_0, 1]$ and $(t,x) \in C_R$.

Indeed, for q_0 from Lemma 2 and $q \in [q_0, 1]$, if $u \in \mathbb{U}^{t,M}_q$, then $u(1+\cdot,\cdot) \in \mathbb{U}^{t,m}_q$, and $u(1,x) \geq m_0$, for $|x| \leq R$, where m_0 is taken from Lemma 2, and it only remains to use Lemma 1. □

The following lemma asserts that "if you win locally in time, you win globally".

4. Lemma. *Let $\mathcal{L} \in \mathbb{L}^{t,\infty}_{\delta,K}$, let Γ be an open subset of C_1, and let nonpositive $f, g \in L_{d+1}(C_1)$, be such that*

$$g = gI_\Gamma.$$

Denote by u and v the $\overset{0}{W}{}^{1,2}_{d+1}(C_1)$-solutions of the equations

$$\mathcal{L}u = f, \quad \mathcal{L}v = g.$$

Suppose that for any point, $(t_0, x_0) \in \Gamma$, there exist numbers τ, $\rho > 0$ and a point $(t_1, x_1) \in C_1$, such that

$$(t_0, x_0) \in C_{\tau,\rho}(t_1, x_1) \subset C_1$$

and for $u^1, v^1 \in \overset{0}{W}{}^{1,2}_{d+1}\big(C_{\tau,\rho}(t_1, x_1)\big)$ defined as the solutions of the equations

$$\mathcal{L}u^1 = f, \quad \mathcal{L}v^1 = g,$$

we have $u^1(t_0, x_0) \geq v^1(t_0, x_0)$. Then $u \geq v$ in C_1.

Proof. By subtracting a constant $\chi > 0$ from f and then passing to the limit as $\chi \downarrow 0$, we see that we may assume that f is separated from zero. In that case

$$u^1(t_0, x_0) > 0 \tag{4}$$

(apply Theorem 3.2.8 to $-u^1$ in place of u). The maximum principle shows that, if for fixed $\mathcal{L}$, Γ, and f the conditions of the lemma are satisfied for a function g, they are also satisfied for gI_{Γ^1}, where Γ^1 is any measurable subset of C_1. By Theorem 3.0.1, $v_n \to v$ if $v_n \in \overset{0}{W}{}^{1,2}_{d+1}(C_1)$ are solutions of $\mathcal{L}v_n = gI_{\Gamma^n}$ with $\Gamma^n \uparrow \Gamma$. It follows that it suffices to concentrate on the case that $g = 0$ outside a closed subset Γ^1 of Γ and it suffices to prove that $v \le (1+\varepsilon)u$ on $\bar{C}_1$ for any $\varepsilon > 0$.

Assume the contrary. Then for an $\varepsilon > 0$

$$\gamma := \max\big(v - (1+\varepsilon)u,\ \bar{C}_1\big) > 0.$$

If we apply the maximum principle to $v - (1+\varepsilon)u$ considered in $C_1 \setminus \Gamma^1$ then we see that the above maximum is attained at a point $(t_0, x_0) \in \Gamma^1$.

Let τ, $\rho > 0$ and (t_1, x_1) be as in the formulation of the lemma. Observe that

$$\mathcal{L}(u - u^1) = \mathcal{L}(v - v^1) = 0$$

in $C_{\tau,\rho}(t_1, x_1)$ (a.e.) and, on $\partial' C_{\tau,\rho}(t_1, x_1)$ we have

$$v - v^1 = v \le \gamma + (1+\varepsilon)u = \gamma + (1+\varepsilon)(u - u^1).$$

By the maximum principle (note $\mathcal{L}\gamma \le 0$),

$$v - v^1 \le \gamma + (1+\varepsilon)(u - u^1)$$

at (t_0, x_0). Hence, at (t_0, x_0) (see (4))

$$\gamma = v - (1+\varepsilon)u \le \gamma + v^1 - (1+\varepsilon)u^1 < \gamma + v^1 - u^1 \le \gamma.$$

We have arrived at $\gamma < \gamma$ which is false. The lemma is proved. □

Next comes the lemma on "running ink spots under the hot wind". It is, actually, a part of measure theory and we choose to defer its proof to Appendix B. Let Γ be a measurable subset of C_1 and let

$$q, \eta, \zeta \in (0,1).$$

For $T, R > 0$ introduce

$$C^o_{T,R} = (0,T) \times B_R, \quad C^o_{T,R}(t,x) = (t,x) + C^o_{T,R}, \quad C^o_R(t,x) = C^o_{R^2,R}(t,x).$$

We also introduce $\mathcal{B} = \mathcal{B}(\Gamma, q)$ as the family of *open* cylinders Q of type $C^o_\rho(t_0, x_0)$ such that

$$Q \subset C_1 \quad \text{and} \quad |Q \cap \Gamma| \ge q|Q|.$$

For $Q = C^o_\rho(t_0, x_0) \in \mathcal{B}$ define

$$Q' = (t_0, x_0) - C^o_{\eta^{-1}\rho^2,\rho}, \quad Q'' = \big(t_0 - \eta^{-1}\rho^2, x_0\big) + C^o_{\eta^{-1}\rho^2\zeta^2,\rho\zeta}.$$

Imagine that the t-axis is pointed up vertically. Then Q' is adjacent to Q from below, the two cylinders have a common base, and along the t-axis Q' is η^{-1} times longer than Q. The possibility that Q' sticks out of C_1 is not excluded. The cylinder Q'' is obtained by contracting Q' to the center of its lower base with the contraction factor ζ^{-2} for the t-axis and ζ^{-1} for the spatial axes.

5. REMARK. If $Q = C^o_\rho(t_0, x_0)$, then the shortest distance between Q and Q'' along the t axis is

$$\eta^{-1}\rho^2 - \eta^{-1}\rho^2\zeta^2 = \eta^{-1}\rho^2(1-\zeta^2), \tag{5}$$

which is bigger than $2\rho^2$ if

$$\zeta^2 \le 1 - 2\eta. \tag{6}$$

Finally, define

$$\Gamma'' = \bigcup_{Q\in\mathcal{B}} Q''.$$

Observe that for $Q \in \mathcal{B}$ the set Q'' is open. Hence, Γ'' is open and measurable.

6. LEMMA. *If* $|\Gamma| \le q|C_1|$, *then*

$$|\Gamma''| \ge \Big(1 - \frac{1-q}{3^{d+1}}\Big)^{-1}(1+\eta)^{-1}\zeta^{d+2}|\Gamma|.$$

3. Proof of Theorems 1.1 and 1.2

Proof of Theorem 1.1. We follow the exposition in [**61**] and [**70**] originated in [**84**] and [**100**].

Observe that, obviously, in suffices to prove the theorem for $R \in [0,1)$ such that $R^2 \ge 1/2$. We fix such an R and set

$$\mu(q) = \inf\big\{M(0,x,q) : |x| \le R\big\}, \quad \nu(q) = \inf\big\{m(0,x,q) : |x| \le R\big\}.$$

By Lemma 2.2 and Corollary 2.3 there is a $q_0 \in (0,1)$ and $\mu_0 > 0$ such that

$$\mu(q), \nu(q) \ge \mu_0$$

for $q \in [q_0, 1]$.

Next, we take some $0 < q' < q'' < 1$ and try to relate $\mu(q')$ to $\mu(q'')$. More specifically, we set

$$\xi := \big(1 - (1-q_0)3^{-d-1}\big)^{-1}(1+\eta)^{-1}\zeta^{d+2},$$

and find and fix $\eta = \eta(q_0) \in (0,1)$ and $\zeta = \zeta(q_0) \in (0,1)$ such that (2.6) is satisfied and $\xi = \xi(q_0) > 1$. We will be interested in $0 < q' < q'' < 1$ such that

$$(1+\xi)q' \ge 2q''. \tag{1}$$

Take a $u \in \mathbb{U}^{t,M}_{q'}$, let $\mathcal{L}$ be the corresponding operator, and define

$$\Gamma = C_1(1,0) \cap \{\mathcal{L}u \le -1\}.$$

In the construction before Lemma 2.6 we replace C_1 by $C_1(1,0)$, keep all other notation, and from the chosen Γ, q_0, η, and ζ we build up the set Γ''. There are two cases:

(i) $\big|\Gamma'' \setminus C_1(1,0)\big| \le (q''-q')|C_1|$,

(ii) $\big|\Gamma'' \setminus C_1(1,0)\big| > (q''-q')|C_1|$.

Case (i). Our goal is to show that

$$u(0,x) \ge \min\big(\mu_0, N_0^{-1}\mu(q'')\big), \quad |x| \le R, \tag{2}$$

where N_0 depends only on d, δ, K, and R.

Observe that, if $|\Gamma| \geq q_0|C_1|$, by definition $u(0,x) \geq \mu(q_0) \geq \mu_0$ for $|x| \leq R$. Hence, we may assume that

$$|\Gamma| < q_0|C_1|.$$

With the goal of using Lemma 2.4 denote by $\tilde{u}$ and $\tilde{v}$ the $\overset{0}{W}{}^{1,2}_{d+1}(C_{2,1})$-solutions of

$$\mathcal{L}\tilde{u} = -I_\Gamma, \quad \mathcal{L}\tilde{v} = -I_{\Gamma_0},$$

where

$$\Gamma_0 = \Gamma'' \cap C_1^o(1,0).$$

Since the coefficients of $\mathcal{L}$ are infinitely differentiable, such solutions exist and by the maximum principle $u \geq \tilde{u}$. Then we are going to estimate $\tilde{v}$ from below and after that use Lemma 2.4 to estimate from below $\tilde{u}$ in terms of $\tilde{v}$.

Notice that by definition and Lemma 2.6

$$q'|C_1| \leq |\Gamma| \leq \xi^{-1}|\Gamma''|.$$

Moreover, by assumption

$$|\Gamma''| = \big|\Gamma'' \setminus C_1(1,0)\big| + |\Gamma_0| \leq (q''-q')|C_1| + |\Gamma_0|.$$

Due to (1), it follows that

$$|\Gamma_0| \geq q''|C_1|,$$

so that by Lemma 1.4

$$\tilde{v}(0,x) \geq \mu(q''), \quad |x| \leq R.$$

We thus have estimated $\tilde{v}$ from below. As we have mentioned above, $u \geq \tilde{u}$ and we plan to estimate $\tilde{u}$ from below in terms of $\tilde{v}$. Here we apply Lemma 2.4.

If $(t_0,x_0) \in \Gamma_0$, there exists $Q \in \mathcal{B}$ for which $(t_0,x_0) \in Q''$. Define (t_1,x_1) and τ, ρ from the equation

$$C^o_{\tau,\rho}(t_1,x_1) = Q_1 := (Q' \cup Q) \cap C^o_{2,1},$$

so that

$$Q = C^o_\rho\big(\tau + t_1 - \rho^2, x_1\big).$$

Now, let $\hat{u}$ and $\hat{v}$ be the $\overset{0}{W}{}^{1,2}_{d+1}(Q_1)$-solutions of

$$\mathcal{L}\hat{u} = -N_0 I_\Gamma, \quad \mathcal{L}\hat{v} = -I_{\Gamma_0},$$

where the constant N_0, depending only on d, δ, K, and R, will be specified later in such a way that

$$\hat{u}(t_0,x_0) \geq \hat{v}(t_0,x_0). \tag{3}$$

If we can do this, then by Lemma 2.4 we have $N_0\tilde{u} \geq \tilde{v}$ on $C_{2,1}$ and

$$u(0,x) \geq \tilde{u}(0,x) \geq N_0^{-1}\tilde{v}(0,x) \geq N_0^{-1}\mu(q''), \quad |x| \leq R. \tag{4}$$

Observe that the height of the cylinder Q_1 is less than the height of $Q' \cup Q$, which is $\rho^2 + \eta^{-1}\rho^2$. It follows that

$$t_1 + \tau - t_0 \leq \tau \leq \rho^2 + \eta^{-1}\rho^2. \tag{5}$$

By the maximum principle $\hat{v}(t,x) \leq t_1 + \tau - t$, so that, owing to (5),

$$\hat{v}(t_0, x_0) \le (1 + \eta^{-1})\rho^2. \tag{6}$$

Furthermore, by the choice of Q we have $q_0|Q| \le |\Gamma \cap Q|$. This relation is preserved under the parabolic scaling $x \to \rho^{-1}x$, $t \to \rho^{-2}t$ which transforms Q into a cylinder $\check{Q}$ congruent to C_1 and $\hat{u}$ into a function $\check{u}$, that satisfies $\check{\mathcal{L}}\check{u} = -N_0\rho^2 I_{\check{\Gamma}}$ in $\check{Q}$ (a.e.) with an $\check{\mathcal{L}} \in \mathbb{L}_{\delta,K}^{t,\infty}$ and $\check{\Gamma}$ being the image of Γ. It follows by the definition of $\nu(q_0)$ that, if (t, x) belongs to the lower base of $\check{Q}$ and is at a distance less than R of the lower base's center, then

$$\check{u}(t, x) \ge N_0\rho^2 \inf_{|y|\le R} m(0, y, q_0) \ge N_0\rho^2\nu(q_0) \ge N_0\rho^2\mu_0.$$

Consequently, we have $\hat{u}(\tau + t_t - \rho^2, x) \ge N_0\rho^2\mu_0$ for $|x - x_1| \le \rho R$ (that is in a ball on the bottom of Q).

Furthermore, by Remark 2.5, since $(t_0, x_0) \in Q''$ $(\subset Q')$, the distance from (t_0, x_0) to the bottom of Q is bigger than $2\rho^2$. On the other hand, the said distance is less than the hight of Q', which is $\eta^{-1}\rho^2$. Since also $|x_0 - x_1| \le (1 - \zeta)\rho$, it follows by Lemma 2.1 that,

$$\hat{u}(t_0, x_0) \ge \alpha N_0\rho^2\mu_0,$$

where $\alpha > 0$ depends only on d, δ, K, and R.

By taking

$$N_0 = \alpha^{-1}(1 + \eta^{-1})\mu_0^{-1}$$

and recalling (6) we come to (3) and thus (4) and (2) are established in case (i). Observe that N_0 indeed depends only on d, δ, K, and R (by the way, recall that $\eta = \eta(q_0)$).

Case (ii). Here the goal is to prove that

$$u(0, x) \ge \mu_0\varepsilon\eta^n(q'' - q')^n, \quad |x| \le R, \tag{7}$$

where $\varepsilon > 0$ and $n \ge 1$ depend only on δ, K, d, and R.

First we claim that for some $(t, x) \in \Gamma''$ it holds that $t < q' - q'' + 1$. Indeed, otherwise $\Gamma'' \setminus C_1(1, 0) \subset C_{q''-q',1}(q' - q'' + 1, 0)$ and $|\Gamma'' \setminus C_1(1, 0)| \le (q'' - q')|C_1|$. It follows that there is a cylinder

$$Q = C_\rho^o(t_0, x_0) \in \mathcal{B}$$

such that Q' contains points in the half-space $t < q' - q'' + 1$. Since $q' < q''$, we have $q' - q'' + 1 < 1$, and since Q' is adjacent to $Q \subset C_1(1, 0)$, this implies that the height of Q' is at least $q'' - q'$, that is,

$$\rho^2\eta^{-1} \ge q'' - q', \quad \rho^2 \ge \eta(q'' - q'). \tag{8}$$

On the other hand, $Q \subset C_1(1, 0)$, $t_0 > 1$, and $\rho < 1$.

Moreover, by definition, $|\Gamma \cap Q| \ge q_0|Q|$ and by using the definition of M and scalings we see that

$$u\big(t_0 - \rho^2, x\big) \ge \mu_0\rho^2, \quad |x - x_0| \le \rho R. \tag{9}$$

If $t_0 - \rho^2 \ge 1/4$, then (9), by Lemma 2.1, implies that

$$u(0, x) \ge \mu_0\varepsilon\rho^n, \quad |x| \le R,$$

where $\varepsilon > 0$ and $n \geq 1$ depend only on δ, K, d, and R. In particular (see (8)), we come to (7).

However, if $t_0 - \rho^2 < 1/4$, then $\rho^2 \geq 3/4$, $u(t_0, x) \geq (3/4)\mu_0$ for $|x - x_0| \leq R\sqrt{3/4}$, and $u(0, x) \geq \mu_0\varepsilon$ by Lemma 2.1 and (7) holds again.

By combining the two cases (i) and (ii) we conclude that

$$u(0,x) \geq \min\big(\mu_0, N_0^{-1}\mu(q''), \mu_0\varepsilon\eta^n(q''-q')^n\big), \quad |x| \leq R,$$

and since $u \in \mathbb{U}^{t,M}_{q'}$ is arbitrary

$$\mu(q') \geq \min\big(\mu_0, N_0^{-1}\mu(q''), \mu_0\varepsilon\eta^n(q''-q')^n\big), \tag{10}$$

whenever (1) holds.

After introducing

$$\hat{\mu}(q) = \min\big(\mu_0, \mu(q)\big)$$

(10) yields that there exist $\varepsilon_0 \in (0,1)$ and $n_0 \geq 1$ depending only on δ, K, d, and R, such that for any $0 < q' < q'' < 1$ satisfying $(1+\xi)q' \geq 2q''$ we have

$$\hat{\mu}(q') \geq \varepsilon_0 \min\big((q''-q')^{n_0}, \hat{\mu}(q'')\big). \tag{11}$$

We also know that $\hat{\mu}(q) \geq \mu_0 > 0$ for $q \geq q_0$.

We may certainly assume that $\varepsilon_0 \leq \bar{\varepsilon} := 2/(1+\xi)$ (recall that $\xi > 1$) and we claim that for $q_k = \bar{\varepsilon}^k q_0$, $k = 0, 1, 2, \dots$, we have

$$\hat{\mu}(q_k) \geq \varepsilon_0^{kn_0}\chi, \quad \chi := \min\big(\mu_0, q_0^{n_0}(1-\bar{\varepsilon})^{n_0}\big). \tag{12}$$

To prove the claim we use induction. If $k = 0$, (12) is obvious. If it is true for a k, then $q_k - q_{k+1} = \bar{\varepsilon}^k q_0(1-\bar{\varepsilon})$,

$$(q_k - q_{k+1})^{n_0} = \bar{\varepsilon}^{kn_0} q_0^{n_0}(1-\bar{\varepsilon})^{n_0} \geq \varepsilon_0^{kn_0}\chi,$$

so that by (11) and the fact that $(1+\xi)q_{k+1} = 2q_k$

$$\hat{\mu}(q_{k+1}) \geq \varepsilon_0 \min\big(\varepsilon_0^{kn_0}\chi, \hat{\mu}(q_k)\big) \geq \varepsilon_0\varepsilon_0^{kn_0}\chi \geq \varepsilon_0^{(k+1)n_0}\chi.$$

This proves (12) and shows that, if we define $r > 1$ so that $\varepsilon_0^{n_0} = \bar{\varepsilon}^r$, then $\hat{\mu}(q_k) \geq Nq_k^r$ with $r, N > 0$ depending only on δ, K, d, and R. By observing that $\hat{\mu}$ is an increasing function we obtain that $\hat{\mu}(q) \geq Nq^r$, $\mu(q) \geq Nq^r$ for $q \leq 1$. This yields (1.7) with $\gamma = 1/r$ and proves the theorem. □

Proof of Theorem 1.2. This theorem is a rather simple consequence of Theorem 1.1 and Lemma 1.4. Indeed, take $u \in \mathbb{U}^{t,m}_q$, the corresponding operator $\mathcal{L} \in \mathbb{L}^{t,\infty}_{\delta,K}$, $t_0 \in [0, qR]$, and introduce a piecewise linear continuous function $f(t)$ such that

$$f\big([t_0, q + t_0/2]\big) = [0,1], \quad f\big([q + t_0/2, 1]\big) = [1,2].$$

Obviously the first derivatives of f and of its inverse f^{-1} are controlled by R and q uniformly with respect to $t_0 \in [0, qR]$. Since the portion of the set $\{\mathcal{L}u \leq -1\}$ in C_1 is greater than q, the image of $\{\mathcal{L}u \leq -1\}$ under the mapping $F : (t,x) \to \big(f(t), x\big)$ will occupy at least $(1-R)q/2$ portion of $C_1(1,0)$.

Furthermore, introduce a function and an operator by

$$\tilde{u}(t,x) = \big(1 - (q+t_0)/2\big)^{-1} u\big(F^{-1}(t,x)\big), \quad \tilde{\mathcal{L}}\phi = \Big[(f')^{-1}\mathcal{L}\{\phi(F)\}\Big](F^{-1}).$$

As is easy to see $\tilde{\mathcal{L}}$ is of type (1.1) with coefficients satisfying (1.2), where one should replace δ and K with new ones, but depending only on the original K, δ, d, q, and R. Also notice that $f' = \big(1-(q+t_0)/2\big)^{-1}$ for $t \in [(q+t_0/2, 1]$, which implies that the set

$$\{(t,x) \in C_1(1,0) : \tilde{\mathcal{L}}\tilde{u}(t,x) \le -1\}$$

coincides with

$$\{(t,x) \in C_1(1,0) : (\mathcal{L}u)\big(F^{-1}(t,x)\big) \le -1\}.$$

By the above this set occupies at least $(1-R)q/2$ portion of $C_1(1,0)$, so that by Theorem 1.1, for $|x| \le R$,

$$\big(1-(1+R)q/2\big)^{-1}u(t_0,x) \ge \big(1-(q+t_0)/2\big)^{-1}u(t_0,x) = \tilde{u}(0,x) \ge m_0,$$

where m_0 depends only on d, K, δ, q, and R. This, obviously, proves the theorem. □

4. Fang-Hua Lin's type estimates for elliptic and parabolic equations

In the stated form Theorem 1 below first appeared in [**70**] without really a proof. The complete proof is given in [**73**]. Recall that the sets of operators $\mathbb{L}^t_{\delta,K}$ and $\mathbb{L}_{\delta,K}$ are introduced in Section 1 before Theorems 1.1 and 1.6, respectively.

1. THEOREM. *Let $w \in W^{1,2}_{d+1,\mathrm{loc}}(C_{2,1}) \cap C(\bar{C}_{2,1})$. Then there are constants $\gamma \in (0,1]$ and N, depending only on δ, K, and d, such that for any $\gamma \in (0,\gamma_0]$ and $\mathcal{L} \in \mathbb{L}^t_{\delta,K}$ we have*

$$\int_{C_{1,1}(1,0)} |D^2 w|^\gamma \, dxdt \le N\Big(\int_{C_{2,1}} |\mathcal{L}w|^{d+1}\, dxdt\Big)^{\gamma/(d+1)} + N \sup_{\partial' C_{2,1}} |w|^\gamma. \quad (1)$$

The proof of Theorem 1 consists of several steps. We first draw the following conclusion from Theorem 1.1.

2. LEMMA. *Let $u \in W^{1,2}_{d+1}(C_{2,1})$ and assume that $u \ge 0$ on $\partial' C_{2,1}$ and there exists an operator $\mathcal{L} \in \mathbb{L}^t_{\delta,K}$ such that $\mathcal{L}u \le 0$ in $C_{2,1}$ (a.e.). Let γ' be the constant called γ in Theorem 1.1 when $R = 0$ and set*

$$\gamma_0 = \gamma_0(d,\delta,K) = \gamma'/2.$$

Then, for any $\gamma \in (0,\gamma_0]$, we have

$$\int_{C_{1,1}(1,0)} |\mathcal{L}u|^\gamma \, dxdt \le N u^\gamma(0,0), \quad (2)$$

where $N = N(\delta,d,K)$.

Proof. First, observe that for any $\lambda > 0$

$$\big|C_1(1,0) \cap \{(t,x) : -\mathcal{L}u(t,x) \ge \lambda\}\big| \le N\lambda^{-2\gamma_0}u^{2\gamma_0}(0,0). \quad (3)$$

Indeed, (3) follows from (1.7) if one considers u/λ in place of u.

Next, the left-hand side of (2) equals

$$\int_0^\infty \big|C_1(1,0)\cap\{-\mathcal{L}u\ge\lambda^{1/\gamma}\}\big|\,d\lambda$$

$$\le\int_0^\mu \big|C_1(1,0)\big|\,d\lambda+Nu^{2\gamma_0}(0,0)\int_\mu^\infty \lambda^{-2\gamma_0/\gamma}\,d\lambda,$$

where $\mu=u^\gamma(0,0)$. Upon computing the last integral we arrive at (2). The lemma is proved. □

For elliptic equations we have the following version of Lemma 2 which is obtained as that lemma with the help of Theorem 1.6.

3. LEMMA. *Let $u\in W^2_d(B_1)$ and assume that $u\ge0$ on ∂B_1 and there exists an operator $\mathcal{L}\in\mathbb{L}_{\delta,K}$ such that $\mathcal{L}u\le0$ in B_1 (a.e.). Then for any $\gamma\in(0,\gamma_0]$ we have*

$$\int_{B_1}|\mathcal{L}u|^\gamma\,dx\le Nu^\gamma(0),$$

where $N=N(\delta,d,K)$.

Here is a useful generalization of Lemma 2.

4. LEMMA. *There is a constant N, depending only on δ, K, and d, such that, if $\gamma\in(0,\gamma_0]$, $\mathcal{L}\in\mathbb{L}^t_{\delta,K}$, $u\in W^{1,2}_{d+1,\mathrm{loc}}(C_{2,1})\cap C(\bar C_{2,1})$ and $\mathcal{L}u=g-f$ in $C_{2,1}$ (a.e.) with $f\ge0$, then we have*

$$\int_{C_1(1,0)}|f|^\gamma\,dxdt\le N\Big(\int_{C_{2,1}}|g_+|^{d+1}\,dxdt\Big)^{\gamma/(d+1)}$$

$$+N\big|u_+(0,0)\big|^\gamma+N\sup_{\partial' C_{2,1}}|u_-|^\gamma. \tag{4}$$

Proof. First we reduce the general case to the one in which $u\in W^{1,2}_{d+1}(C_{2,1})$. To do that we introduce "shifted and dilated" u, that is for $\varepsilon\in[0,1)$ we define

$$u_\varepsilon(t,x)=u(\varepsilon^2t+1-\varepsilon^2,\varepsilon x).$$

Obviously, $u_\varepsilon\in C(\bar C_{2,1})\cap W^{1,2}_{d+1}(C_{2,1})$. We also modify the coefficients of $\mathcal{L}$ in such a way that the new g and f are just shifted and dilated original g and f, respectively, times ε^2. If (4) holds for u_ε, then we obtain it as is by letting $\varepsilon\uparrow1$ by the monotone convergence theorem owing to the continuity of u in $\bar C_{2,1}$. By the way, we do not assume that the integral in the right-hand side of (4) is finite. Thus indeed we may concentrate on $u\in W^{1,2}_{d+1}(C_{2,1})$.

Then observe that if the integral in the right-hand side of (4) is infinite, we have nothing to prove. Therefore, we may assume that it is finite. Then $g\in L_{d+1}(C_{2,1})$ since $g_+\ge g\ge\mathcal{L}u$. It follows that $f\in L_{d+1}(C_{2,1})$ as well.

Now take an operator $\mathcal{L}'\in\mathbb{L}^{t,\infty}_{\delta,K}$ and introduce

$$f'=f+\big[(\mathcal{L}'-\mathcal{L})u\big]_-,\quad g'=g+\big[(\mathcal{L}'-\mathcal{L})u\big]_+,$$

so that $\mathcal{L}'u=g'-f'$, $f',g'\in L_{d+1}(C_{2,1})$, and $f'\ge0$. Obviously, if (4) were true with f',g' in place of f,g for any $\mathcal{L}'$, then by approximating $\mathcal{L}$ by operators $\mathcal{L}'$ we would obtain (4) in its original form.

Therefore, in the rest of the proof without losing generality we assume that $\mathcal{L} \in \mathbb{L}^{t,\infty}_{\delta,K}$ and introduce functions v and w as $W^{1,2}_{d+1}(C_{2,1})$-solutions of

$$\mathcal{L}v = -f, \quad \mathcal{L}w = -g$$

with zero condition on $\partial' C_{2,1}$ for v and with condition $w = -u$ on $\partial' C_{2,1}$. The existence and uniqueness of v and w is a classical result (see, for instance, [**89**]).

Clearly, $v = u - w$ and by the maximum principle $v \geq 0$. By Lemma 2 the left-hand side of (4) is less than a constant times

$$v^\gamma(0,0) \leq u^\gamma_+(0,0) + w^\gamma_-(0,0).$$

After that it only remains to use the parabolic Aleksandrov estimate. The lemma is proved. □

For elliptic operators Lemma 4 becomes the following.

5. LEMMA. *There is a constant N, depending only on δ, K, and d, such that, if $\gamma \in (0, \gamma_0]$, $\mathcal{L} \in \mathbb{L}_{\delta,K}$, $u \in W^2_{d,\mathrm{loc}}(B_1) \cap C(\bar{B}_1)$, and $\mathcal{L}u = g - f$ in B_1 (a.e.) with $f \geq 0$, then we have*

$$\int_{B_1} |f|^\gamma \, dx \leq N\Big(\int_{B_1} |g_+|^d \, dx\Big)^{\gamma/d} + N\big|u_+(0)\big|^\gamma + N \sup_{\partial B_1} |u_-|^\gamma.$$

The proof is based on Lemma 3 and consists of repeating the proof of Lemma 4 with obvious changes. Of course, at the last step one applies the original Aleksandrov estimate rather than its parabolic version.

Proof of Theorem 1. Introduce $h = \mathcal{L}u$, take an operator $\mathcal{L}' \in \mathbb{L}^t_{\delta/2,K}$, and observe that

$$\mathcal{L}'u = g - f, \quad g := h + 2\big[(\mathcal{L}' - \mathcal{L})u\big]_+, \quad f := \big|(\mathcal{L}' - \mathcal{L})u\big|.$$

According to (4) and the parabolic Aleksandrov estimate

$$\int_{C_{1,1}(1,0)} \big|(\mathcal{L}' - \mathcal{L})u\big|^\gamma \, dxdt \leq N\big\|\big[(\mathcal{L}' - \mathcal{L})u\big]_+\big\|^\gamma_{L_{d+1}(C_{2,1})}$$
$$+ N\|h\|^\gamma_{L_{d+1}(C_{2,1})} + N \sup_{\partial' C_{2,1}} |u|^\gamma. \tag{5}$$

We now use the arbitrariness of $\mathcal{L}'$. Obviously, there exists an $\varepsilon = \varepsilon(\delta, d) > 0$ and an operator $\mathcal{L}' \in \mathbb{L}^t_{\delta/2,K}$ with lower order coefficients coinciding with the ones of $\mathcal{L}$ and such that

$$\mathcal{L}'u = \mathcal{L}u - \varepsilon|D^2u|.$$

With such an operator (5) becomes (1). The theorem is proved. □

The reader understands that the following result, basically belonging to Fang-Hua Lin [**90**], is obtained by mimicking the proof of Theorem 1 and using Lemma 5 instead of Lemma 4.

6. THEOREM. *There is a constant N, depending only on δ, K, and d, such that for any $\gamma \in (0, \gamma_0]$, $\mathcal{L} \in \mathbb{L}_{\delta,K}$, and $u \in W^2_{d,\mathrm{loc}}(B_1) \cap C(\bar{B}_1)$ we have*

$$\int_{B_1} |D^2u|^\gamma \, dx \leq N\Big(\int_{B_1} |\mathcal{L}u|^d \, dx\Big)^{\gamma/d} + N \sup_{\partial B_1} |u|^\gamma. \tag{6}$$

Introduce

$$\mathbb{R}^d_+ = \{x = (x^1, x') \in \mathbb{R}^d : x^1 > 0, x' \in \mathbb{R}^{d-1}\}, \quad B_r^+ = B_r \cap \mathbb{R}^d_+.$$

By the odd continuation of u across the hyperplane $\{x^1 = 0\}$ we get the following.

7. COROLLARY. *Assume that $u \in W^2_d(B_r^+) \cap C(\bar{B}_1^+)$ for any $r < 1$ and $u = 0$ on $\{x^1 = 0\} \cap B_1$. Then with the same $\mathcal{L}$ and γ as in Theorem 6 and N, depending only on δ, K, and d, we have*

$$\int_{B_1^+} |D^2 u|^\gamma \, dx \le N\Big(\int_{B_1^+} |\mathcal{L}u|^d \, dx\Big)^{\gamma/d} + N \sup_{\partial B_1^+} |u|^\gamma.$$

8. REMARK. The next result is stronger than Theorem 1 and looks like the right parabolic counterpart of Theorem 6. Theorem 9 below is one of the cornerstones of the L_p-theory of fully nonlinear parabolic equations. This theorem in case $K = 0$ first appeared in [**30**] with the proof based on Theorem 1. General case is given in [**73**]. We note that a statement similar to Theorem 9 for viscosity solutions of fully nonlinear parabolic equations can be found in [**113**], where the reader is referred to [**114**] for the proof. In it, however, the domain on the right in (7) is wider than C_1. It is also worth noting that a version of Theorem 9 appeared in [**39**], where on the left of (7) one has $C_{1,1/2}$ in place of C_1.

9. THEOREM. *Let $u \in C(\bar{C}_1) \cap W^{1,2}_{d+1,\text{loc}}(C_1)$. Then there is a constant N, depending only on δ, d, and K, such that for any $\gamma \in (0, \gamma_0]$ and $\mathcal{L} \in \mathbb{L}^t_{\delta,K}$ we have*

$$\int_{C_1} |D^2 u|^\gamma \, dxdt \le N \sup_{\partial' C_1} |u|^\gamma + N\Big(\int_{C_1} |\mathcal{L}u|^{d+1} \, dxdt\Big)^{\gamma/(d+1)}. \tag{7}$$

Proof. First as in the proof of Lemma 6.1.7 one reduces the general situation to the one in which $u \in W^{1,2}_{d+1}(C_1)$.

Then we may also assume that the coefficients of $\mathcal{L}$ are infinitely differentiable in $\mathbb{R}^{d+1}$. Now set $f = \mathcal{L}u$ in C_1 and extend $f(t,x)$ for $t \le 0$ as zero. Also set $u(t,x) = u(-t,x)$ for $t \le 0$. Observe that the new u belongs to $W^{1,2}_{d+1}\big((-1,1) \times B_1\big)$. After that define $v(t,x)$ as a unique $W^{1,2}_{d+1,\text{loc}}\big((-1,1) \times B_1\big) \cap C\big([-1,1] \times \bar{B}_1\big)$-solution of $\mathcal{L}v = f$ with terminal and lateral conditions being u. The existence and uniqueness of such a solution is a classical result (see, for instance, Theorem 7.17 of [**89**]). By uniqueness $v = u$ in C_1, so that owing to Theorem 1,

$$\begin{aligned}\int_{C_1} |D^2 u|^\gamma \, dxdt &= \int_{C_1} |D^2 v|^\gamma \, dxdt \\ &\le N\Big(\int_{(-1,1)\times B_1} |f|^{d+1} \, dxdt\Big)^{\gamma/(d+1)} + N \sup_{\partial'(-1,1)\times B_1} |v|^\gamma \\ &= N\Big(\int_{C_1} |f|^{d+1} \, dxdt\Big)^{\gamma/(d+1)} + N \sup_{\partial' C_1} |u|^\gamma.\end{aligned}$$

The theorem is proved. □

Now we pass to estimating Du.

10. THEOREM. *There is a constant N, depending only on δ, K, and d, such that, if $\gamma\in(0,\gamma_0]$, $\mathcal{L}\in\mathbb{L}^t_{\delta,K}$, and $u\in W^{1,2}_{d+1,\mathrm{loc}}(C_1)\cap C(\bar C_1)$, then we have*

$$\int_{C_1}|Du|^{2\gamma}\,dxdt\le N\Big(\int_{C_1}|\mathcal{L}u|^{d+1}\,dxdt\Big)^{2\gamma/(d+1)}+N\sup_{\partial' C_1}|u|^{2\gamma}. \tag{8}$$

Proof. Let $\mathcal{L}\in\mathbb{L}^t_{\delta,K}$ and $v\in C^{1,2}(\bar C_{2,1})$. Observe that

$$\mathcal{L}(-v^2)=g-f,\quad g:=-2v\mathcal{L}v+cv^2,\quad f=2a^{ij}(D_iv)D_jv. \tag{9}$$

By Lemma 4

$$\int_{C_{2,1}}|Dv|^{2\gamma}\,dx\,dt\le N\sup_{C_{2,1}}|v|^{\gamma}\Big(\int_{C_{2,1}}|\mathcal{L}v|^{d+1}\,dx\,dt\Big)^{\gamma/(d+1)}+N\sup_{C_{2,1}}|v|^{2\gamma}.$$

The inequality $ab\le a^2+b^2$ and the parabolic Aleksandrov estimate allow us to conclude that

$$\int_{C_{1,1}(1,0)}|Dv|^{2\gamma}\,dxdt\le N\Big(\int_{C_{2,1}}|\mathcal{L}v|^{d+1}\,dxdt\Big)^{2\gamma/(d+1)}+N\sup_{\partial' C_{2,1}}|v|^{2\gamma}.$$

By continuity we extend this estimate to $v\in W^{1,2}_{d+1}(C_{2,1})$ and then to $v\in W^{1,2}_{d+1,\mathrm{loc}}(C_{2,1})\cap C(\bar C_{2,1})$ as in the proof of Lemma 4. After that we get (8) by the same "trick" as in the proof of Theorem 9. The theorem is proved. □

One more time by the odd continuation of u across the hyperplane $\{x^1=0\}$ we get the following.

11. COROLLARY. *Introduce*

$$C_r^+=(0,r^2)\times B_r^+$$

and assume that $u\in W^{1,2}_{d+1}(C_r^+)\cap C(\bar C_1^+)$ for any $r<1$ and $u=0$ on $\{x^1=0\}\cap C_1$. Then with the same $\mathcal{L}$ and γ as in Theorem 9 and N, depending only on δ, K, and d, we have

$$\int_{C_1^+}(|D^2u|^{\gamma}+|Du|^{\gamma})\,dx\le N\Big(\int_{C_1^+}|\mathcal{L}u|^{d+1}\,dx\Big)^{\gamma/(d+1)}+N\sup_{\partial' C_1^+}|u|^{\gamma}.$$

We also have (9) for elliptic operators. Therefore, as above, Lemma 5 yields the following result of L.C. Evans [**35**].

12. THEOREM. *There is a constant N, depending only on δ, K, and d, such that, if $\gamma\in(0,\gamma_0]$, $\mathcal{L}\in\mathbb{L}_{\delta,K}$, and $u\in W^2_{d,\mathrm{loc}}(B_1)\cap C(\bar B_1)$, then we have*

$$\int_{B_1}|Du|^{2\gamma}\,dx\le N\sup_{\partial B_1}|u|^{2\gamma}+N\Big(\int_{B_1}|\mathcal{L}u|^{d}\,dx\Big)^{2\gamma/d}.$$

Similarly to Corollary 7 we have

13. COROLLARY. *Assume that $u\in W^2_d(B_r^+)\cap C(\bar B_1^+)$ for any $r<1$ and $u=0$ on $\{x^1=0\}\cap B_1$. Then with the same $\mathcal{L}$ and γ, and N as in Theorem 12 we have*

$$\int_{B_1^+}|Du|^{2\gamma}\,dx\le N\Big(\int_{B_1^+}|\mathcal{L}u|^{d}\,dx\Big)^{2\gamma/d}+N\sup_{\partial B_1^+}|u|^{2\gamma}.$$

5. Estimating functions that are $\mathcal{L}$-harmonic outside small sets

In this and the next section of the chapter we basically follow the exposition in [**84**] and [**61**].

Recall that the sets of operators $\mathbb{L}^t_{\delta,K}$ and $\mathbb{L}_{\delta,K}$ are introduced in Section 1 before Theorems 1.1 and 1.6, respectively. Using the properties of the functions m and M from Section 1, we now derive some properties of a function $p(t,x,q)$ which is defined as follows. Let $q \in [0,1)$. Introduce $\mathbb{U}^{t,p}_q$ as the set of functions $u \in C^{1,2}(C_1)$ vanishing on $\partial' C_1$ such that for each of them one can find $\mathcal{L} \in \mathbb{L}^{t,\infty}_{\delta,K}$ for which $\mathcal{L}u \leq 0$ on C_1 and

$$\big|C_1 \cap \{u \geq 1\}\big| \geq q|C_1|.$$

It is intuitively clear that if we fix a closed set $\Gamma \subset C_1$ and $\mathcal{L} \in \mathbb{L}^{t,\infty}_{\delta,K}$, then the smallest u satisfying $\mathcal{L}u \leq 0$ in $C_1 \setminus \Gamma$ such that $u = 0$ on $\partial' C_1$ and $u \geq 1$ on Γ should be such that $\mathcal{L}u = 0$ in $C_1 \setminus \Gamma$, $u = 0$ on $\partial' C_1$, and $u = 1$ on the rest of the parabolic boundary of $C_1 \setminus \Gamma$. Such functions are called $\mathcal{L}$-harmonic (or $\mathcal{L}$-caloric) in $C_1 \setminus \Gamma$, and our goal in this section is to estimate them from below in terms of the measure of Γ.

In this connection define

$$p(t,x,q) := \inf\big\{u(t,x) : u \in \mathbb{U}^{t,p}_q\big\}, \quad (t,x) \in C_{1,1}, \quad q \in [0,1).$$

The main properties of p are listed in the following theorem.

1. THEOREM. *If $(t,x) \in C_1$ and $q \in [0,1)$, then $1 \geq p(t,x,q) \geq m(t,x,q)$. Specifically, for any $q \in (0,1)$ and $\kappa \in [0,1)$ there exists a constant $p_0 > 0$ which depends only on d, K, δ, q, and κ such that $p(t,x,q) \geq p_0$ if $|x| \leq \kappa$ and $0 \leq t \leq \kappa q$. Moreover, for any $\kappa \in [0,1)$,*

$$\inf\big\{p(t,x,q) \,:\, |x| \leq \kappa, \quad 0 \leq t \leq \kappa\big\} \to 1 \tag{1}$$

as $q \uparrow 1$.

Proof. We prove that $1 \geq p$ in the next theorem. The reader can easily check that no vicious circle is caused by this delay. To show that $p \geq m$, we take a $u \in \mathbb{U}^{t,p}_q$, the corresponding operator $\mathcal{L}$, set

$$\Gamma = C_1 \cap \{u \geq 1\},$$

and denote by v the $\overset{0}{W}{}^{1,2}_{d+1}(C_1)$-solution of the equation

$$\mathcal{L}v = -I_\Gamma.$$

We have $\mathcal{L}t \geq 1$ and, by the maximum principle, $v \leq 1 - t \leq 1$ on C_1. The same principle applied to $Q = C_1 \setminus \Gamma$ implies that $u \geq v$ on $\bar{Q}$ and also, obviously, on Γ. We conclude that $u \geq v$ on C_1. By Lemma 1.4, $v \geq m(q)$ and $u \geq m(q)$, and the arbitrariness of $u \in \mathbb{U}^{t,p}_q$ proves that $p \geq m$ indeed.

To prove (1), take $\kappa \in [0,1)$, $q \in (0,1)$, let (t_0,x_0) be a point for which $|x_0| \leq \kappa$ and $0 \leq t_0 \leq \kappa$. Also take $u \in \mathbb{U}^{t,p}_q$ and $\mathcal{L}$ corresponding to u. Assume that $u(t_0,x_0) < 1$, denote $\sigma = 1 - \kappa$, and introduce

$$v(t,x) = 1 - u(t,x) - 2\sigma^{-1}(t - t_0) - 2\sigma^{-2}|x - x_0|^2.$$

Next, define

$$Q = \{(t,x) : t > t_0,\ v(t,x) > 0\} \cap C_1.$$

Since $u \geq 0$, we have $v(t,x) < 0$ if either

$$t \geq t_0 + \sigma \quad \text{or} \quad |x - x_0| \geq \sigma.$$

Therefore, $\partial' Q \subset C_1$ and $v = 0$ on $\partial' Q$. Furthermore, $\mathcal{L}v \geq -N = -N(K, d, \kappa)$ in Q (a.e.). By Theorem 3.0.1,

$$v \leq N|Q|^{1/(d+1)},$$

in $\bar{Q}$ and, in particular, at (t_0, x_0), where $N = (K, d, \kappa, \delta)$. By adding that $Q \subset C_1 \setminus \{u \geq 1\}$, we get that

$$1 - u(t_0, x_0) \leq N|Q|^{1/(d+1)},$$

$$u(t_0, x_0) \geq 1 - N|C_1|^{1/(d+1)}(1-q)^{1/(d+1)}.$$

We obtained this estimate of $u(t_0, x_0)$ assuming that $u(t_0, x_0) < 1$. However, if $u(t_0, x_0) \geq 1$, the estimate is trivial. We now take inf over $u \in \mathbb{U}_q^{t,p}$, $|x| \leq \kappa$, and $0 \leq t_0 \leq \kappa$, and let $q \uparrow 1$. Then we come to (1). The theorem is proved. □

The following further properties of p are sometimes useful.

2. THEOREM. *Let $\kappa \in (0,1]$, $u \in W^{1,2}_{d+1}(C_{\kappa,1})$, $\mathcal{L} \in \mathbb{L}^t_{\delta,K}$. Assume that $\mathcal{L}u \leq 0$ on $C_{\kappa,1}$ (a.e.) and $u \geq 0$ on $\partial' C_{\kappa,1}$. Also, let $q \in [0,1)$ and*

$$\big|C_{\kappa,1} \cap \{u \geq 1\}\big| \geq q|C_{\kappa,1}|.$$

Then $u(t,x) \geq p(t,x,\kappa q)$ on $C_{\kappa,1}$.

Proof. Fulfilling the promise given in the proof of Theorem 1, we first observe that in case $u \equiv 1$, by the current theorem, one gets that $1 \geq p$.

Then, we prove the theorem in case $\mathcal{L} \in \mathbb{L}^{t,\infty}_{\delta,K}$ and $u = 0$ on $\partial' C_{\kappa,1}$. Denote $\mathcal{L}u = f$ and find $f_n \in C_0^\infty(C_{\kappa,1})$ such that $f_n \leq 0$ and $f_n \to f$ in $L_{d+1}(C_{\kappa,1})$. Next introduce u_n as $C^{1,2}(C_1)$-solution of $\mathcal{L}u_n = f_n$ in C_1 vanishing on $\partial' C_1$. Obviously, $u_n = 0$ in $C_1 \setminus C_{\kappa,1}$. By Theorem 3.0.1 we have $u_n \to u$ uniformly on $\bar{C}_{\kappa,1}$. Therefore, for any $\varepsilon \in (0,1)$ and sufficiently large n, we have

$$\big|C_1 \cap \{u_n \geq \varepsilon\}\big| = \big|C_{\kappa,1} \cap \{u_n \geq \varepsilon\}\big|$$

$$> \big|C_{\kappa,1} \cap \{u \geq 1\}\big| \geq q|C_{\kappa,1}| = \kappa q|C_1|.$$

It follows by definition that $u_n \varepsilon^{-1} \geq p(\kappa q)$ on C_1. As $n \to \infty$ and $\varepsilon \uparrow 1$, the desired result follows.

To drop the assumption that u vanishes on $\partial' C_1$ take $\mathcal{L} \in \mathbb{L}^{t,\infty}_{\delta,K}$ and u as in the statement of the theorem, and for $\varepsilon \in (0,1)$, introduce closed sets

$$\Gamma^\varepsilon = \{(t,x) \in \bar{C}_{\kappa,1} : u(t,x) \geq \varepsilon\}.$$

It is not hard to see that, for $\varepsilon > 0$ sufficiently close to 1, we have

$$|\Gamma^\varepsilon| > q|C_{\kappa,1}| \tag{2}$$

in both cases: if $u \geq 1$ on $C_{\kappa,1}$, or, on the contrary, $u < 1$ on a subset of $C_{\kappa,1}$.

Fix such an $\varepsilon \in (0,1)$ and, for $n = 1, 2, ...$, introduce $v_n \in \overset{0}{W}{}^{1,2}_{d+1}(C_{\kappa,1})$ by the equations

$$\mathcal{L}v_n + nI_{\Gamma^\varepsilon}(1 - v_n) = 0.$$

Since $\mathcal{L}1 \leq 0$, we have $\mathcal{L}1 + nI_{\Gamma^\varepsilon}(1-1) \leq 0$ and, again by the maximum principle, $v_n \leq 1$, $\mathcal{L}v_n \leq 0$, $v_n \geq 0$ on $C_{\kappa,1}$, and $\mathcal{L}v_n = 0$ on $C_{\kappa,1} \setminus \Gamma^\varepsilon$. The last equation and the maximum principle applied to the domain $C^o_{\kappa,1} \setminus \Gamma^\varepsilon$, implies that $v_n \leq \varepsilon^{-1}u$ on $\bar{C}_{\kappa,1}$.

It follows from the fact that in $C_{\kappa,1}$ (a.e.)

$$\mathcal{L}\big(v_n/n\big) + I_{\Gamma^\varepsilon}(1 - v_n) = 0,$$

by Corollary 1.5 (or by Corollary 4.2.12), that $v_n \to 1$ in measure on Γ^ε. In particular,

$$\big|C_{\kappa,1} \cap \{v_n \geq \varepsilon\}\big| \geq q|C_{\kappa,1}|$$

(see (2)) for sufficiently large n. By the first part of the proof, this allows us to conclude that $v_n\varepsilon^{-1} \geq p(\kappa q)$ on $C_{\kappa,1}$ for large n. Hence,

$$u \geq \varlimsup_{n\to\infty} \varepsilon v_n \geq \varepsilon^2 p(\kappa q)$$

which implies that $u \geq p(\kappa q)$ on $C_{\kappa,1}$ after setting $\varepsilon \uparrow 1$.

It only remains to drop the assumption that $\mathcal{L} \in \mathbb{L}^{t,\infty}_{\delta,K}$. Take a sequence $\mathcal{L}_n \in \mathbb{L}^{t,\infty}_{\delta,K}$ whose coefficients converge to the corresponding coefficients of $\mathcal{L}$ (a.e.). Introduce $w_n \in \overset{0}{W}{}^{1,2}_{d+1}(C_{\kappa,1})$ by solving the equation

$$\mathcal{L}_n w_n = (\mathcal{L}_n - \mathcal{L})u$$

in $C_{\kappa,1}$. By the Aleksandrov estimates $w_n \to 0$ uniformly on $C_{\kappa,1}$ and, for a sequence $\varepsilon_n \downarrow 0$, it holds that $\varepsilon_n - w_n \geq 0$,

$$\mathcal{L}_n(u - w_n + \varepsilon_n) \leq \mathcal{L}_n(u - w_n) = \mathcal{L}u \leq 0$$

on $C_{\kappa,1}$ (a.e.), and $u - w_n + \varepsilon_n \geq 1$ on $C_{\kappa,1} \cap \{u \geq 1\}$. Since $\mathcal{L}_n \in \mathbb{L}^{t,\infty}_{\delta,K}$, by the above result, $u - w_n + \varepsilon_n \geq p(\kappa q)$ on $C_{\kappa,1}$. It only remains to let $n \to \infty$, The theorem is proved. □

3. LEMMA. *Let $\kappa \in (0,1]$, $u \in W^{1,2}_{d+1}(C_{\kappa,1})$, $\mathcal{L} \in \mathbb{L}^{t}_{\delta,K}$. Assume that $\mathcal{L}u \geq 0$ on $C_{\kappa,1}$ (a.e.) and*

$$\big|C_{\kappa,1} \cap \{u \leq \zeta\}\big| \geq q|C_{\kappa,1}|,$$

where $\zeta \in (-\infty,\infty)$ and $q \in [0,1)$. Also, let one of the following two conditions hold:

(a) $c \equiv 0$ or

(b) $\max(u, \bar{C}_{\kappa,1}) \geq 0$. Then

$$u(t,x) \leq p(t,x,\kappa q)\zeta + \big[1 - p(t,x,\kappa q)\big]\max(u, \bar{C}_{\kappa,1}) \tag{3}$$

on $C_{\kappa,1}$.

Proof. Define M as the maximum in (3). If $\zeta \geq M$, (3) is obvious. In the remaining case $\zeta < M$ and we introduce

$$v = (M - u)(M - \zeta)^{-1}.$$

Observe that in both cases (a) and (b) we have $\mathcal{L}v \leq 0$ on $C_{\kappa,1}$. In addition, $\{u \leq \zeta\} = \{v \geq 1\}$ and $v \geq 0$ on $C_{\kappa,1}$. Now Theorem 2 implies that $v \geq p(\kappa q)$ on $C_{\kappa,1}$ which is equivalent to (3). The lemma is proved. □

Next we extend the obtained results to elliptic operators. Recall that the sets of operators $\mathbb{L}_{\delta,K}$ and $\mathbb{L}^{\infty}_{\delta,K}$ are introduced in Section 1 before Theorem 1.6. For $q \in [0, 1)$, we write $u \in \mathbb{U}^p_q$ if $u \in C^2(B_1)$, $u = 0$ on ∂B_1 and there exists an operator $\mathcal{L} \in \mathbb{L}^{\infty}_{\delta,K}$ such that $\mathcal{L}u \leq 0$ in B_1 and

$$\big|B_1 \cap \{u \geq 1\}\big| \geq q|B_1|.$$

Define

$$p(x, q) = \inf \big\{u(x) : u \in \mathbb{U}^p_q\big\}, \quad x \in B_1, \quad q \in [0, 1).$$

4. THEOREM. *(a)* $1 \geq p(x, q) \geq p(0, x, q)$ *and* $p(x, q) \to 1$ *as* $q \uparrow 1$ *uniformly on any closed subset of* B_1.

(b) If $q \in [0, 1)$, $u \in W^2_d(B_1)$, $\mathcal{L} \in \mathbb{L}_{\delta,K}$, $\mathcal{L}u \leq 0$ *in* B_1 *(a.e.),* $u \geq 0$ *on* ∂B_1, *and*

$$\big|B_1 \cap \{u \geq 1\}\big| \geq q|B_1|,$$

then $u(x) \geq p(x, q)$ *on* B_1.

(c) If $q \in [0, 1)$, $u \in W^2_d(B_1)$, $\mathcal{L} \in \mathbb{L}_{\delta,K}$, $\mathcal{L}u \geq 0$ *in* B_1 *(a.e.),* $\zeta \in (-\infty, \infty)$,

$$\big|B_1 \cap \{u \leq \zeta\}\big| \geq q|B_1|,$$

and one of the following two conditions holds: $c \equiv 0$ *or* $\max(u, \bar{B}_1) \geq 0$, *then*

$$u(x) \leq p(x, q)\zeta + \big[1 - p(x, q)\big] \max(u, \bar{B}_1)$$

on B_1.

Proof. Theorem 2 and the fact that $C^2(B_1)$ is just a subset of $C^{1,2}(C_1)$ of functions independent of t readily prove that $p(x, q) \geq p(0, x, q)$. The second assertion in (a) follows from Theorem 1. One proves the remaining assertions by mimicking the proofs of Lemma 3 and Theorem 2. The theorem is proved. □

Sometimes the following result is useful.

5. THEOREM. *Let* Ω *be a bounded connected domain and let* $u \in W^2_{d,\mathrm{loc}}(\Omega) \cap C(\bar{\Omega})$ *be such that* $\mathcal{L}u \geq 0$ *in* Ω *(a.e.), where* $\mathcal{L} \in \mathbb{L}_{\delta,K}$. *Let* $\zeta \leq M := \max(u, \bar{\Omega})$ *and assume that*

$$q := \big|\{u \leq \zeta\}\big| > 0.$$

Finally, assume that either $c \equiv 0$ *or* $M \geq 0$. *Then for any* $\varepsilon > 0$ *there exists* $\gamma \in (0, 1)$, *depending only on* $\Omega, \varepsilon, q, \delta, K$, *and* d, *such that*

$$u(x) - \zeta \leq \gamma(M - \zeta),$$

whenever $\mathrm{dist}\,(x, \Omega^c) \geq \varepsilon$.

Proof. The assertion of the theorem becomes stronger if we decrease ε. Therefore, we take $\varepsilon \le \varepsilon_0$, where $\varepsilon_0 > 0$ is so small that $\Omega^{2\varepsilon}$ is connected and

$$|\Omega \setminus \Omega^{3\varepsilon_0}| \le q/2.$$

Then, for $\Gamma := \{x \in \Omega^{3\varepsilon} : u(x) \le \zeta\}$ we have $|\Gamma| \ge q/2$ and

$$\int_{\Omega^{2\varepsilon}} I_{|x-x_0|<\varepsilon} I_\Gamma(x)\,dx_0 = |B_\varepsilon| I_\Gamma(x).$$

It follows that

$$\int_{\Omega^{2\varepsilon}} \Big(\int_{\Omega^{3\varepsilon}} I_{|x-x_0|<\varepsilon} I_\Gamma(x)\,dx \Big)\,dx_0 = |B_\varepsilon|\,|\Gamma| \ge (q/2)|B_\varepsilon|,$$

and there exists a point $x_0 \in \Omega^{2\varepsilon}$ such that

$$\big|B_\varepsilon(x_0) \cap \{u \le \zeta\}\big| \ge q_1 |B_\varepsilon|,$$

where $q_1 = q\big(2|\Omega|\big)^{-1}$.

By Theorem 4, applied to $B_{2\varepsilon}$ in place of B_1, and by the properties of $p(x,q)$, we have in $B_\varepsilon(x_0)$ that

$$u - \zeta \le \gamma_0 \sup_{B_{2\varepsilon}(x_0)} (u-\zeta) \le \gamma_0(M-\zeta), \quad u \le \zeta + \gamma_0(M-\zeta) =: \zeta_1,$$

where $\gamma_0 \in (0,1)$ depends only on q_1, δ, K, and d.

Now, if $x_1 \in \Omega^{2\varepsilon}$ and $|x_1 - x_0| \le \varepsilon$, then $|B_\varepsilon(x_1) \cap B_\varepsilon(x_0)|$ is bounded from below by a $q_2|B_\varepsilon|$ where $q_2 > 0$ depends only on d. The same is true for $\big|B_\varepsilon(x_1) \cap \{u \le \zeta_1\}\big|$. Hence, by the same token as above,

$$u \le \zeta_1 + \gamma_1(M - \zeta_1) = \zeta + \big[\gamma_1(1-\gamma_0) + \gamma_0\big](M - \zeta)$$

in $B_\varepsilon(x_1)$. One continues in an obvious way and gets the desired result in finitely many steps, albeit with 2ε in place of ε, owing to the boundedness and connectivity of $\Omega^{2\varepsilon}$. The theorem is proved. □

Sometimes the following result is useful in investigating the modulus of continuity of solutions near a quite irregular boundary.

6. THEOREM. *Let $u \in W^2_d(B_1)$ and $\mathcal{L} \in \mathbb{L}_{\delta,K}$. Assume that $u \ge 0$, $\mathcal{L}u \le 0$ in B_1 (a.e.), and*

$$\big|B_1 \cap \{u \ge 1\}\big| \ge q|B_1|,$$

where $q \in (0,1)$. Then, for any $R \in [0,1)$ there exist constants $N \in (0,\infty)$ and $\gamma \in (0,1)$, depending only on K, δ, d, and R, such that for all $x \in B_R$ we have

$$Nu(x) \ge q^{1/\gamma}. \tag{4}$$

In particular, $Np(x,q) \ge q^{1/\gamma}$.

Proof. Due to Theorem 1.1, it suffices to prove that $2u(x) \ge M(0,x,q)$. To do that we repeat the proof of part of the first assertion of Theorem 1. Set

$$\Gamma = B_1 \cap \{u \ge 1\},$$

and, first assuming that the coefficients of $\mathcal{L}$ are smooth, introduce v as a $\overset{0}{W}{}^{1,2}_d(C_{2,1})$-solution of the equation

$$\partial v + \mathcal{L}v = -I_{[1,2]\times\Gamma}.$$

By the maximum principle $v \leq 2 - t \leq 2$ on $C_{2,1}$. Obviously $2u \geq v$ on $[1,2] \times \Gamma$. By the maximum principle, applied in the domain $Q = C^o_{2,1} \setminus [1,2] \times \Gamma$ we have $2u \geq v$ in $\bar{Q}$. Thus, $2u \geq v$ in $\bar{C}_{2,1}$ and, since by Lemma 1.3, $v(0,x) \geq M(0,x,q)$, we have $2u(x) \geq M(0,x,q)$ indeed.

The passage to the general $\mathcal{L}$ is done by approximation along the lines in the proof of Theorem 2. The theorem is proved. □

6. The Harnack inequality and the interior Hölder continuity for $\mathcal{L}$-harmonic functions

Let $K \in [0,\infty)$ and $\delta \in (0,1]$ be some constants. Let

$$\mathcal{L} = \partial_t + a^{ij}(t,x)D_{ij} + b^i(t,x)D_i - c(t,x) \tag{1}$$

be an operator whose measurable coefficients are given on $\mathbb{R}^{d+1}$ and are such that

$$a^{ij} = a^{ji}, \quad \delta|\lambda|^2 \leq a^{rs}\lambda^r\lambda^s \leq \delta^{-1}|\lambda|^2, \quad |b^i| \leq K, \quad |c| \leq K \tag{2}$$

for all $(t,x) \in \mathbb{R}^{d+1}$, $\lambda \in \mathbb{R}^d$, $i,j = 1,...,d$.

Harnack's inequality, dealt with first in this section, says that, if $u \geq 0$ satisfies $\mathcal{L}u = 0$ in a domain Q and $u \geq 1$ at a point $(t_0,x_0) \in Q$, u cannot be arbitrarily small at a point $(t,x) \in Q$ which is sufficiently close to (t_0,x_0) but such that $t \leq t_0 - \varepsilon$ where $\varepsilon > 0$. We emphasize that the condition $t \leq t_0 - \varepsilon$ is necessary. For instance, (see Moser [**93**]), $u(t,x) = \exp(\lambda x - \lambda^2 t)$ satisfies $u_t + u_{xx} = 0$ on $\mathbb{R}^2$ and $u(0,\,0) = 1$ but $u(0,\,-\varepsilon) \to 0$ as $\lambda \to \infty$ for any $\varepsilon > 0$.

The Harnack inequality admits interpretations in the framework of potential theory or probability theory; both interpretations look very much alike. Arguing somewhat vaguely, let $\mathcal{L}$ and Q be regular enough so that for any sufficiently smooth function φ given on $\partial' Q$ the equation $\mathcal{L}u = 0$ in Q with boundary condition $u = \varphi$ on $\partial' Q$, admits a sufficiently smooth solution. In that case we have the mapping

$$\varphi \to u(t_0,x_0)$$

defined for any $(t_0,x_0) \in Q$. By the maximum principle, this mapping is a linear bounded functional defined on a dense subset of $C(\overline{\partial' Q})$.

By the Riesz representation theorem, there exists a measure μ_{t_0,x_0} concentrated on $\overline{\partial' Q}$ such that

$$u(t_0,x_0) = \int \varphi(t,x)\mu_{t_0,x_0}(dtdx).$$

The maximum principle implies that μ_{t_0,x_0} is nonnegative. Now it is easy to see that the Harnack inequality means exactly that

$$\mu_{t_0,x_0} \leq N\mu_{t,x}$$

if (t,x) is sufficiently close to (t_0,x_0) and such that $t \leq t_0 - \varepsilon$, where $\varepsilon > 0$ and N is a constant.

It is important to keep in mind that, if $\mu_{t,x}$ is constructed from a different operator, the above inequality can only hold with a constant N depending on the regularity of the coefficients of both operators (see Safonov [**101**] or Fabes and Kenig [**40**]).

Here is the statement of the Harnack inequality.

1. THEOREM. *Let $\theta > 1$, let $R \leq 2$, and let $u \in W^{1,2}_{d+1}(C_{\theta R^2,R})$ be such that $u \geq 0$ in $C_{\theta R^2,R}$ and $\mathcal{L}u = 0$ in $C_{\theta R^2,R}$ (a.e.). Then there exists a constant N, which depends only on θ, δ, K, and d, such that*

$$u(R^2, 0) \leq N u(0, x)$$

whenever $|x| \leq R/2$. Moreover, if $(1-\theta)^{-1}$, δ^{-1}, and K vary within a bounded range, then N also varies within a bounded range. Finally, if $b \equiv 0$ and $c \equiv 0$, then our assertions hold true for $R > 2$ as well.

Proof. We follow the arguments in [**84**], that in turn are based on some ideas in E.M. Landis [**88**]. First use scaling by introducing $\tilde{u}(t,x) = u(R^2t/4, Rx/2)$. Simple computations show that $\tilde{u}$ in $C_{4\theta,2}$ satisfies an equation $\tilde{\mathcal{L}}\tilde{u} = 0$, with $\tilde{\mathcal{L}}$ of type (1) and with the coefficients satisfying (2) in both cases $R \leq 2$ or $K = 0, R > 2$. It follows that without loss of generality we may concentrate on $R = 2$. Then the case of general $\theta > 1$ is reduced to that of $\theta \geq 2$ by appropriate change of the time variable $t \to \tau(t)$. One more observation that the best constant N is obviously increasing in θ, allows us to restrict our attention to the case of $\theta = 2$. If $\theta = R = 2$, the best possible N is decreasing in δ and increasing in K. Therefore, the second assertion of the theorem follows from the first one and this argument.

Now, we deal with the first assertion, in case $\theta = R = 2$ and, to exclude a trivial situation, additionally assume that

$$u(4,0) > 0.$$

Furthermore, the possibility of replacing $u(t,x)$ by $e^{Kt}u(t,x)$ convinces us that we may as well assume that that

$$c \geq 0.$$

In that case conditions (1.2) hold. For $\kappa = 1/5$, we take ε and n from Lemma 2.1 (oblique cylinder lemma) and, having in mind Theorem 5.1, find $q \in (0,1)$, close to 1, for which

$$1 > p(0,0,q)2^{-1} + \big[1 - p(0,0,q)\big]2^n. \tag{3}$$

Next, for $r \in [0,1)$, introduce

$$\nu(r) = u(4,\,0)(1-r)^{-n}, \quad n(r) = \max\{u, \bar{C}_r(4,0)\} \quad (\bar{C}_0(4,0) := \{(4,0)\}),$$

and define r_0 as the greatest number in $[0,1)$ satisfying

$$n(r) = \nu(r).$$

Such a number does exist because $n(0) = \nu(0)$, $\nu(r) \to \infty$ as $r \uparrow 1$, and $n(r)$ is bounded. Choose $(t_0, x_0) \in \bar{C}_{r_0}(4,0)$ such that $\nu(r_0) = u(t_0,x_0)$ and consider the cylinder

$$Q := \Big\{(t,x) \,:\, 0 < t - t_0 < \frac{(1-r_0)^2}{4}, \quad |x - x_0| < \frac{1-r_0}{2}\Big\}.$$

As is easy to see $\bar{Q} \subset \bar{C}_{r_1}(4,0)$, where $r_1 = (1+r_0)/2$. By the definition of r_0, this implies that in $\bar{Q}$

$$u < \nu(r_1) = u(4,0)\Big(\frac{1-r_0}{2}\Big)^{-n} = 2^n\nu(r_0).$$

We claim that owing to this and (3),

$$\big|Q \cap \{u > \nu(r_0)/2\}\big| \geq (1-q)|Q|. \tag{4}$$

To argue by contradiction, assume (4) is false. Then dilate Q to the dimensions of C_1 and shift the result so that it coincides with C_1. This will bring u into a function $\tilde{u}$ on C_1 satisfying an equation $\tilde{\mathcal{L}}\tilde{u} = 0$ with an operator $\tilde{\mathcal{L}}$ of type (1) whose coefficients satisfy (1.2). It is clear that our assumption concerning (4) leads to

$$\big|C_1 \cap \{\tilde{u} > \nu(r_0)/2\}\big| < (1-q)|C_1|$$

and

$$\big|C_1 \cap \{\tilde{u} \leq \nu(r_0)/2\}\big| > q|C_1|.$$

It follows, by Lemma 5.3 with $\zeta = \nu(r_0)/2$, that

$$\begin{aligned}\nu(r_0) &= u(t_0, x_0) = \tilde{u}(0,0)\\ &\leq p(0,0,q)\nu(r_0)2^{-1} + \big[1 - p(0,0,q)\big]\max(\tilde{u}, \bar{C}_1)\\ &= p(0,0,q)\nu(r_0)2^{-1} + \big[1 - p(0,0,q)\big]\max(u, Q)\\ &\leq p(0,0,q)\nu(r_0)2^{-1} + \big[1 - p(0,0,q)\big]2^n\nu(r_0).\end{aligned}$$

However, this contradicts (3) and proves (4). We now apply Theorems 5.2 and 5.1 bearing in mind the above manipulations with Q and u and get that

$$u(t_0, x) \geq p_0\nu(r_0)2^{-1}$$

if $|x - x_0| \leq (1 - r_0)4^{-1}$, where $p_0 = p_0(d, \delta, K, q) > 0$. After that it only remains to apply Lemma 2.1 to conclude that for $|x| \leq 1$ we have

$$u(0,x) \geq \frac{1}{2}p_0\nu(r_0)\varepsilon\Big(\frac{1-r_0}{4}\Big)^n = 2^{-2n-1}p_0\varepsilon u(4,0).$$

The theorem is proved. □

In the next part of the section we deal with Hölder norm estimates for solutions of $\mathcal{L}u = f$. Remember that, if $z_1 = (t_1, x_1)$ and $z_2 = (t_2, x_2)$, we define

$$\rho(z_1, z_2) = |x_1 - x_2| + |t_1 - t_2|^{1/2} \tag{5}$$

and call $\rho(z_1, z_2)$ the parabolic distance between z_1 and z_2.

2. LEMMA. *Let $r \in (0, K^{-1} \wedge 1]$ and $u \in W^{1,2}_{d+1}(C_{2r})$. Assume that $\mathcal{L}u = 0$ and $c = 0$ on C_{2r} (a.e.). Then there exist constants $N(d, \delta) < \infty$ and*

$$\alpha_0 = \alpha_0(\delta, d) \in (0,1),$$

such that, for any $\alpha \in (0, \alpha_0]$ and $z_1, z_2 \in C_r$, we have

$$\big|u(z_1) - u(z_2)\big| \leq N r^{-\alpha}\rho^\alpha(z_1, z_2)\sup\big(|u|, C_{2r}\big). \tag{6}$$

Furthermore, $\sup(|u|, C_{2r})$ in (6) can be replaced by $\mathrm{osc}(u, C_{2r})$, where we use the notation

$$\mathrm{osc}(g, \Gamma) = \operatorname*{osc}_\Gamma g = \sup_\Gamma g - \inf_\Gamma g.$$

Finally, if, in addition, $b \equiv 0$, then (6) is also true for $r > K^{-1} \wedge 1$.

Proof. Parabolic scalings $(t,x) \to (r^{-2}t, r^{-1}x)$ allow us to reduce the general case to the one where $r = 1$, $K \le 1$. For $r = 1$ we observe that condition (1.2) is satisfied ($c \equiv 0$) and we adapt De Giorgi's argument from [**27**], relying on Lemma 5.3 and Theorem 5.1.

For $R \in (0, 2]$, set

$$w(R) = \operatorname{osc}(u, C_R).$$

In case $R \le 1/2$ and

$$\Big|C_{2R} \cap \Big\{u \le \frac{1}{2}\big(m(R) + M(R)\big)\Big\}\Big| \ge \frac{1}{2}|C_{2R}|, \tag{7}$$

Lemma 5.3 implies that on C_{2R} we have

$$u(t,x) \le p\Big(\frac{t}{4R^2}, \frac{x}{2R}, \frac{1}{2}\Big)\frac{1}{2}\big(m(R) + M(R)\big)$$
$$+ \Big[1 - p\Big(\frac{t}{4R^2}, \frac{x}{2R}, \frac{1}{2}\Big)\Big] M(2R).$$

By comparing the sup's of the left and right-hand sides over C_R we take into account that $m(R) + M(R) \le 2M(2R)$, so that the right-hand side increases if we replace p with a smaller number. We also observe that

$$p_0 := \inf\big\{p(t,x,1/2) : 0 \le t \le 1/4, \quad |x| \le 1/2\big\} > 0$$

in light of Theorem 5.1, where $p_0 = p_0(d, \delta)$. Then we conclude that

$$M(R) \le p_0 \frac{1}{2}\big(m(R) + M(R)\big) + (1 - p_0)M(2R).$$

Adding to this the obvious inequality

$$(p_0 - 1)m(R) \le (p_0 - 1)m(2R),$$

we get

$$\big(1 - \frac{p_0}{2}\big)w(R) \le (1 - p_0)w(2R), \quad w(R) \le \varepsilon w(2R), \tag{8}$$

where $\varepsilon < 1$, $\varepsilon = \varepsilon(d, K, \delta)$. We may, certainly, assume that $\varepsilon > 1/2$.

We have proved (8) assuming that (7) is true. However if (7) is false, then $-u$ satisfies an inequality similar to (7) and this leads to (8) again.

Therefore, $w(R) \le \varepsilon w(2R)$ for all $R \le 1/2$. Iterations then yield

$$w(R) \le \varepsilon^2 w(4R) \quad \text{for} \quad R \le 1/4, ..., w(R) \le \varepsilon^n w(2^n R) \quad \text{for} \quad R \le 2^{-n}.$$

If $R \le 1/2$ and we take $n := \lfloor -\log_2 R \rfloor$, then $R \le 2^{-n}$ and

$$w(R) \le \varepsilon^n w(2^n R) \le \varepsilon^{-1} R^\alpha w(1) \le 2\varepsilon^{-1} R^\alpha \sup\big(|u|, C_1\big),$$

where $\alpha = -\log_2 \varepsilon \in (0,1)$. This provides an estimate of the oscillation of u in any C_R with $R \le 1/2$. The same estimate obviously holds for the oscillation of u in any $C_R(t,x) \subset C_2$ as long as $R \le 1/2$.

Now take $z_1 = (t_1, x_1), z_2 = (t_2, x_2) \in C_1$ such that $R := \rho(z_1, z_2) \le 1/2$ and define

$$t = t_1 \wedge t_2, \quad x = (x_1 + x_2)/2.$$

Then we have $z_i \in \bar{C}_R(t,x)$, $i = 1,2$, and

$$\big|u(z_1) - u(z_2)\big| \le 2\varepsilon^{-1} R^{\alpha} \sup\big(|u|, C_1(t,x)\big)$$

$$\le 2\varepsilon^{-1}\rho^{\alpha}(z_1,z_2)\sup\big(|u|, C_2\big).$$

In the case that $\rho(z_1,z_2) \ge 1/2$ we have

$$\big|u(z_1) - u(z_2)\big| \le 2\sup\big(|u|, C_2\big)$$

$$\le 2^{1+\alpha}\rho^{\alpha}(z_1,z_2)\sup\big(|u|, C_2\big).$$

Thus, $N = 2^{1+\alpha} + 2\varepsilon^{-1}$ in (6) is always a good choice with $r = 1$ and $\alpha = \alpha(\delta, d)$ found above. One can take any smaller α as well since $\rho(z_1,z_2) \le N(d)r$. The lemma is proved. □

3. Corollary (two-sided Liouville theorem). *If $c \equiv 0$, $b \equiv 0$, $u \in W^{1,2}_{d+1}(C_r)$ for all $r > 0$, $\mathcal{L}u = 0$ almost everywhere in $\{t > 0\} \times \mathbb{R}^d$, and u is bounded on this set, then $u =$ const.*

Indeed, as $r \to \infty$, (6) implies that u=const. By the way, observe that, if we are only given that $u \ge 0$, u may be unbounded, as the example of the one-dimensional equations $\partial_t u + D^2 u = 0$ and its solution $u(t,x) = \exp(x - t)$ shows.

4. Corollary. *Under the assumptions of Lemma 2 there exists a constant $N = N(d,\delta) < \infty$ such that, for any $\alpha \in (0,\alpha_0]$, $\rho \in (0,2r)$, and $z_1, z_2 \in \bar{C}_\rho$, we have*

$$\big|u(z_1) - u(z_2)\big| \le N(2r - \rho)^{-\alpha}\rho^{\alpha}(z_1,z_2)\sup\big(|u|, C_{2r}\big). \tag{9}$$

Furthermore, $\sup\big(|u|, C_{2r}\big)$ in (9) can be replaced with $\operatorname{osc}(u, C_{2r})$.

Finally, if, in addition, $b \equiv 0$, then (9) is also true for $r > K^{-1} \wedge 1$.

Indeed, if $\rho(z_1,z_2) \le (2r - \rho)/4 := r_0$, then

$$z_1, z_2 \in \bar{C}_{r_0}(t,x) \subset \bar{C}_{4r_0}(t,x) \subset \bar{C}_{2r},$$

where, for $z_1 = (t_1, x_1)$ and $z_2 = (t_2, x_2)$, $t = t_1 \wedge t_2$ and $x = (x_1 + x_2)/2$. In that case (9) follows from (6) with the same N and α. However if $\rho(z_1,z_2) \ge (2r-\rho)/4$, (9) obviously holds with any $\alpha > 0$ and $N = 2$.

Lemma 2 admits an almost trivial generalization. Recall that α_0 is introduced in Lemma 2.

5. Theorem. *Let $r \in (0, K^{-1} \wedge 1]$ and $u \in W^{1,2}_{d+1}(C_{2r})$. Assume that $c \equiv 0$. Then there exists a constant N, which depends only on d and δ, such that*

$$\big|u(z_1) - u(z_2)\big| \le N\big(r^{-\alpha}\rho^{\alpha}(z_1,z_2)\sup_{C_{2r}}|u| + r^{d/(d+1)}\|\mathcal{L}u\|_{L_{d+1}(C_{2r})}\big) \tag{10}$$

for z_1, $z_2 \in C_r$ and $\alpha \in (0,\alpha_0]$. If, in addition, $b \equiv 0$, then (10) is also true for $r > K^{-1} \wedge 1$.

Proof. Parabolic scalings allow us to only concentrate on the case when $r = 1$ and $K \le 1$. Also, the form of (10) allows us to approximate $\mathcal{L}$ having measurable coefficients by the operators with smooth coefficients. Hence, we may assume that $r = 1$, $K \le 1$, and $\mathcal{L}$ has infinitely differentiable coefficients.

Let $h, w \in W^{1,2}_{d+1}(C_2)$ be such that $\mathcal{L}h = 0$, $\mathcal{L}w = \mathcal{L}u$ in C_2 (a.e.), $h = u$, $w = 0$ on $\partial' C_2$. Then, by uniqueness, $u = h + w$. Observe that Lemma 2 is applicable to h and since, by the maximum principle,

$$\sup\big(|h|, C_2\big) \le \sup\big(|h|, \partial' C_2\big) \le \sup\big(|u|, C_2\big).$$

it will only remain to use Theorem 3.0.1 implying that in C_2

$$|w| \le N\|\mathcal{L}w\|_{L_{d+1}(C_2)} = N\|\mathcal{L}u\|_{L_{d+1}(C_2)}.$$

The theorem is proved. □

Here is a version of Theorem 5 which sometimes is slightly more convenient.

6. THEOREM. *Let $r \in (0,1]$ and $u \in W^{1,2}_{d+1}(C_{2r})$. Assume that $c \equiv 0$. Then there exists a constant N, which depends only on d, K, and δ, such that*

$$\big|u(z_1) - u(z_2)\big| \le N r^{-\beta} \rho^{\beta}(z_1, z_2)\big(\sup_{C_{2r}} |u| + r^{d/(d+1)}\|\mathcal{L}u\|_{L_{d+1}(C_{2r})}\big) \tag{11}$$

for z_1, $z_2 \in C_r$, where

$$\beta = \frac{\alpha d}{\alpha(d+1) + d}$$

and $\alpha = \alpha_0(\delta, d)$ is the constant from Theorem 5 (or Lemma 2).

Proof. Fix z_1, $z_2 \in C_r$. Since there is the sup norm of u on the right, it suffices to prove (11) assuming that

$$\xi := \Big(\frac{r}{\rho(z_1, z_2)}\Big)^{\beta/\alpha} \ge 4.$$

Then set

$$\bar r = \xi \rho(z_1, z_2).$$

If $z_i = (t_i, x_i)$, $i = 1, 2$, without losing generality we may assume that $t_1 \le t_2$. Then for

$$|x_1| + \bar r \le 2r \quad \text{and} \quad t_1 + \bar r^2 \le 4r^2 \tag{12}$$

we have

$$z_1, z_2 \in \bar C_{\bar r/4}(z_1) \subset \bar C_{\bar r}(z_1) \subset \bar C_{2r}. \tag{13}$$

Since $z_1 \in \bar C_r$, we have $|x_1| \le r$ and $t_1 \le r^2$ and, for any of the inequalities (12) to go wrong, we have to have $\bar r \ge r$, that is,

$$\Big(\frac{r}{\rho(z_1, z_2)}\Big)^{\beta/\alpha - 1} \ge 1,$$

which is only possible if $\rho(z_1, z_2) \ge r$ when (11) holds trivially with $N = 2$. Therefore, in what follows we assume (13) and that $\bar r \le r$.

Next, if $\bar{r} \geq K^{-1}$, then, since $r \leq 1$,

$$\rho^{1-\beta/\alpha}(z_1, z_2) \geq K^{-1}$$

and $\rho(z_1, z_2)$ is bounded from below. In that case again (11) is trivial.

Finally, if (13) holds, $\bar{r} \leq r$, and $\bar{r} \leq K^{-1}$, by Theorem 5 applied to $C_{\bar{r}}(z_1)$ in place of C_r we obtain

$$\big|u(z_1) - u(z_2)\big| \leq N\big(\bar{r}^{-\alpha}\rho^{\alpha}(z_1, z_2) \sup_{C_{2r}} |u| + \bar{r}^{d/(d+1)} \|\mathcal{L}u\|_{L_{d+1}(C_{2r})}\big),$$

where the right-hand side is transformed to that of (11) by simple arithmetics. The theorem is proved. □

Much later in Section 13.12 we will need one more consequence of Lemma 2. Recall some notation introduced earlier and introduce some new ones:

$$\mathbb{R}^d_+ = \{x = (x^1, x') : x^1 > 0, x' \in \mathbb{R}^{d-1}\},$$

$$B_r^+ = B_r \cap \mathbb{R}^d_+, \quad \partial_{x^1} B_r^+ = \bar{B}_r \cap \partial\mathbb{R}^d_+, \tag{14}$$

$$C_{\tau,r}^+ = [0, \tau) \times B_r^+, \quad \partial_{x^1} C_{\tau,r}^+ = [0, \tau) \times \partial_{x^1} B_r^+, \tag{15}$$

$$C_r^+ = C_{r^2,r}^+, \quad \partial_{x^1} C_r^+ = \partial_{x^1} C_{r^2,r}^+.$$

7. LEMMA. *Let $R \in (0, \infty)$ and $u \in W^{1,2}_{d+1,\mathrm{loc}}(C_{2R}^+) \cap C(\bar{C}_{2R}^+)$. Assume that $b = 0$, $c = 0$, $\mathcal{L}u = 0$ in C_{2R} (a.e.), and $u = 0$ on $\partial_{x^1} C_{2R}^+$. Then with α_0 from Lemma 2, $\alpha \in (0, \alpha_0]$, and $N = N(d, \delta)$ we have*

$$\big|u(z_1) - u(z_2)\big| \leq N R^{-\alpha} \rho^{\alpha}(z_1, z_2) \sup\big(|u|, C_{2R}^+\big), \tag{16}$$

whenever z_1, $z_2 \in C_R^+$.

Proof. As always, it suffices to concentrate on $R = 1$. First we claim that there is a constant $N = N(d, \delta)$ such that in $C_{3/2}^+$ we have

$$\big|u(t, x)\big| \leq N x^1 \sup\big(|u|, C_2^+\big). \tag{17}$$

To prove (17) fix $(0, x_0') \in B_{3/2}$ and choose $\gamma = \gamma(\delta, d) > 0$ so that the function

$$\psi(x) = (1/2)^{-\gamma} - \big|x - (-1/2, x_0')\big|^{-\gamma}$$

would satisfy $\mathcal{L}\psi \leq -1$ in C_2^+ (see the proof of Lemma 5.5.7). After that set

$$\Psi(t, x) = (4 - t)^{-1}\psi(x), \quad G = \big\{(t, x) : \Psi(t, x) < 1\big\} \cap C_2^+.$$

As is easy to check, $\mathcal{L}\Psi \leq 0$ in G. Furthermore, if $(t, x) \in \partial' G$, then either $\Psi(t, x) = 1$, or $x^1 = 0$, when $u(t, x) = 0$, or else $|x| = 2$, in which case

$$4\Psi(t, x) \geq \psi\big(0, 2x_0'/|x_0'|\big) \geq (1/2)^{-\gamma}$$

$$-\big(1/4 + \big(2 - |x_0'|\big)^2\big)^{-\gamma/2} \geq (1/2)^{-\gamma} - (1/2)^{-\gamma/2}.$$

It follows that

$$|u| \leq N\Psi \sup\big(|u|, C_2^+\big)$$

on $\partial' G$ and, by the maximum principle, in G, and, since $\Psi \geq 1$ outside G, in C_2^+ as well. After noting that for $(t_0, x_0) \in C_{3/2}^+$ we have

$$\Psi(t_0, x_0) \leq \psi(x_0) = (1/2)^{-\gamma} - (x_0^1 + 1/2)^{-\gamma} \leq N x_0^1,$$

we obtain (17).

Now take $z_1 = (t_1, x_1), z_2 = (t_2, x_2) \in C_1^+$, assume that $t_1 \leq t_2$ and consider two cases:

(a) $x_1^1 > 4\rho(z_1, z_2)$,

(b) $x_1^1 \leq 4\rho(z_1, z_2)$.

In case (a) set $r = (1/4)x_1^1$ and note that

$$z_1, z_2 \in C_r(z_1) \subset C_{2r}(z_1) \subset C_{3/2}^+.$$

By Lemma 2

$$\big|u(z_1) - u(z_2)\big| \leq N r^{-\alpha} \rho^\alpha(z_1, z_2) \sup\big(|u|, C_{2r}(z_1)\big). \tag{18}$$

Upon observing that, if $z = (t, x) \in C_{2r}(z_1)$, then $x^1 \leq x_1^1 + 2r = 6r$ and using (17) we conclude from (18) that

$$\begin{aligned}\big|u(z_1) - u(z_2)\big| &\leq N r^{-\alpha} \rho^\alpha(z_1, z_2) r \sup(|u|, C_2)\\ &\leq N \rho^\alpha(z_1, z_2) \sup(|u|, C_2).\end{aligned}$$

After that it only remains to observe that in case (b) in light of (17)

$$\begin{aligned}\big|u(z_1) - u(z_2)\big| &\leq \big|u(z_1)\big| + \big|u(z_2)\big| \leq N\big(x_1^1 + |x_1 - x_2|\big) \sup(|u|, C_2)\\ &\leq N\rho(z_1, z_2) \sup(|u|, C_2) \leq N\rho^\alpha(z_1, z_2) \sup(|u|, C_2).\end{aligned}$$

The lemma is proved. □

We now extend the above results to elliptic equations. We say that an operator $\mathcal{L}$ is of type (19) if

$$\mathcal{L} = a^{ij}(x) D_{ij} + b^i(x) D_i - c(x), \tag{19}$$

where measurable a^{ij}, b^i, c are defined on $\mathbb{R}^d$, satisfy (2) for all x, $\lambda \in \mathbb{R}^d$, $i, j = 1, ..., d$, and $c \geq 0$.

8. Theorem. *Let $R \in (0, 2]$ and let $u \in W_d^{1,2}(B_R)$ be such that $u \geq 0$ in B_R and $\mathcal{L}u = 0$ in B_R (a.e.), where $\mathcal{L}$ is of type (19). Then there exists a constant N, which depends only on d, K, and δ, such that*

$$u(0) \leq N u(x)$$

for $|x| \leq R/2$. Moreover, if δ^{-1} and K vary in a bounded range, then N also varies in a bounded range. Finally, if $b \equiv 0$ and $c \equiv 0$, then our assertions are also true for $R > 2$.

Proof. Our assertion follows directly from Theorem 1 if $u \in C^2(B_R)$. To see this, it suffices to take $\theta = 2$ and $\partial_t + \mathcal{L}$ in place of $\mathcal{L}$ in Theorem 1. In the general case, find a sequences $\varphi_n \in C^3(B_R)$ and $\mathcal{L}_n$ of type (19) with smooth coefficients such that $\varphi_n \geq 0$ in B_R, $\varphi_n \to u$ uniformly on ∂B_R, and the coefficients of $\mathcal{L}_n$ converge to the corresponding coefficients of $\mathcal{L}$ in B_R (a.e.). Define $u_n \in C^2(B_R)$ as solutions of $\mathcal{L}_n u_n = 0$ with boundary conditions $u_n = \varphi_n$ on ∂B_R. Then $\mathcal{L}_n(u_n - u) = (\mathcal{L} - \mathcal{L}_n)u$ in B_R (a.e.) and by Theorem 3.0.2, $u_n \to u$ uniformly on $\bar{B}_R$. Since the assertion of the theorem holds for u_n, it is also holds for u. The theorem is proved. □

9. REMARK. Safonov ([**105**]) extended the elliptic Harnack inequality to the case of $|b| \in L_d$.

10. COROLLARY (one-sided Liouville's theorem). *If $b \equiv 0$, $c \equiv 0$, $u \in W^2_d(B_R)$, for all R, $u \geq 0$ in $\mathbb{R}^d$, $\mathcal{L}u = 0$ on $\mathbb{R}^d$ (a.e.), where $\mathcal{L}$ is of type* (19), *then u=const.*

Indeed, replacing u by $u - \inf(u, \mathbb{R}^d)$, reduces the general case to the one that $\inf(u, \mathbb{R}^d) = 0$. Then choose $x_n \in E$ so that $u(x_n) \to 0$. By Theorem 8, $0 \leq u(0) \leq Nu(x_n)$, where N is independent of n. It follows that $u(0) = 0$ and, since one can take any point as the origin, $u(x) = 0$ for all x.

11. THEOREM. *Let $r \in (0, K^{-1} \wedge 1]$, let $u \in W^2_d(B_{2r})$, and let $c \equiv 0$. Then there exist a constant $N = N(d, \delta)$ such that, for any operator $\mathcal{L}$ of type* (19) *and $\alpha \in (0, \alpha_0]$,*

$$\big|u(x_1) - u(x_2)\big| \leq N\big(r^{-\alpha}|x_1 - x_2|^{\alpha} \sup_{B_{2r}} |u| + r\|\mathcal{L}u\|_{L_d(B_{2r})}\big) \tag{20}$$

whenever $x_1, x_2 \in \bar{B}_r$. If, in addition, $b \equiv 0$, then (20) *also holds for $r > K^{-1} \wedge 1$.*

Proof. Scalings allow us to assume that $r = 1$ and $K \leq 1$. In case $r = 1$, $u \in C^2(B_{2r})$, and $\mathcal{L}u = 0$, (20) follows directly from Lemma 2 if we take $z_1 = (0, x_1), z_2 = (0, x_2)$, and take $\partial_t + \mathcal{L}$ in place of $\mathcal{L}$. Next, following the argument in the proof of Theorem 8, we get (20) in the case that $u \in W^2_d(B_{2r})$ and $\mathcal{L}u = 0$. The general case is dealt with as in Theorem 5, where one applies Theorem 3.0.2 instead of Theorem 3.0.1. The theorem is proved. □

Here is an elliptic counterpart of Theorem 6.

12. THEOREM. *Let $r \in (0, 1]$ and $u \in W^2_d(B_{2r})$. Assume that $c \equiv 0$. Then there exists a constant $N = N(d, \delta, K)$ such that, for any operator $\mathcal{L}$ of type* (19), *we have*

$$\big|u(x_1) - u(x_2)\big| \leq Nr^{-\beta}|x_1 - x_2|^{\beta}\big(\sup_{B_{2r}} |u| + r\|\mathcal{L}u\|_{L_d(B_{2r})}\big) \tag{21}$$

for x_1, $x_2 \in \bar{B}_r$, where

$$\beta = \frac{\alpha}{\alpha + 1}$$

and $\alpha = \alpha_0(\delta, d)$ is the constant from Theorem 11 *(or Lemma* 2*).*

Proof. Fix x_1, $x_2 \in \bar{B}_r$. Since there is the sup norm of u on the right, it suffices to prove (21) if

$$\xi := \Big(\frac{r}{|x_1 - x_2|}\Big)^{\beta/\alpha} \geq 4.$$

Then set

$$\bar{r} = \xi|x_1 - x_2|,$$

and observe that

$$x_1, x_2 \in \bar{B}_{\bar{r}/4}(x_1) \subset \bar{B}_{\bar{r}}(x_1) \subset \bar{C}_{2r}, \tag{22}$$

where the last inclusion holds provided

$$|x_1| + \bar{r} \leq 2r.$$

Since $x_1 \in \bar{B}_r$, we have $|x_1| \le r$ and, for the above inequality to fail, we have to have $\bar{r} \ge r$, that is,

$$\Big(\frac{r}{|x_1 - x_2|}\Big)^{\beta/\alpha - 1} \ge 1,$$

which is only possible if $|x_1 - x_2| \ge r$ when (21) holds trivially with $N = 2$. Therefore, in what follows we assume (22) and that $\bar{r} \le r$ (≤ 1).

Next, if $\bar{r} \ge K^{-1}$, then, since $r \le 1$,

$$|x_1 - x_2|^{1-\beta/\alpha} \ge K^{-1}$$

and $|x_1 - x_2|$ is bounded from below. In that case again (21) is trivial.

Finally, if (22) holds, $\bar{r} \le r$, and $\bar{r} \le K^{-1}$, by Theorem 5 applied to $B_{\bar{r}}(x_1)$ in place of B_r we obtain

$$\big|u(x_1) - u(x_2)\big| \le N\big(\bar{r}^{-\alpha}|x_1 - x_2|^\alpha \sup_{B_{2r}} |u| + \bar{r}\|\mathcal{L}u\|_{L_d(B_{2r})}\big),$$

where the right-hand side is transformed to that of (21) by simple arithmetics. The theorem is proved. □

7. On the global modulus of continuity of solutions of nonlinear elliptic equations

Let Ω be a bounded domain satisfying the exterior ball condition and let $\delta \in (0, 1]$ and $K_0 \in [0, \infty)$ be fixed constants.

Here we estimate the global modulus of continuity of solutions. This will be used in many proofs involving passing to the limit.

1. Theorem. *Suppose that we are given a Borel measurable function $H(\mathsf{u}, x)$,*

$$\mathsf{u} = (\mathsf{u}', \mathsf{u}''), \quad \mathsf{u}' = \big(\mathsf{u}'_0, \mathsf{u}'_1, ..., \mathsf{u}'_d\big) \in \mathbb{R}^{d+1}, \quad \mathsf{u}'' \in \mathbb{S}, \quad x \in \mathbb{R}^d,$$

such that

$$\big|H(\mathsf{u}', 0, x)\big| \le K_0|\mathsf{u}'| + \bar{H}(x) \tag{1}$$

for all $\mathsf{u}' \in \mathbb{R}^{d+1}$ and $x \in \mathbb{R}^d$, where $\bar{H} \in L_d(\Omega)$. Assume that H is Lipschitz continuous with respect to u'' and $D_{\mathsf{u}''}H \in \mathbb{S}_\delta$ at all point of its differentiability with respect to u''. Let $u \in W^2_{d,\text{loc}}(\Omega) \cap C(\bar{\Omega})$ satisfy

$$H[u](x) := H\big(u(x), Du(x), D^2u(x), x\big) = 0$$

in Ω (a.e.). Then there exists a function $\bar{\omega}(\tau)$, depending only on

$$\|u\|_{C(\Omega)}, \quad \|\bar{H}\|_{L_d(\Omega)},$$

$\tau \ge 0$, δ, d, K_0, $\rho_{\text{ext}}(\Omega)$, and the modulus of continuity of u restricted to $\partial\Omega$, such that $\bar{\omega}(\tau) \to 0$ as $\tau \downarrow 0$ and

$$\big|u(x) - u(y)\big| \le \bar{\omega}\big(|x - y|\big)$$

for any $x, y \in \bar{\Omega}$.

Proof. By Lemma 4.1.6 in which we ignore the dependence on q we have the following.

2. LEMMA. *Let $u \in W^2_{d,\text{loc}}(\Omega)$. Then there exist an $\mathbb{S}_\delta$-valued function $a(x)$, $\mathbb{R}^d$-valued functions $b(x)$, and real-valued function $f(x)$, such that they are measurable,*

$$|b| \leq K_0, \quad |f| \leq \bar{H} + K_0|u|,$$

and in Ω (a.e.)

$$a^{ij}D_{ij}u + b^iD_iu + f = H[u]. \tag{2}$$

Next, take $x_1, x_2 \in \bar{\Omega}$ and introduce

$$\rho_i = \text{dist}\,(x_i, \partial\Omega), \quad \rho = |x_1 - x_2|.$$

Owing to the fact that we allow $\bar{\omega}$ to depend on $\sup_\Omega |u|$, while estimating $|u(x_1) - u(x_2)|$, we may concentrate on the case that

$$\rho \leq K_0^{-2} \wedge \big(\rho^2_{\text{ext}}(\Omega)/4\big). \tag{3}$$

In this situation consider two subcases:

(a) $\rho \geq \rho_1^2 \wedge \rho_2^2$;

(b) $\rho < \rho_1^2 \wedge \rho_2^2$.

In case (a) similarly to the proof of Lemma 5.5.8 let $\bar{x}_i$ be a closest point to x_i on $\partial\Omega$. Then, for each $i = 1, 2$, the closures of the two balls has only one common point $\bar{x}_i$: one is the ball centered at x_i with radius ρ_i and the other is the ball of radius $\rho_{\text{ext}}(\Omega)$, which lies outside Ω and is centered, say at y_i. Then, for each $i = 1, 2$, all three points

$$x_i \in \bar{\Omega}, \quad \bar{x}_i \in \partial\Omega, \quad y_i \notin \bar{\Omega}$$

lie on the same straight line and $|\bar{x}_i - y_i| = \rho_{\text{ext}}(\Omega)$ (≤ 1 by assumption).

Observe that $|\rho_1 - \rho_2| \leq \rho < \sqrt{\rho}$ and (case (a)) $\rho_1, \rho_2 < 2\sqrt{\rho}$, so that

$$x_i \in B_{\rho_{\text{ext}}(\Omega)+2\sqrt{\rho}}(y_i), \quad i = 1, 2.$$

Introduce

$$N_i = \sup\big\{|u(y) - u(\bar{x}_i)| \,: y \in \partial\Omega \cap B_{\rho_{\text{ext}}(\Omega)+2\sqrt{\rho}}(y_i)\big\}, \quad i = 1, 2.$$

Then take ψ from Lemma 6.6.17 and set

$$\phi(x) = N' - N''\rho^{-1/2}\psi(x - y_1),$$

where

$$N' = u(\bar{x}_1) + N_1,$$

and N'', depending only on d, δ, $\rho_{\text{ext}}(\Omega)$, K_0, and $\|u\|_{C(\Omega)}$, is chosen so that, for $i = 1, 2$,

$$-N''\rho^{-1/2}\psi(x - y_i) \geq \big|u(x)\big| \quad \text{on} \quad \Omega \cap \partial B_{\rho_{\text{ext}}(\Omega)+2\sqrt{\rho}}(y_i),$$

which is possible since on $\partial B_{\rho_{\text{ext}}(\Omega)+2\sqrt{\rho}}(y_i)$ we have (see (3))

$$-\psi(x - y_i) = \rho^{-\beta}_{\text{ext}}(\Omega) - \big|\rho_{\text{ext}}(\Omega) + 2\sqrt{\rho}\big|^{-\beta} \geq \sqrt{\rho}\gamma,$$

where $\gamma = \gamma\big(\beta, \rho_{\text{ext}}(\Omega)\big) > 0$.

Now, by Lemmas 2 and 6.6.17 (with $c \equiv 0$ in the latter) the function $v := u - \phi$ satisfies

$$a^{kr}D_{kr}v + b^kD_kv \geq -\bar{H} - K_0|u|$$

in $\Omega \cap B_{\rho_{\text{ext}}(\Omega)+2\sqrt{\rho}}(y_1)$ (a.e.) for appropriate a and b. Moreover, $v \le 0$ on the boundary of

$$\Omega \cap B_{\rho_{\text{ext}}(\Omega)+2\sqrt{\rho}}(y_1).$$

Before applying Theorem 3.0.2 in this situation we estimate the size of $\Omega \cap B_{\rho_{\text{ext}}}(\Omega)+2\sqrt{\rho}(y_1)$. Since $2\sqrt{\rho} \le \rho_{\text{ext}}(\Omega)$ (see (3)), it follows from Definition 5.5.3 that

$$\Omega \cap B_{\rho_{\text{ext}}(\Omega)+2\sqrt{\rho}}(y_1) \subset B_{3\sqrt{2\rho_{\text{ext}}(\Omega)}\rho^{1/4}}(\bar{x}_1). \tag{4}$$

Hence, in light of Theorem 3.0.2, we have

$$u(x_1) \le u(\bar{x}_1) + N_1 - N''\rho^{-1/2}\psi(x_1 - y_1) + N\rho^{1/4},$$

where and below N depends only on δ, d, K_0, $\rho_{\text{ext}}(\Omega)$, $\|u\|_{C(\Omega)}$, $\|\bar{H}\|_{L_d(\Omega)}$. Similarly, one proves that

$$u(x_2) \ge u(\bar{x}_2) - N_2 + N''\rho^{-1/2}\psi(x_2 - y_2) - N\rho^{1/4}.$$

Next, observe that $|\bar{x}_i - x_i| = \rho_i \le 2\sqrt{\rho}$ and and $|\bar{x}_1 - \bar{x}_2| \le \rho_1 + \rho_2 + \rho \le 5\sqrt{\rho}$. It follows that

$$|\psi(x_i - y_i)| = |\psi(x_i - y_i) - \psi(\bar{x}_i - y_i)| \le N\sqrt{\rho},$$

$$|u(\bar{x}_1) - u(\bar{x}_2)| \le \omega(5\sqrt{\rho}),$$

where ω is the modulus of continuity of u restricted to $\partial\Omega$. Finally, owing to (4)

$$N_1 \le \omega\Big(3\sqrt{2\rho_{\text{ext}}(\Omega)}\rho^{1/4}\Big)$$

and the same estimate holds for N_2. We conclude that

$$u(x_1) - u(x_2) \le N\rho^{1/4} + \omega(5\sqrt{\rho}) + 2\omega\Big(3\sqrt{2\rho_{\text{ext}}(\Omega)}\rho^{1/4}\Big).$$

By interchanging x_1 and x_2 we get the desired estimate in case (a).

In case (b) we apply Theorem 6.11 with $r = \sqrt{\rho}/2$ to $B_{\sqrt{\rho}}(x_1)$ in place of B_{2r}. We need to have $r \le K_0^{-1} \wedge 1$, that is $\rho \le 4\big(K_0^{-2} \wedge 1\big)$, which is guaranteed by (3). Note also that $\sqrt{\rho} \le \rho_1$, so that $B_{\sqrt{\rho}}(x_1) \subset \Omega$ and

$$|x_1 - x_2| = \rho \le \sqrt{\rho}/2 = r,$$

since in light of (3) and the fact that $\rho_{\text{ext}}(\Omega) \le 1$, we have $\rho \le 1/4$.

By Theorem 6.11

$$\big|u(x_1) - u(x_2)\big| \le Nr^{-\alpha}\rho^{\alpha} \sup_{\Omega} |u| + Nr\|\bar{H}\|_{L_d(\Omega)} \le N\rho^{\alpha/2}.$$

We have estimated $\big|u(x_1) - u(x_2)\big|$ in both cases (a) and (b), and the theorem is proved. □

8. On the global modulus of continuity of solutions of nonlinear parabolic equations

Let Ω be a bounded domain satisfying the exterior ball condition and let $\delta \in (0,1]$, $T, K_0 \in (0,\infty)$ be fixed constants and set

$$\Pi = [0,T) \times \Omega.$$

1. THEOREM. *Suppose that we are given a Borel measurable function* $H(\mathsf{u},t,x)$,

$$\mathsf{u} = (\mathsf{u}',\mathsf{u}''), \quad \mathsf{u}' = (\mathsf{u}'_0,\mathsf{u}'_1,...,\mathsf{u}'_d) \in \mathbb{R}^{d+1}, \quad \mathsf{u}'' \in \mathbb{S}, \quad t \in \mathbb{R}, \quad x \in \mathbb{R}^d,$$

satisfying

$$\big|H(\mathsf{u}',0,t,x)\big| \le K_0|\mathsf{u}'| + \bar{H}(t,x) \tag{1}$$

for all $\mathsf{u}', (t,x) \in \mathbb{R}^{d+1}$, *where* $\bar{H} \in L_{d+1}(\Pi)$. *Assume that* H *is Lipschitz continuous with respect to* u'' *and* $D_{\mathsf{u}''}H \in \mathbb{S}_\delta$ *at all point of its differentiability with respect to* u''. *Let* $u \in W^{1,2}_{d+1,\text{loc}}(\Pi) \cap C(\bar{\Pi})$ *satisfy*

$$\partial_t v(t,x) + H[v](t,x) = 0 \tag{2}$$

in Π *(a.e.). Then there exists a function* $\bar{\omega}(\tau)$, *depending only on* τ,

$$\|u\|_{C(\Pi)}, \quad \|\bar{H}\|_{L_{d+1}(\Pi)},$$

δ, d, K_0, $\rho_{\text{ext}}(\Omega)$, *and the modulus* ω *of continuity of* u *restricted to* $\partial'\Pi$, *such that* $\bar{\omega}(\tau) \to 0$ *as* $\tau \downarrow 0$ *and*

$$\big|u(t,x) - u(s,y)\big| \le \bar{\omega}\big[\rho\big((t,x),(s,y)\big)\big]$$

for any $x, y \in \bar{\Omega}$ *and* $s, t \in [0,T]$.

For the proof we need a few auxiliary results.

2. LEMMA. *For a constant* $B \ge 0$ *and* $t > 0, x \ge 0$ *introduce*

$$\xi(t,x) = \frac{1}{2\sqrt{\pi t\delta}} \int_{-\infty}^{x} e^{-By-B^2t\delta}\Big(e^{-y^2/(4t\delta)} - e^{-(2x-y)^2/(4t\delta)}\Big)\,dy.$$

For $t = 0$, $x \ge 0$ *set* $\xi(0,x) = 1$.

Then

(i) the function ξ *is continuous in* $[0,\infty) \times [0,\infty) \setminus \{(0,0)\}$, $\xi(t,0) = 0$ *for* $t > 0$, *it is an increasing function of* x, *and* $0 \le \xi \le 1$,

(ii) the function ξ *is a decreasing function of* t, *concave with respect to* x, *and in* $(0,\infty) \times [0,\infty)$ *we have*

$$\partial_t \xi = \delta D^2\xi + 2\delta B D\xi, \tag{3}$$

(iii) for $t > 0$ *and* $x \ge 0$ *we have*

$$|D\xi(t,x)| \le \frac{1}{\sqrt{\pi t\delta}} + 2B.$$

Proof. (i) The continuity of ξ is obvious. Next, for $t > 0$ we have

$$D\xi(t,x) = \frac{1}{2\sqrt{\pi}(t\delta)^{3/2}} \int_{-\infty}^{x} e^{-By-B^2t\delta}(2x-y)e^{-(2x-y)^2/(4t\delta)}\,dy,$$

which shows, in particular, that $D\xi > 0$. Also, obviously, for $t > 0$ we have $\xi(t,0) = 0$ and

$$\lim_{x\to\infty}\xi(t,x)=\frac{1}{2\sqrt{\pi t\delta}}\int_{-\infty}^{\infty}e^{-By-B^2t\delta-y^2/(4t\delta)}\,dy$$

$$=\frac{1}{2\sqrt{\pi t\delta}}\int_{-\infty}^{\infty}e^{-(y+2Bt\delta)^2/(4t\delta)}\,dy=1,$$

so that $\xi\le 1$.

(ii) To deal with $\partial_t\xi$ recall how the fundamental solution of $\partial_t u=\delta D^2 u$ looks to get that, for $t>0$,

$$\partial_t\xi=\frac{\delta}{2\sqrt{\pi t\delta}}\int_{-\infty}^{x}e^{-By-B^2t\delta}D_y^2\Big(e^{-y^2/(4t\delta)}-e^{-(2x-y)^2/(4t\delta)}\Big)\,dy-B^2\delta\xi.$$

Here in the first term on the right we integrate by parts and obtain that it equals

$$\frac{\delta B}{2\sqrt{\pi t\delta}}\int_{-\infty}^{x}e^{-By-B^2t\delta}D_y\Big(e^{-y^2/(4t\delta)}-e^{-(2x-y)^2/(4t\delta)}\Big)\,dy$$

$$-\frac{x\delta}{2\sqrt{\pi}(t\delta)^{3/2}}e^{-Bx-B^2t\delta-x^2/(4t\delta)}=-\frac{x\delta}{2\sqrt{\pi}(t\delta)^{3/2}}e^{-Bx-B^2t\delta-x^2/(4t\delta)}+B^2\delta\xi.$$

Hence,

$$\partial_t\xi=-\frac{x\delta}{2\sqrt{\pi}(t\delta)^{3/2}}e^{-Bx-B^2t\delta-x^2/(4t\delta)}.\tag{4}$$

This expression is negative and ξ decreases in t indeed.

Next,

$$D\xi=-\frac{1}{2\sqrt{\pi t\delta}}\int_{-\infty}^{x}e^{-By-B^2t\delta}D_x e^{-(2x-y)^2/(4t\delta)}\,dy,$$

$$D^2\xi=\frac{x}{2\sqrt{\pi}(t\delta)^{3/2}}e^{-Bx-B^2t\delta-x^2/(4t\delta)}$$

$$-\frac{1}{2\sqrt{\pi t\delta}}\int_{-\infty}^{x}e^{-By-B^2t\delta}D_x^2 e^{-(2x-y)^2/(4t\delta)}\,dy.$$

In the last integral D_x^2 can be replaced with $-2D_yD_x$, and then integrating by parts once we see that

$$D^2\xi=-\frac{x}{2\sqrt{\pi}(t\delta)^{3/2}}e^{-Bx-B^2t\delta-x^2/(4t\delta)}-2BD\xi.$$

This implies, in the first place, that $D^2\xi<0$ and ξ is concave in x, and also combined with with (4) leads to (3).

To prove assertion (iii) observe that $0<D\xi(t,x)\le D\xi(t,0)$ since ξ is a concave increasing function of x. After that it only remains to note that

$$D\xi(t,0)=\frac{1}{2\sqrt{\pi}(t\delta)^{3/2}}\int_0^{\infty}ye^{-(y-2B\delta t)^2/(4t\delta)}\,dy=I_1+I_2,$$

where

$$I_1=\frac{1}{2\sqrt{\pi}(t\delta)^{3/2}}\int_0^{\infty}(y-2B\delta t)e^{-(y-2B\delta t)^2/(4t\delta)}\,dy=\frac{1}{\sqrt{\pi t\delta}}e^{-B^2t\delta},$$

$$I_2 \leq \frac{B}{\sqrt{\pi t \delta}} \int_{-\infty}^{\infty} e^{-(y-2B\delta t)^2/(4t\delta)}\, dy = 2B.$$

The lemma is proved. □

3. Lemma. *For $\mu \geq K_0$ and $(t,x) \in \mathbb{R}^{d+1}$ set*

$$\zeta(t,x) = e^{\alpha\mu^2 t} \cosh \mu|x|.$$

Then there exists $\alpha = \alpha(\delta, d) > 0$ such that

$$\partial_t \zeta(t,x) \geq a^{ij} D_{ij}\zeta(t,x) + b^i D_i \zeta(t,x) \tag{5}$$

whenever $a = (a^{ij}) \in \mathbb{S}_\delta$ and $|b| \leq K_0$, $b = (b^i)$.

Proof. Due to the continuity of $D^2\zeta$, it suffices to prove (5) for $x \neq 0$. Set $\lambda = x/|x|$ and observe that the right-hand side of (5) multiplied by $e^{-\alpha\mu^2 t}$ equals

$$\mu^2(a\lambda,\lambda)\cosh\mu|x| + \frac{\mu}{|x|}\big[\operatorname{tr} a - (a\lambda,\lambda)\big]\sinh\mu|x| + \mu(b,\lambda)\sinh\mu|x|. \tag{6}$$

We use the inequalities $\sinh\mu|x| \leq \cosh\mu|x|$ and $\sinh\mu|x| \leq \mu|x|\cosh\mu|x|$ and conclude that (6) is less than

$$\big[\mu^2 \operatorname{tr} a + \mu K_0\big]\cosh\mu|x| \leq \mu^2[\delta^{-1}d + 1]\cosh\mu|x|.$$

Now (5) follows if $\alpha = \delta^{-1}d + 1$. The lemma is proved. □

In the remaining part of the section the assumptions of Theorem 1 are supposed to be satisfied.

4. Lemma. *Let $z_0 = (t_0, x_0) \in \Pi$ and let $\bar{x}_0 \in \partial\Omega$ satisfy $\rho_\Omega(x_0) = |x_0 - \bar{x}_0|$. Then*

$$\big|u(t_0,x_0) - u(t_0,\bar{x}_0)\big| \leq \hat{\omega}\big(|x_0 - \bar{x}_0|\big),$$

where the function $\hat{\omega}(\tau)$, $\tau \geq 0$, depending only on τ, d, δ, $\rho_{\text{ext}}(\Omega)$, K_0,

$$\|u\|_{C(\Pi)}, \quad \|\bar{H}\|_{L_{d+1}(\Pi)},$$

and the modulus $\omega(\tau)$ of continuity of u on $\partial'\Pi$, is such that $\hat{\omega}(\tau) \to 0$ as $\tau \downarrow 0$.

Proof. Owing to the fact that we allow $\bar{\omega}$ to depend on $\sup_\Pi |u|$, while estimating $|u(t_0,x_0) - u(t_0,\bar{x}_0)|$, we may concentrate on the case that

$$\rho_\Omega(x_0) = |x_0 - \bar{x}_0| < \rho^2_{\text{ext}}(\Omega).$$

Then set $r = \rho_\Omega^{1/2}(x_0)$, note that $\rho_\Omega(x_0) < r \leq \rho_{\text{ext}}(\Omega)$, and consider two cases:

(a) $r^2 \leq T - t_0$,

(b) $r^2 > T - t_0$.

In case (a) let $y_0 \in \mathbb{R}^d$ be the point outside Ω on the line passing through x_0 and $\bar{x}_0$ at the distance $\rho_{\text{ext}}(\Omega)$ from $\bar{x}_0$. Define

$$\Sigma = \Omega \cap B_{\rho_{\text{ext}}(\Omega)+r}(y_0), \quad \Gamma = (t_0, t_0 + r^2) \times \Sigma.$$

Take the number $B = B\big(K_0, \delta, d, \rho_{\text{ext}}(\Omega)\big) \geq 0$ in Lemma 2 in such a way that for

$$\eta(t,x) = \xi\big(t_0 + r^2 - t, |x - y_0| - \rho_{\text{ext}}(\Omega)\big)$$

we have

$$\partial_t \eta + a^{ij} D_{ij}\eta + b^i D_i \eta \leq 0 \tag{7}$$

in $(0, t_0 + r^2) \times B^c_{\rho_{\text{ext}}(\Omega)}(y_0)$, whenever $(a^{ij}) \in \mathbb{S}_\delta$ and $|(b^i)| \le K_0$. This is possible because the left-hand side of (7) is (skipping obvious values of the arguments)

$$-\partial_t \xi + (D^2\xi) a^{ij} \frac{(x^i - y_0^i)(x^j - y_0^j)}{|x - y_0|^2} + (D\xi) b^i \frac{x^i - y_0^i}{|x - y_0|}$$

$$+(D\xi)\frac{1}{|x - y_0|}\Big[\operatorname{tr}(a^{ij}) - a^{ij}\frac{(x^i - y_0^i)(x^j - y_0^j)}{|x - y_0|^2}\Big],$$

which owing to the properties of ξ is less than

$$-\partial_t \xi + \delta D^2 \xi + \big[K_0 + \delta^{-1}\rho_{\text{ext}}^{-1}(\Omega) d\big] D\xi.$$

Next, introduce

$$\phi(t, x) = u(t_0, \bar{x}_0) + N_0 + N_1 \eta(t, x) - N_2 r^{-1} \psi(x - y_0),$$

where ψ is taken from Lemma 6.6.17,

$$N_0 = \sup\big\{|u(z) - u(t_0, \bar{x}_0)| : z \in (\partial'\Pi) \cap \partial'\Gamma\big\},$$

$$N_1 = 2 \sup_{\Pi} u\,,$$

and N_2, depending only on d, δ, K_0, $\|u\|_{C(\Pi)}$, and $\rho_{\text{ext}}(\Omega)$, is chosen so that

$$-N_2 r^{-1} \psi(x - y_0) \ge |u(t, x) - u(t_0, \bar{x}_0)| \text{ on } \Pi \cap (0, T) \times \partial B_{\rho_{\text{ext}}(\Omega) + r}(y_0),$$

which is possible by the same reasons as in the proof of Theorem 7.1.

Observe that by construction

$$\phi \ge u \quad \text{on} \quad \partial'\Gamma. \tag{8}$$

Furthermore,

$$\partial_t \phi + a^{ij} D_{ij} \phi + b^i D_i \phi \le 0$$

in Γ, whenever $(a^{ij}) \in \mathbb{S}_\delta$ and $|(b^i)| \le K_0$.

Notice that $\Gamma \subset \Pi$, and by Definition 5.5.3 (recall that $r \le \rho_{\text{ext}}(\Omega)$) we have

$$\Sigma = \Omega \cap B_{\rho_{\text{ext}}(\Omega) + r}(y_0) \subset B_{3\sqrt{r \rho_{\text{ext}}(\Omega)}}(\bar{x}_0). \tag{9}$$

It follows in light of the parabolic version of Lemma 7.2 and Theorem 3.0.1 that

$$u(t_0, x_0) \le \phi(t_0, x_0) + N r^{d/(2d+2)} \|\bar{H}, u\|_{L_{d+1}(\Gamma)},$$

$$u(t_0, x_0) - u(t_0, \bar{x}_0) \le N_0 + N_1 \eta(t_0, x_0) - N_2 r^{-1} \psi(x_0 - y_0)$$

$$+ N r^{d/(2d+2)} \|\bar{H}\|_{L_{d+1}(\Pi)} + N r^{d/(2d+2)} \|u\|_{C(\Pi)}.$$

The reader might have noticed a flaw in the above argument caused by the fact that ϕ is not continuous in $\bar{\Gamma}$. To avoid this flaw it suffices to replace $|x - y_0|$ in the definition of η with $\varepsilon + |x - y_0|$, where $\varepsilon > 0$, and then let $\varepsilon \downarrow 0$.

By comparing $-\phi + 2u(t_0, \bar{x}_0)$ and u we get a complementary inequality and hence

$$\big|u(t_0, x_0) - u(t_0, \bar{x}_0)\big| \leq N_0 + N_1\eta(t_0, x_0) - N_2 r^{-1}\psi(x_0 - y_0) + N r^{d/(2d+2)}.$$

Owing to (9) we have $N_0 \leq \omega\big(3\sqrt{r\rho_{\text{ext}}(\Omega)} + r\big)$. Also

$$\big|\psi(x_0 - y_0)\big| \leq N|x_0 - \bar{x}_0| = Nr^2.$$

Therefore, using also Lemma 2 we conclude

$$\eta(t_0, x_0) = \xi\big(r^2, \rho_\Omega(x_0)\big) \leq (\pi\delta)^{-1/2} r + 2Br^2,$$

and

$$\big|u(t_0, x_0) - u(t_0, \bar{x}_0)\big| \leq \omega\big(3\sqrt{r\rho_{\text{ext}}(\Omega)} + r\big) + Nr + Nr^{d/(2d+2)}. \tag{10}$$

In case (b) set

$$\Gamma = (t_0, T) \times \Sigma, \quad \phi(t, x) = u(t_0, \bar{x}_0) + N_0 - N_2 r^{-1}\psi(x - y_0),$$

where N_0 and N_2 are defined in the same way as above. Then we again obtain (8) and (10). After that it only remains to recall that $r = |x_0 - \bar{x}_0|^{1/2}$. The lemma is proved. □

5. LEMMA. *Let* $z_0 = (t_0, x_0) \in \Pi$. *Then*

$$\big|u(T, x_0) - u(t_0, x_0)\big| \leq \breve{\omega}(T - t_0),$$

where the function $\breve{\omega}(\tau)$, $\tau \geq 0$, *depending only on* τ, d, δ, $\rho_{\text{ext}}(\Omega)$, K_0,

$$\|u\|_{C(\Pi)}, \quad \|\bar{H}\|_{L_{d+1}(\Pi)},$$

and the modulus $\omega(\tau)$ *of continuity of* u *on* $\partial'\Pi$, *is such that* $\breve{\omega}(\tau) \to 0$ *as* $\tau \downarrow 0$.

Proof. Take α from Lemma 3. Again it suffices to concentrate on the case that

$$r := (T - t_0)^{1/4} \leq (2\alpha K_0)^{-1/3}. \tag{11}$$

Furthermore, if $\rho_\Omega(x_0) \leq r$, then by Lemma 4 and in its notation we have

$$\begin{aligned}\big|u(T, x_0) - u(t_0, x_0)\big| &\leq \omega\big(|x_0 - \bar{x}_0|\big) + \hat{\omega}\big(|x_0 - \bar{x}_0|\big) + \big|u(T, \bar{x}_0) - u(t_0, \bar{x}_0)\big| \\ &\leq \omega(r) + \hat{\omega}(r) + \omega(T - t_0).\end{aligned}$$

Therefore, in the remaining part of the proof we assume that

$$\rho_\Omega(x_0) \geq r,$$

so that $B_r(x_0) \subset \Omega$. Then set

$$\mu = (2\alpha r^3)^{-1},$$

which is greater than K_0 owing to (11), and introduce

$$\Gamma = (t_0, T) \times B_r(x_0), \quad \phi(t, x) = N_0 + N_1 e^{\alpha|\mu|^2(T-t)} \frac{\cosh \mu|x - x_0|}{\cosh \mu r},$$

where

$$N_0 = \sup_{B_r(x_0)} \big|u(T, x) - u(T, x_0)\big|, \quad N_1 = 2\sup_{\Pi} |u|.$$

Obviously $\phi \geq u - u(T, x_0)$ on $\partial'\Gamma$, so that, by taking into account Lemma 3, the parabolic version of equation (7.2), and Theorem 3.0.1, we get that

$$u(t_0, x_0) - u(T, x_0) \leq \phi(t_0, x_0) + N r^{d/(d+1)} \|\bar{H}, u\|_{L_{d+1}(\Gamma)}.$$

As usual changing signs we obtain

$$\big|u(t_0, x_0) - u(T, x_0)\big| \leq \phi(t_0, x_0) + N r^{d/(d+1)}.$$

Since $N_0 \leq \omega(r)$, we have

$$\phi(t_0, x_0) \leq \omega(r) + N \frac{e^{\alpha\mu^2 r^4}}{\cosh \mu r} \leq \omega(r) + N \exp(\alpha\mu^2 r^4 - \mu r).$$

Upon observing that

$$\alpha\mu^2 r^4 - \mu r = -(4\alpha r^2)^{-1}$$

and recalling what r is we finish the proof. The lemma is proved. □

Proof of Theorem 1. Take $z_1 = (t_1, x_1), z_2 = (t_2, x_2) \in \Pi$, assume that $t_1 \leq t_2$, and set

$$\rho = \rho(z_1, z_2), \quad r = \rho^{1/2}.$$

Observe that it suffices to estimate $\big|u(z_1) - u(z_2)\big|$ only if $\rho < K_0^{-2} \wedge 1$ when $K_0^{-1} \wedge 1 > r > \rho$ and

$$z_1, z_2 \subset C_r(z_1).$$

Then first assume that $C_{2r}(z_1) \subset \Pi$. In that case thanks to Theorem 6.5 and (7.2) we have

$$\big|u(z_1) - u(z_2)\big| \leq N r^\alpha + N r^{d/(d+1)}, \tag{12}$$

where α is taken from Theorem 6.5.

However, if $C_{2r}(z_1) \not\subset \Pi$ only two things may happen:

(a) $2r > \rho_\Omega(x_1)$,

(b) $t_1 + 4r^2 > T$.

In case (a) we have

$$\rho_\Omega(x_2) \leq \rho_\Omega(x_1) + |x_1 - x_2| \leq 2r + \rho \leq 3r,$$

and if we denote $\bar{x}_i$ some points on $\partial\Omega$ such that $\rho_\Omega(x_i) = |x_i - \bar{x}_i|$, $i = 1, 2$, then by Lemma 4

$$\big|u(t_1, x_1) - u(t_1, \bar{x}_1)\big| \leq \hat{\omega}\big(|x_1 - \bar{x}_1|\big) \leq \hat{\omega}(2r), \quad \big|u(t_2, x_2) - u(t_2, \bar{x}_2)\big| \leq \hat{\omega}(3r).$$

Furthermore,

$$|\bar{x}_1 - \bar{x}_2| \leq \rho_\Omega(x_1) + \rho_\Omega(x_2) + |x_1 - x_2| \leq 5r, \quad |t_1 - t_2|^{1/2} \leq \rho \leq r.$$

It follows that

$$\big|u(z_1) - u(z_2)\big| \leq 2\hat{\omega}(3r) + \big|u(t_1, \bar{x}_1) - u(t_2, \bar{x}_2)\big| \leq 2\hat{\omega}(3r) + \omega(6r). \tag{13}$$

Finally, in case (b)

$$T - t_2 \leq T - t_1 \leq 4r^2$$

and by Lemma 5

$$\big|u(t_1, x_1) - u(T, x_1)\big| \leq \check{\omega}(4r^2), \quad \big|u(t_2, x_2) - u(T, x_2)\big| \leq \check{\omega}(4r^2).$$

By recalling that $|x_1 - x_2| \leq r$ we get

$$\big|u(z_1) - u(z_2)\big| \leq 2\breve{\omega}(4r^2) + \omega(r). \tag{14}$$

Thus, one of the estimates (12)-(14) always holds whenever $\rho < K_0^{-2} \wedge 1$, and the theorem is proved. □.

CHAPTER 10

Solvability in $W^2_{p,\text{loc}}$ of fully nonlinear elliptic equations

1. Main results

In this chapter we consider elliptic equations

$$H[v](x) := H\big(v(x), Dv(x), D^2v(x), x\big) = 0 \tag{1}$$

in subdomains of $\mathbb{R}^d$, where $H(\mathsf{u}, x)$ is a function given for $x \in \mathbb{R}^d$ and $\mathsf{u} = (\mathsf{u}', \mathsf{u}'')$,

$$\mathsf{u}' = (\mathsf{u}'_0, \mathsf{u}'_1, ..., \mathsf{u}'_d) \in \mathbb{R}^{d+1}, \quad \mathsf{u}'' \in \mathbb{S}.$$

Recall that $\mathbb{S}$ is the set of symmetric $d \times d$ matrices and, for $\delta \in (0, 1]$,

$$\mathbb{S}_\delta = \big\{a \in \mathbb{S} : \delta|\xi|^2 \le a^{ij}\xi^i\xi^j \le \delta^{-1}|\xi|^2, \quad \forall \xi \in \mathbb{R}^d\big\}.$$

Let Ω be an open bounded subset of $\mathbb{R}^d$ satisfying the exterior ball condition. Fix

$$p > d \quad \text{and a measurable function} \quad \bar{G} \ge 0 \quad \text{on} \quad \mathbb{R}^d.$$

Also fix some constants $K_0, K_F \in [0, \infty)$, $\delta \in (0, 1]$.

Quite often we deal with

$$\Omega^\rho = \big\{x \in \Omega : \text{dist}\,(x, \partial\Omega) > \rho\big\},$$

where $\rho > 0$ is a given number. Obviously, $\Omega^\rho \ne \emptyset$ only if $\rho < \rho_{\text{int}}(\Omega)$. However, mentioning this restriction many times looks awkward and quite often we avoid doing that by defining the norms of functions in function spaces over empty domains as being zero.

1. Example. The reader will see that, if $d = 3$ and $f, \bar{G} \in L_p(\Omega)$, the equation $\big(a \wedge b = \min(a, b)\big)$

$$H(D^2u, x) := \bar{G}(x) \wedge |D_{12}u| + \bar{G}(x) \wedge |D_{23}u| + \bar{G}(x) \wedge |D_{31}u|$$
$$+\Delta u - f(x) = 0 \tag{2}$$

satisfies all assumptions listed below. Observe that H in (2) is neither convex nor concave with respect to D^2u. Also note that we can replace Δu with $a^{ij}(x)D_{ij}u$ if $a(x) = (a^{ij}(x))$ is an $\mathbb{S}_\delta$-valued VMO-function such that $a(x) \ge (\delta^{ij})$.

2. Example. For $\tau > 0$ take

$$H(\mathsf{u}) = \Big(1 + \tau \cos\sqrt{|\ln|\mathsf{u}''||}\Big)\text{tr}\,\mathsf{u}'',$$

where for $\mathsf{u}'' \in \mathbb{S}$ by $|\mathsf{u}''|$ we mean $\text{tr}^{1/2}(\mathsf{u}''\mathsf{u}'')$, and choose τ so small that $D_{\mathsf{u}''}H \in \mathbb{S}_\delta$ for a $\delta \in (0, 1]$. Then again H is neither convex nor concave with respect to u'' and our assumptions are satisfied perhaps with a further reduced τ and $\bar{F}(\mathsf{u}'') = \text{tr}\,\mathsf{u}''$.

3. ASSUMPTION. The function $H(\mathsf{u}, x)$ is Borel measurable with respect to (u', x).

The following assumptions contain parameters $\hat{\theta}, \theta \in (0, 1]$ which are specified later in our results.

4. ASSUMPTION. There are Borel measurable functions $F(\mathsf{u}, x) = F\big(\mathsf{u}_0', \mathsf{u}'', x\big)$ and $G(\mathsf{u}, x)$ such that

$$H = F + G.$$

Furthermore, for all $\mathsf{u}'' \in \mathbb{S}$, $\mathsf{u}' \in \mathbb{R}^{d+1}$, and $x \in \mathbb{R}^d$ we have

$$\big|G(\mathsf{u}, x)\big| \leq \hat{\theta}|\mathsf{u}''| + K_0|\mathsf{u}'| + \bar{G}(x), \quad F(0, x) \equiv 0. \tag{3}$$

5. EXAMPLE. One can take $F\big(\mathsf{u}_0', \mathsf{u}'', x\big) = H(0, \mathsf{u}'', x)$ and $G = H - F$ to satisfy the first part of Assumption 4. Since we require that $F(0, x) = 0$, one can then take $F\big(\mathsf{u}_0', \mathsf{u}'', x\big) = H(0, \mathsf{u}'', x) - H(0, x)$ and $G = H - F$. However, we are not bound by this option.

Observe that Assumption 4 is satisfied in Example 1 with $F(u, D^2u) = \Delta u$ and $G = H - F$.

Recall that

$$B_r(x) = \big\{y \in \mathbb{R}^d : |x - y| < r\big\}, \quad B_r = B_r(0)$$

and that for measurable $\Gamma \subset \mathbb{R}^d$ we denote by $|\Gamma|$ the volume of Γ.

The combination of Assumptions 4 and 6 we call "a relaxed convexity assumption on H and an almost VMO condition on F". Recall that Lipschitz continuous functions are almost everywhere differentiable.

6. ASSUMPTION. (i) The function F is Lipschitz continuous with respect to u'' with Lipschitz constant K_F.

Moreover, there exist $R_0 \in (0, 1]$ and $\tau_0 \in [0, \infty)$ such that, if $r \in (0, R_0]$, $z \in \Omega$, $B_r(z) \subset \Omega$, and $\mathsf{u}_0' \in \mathbb{R}$, then one can find a *convex* function $\bar{F}(\mathsf{u}'') = \bar{F}_{z,r,\mathsf{u}_0'}(\mathsf{u}'')$ (independent of x) for which

(ii) We have $\bar{F}(0) = 0$ and $D_{\mathsf{u}''}\bar{F} \in \mathbb{S}_\delta$ at all points of differentiability of $\bar{F}$;

(iii) For any $\mathsf{u}'' \in \mathbb{S}$ with $|\mathsf{u}''| = 1$, we have

$$\int_{B_r(z)} \sup_{\tau > \tau_0} \tau^{-1}\big|F\big(\mathsf{u}_0', \tau\mathsf{u}'', x\big) - \bar{F}(\tau\mathsf{u}'')\big|\, dx \leq \theta\big|B_r(z)\big|; \tag{4}$$

(iv) There exists a continuous increasing function $\omega_F(\tau)$, $\tau \geq 0$, such that $\omega_F(0) = 0$ and for any $\mathsf{u}_0', \mathsf{v}_0' \in \mathbb{R}$, $x \in \Omega$, and $\mathsf{u}'' \in \mathbb{S}$ we have

$$\big|F\big(\mathsf{u}_0', \mathsf{u}'', x\big) - F\big(\mathsf{v}_0', \mathsf{u}'', x\big)\big| \leq \omega_F\big(|\mathsf{u}_0' - \mathsf{v}_0'|\big)|\mathsf{u}''|.$$

7. REMARK. It is useful to note that Assumptions 4 and 6 (iv) imply that $F(\mathsf{u}_0', 0, x) = 0$ for any $\mathsf{u}_0' \in \mathbb{R}$ and $x \in \Omega$.

8. REMARK. Assumptions 4 and 6 (iv) imply

$$\big|H(\mathsf{u}', 0, x)\big| \leq K_0|\mathsf{u}'| + \bar{G}(x) \quad \forall \mathsf{u}', x. \tag{5}$$

Note that Assumption 6 is obviously satisfied in Examples 1 and 2 if we take $F(u, D^2u, x) = \Delta u$ and choose τ appropriately in Example 2.

Next, we introduce the following definition.

9. DEFINITION. For a function $u \in C(\bar{\Omega})$ set

$$\omega_u(\Omega, \rho) = \sup\{|u(x_1) - u(x_2)| : x_1, x_2 \in \Omega, |x_1 - x_2| \leq \rho\},$$

$$\omega_{F,u,\Omega}(\rho) = \omega_F(\omega_u(\Omega, \rho)),$$

and in the formulations of a theorem, lemma,... let us say that a certain constant depends only on A,B,..., and the function $\omega_{F,u,\Omega}$ if it depends only on A,B,..., and on the maximal solution of an inequality like $N_0\omega_{F,u,\Omega}(\rho) \leq 1/2$, where the range of ρ and the value of N_0 depending only on A,B,... could be always traced down in our arguments.

We now come to the main a priori estimate in Sobolev spaces. Observe that in Theorem 10 we do not even assume that equation (1) is elliptic. Note also that the standing assumption that Ω satisfies the exterior ball condition, actually, is not needed in Theorem 10.

10. THEOREM. *Under the above assumptions there exist constants $\hat{\theta}, \theta \in (0, 1]$, depending only on d, p, δ, and K_F, such that, if Assumptions* 6 *and* 4 *are satisfied with these θ and $\hat{\theta}$, respectively, then, for any $u \in W^2_{p,\mathrm{loc}}(\Omega) \cap C(\bar{\Omega})$ that satisfies* (1) *in Ω (a.e.) and $\rho > 0$ (such that $\Omega^\rho \neq \emptyset$), we have*

$$\|u\|_{W^2_p(\Omega^\rho)} \leq N\|\bar{G}\|_{L_p(\Omega)} + N\rho^{-2}\|u\|_{C(\Omega)} + N\tau_0, \tag{6}$$

where the constants N depend only on K_0, K_F, d, p, δ, R_0, $\mathrm{diam}(\Omega)$, *and the function $\omega_{F,u,\Omega}$.*

We prove this theorem in Section 5.

Generally, having a priori estimates does not guarantee that there are existence results. Say, if $d = 1$ and the equation is

$$u'' - (2u'')I_{|u''|\leq 5} - 10u = 1 \quad \text{in} \quad (-1, 1),$$

a priori W^2_p-estimates are trivial since $|(2u'')I_{|u''|\leq 5}| \leq 10$. However, the solvability is quite questionable.

To have the solvability we need ellipticity and more regularity of H.

11. ASSUMPTION. The function $H(\mathsf{u}, x)$ is continuous in u for any x, is Lipschitz continuous with respect to u'', $D_{\mathsf{u}''}H \in \mathbb{S}_\delta$ at all points of differentiability of H with respect to u''.

12. ASSUMPTION. For all values of the arguments,

$$H(\mathsf{u}', 0, x)\,\mathrm{sign}\,\mathsf{u}'_0 \leq K_0|[\mathsf{u}']| + \bar{G}(x) \quad (\mathrm{sign}\,0 := \pm 1) \tag{7}$$

(regarding the notation $[\mathsf{u}']$ see (4.1.15)).

Assumption 11 is satisfied in Example 1 because $H_{\mathsf{u}''_{ii}} = 1$, $|H_{\mathsf{u}''_{13}}| = |H_{\mathsf{u}''_{31}}| \leq 1/2$, $|H_{\mathsf{u}''_{12}}| = |H_{\mathsf{u}''_{21}}| \leq 1/2$, $|H_{\mathsf{u}''_{23}}| = |H_{\mathsf{u}''_{32}}| \leq 1/2$ (see Remark 4.1.10).

13. REMARK. We draw the reader's attention to the fact that we do not assume that H is a nonincreasing function of u'_0 (cf. Remark 6.6.8).

Here is one of the main results of the chapter concerning the solvability of (1) in Sobolev spaces. We fix a function

$$g \in C(\partial\Omega).$$

14. THEOREM. *There exist constants $\hat\theta,\theta\in(0,1]$, depending only on d, p, δ, and K_F, which are, generally, smaller than $\hat\theta,\theta$ from Theorem 10 and such that, if Assumptions 6 and 4 are satisfied with these θ and $\hat\theta$, respectively, and Assumptions 11 and 12 are also satisfied and $\bar G\in L_p(\Omega)$, then there exists $u\in W^2_{p,\text{loc}}(\Omega)\cap C(\bar\Omega)$ satisfying (1) in Ω (a.e.) and such that $u=g$ on $\partial\Omega$.*

15. REMARK. Since none of characteristics of Ω, apart from $\rho_{\text{int}}(\Omega)$ and $\operatorname{diam}(\Omega)$ enters Theorem 10 and Lemma 6.6.10, one can use Theorem 14 to prove the solvability in much worst domains than those satisfying the exterior ball condition. Usually one does it by approximating from inside a given domain, say with smooth ones. For instance, it would suffice to have

$$\lim_{\rho\downarrow 0}\inf_{x\in\partial\Omega}\frac{|B_\rho(x)\cap\Omega^c|}{\rho^d}>0, \tag{8}$$

see Theorem 3.1 of [**103**].

We are not pursuing this path and leave it to the interested reader.

We give the proof of Theorem 14 in Section 5 after some preparations. As a simple corollary of this theorem we obtain the following in which Assumption 6 is not used.

16. THEOREM. *Suppose that Assumptions 3, 11, and 12 are satisfied, $\bar G\in L_p(\Omega)$, and (5) holds true. Let $P(\mathsf{u}'')$ be a convex function on $\mathbb{S}$ such that $D_{\mathsf{u}''}P\in\mathbb{S}_{\delta'}$ at each point of its differentiability, where $\delta'\in(0,\delta]$. Also assume that for any $a\in\mathbb{S}_\delta$ and $\mathsf{u}''\in\mathbb{S}$ we have*

$$a^{ij}\mathsf{u}''_{ij}\le P(\mathsf{u}'')+K,$$

where K is a constant. Then the equation

$$\max\big(H[u],P[u]\big)=0$$

in Ω (a.e.) with boundary condition $u=g$ on $\partial\Omega$ has a solution $u\in W^2_{p,\text{loc}}(\Omega)\cap C(\bar\Omega)$.

Proof. Introduce

$$\hat H(\mathsf{u},x)=\max\big(H(\mathsf{u},x),P(\mathsf{u}'')\big),\quad \hat F(\mathsf{u}'',x)=P(\mathsf{u}'')-P(0),\quad \hat G=\hat H-\hat F.$$

Obviously Assumptions 3, 6, 11, and 12 are satisfied for $\hat H$, $\hat F$, and $\hat F$ in place of H, F, and $\bar F$, respectively, with a K_F, $\tau_0=\theta=0$, the same K_0, δ' in place of δ, and $\bar G+|P(0)|$ in place of $\bar G$. Finally, for any u,x,

$$\hat G(\mathsf{u},x)=\max\big(H(\mathsf{u},x)-P(\mathsf{u}'')+P(0),P(0)\big)\ge P(0),$$

where for an $a\in\mathbb{S}_\delta$

$$H(\mathsf{u},x)-P(\mathsf{u}'')=H(\mathsf{u},x)-H(\mathsf{u}',0,x)-P(\mathsf{u}'')+H(\mathsf{u}',0,x)$$
$$=a^{ij}\mathsf{u}''_{ij}-P(\mathsf{u}'')+H(\mathsf{u}',0,x)\le K+H(\mathsf{u}',0,x),$$

which shows that Assumption 4 is also satisfied with $\hat\theta=0$ and $\bar G+K+|P(0)|$ in place of $\bar G$.

Hence, Theorem 14 is applicable and our theorem is proved. □

One more solvability result, in which there is no monotonicity assumption on the dependence of H on u'_0, is the following.

17. THEOREM. *There exist constants $\hat\theta,\theta\in(0,1]$, depending only on d, p, δ, and K_F, such that, if Assumptions* 6 *and* 4 *are satisfied with these θ and $\hat\theta$, respectively, and Assumptions* 11 *and* 6.6.4 *are also satisfied with $M_0\geq 0,\varepsilon>0$, and $|g|\leq M_0$, and $\bar G\in L_p(\Omega)$, then there exists $u\in W^2_{p,\mathrm{loc}}(\Omega)\cap C(\bar\Omega)$ satisfying* (1) *in Ω (a.e.) and such that $u=g$ on $\partial\Omega$.*

We prove this theorem in Section 5.

18. REMARK. Similarly to Remark 6.6.14 we observe that generally there is no uniqueness in Theorems 14 and 17. For instance, in the one-dimensional case the (quasilinear) equation

$$D^2u+\sqrt{12|Du|}=0$$

for $x\in(-1,1)$ with zero boundary data has two solutions: one is identically equal to zero and the other one is $1-|x|^3$.

Another example is given by the (semilinear) equation

$$D^2u+2u(1+\sin^2 x+u^2)^{-1}=0$$

on $(-\pi/2,\pi/2)$ with zero boundary condition. Again there are two solutions: one is $\cos x$ and the other one is identically equal to zero.

To have uniqueness we need additional assumptions (see Section 4.1:2).

19. REMARK. The parameter θ in Theorems 14 and 17 depends on p and we cannot guarantee that it stays bounded away from zero for all $p>d$. Our arguments are only valid if we take θ sufficiently small and, as $p\to\infty$, θ should go to zero.

20. REMARK. For measurable $\Gamma\subset\mathbb{R}^d$ with nonzero Lebesgue measure and locally summable f denote

$$⨍_\Gamma f(x)\,dx=\frac{1}{|\Gamma|}\int_\Gamma f(x)\,dx$$

(recall that $|\Gamma|$ is the volume of Γ). Then for $z\in\Omega$, $r>0$, and $\mathsf{u}_0'\in\mathbb{R}$ introduce

$$\hat F(\mathsf{u}'')=\hat F_{z,r,\mathsf{u}_0'}(\mathsf{u}'')=⨍_{B_r(z)}F\big(\mathsf{u}_0',\mathsf{u}'',x\big)\,dx.$$

Observe that

$$\big|F\big(\mathsf{u}_0',\tau\mathsf{u}'',x\big)-\hat F(\tau\mathsf{u}'')\big|\leq\big|F\big(\mathsf{u}_0',\tau\mathsf{u}'',x\big)-\bar F(\tau\mathsf{u}'')\big|$$
$$+⨍_{B_r(z)}\big|F\big(\mathsf{u}_0',\tau\mathsf{u}'',y\big)-\bar F(\tau\mathsf{u}'')\big|\,dy,$$

which implies that

$$⨍_{B_r(z)}\sup_{\tau>\tau_0}\tau^{-1}\big|F\big(\mathsf{u}_0',\tau\mathsf{u}'',x\big)-\hat F(\tau\mathsf{u}'')\big|\,dx\leq 2\theta, \tag{9}$$

if $|u''|=1$, $B_r(z)\subset\Omega$ and $r\leq R_0$. Thus, one can be tempted to always take $\hat F$ as $\bar F$. However, there is no guarantee that $\hat F(\mathsf{u}'')$ is convex in u''.

21. REMARK. Under the above assumptions the function

$$F(\mathsf{u}'',x)=H(0,\mathsf{u}'',x)-H(0,x)$$

does not necessarily satisfy Assumption 6, so that this choice of F and G in Example 5 may not be optimal. The simplest example in case $d = 2$ is given by

$$H(0, \mathsf{u}'', x) = G(x) \wedge |\mathsf{u}''_{11}| + 2\mathsf{u}''_{11} + \mathsf{u}''_{22},$$

where $G(x) = (x^1)^{-\alpha}$ for $x^1 > 0$ and $G(x) = 0$ for $x^1 \leq 0$ with a small $\alpha > 0$, so that G is summable to a high power.

Indeed, assume that $0 \in \Omega$. Then for $z = 0$, small $r > 0$, and $\mathsf{u}''_{11} = 1$ the left-hand side of (9) (with $H(0, \mathsf{u}'', x) - H(0, x)$ in place of $F(\mathsf{u}'', x)$) becomes

$$\fint_{B_r} \sup_{\tau > \tau_0} \Big| 1 \wedge \big(G(x)/\tau\big) - \fint_{B_r} 1 \wedge \big(G(y)/\tau\big)\, dy \Big|\, dx$$

$$\geq \fint_{B_r} \Big| 1 \wedge \big(G(x)/\tau_0\big) - \fint_{B_r} 1 \wedge \big(G(y)/\tau_0\big)\, dy \Big|\, dx,$$

which for $r \leq \tau_0^{1/\alpha}$ equals

$$\fint_{B_r} \Big| I_{x^1 > 0} - \fint_{B_r} I_{y^1 > 0}\, dy \Big|\, dx = \fint_{B_r} |I_{x^1 > 0} - 1/2|\, dx = 1/2.$$

Hence, (4) cannot be satisfied with small θ and the natural choice for F in this example is, of course, $2\mathsf{u}''_{11} + \mathsf{u}''_{22}$.

22. REMARK. In condition (4) no restriction is imposed on $F\big(\mathsf{u}'_0, \mathsf{u}'', x\big)$ for $|\mathsf{u}''| \leq \tau_0$. But even for large $|\mathsf{u}''|$ the function F satisfying Assumption 6 need not be even locally convex. For instance

$$F(D^2u, x) = \Big(1 + \varepsilon \sin\big(\ln |D^2 u|\big)\Big)\, a^{ij}(x) D_{ij} u$$

satisfies Assumption 6 with $\tau_0 = 0$ and any $\theta > 0$ given in advance if $a^{ij} \in VMO$, $(a^{ij}) \in \mathbb{S}_\delta$, and $\varepsilon > 0$ is small enough.

Indeed, it suffices to take sufficiently small R_0 and set $\bar{F}_{z,r}(\mathsf{u}'') = \mathsf{u}''_{ij} \bar{a}^{ij}_{z,r}$, where $\bar{a}^{ij}_{z,r}$ are the averages of a^{ij} over $B_r(z)$. Moreover, since $\theta > 0$, a^{ij} need not be exactly in VMO but, as we say, almost in VMO.

23. REMARK. A somewhat more transparent condition than Assumption 6 can be stated in terms of

$$\hat{F}\big(\mathsf{u}'_0, \mathsf{u}'', x\big) := \lim_{\tau \to \infty} \frac{1}{\tau} F\big(\mathsf{u}'_0, \tau \mathsf{u}'', x\big)$$

provided that the limit exists. Then requiring that $\hat{F}$ be convex in u'', $D_{\mathsf{u}''} \hat{F} \in \mathbb{S}_\delta$, and $\hat{F}(\mathsf{u}'_0, \mathsf{u}'', \cdot) \in VMO$ for any $\mathsf{u}'_0, \mathsf{u}''$ seems to be enough to satisfy Assumption 6 (perhaps, modulo some kind of uniformity).

However, observe that $\hat{F}$ does not exist for F from Remark 22. One way to remedy the situation is to absorb $\varepsilon \sin\big(\ln |D^2 u|\big) a^{ij}(x) D_{ij} u$ into G. But, generally, requiring the above limit to exist is an additional, rather strong, and unnecessary assumption.

Yet it is worth mentioning [**97**], which shows that for $H = F(\mathsf{u}'') - f(x)$, any viscosity solution of $H[u] = 0$ is in $W^2_{p,\mathrm{loc}}$ with a corresponding estimate provided that $\hat{F}$ exists, is convex, and $f \in L_p$, $p > d$. We show that in such situations there exists a unique $W^2_{p,\mathrm{loc}}$-solution admitting such estimates.

In addition to the examples presented above we give two more examples related to the stochastic differential games and the Isaacs equations.

24. EXAMPLE. Let A and B be some countable sets and assume that for $\alpha \in A$, $\beta \in B$, $\mathsf{u}_0' \in \mathbb{R}$, and $x \in \mathbb{R}^d$ we are given functions

$$a^{\alpha}(\mathsf{u}_0', x), \quad b^{\alpha\beta}(\mathsf{u}_0', x), \quad c^{\alpha\beta}(\mathsf{u}_0', x), \quad f^{\alpha\beta}(\mathsf{u}_0', x)$$

with values in

$$\mathbb{S}_\delta, \quad \mathbb{R}^d, \quad [0, \infty), \quad \mathbb{R},$$

respectively. Assume that these functions are measurable in x, a^{α}, $b^{\alpha\beta}$, and $c^{\alpha\beta}$ are bounded, they are continuous with respect to u_0' uniformly with respect to α, β, x, and

$$\bar{G} := \sup_{\mathsf{u}_0', \alpha, \beta} \left| f^{\alpha\beta}(\mathsf{u}_0', \cdot) \right| \in L_p(\Omega).$$

Consider the following Isaacs equation

$$H(v, Dv, D^2 v, x) = 0, \tag{10}$$

where

$$H(\mathsf{u}, x) := \inf_{\beta \in B} \sup_{\alpha \in A} \Big[\sum_{i,j=1}^{d} a_{ij}^{\alpha}(\mathsf{u}_0', x) \mathsf{u}_{ij}'' + \sum_{i=1}^{d} b_i^{\alpha\beta}(\mathsf{u}_0', x) \mathsf{u}_i'$$

$$- c^{\alpha\beta}(\mathsf{u}_0', x) \mathsf{u}_0' + f^{\alpha\beta}(\mathsf{u}_0', x) \Big].$$

Our measurability, boundedness, and countability assumptions guarantee that H is measurable in x and Lipschitz continuous in u''. One can also easily check that at all points of differentiability $D_{\mathsf{u}''} H \in \mathbb{S}_\delta$ (see Corollary 4.1.4). Next assume that there is an $R_0 \in (0, \infty)$ such that for any $z \in \Omega$, $r \in (0, R_0]$, and $\mathsf{u}_0' \in \mathbb{R}$ one can find $\bar{a}^{\alpha} \in \mathbb{S}_\delta$ (independent of x) such that

$$\fint_{B_r(z)} \sup_{\alpha \in A} \left| a^{\alpha}(\mathsf{u}_0', x) - \bar{a}^{\alpha} \right| dx \le \theta,$$

where θ is taken from Theorem 14 (with $K_F = \delta^{-1}\sqrt{d}$).

Then we claim that there exists $u \in W^2_{p,\mathrm{loc}}(\Omega) \cap C(\bar{\Omega})$ satisfying (10) in Ω (a.e.) and such that $u = g$ on $\partial\Omega$. Furthermore, (6) holds with $\tau_0 = 0$.

To prove the claim introduce

$$F(\mathsf{u}_0', \mathsf{u}'', x) = \sup_{\alpha \in A} \sum_{i,j=1}^{d} a_{ij}^{\alpha}(\mathsf{u}_0', x) \mathsf{u}_{ij}'', \quad G = H - F.$$

Notice that Assumption 6 is satisfied with $\tau_0 = 0$ and

$$\bar{F}(\mathsf{u}'') := \sup_{\alpha \in A} \sum_{i,j=1}^{d} \bar{a}_{ij}^{\alpha} \, \mathsf{u}_{ij}''$$

because these functions are convex, positive homogeneous of degree one with respect to u'' and, for $|\mathsf{u}''|=1$,

$$\fint_{B_r(z)} \big|F(\mathsf{u}_0',\mathsf{u}'',x)-\bar F(\mathsf{u}'')\big|\,dx \le \fint_{B_r(z)} \sup_{\alpha\in A}\Big|\sum_{i,j=1}^{d}\big[a^\alpha_{ij}(\mathsf{u}_0',x)-\bar a^\alpha\big]\mathsf{u}''_{ij}\Big|\,dx$$

$$\le \fint_{B_r(z)} \sup_{\alpha\in A}\big|a^\alpha(\mathsf{u}_0',x)-\bar a^\alpha\big|\,dx \le \theta.$$

On can easily check that the remaining item (iv) in Assumption 6 and Assumptions 4 (with $\hat\theta=0$) are satisfied as well. Assumption 12 is satisfied because $c\ge 0$.

As a result we have a solvability theorem for (10), which covers (apart from the restriction on p), as A and B are singletons, the first result about solvability of linear elliptic equations with VMO coefficients obtained by Chiarenza, Frasca, and Longo in their pioneering paper [**20**]. In this singleton case we in addition treat semilinear equations.

25. Example. Let A and B be some countable sets and assume that for $\alpha\in A$, $\beta\in B$, $x\in\mathbb{R}^d$, and $\mathsf{u}'\in\mathbb{R}^{d+1}$ we are given an $\mathbb{S}_\delta$-valued function $a^\alpha(\mathsf{u}_0',x)$ and a real-valued function $b^{\alpha\beta}(\mathsf{u}',x)$. Assume that these functions are measurable in x, a^α and $b^{\alpha\beta}$ are continuous with respect to u' uniformly with respect to α,β,x, and

$$\big|b^{\alpha\beta}(\mathsf{u}',x)\big|\le K_0\big|(\mathsf{u}_1',...,\mathsf{u}_d')\big|+\bar G(x),$$

where $\bar G\in L_p(\Omega)$. Also assume that a^α satisfies the condition from Example 24.

Consider equation (10), where

$$H(\mathsf{u},x):=\inf_{\beta\in B}\sup_{\alpha\in A}\Big[\sum_{i,j=1}^{d}a^\alpha_{ij}(\mathsf{u}_0',x)\mathsf{u}''_{ij}+b^{\alpha\beta}(\mathsf{u}',x)\Big].$$

As in Example 24 one easily sees that Theorem 14 is applicable.

26. Remark. In the literature, the interior W^2_p, $p>d$, estimates like (6) with $\tau_0=0$ and $\Omega=B_1$ for viscosity solutions of a class of fully nonlinear uniformly elliptic equations of the form

$$H(D^2u,x)=f(x)$$

were first obtained by Caffarelli in [**12**] (see also [**14**]). His proof is based on an ingenious application of the Aleksandrov–Bakel'man–Pucci a priori estimate, the Krylov–Safonov Harnack inequality, and a covering result a version of which can be found in [**84**] and [**100**]. Our results are based on technically less sophisticated ideas and results from [**75**], which uses the Evans-Krylov, Fang-Hua Lin, and Fefferman–Stein theorems. By exploiting a weak reverse Hölder's inequality, the result of [**12**] was sharpened by Escauriaza in [**32**], who obtained the interior W^2_p-estimate for the same equations allowing $p>d-\varepsilon$, with a small constant $\varepsilon>0$ depending only on the ellipticity constant and d.

The above cited works [**12**], [**14**], and [**32**] are quite remarkable in one respect–they do not suppose that H is convex or concave in D^2u and relate to any viscosity solution. The assumptions in [**12**] and [**14**] are quite different from ours. One of these assumptions is that the equations $H(D^2u,x_0)=0$ admit $C^2_{\mathrm{loc}}(B_r(x_0))$-solutions for any $B_r(x_0)\subset B_1$ and any continuous boundary data. Until now we only know that, generally, this assumption is satisfied if H is convex or concave with respect to u''. Paper [**109**] and the references there present a few exceptions. Another condition is stated in terms of

$$\beta(x, x_0) = \sup_{\mathsf{u}'' \in \mathbb{S}} \frac{|H(\mathsf{u}'', x) - H(\mathsf{u}'', x_0)|}{|\mathsf{u}''|}.$$

It is assumed that for all $B_r(x_0) \subset B_1$

$$\fint_{B_r(x_0)} \beta^d(x, x_0)\, dx \leq \theta^d, \tag{11}$$

where θ is sufficiently small.

Since condition (11) is repeated in very many papers, we want to compare (11) and (4) taking $d = 3$, $p > 3$, $G \in L_p(\mathbb{R}^d)$, $G \geq 0$, and

$$H(\mathsf{u}'', x) = G(x) \wedge |\mathsf{u}''_{11}| + 2\mathsf{u}''_{11} + \mathsf{u}''_{22} + \mathsf{u}''_{33}.$$

If we take $F(\mathsf{u}'', x) = 2\mathsf{u}''_{11} + \mathsf{u}''_{22} + \mathsf{u}''_{33}$, then *all* our conditions are automatically satisfied and we have a priori estimates and *solvability*.

At the same time, note that $H(\mathsf{u}'', x_0)$ is neither concave nor convex with respect to u'' and it is not clear if the solvability assumption in [**14**] is satisfied.

Also, as is easy to see, condition (11) implies that, if $B_r(x_0) \subset B_1$ and, say $G \leq 1$, then (take $|\mathsf{u}''| = 1$ in the definition of $\beta(x, x_0)$)

$$\Big| \fint_{B_r(x_0)} G(x)\, dx - G(x_0) \Big| \leq \theta. \tag{12}$$

In particular, G should lie between two continuous functions, which collapse as $\theta \downarrow 0$. This condition is reminiscent of the usual condition in the L_p-theory of *linear* elliptic equations when one requires the leading coefficients be not necessarily continuous but be sufficiently uniformly close to continuous ones. We just repeat that our conditions, one of which may look like (11), are automatically satisfied even without anything like (12).

A number of existence results of $W^2_{p,\text{loc}}$-solutions and a priory estimates in $W^2_{p,\text{loc}}$ obtained by means of the theory of viscosity solutions can be found in [**13**], [**14**], and [**15**]. In all of them H is supposed to be Lipschitz continuous in u uniformly with respect to x and satisfy a continuity assumption with respect to x similar to (11). Note that these assumptions exclude, for instance, Example 25. On the other hand, our results do not cover those from [**13**], [**14**], and [**15**] either, in particular, just because $p < d$ is allowed there.

Note that in Theorem 4.2 of [**116**] one more interior estimate of type (6) is obtained under the assumptions that H is convex in u'', Lipschitz continuous in u and satisfies a condition similar to (11). On the other hand, again some values of $p < d$ are allowed. Finally, in [**15**] and [**116**] the function H is assumed to be nonincreasing with respect to u'_0 unlike in our Theorem 14.

2. A priori estimates in $C^{2+\alpha}_{\text{loc}}$ for equations $F(D^2u) = 0$

If $\alpha \in (0,1)$ and Q is a domain in $\mathbb{R}^d$ and u is a function on Q, we define

$$[u]_{C^\alpha(Q)} = \sup_{\substack{x,y\in Q\\ x\neq y}} \frac{|u(x)-u(y)|}{|x-y|^\alpha}$$

and by $C^{2+\alpha}(Q)$ we mean the subset of $C^2(Q)$ consisting of functions u for which

$$\|u\|_{C^{2+\alpha}(Q)} = \|u\|_{C^2(Q)} + [D^2u]_{C^\alpha(Q)} < \infty.$$

One knows that, if Q is regular enough, the norm $\|u\|_{C^{2+\alpha}(Q)}$ is equivalent to

$$\|u\|_{C^{2+\alpha}(Q)} = \|u\|_{C(Q)} + [D^2u]_{C^\alpha(Q)} < \infty.$$

As usual, by $C^{2+\alpha}_{\text{loc}}(Q)$ we mean the set of functions which belong to $C^{2+\alpha}(Q')$ for any bounded domain $Q' \subset \bar{Q}' \subset Q$.

The following fundamental result is usually attributed to L.C. Evans and the author. Let $F(\mathsf{u}'')$ be a *convex* function defined for $\mathsf{u}'' \in \mathbb{S}$ such that at all points of its differentiability we have

$$D_{\mathsf{u}''}F(\mathsf{u}'') \in \mathbb{S}_\delta,$$

where $\delta \in (0,1]$ is a fixed number.

1. THEOREM. *There exists a constant*

$$\alpha_0 = \alpha_0(\delta, d) \in (0,1)$$

such that for any $R \in (0,\infty)$ and $u \in C^{1,1}(B_R)$, satisfying $F(D^2u) = 0$ (a.e.) in B_R, we have $u \in C^{2+\alpha_0}_{\text{loc}}(B_R)$. Furthermore, for any $r \in (0,R)$ and $\alpha \in (0,\alpha_0]$,

$$[D^2u]_{C^\alpha(B_r)} \le N\frac{1}{(R-r)^{2+\alpha}} \sup_{\partial B_R} |u - l|, \tag{1}$$

where $l = l(x)$ is an arbitrary quadratic *function and $N = N(\delta, d, \alpha) < \infty$.*

To prove the theorem we need a few auxiliary results, but first we observe that we may take $l \equiv 0$ and assume that

$$F(0) = 0. \tag{2}$$

Indeed, take an arbitrary quadratic function $l(x)$ and introduce

$$\hat{F}(\mathsf{u}'') = F\big(\mathsf{u}'' + D^2l + tD^2|x|^2\big),$$

where t is chosen so that

$$\hat{F}(0) = F\big(D^2l + tD^2|x|^2\big) = 0.$$

The latter is possible since the derivative with respect to t of the middle term is $2\,\text{tr}\, D_{\mathsf{u}''}F$ which is between $2\delta d$ and $2\delta^{-1}d$. Then, under the assumption that the theorem is true if $F(0) = 0$ and $l \equiv 0$, we apply it to $v(x) = u(x) - t\big(|x|^2 - R^2\big) - l(x)$ and $\hat{F}$ in place of u and F, respectively, to get

$$[D^2u]_{C^\alpha(B_r)} = [D^2v]_{C^\alpha(B_r)} \le N\frac{1}{(R-r)^{2+\alpha}} \sup_{\partial B_R} |v|$$

$$= N\frac{1}{(R-r)^{2+\alpha}} \sup_{\partial B_R} |u - l|.$$

Therefore, below we always take $l \equiv 0$, assume that $u \in C^{1,1}(B_R)$, $F(D^2u) = 0$ (a.e.) in B_R, and (2) holds. However, in the first auxiliary result the fact that $F(D^2u) = 0$ (a.e.) in B_R is not used.

Observe that for almost any $x_0 \in B_R$, as $x \to x_0$,

$$\begin{aligned} u(x) &= u(x_0) + D_i u(x_0)(x^i - x_0^i) \\ &\quad + (1/2)D_{ij}u(x_0)(x^i - x_0^i)(x^j - x_0^j) + o(|x - x_0|^2). \end{aligned} \tag{3}$$

Let U be a Borel subset of these points having full measure.

Recall that for $\lambda \in \mathbb{R}^d$ and $h > 0$

$$\Delta_{h,\lambda}u(x) := \frac{u(x + h\lambda) - 2u(x) + u(x - h\lambda)}{h^2}.$$

2. LEMMA. *For any unit $\lambda \in \mathbb{R}^d$ we have $\Delta_{h,\lambda}u \to D_{(\lambda)(\lambda)}u$ on U as $h \downarrow 0$ and for any ball $B \subset B_R$ we have*

$$\sup_{B \cap U} D_{(\lambda)(\lambda)}u = \lim_{h \downarrow 0} \sup_{B^h} \Delta_{h,\lambda}u, \tag{4}$$

$$\inf_{B \cap U} D_{(\lambda)(\lambda)}u = \lim_{h \downarrow 0} \inf_{B^h} \Delta_{h,\lambda}u. \tag{5}$$

Proof. Owing to (3) we have $\Delta_{h,\lambda}u = D_{(\lambda)(\lambda)}u + o(1)$ at any given point of U as $h \downarrow 0$. This proves the first assertion and also shows that (4) holds with $\leq$ in place of $=$.

To prove the converse inequality, fix a unit λ and for $x_0 \in B_R$ denote by $\ell(x_0)$ the intersection of the line $\{x_0 + t\lambda : t \in \mathbb{R}\}$ with B_R. We call a point $x_0 \in B_R$ admissible if $\ell(x_0) \cap U$ has full (one-dimensional) measure.

By Fubini's theorem the set of admissible points has full measure. Furthermore, on each line $\ell(x_0)$ the function u is of class $C^{1,1}$ and hence it is twice differentiable in the sense of Taylor's formula almost everywhere on the line. If x_0 is admissible, then by comparing (3) with the Taylor's expansion mentioned above we conclude that

$$u_{tt}(x_0 + t\lambda) := \frac{\partial^2 u(x_0 + t\lambda)}{(\partial t)^2} = D_{(\lambda)(\lambda)}u(x_0 + t\lambda)$$

on $\ell(x_0)$ almost everywhere. Since, for $x_0 \in B^h$, it holds that

$$u(x_0 + h\lambda) - 2u(x_0) + u(x_0 - h\lambda) = \int_{-h}^{h} (h - |r|)u_{tt}(x_0 + t\lambda)\, dt,$$

it follows that, for admissible $x_0 \in B^h$, $\Delta_{h,\lambda}u(x_0)$ is dominated by the left-hand side of (4). To finish proving (4), it only remains to add that $\Delta_{h,\lambda}u$ is a continuous function. Equation (5) is proved by replacing u with $-u$ in (4). The lemma is proved. □

3. LEMMA. *Let U, V, W be $d \times d$ symmetric matrices. Then there exist $a \in \mathbb{S}_\delta$ such that*

$$F(U) - 2F(V) + F(W) \geq a^{ij}(U - 2V + W)_{ij}. \tag{6}$$

Proof. Since F is differentiable almost everywhere we may assume that it is differentiable at V. Then observe that owing to the convexity of F

$$F(U) - F(V) = \int_0^1 \frac{\partial}{\partial t} F(tU + (1-t)V)\,dt$$

$$\geq \int_0^1 \frac{\partial}{\partial t} F\big(tU + (1-t)V\big)\big|_{t=0}\,dt = D_{\mathsf{u}''_{ij}} F(V)(U_{ij} - V_{ij}).$$

Similarly,

$$F(W) - F(V) \geq D_{\mathsf{u}''_{ij}} F(V)(W_{ij} - V_{ij}),$$

and the lemma is proved. □

4. LEMMA. *For any constant $\nu \in (0,1)$ there exists a constant*

$$\beta = \beta(\nu, d, \delta) \in (0,1)$$

such that if λ is a unit vector in $\mathbb{R}^d$, $2r \leq R$, $\omega > 0$, and

$$\big|\{x \in B_{2r} : D_{(\lambda)(\lambda)}u - \inf_{U \cap B_{2r}} D_{(\lambda)(\lambda)}u > \omega\}\big| \geq \nu |B_{2r}|, \tag{7}$$

then

$$\inf_{U \cap B_r} D_{(\lambda)(\lambda)}u - \inf_{U \cap B_{2r}} D_{(\lambda)(\lambda)}u \geq \beta\omega. \tag{8}$$

Proof. By (7) and Lemma 2 for all sufficiently small $h > 0$ we have

$$\big|\{x \in B^h_{2r} : \Delta_{h,\lambda}u - \inf_{B^h_{2r}} \Delta_{h,\lambda}u > \omega'\}\big| \geq \nu' |B^h_{2r}|, \tag{9}$$

where $\omega' = \omega/2$, $\nu' = \nu/2$. Furthermore, Lemma 3 (combined with Lemma 4.1.5) implies that for an $\mathbb{S}_\delta$-valued measurable function a (and $h \leq r/4$)

$$a_{ij} D_{ij} \Delta_{h,\lambda} u \leq 0$$

almost everywhere in B^h_{2r}. The same inequality is true if we replace $\Delta_{h,\lambda}u$ with

$$\Delta_{h,\lambda}u - \inf_{B^h_{2r}} \Delta_{h,\lambda}u \quad (\geq 0).$$

It follows by Theorem 9.5.4 (a) and Theorem 9.5.1 that there exists a constant $\beta = \beta(\nu, d, \delta) \in (0,1)$ such that

$$\inf_{B_r} \Delta_{h,\lambda}u - \inf_{B^h_{2r}} \Delta_{h,\lambda}u \geq \beta\omega.$$

By letting $h \downarrow 0$ we get (8) and the lemma is proved. □

5. REMARK. Lemma 4 with the proof not using finite differences can be found in [**104**]. Below we follow even closer the exposition in [**104**], that is shorter than the original arguments of Evans and Krylov, which have their own advantage of being more intuitively clear. They reveal themselves in the original form only in the proof of Theorem 13.7.1. It is also worth saying that another approach to proving Theorem 1 is given in [**16**] and extended to integro-differential operators in [**18**].

Define

$$\omega(r, x) = \max_{|\lambda|=1} \operatorname*{osc}_{U \cap B_r(x)} D_{(\lambda)(\lambda)}u, \quad \omega(r) = \omega(r, 0).$$

6. LEMMA. *For unit $\lambda \in \mathbb{R}^d$, $\mu,\nu \in (0,1)$, and $r \in (0,R]$ introduce*

$$\Gamma_{\lambda,\mu,r} = \big\{x \in U \cap B_r : D_{(\lambda)(\lambda)}u(x) - \inf_{U\cap B_r} D_{(\lambda)(\lambda)}u > \mu\omega(r)\big\},$$

$$\Sigma_{\mu,\nu,r} = \big\{\lambda \in \partial B_1 : |\Gamma_{\lambda,\mu,r}| > \nu |B_r|\big\}.$$

Then there exist $\mu = \mu(d,\delta), \nu = \nu(d,\delta) \in (0,1)$ such that $|\Sigma_{\mu,\nu,r}| \geq \nu|\partial B_1|$ (where this time $|\cdot|$ stands for the surface measure).

Proof. For any $x \in U\cap B_r$ and unit λ, if

$$D_{(\lambda)(\lambda)}u(x) \geq (1/2)\Big[\sup_{U\cap B_r} D_{(\lambda)(\lambda)}u + \inf_{U\cap B_r} D_{(\lambda)(\lambda)}u\Big],$$

then

$$D_{(\lambda)(\lambda)}u(x) - \inf_{U\cap B_r} D_{(\lambda)(\lambda)}u \geq (1/2) \operatorname*{osc}_{U\cap B_r} D_{(\lambda)(\lambda)}u.$$

In the remaining case

$$\sup_{U\cap B_r} D_{(\lambda)(\lambda)}u - D_{(\lambda)(\lambda)}u(x) > (1/2) \operatorname*{osc}_{U\cap B_r} D_{(\lambda)(\lambda)}u.$$

Thus,

$$\sup_{y\in U\cap B_r} \big|D_{(\lambda)(\lambda)}u(x) - D_{(\lambda)(\lambda)}u(y)\big| \geq (1/2) \operatorname*{osc}_{U\cap B_r} D_{(\lambda)(\lambda)}u.$$

By taking the supremums over $|\lambda|=1$, we conclude that for each $x \in U\cap B_r$ there is a $y \in U \cap B_r$ such that

$$\max_{|\lambda|=1} \big|D_{(\lambda)(\lambda)}u(x) - D_{(\lambda)(\lambda)}u(y)\big| \geq (1/3)\omega(r). \tag{10}$$

Next, there are two cases: either there exists $\lambda \in \partial B_1$ such that

$$D_{(\lambda)(\lambda)}u(x) - D_{(\lambda)(\lambda)}u(y) \geq (1/3)\omega(r),$$

or the opposite inequality holds for all $\lambda \in \partial B_1$, but

$$-\min_{|\lambda|=1}\big[D_{(\lambda)(\lambda)}u(x) - D_{(\lambda)(\lambda)}u(y)\big] \geq (1/3)\omega(r).$$

In the latter case we observe that there exists an $a \in \mathbb{S}_\delta$ such that $a_{ij}\big[D_{ij}u(x) - D_{ij}u(y)\big] = 0$, which shows that

$$0 \leq \frac{d-1}{\delta}\max_{|\lambda|=1}\big[D_{(\lambda)(\lambda)}u(x) - D_{(\lambda)(\lambda)}u(y)\big] + \delta \min_{|\lambda|=1}\big[D_{(\lambda)(\lambda)}u(x) - D_{(\lambda)(\lambda)}u(y)\big]$$

$$\leq \frac{d-1}{\delta}\max_{|\lambda|=1}\big[D_{(\lambda)(\lambda)}u(x) - D_{(\lambda)(\lambda)}u(y)\big] - (\delta/3)\mu(r),$$

and along with the former case leads to the conclusion that there is a $\mu = \mu(d,\delta) \in (0,1)$ and a unit $\lambda = \lambda(x,y)$ such that

$$D_{(\lambda)(\lambda)}u(x) - D_{(\lambda)(\lambda)}u(y) \geq 2\mu\omega(r). \tag{11}$$

Moreover, for $\lambda \in \partial B_1$ the absolute value of the ratio

$$\frac{1}{\omega(r)}\big[D_{(\lambda)(\lambda)}u(x) - D_{(\lambda)(\lambda)}u(y)\big]$$

is less than one and therefore has a modulus of continuity with respect to λ depending only on d. It follows that (11) with μ in place of 2μ also holds in a neighborhood of $\lambda(x,y)$ of a fixed size.

Since

$$D_{(\lambda)(\lambda)}u(y) \geq \inf_{U\cap B_r} D_{(\lambda)(\lambda)}u$$

it follows that, for any $x \in U \cap B_r$, we have

$$\int_{\partial B_1} I_{\Gamma_{\lambda,\mu,r}}(x)\, d\sigma_\lambda \geq 2\nu|\partial B_1|, \tag{12}$$

where σ_λ is the surface measure on ∂B_1 and $\nu = \nu(d,\delta) \in (0,1)$. By integrating through (12) with respect to x and using Fubini's theorem, we obtain

$$\int_{\partial B_1} |\Gamma_{\lambda,\mu,r}|\, d\sigma_\lambda \geq 2\nu|\partial B_1|\cdot|B_r|.$$

Due to the fact that $|\Gamma_{\lambda,\mu,r}| \leq |B_r|$, the left-hand side is less than

$$|B_r|\cdot|\Sigma_{\mu,\nu,r}| + \nu|\partial B_1|\cdot|B_r|,$$

which yields the desired result. The lemma is proved. □

7. LEMMA. *Define*

$$\omega^*(r) = \int_{\partial B_1} \operatorname*{osc}_{U\cap B_r} D_{(\lambda)(\lambda)}u\, d\sigma_\lambda.$$

Then there is a constant $N = N(d)$ such that $\omega \leq N\omega^*$ and $\omega^* \leq N\omega$ for $r \leq R$.

This result follows easily from the observation that

$$\frac{\operatorname{osc}_{U\cap B_r} D_{(\lambda)(\lambda)}u}{\omega(r)}$$

is a Lipschitz continuous function of $\lambda \in \partial B_1$ with Lipschitz constant depending only on d.

8. LEMMA. *There are constants $\alpha_0 = \alpha_0(\delta,d) \in (0,1)$ and $N = N(d,\delta) \in (0,\infty)$ such that for $r \leq R$ and $\alpha \in (0,\alpha_0]$*

$$\omega(r) \leq N\Big(\frac{r}{R}\Big)^\alpha \omega(R). \tag{13}$$

Proof. By Lemma 7 it suffices to prove (13) with ω^* in place of ω. Furthermore, since $r/R \leq 1$, it suffices to prove (13) with $\alpha = \alpha_0$.

In light of Lemmas 6 and 4 for

$$\lambda \in \Sigma_{\mu,\nu,2r}$$

inequality (8) holds with $\omega = \mu\omega(2r)$, provided that $r \leq R/2$. Observe that

$\beta\mu\omega(2r) \geq \mu'\omega^*(2r)$, where $\mu' = \mu'(d, \delta) \in (0, 1)$. In particular, for those λ

$$\underset{U\cap B_{2r}}{\text{osc}} D_{(\lambda)(\lambda)}u - \underset{U\cap B_r}{\text{osc}} D_{(\lambda)(\lambda)}u$$

$$\geq \inf_{U\cap B_r} D_{(\lambda)(\lambda)}u - \inf_{U\cap B_{2r}} D_{(\lambda)(\lambda)}u \geq \mu'\omega^*(2r).$$

It follows that

$$\omega^*(2r) - \omega^*(r) = \int_{\partial B_1} \big[\underset{U\cap B_{2r}}{\text{osc}} D_{(\lambda)(\lambda)}u - \underset{U\cap B_r}{\text{osc}} D_{(\lambda)(\lambda)}u\big]\, d\sigma_\lambda$$

$$\geq |\Sigma_{\mu,\nu,2r}|\mu'\omega^*(2r) \geq \gamma\omega^*(2r),$$

where $\gamma = \gamma(d, \delta) \in (0, 1)$. Hence, $\omega^*(r) \leq (1 - \gamma)\omega^*(2r)$. Now by repeating an iterative argument on page 190 we conclude that (13) holds with ω^* in place of ω and with $\alpha_0 = \log_2(1 + \gamma)$ if $r \leq R/2$. For the remaining range of r and R it holds because of the monotonicity of ω^*. The lemma is proved. □

Proof of Theorem 1. Lemma 8 is applicable not only to the balls centered at the origin but at any other point in B_R as well, as long as these balls belong to B_R. Then we see, in particular, that $D_{(\lambda)(\lambda)}u$ are uniformly continuous on $U \cap B_{R'}$ with any $R' < R$. By calculus they admit a unique continuation on $B_{R'}$ and these continuations are second-order derivative of u. Furthermore, by enclosing $x, y \in B_r$ into a ball of radius $|x - y|$, finding the largest concentric ball still lying in B_R, and using Lemma 8 we easily see that, for $r < R$,

$$M(r) := [D^2u]_{C^\alpha(B_r)} \leq N\frac{1}{(R-r)^\alpha}\omega(R) \leq N\frac{1}{(R-r)^\alpha} \sup_{B_R, |\lambda|=1} |D_{(\lambda)(\lambda)}u|. \tag{14}$$

In particular, $M(r) < \infty$ for $r < R$.

To convert the right-hand side of (14) to the one in (1), observe that, for any $r < R$, unit $\lambda \in \mathbb{R}^d$, $h < r/2$, $x \in B_r$, $x' \in B_{r-h}$, such that x' lies on the same diameter as x and has the distance to x equal to h, we have

$$|D_{(\lambda)(\lambda)}u(x)| \leq |D_{(\lambda)(\lambda)}u(x) - \Delta_{h,\lambda}u(x')| + \frac{4}{h^2}\sup_{B_R}|u|$$

$$\leq (2h)^\alpha[D_{(\lambda)(\lambda)}u]_{C^\alpha(B_r)} + \frac{4}{h^2}\sup_{B_R}|u|.$$

It follows that if $r_n < r_{n+1} < R$ and $h < r_{n+1}/2$, then

$$M(r_n) \leq N\frac{1}{(r_{n+1} - r_n)^\alpha} \sup_{B_{r_{n+1}}, |\lambda|=1} |D_{(\lambda)(\lambda)}u|$$

$$\leq N_0^\alpha \frac{h^\alpha}{(r_{n+1} - r_n)^\alpha} M(r_{n+1}) + N\frac{1}{(r_{n+1} - r_n)^\alpha h^2}\sup_{B_R}|u|. \tag{15}$$

Without losing generality, we assume that $N_0 \geq 2$ and we take here $h = (r_{n+1} - r_n)/N_0$ $(< r_{n+1}/2)$. Then we get

$$M(r_n) \leq 2^{-\alpha}M(r_{n+1}) + N\frac{1}{(r_{n+1} - r_n)^{2+\alpha}}\sup_{B_R}|u|.$$

Now we choose $r_1 = r$ and for $n \geq 1$ define

$$r_{n+1} = r + (R - r)\kappa \sum_{k=1}^{n} \frac{1}{k^2},$$

where κ^{-1} is the value of the above series with ∞ in place of n, so that $r_n \uparrow R$. Then we conclude

$$\sum_{n=1}^{\infty} 2^{-n\alpha} M\big(r(n)\big) \leq \sum_{n=2}^{\infty} 2^{-n\alpha} M\big(r(n)\big) + N \frac{1}{(R-r)^{2+\alpha}} \sup_{B_R} |u| \sum_{n=1}^{\infty} n^2 2^{-n\alpha}.$$

To ensure that the above series converge we replace R with a smaller number R' and use the fact that $M(R') < \infty$. Then we cancel like terms and let $R' \uparrow R$ to get (1) with $l \equiv 0$ and B_R in place of ∂B_R. After that we observe that (recall (2))

$$0 = F(D^2 u) - F(0) = a_{ij} D_{ij} u$$

in B_R (a.e.) with a measurable $\mathbb{S}_\delta$-valued matrix (a_{ij}). Then the maximum principle shows that

$$\sup_{B_R} |u| = \sup_{\partial B_R} |u|.$$

The theorem is proved. □

9. COROLLARY. *Let $u \in C^{1,1}_{\mathrm{loc}}(\mathbb{R}^d)$ satisfy $F(D^2 u) = 0$ in $\mathbb{R}^d$ (a.e.). Assume that*

$$\varliminf_{|x| \to \infty} \frac{u(x)}{|x|^{2+\alpha_0}} = 0.$$

Then u is a quadratic function.

3. Solvability in $C^{2+\alpha}_{\mathrm{loc}}$ for equations $F(D^2 v) = 0$ and estimates of sharp functions of $D^2 u$

Let $F(\mathsf{u}'')$ be a *convex* function defined for $\mathsf{u}'' \in \mathbb{S}$ such that at all points of its differentiability we have

$$D_{\mathsf{u}''} F(\mathsf{u}'') \in \mathbb{S}_\delta,$$

where $\delta \in (0, 1]$ is a fixed number. Let $\alpha_0 = \alpha_0(\delta, d) \in (0, 1)$ be the constant from Theorem 2.1.

1. THEOREM. *For any $g \in C(\partial B_2)$ there exists a unique $v \in C(\bar{B}_2) \cap C^{2+\alpha_0}_{\mathrm{loc}}(B_2)$ satisfying*

$$F(D^2 v) = 0 \quad \text{in} \quad B_2, \quad v = g \quad \text{on} \quad \partial B_2. \tag{1}$$

Furthermore,

$$\big|D^2 v(x) - D^2 v(y)\big| \leq N |x - y|^{\alpha_0} \sup_{\partial B_2} |g - p|$$

as long as $x, y \in B_1$, where p is an arbitrary polynomial of degree 2 on $\mathbb{R}^d$ and N depends only on δ and d.

Proof. As usual, uniqueness follows from the maximum principle. Furthermore, replacing $F(\mathsf{u}'')$ with $F(\mathsf{u}''+D^2p)$ and g with $g-p$ shows that it suffices to concentrate on $p\equiv 0$. Also, in view of Theorem 2.1, it suffices to prove the existence of $v\in C(\bar B_2)\cap C^{1,1}_{\text{loc}}(B_2)$ satisfying (1).

Recall that $\bar\delta$ is introduced in Remark 6.2.2, and set

$$\bar\alpha=\bar\alpha(\delta,d)=\alpha_0\big(\bar\delta(d,\delta),d\big)\quad(\in(0,1)). \tag{2}$$

First assume that $g\in C^2(\mathbb{R}^d)$. Take P from (6.2.4), introduce $P[u](x)=P\big(D^2u(x)\big)$, and for $K>0$ consider the equation

$$\max\big(F[u],P[u]-K\big)=0$$

in B_2 (a.e.) with boundary condition $u=g$ on ∂B_2. Observe that $|F(0)|\le \bar H+0|\mathsf{u}'|$, where $\bar H=|F(0)|$. Hence, by Theorem 6.6.20 and Remark 6.6.18, this problem has a unique solution $u_K\in C(\bar B_2)\cap C^{1,1}_{\text{loc}}(B_2)$.

By Theorem 2.1 $u_K\in C^{2+\bar\alpha}_{\text{loc}}(B_2)$ and for any $r\in(0,2)$,

$$[D^2u_K]_{C^{\bar\alpha}(B_r)}\le N(\delta,d)\frac{1}{(2-r)^{2+\bar\alpha}}\sup_{\partial B_2}|g|. \tag{3}$$

In light of Remark 6.6.19 and the Arzelà-Ascoli theorem, we see that the set $\{u_K:K>0\}$ is precompact in $C(\bar B_2)$. Interpolation theorems and (3), tell us that the set $\{u_K:K>0\}$ is precompact in $C^2(B_r)$ with any $r<R$. Therefore, there is a sequence $K_n\to\infty$ such that u_{K_n}, Du_{K_n}, and $D^2u_{K_n}$ converge uniformly in $\bar B_2$, $C(B_r)$, and $C(B_r)$ for any $r<2$, respectively, to a function v, Dv, and D^2v, respectively. Obviously, $v\in C(\bar B_2)\cap C^2_{\text{loc}}(B_2)$ and v satisfies (1). Also (3) holds with v in place of u_K. This proves the theorem if $g\in C^2(\mathbb{R}^d)$.

In the case of general $g\in C(\partial B_2)$ we approximate it in $C(\partial B_2)$-norm by $g_n\in C^3(\mathbb{R}^d)$. Then we obtain the functions v_n corresponding to g_n. By the maximum principle

$$\sup_{\bar B_2}|v_n-v_m|\le\sup_{\partial B_2}|g_n-g_m|.$$

Hence, there is $v\in C(\bar B_2)$ such that $v_n\to v$ uniformly in $\bar B_2$. Of course, $v=g$ on ∂B_2. Furthermore, estimate (3) with v_n and g_n in place of u_K and g yields a similar estimate for v, which shows that $v\in C^{2+\alpha}_{\text{loc}}(B_2)$ and proves the theorem. □

Below in this section we fix

$$\alpha\in(0,\alpha_0].$$

2. LEMMA. *Let $r\in(0,\infty)$, $\nu\ge 2$ and let $\phi\in C(\partial B_{\nu r})$. Then there exists a unique $v\in C(\bar B_{\nu r})\cap C^{2+\alpha}_{\text{loc}}(B_{\nu r})$ such that*

$$F(D^2v)=0\quad in\quad B_{\nu r},\quad v=\phi\quad on\quad \partial B_{\nu r}.$$

Furthermore,

$$\fint_{B_r}\fint_{B_r}|D^2v(x)-D^2v(y)|\,dxdy\le N(d,\alpha,\delta)\nu^{-2-\alpha}r^{-2}\sup_{\partial B_{\nu r}}|\phi|.$$

Proof. Scalings show that it suffices to concentrate on $r=2/\nu$. In that case the existence of solution follows from Theorem 1, which also implies that for $x,y\in B_{2/\nu}\subset B_1$

$$|D^2v(x)-D^2v(y)|\le N\nu^{-\alpha}\sup_{\partial B_2}|\phi|.$$

It only remains to observe that

$$\fint_{B_{2/\nu}} \fint_{B_{2/\nu}} \left|D^2 v(x) - D^2 v(y)\right| dxdy \le \sup_{x,y \in B_{2/\nu}} \left|D^2 v(x) - D^2 v(y)\right|.$$

The lemma is proved. □

Recall that $\gamma_0 = \gamma_0(d, \delta, K)$ is introduced in Lemma 9.4.2 and set

$$\bar{\gamma} = \bar{\gamma}(d, \delta) := \gamma_0(d, \delta, 0) \quad (\in (0,1)). \tag{4}$$

3. LEMMA. *Let $r \in (0, \infty)$ and $\nu \in [2, \infty)$. Then for any $u \in W^2_d(B_{\nu r})$ and $\gamma \in (0, \bar{\gamma}]$ we have*

$$\Big(\fint_{B_r} \fint_{B_r} \left|D^2 u(x) - D^2 u(z)\right|^{\gamma} dxdy \Big)^{1/\gamma}$$

$$\le N \nu^{d/\gamma} \Big(\fint_{B_{\nu r}} \left|F[u]\right|^d dx \Big)^{1/d} + N \nu^{-\alpha} \Big(\fint_{B_{\nu r}} |D^2 u|^d \, dx \Big)^{1/d}, \tag{5}$$

where N depends only on d and δ.

Proof. Define v to be a unique $C(\bar{B}_{\nu r}) \cap C^{2+\alpha}_{\mathrm{loc}}(B_{\nu r})$-solution of the equation $F[v] = 0$ in $B_{\nu r}$ with boundary condition $v = u$ on $\partial B_{\nu r}$. Such a function exists by Lemma 2. Furthermore, $v(x) - b^i x^i - c$ satisfies the same equation for any constants b^i, c. Hence by Lemma 2 and Hölder's inequality

$$I_r := \Big(\fint_{B_r} \fint_{B_r} \left|D^2 v(x) - D^2 v(y)\right|^{\gamma} dxdy \Big)^{1/\gamma}$$

$$\le N \nu^{-2-\alpha} r^{-2} \sup_{x \in \partial B_{\nu r}} \left|u(x) - (D_i u)_{B_{\nu r}} x^i - u_{B_{\nu r}}\right|.$$

By Poincaré's inequality (see, for instance, Lemma 2.1 in [**70**]) the last supremum is dominated by a constant times

$$\nu^2 r^2 \Big(\fint_{B_{\nu r}} |D^2 u|^d \, dx \Big)^{1/d}.$$

It follows that

$$I_r \le N \nu^{-\alpha} \Big(\fint_{B_{\nu r}} |D^2 u|^d \, dx \Big)^{1/d}. \tag{6}$$

Next, the function $w = u - v$ is of class $C(\bar{B}_{\nu r}) \cap W^2_{d,\mathrm{loc}}(B_{\nu r})$ and for an operator $\mathcal{L} \in \mathbb{L}_{\delta,0}$ we have

$$F[u] - F[v] = \mathcal{L}(u - v), \quad \mathcal{L} w = F[u]$$

in $B_{\nu r}$ (a.e.). Moreover, $w = 0$ on $\partial B_{\nu r}$. Therefore, by Theorem 9.4.6 there exists $N = N(d, \delta) < \infty$ such that

$$\fint_{B_r} |D^2 w|^{\gamma} \, dx \le \nu^d \fint_{B_{\nu r}} |D^2 w|^{\gamma} \, dx \le N \nu^d \Big(\fint_{B_{\nu r}} \left|F[u]\right|^d dx \Big)^{\gamma/d}.$$

In fact, Theorem 9.4.6 provides the above estimate only for $\nu r = 1$. The general case is taken care of by scalings. Upon combining this result with (6) we come to (5) and the lemma is proved. □

4. A priori estimates in $W^2_{p,\text{loc}}$

Here we suppose that Assumptions 1.4 and 1.6 are satisfied. Thus, we suppose that all assumptions stated before Theorem 1.10 are satisfied.

First we derive the following.

1. LEMMA. *For any $q \in [1,\infty)$ and $\mu > 0$ there is a $\theta = \theta(d,\delta,K_F,\mu,q) > 0$ such that, if Assumption 1.6 is satisfied with this θ, then for any $\mathsf{u}_0' \in \mathbb{R}$, $r \in (0,R_0]$ and $z \in \Omega$ such that $B_r(z) \subset \Omega$ we have*

$$I(r,q,z) := \fint_{B_r(z)} \sup_{\substack{\mathsf{u}''\in\mathbb{S},\\ |\mathsf{u}''|>\tau_0}} \frac{\big|F(\mathsf{u}_0',\mathsf{u}'',x) - \bar F(\mathsf{u}'')\big|^q}{|\mathsf{u}''|^q}\,dx \le \mu^q,$$

where $\bar F = \bar F_{z,r,\mathsf{u}_0'}$.

Proof. The variable u_0' in this statement is just a parameter and we drop it. Then observe that $\big|F(\mathsf{u}'',x)\big| = \big|F(\mathsf{u}'',x) - F(0,x)\big| \le K_F|\mathsf{u}''|$ and $\big|\bar F(\mathsf{u}'')\big| \le \delta^{-1}|\mathsf{u}''|$, so that $I(r,q,z) \le N(\delta,K_F,q)I(r,1,z)$ and we may assume that $q=1$.

Next, the functions $\tau^{-1}F(\tau\mathsf{u}'',x)$ and $\tau^{-1}\bar F(\tau\mathsf{u}'')$ are Lipschitz continuous with respect to u'' with constants depending only on δ and K_F. Therefore, there exist points $\mathsf{u}''(1),...,\mathsf{u}''(n)$ with $n = n(\mu,d,\delta,K_F)$, such that $\big|\mathsf{u}''(k)\big| = 1$ and for any $\mathsf{u}'' \in \mathbb{S}$ with $|\mathsf{u}''| = 1$ there exists a k such that

$$\big|\tau^{-1}F(\tau\mathsf{u}'',x) - \tau^{-1}F\big(\tau\mathsf{u}''(k),x\big)\big| \le \mu/4,\ \big|\tau^{-1}\bar F(\tau\mathsf{u}'') - \tau^{-1}\bar F\big(\tau\mathsf{u}''(k)\big)\big| \le \mu/4$$

for all $\tau > 0$. Also note that setting $\tau = |\mathsf{u}''|$ we get

$$\sup_{\mathsf{u}''\in\mathbb{S},|\mathsf{u}''|>\tau_0} \frac{\big|F(\mathsf{u}'',x) - \bar F(\mathsf{u}'')\big|}{|\mathsf{u}''|} = \sup_{\mathsf{u}'':|\mathsf{u}''|=1}\ \sup_{\tau>\tau_0} \tau^{-1}\big|F(\tau\mathsf{u}'',x) - \bar F(\tau\mathsf{u}'')\big|$$

$$\le \sum_{k=1}^n \sup_{\tau>\tau_0} \tau^{-1}\big|F\big(\tau\mathsf{u}''(k),x\big) - \bar F\big(\tau\mathsf{u}''(k)\big)\big| + \mu/2.$$

After that it is seen that our assertion is true with $q = 1$ for $\theta(d,\delta,K_F,\mu,1) = \mu/(2n)$. The lemma is proved. □

Recall that α_0 and $\bar\gamma$ are introduced in Theorem 2.1 and (3.4), respectively.

2. LEMMA. *Let $r \in (0,\infty)$ and $\nu \ge 2$ be such that $\nu r \le R_0$ and $\Omega^{\nu r} \ne \emptyset$. Take*

$$\mu \in (0,\infty), \quad \beta \in (1,\infty),$$

and suppose that Assumption 1.6 is satisfied with $\theta = \theta(d,\delta,K_F,\mu,\beta d)$ (see Lemma 1). Take $\gamma \in (0,\bar\gamma]$, $\alpha \in (0,\alpha_0]$, a function $u \in W^2_d(\Omega)$, and for $x_0 \in \Omega^{\nu r}$ denote

$$I_r(x_0) = \Big(\fint_{B_r(x_0)}\fint_{B_r(x_0)} \big|D^2u(x_1) - D^2u(x_2)\big|^\gamma\,dx_1dx_2\Big)^{1/\gamma}.$$

Then for $x_0 \in \Omega^{\nu r}$

$$I_r(x_0) \le N\nu^{d/\gamma}\Big(\fint_{B_{\nu r}(x_0)} \big|F[u]\big|^d\,dx\Big)^{1/d} + N\tau_0\nu^{d/\gamma}$$

$$+ N\Big[\big(\mu + \omega_{F,u,\Omega}(\nu r)\big)\nu^{d/\gamma} + \nu^{-\alpha}\Big]\Big(\fint_{B_{\nu r}(x_0)} |D^2u|^{\beta' d}\,dx\Big)^{1/(\beta' d)}, \tag{1}$$

where $\beta' = \beta/(\beta-1)$ *and* N *depends only on* d, K_F, α, *and* δ.

Proof. Set $\rho := \nu r$. Since $\rho \le R_0$, $\bar F = \bar F_{x_0,\rho,u(x_0)}$ is well defined and by Lemma 3.3

$$I_r(x_0) \le N\nu^{d/\gamma}\Big(\fint_{B_\rho(x_0)} \big|\bar F[u]\big|^d\,dx\Big)^{1/d} + N\nu^{-\alpha}\Big(\fint_{B_\rho(x_0)} |D^2u|^d\,dx\Big)^{1/d}. \tag{2}$$

By setting $\hat F[u](x) = F\big(u(x_0), D^2u(x), x\big)$ we find

$$\fint_{B_\rho(x_0)} \big|\bar F[u]\big|^d\,dx \le N \fint_{B_\rho(x_0)} \big|F[u]\big|^d\,dy + NJ_1 + NJ_2,$$

where

$$J_1 = \fint_{B_\rho(x_0)} \big|\hat F[u] - \bar F[u]\big|^d\,dx$$

is dominated by (cf. Remark 1.7)

$$\fint_{B_\rho(x_0)} I_{|D^2u|>\tau_0}\frac{\big|\hat F[u] - \bar F[u]\big|^d}{|D^2u|^d}|D^2u|^d\,dx + N\tau_0^d,$$

which in turn owing to Lemma 1 and Hölder's inequality is less than

$$\mu^d\Big(\fint_{B_\rho(x_0)} |D^2u|^{\beta' d}\,dx\Big)^{1/\beta'} + N\tau_0^d,$$

and

$$J_2 = \fint_{B_\rho(x_0)} \big|\hat F[u] - F[u]\big|^d\,dx \le \omega_F^d\big(\operatorname*{osc}_{B_\rho(x_0)} u\big) \fint_{B_\rho(x_0)} |D^2u|^d\,dx.$$

It follows that

$$\Big(\fint_{B_\rho(x_0)} \big|\bar F[u]\big|^d\,dy\Big)^{1/d} \le N\Big(\fint_{B_\rho(x_0)} \big|F[u]\big|^d\,dy\Big)^{1/d}$$

$$+N\mu\Big(\fint_{B_\rho(x_0)} |D^2u|^{\beta' d}\,dy\Big)^{1/(\beta' d)} + N\tau_0 + N\omega_{F,u,\Omega}(\rho)\Big(\fint_{B_\rho(x_0)} |D^2u|^d\,dx\Big)^{1/d}.$$

This and (2) yield (1) since

$$\Big(\fint_{B_\rho(x_0)} |D^2u|^d\,dx\Big)^{1/d} \le \Big(\fint_{B_\rho(x_0)} |D^2u|^{\beta' d}\,dx\Big)^{1/\beta' d}$$

by Hölder's inequality. The lemma is proved. □

3. LEMMA. *Set $\gamma = \bar{\gamma}$, recall that $p > d$, and take $R \in (0,1]$ and $u \in W^2_p(B_{2R})$. Assume that $B_{2R} \subset \Omega$. Then there exist constants $\hat{\theta}, \theta \in (0,1]$, depending only on d, p, δ, and K_F, such that, if Assumptions 1.4 and 1.6 are satisfied with these $\hat{\theta}$ and θ, respectively, then there is a constant N, depending only on R_0, d, p, K_0, K_F, δ, and the function $\omega_{F,u,B_{2R}}$, such that*

$$\|D^2u\|_{L_p(B_R)} \le N\big\|H[u]\big\|_{L_p(B_{2R})} + N\|\bar{G}\|_{L_p(B_{2R})} + N\tau_0 R^{d/p}$$

$$+ NR^{d/p-d/\gamma}\big\|\,|D^2u|^\gamma\big\|^{1/\gamma}_{L_1(B_{2R})} + N\|u\|_{L_p(B_{2R})}, \tag{3}$$

$$\|D^2u\|_{L_p(B_R)} \le N\tau_0 R^{d/p} + NR^{d/p-2}\sup_{B_{2R}}|u|$$

$$+ N\big(\big\|H[u]\big\|_{L_p(B_{2R})} + \|\bar{G}\|_{L_p(B_{2R})}\big). \tag{4}$$

Proof. For $\rho > 0$ and $x \in \mathbb{R}^d$ introduce

$$h^{\#}_{\gamma,\rho}(x) = \sup_{\substack{r\in(0,\rho],\\ B_r(x_0)\ni x}} \Big(\fint_{B_r(x_0)}\fint_{B_r(x_0)} \big|h(x_1)-h(x_2)\big|^\gamma\,dx_1dx_2\Big)^{1/\gamma},$$

$$\mathbb{M}h(x) = \sup_{\substack{r>0,\\ B_r(x_0)\ni x}} \fint_{B_r(x_0)} |h(y)|\,dy, \tag{5}$$

whenever these definitions make sense.

Take $\varepsilon \in (0,1]$ to be specified later and take $R_1 < R_2 \le 2R$ such that

$$R_2 - R_1 \le \varepsilon R_0, \quad R_2 \le 2R_1. \tag{6}$$

Then take $\nu \ge 2$ and set

$$r_0 = (R_2 - R_1)/(\nu+1).$$

Next, take x, x_0, and $r > 0$ such that

$$r \le r_0, \quad x \in B_{R_1}, \quad x \in B_r(x_0)$$

and observe that, since $R_2 - \nu r_0 = R_1 + r_0$, we have $x_0 \in B_{R_2-\nu r_0}$ and $B_{\nu r}(x_0) \subset B_{R_2}$. Also $\nu r \le \nu r_0 \le R_0$. Therefore, by Lemma 2 applied to $\Omega = B_{R_2}$ and $\alpha = \alpha_0$ we have (note x_0 on the left and x on the right)

$$I_r(x_0) \le N\nu^{d/\gamma}\mathbb{M}^{1/d}\big(\big|F[u]\big|^d I_{B_{R_2}}\big)(x) + N\tau_0\nu^{d/\gamma}$$

$$+ N\Big[\big(\mu + \omega_{F,u,B_{2R}}(\nu r_0)\big)\nu^{d/\gamma} + \nu^{-\alpha}\Big]\mathbb{M}^{1/(\beta'd)}\big(|D^2u|^{\beta'd}I_{B_{R_2}}\big)(x)$$

with N depending only on d, K_F, and δ. It follows that in B_{R_1}

$$(D^2u)^{\#}_{\gamma,r_0} \le N\nu^{d/\gamma}\mathbb{M}^{1/d}\big(\big|F[u]\big|^d I_{B_{R_2}}\big) + N\tau_0\nu^{d/\gamma}$$

$$+ N\Big[\big(\mu + \omega_{F,u,B_{2R}}(\varepsilon R_0)\big)\nu^{d/\gamma} + \nu^{-\alpha}\Big]\mathbb{M}^{1/(\beta'd)}\big(|D^2u|^{\beta'd}I_{B_{R_2}}\big).$$

By Theorem C.2.6 with

$$\kappa = r_0/R_1 \le 1/3, \quad \chi_1 = (d+2)/\gamma, \quad \chi_2 = d/\gamma - d/p$$

and the Hardy-Littlewood maximal function theorem, by taking β so that $p > \beta' d$, we obtain

$$\begin{aligned}\|D^2u\|_{L_p(B_{R_1})} &\le N\nu^{d/\gamma}\big\|F[u]\big\|_{L_p(B_{R_2})} + N\tau_0\nu^{d/\gamma}R_1^{d/p} \\ &+ \Big[N\big(\mu + \omega_{F,u,B_{2R}}(\varepsilon R_0)\big)\nu^{d/\gamma} + N_0\nu^{-\alpha})\Big]\|D^2u\|_{L_p(B_{R_2})} \\ &+ N\nu^{\chi_1}(R_2 - R_1)^{-\chi_1}R_1^{-\chi_2+\chi_1}\big\|\,|D^2u|^\gamma\big\|^{1/\gamma}_{L_1(B_{2R})}, \qquad (7)\end{aligned}$$

where and in a few lines below the constants N, N_i depend only on d, p, K_F, and δ. Somewhat later they will also depend on K_0, and the moment when they depend on the data as in the statement of the lemma will be specifically noted. Now we take and fix $\nu \ge 2$ so that

$$N_0\nu^{-\alpha} \le 1/8.$$

Then (7) becomes

$$\begin{aligned}\|D^2u\|_{L_p(B_{R_1})} &\le N_1\big\|F[u]\big\|_{L_p(B_{R_2})} + N\tau_0R_1^{d/p} \\ &+ \Big[N_2\big(\mu + \omega_{F,u,B_{2R}}(\varepsilon R_0)\big) + 1/8\Big]\|D^2u\|_{L_p(B_{R_2})} \\ &+ N(R_2 - R_1)^{-\chi_1}R_1^{-\chi_2+\chi_1}\big\|\,|D^2u|^\gamma\big\|^{1/\gamma}_{L_1(B_{2R})}. \qquad (8)\end{aligned}$$

Next, we use the fact that

$$\big|F[u]\big| \le \big|H[u]\big| + K_0|u| + K_0|Du| + \bar{G} + \hat{\theta}|D^2u|$$

and that by interpolation inequalities

$$K_0N_1\|Du\|_{L_p(B_{R_2})} \le (1/8)\|D^2u\|_{L_p(B_{R_2})} + N\|u\|_{L_p(B_{R_2})}.$$

Then we take $\hat{\theta}$ and μ so small that

$$N_1\hat{\theta} \le 1/8, \quad N_2\mu \le 1/8,$$

and, finally, take the largest $\varepsilon \le 1$ such that

$$N_2\omega_{F,u,B_{2R}}(\varepsilon R_0) \le 1/8.$$

This ε, which depends only on d, p, K_F, R_0, the function $\omega_{F,u,B_{2R}}$, and δ, will appear later in our arguments and this is the way how the constant N in the statement of the lemma depends on $\omega_{F,u,B_{2R}}$.

We require Assumptions 1.4 and 1.6 be satisfied with the above chosen $\hat{\theta}$ and $\theta = \theta(d, \delta, K_F, \mu, \beta d)$ (see Lemma 1), respectively. By combining the above, we get

$$\begin{aligned}\|D^2u\|_{L_p(B_{R_1})} &\le N\big\|H[u]\big\|_{L_p(B_{R_2})} + N\tau_0R^{d/p} + (5/8)\|D^2u\|_{L_p(B_{R_2})} \\ &+ N(R_2 - R_1)^{-\chi_1}R_1^{-\chi_2+\chi_1}\big\|\,|D^2u|^\gamma\big\|^{1/\gamma}_{L_1(B_{2R})} + N\|u\|_{L_p(B_{2R})} + N\|\bar{G}\|_{L_p(B_{2R})}.\end{aligned}$$

Now we are going to iterate this estimate by defining $R_1 = R$ and for $k \ge 1$

$$R_{k+1} = R_k + cR(n_0 + k)^{-2},$$

where the constant $c = O(n_0)$ is chosen so that $R_k \uparrow 2R$ as $k \to \infty$, that is

$$c\sum_{k=1}^{\infty}(n_0 + k)^{-2} = 1,$$

and $n_0 > 0$ is chosen so that for $k \geq 1$

$$R_{k+1} - R_k = cR(n_0 + k)^{-2} \leq Rcn_0^{-2} \leq R \leq R_k,$$

which is satisfied if n_0 is just an appropriate absolute constant, and

$$R_{k+1} - R_k = cR(n_0 + k)^{-2} \leq cn_0^{-2} \leq \varepsilon R_0$$

(this time we need $n_0^{-1} = o(\varepsilon R_0)$ as $\varepsilon R_0 \to 0$). Also observe that $R \leq R_k \leq 2R$ and

$$(R_{k+1} - R_k)^{-\chi_1} R_k^{-\chi_2+\chi_1} \leq N(n_0 + k)^{2\chi_1} R^{-\chi_2}.$$

Then for $k \geq 1$ we get

$$\|D^2u\|_{L_p(B_{R_k})} \leq N\big\|H[u]\big\|_{L_p(B_{R_{k+1}})} + N\tau_0 R^{d/p} + (5/8)\|D^2u\|_{L_p(B_{R_{k+1}})}$$

$$+N(n_0 + k)^{2\chi_1} R^{-\chi_2}\big\|\,|D^2u|^\gamma\big\|^{1/\gamma}_{L_1(B_{2R})} + N\|u\|_{L_p(B_{2R})} + N\|\bar{G}\|_{L_p(B_{2R})},$$

where and below the constants N are as in the statement of the lemma. We multiply both parts of this inequality by $(5/8)^k$ and sum up the results over $k = 1, 2, \ldots$. Then we cancel the like terms

$$\sum_{k=2}^{\infty}(5/8)^k\|D^2u\|_{L_p(B_{R_k})},$$

which are finite since $u \in W^2_p(B_{2R})$, and finally take into account that

$$\sum_{k=2}^{\infty}(5/8)^k(n_0 + k)^{2\chi_1} \leq Nn_0^{2\chi_1}\sum_{k=2}^{\infty}(5/8)^k + N\sum_{k=2}^{\infty}(5/8)^k k^{2\chi_1} \leq N.$$

Then we come to (3).

Next, by using equation (9.7.2) and performing scaling in Theorem 9.4.6 (here we need $R \leq 1$), using Hölder's inequality (to go from d to p), and denoting

$$I = \|\bar{G}\|_{L_p(B_{2R})} + \big\|H[u]\big\|_{L_p(B_{2R})}$$

we infer that in (3)

$$\big\|\,|D^2u|^\gamma\big\|^{1/\gamma}_{L_1(B_{2R})} \leq NR^{\chi_2}\Big(\|\bar{G} + K_0|u|\,\|_{L_p(B_{2R})} + \big\|H[u]\big\|_{L_p(B_{2R})}\Big)$$

$$+ NR^{\chi_3}\sup_{B_{2R}}|u| \leq NR^{\chi_2}I + NR^{\chi_3}\sup_{B_{2R}}|u|,$$

where $\chi_3 = d/\gamma - 2$. After that it suffices to roughly estimate $\|u\|_{L_p(B_{2R})}$ in (3) by the last supremum above. The lemma is proved. □

Until this moment we followed a general framework laid out in [**70**] and [**30**], where we also used the localization techniques based on partitions of unity. In particular, we considered $H[\zeta u]$ and were able to relate it (due to certain assumptions) to $\zeta H[u]$ plus unimportant terms. In our present situation this localization does not work and instead we rely on the following simple fact.

The following lemma will allow us to get interior W^2_p-estimates in domains by using Lemma 3.

4. LEMMA. *Let Ω be just a bounded domain in $\mathbb{R}^d$ and $\rho > 0$ be such that $\Omega^{3\rho} \neq \emptyset$, and let f and h be nonnegative functions given in Ω. Assume that for any $z \in \Omega^{2\rho}$ we have*

$$\int_{B_\rho(z)} f(x)\,dx \le \int_{B_{2\rho}(z)} h(x)\,dx + N_1\rho^d + N_2\rho^{d-2p}, \tag{9}$$

where N_1 and N_2 are fixed constants ≥ 0. Then

$$\int_{\Omega^{3\rho}} f(x)\,dx \le 2^d \int_\Omega h(x)\,dx + NN_1 + NN_2\rho^{-2p},$$

where the constants N depend only on d and $|\Omega|$.

Proof. We integrate (9) with respect to z over $\Omega^{2\rho}$ and observe that

$$\int_{\Omega^{2\rho}} I_{B_\rho(z)}(x)\,dz = \int_{\Omega^{2\rho}} I_{B_\rho(x)}(z)\,dz = \big|B_\rho(x) \cap \Omega^{2\rho}\big|.$$

For $x \in \Omega^{3\rho}$, $B_\rho(x) \subset \Omega^{2\rho}$, and $\big|B_\rho(x) \cap \Omega^{2\rho}\big| = \rho^d|B_1|$. On the other hand,

$$\int_{\Omega^{2\rho}} I_{B_{2\rho}(z)}(x)\,dz = \big|B_{2\rho}(x) \cap \Omega^{2\rho}\big| \le 2^d\rho^d|B_1|I_\Omega(x).$$

This easily yields the desired result. The lemma is proved. □

5. Proof of Theorems 1.10, 1.14, and 1.17

Proof of Theorem 1.10. We take the constants $\hat\theta, \theta \in (0,1]$ from Lemma 4.3. By that lemma, if $\rho \in (0,1]$ and $\Omega^{2\rho} \neq \emptyset$ and $z \in \Omega^{2\rho}$, we have $\bar B_{2\rho}(z) \subset \Omega$ and

$$\|D^2u\|^p_{L_p(B_\rho(z))} \le N\tau_0^p|\rho|^d + N\rho^{d-2p} \sup_{B_{2\rho}(z)} |u|^p + N\|\bar G\|^p_{L_p(B_{2\rho}(z))},$$

and by Lemma 4.4 for $0 < 3\rho < \rho_{\mathrm{int}}(\Omega) \wedge 3$

$$\int_{\Omega^{3\rho}} \big|D^2u(x)\big|^p\,dx \le N\int_\Omega \big|\bar G(x)\big|^p\,dx + N\tau_0^p + N\rho^{-2p}\sup_\Omega |u|^p. \tag{1}$$

If $3\rho \ge \rho_{\mathrm{int}}(\Omega) \wedge 3$ and still $\Omega^{3\rho} \neq \emptyset$, we have $3\rho < \rho_{\mathrm{int}}(\Omega)$, hence $\rho \ge 1$, $\rho \le \mathrm{diam}(\Omega)$, and (1) follows because $\Omega^{3\rho} \subset \Omega^3$ and $3^{-2p} \le 3^{-2p}\rho^{-2p}\mathrm{diam}^{2p}(\Omega)$. Using interpolation inequalities also allows us to estimate the $L_p(\Omega^{3\rho})$-norm of Du. The theorem is proved. □

Proof of Theorem 1.17. It is based on Theorem 6.6.5 and, as there, we take P from (6.2.4) and set $P[u](x) = P(D^2u(x))$.

Step 1. Assume that $g \in C^{1,1}(\mathbb{R}^d)$ and there exist a constant N_0 and a *constant* $\bar H$ such that, for all x, u',

$$\big|H(\mathsf{u}', 0, x)\big| \le N_0|\mathsf{u}'| + \bar H. \tag{2}$$

We need this assumption in addition to a weaker property (1.5) in order to be able to apply Theorem 6.6.5. By that theorem for any $K > 0$ there exists a solution $v_K \in W^2_{p,\mathrm{loc}}(\Omega) \cap C(\bar{\Omega})$, $p > 1$, of the equation

$$\max\big(H[v_K], P[v_K] - K\big) = 0 \tag{3}$$

in Ω (a.e.) and $v_K = g$ on $\partial\Omega$, and

$$\sup_{\Omega} |v_K| \le M_0. \tag{4}$$

Next, we want to use Theorems 9.7.1 and 1.10. Set

$$H_K(\mathsf{u}, x) = \max\big(H(\mathsf{u}, x), P(\mathsf{u}'') - K\big),$$

$$F_K\big(\mathsf{u}'_0, \mathsf{u}'', x\big) = \max\big(F\big(\mathsf{u}'_0, \mathsf{u}'', x\big), P(\mathsf{u}'') - K\big), \quad G_K = H_K - F_K.$$

$$\bar{F}_{K,z,r,\mathsf{u}'_0}(\mathsf{u}'') = \max\big(\bar{F}_{z,r,\mathsf{u}'_0}(\mathsf{u}''), P(\mathsf{u}'') - K\big).$$

Then condition (1.3) is satisfied for G_K in place of G just because $|G_K| \le |G|$. Assumption 6.6.4 is also easily checked.

By Remark 6.2.2, H_K satisfies Assumption 1.11 with $\bar{\delta} = \bar{\delta}(d, \delta) \in (0, \delta]$ in place of δ. Also Assumptions 1.6 (i) and (ii) are satisfied with new K_F and δ depending only on d and the original K_F and δ. Finally, Assumptions 1.6 (iii) and (iv) are satisfied with the same θ and ω_F because, for instance,

$$\big|F_K\big(\mathsf{u}'_0, \mathsf{u}'', x\big) - \bar{F}_{K,z,r,\mathsf{u}'_0}(\mathsf{u}'')\big| \le \big|F\big(\mathsf{u}'_0, \mathsf{u}'', x\big) - \bar{F}_{z,r,\mathsf{u}'_0}(\mathsf{u}'')\big|.$$

This allows us to use the results proved in Theorem 1.10 for v_K, but first we recall Remark 1.8 and combine it with (4) and Theorem 9.7.1 to conclude that the family of the functions v_K is equicontinuous in $\bar{\Omega}$, and, in particular, the functions $\omega_{F,v_K,\Omega}$ are majorated by a continuous function independent of K and vanishing at zero.

As an important observation tacitly used later we point out that in the above argument guaranteeing the precompactness of $\{v_K : K \ge 0\}$ in $C(\bar{\Omega})$, in what concerns g, only the supremum of $|g|$ and the modulus of continuity of g on $\partial\Omega$ were involved. This is important when we will need to approximate our general $g \in C(\partial\Omega)$ with the ones of class $C^{1,1}(\mathbb{R}^d)$.

Then by Theorem 1.10 we obtain that, provided that an appropriate choice of $\hat{\theta}$ and θ was made, for any $\rho > 0$ such that $\Omega^\rho \neq \emptyset$

$$\sup_{K \ge 1} \|v_K\|_{W^2_p(\Omega^\rho)} < \infty. \tag{5}$$

We now let $K \to \infty$. In light of the precompactness of $\{v_K\}$ and estimate (5) there is a sequence $K_n \to \infty$ as $n \to \infty$ and $v \in W^2_{p,\mathrm{loc}}(\Omega)$ such that $v_{K_n} \to v$ weakly in $W^2_p(\Omega^\rho)$ for any ρ and $v_{K_n} \to v$ uniformly in $\bar{\Omega}$. Then, of course, $v \in C(\bar{\Omega})$ and $v = g$ on $\partial\Omega$. Embedding theorems and the fact that (5) holds with $p > d$ also imply that $Dv_{K_n} \to Dv$ locally uniformly in Ω.

Next, for $m = 1, 2, \ldots$ define

$$H^m(\mathsf{u}'', x) = \sup_{n\geq m} \max\big(H\big(v_{K_n}(x), Dv_{K_n}(x), \mathsf{u}'', x\big), P(\mathsf{u}'') - K_n\big).$$

Observe that $H^m(\mathsf{u}'', x)$ are Lipschitz continuous in u'' and at all points of differentiability satisfy $D_{\mathsf{u}''}H^m \in \mathbb{S}_{\bar\delta}$. Also

$$\big|H^m(0, x)\big| \leq K_0 \max_{n\geq m}\big(\big|v_{K_n}(x)\big| + \big|Dv_{K_n}(x)\big|\big) + \bar G(x),$$

which is in $L_{p,\mathrm{loc}}(\Omega)$. Therefore, the operators $H^m[u]$ fit into the scheme of Section 4.2. Furthermore, for $n \geq m$ obviously

$$H^m(v_{K_n}, x) \geq 0$$

in Ω (a.e.). By Theorems 4.2.15 and 4.2.6 we conclude that for any m

$$\sup_{n\geq m} \max\big[H\big(v_{K_n}(x), Dv_{K_n}(x), D^2v(x), x\big), P\big(D^2v(x)\big) - K_n\big] \geq 0 \tag{6}$$

in Ω (a.e.). We fix x at which (6) holds for all m (that is, we fix almost any x) and since $H(\mathsf{u}', \mathsf{u}'', x)$ is continuous in u', we have that

$$\big|H\big(v_{K_n}(x), Dv_{K_n}(x), D^2v(x), x\big) - H\big(v(x), Dv(x), D^2v(x), x\big)\big| \to 0$$

as $n \to \infty$. Then in light of (6)

$$\max\big[H\big(v(x), Dv(x), D^2v(x), x\big), P\big(D^2v(x)\big) - K_m\big] \geq o(1),$$

which for $m \to \infty$ yields

$$H\big(v(x), Dv(x), D^2v(x), x\big) = H[v](x) \geq 0.$$

The inequality $H[v] \leq 0$ is proved similarly starting from the function

$$\inf_{n\geq m} \max\big[H\big(v_{K_n}(x), Dv_{K_n}(x), \mathsf{u}'', x\big), P(\mathsf{u}'') - K_n\big].$$

This proves the theorem if $g \in C^{1,1}(\mathbb{R}^d)$ and (2) holds.

Step 2. Assume that $g \in C^{1,1}(\mathbb{R}^d)$ and abandon (2). Let $\eta(t) = t$ for $|t| \leq 1$ and $\eta(t) = \operatorname{sign} t$ for $|t| \geq 1$. For $\kappa > 0$ define $\eta(\kappa, t) = \kappa\eta(t/\kappa)$,

$$H^n(\mathsf{u}, x) = H(\mathsf{u}, x) - H(\mathsf{u}', 0, x) + \eta\left(n, H(\mathsf{u}', 0, x)\right),$$

$$F^n(\mathsf{u}_0', \mathsf{u}'', x) = F(\mathsf{u}_0', \mathsf{u}'', x), \quad G^n = H^n - F^n.$$

Observe that

$$\big|H^n(\mathsf{u}', 0, x)\big| \leq n,$$

so that (2) is satisfied.

In light of (1.3)

$$\big|G^n(\mathsf{u}, x)\big| = \big|G(\mathsf{u}, x) - G(\mathsf{u}', 0, x) + \eta\left(n, G(\mathsf{u}', 0, x)\right)\big|$$

$$\leq \big|G(\mathsf{u}, x)\big| + \big|G(\mathsf{u}', 0, x)\big| \leq \hat\theta|\mathsf{u}''| + 2K_0|\mathsf{u}'| + 2\bar G(x), \tag{7}$$

and (1.3) is satisfied for G^n in place of G with slightly modified K_0 and $\bar G$. Thus Assumption 1.4 is satisfied for H^n.

Owing to (6.6.4), for $\mathsf{u}' = (M_0 + \varepsilon_0, \varepsilon')$ with $\varepsilon' = (\varepsilon_1, ..., \varepsilon_d)$, since $\eta(\kappa, t)$ has the sign of t, we have

$$-H^n(\mathsf{u}', 0, x) = -\eta\,(n, H(\mathsf{u}', 0, x)) \geq 0.$$

Similarly, $H^n(-\mathsf{u}', 0, x) \leq 0$, which finishes checking Assumption 6.6.4. for H^n.

Assumption 1.6 is satisfied because we did not change F and Assumption 1.11 is satisfied because $D_{\mathsf{u}''}H^n = D_{\mathsf{u}''}H$.

Hence, with the same $\hat{\theta}$ and θ as in Step 1, for any n there exists $v^n \in W^2_{p,\mathrm{loc}}(\Omega) \cap C(\bar{\Omega})$ satisfying

$$H^n[v^n] = 0 \tag{8}$$

in Ω (a.e.) and such that $v^n = g$ on $\partial\Omega$. The derivation of (5) shows that for any $\rho > 0$ such that $\Omega^\rho \neq \emptyset$

$$\sup_n \|v^n\|_{W^2_p(\Omega^\rho)} < \infty. \tag{9}$$

Then by repeating what is said after (5) we find a subsequence $v^{n'}$ and a function $v \in W^2_{p,\mathrm{loc}}(\Omega)$ such that $v^{n'} \to v$ weakly in $W^2_p(\Omega^\rho)$ for any ρ and $v^{n'} \to v$ uniformly in $\bar{\Omega}$. Then, of course, $v = g$ on $\partial\Omega$. Embedding theorems and the fact that (5) holds with $p > d$ also imply that $Dv^{n'} \to Dv$ locally uniformly in Ω.

Next, for large $m = 1, 2, ...$ define

$$\hat{H}^m(\mathsf{u}'', x) = \sup_{n' \geq m} H^{n'}\big(v^{n'}(x), Dv^{n'}(x), \mathsf{u}'', x\big).$$

Observe that $\hat{H}^m(\mathsf{u}'', x)$ are Lipschitz continuous in u'' and at all points of differentiability satisfy $D_{\mathsf{u}''}\hat{H}^m \in \mathbb{S}_\delta$. Also (7) shows that

$$\big|\hat{H}^m(0, x)\big| \leq 2K_0 \max_{n' \geq m}\big(\big|v^{n'}(x)\big| + \big|Dv^{n'}(x)\big|\big) + 2\bar{G}(x),$$

which is in $L_{p,\mathrm{loc}}(\Omega)$. Therefore, the operators $\hat{H}^m[u]$ fit into the scheme of Section 4.2. Furthermore, for $n' \geq m$ obviously

$$\hat{H}^m(D^2 v^{n'}, x) \geq 0$$

in Ω (a.e.). By Theorem 4.2.15 and 4.2.6 we conclude that for any m in Ω (a.e.)

$$\hat{H}^m(D^2 v, x) \geq 0. \tag{10}$$

Next, note that, for any u'' and $x \in \Omega$, for all large n we have

$$H^n\big(v^n(x), Dv^n(x), \mathsf{u}'', x\big) = H\big(v^n(x), Dv^n(x), \mathsf{u}'', x\big)$$

because the sequences $v^n(x), Dv^n(x)$ are bounded. It follows from the continuity of $H(\mathsf{u}, x)$ with respect to u that for any u'' and $x \in \Omega$

$$\hat{H}^m(\mathsf{u}'', x) \to H\big(v(x), Dv(x), \mathsf{u}'', x\big).$$

Now we infer from (10) that $H[v] \geq 0$ in Ω (a.e.).

The fact that $H[v] \leq 0$ in Ω (a.e.) is proved similarly starting with the function

$$\inf_{n' \geq m} H^{n'}\big(v^{n'}(x), Dv^{n'}(x), \mathsf{u}'', x\big).$$

This yields the desired result if $g \in C^{1,1}(\mathbb{R}^d)$. The final step where one approximates $g \in C(\partial\Omega)$ by polynomials and passes to the limit is much simpler than the argument in Step 2 and is left to the reader as a routine exercise. The theorem is proved. □

Proof of Theorem 1.14. It suffices to literally repeat the above proof using Theorem 6.6.9 in place of Theorem 6.6.5. □

CHAPTER 11

Nonlinear elliptic equations in $C^{2+\alpha}_{\mathrm{loc}}(\Omega) \cap C(\bar{\Omega})$

The goal of this short chapter is to prepare some necessary tools for developing the Sobolev space theory of *parabolic* equations in $W^{1,2}_{p,\mathrm{loc}}$ and later on the theory of solvability in global Sobolev spaces of nonlinear elliptic and parabolic equations. We concentrate on *only* the facts which we need for our development of the Sobolev space or viscosity solutions theory of fully nonlinear equations leaving aside a rich and well-developed Hölder space theory, which the reader can find in [**44**], [**45**], [**61**], [**66**], [**89**], [**103**], [**104**], and in the references therein.

1. Main results

We consider elliptic equations

$$H[v](x) := H\big(v(x), Dv(x), D^2v(x), x\big) = 0 \tag{1}$$

in subdomains of $\mathbb{R}^d$ and deal with their solvability in $C^{2+\alpha}_{\mathrm{loc}}(\Omega) \cap C(\bar{\Omega})$, where Ω is an open bounded subset of $\mathbb{R}^d$ satisfying the exterior ball condition.

1. Assumption. The function $H(\mathsf{u}, x)$, $\mathsf{u} = (\mathsf{u}', \mathsf{u}'')$, $\mathsf{u}' = \big(\mathsf{u}'_0, ..., \mathsf{u}'_d\big) \in \mathbb{R}^{d+1}$, $\mathsf{u}'' \in \mathbb{S}$, $x \in \mathbb{R}^d$, is *convex* with respect to u'' and at all points of its differentiability with respect to u'' we have

$$D_{\mathsf{u}''}H \in \mathbb{S}_\delta,$$

where $\delta \in (0, 1]$ is a fixed constant.

2. Assumption. There exists a (finite) function $K_R \geq 0$ defined for $R \in [0, \infty)$, such that, for $R \geq 0$ and (u, x) and (v, y) satisfying

$$|\mathsf{u}' - \mathsf{v}'|, |x - y| \leq 1 \quad \text{and} \quad |\mathsf{u}'|, |\mathsf{v}'| \leq R,$$

we have

$$\big|H(\mathsf{u}', \mathsf{u}'', x) - H(\mathsf{v}', \mathsf{u}'', y)\big| \leq K_0|\mathsf{u}' - \mathsf{v}'|^\alpha + K_R\big(1 + |\mathsf{u}''|\big)|x - y|^\alpha,$$

where, for $\alpha_0 = \alpha_0(\delta, d)$ introduced in Theorem 10.2.1,

$$\alpha \in (0, \alpha_0). \tag{2}$$

3. Assumption. The function $H(\mathsf{u}, x)$ is nonincreasing with respect to u'_0 and $H(0, x)$ is bounded.

4. Theorem. *Let $g \in C(\partial\Omega)$. Then under the above assumptions there exists $u \in C^{2+\alpha}_{\mathrm{loc}}(\Omega) \cap C(\bar{\Omega})$ satisfying* (1) *in Ω and equal g on $\partial\Omega$. Furthermore, for this solution for any $\rho > 0$ there exists a constant N, depending only on d, ρ, δ, α, Ω,* $\sup_{\partial\Omega}|g|$, $\sup\big|H(0, x)\big|$, *and the function K_R, such that*

$$[D^2u]_{C^\alpha(\Omega^\rho)} \leq N. \tag{3}$$

5. REMARK. On the one hand, Theorem 4 generalizes the corresponding results in [**45**] and [**61**], because no differentiability or Lipschitz continuity of H with respect to (u',x) is assumed. On the other hand, the methods in [**45**] and [**61**] lead there to results for H growing quadratically with respect to $[\mathsf{u}']$. Basically, in the case of *linear equations* (when H is linear in u) the nonlinear results from [**45**] and [**61**] require excessive smoothness of the coefficients than necessary, whereas Theorem 4 in that regard is the optimal result (apart from the restriction on α).

Theorem 4 has much in common with the corresponding result of [**104**] and an earlier result Theorem 1.1 of [**103**], but will never be used later in this book in its full generality. We only use its particular case for the Laplace equation in a half ball in the first proof of Theorem 13.4.1 and in the proof of the following weaker version of Theorem 4 in which estimate (3) is better elaborated. Theorem 6 (below) is a particular case of the results in [**103**] or [**104**].

It is also worth noting that in Theorem 1.1 of [**103**] condition (10.1.8) rather than the exterior ball condition is imposed.

6. THEOREM. *Let $R\in(0,\infty)$, and let $\Omega=B_R$. Suppose that $H(\mathsf{u},x)=H(\mathsf{u}'')+f(x)$ and $H(\mathsf{u}'')$ satisfies Assumption 1 whereas $f\in C^{\alpha}(B_R)$. Let $g\in C(\partial B_R)$. Then the solution from Theorem 4 is unique, it belongs to $C^{2+\alpha}_{\text{loc}}(\Omega)\cap C(\bar\Omega)$, and there exists a constant N, depending only on δ, d, and α, such that for any $\rho\in(0,R)$*

$$[D^2u]_{C^{\alpha}(B_\rho)}\le N[f]_{C^{\alpha}(B_R)}+N(R-\rho)^{-2-\alpha}\sup_{\partial B_R}|g-p|, \tag{4}$$

where p is any polynomial of degree 2 on $\mathbb{R}^d$. Furthermore, there exists $\nu=\nu(d,\delta,\alpha)\ge 2$ such that, if $B_{\nu r}(x_0)\subset B_R$ (for some r,x_0), then there exists a polynomial p of degree 2 for which the inequality

$$\sup_{B_r(x_0)}|u-p|\le N[f]_{C^{\alpha}(B_{\nu r}(x_0))}r^{2+\alpha}+(1/2)\nu^{-(2+\alpha)}\sup_{B_{\nu r}(x_0)}|u-q| \tag{5}$$

holds for any quadratic polynomial q.

Theorem 6 turns out to be an indispensable tool in Chapter 12 devoted to *parabolic* equations.

7. REMARK. From the proof of Theorem 6 it will be clear that its second assertion holds each time when there is $u\in C^{2+\alpha}(B_{\nu r}(x_0))$ satisfying (1) in $B_{\nu r}(x_0)$ rather than in B_R.

2. Characterizations of the spaces $C^{s+\alpha}$

In this section $R\in(0,\infty)$ and $\alpha\in(0,1)$ are fixed. The spaces $C^{2+\alpha}(\Omega)$ are introduced on page 216. Similarly $C^{1+\alpha}(\Omega)$ is the subset of $C^1(\Omega)$ consisting of functions u such that

$$\|u\|_{C^{1+\alpha}(\Omega)}=\|u\|_{C^1(\Omega)}+[Du]_{C^{\alpha}(\Omega)}<\infty.$$

We will discuss two types of characterizations of $C^{s+\alpha}$: by using finite-differences and by using polynomial approximation, the latter being the most important tool in the later development. This tool was introduced into the theory of fully nonlinear equations by Safonov (see [**103**], [**104**]). In what concerns the finite-difference characterization, its long and very interesting history can be found in [**46**], which along with [**47**] is the source of the results but not the proofs in this part of the section. In many proofs we use the methods borrowed from [**103**] and [**104**].

In order to treat the spaces $C^{1+\alpha}$ and $C^{2+\alpha}$ in a unified manner for unit $l \in \mathbb{R}^d$ and $h > 0$ introduce

$$\Delta^{(1)}_{h,l} = \delta_{h,l}, \quad \Delta^{(2)}_{h,l} = \Delta_{h,l}, \quad D^1_l = D_{(l)} = l^i D_i, \quad D^2_l = (D_{(l)})^2 = D_{(l)(l)}.$$

1. LEMMA. *Let $K \in [0,\infty)$, $\nu, \eta \in [1,\infty)$, $s \in \{1,2\}$, and let Ω be an open bounded convex domain in $\mathbb{R}^d$ such that*

$$\operatorname{diam}\Omega \leq \eta \rho_{\text{int}}(\Omega). \tag{1}$$

Let u be a continuous function on Ω such that for any $h > 0$, and $x_0 \in \Omega$, satisfying $B_{2\nu h}(x_0) \subset \Omega$, and any unit $l \in \mathbb{R}^d$, it holds that

$$\operatorname*{osc}_{B_h(x_0)} \Delta^{(s)}_{h,l} u \leq K h^{\alpha}. \tag{2}$$

Then $u \in C^{s+\alpha}(\Omega)$ and

$$[D^s u]_{C^{\alpha}(\Omega)} \leq N(d,\alpha,\nu,\eta) K. \tag{3}$$

Proof. We start by noticing that, as is easy to see, (2) and (3) are scale invariant in the sense that, of course, scalings change the constant K in (2) but they also change the constant K in (3) in the same way. Furthermore, scalings do not affect the ratio

$$\frac{\rho_{\text{int}}(\Omega)}{\operatorname{diam}\Omega}. \tag{4}$$

Therefore, without loss of generality we assume that $\rho_{\text{int}}(\Omega) = 1$ and the open unit ball B_1 lies in Ω. In that case also $\operatorname{diam}\Omega \leq \eta$.

Next, notice that for any $x \in \Omega$, the smallest open convex set containing B_1 and x is either B_1 or an ice cream cone-like domain $I(x)$ belonging to Ω. The openings of these "cones" depend only on $|x|$ and are obviously bounded from below since $|x| \leq \eta$ on Ω. It follows easily that there exists a $\theta_0 \in [1,\infty)$, depending only on η, such that for any $x \in \Omega$ and $\rho \in (0,1]$ there is an x_ρ on the ray tx, $t \in [0,1]$, satisfying

$$|x_\rho - x| \leq \theta_0 \rho \quad \text{and} \quad B_\rho(x_\rho) \subset \Omega. \tag{5}$$

To find such an x_ρ, it suffices to take B_ρ and to move it along the ray tx, $t \geq 0$, until it either covers x (if $x \in B_1$) or touches the boundary of $I(x)$. In the former case $|x_\rho - x| \leq \rho$ and in the latter case $|x_\rho - x| \leq \rho/\sin|\phi|$, where ϕ is the angle between x and any tangent line to B_1 passing through x.

We continue under the additional assumption that $u \in C^{s+\alpha}(\Omega)$. Set

$$K_0 = \sup_{|l|=1} \sup_{\substack{x,y\in\Omega \\ |x-y|\leq\theta_0}} \frac{|D^s_l u(x) - D^s_l u(y)|}{|x-y|^{\alpha}} \quad (<\infty).$$

Take $x, y \in \Omega$ such that

$$|x - y| \leq \theta_0,$$

take $\varepsilon\in(0,1)$, to be specified later, and introduce

$$\rho=\varepsilon|x-y|\theta_0^{-1},\quad x_0=x_\rho,\quad y_0=y_\rho,\quad h=\rho/(2\nu).$$

Clearly, $|x_0-y_0|\le|x-y|$, $|x-x_0|,|y-y_0|\le\theta_0\rho$.

Next, we cover the straight segment connecting x_0 and y_0 by a chain of open balls of radius ρ such that their centers lie on the segment, x_0 and y_0 are centers, and the distance between the centers of consecutive balls is less than or equal to h. We will need not more than n balls, where

$$n\le 1+\frac{|x_0-y_0|}{h}=1+\frac{|x_0-y_0|}{(2\nu\theta_0)^{-1}\varepsilon|x-y|}\le 1+2\nu\theta_0\varepsilon^{-1}.$$

Denote by $z_1,\dots,z_n$ the centers of the consecutive balls and observe that in light of (5) and the convexity of Ω we have $B_{2\nu h}(z_k)\subset\Omega$ for any k.

Now take a unit $l\in\mathbb{R}^d$ and note that the points $x_0,x_0\pm hl$ are in Ω, and by the mean value theorem and by the definition of K_0

$$\big|D_l^s u(x_0)-\Delta_{h,l}^{(s)}u(x_0)\big|\le K_0h^\alpha. \tag{6}$$

Again by the definition of K_0

$$\begin{aligned}\big|D_l^s u(x)-\Delta_{h,l}^{(s)}u(x_0)\big|&\le\big|D_l^s u(x)-D_l^s u(x_0)\big|+K_0h^\alpha\\&\le K_0(\theta_0\rho)^\alpha+K_0h^\alpha\\&\le N_0\varepsilon^\alpha K_0|x-y|^\alpha,\end{aligned}\tag{7}$$

where $N_0=1+(2\nu\theta_0)^{-\alpha}$.

Similarly,

$$\big|D_l^s u(y)-\Delta_{h,l}^{(s)}u(y_0)\big|\le N_0\varepsilon^\alpha K_0|x-y|^\alpha. \tag{8}$$

It follows that

$$\big|D_l^s u(x)-D_l^s u(y)\big|\le 2N_0\varepsilon^\alpha K_0|x-y|^\alpha+\big|\Delta_{h,l}^{(s)}u(x_0)-\Delta_{h,l}^{(s)}u(y_0)\big|. \tag{9}$$

Next, we choose and fix ε so that $N_0\varepsilon^\alpha\le 1/4$ and write

$$\big|\Delta_{h,l}^{(s)}u(x_0)-\Delta_{h,l}^{(s)}u(y_0)\big|\le\sum_{k=1}^{n-1}\big|\Delta_{h,l}^{(s)}u(z_k)-\Delta_{h,l}^{(s)}u(z_{k+1})\big|.$$

Here $z_{k+1}\in\bar B_h(z_k)$ and $B_{2\nu h}(z_k)\subset B_1$, so that by assumption

$$\big|\Delta_{h,l}^{(s)}u(z_k)-\Delta_{h,l}^{(s)}u(z_{k+1})\big|\le Kh^\alpha,$$

implying that

$$\big|\Delta_{h,l}^{(s)}u(x_0)-\Delta_{h,l}^{(s)}u(y_0)\big|\le nKh^\alpha=NK|x-y|^\alpha.$$

In combination with (9) this yields

$$\big|D_l^s u(x)-D_l^s u(y)\big|\le(1/2)K_0|x-y|^\alpha+NK|x-y|^\alpha.$$

The arbitrariness of x, y, and l allows us to conclude that

$$K_0 \le (1/2)K_0 + N_1 K, \quad K_0 \le N.$$

This is equivalent to (3) and proves the lemma in our particular case.

Now we abandon the additional assumption that $u \in C^{s+\alpha}(\Omega)$. Take a non-negative $\zeta \in C_0^\infty(\mathbb{R}^d)$ with unit integral and support in B_1 and for $\gamma > 0$ define $\zeta_\gamma(x) = \gamma^{-d}\zeta(x/\gamma)$ and

$$u^{(\gamma)} = u * \zeta_\gamma.$$

Then $u^{(\gamma)} \to u$ uniformly on compact subsets of Ω as $\gamma \downarrow 0$. Furthermore, $u^{(\gamma)}$ is infinitely differentiable with any derivative bounded in $\Omega^{2\gamma}$ if $2\gamma < \rho_{\text{int}}(\Omega)$. Next, take $\varepsilon \in (0, \rho_{\text{int}}(\Omega))$ and $2\gamma < \varepsilon$ and observe that $u^{(\gamma)} \in C^{2+\alpha}(\Omega^\varepsilon)$. Furthermore, if $B_{2\nu h}(x_0) \subset \Omega^\varepsilon$, then $B_{2\nu h}(x_0 + \gamma z) \subset \Omega$ for any $z \in B_1$ and

$$\operatorname*{osc}_{B_h(x_0)} \Delta_{h,l}^{(s)} u^{(\gamma)} \le \sup_{z \in B_1} \operatorname*{osc}_{B_h(x_0+\gamma z)} \Delta_{h,l}^{(s)} u \le K h^\alpha.$$

Also Ω^ε is a convex domain with the corresponding ratio (4) as close to (4) as we wish on the account of taking ε small enough and by the first part of the proof, for ε small enough and $2\gamma < \varepsilon$,

$$\big[D^s u^{(\gamma)}\big]_{C^\alpha(\Omega^\varepsilon)} \le N(d, \alpha, \nu, \eta)K. \tag{10}$$

It follows that $D^s u^{(1/n)}$ are uniformly continuous in any closed subset of Ω for all large n. They are also bounded on such subsets, since $D_l^s u^{(1/n)}$ admit a good approximation by $\Delta_{h,l}^{(s)} u^{(1/n)}$ (see (6)) which are bounded on such subsets. Then there is a subsequence of $D^s u^{(1/n)}$ which converges uniformly on any closed subset of Ω. Since $u_n \to u$, from Calculus we know that $D^s u$ exists, is continuous, and the whole sequence $D^s u^{(1/n)}$ converges to $D^s u$ uniformly on any closed subset of Ω. Passing to the limit in (10) as first $\gamma = 1/n \to 0$ and then $\varepsilon \downarrow 0$ presents no problem and yields (3). Mean value theorems allow us to also estimate the magnitudes of the lower order derivatives and $|u|$, proving that $u \in C^{s+\alpha}(\Omega)$. The lemma is proved. □

Somewhat down the road we will need a simple fact that is closely related to the above arguments.

2. Remark. If Ω is as in Lemma 1 and $u \in C^{s+\alpha}(\Omega)$, then there exists a constant $N = N(\eta, \alpha)$ such that for any unit $l \in \mathbb{R}^d$ and $\varepsilon \in (0, \rho_{\text{int}}(\Omega))$ we have

$$\sup_{x,y\in\Omega} \big|D_l^s u(x) - D_l^s u(y)\big| \le N\varepsilon^\alpha \big[D_l^s u\big]_{C^\alpha(\Omega)} + \sup_{x,y\in\Omega^\varepsilon} \big|\Delta_{\varepsilon,l}^{(s)} u(x) - \Delta_{\varepsilon,l}^{(s)} u(y)\big|. \tag{11}$$

Indeed, as in the proof of Lemma 1 we may assume that $\rho_{\text{int}}(\Omega) = 1$. Then take $x, y \in \Omega$, define x_0, y_0 according to (5) with $\rho = \varepsilon$ and observe that (6), (7), (8), and (9) are valid with

$$h = \varepsilon, \quad K_0 = \big[D_l^s u\big]_{C^\alpha(\Omega)},$$

and $K_0\varepsilon^\alpha(1 + \theta_0^\alpha)$ in place of $N_0\varepsilon^\alpha K_0|x - y|^\alpha$. This yields (11).

We single out one more piece of the proof of Lemma 1, which yields interpolation inequalities we use a few times in the future.

3. REMARK. If Ω is as in Lemma 1 and $u\in C^{s+\alpha}(\Omega)$, then there exists a constant $N=N(d,\eta,\alpha)$ such that for any $\varepsilon\in(0,1]$

$$|D^s u|_{C(\Omega)} \le N\varepsilon^\alpha \rho^\alpha_{\text{int}}(\Omega)[D^s u]_{C^\alpha(\Omega)} + N\varepsilon^{-s}\rho^{-s}_{\text{int}}(\Omega)\|u\|_{C(\Omega)}. \tag{12}$$

Indeed, again we may assume that $\rho_{\text{int}}(\Omega)=1$. Then take $x\in\Omega$ define x_0 according to (5) with $\rho=\varepsilon$ and observe that similarly to (7)

$$\left|D^s_l u(x)-\Delta^{(s)}_{\varepsilon,l}u(x_0)\right| \le K_0\varepsilon^\alpha\left(1+\theta_0^\alpha\right)$$

It follows that

$$\left|D^s_l u(x)\right| \le N\varepsilon^\alpha[D^s u]_{C^\alpha(\Omega)} + \left|\Delta^{(s)}_{\varepsilon,l}u(x_0)\right|$$

$$\le N\varepsilon^\alpha[D^s u]_{C^\alpha(\Omega)} + 4\varepsilon^{-s}\|u\|_{C(\Omega)},$$

which yields (12) since we can use arbitrary unit vectors $l\in\mathbb{R}^d$.

Denote by $\mathcal{P}_s$ the set of polynomials of degree s on $\mathbb{R}^d$.

4. THEOREM. *Let $K\in[0,\infty)$, $\nu,\eta\in[1,\infty)$, $s\in\{1,2\}$, and let Ω be an open bounded convex domain in $\mathbb{R}^d$ such that (1) holds.*

Let u be a function on Ω such that for any ball $B_{\nu r}(x_0)\subset\Omega$ there exists $p\in\mathcal{P}_s$ for which

$$\sup_{B_r(x_0)}|u-p| \le Kr^{s+\alpha}. \tag{13}$$

Then $u\in C^{s+\alpha}(\Omega)$ and

$$[D^s u]_{C^\alpha(\Omega)} \le N(d,\alpha,\nu,\eta)K. \tag{14}$$

Proof. First, observe that u is continuous in Ω. Indeed, this follows from the fact that, if $x_0\in\Omega$, then

$$\varlimsup_{x\to x_0}\left|u(x)-u(x_0)\right| \le 2\sup_{B_r(x_0)}|u-p| \le Kr^{s+\alpha}$$

for any sufficiently small r and an appropriate polynomial p.

Next, take $h>0$ and $x_0\in\Omega$ such that $B_{2\nu h}(x_0)\subset\Omega$ and observe that for any $x,y\in B_h(x_0)$ and unit $l\in\mathbb{R}^d$ we have

$$\left|\Delta^{(s)}_{h,l}u(x)-\Delta^{(s)}_{h,l}u(y)\right| \le \left|\Delta^{(s)}_{h,l}(u-p)(x)\right| + \left|\Delta^{(s)}_{h,l}(u-p)(x)\right|,$$

where p is any polynomial of degree s and we have used the fact that the sth order differences of such polynomials are constant. If p is chosen according to (13) with $r=2h$, then we see that

$$\left|\Delta^{(s)}_{h,l}u(x)-\Delta^{(s)}_{h,l}u(y)\right| \le 8K2^{s+\alpha}h^\alpha,$$

and it only remains to refer to Lemma 1. The theorem is proved. □

5. REMARK. Theorem 4 and similar results started to be widely used in the theory of linear and fully nonlinear equations after Safonov [**103**] showed the usefulness of them. He himself proves (14) assuming (13) *and that* $u\in C^{s+\alpha}(\Omega)$ in a very short way different from ours. His assumption that $u\in C^{s+\alpha}(\Omega)$ was justified after the full scale of a priori estimates in domains was developed and the method of continuity could be applied to provide the existence of sufficiently smooth solutions.

In the framework of the $C^{2+\alpha}$-theory this is quite natural. But in the framework of the general W^2_p-theory this turns out to be an unnecessary digression. Still we could not avoid doing this digression altogether albeit in the simplest case in Chapter 13 where we treat the simplest equations $F(D^2u,x)=f$ in half balls.

Speaking about the method of continuity, it could be appropriate at this moment to point out that the method of continuity does not work in the setting of the W^2_p-theory.

The proof of Theorem 4 and Remark 2 immediately lead to the following.

6. REMARK. If Ω is as in Lemma 1 (and Theorem 4) and $u\in C^{s+\alpha}(\Omega)$, then there exists a constant $N=N(\eta,\alpha)$ such that for any unit $l\in\mathbb{R}^d$, $\varepsilon\in\big(0,\rho_{\text{int}}(\Omega)\big)$, and $p\in\mathcal{P}_s$ we have

$$\sup_{x,y\in\Omega}\big|D^s_lu(x)-D^s_lu(y)\big|\le N\varepsilon^\alpha\big[D^s_lu\big]_{C^\alpha(\Omega)}+8\varepsilon^{-s}\sup_\Omega|u-p|,$$

which after replacing ε with $\varepsilon\rho_{\text{int}}(\Omega)$ says that for any $\varepsilon\in(0,1]$

$$\operatorname*{osc}_\Omega D^s_lu\le N\rho^\alpha_{\text{int}}(\Omega)\varepsilon^\alpha\big[D^s_lu\big]_{C^\alpha(\Omega)}+8\rho^{-s}_{\text{int}}(\Omega)\varepsilon^{-s}\sup_\Omega|u-p|.$$

7. THEOREM. *Let $K\in[0,\infty)$, $\varepsilon\in[0,1)$, $s\in\{1,2\}$, and $\nu\in[1,\infty)$. Let $\eta\in[1,\infty)$ and let Ω be an open bounded convex domain in $\mathbb{R}^d$ such that (1) holds. Let $\mathcal{Q}$ be a linear subset of $\mathcal{P}_s$. Finally, let u be a function on Ω such that for any ball $B_{\nu r}(x)\subset\Omega$ there exists $p\in\mathcal{Q}$ for which the inequality*

$$\sup_{B_r(x)}|u-p|\le Kr^{s+\alpha}+\varepsilon\nu^{-s-\alpha}\sup_{B_{\nu r}(x)}|u-q|$$

holds for any $q\in\mathcal{Q}$.

Then, for any $\rho\in(0,\rho_{\text{int}}(\Omega))$, $u\in C^{s+\alpha}(\Omega^\rho)$ and for any $q\in\mathcal{Q}$

$$[D^su]_{C^\alpha(\Omega^\rho)}\le NK+N\varepsilon\rho^{-s-\alpha}\sup_\Omega|u-q|,\tag{15}$$

where the constants N depend only on $d,\nu,\varepsilon,\alpha,\eta$.

Proof. By considering $u-q$ in place of u and taking into account that the set $\mathcal{Q}$ is linear, we see that it suffices to prove (15) with $q=0$. In case $\nu=1$ (which will never occur in our applications in which ν is large) (13) is satisfied with $K/(1-\varepsilon)$ in place of K, so that one can delete the last term in (15). Therefore we assume that $\nu>1$.

If $B_r(x)\subset\Omega$ define

$$\omega_r(x)=r^{-s-\alpha}\inf_{p\in\mathcal{Q}}\sup_{B_r(x)}|u-p|.$$

Observe that if $n\in\{1,2,...\}$ and r,x are such that $B_{\nu^nr}(x)\subset\Omega$, then by assumption

$$\omega_{\nu^{n-1}r}(x)\le K+\varepsilon\omega_{\nu^nr}(x).$$

Hence,

$$\omega_r(x)\le(1-\varepsilon)^{-1}K+\varepsilon\omega_{\nu^{n(x)}r}(x),$$

where $n(x)$ is the largest integer n for which $B_{\nu^nr}(x)\subset\Omega$. We have that $B_{\nu^{n(x)+1}r}(x)\not\subset\Omega$, so that

$$\rho_\Omega(x) \le \nu^{n(x)+1} r, \quad \nu^{n(x)} r \ge \rho_\Omega(x)/\nu,$$

$$\omega_{\nu^{n(x)}r}(x) \le \nu^{s+\alpha} \rho_\Omega^{-s-\alpha}(x) \sup_\Omega |u|,$$

$$\omega_r(x) \le (1-\varepsilon)^{-1} K + \nu^{s+\alpha} \varepsilon \rho_\Omega^{-s-\alpha}(x) \sup_\Omega |u|. \tag{16}$$

Now, if $B_{\nu r}(x) \subset \Omega^\rho$, then $\rho_\Omega(x) \ge \rho$ and in light of (16) we have

$$\omega_r(x) \le (1-\varepsilon)^{-1} K + \nu^{s+\alpha} \varepsilon \rho^{-s-\alpha} \sup_\Omega |u|.$$

To finish the proof, it only remains to apply Theorem 4. The theorem is proved. □

Much later in Section 16.3 we will need a different characterization of C^α and $C^{1+\alpha}$ before which we state a simple geometric lemma.

8. LEMMA. *If $r \in (0, 1/4]$, $x, y \in B_R$, $|y| \le |x|$, and $|x-y| = rR$, then there exists $y^0 \in B_R$ for which*

$$|y^0 - y| \le 8r^2 R, \quad |y^0 - x| \le |y - x|,$$

and the length of the straight segment passing through y^0 with one end at x and the other on ∂B_R is at least $2|y^0 - x|$.

Proof. Scalings show that we may assume $R = 1$. In that case take $x, y \in B_1$ such that $|x-y| = r \le 1/4$. It suffices to concentrate on the plane passing trough the origin and x, y. In other words we may assume that $d = 2$. Obviously, if

$$|2(y-x)+x| = |2y - x| \le |x|, \tag{17}$$

then the straight segment starting at x, passing through y, and stopped when it reaches ∂B_1 has the length at least $2|y-x|$. Therefore, $y^0 = y$ is a right choice.

Generally, if $|y| \le 1/4$, one can again take $y^0 = y$. In the remaining case we have

$$|2y - x| > |x|, \quad |x| \ge |y| > 1/4.$$

It follows that $4|y|^2 - 4(x,y) > 0$, so that

$$s^0 := \frac{(x,y)}{|y|^2} \le 1, \quad |2y^0 - x| = |x|,$$

where $y^0 := s^0 y$.

An easy computation also shows that s^0 is the minimum point of $|sy - x|^2$. Therefore $|y^0 - x| \le |y - x|$.

Next, $2(x, y) = |x|^2 + |y|^2 - r^2 \geq 2|y|^2 - r^2$, so that

$$s^0 \geq 1 - r^2/(2|y|^2), \quad 1 - s^0 \leq r^2/(2|y|^2),$$

$$|y^0 - y| = (1 - s^0)\,|y| \leq r^2/(2|y|) \leq 8r^2.$$

Now it only remains to note that the new couple x, y^0 satisfies (17). The lemma is proved. □

The characterization we have in mind is also based on the following.

9. LEMMA. *Let $u, v \in C^\alpha(B_1)$ and let K_u, K_v be some constants. Take $\varepsilon \in (0, 1/2]$ and assume that for any unit $l \in \mathbb{R}^d$, and $h > 0$ we have*

$$\big|(T_{h,l} - 1)v(x)\big| \leq K_v h^\alpha, \tag{18}$$

$$\big|(T_{h,l} - 1)^2 u(x)\big| \leq K_u h^\alpha, \tag{19}$$

whenever

$$x, x + 2hl \in \bar{B}_{1-h\varepsilon}.$$

Then

$$[v]_{C^\alpha(B_1)} \leq 2\varepsilon^\alpha [v]_{C^\alpha(B_1)} + K_v + 16\varepsilon^{-\alpha} \operatorname*{osc}_{B_1} v, \tag{20}$$

$$[u]_{C^\alpha(B_1)} \leq \big[2\varepsilon^\alpha + 2^{\alpha-1}\big][u]_{C^\alpha(B_1)} + K_u + 16\varepsilon^{-\alpha} \operatorname*{osc}_{B_1} u. \tag{21}$$

Proof. First note that

$$\begin{aligned} 2(T_{h,l} - 1) &= T_{2h,l} - 1 - \big[T_{2h,l} - 1 - 2(T_{h,l} - 1)\big] \\ &= T_{2h,l} - 1 - (T_{h,l} - 1)^2. \end{aligned}$$

Hence under condition (19)

$$\big|(T_{h,l} - 1)u(x)\big| \leq \big[2^{\alpha-1}[u]_{C^\alpha(B_1)} + K_u\big]h^\alpha.$$

It follows that if (20) is true, then

$$[u]_{C^\alpha(B_1)} \leq 2\varepsilon^\alpha [u]_{C^\alpha(B_1)} + 2^{\alpha-1}[u]_{C^\alpha(B_1)} + K_u + 16\varepsilon^{-\alpha} \operatorname*{osc}_{B_1} u,$$

which is (21).

Thus, it suffices to concentrate on (20). We define

$$\bar{h} = \bar{h}(\varepsilon) = \varepsilon/16,$$

fix $h \in (0, \bar{h}]$, and take a unit l and

$$x, y := x + hl \in \bar{B}_{1-h\varepsilon}. \tag{22}$$

Consider two cases:

a) $|y| \leq |x|$,

b) $|x| \leq |y|$.

Generally if (22) holds, then $h = |x - y| =: r(1 - h\varepsilon)$ with $r \leq \bar{h}/(1 - h\varepsilon) < 1/4$. Therefore, in the situation of a), in light of Lemma 8, there exists a point $y^0 \in B_{1-h\varepsilon}$

such that

$$|y^0-y|\le 8h^2(1-h\varepsilon)^{-1}\le 16h^2\le \varepsilon h,\quad x,x+2h_0l_0\in \bar B_{1-h\varepsilon}\subset \bar B_{1-h_0\varepsilon},$$

where $l_0=(y^0-x)/|y^0-x|$, $h_0=|y^0-x|$ $(\le h)$, and the last two inequalities hold since $\varepsilon\le 1/2$ and $h\le \bar h=\bar h(\varepsilon)$. Due to (18), in that case we have

$$|T_{h_0,l_0}v(x)-v(x)|\le K_v h_0^\alpha. \tag{23}$$

Furthermore, observe that

$$\begin{aligned}\big|T_{h_0,l_0}v(x)-T_{h,l}v(x)\big| &\le [v]_{C^\alpha(B_1)}\big|x+h_0l_0-(x+hl)\big|^\alpha\\ &=[v]_{C^\alpha(B_1)}|y^0-y|^\alpha\\ &\le [v]_{C^\alpha(B_1)}(\varepsilon h)^\alpha.\end{aligned}$$

Hence,

$$\big|v(y)-v(x)\big|=\big|T_{h,l}v(x)-v(x)\big|\le\big[\varepsilon^\alpha[v]_{C^\alpha(B_1)}+K_v\big]h^\alpha. \tag{24}$$

In case b) we again obtain (24) by interchanging the roles of x and y in the above argument. Thus, (24) always holds if $\varepsilon\in(0,1/2]$, $x,x+hl\in B_{1-h\varepsilon}$ and $h\le\bar h(\varepsilon)$.

In the general case that $x,y:=x+hl\in\bar B_1$ and $h\le\bar h(\varepsilon)$ define $x_\varepsilon=(1-h\varepsilon)x$ and $y_\varepsilon=(1-h\varepsilon)y$. Then $x_\varepsilon,y_\varepsilon\in\bar B_{1-h\varepsilon}$, $y_\varepsilon=x_\varepsilon+h_\varepsilon l$, where $h_\varepsilon=(1-h\varepsilon)h\le\bar h(\varepsilon)$, so that (24) holds with $x_\varepsilon,y_\varepsilon,h_\varepsilon$ in place of x,y,h, respectively. Furthermore,

$$\begin{aligned}\big|T_{h_\varepsilon,l}v(x_\varepsilon)-T_{h,l}v(x)\big| &\le [v]_{C^\alpha(B_1)}\big|x_\varepsilon+h_\varepsilon l-(x+hl)\big|^\alpha\\ &=[v]_{C^\alpha(B_1)}|x+hl|^\alpha(h\varepsilon)^\alpha,\end{aligned}$$

so that owing to (24),

$$\big|v(y)-v(x)\big|\le\big[2\varepsilon^\alpha[v]_{C^\alpha(B_1)}+K_v\big]h^\alpha. \tag{25}$$

Also observe that if $x,y=x+hl\in\bar B_1$ and $h\ge\bar h(\varepsilon)$, then

$$\big|v(y)-v(x)\big|\le \bar h^{-\alpha}h^\alpha\operatorname*{osc}_{B_1}v. \tag{26}$$

Upon combining (25) and (26) we conclude that, for any $\varepsilon\in(0,1/2]$ and $x,y\in\bar B_1$, $x\ne y$, we have

$$|x-y|^{-\alpha}\big|v(x)-v(y)\big|\le\bar h^{-\alpha}(\varepsilon)\operatorname*{osc}_{B_1}v+2\varepsilon^\alpha[v]_{C^\alpha(B_1)}+K_v,$$

which is even a notch stronger than (20). The lemma is proved. □

10. THEOREM. *Let $R\in(0,\infty)$, $u\in C^{1+\alpha}(B_R)$, $v\in C^\alpha(B_R)$ and assume that there exists a constant $K\in[0,\infty)$ such that for any unit $l,l_1\in\mathbb{R}^d$, $0\le h_1\le h$, we have*

$$\big|(T_{h,l}-1)^2(T_{h_1,l_1}-1)u(x)\big|\le Kh^{1+\alpha}, \tag{27}$$

whenever

$$x,x+h_1l_1,x+2hl,x+h_1l_1+2hl\in\bar B_R, \tag{28}$$

and

$$\big|(T_{h,l}-1)^2 v(x)\big| \le K h^{\alpha}, \tag{29}$$

whenever

$$x, x+2hl \in \bar{B}_R. \tag{30}$$

Then there exists a constant N depending only on d and α such that

$$[Du]_{C^{\alpha}(B_R)} \le N\big(K + R^{-1-\alpha} \operatorname*{osc}_{B_R} u\big), \quad [v]_{C^{\alpha}(B_R)} \le N\big(K + R^{-\alpha} \operatorname*{osc}_{B_R} u\big). \tag{31}$$

Proof. As usual, scalings allow us to assume that $R = 1$. In that case first observe that, if x, l, l_1, h, h_1 are as in (28), then

$$\big|(T_{h,l}-1)^2 D_{(l_1)}u(x) - (T_{h,l}-1)^2 h_1^{-1}(T_{h_1,l_1}-1)u(x)\big| \le 4[D_{(l_1)}u]_{C^{\alpha}(B_1)} h_1^{\alpha}.$$

We choose h_1 so that

$$h_1 = \varepsilon h,$$

where $\varepsilon \in (0, 1/2]$ will be specified later, fix l_1, and set $w = D_{(l_1)}u$. Then, for any unit $l \in \mathbb{R}^d$ and $h > 0$, we have

$$\big|(T_{h,l}-1)^2 w(x)\big| \le K_1 h^{\alpha}, \tag{32}$$

where

$$K_1 = 4\varepsilon^{\alpha}[w]_{C^{\alpha}(B_1)} + \varepsilon^{-1}K,$$

provided that

$$x, x+2hl \in \bar{B}_{1-h\varepsilon}. \tag{33}$$

By Lemma 9

$$[w]_{C^{\alpha}(B_1)} \le (6\varepsilon^{\alpha} + 2^{\alpha-1})[w]_{C^{\alpha}(B_1)} + \varepsilon^{-1}K + 16\varepsilon^{-\alpha} \operatorname*{osc}_{B_1} w.$$

We take $\varepsilon = \varepsilon(\alpha) \in (0, 1/2]$ so that $6\varepsilon^{\alpha} + 2^{\alpha-1} \le 2^{(\alpha-1)/2}$ (< 1) and conclude

$$[w]_{C^{\alpha}(B_1)} \le N\ (K + \operatorname*{osc}_{B_1} w).$$

Since this holds for any unit l_1 with $w = D_{(l_1)}u$, we have

$$[u]_{C^{1+\alpha}(B_1)} \le N(K + \operatorname*{osc}_{B_1} Du),$$

and using the interpolation inequalities allowing to estimate $\operatorname{osc}_{B_1} Du$ through $[Du]_{C^{\alpha}(B_1)}$ and $\operatorname{osc}_{B_1} u$ brings the proof of the first estimate in (31) to an end.

The second estimate in (31) follows immediately from Lemma 9 (in case $R = 1$) when in (21) we take $\varepsilon \in (0, 1/2]$ such that $2\varepsilon^{\alpha} + 2^{\alpha-1} \le 2^{(\alpha-1)/2}$. The theorem is proved. □

3. Proof of Theorems 1.4 and 1.6

To prove Theorem 1.4 we need one more lemma.

1. LEMMA. *Let the assumptions of Theorem 1.4 be satisfied, let $\rho_0\in(0,1]$, and let $u\in C^{1,1}_{\text{loc}}(B_{2\rho_0})$ be a solution of (1.1) in $B_{2\rho_0}$ (a.e.). Let N' and N'' be constants such that*

$$|u,Du|_{C(B_{2\rho_0})}\le N',\quad [u,Du]_{C^\alpha(B_{2\rho_0})}\le N'',\tag{1}$$

and assume that

$$\operatorname*{osc}_{B_{2\rho_0}} u,\ \operatorname*{osc}_{B_{2\rho_0}} Du\le 1.\tag{2}$$

Then $u\in C^{2+\alpha}_{\text{loc}}(B_{2\rho_0})$ and

$$[D^2u]_{C^\alpha(B_{\rho_0})}\le N\big(1+\rho_0^{-2-\alpha}\sup_{B_{2\rho_0}}|u|\big),\tag{3}$$

where N depends only on $K_0(N'')^\alpha$, d, δ, α, and $K_{N'}$.

Proof. Take an $R\in(0,2\rho_0]$, a ball $B_\rho(x_0)\subset B_R$, and call v the solution of

$$H\big(u(x_0),Du(x_0),D^2v(x),x_0\big)=0$$

in $B_\rho(x_0)$ with boundary condition $v=u$ on $\partial B_\rho(x_0)$. By a scaled version of Theorem 10.3.1 such a solution of class $C\big(\bar B_\rho(x_0)\big)\cap C^{2+\alpha_0}_{\text{loc}}\big(B_\rho(x_0)\big)$ does exist and

$$\big|D^2v(x)-D^2v(y)\big|\le N|x-y|^{\alpha_0}\rho^{-2-\alpha_0}\sup_{\partial B_\rho(x_0)}|u-p|$$

if $x,y\in B_{\rho/2}(x_0)$, where p is an arbitrary polynomial of degree 2 on $\mathbb{R}^d$ and N depends only on δ,d. Therefore, if T is the Taylor polynomial of degree 2 of v at x_0, then on $B_r(x_0)$, for any $r\le\rho/2$, we have

$$|u-T|\le|u-v|+|v-T|\le|u-v|+Nr^{2+\alpha_0}\rho^{-2-\alpha_0}\sup_{\partial B_\rho(x_0)}|u-p|.\tag{4}$$

Furthermore, in $B_\rho(x_0)$

$$0=H\big(u(x),Du(x),D^2u(x),x\big)-H\big(u(x_0),Du(x_0),D^2v(x),x_0\big)=I_1+I_2,$$

where

$$\begin{aligned}I_1&=H(u(x_0),Du(x_0),D^2u(x),x_0)-H(u(x_0),Du(x_0),D^2v(x),x_0)\\&=a^{ij}(x)D_{ij}(u-v)(x)\end{aligned}$$

(a.e.) with an $\mathbb{S}_\delta$-valued (a^{ij}) and (due to Assumption 1.2 and (2))

$$\begin{aligned}|I_2|&=\big|H\big(u(x),Du(x),D^2u(x),x\big)-H\big(u(x_0),Du(x_0),D^2u(x),x_0\big)\big|\\&\le K\big(\big|u(x)-u(x_0)\big|^\alpha+\big|Du(x)-Du(x_0)\big|^\alpha\big)+K_{N'}(1+M_R)\rho^\alpha\\&\le 2K(N'')^\alpha\rho^{\alpha^2}+K_{N'}(1+M_R)\rho^\alpha\\&\le N(1+M_R)\rho^{\alpha^2},\end{aligned}\tag{5}$$

where

$$M_R = \sup_{B_R} |D^2 u|.$$

Hence,

$$\left|a^{ij}(x)D_{ij}(u-v)\right| \le N(1+M_R)\rho^{\alpha^2}$$

in $B_\rho(x_0)$ and $u - v = 0$ on $\partial B_\rho(x_0)$. It follows that

$$|u-v| \le N(1+M_R)\rho^{2+\alpha^2},$$

which along with (4) shows that, for any $r \le \rho/2$, we have on $B_r(x_0)$ that

$$|u-T| \le N(1+M_R)\rho^{2+\alpha^2} + N_1 r^{2+\alpha_0}\rho^{-2-\alpha_0} \sup_{\partial B_\rho(x_0)} |u-p|.$$

We set $r = \rho/\nu$, where $\nu \ge 2$ is chosen so that (recall (1.2))

$$N_1 \nu^{-2-\alpha_0} \le (1/2)\nu^{-2-\alpha^2}. \tag{6}$$

Then we find that, if $B_{\nu r}(x_0) \subset B_R$, then there exists a polynomial p $(= T)$ of degree 2 such that the inequality

$$\sup_{B_r(x_0)} |u-p| \le N\big(1 + \sup_{B_R}|D^2 u|\big)r^{2+\alpha^2} + (1/2)\nu^{-(2+\alpha^2)} \sup_{B_{\nu r}(x_0)} |u-q|$$

holds for any quadratic polynomial q.

It follows by Theorem 2.7 that for any $\rho < R$ we have

$$[D^2 u]_{C^{\alpha^2}(B_\rho)} \le N_2\big(1 + \sup_{B_R}|D^2 u|\big) + N(R-\rho)^{-2-\alpha^2} \sup_{B_R}|u|. \tag{7}$$

Since, R is any number $\le 2\rho_0$, in particular, we get that

$$u \in C^{2+\alpha^2}_{\text{loc}}(B_{2\rho_0}).$$

By Remark 2.3

$$N_2 \sup_{B_R}|D^2 u| \le (1/4)R^{\alpha^2}[D^2 u]_{C^{\alpha^2}(B_R)} + NR^{-2}\sup_{B_R}|u|,$$

which along with (7) (and $R \le 2\rho_0 \le 2$) show that

$$[D^2 u]_{C^{\alpha^2}(B_\rho)} \le N + N(R-\rho)^{-2-\alpha^2} \sup_{B_R}|u| \tag{8}$$

for any $\rho < R \le 2\rho_0$.

By taking $R = 2\rho_0$, and $\rho = (3/2)\rho_0$ and using again interpolation, we conclude that

$$|D^2 u|_{C(B_{(3/2)\rho_0})} \le N\rho_0^{\alpha^2} + N\rho_0^{-2} \sup_{B_{2\rho_0}}|u| \le N + N\rho_0^{-2} \sup_{B_{2\rho_0}}|u| =: J(2\rho_0).$$

Also

$$|Du|_{C(B_{(3/2)\rho_0})} \le \rho_0 |D^2 u|_{C(B_{(3/2)\rho_0})} + N(d)\rho_0^{-1} \sup_{B_{(3/2)\rho_0}}|u| \le NJ(2\rho_0).$$

With these estimates of $D^2 u$ and Du at hand we come back to (5) and see that, if $B_\rho(x_0) \subset B_R \subset B_{(3/2)\rho_0}$, then

$$|I_2| \le N\big(J(2\rho_0) + M_R\big)\rho^\alpha.$$

This allows us to repeat with obvious changes what follows (5) replacing α^2 with α and paying attention that the solvability of the modified (6) crucially depends on the assumption that $\alpha<\alpha_0$. Then we come to the new version of (8), which implies (3). The lemma is proved. □

Proof of Theorems 1.4. First assume that $g\in C^{1,1}(\mathbb{R}^d)$. Observe that

$$\big|H(\mathsf{u}',0,x)\big|\le\big|H(\mathsf{u}',0,x)-H(0,x)\big|+\big|H(0,x)\big|$$

$$\le K_0\big(|\mathsf{u}'|+1\big)+\sup_x\big|H(0,x)\big|.$$

This shows that Assumption 6.6.2 (ii) is satisfied. Its requirements (i) and (iii) are also, obviously satisfied. Take P from (6.2.4) and for $K>0$ set

$$H_K(\mathsf{u},x)=\max\big(H(\mathsf{u},x),P(\mathsf{u}'')-K\big).$$

Then by Theorem 6.6.20, for any $K>0$, there exists $u_K\in C(\bar\Omega)\cap C^{0,1}(\Omega)\cap C^{1,1}_{\text{loc}}(\Omega)$ such that $u_K=g$ on $\partial\Omega$ and

$$H_K[u_K]=0\quad\text{in}\quad\Omega\text{ (a.e.)}.\tag{9}$$

Now we want to apply Lemma 1 to u_K and (9) in place of u and (1.1), respectively, after shifting the origin appropriately and obtain interior estimates of $C^{2+\alpha}$-norms of u_K independent of K. But first we have to ensure that the shifted analogs of (1) hold with N' and N'' independent of K if K is sufficiently large, say

$$K\ge\sup_x\big|H(0,x)\big|.\tag{10}$$

This we do by using Theorem 10.1.10 applied to H_K. For K fixed, introduce

$$F(\mathsf{u}'',x)=H_K(0,\mathsf{u}'',x)-H(0,x),\quad G(\mathsf{u},x)=H_K(\mathsf{u},x)-F(\mathsf{u}'',x),$$

$$\bar F_{r,z}(\mathsf{u}'')=F(\mathsf{u}'',z),\quad r>0,\ z\in\mathbb{R}^d.$$

In light of Assumption 1.2 and (10) we have $F(0,x)\equiv0$ and

$$\big|G(\mathsf{u},x)\big|\le\big|H(\mathsf{u},x)-H(0,\mathsf{u}'',x)\big|+\big|H(0,x)\big|$$

$$\le K_0\big(|\mathsf{u}'|+1\big)+\big|H(0,x)\big|$$

$$=:K_0|\mathsf{u}'|+\bar G(x),$$

which shows that Assumption 10.1.4 is satisfied with $\hat\theta=0$. Also

$$\big|F(\mathsf{u}'',x)-\bar F_{r,z}(\mathsf{u}'')\big|=\big|H_K(0,\mathsf{u}'',x)-H(0,x)-H_K(0,\mathsf{u}'',z)+H(0,z)\big|$$

$$\le K_0\big(2+|\mathsf{u}''|\big)|x-z|^\alpha$$

if $|x-z|\le1$, which implies that, for any $\theta>0$, there exists R_0, such that (10.1.4) holds as long as $r\le R_0$, $|\mathsf{u}''|=1$, and $\tau_0=1$. Other requirements in Assumption 10.1.6 are satisfied as well.

The remaining assumption, Assumption 10.1.3, is obviously satisfied and thus all assumptions of Theorem 10.1.10 are satisfied with parameters involved independent of K. Note that, since F is independent of u_0', we have $\omega_F\equiv0$. Furthermore, thanks to Remark 10.1.8 and the fact that $H(\mathsf{u},x)$ is a nonincreasing function of

u_0' by assumption, Assumption 10.1.12 is satisfied. Hence by Theorem 10.1.10 and Lemma 6.6.10, for any $p > d$ and $\rho < \rho_{\text{int}}(\Omega)$, we have

$$\|u_K\|_{W^2_p(\Omega^\rho)} \le N\rho^{-2}\big(\|\bar{G}\|_{L_p(\Omega)} + \|g\|_{C(\partial\Omega)}\big) + N = M(\rho),$$

$$\sup_{\Omega}|u| \le N\|\bar{G}\|_{L_d(\Omega)} + \|g\|_{C(\partial\Omega)},$$

where the constants N depend only on d, δ, α, p, K_0, and $\operatorname{diam}(\Omega)$. By adding here Theorem 9.7.1, we see that the family $\{u_K\}$ is uniformly continuous and uniformly bounded in $\bar{\Omega}$. It is important for the future that the $C^{1,1}(\mathbb{R}^d)$-norm of g is not involved in the last statement but rather its modulus of continuity on $\partial\Omega$.

By embedding theorems, for $p = p(d,\alpha)$ large enough, if $B_{2\rho_0}(x_0) \subset \Omega^{2\rho}$, then

$$|u_K, Du_K|_{C(B_{2\rho_0}(x_0))}, [u_K, Du_K]_{C^\alpha(B_{2\rho_0}(x_0))} \le N(\Omega^\rho, d, \alpha)M(\rho), \tag{11}$$

If we restrict ρ_0 to also satisfy

$$N(\Omega^\rho, d, \alpha)M(\rho)(2\rho_0)^\alpha \le 1,$$

then

$$\operatorname*{osc}_{B_{2\rho_0}(x_0)} u_K, \quad \operatorname*{osc}_{B_{2\rho_0}(x_0)} Du_K \le 1.$$

It follows, owing to Lemma 1 applied to u_K and $B_{2\rho_0}(x_0)$ in place of u and $B_{2\rho_0}$, that the $C^{2+\alpha}\big(B_{\rho_0}(x_0)\big)$-norms of u_K are bounded by a constant N which depends only on d, ρ, δ, α, Ω, $\sup_{\partial\Omega}|g|$, $\sup\big|H(0,x)\big|$, the function K_R, and ρ_0.

Due to the arbitrariness in the choice of $B_{2\rho_0}(x_0)$, the statement of the theorem holds true with u_K in place of u. Upon combining this fact with the uniform boundedness and uniform continuity of the family $\{u_K\}$, choosing an appropriate subsequence $u_{K(n)}$, $K(n) \to \infty$, and passing to the limit in (9) along this subsequence, we easily complete the proof in case $g \in C^{1,1}(\mathbb{R}^d)$.

The general case is proved by uniform approximating $g \in C(\partial\Omega)$ by polynomials and taking into account that by Theorem 9.7.1 the solutions with approximating data are all uniformly bounded and uniformly continuous and, by the above particular case, their interior $C^{2+\alpha}$-norms are also bounded.

The theorem is proved. □

Proof of Theorem 1.6. The uniqueness statement is a simple consequence of the maximum principle. In light of Theorem 1.4 it only remains to prove that (1.4) and (1.5) hold for any solution $u \in C^2_{\text{loc}}(B_R) \cap C(\bar{B}_R)$ of

$$H(D^2u) + f = 0$$

in B_R with boundary condition $u = g$.

We will just repeat part of the proof of Lemma 1. Take a ball $B_\rho(x_0) \subset B_R$ and call v the solution of

$$H\big(D^2v(x)\big) + f(x_0) = 0$$

in $B_\rho(x_0)$ with boundary condition $v = u$ on $\partial B_\rho(x_0)$. By a scaled version of Theorem 10.3.1 such a solution of class $C\big(\bar{B}_\rho(x_0)\big) \cap C^{2+\alpha_0}_{\text{loc}}\big(B_\rho(x_0)\big)$ does exist and

$$\big|D^2v(x) - D^2v(y)\big| \le N|x-y|^{\alpha_0}\rho^{-2-\alpha_0} \sup_{\partial B_\rho(x_0)} |u - p|$$

if $x, y \in B_{\rho/2}(x_0)$, where p is an arbitrary polynomial of degree 2 on $\mathbb{R}^d$ and N depends only on δ, d. Therefore, if T is the Taylor polynomial of degree 2 of v at x_0, then on $B_r(x_0)$, for any $r \le \rho/2$, we have

$$|u - T| \le |u - v| + |v - T| \le |u - v| + N r^{2+\alpha_0} \rho^{-2-\alpha_0} \sup_{\partial B_\rho(x_0)} |u - p|. \tag{12}$$

Furthermore, in $B_\rho(x_0)$

$$0 = H\big(D^2 u(x)\big) + f(x) - H\big(D^2 v(x)\big) - f(x_0)$$

$$= a^{ij}(x) D_{ij}(u - v)(x) + h(x)\rho^\alpha,$$

where (a^{ij}) is $\mathbb{S}_\delta$-valued and

$$|h(x)| = \big|f(x) - f(x_0)\big|\rho^{-\alpha} \le [f]_{C^\alpha(B_R)}.$$

It follows that $|u - v| \le N(d, \delta)[f]_{C^\alpha(B_R)}\rho^{2+\alpha}$ which combined with (12) yields that for any $r \le \rho/2$

$$|u - T| \le N[f]_{C^\alpha(B_R)}\rho^{2+\alpha} + N_1 r^{2+\alpha_0}\rho^{-2-\alpha_0} \sup_{\partial B_\rho(x_0)} |u - p|$$

on $B_r(x_0)$. We set $r = \rho/\nu$, where $\nu \ge 2$ is chosen so that (recall that $\alpha < \alpha_0$)

$$N_1 \nu^{-2-\alpha_0} \le (1/2)\nu^{-2-\alpha}.$$

Then we find that, if $B_{\nu r}(x_0) \subset B_R$, then there exists a polynomial p $(= T)$ of degree 2 such that (1.5) holds for any quadratic polynomial q.

After that it only remains to apply Theorem 2.7. $\square$

CHAPTER 12

Solvability in $W^{1,2}_{p,\mathrm{loc}}$ of fully nonlinear parabolic equations

1. Main results

In this chapter we consider parabolic equations

$$\partial_t v(t,x) + H[v](t,x) = 0, \tag{1}$$

where

$$H[v](t,x) = H\big(v(t,x), Dv(t,x), D^2v(t,x), t, x\big),$$

in subdomains of $\mathbb{R}^{d+1} = \big\{(t,x) : t \in \mathbb{R}, x \in \mathbb{R}^d\big\}$. Recall that $\mathbb{S}$ is the set of symmetric $d \times d$ matrices and, for $\delta \in (0,1]$,

$$\mathbb{S}_\delta = \big\{a \in \mathbb{S} : \delta|\xi|^2 \le a^{ij}\xi^i\xi^j \le \delta^{-1}|\xi|^2, \quad \forall \xi \in \mathbb{R}^d\big\}.$$

Let Ω be an open bounded subset of $\mathbb{R}^d$ satisfying the exterior ball condition. Fix $T \in (0,\infty)$ and set

$$\Pi = [0,T) \times \Omega.$$

Fix

$$p > d+1 \quad \text{and a measurable function} \quad \bar{G} \ge 0 \quad \text{on} \quad \mathbb{R}^{d+1}.$$

Also fix some constants $K_0, K_F \in [0,\infty)$, and $\delta \in (0,1]$.

1. Example. As in Section 10.1 the reader will see that, if $d = 3$ and $f, \bar{G} \in L_p(\Pi)$, the equation ($a \wedge b = \min(a,b)$)

$$\partial_t u(t,x) + H(D^2u,t,x) := \bar{G}(t,x) \wedge |D_{12}u| + \bar{G}(t,x) \wedge |D_{23}u| + \bar{G}(t,x) \wedge |D_{31}u|$$

$$+\Delta u - f(t,x) + \partial_t u(t,x) = 0 \tag{2}$$

satisfies all assumptions listed below (see also Remark 11). Observe that H in (2) is neither convex nor concave with respect to D^2u.

2. Assumption. The function $H(\mathsf{u},t,x)$,

$$u = (\mathsf{u}', \mathsf{u}''), \quad \mathsf{u}' = \big(\mathsf{u}'_0, \mathsf{u}'_1, ..., \mathsf{u}'_d\big) \in \mathbb{R}^{d+1}, \quad \mathsf{u}'' \in \mathbb{S}, \quad (t,x) \in \mathbb{R}^{d+1},$$

is Borel measurable with respect to (u', t, x).

The following assumptions contain parameters $\hat{\theta}, \theta \in (0,1]$ which are specified later in our results.

3. ASSUMPTION. There are two Borel measurable functions

$$F(u,t,x)=F\big(\mathsf{u}_0',\mathsf{u}'',t,x\big),\quad G(\mathsf{u},t,x)$$

such that

$$H=F+G.$$

For all $\mathsf{u}''\in\mathbb{S}$, $\mathsf{u}'\in\mathbb{R}^{d+1}$, and $(t,x)\in\mathbb{R}^{d+1}$ we have

$$\big|G(\mathsf{u}',\mathsf{u}'',t,x)\big|\le\hat\theta|\mathsf{u}''|+K_0|\mathsf{u}'|+\bar G(t,x),\quad F(0,t,x)\equiv 0.$$

Examples of decomposition $H=F+G$ can be found in Example 10.1.5. Observe that Assumption 3 is satisfied in Example 1 with $F(D^2u)=\Delta u$ and $G=H-F$.

Recall that

$$C_r(t,x)=[t,t+r^2)\times B_r(x),\quad C_r=C_r(0),$$

and for measurable $\Gamma\subset\mathbb{R}^{d+1}$ we denote by $|\Gamma|$ the volume of Γ.

4. ASSUMPTION. (i) The function F is Lipschitz continuous with respect to u'' with Lipschitz constant K_F.

Moreover, there exist $R_0\in(0,1]$ and $\tau_0\in[0,\infty)$ such that, if $C_r(z)\subset\Pi$, $r\in(0,R_0]$, $z\in\mathbb{R}^{d+1}$, and $\mathsf{u}_0'\in\mathbb{R}$, then one can find a *convex* function $\bar F(\mathsf{u}'')=\bar F_{z,r,\mathsf{u}_0'}(\mathsf{u}'')$ (depending only on u'') for which

(ii) We have $\bar F(0)=0$ and $D_{\mathsf{u}''}\bar F\in\mathbb{S}_\delta$ at all points of differentiability of $\bar F$;

(iii) For any $\mathsf{u}''\in\mathbb{S}$ with $|\mathsf{u}''|=1$, we have

$$\int_{C_r(z)}\sup_{\tau>\tau_0}\tau^{-1}\big|F\big(\mathsf{u}_0',\tau\mathsf{u}'',t,x\big)-\bar F(\tau\mathsf{u}'')\big|\,dtdx\le\theta\big|C_r(z)\big|;\tag{3}$$

(iv) There exists a continuous increasing function $\omega_F(\tau)$, $\tau\ge 0$, such that $\omega_F(0)=0$ and for any $\mathsf{u}_0',\mathsf{v}_0'\in\mathbb{R}$, $(t,x)\in\Pi$, and $\mathsf{u}''\in\mathbb{S}$ we have

$$\big|F\big(\mathsf{u}_0',\mathsf{u}'',t,x\big)-F\big(\mathsf{v}_0',\mathsf{u}'',t,x\big)\big|\le\omega_F\big(|\mathsf{u}_0'-\mathsf{v}_0'|\big)|\mathsf{u}''|.$$

5. REMARK. Assumptions 3 and 4 (iv) imply that

$$\big|H(\mathsf{u}',0,t,x)\big|\le K_0|\mathsf{u}'|+\bar G(t,x)\quad\forall\mathsf{u}',(t,x)\in\mathbb{R}^{d+1}.\tag{4}$$

Note that Assumption 4 is obviously satisfied in Example 1 for the choice $F(u,D^2u,t,x)=\Delta u$.

Next, we introduce a definition, similar to Definition 10.1.9.

6. DEFINITION. For a function $u\in C(\bar\Pi)$ set

$$\omega_u(\Pi,\rho)=\sup\big\{|u(z_1)-u(z_2)|:z_1,z_2\in\Pi,\rho(z_1,z_2)\le\rho\big\},$$

$$\omega_{F,u,\Pi}(\rho)=\omega_F\big(\omega_u(\Pi,\rho)\big),$$

and in the formulations of a theorem, lemma,... let us say that a certain constant depends only on A,B,..., and the function $\omega_{F,u,\Pi}$ if it depends only on A,B,..., and on the maximal solution of the inequality $N_0\omega_{F,u,\Pi}(\rho)\le 1/2$, where the range of ρ and the value of N_0 depending only on A,B,... could be always found out from our arguments.

We now come to the main a priori estimate in Sobolev spaces. Observe that in Theorem 7 we do not even assume that equation (1) is parabolic. For $\rho > 0$ we denote

$$\Pi^\rho = [0, T - \rho^2) \times \Omega^\rho.$$

7. THEOREM. *There exist constants $\theta, \hat{\theta} \in (0,1]$, depending only on d, p, δ, and K_F, such that, if Assumptions 4 and 3 are satisfied with these θ and $\hat{\theta}$, respectively, then for any $u \in W^{1,2}_{p,\mathrm{loc}}(\Pi) \cap C(\bar{\Pi})$ that satisfies (1) in Π (a.e.) and $\rho > 0$, such that $\Pi^\rho \neq \emptyset$, we have*

$$\|u\|_{W^{1,2}_p(\Pi^\rho)} \leq N\|\bar{G}\|_{L_p(\Pi)} + N\rho^{-2}\|u\|_{C(\Pi)} + N\tau_0, \tag{5}$$

where the constants N depend only on K_0, K_F, d, p, δ, R_0, $\operatorname{diam}(\Omega)$, T, and the function $\omega_{F,u,\Pi}$.

We prove this theorem in Section 5.

Generally, having a priori estimates does not guarantee that there are existence results. Say, if $d = 1$ and the equation is

$$\partial_t u + u'' - (2u'')I_{|u''| \leq 5} - 10u = 1 \quad \text{in} \quad (0,T) \times (-1,1),$$

a priori $W^{1,2}_p$-estimates are trivial since $\big|(2u'')I_{|u''|\leq 5}\big| \leq 10$. However, the solvability is quite questionable.

To have the solvability we need ellipticity and more regularity of H.

8. ASSUMPTION. For any $(t,x) \in \mathbb{R}^{d+1}$, the function $H(\mathsf{u},t,x)$ is continuous with respect to u, is Lipschitz continuous with respect to u'', and $D_{\mathsf{u}''}H \in \mathbb{S}_\delta$ at all points of differentiability of H with respect to u''.

We can now state another a priori estimate.

9. LEMMA. *Suppose that (4) and Assumption 8 are satisfied. Then for any $v \in W^{1,2}_{d+1,\mathrm{loc}}(\Pi) \cap C(\bar{\Pi})$ that satisfies (1) in Π (a.e.) we have*

$$\sup_{\Pi} |v| \leq Ne^{K_0 T}\big(\|\bar{G}\|_{L_{d+1}(\Pi)} + \sup_{\partial'\Pi}|v|\big), \tag{6}$$

where N depends only on d, δ, K_0, and $\operatorname{diam}(\Omega)$.

Proof. By Remark 6.6.16, obviously applicable in the parabolic case,

$$\partial_t v + a^{ij}D_{ij}v + b^i D_i v + cv + \tau\bar{G} = 0 \tag{7}$$

in Π (a.e.) for certain measurable $\mathbb{S}_\delta$-valued (a^{ij}), measurable $\mathbb{R}^d$-valued b, and measurable real-valued τ and c, such that $|b|, |c| \leq K_0$, and $|\tau| \leq 1$. Hence $w = ve^{K_0(t-T)}$ satisfies

$$\partial_t w + a^{ij}D_{ij}w + b^i D_i w + (c - K_0)w + \tau e^{K_0(t-T)}\bar{G} = 0,$$

which yields (6) by Theorem 3.0.1 since $c - K_0 \leq 0$. The lemma is proved. □

Observe that Assumption 8 is satisfied in Example 1 because $H_{\mathsf{u}''_{ii}} = 1$, $|H_{\mathsf{u}''_{13}}| = |H_{\mathsf{u}''_{31}}| \leq 1/2$, $|H_{\mathsf{u}''_{12}}| = |H_{\mathsf{u}''_{21}}| \leq 1/2$, $|H_{\mathsf{u}''_{23}}| = |H_{\mathsf{u}''_{32}}| \leq 1/2$.

Here is the main result of the chapter concerning the solvability of (1) in Sobolev spaces.

10. THEOREM. *Let $p > d+2$. There exist constants $\theta, \hat{\theta} \in (0,1]$, depending only on d, p, δ, and K_F, such that, if Assumptions 4 and 3 are satisfied with these θ and $\hat{\theta}$, respectively, and Assumption 8 is satisfied, and $\bar{G} \in L_p(\Pi)$, then for any*

$g \in C(\overline{\partial' \Pi})$ *there exists*

$$u \in \bigcap_{\rho>0} W^{1,2}_p(\Pi^\rho) \cap C(\bar{\Pi})$$

satisfying (1) *in* Π (*a.e.*) *and such that* $u = g$ *on* $\partial' \Pi$.

11. REMARK. The reader will see from the proof of Theorem 10 that if H is independent of $[\mathsf{u}']$ then the condition $p > d+2$ can be relaxed to $p > d+1$. The same applies to Theorem 13 below.

12. REMARK. Theorem 8.4 of [**25**] is quite similar to Theorem 10 and is proved for Ω satisfying the exterior cone, rather than the ball condition. The latter condition can be also relaxed further in Theorem 10, as we discussed it in Remark 10.1.15. Also in Theorem 8.4 of [**25**] the case $p < d+2$ is allowed. The results of [**25**] are build in part on the earlier results in [**113**], [**114**], and [**115**].

On the other hand, H is assumed in [**25**] to be either convex or concave in u'' and satisfy the continuity requirement in (t, x) similar to (10.1.11) (although $|\mathsf{u}''|$ in the denominator of (10.1.11) is replaced with $1 + |\mathsf{u}''|$). We discussed the relation of such conditions to our almost VMO conditions in Remark 10.1.26. Another assumption in Theorem 8.4 of [**25**], absent in Theorem 10, is that $H(\mathsf{u}, t, x)$ is Lipschitz continuous with respect to $[\mathsf{u}'] = (\mathsf{u}'_1, ..., \mathsf{u}'_d)$ and is uniformly continuous with respect to u'_0 uniformly with respect to all other arguments. In case $d = 1$ this excludes, for instance, the equation

$$\partial_t u(t,x) + a(u(t,x), t, x) D^2 u(t,x) + f(u(t,x), t, x) = 0.$$

Done with the discussion, note that obviously, the arguments in the proof of Theorem 10.1.16 are valid in the parabolic case as well and allow us to derive from Theorem 10 the following.

13. THEOREM. *Suppose that Assumptions* 2 *and* 8 *are satisfied,* $p > d+2$, $\bar{G} \in L_p(\Pi)$, *and* (4) *holds true. Let* $P(\mathsf{u}'')$ *be a convex function on* $\mathbb{S}$ *such that at each point of its differentiability* $D_{\mathsf{u}''} P \in \mathbb{S}_{\delta'}$, *where* $\delta' \in (0, \delta]$. *Also assume that for any* $a \in \mathbb{S}_\delta$ *and* $\mathsf{u}'' \in \mathbb{S}$ *we have*

$$a^{ij} \mathsf{u}''_{ij} \leq P(\mathsf{u}'') + K,$$

where K *is a constant. Then, for any* $g \in C(\overline{\partial' \Pi})$, *the equation*

$$\partial_t u + \max\big(H[u], P[u]\big) = 0$$

in Π (*a.e.*) *with boundary condition* $u = g$ *on* $\partial' \Pi$ *has a solution*

$$u \in \bigcap_{\rho>0} W^{1,2}_p(\Pi^\rho) \cap C(\bar{\Pi}).$$

14. REMARK. Similarly to Remark 6.6.14 we observe that generally there is no uniqueness in Theorem 10. For instance, in the one-dimensional case the (quasilinear) equation

$$\partial_t u + D^2 u + (1-t)\sqrt{12|Du|} + 2\sqrt{(1-|x|^3)u} = 0$$

in $\Pi = [0, 1) \times (-1, 1)$ with zero boundary data on $\partial' \Pi$ has two solutions: one is identically equal to zero and the other one is $(1-t)^2(1-|x|^3)$.

Uniqueness of solutions can be investigated on the basis of the results in Subsection 4.1:1.

We prove Theorem 10 in Section 5 after some preparations.

15. REMARK. The parameter θ in Theorem 10 depends on p and we cannot guarantee that it stays bounded away from zero for all $p > d+2$. Our arguments are only valid if we take θ sufficiently small and, as $p \to \infty$, θ should go to zero.

16. REMARK. For measurable $\Gamma \subset \mathbb{R}^{d+1}$ with nonzero Lebesgue measure and locally summable f denote

$$\fint_{\Gamma} f(t,x)\,dxdt = \frac{1}{|\Gamma|}\int_{\Gamma} f(t,x)\,dxdt$$

(recall that $|\Gamma|$ is the volume of Γ). Then for $z \in \Pi$, $r > 0$, and $\mathsf{u}_0' \in \mathbb{R}$ introduce

$$\hat{F}(\mathsf{u}'') = \hat{F}_{z,r,\mathsf{u}_0'}(\mathsf{u}'') = \fint_{C_r(z)} F(\mathsf{u}_0', \mathsf{u}'', t, x)\,dxdt.$$

As in Remark 10.1.20 it is very tempting to always use this $\hat{F}(\mathsf{u}'')$ as $\bar{F}$ in Assumption 4. However, there is no guarantee that $\hat{F}(\mathsf{u}'')$ is convex in u''.

17. REMARK. Similarly to Remark 10.1.21, under the above assumptions, the function

$$F(\mathsf{u}'', t, x) = H(0, \mathsf{u}'', t, x) - H(0, t, x)$$

does not necessarily satisfy Assumption 4, so that this choice of F may not be optimal.

18. REMARK. As in Remark 10.1.22 one observes that in condition (3) no restriction is imposed on $F(\mathsf{u}_0', \mathsf{u}'', t, x)$ for $|\mathsf{u}''| \leq \tau_0$. But even for large $|\mathsf{u}''|$ the function F satisfying Assumption 4 need not be even locally convex.

19. REMARK. A somewhat more transparent condition than Assumption 4 can be stated in terms of

$$\hat{F}\big(\mathsf{u}_0', \mathsf{u}'', t, x\big) := \lim_{\tau\to\infty} \frac{1}{\tau} F\big(\mathsf{u}_0', t\mathsf{u}'', t, x\big)$$

provided that the limit exists. Then requiring that $\hat{F}$ be convex in u'', $D_{\mathsf{u}''}\hat{F} \in \mathbb{S}_\delta$, and $\hat{F}\big(\mathsf{u}_0', \mathsf{u}'', \cdot\big) \in VMO$ for any $\mathsf{u}_0', \mathsf{u}''$ seems to be enough to satisfy Assumption 4 (perhaps, modulo some kind of uniformity).

However, requiring the above limit to exist is an additional, rather strong, and unnecessary assumption. For instance, it is not satisfied in Example 10.1.2.

In addition to the examples presented above we give one more related to stochastic differential games and the Isaacs equations.

20. EXAMPLE. Let A and B be some countable sets and assume that for $\alpha \in A$, $\beta \in B$, $\mathsf{u}_0' \in \mathbb{R}$, and $(t,x) \in \mathbb{R}^{d+1}$ we are given functions

$$a^{\alpha}\big(\mathsf{u}_0', t, x\big), \quad b^{\alpha\beta}\big(\mathsf{u}_0', t, x\big), \quad c^{\alpha\beta}\big(\mathsf{u}_0', t, x\big), \quad f^{\alpha\beta}\big(\mathsf{u}_0', t, x\big)$$

with values in

$$\mathbb{S}_\delta, \quad \mathbb{R}^d, \quad \mathbb{R}, \quad \mathbb{R},$$

respectively. Assume that these functions are measurable in (t,x), a^{α}, $b^{\alpha\beta}$, and $c^{\alpha\beta}$ are bounded and continuous with respect to u_0' uniformly with respect to $\alpha, \beta, (t,x)$, and

$$\bar{G} := \sup_{\mathsf{u}_0', \alpha, \beta} |f^{\alpha\beta}(\mathsf{u}_0', \cdot)| \in L_p(\Pi).$$

Consider the following Isaacs equation

$$\partial_t v + H(v, Dv, D^2 v, t, x) = 0,$$

where

$$H(\mathsf{u}, t, x) := \inf_{\beta\in B} \sup_{\alpha\in A} \Big[\sum_{i,j=1}^{d} a^{\alpha}_{ij}(\mathsf{u}'_0, t, x)\mathsf{u}''_{ij} + \sum_{i=1}^{d} b^{\alpha\beta}_{i}(\mathsf{u}'_0, t, x)\mathsf{u}'_i$$

$$-c^{\alpha\beta}(\mathsf{u}'_0, t, x)\mathsf{u}'_0 + f^{\alpha\beta}(\mathsf{u}'_0, t, x)\Big]. \quad (8)$$

Next assume that there is an $R_0 \in (0,\infty)$ such that for any point $z \in \Pi$, $r \in (0, R_0]$, and $\mathsf{u}'_0 \in \mathbb{R}$ one can find $\bar{a}^{\alpha} \in \mathbb{S}_\delta$ such that

$$⨍_{C_r(z)} \sup_{\alpha\in A} \big|a^{\alpha}(\mathsf{u}'_0, t, x) - \bar{a}^{\alpha}\big|\, dxdt \le \theta,$$

where θ is taken from Theorem 10.

Then, as in Example 10.1.24, it turns out that the assertions of Theorem 10 hold true and estimate (5) holds with $\tau_0 = 0$.

21. REMARK. The solvability in $W^{1,2}_{p,\mathrm{loc}}(\Pi) \cap C(\bar{\Pi})$ of equation (1) with H from (8) was proved in Theorem 9.1 of [**25**] under the additional condition that a^α are independent of u'_0 and are uniformly continuous with respect to (t, x) uniformly with respect to α. On the other hand, again some $p < d + 2$ are allowed and Ω is a domain satisfying the exterior cone condition.

In the case of linear equations Theorem 10 contains (apart from the restrictions on p and Ω) the corresponding result of the linear theory for equations whose main coefficients are almost in VMO (cf. [**10**]), whereas Theorem 9.1 of [**25**] reduces to the usual existence theorem for equations with continuous main coefficients.

2. A priori estimates in $C^{2+\alpha}_{\mathrm{loc}}$ for equations $\partial_t u + F(D^2 u) = 0$

The results of this section are either close or coincide with the corresponding results in [**89**]. The proofs, however, are quite different and are based on some estimates obtained for the elliptic equations.

Let $F(\mathsf{u}'')$ be a *convex* function defined for $\mathsf{u}'' \in \mathbb{S}$ such that at all points of its differentiability we have

$$D_{u''}F(\mathsf{u}'') \in \mathbb{S}_\delta,$$

where $\delta \in (0, 1]$ is a fixed number. Also assume that

$$F(0) = 0. \quad (1)$$

In this section we denote by α_0 the smallest of the constants called α_0 in Lemma 9.6.2 and Theorem 10.2.1 and fix

$$\alpha \in (0, \alpha_0).$$

1. THEOREM. *For any $R \in (0,\infty)$ and $u \in W^{1,2}_\infty(C_R)$, satisfying $\partial_t u + F(D^2 u) = 0$ (a.e.) in C_R, we have $u \in C^{2+\alpha}_{\mathrm{loc}}(C_R)$. Furthermore, for $r \in (0, R)$,*

$$[D^2 u]_{C^\alpha(C_r)} \le N \frac{1}{(R - r)^{2+\alpha}} \sup_{\partial' C_R} |u|, \quad (2)$$

where $N = N(\delta, d, \alpha) < \infty$.

2. REMARK. Unlike in the elliptic case, (2) does not hold if we drop condition (1). Indeed, in the future we will see that the equation $\partial_t u + F(D^2u) = 0$ with any continuous boundary condition on $\partial' C_R$ has a unique classical solution. But if the boundary data is zero, then (2) implies that D^2u is constant and then the equation implies that $\partial_t u$ is constant, so that $u(t,x) = \beta t + l(x)$, where l is a quadratic polynomial. Then the boundary data are only satisfied if $\beta = 0$ and $l \equiv 0$, but $u \equiv 0$ is not a solution if (1) does not hold.

Also (2) does not hold if we replace u on the right with $v = u - \big(\beta t + l(x)\big)$, where β is a constant and l is a quadratic function of x, because v satisfies

$$\partial_t v + \hat{F}(D^2v) = 0,$$

where $\hat{F}(\mathsf{u}'') = \beta + F(\mathsf{u}'' + D^2 l)$, and $\hat{F}(0)$ is generally non zero. Of course, $\hat{F}(0) = 0$ if $\beta = 0$ and l is linear.

There is, however, one more very important exception when $\beta t + l(x)$ is the Taylor polynomial of u at a point $z_0 = (t_0, x_0) \in C_R$:

$$\beta t + l(x) = (t - t_0)\partial_t u(z_0) + (1/2)(x^i - x_0^i)(x^j - x_0^j)D_{ij}u(z_0)$$

$$+ (x^i - x_0^i)D_i u(z_0) + u(z_0).$$

In this case $\hat{F}(0) = \big[\partial_t u + F(D^2u)\big](z_0) = 0$.

To prove the theorem we need two auxiliary results.

3. LEMMA. *For any $R \in (0,\infty)$ and $u \in W^{1,2}_\infty(C_R)$, satisfying $\partial_t u + F(D^2u) = 0$ in C_R (a.e.), we have $\partial_t u \in C^{\alpha_0}_{\text{loc}}(C_R)$. Furthermore, for $r \in (0,R)$ and $\beta \in (0,\alpha_0]$,*

$$[\partial_t u]_{C^\beta(C_r)} \leq N \frac{1}{(R-r)^{2+\beta}} \sup_{\partial' C_R} |u|, \tag{3}$$

where $N = N(\delta, d, \beta)$.

Proof. Take $h \in (0, R^2/2]$ and define

$$u_h(t,x) = \frac{u(t+h,x) - u(t,x)}{h}.$$

Then one sees that u_h satisfies

$$\partial_t u_h + a^{ij} D_{ij} u_h = 0$$

in $C_{R^2-h,R}$ (a.e.), where (a^{ij}) is a certain $\mathbb{S}_\delta$-valued measurable function on $C_{R^2-h,R}$. Then take $0 < \rho_0 < \rho_1 < R$ and $z_1 = (t_1, x_1), z_2 = (t_2, x_2) \in C_{\rho_0}$, such that $t_1 \leq t_2$, and consider two cases

(a) $2\rho(z_1, z_2) \geq \rho_1 - \rho_0$,

(b) $2\rho(z_1, z_2) < \rho_1 - \rho_0$.

In case (a) we have

$$\frac{\big|u_h(z_1) - u_h(z_2)\big|}{\rho^\beta(z_1,z_2)} \leq \frac{2\cdot 2^\beta}{|\rho_1 - \rho_0|^\beta} \sup\big(|u_h|, C_{\rho_1}\big).$$

In case (b) define $r = (\rho_1 - \rho_0)/2$ and note that, as is easy to see,

$$z_1, z_2 \in C_r(z_1) \subset C_{2r}(z_1) \subset C_{\rho_1}.$$

It follows by Lemma 9.6.2 that for all h small enough (such that $C_{\rho_1} \subset C_{R^2-h,R}$) and any $\beta \in (0, \alpha_0]$

$$\frac{\big|u_h(z_1) - u_h(z_2)\big|}{\rho^\beta(z_1, z_2)} \leq \frac{N}{|\rho_1 - \rho_0|^\beta} \sup\big(|u_h|, C_{\rho_1}\big),$$

where and below by N we denote various constants depending only on d, β, and δ.

Thus, for $0 < \rho_0 < \rho_1 < R$ and all sufficiently small h we have

$$[u_h]_{C^\beta(C_{\rho_0})} \leq \frac{N}{|\rho_1 - \rho_0|^\beta} \sup\big(|u_h|, C_{\rho_1}\big). \tag{4}$$

Observe that u_h are bounded in C_{ρ_1} uniformly with respect to h for small h since $\partial_t u$ is bounded in C_R. Then, as $h \downarrow 0$, (4) shows that $\partial_t u \in C^\beta(C_{\rho_0})$ and the arbitrariness of ρ_0 (and $\beta \in (0, \alpha_0]$) implies that $\partial_t u \in C^{\alpha_0}_{\mathrm{loc}}(C_R)$ and

$$[\partial_t u]_{C^\beta(C_{\rho_0})} \leq \frac{N}{|\rho_1 - \rho_0|^\beta} \sup\big(|\partial_t u|, C_{\rho_1}\big). \tag{5}$$

Now the only thing remaining to be proved is (3). We will derive it from (5) by using interpolation inequalities very similarly to what is done after (10.2.14). Observe that, for $h \in (0, \rho_0)$ and $z = (t, x), z' = (t', x) \in C_{\rho_0}$, such $|t - t'| = h^2$, we have

$$\big|\partial_t u(z)\big| \leq \Big|\partial_t u(z) - \frac{u(z) - u(z')}{t - t'}\Big| + \frac{2}{h^2} \sup\big(|u|, C_{\rho_1}\big)$$

$$\leq h^\beta [\partial_t u]_{C^\beta(C_{\rho_1})} + \frac{2}{h^2} \sup\big(|u|, C_{\rho_1}\big).$$

By using the notation

$$M(\rho) := [\partial_t u]_{C^\beta(C_\rho)},$$

we infer that, if $\rho_n < \rho_{n+1} < R$ and $h \in (0, \rho_{n+1})$, then

$$M(\rho_n) \leq \frac{N}{|\rho_{n+1} - \rho_n|^\beta} \sup\big(|\partial_t u|, C_{\rho_{n+1}}\big)$$

$$\leq N_0^\beta \frac{h^\beta}{(\rho_{n+1} - \rho_n)^\beta} M(\rho_{n+1}) + N \frac{1}{(\rho_{n+1} - \rho_n)^\beta h^2} \sup\big(|u|, C_R\big).$$

By repeating further the arguments following (10.2.15), we arrive at (3) with C_R in place of $\partial' C_R$. After that it only remains to use the maximum principle and the fact that, owing to (1)

$$0 = \partial_t u + F(D^2 u) - F(0) = \partial_t u + a^{ij} D_{ij} u \tag{6}$$

in C_R (a.e.), where $a = (a^{ij})$ is an $\mathbb{S}_\delta$-valued measurable function. The lemma is proved. □

The next is a kind of interpolation theorem.

4. LEMMA. *Let $\beta \in (0,1]$, $\eta \in [1,\infty)$, and let Ω be an open bounded convex domain in $\mathbb{R}^d$ such that (11.2.1) holds. Take $T \le \eta\rho^2_{\mathrm{int}}(\Omega)$ and assume that in $\Sigma = [0,T]\times\Omega$ we are given a bounded measurable function u such that, for any $t \in [0,T]$, $u(t,\cdot)$ is twice continuously differentiable in Ω and*

$$[D^2u(t,\cdot)]_{C^\beta(\Omega)} \le N_0, \tag{7}$$

where N_0 is a fixed constant. Also suppose that, for any $x \in \Omega$, the function $u(t,x)$ is absolutely continuous with respect to t and its derivative $\partial_t u$ admits a modification still denoted by $\partial_t u$ such that, for almost any $t \in (0,T)$,

$$[\partial_t u(t,\cdot)]_{C^\beta(\Omega)} \le N_0. \tag{8}$$

Then D^2u is continuous in Σ and

$$[D^2u]_{C^\beta(\Sigma)} \le N(1+T)N_0,$$

where N depends only on d, β, and η.

Proof. By using parabolic scaling we reduce the general case to the one with $\rho_{\mathrm{int}}(\Omega) = 1$. Then, if $T < 1$, one can extend $u(t,x)$ beyond Σ by setting $u(t,x) = u(T,x)$ for $t > T$ preserving all required properties. On the other hand, if $1 \le T \le \eta$, one can split the interval $(t_1,t_2) \subset (0,T)$ into several subintervals and estimate the increment of $D^2u(s,x)$ over this interval through the sum of the increments over smaller intervals. This shows that without losing generality we may assume that $T = 1$.

In that case, as we know there exists a $\theta_0 \in [1,\infty)$, depending only on η, such that for any $x_0 \in \Omega$ and $\rho \in (0,1]$ there is an x_ρ on the ray tx, $t \in [0,1]$, satisfying (11.2.5), that is such that $|x_0 - x_\rho| \le \theta_0\rho$ and $B_\rho(x_\rho) \subset \Omega$.

Now we take $(t_1,t_2) \subset (0,1)$, $x_0 \in \Omega$, set

$$\rho = \sqrt{t_2 - t_1}, \quad z_i = (t_i, x_0), \quad \hat{z}_i = (t_i, x_\rho),$$

take a unit $l \in \mathbb{R}^d$, and write

$$\big|D_l^2u(z_1) - D_l^2u(z_2)\big| \le I + NN_0\rho^\beta, \tag{9}$$

where

$$I = \big|D_l^2u(\hat{z}_1) - D_l^2u(\hat{z}_2)\big|.$$

Observe that that for $h = \rho/2$ we have $x_\rho \pm hl \in \Omega$, and therefore

$$I \le \big|D_l^2u(\hat{z}_1) - \Delta_{h,l}u(\hat{z}_2)\big| + \big|D_l^2u(\hat{z}_2) - \Delta_{h,l}u(\hat{z}_2)\big|$$

$$\le NN_0h^\beta + |J|,$$

where,

$$J = \Delta_{h,l}u(t_1, x_\rho) - \Delta_{h,l}u(t_2, x_\rho)$$

$$= \int_{t_1}^{t_2} \Delta_{h,l}\partial_t u_K(s, x_\rho)\, ds$$

$$= \frac{1}{h^2}\int_{t_1}^{t_2} \big[\partial_t u(s, x_\rho + hl) - \partial_t u(s, x_\rho)\big]\, ds$$

$$+ \frac{1}{h^2}\int_{t_1}^{t_2} \big[\partial_t u(s, x_\rho - hl) - \partial_t u(s, x_\rho)\big]\, ds.$$

Since $|t_1 - t_2| = 4h^2$, it follows from (8) that $|J| \leq NN_0\rho^\beta$ and $I \leq NN_0\rho^\beta$ which along with (9) yields

$$\big|D_l^2 u(z_1) - D_l^2 u(z_2)\big| \leq NN_0\rho^\beta.$$

The arbitrariness of l shows that D^2u is Hölder continuous with respect to t. That it is also Hölder continuous with respect to x, is given in (7). The lemma is proved. □

Proof of Theorem 1. Observe that, owing to Lemma 3, $\partial_t u(t,\cdot) \in C^\alpha_{\mathrm{loc}}(B_R)$, and for almost any $t \in (0, R^2)$ we have $u(t,\cdot) \in W^2_{\infty,\mathrm{loc}}(B_R)$ and $u(t,\cdot)$ satisfies

$$F[u](t,\cdot) = -\partial_t u(t,\cdot) \tag{10}$$

in B_R (a.e.) with boundary condition $u(t,\cdot) = g(t,\cdot)$ on ∂B_R. Furthermore, $u(t,\cdot) \in C(\bar{B}_R)$. For each of those t and any $\rho \in (0, R)$, there exists by Theorem 11.1.6 a unique $v(t,\cdot) \in C^{2+\alpha}_{\mathrm{loc}}(B_\rho) \cap C(\bar{B}_\rho)$ such that

$$F[v](t,\cdot) = -\partial_t u(t,\cdot)$$

in B_ρ and $v(t,\cdot) = u(t,\cdot)$ on ∂B_ρ. By uniqueness, $v(t,\cdot) = u(t,\cdot)$, $u(t,\cdot) \in C^{2+\alpha}_{\mathrm{loc}}(B_R) \cap C(\bar{B}_R)$ which along with (11.1.4) implies that, for $\rho < R$ and almost all $t < \rho^2$,

$$\big[D^2u(t,\cdot)\big]_{C^\alpha(B_\rho)} \leq N \sup_{s\leq\rho^2} \big[\partial_t u(s,\cdot)\big]_{C^\alpha(B_{(\rho+R)/2})} + N(R-\rho)^{-2-\alpha}\sup_{C_R}|u|.$$

Actually, this estimate holds for all $t \leq \rho^2$ since $u(t,x)$ is continuous in t. In particular, $u(t,\cdot) \in C^{2+\alpha}_{\mathrm{loc}}(B_R)$ for all $t < R^2$.

Also by Lemma 3

$$\sup_{t\leq\rho^2} \big[\partial_t u(t,\cdot)\big]_{C^\alpha(B_{(\rho+R)/2})}(R-\rho)^{2+\alpha}$$

$$\leq 2^{2+\alpha}\sup_{\rho<R}\big[\partial_t u\big]_{C^\alpha(C_\rho)}(R-\rho)^{2+\alpha} \leq N\sup_{C_R}|u|, \tag{11}$$

so that

$$\sup_{t\leq\rho^2}\big[D^2u(t,\cdot)\big]_{C^\alpha(B_\rho)} \leq N(R-\rho)^{-2-\alpha}\sup_{C_R}|u|. \tag{12}$$

Estimates (11) and (12) yield (2) with C_R in place of $\partial' C_R$ by Lemma 4. The argument in the end of the proof of Lemma 3 shows that (2) holds as is. The theorem is proved. □

3. Solvability in $C^{2+\alpha}_{\text{loc}}$ for equations $\partial_t v + F(D^2 v) = 0$ and estimates of the integral oscillations of $D^2 u$

Let $F(\mathsf{u}'')$ and α_0 be the same as in Section 2. The following theorem is a particular case of the results in [**89**]. Again we give a different proof.

1. THEOREM. *For any $\alpha \in (0, \alpha_0)$ and $g \in C(\overline{\partial' C_2})$ there exists a unique $v \in C(\bar{C}_2) \cap C^{2+\alpha}_{\text{loc}}(C_2)$ satisfying*

$$\partial_t v + F(D^2 v) = 0 \quad in \quad C_2, \quad v = g \quad on \quad \partial' C_2. \tag{1}$$

Furthermore,

$$\left|D^2 v(z_1) - D^2 v(z_2)\right| \le N \rho^\alpha(z_1, z_2) \sup_{\partial' C_2} |g|$$

as long as $z_1, z_2 \in C_1$, where N depends only on δ, α, and d.

Proof. As usual uniqueness follows from the maximum principle. To prove the existence, first assume that $g \in W^{1,2}_{\infty,\text{loc}}(\mathbb{R}^{d+1})$. In that case introduce $P[u](x) = P\big(D^2 u(x)\big)$, where P is defined by formula (6.2.4), and for $K > 0$ consider the equation

$$\partial_t u + \max\big(F[u], P[u] - K\big) = 0$$

in C_2 with boundary condition $u = g$ on $\partial' C_2$. Recall that $F(0) = 0$, so that by Theorem 8.3.3, this problem has a unique solution $u_K \in C(\bar{C}_2) \cap W^{1,2}_{\infty,\text{loc}}(C_2)$ and by the maximum principle (cf. (2.6))

$$|u_K| \le \sup_{\partial' C_2} |g|. \tag{2}$$

Recall that $\bar{\alpha}$ is introduced in (10.3.2) and use Theorem 2.1 to conclude that, for any $\beta \in (0, \bar{\alpha})$, $u_K \in C^{2+\beta}_{\text{loc}}(C_2)$ and, for any $r \in (0, 2)$,

$$\big[D^2 u_K\big]_{C^\beta(C_r)} \le N(\delta, \beta, d) \frac{1}{(2-r)^{2+\beta}} \sup_{\partial' C_2} |g|. \tag{3}$$

Also by Theorem 9.8.1 the family $\{u_K, K > 0\}$ is equicontinuous in $\bar{C}_2$. By the Arzelà-Ascoli and interpolation theorems, (2), and (3), we conclude that the set $\{u_K : K \ge 0\}$ is precompact in $C(\bar{C}_2)$ and in $C^2(C_r)$ with any $r < 2$. Therefore, there is a sequence $K_n \to \infty$ such that u_{K_n}, Du_{K_n}, and $D^2 u_{K_n}$ converge uniformly in $\bar{C}_2$, $C(C_r)$, and $C(C_r)$ for any $r < 2$, respectively, to a function v, Dv, and $D^2 v$, respectively. Obviously, $v \in C(\bar{C}_2) \cap C^{2+\beta}_{\text{loc}}(C_2)$ and v satisfies (1). But then by Theorem 2.1 estimate (3) holds with v in place of u_K not only for any $\beta \in (0, \bar{\alpha})$ but also for any $\beta \in (0, \alpha_0)$. This proves the theorem if $g \in W^{1,2}_{\infty,\text{loc}}(\mathbb{R}^{d+1})$.

In the case of general $g \in C\big(\overline{\partial' C_2}\big)$ we approximate it in $C\big(\overline{\partial' C_2}\big)$-norm by $g_n \in W^{1,2}_{\infty,\text{loc}}(\mathbb{R}^{d+1})$. Then we obtain the functions v_n corresponding to g_n. By the maximum principle

$$\sup_{\bar{C}_2} |v_n - v_m| \le \sup_{\partial' C_2} |g_n - g_m|.$$

Hence, there is $v \in C(\bar{C}_2)$ such that $v_n \to v$ uniformly in $\bar{C}_2$. Of course, $v = g$ on $\partial' C_2$. Furthermore, estimate (3) with v_n and g_n in place of u_K and g yields a similar estimate for v, which shows that $v \in C^{2+\alpha}_{\text{loc}}(C_2)$ and proves the theorem. □

Below in this section we fix

$$\alpha \in (0, \alpha_0).$$

2. Lemma. *Let $r \in (0,\infty)$, $\nu \geq 2$ and let $\phi \in C\big(\overline{\partial' C_{\nu r}}\big)$. Then there exists a unique $v \in C(\bar{C}_{\nu r}) \cap C^{2+\alpha}_{\text{loc}}(C_{\nu r})$ such that*

$$\partial_t v + F(D^2 v) = 0 \quad in \quad C_{\nu r}, \quad v = \phi \quad on \quad \partial' C_{\nu r}.$$

Furthermore,

$$\fint_{C_r}\fint_{C_r} \big|D^2 v(z_1) - D^2 v(z_2)\big|\, dz_1 dz_2 \leq N(d,\alpha,\delta)\nu^{-2-\alpha} r^{-2} \sup_{\partial' C_{\nu r}} |\phi|.$$

Proof. Scalings show that it suffices to concentrate on $r = 2/\nu$. In that case the existence of solution follows from Theorem 1, which also implies that for $z_1, z_2 \in C_{2/\nu} \subset C_1$

$$\big|D^2 v(z_1) - D^2 v(z_2)\big| \leq N\nu^{-\alpha} \sup_{\partial' C_2} |\phi|.$$

It only remains to observe that

$$\fint_{C_{2/\nu}}\fint_{C_{2/\nu}} \big|D^2 v(z_1) - D^2 v(z_2)\big|\, dz_1 dz_1 \leq \sup_{z_1,z_2 \in C_{2/\nu}} \big|D^2 v(z_1) - D^2 v(z_2)\big|.$$

The lemma is proved. □

Take $\bar{\gamma} = \bar{\gamma}(d,\delta) \in (0,1)$ from (10.3.4) and take

$$\gamma \in (0, \bar{\gamma}].$$

3. Lemma. *Let $r \in (0,\infty)$ and $\nu \in [2,\infty)$. Then for any $u \in W^{1,2}_{d+1}(C_{\nu r})$ we have*

$$\Big(\fint_{C_r}\fint_{C_r} |D^2 u(z_1) - D^2 u(z_2)|^\gamma\, dz_1 dz_2\Big)^{1/\gamma}$$

$$\leq N\nu^{(d+2)/\gamma}\Big(\fint_{C_{\nu r}} \big|\partial_t u + F[u]\big|^{d+1}\, dz\Big)^{1/(d+1)}$$

$$+N\nu^{-\alpha}\Big(\fint_{C_{\nu r}} |D^2 u|^{d+1}\, dz\Big)^{1/(d+1)}, \tag{4}$$

where N depends only on d, α, and δ.

Proof. Define v to be a unique $C(\bar{C}_{\nu r}) \cap C^{2+\alpha}_{\text{loc}}(C_{\nu r})$-solution of the equation $\partial_t v + F[v] = 0$ in $C_{\nu r}$ with boundary condition $v = u$ on $\partial' C_{\nu r}$. Such a function exists by Lemma 2. Furthermore, $v(x) - b^i x^i - c$ satisfies the same equation for any constant b^i, c. Hence by Lemma 2 and Hölder's inequality

$$I_r := \Big(\fint_{C_r}\fint_{C_r} \big|D^2 v(z_1) - D^2 v(z_2)\big|^\gamma\, dz_1 dz_2\Big)^{1/\gamma}$$

$$\leq N\nu^{-2-\alpha} r^{-2} \sup_{z=(t,x)\in\partial' C_{\nu r}} \big|u(z) - (D_i u)_{C_{\nu r}} x^i - u_{C_{\nu r}}\big|.$$

By Poincaré's inequality (see, for instance, Corollary 5.3 in [**30**]) the last supremum is dominated by a constant times

$$\nu^2 r^2 \Big(\fint_{C_{\nu r}} |D^2 u|^{d+1}\, dz \Big)^{1/(d+1)}.$$

It follows that

$$I_r \le N\nu^{-\alpha} \Big(\fint_{C_{\nu r}} |D^2 u|^{d+1}\, dz \Big)^{1/(d+1)}. \tag{5}$$

Next, the function $w := u - v$ is of class $W^{1,2}_{d+1,\mathrm{loc}}(C_{\nu r}) \cap C(\bar{C}_{\nu r})$ and for an operator $\mathcal{L} \in \mathbb{L}_{\delta,0}$ we have

$$\partial_t u + F[u] = \partial_t u + F[u] - \big(\partial_t v + F[v]\big) = \partial_t w + \mathcal{L} w$$

in $C_{\nu r}$ (a.e.). Moreover, $w = 0$ on $\partial' C_{\nu r}$. Therefore, by a scaled version of Theorem 9.4.9, there exists $N = N(d,\delta) < \infty$ such that

$$\fint_{C_r} |D^2 w|^\gamma \, dz \le \nu^{d+2} \fint_{C_{\nu r}} |D^2 w|^\gamma \, dz$$

$$\le N\nu^{d+2} \Big(\fint_{C_{\nu r}} \big|\partial_t u + F[u]\big|^{d+1}\, dz \Big)^{\gamma/(d+1)},$$

$$\Big(\fint_{C_r} \fint_{C_r} \big|D^2 w(z_1) - D^2 w(z_2)\big|^\gamma \, dz_1 dz_2 \Big)^{1/\gamma}$$

$$\le N\nu^{d+2} \Big(\fint_{C_{\nu r}} \big|\partial_t u + F[u]\big|^{d+1}\, dz \Big)^{\gamma/(d+1)},$$

Upon combining this result with (5) we come to (4) and the lemma is proved. □

4. A priori estimates in $W^{1,2}_{p,\mathrm{loc}}$

Here we suppose that Assumptions 1.3 and 1.4 are satisfied. Thus, we assume that all assumptions stated before Theorem 1.7 are satisfied. Take $\alpha_0 = \alpha_0(\delta, d) \in (0,1)$ from Section 2 and take $\bar\gamma = \bar\gamma(d,\delta) \in (0,1)$ from (10.3.4). Also take

$$\alpha \in (0, \alpha_0), \quad \gamma \in (0, \bar\gamma].$$

First we note the following.

1. LEMMA. *For any $q \in [1,\infty)$ and $\mu > 0$ there is a $\theta = \theta(d, \delta, K_F, \mu, q) > 0$ such that, if Assumption 1.4 is satisfied with this θ, then for any $\mathsf{u}_0' \in \mathbb{R}$, $r \in (0, R_0]$ and $z_0 \in \Pi$ such that $C_r(z_0) \subset \Pi$*

$$\fint_{C_r(z_0)} \sup_{\substack{\mathsf{u}'' \in \mathbb{S}, \\ |\mathsf{u}''| > \tau_0}} \frac{\big|F(\mathsf{u}_0', \mathsf{u}'', z) - \bar F(\mathsf{u}'')\big|^q}{|\mathsf{u}''|^q}\, dz \le \mu^q,$$

where $\bar F = \bar F_{z_0, r, \mathsf{u}_0'}$.

The proof of this lemma is practically identical to that of Lemma 10.4.1.

2. Lemma. *Let $r \in (0,\infty)$ and $\nu \geq 2$ satisfy $\nu r \leq R_0$. Take*

$$\mu \in (0,\infty), \quad \beta \in (1,\infty),$$

and suppose that Assumption 1.4 is satisfied with $\theta = \theta(d,\delta,K_F,\mu,\beta d+\beta)$ (see Lemma 1). Take a function $u \in W^{1,2}_{d+1}(\Pi)$ and for $z_0 \in \Pi$ such that $C_{\nu r}(z_0) \subset \Pi$ (if such z_0's exist) denote

$$I_r(z_0) = \Big(\fint_{C_r(z_0)} \fint_{C_r(z_0)} \big|D^2u(z_1) - D^2u(z_2)\big|^\gamma \, dz_1 dz_2\Big)^{1/\gamma}.$$

Then

$$I_r(z_0) \leq N\nu^{(d+2)/\gamma}\Big(\fint_{C_{\nu r}(z_0)} \big|\partial_t u + F[u]\big|^{d+1}\,dz\Big)^{1/(d+1)} + N\tau_0\nu^{(d+2)/\gamma}$$

$$+N\Big[\big(\mu + \omega_{F,u,\Pi}(\nu r)\big)\nu^{(d+2)/\gamma} + \nu^{-\alpha}\Big]\Big(\fint_{C_{\nu r}(z_0)} |D^2u|^{\beta'(d+1)}\,dz\Big)^{1/(\beta'(d+1))}, \tag{1}$$

where $\beta' = \beta/(\beta-1)$ and N depends only on d, K_F, α, and δ.

The proof of this lemma follows very closely that of Lemma 10.4.2 and is left to the reader.

3. Lemma. *Set $\gamma = \bar\gamma$ and take $p > d+1$, $R \in (0,1]$, and $u \in W^{1,2}_p(C_{2R})$. Assume that $C_{2R} \subset \Pi$. Then there exist constants $\hat\theta, \theta \in (0,1]$, depending only on d, p, δ, and K_F, such that, if Assumptions 1.3 and 1.4 are satisfied with these $\hat\theta$ and θ, respectively, then there is a constant N, depending only on R_0, d, p, K_0, K_F, δ, and the function $\omega_{F,u,C_{2R}}$, such that*

$$\|D^2u\|_{L_p(C_R)} \leq N\big\|\partial_t u + H[u]\big\|_{L_p(C_{2R})} + N\|\bar G\|_{L_p(C_{2R})} + N\tau_0 R^{(d+2)/p} \tag{2}$$

$$+ NR^{(d+2)(1/p-1/\gamma)}\big\|\,|D^2u|^\gamma\big\|^{1/\gamma}_{L_1(C_{2R})} + N\|u\|_{L_p(C_{2R})}, \tag{3}$$

$$\|D^2u\|_{L_p(C_R)} \leq N\tau_0 R^{(d+2)/p} + NR^{(d+2)/p-2}\sup_{C_{2R}}|u| \tag{4}$$

$$+ N\big(\big\|\partial_t u + H[u]\big\|_{L_p(C_{2R})} + \|\bar G\|_{L_p(C_{2R})}\big). \tag{5}$$

Proof. We are going to adapt the proof of Lemma 10.4.3 to the parabolic case at hand. Set

$$\mathbb{R}_+ = (0,\infty).$$

For $\rho > 0$ and $z \in Q := \mathbb{R}_+ \times \mathbb{R}^d$ introduce

$$I_r(h,z) = \Big(\fint_{C_r(z)}\fint_{C_r(z)} \big|h(z_1) - h(z_2)\big|^\gamma\,dz_1dz_2\Big)^{1/\gamma},$$

$$h^{\#}_{Q,\gamma,\rho}(z) = \sup\big\{I_r(h,z_0) : z_0 \in Q, r \in (0,\rho], C_r(z_0) \ni z\big\},$$

$$\mathbb{M}h(z) = \sup_{\substack{r>0, z_0\in Q,\\ C_r(z_0)\ni z}} \fint_{C_r(z_0)} \big|h(\zeta)\big|\,d\zeta, \tag{6}$$

whenever these definitions make sense. Note that $h^{\#}_{Q,\gamma,\rho}$ is well defined in C_R for measurable h defined even only in $C_{R+2\rho}$.

Take $\varepsilon \in (0,1]$ to be specified later and take $R_1 < R_2 \leq 2R$ such that

$$R_2 - R_1 \leq \varepsilon R_0, \quad R_2 \leq 2R_1. \tag{7}$$

Then take $\nu \geq 2$ and set

$$r_0 = (R_2 - R_1)/(\nu + 1).$$

Next, take $z, z_0 \in Q$, and $r > 0$ such that

$$r \leq r_0, \quad z \in C_{R_1}, \quad z \in C_r(z_0),$$

and observe that, since $R_2 - \nu r_0 = R_1 + r_0$, we have $z_0 \in C_{R_2 - \nu r_0}$ and $C_{\nu r}(z_0) \subset C_{R_2}$. Also $\nu r \leq \nu r_0 \leq \varepsilon R_0$. Therefore, by Lemma 2

$$I_r(z_0) \leq N\nu^{(d+2)/\gamma}\mathbb{M}^{1/(d+1)}\big(\big|\partial_t u + F[u]\big|^{d+1} I_{C_{R_2}}\big)(z) + N\tau_0\nu^{(d+2)/\gamma}$$
$$+N\Big[\big(\mu + \omega_{F,u,C_{2R}}(\nu r_0)\big)\nu^{(d+2)/\gamma} + \nu^{-\alpha}\Big]\mathbb{M}^{1/(\beta'(d+1))}\big(|D^2u|^{\beta'(d+1)} I_{C_{R_2}}\big)(z)$$

with N depending only on d, K_F, and δ. It follows that in C_{R_1}

$$(D^2u)^{\#}_{Q,\gamma,r_0} \leq N\nu^{(d+2)/\gamma}\mathbb{M}^{1/(d+1)}\big(\big|\partial_t u + F[u]\big|^{d+1} I_{C_{R_2}}\big) + N\tau_0\nu^{(d+2)/\gamma}$$
$$+N\Big[\big(\mu + \omega_{F,u,C_{2R}}(\varepsilon R_0)\big)\nu^{(d+2)/\gamma} + \nu^{-\alpha}\Big]\mathbb{M}^{1/(\beta'(d+1))}\big(|D^2u|^{\beta'(d+1)} I_{C_{R_2}}\big).$$

By Theorem C.2.4, with

$$\kappa = r_0/R_1 \leq 1/3, \quad \chi_1 = (d+2)/\gamma, \quad \chi_2 = (d+2)(1/\gamma - 1/p)$$

and the Hardy-Littlewood maximal function theorem, by taking β so that $p > \beta'(d+1)$, we obtain

$$\|D^2u\|_{L_p(C_{R_1})} \leq N\nu^{(d+2)/\gamma}\big\|\partial_t u + F[u]\big\|_{L_p(C_{R_2})} + N\tau_0\nu^{(d+2)/\gamma}R_1^{(d+2)/p}$$
$$+ \Big[N\big(\mu + \omega_{F,u,C_{2R}}(\varepsilon R_0)\big)\nu^{(d+2)/\gamma} + N_0\nu^{-\alpha}\Big]\|D^2u\|_{L_p(C_{R_2})}$$
$$+ N\nu^{\chi_1}(R_2 - R_1)^{-\chi_1}R_1^{-\chi_2+\chi_1}\big\|\,|D^2u|^\gamma\big\|^{1/\gamma}_{L_1(C_{2R})}, \tag{8}$$

where the constants N, N_0 depend only on d, p, K_F, and δ.

Now we take and fix $\nu \geq 2$ so that

$$N_0\nu^{-\alpha} \leq 1/8.$$

Then (8) becomes

$$\|D^2u\|_{L_p(C_{R_1})} \leq N\big\|\partial_t u + F[u]\big\|_{L_p(C_{R_2})} + N\tau_0 R_1^{(d+2)/p}$$
$$+ \Big[N\big(\mu + \omega_{F,u,C_{2R}}(\varepsilon R_0)\big) + 1/8\Big]\|D^2u\|_{L_p(C_{R_2})}$$
$$+ N(R_2 - R_1)^{-\chi_1}R_1^{-\chi_2+\chi_1}\big\|\,|D^2u|^\gamma\big\|^{1/\gamma}_{L_1(C_{2R})}.$$

After that the proof of the present lemma goes the same way as the proof of Lemma 10.4.3 almost word for word. One need only replace balls with cylinders, set $\chi_3 = (d+2)/\gamma - 2$, and replace the reference to Theorem 9.4.6 with that to Theorem 9.4.9. The lemma is proved. □

5. Proof of Theorems 1.7 and 1.10

We need a version of Lemma 10.4.4 adapted to cylinders. Recall that Ω^ρ is introduces in the beginning of Section 10.1, recall that a bounded domain Ω in $\mathbb{R}^d$, satisfying the exterior ball condition, and $T \in (0,\infty)$ are fixed and

$$\Pi = [0,T) \times \Omega, \quad \Pi^\rho = [0, T-\rho^2) \times \Omega^\rho.$$

The following is true if Ω is just a bounded domain.

1. LEMMA. *Let f and h be nonnegative functions given in Π. Let $\rho \in (0,1]$ be such that $\Pi^{3\rho} \neq \emptyset$. Assume that for any $z_0 \in \Pi^{2\rho}$ we have*

$$\int_{C_\rho(z_0)} f(z)\,dz \le \int_{C_{2\rho}(z_0)} h(z)\,dz + N_1\rho^{d+2} + N_2\rho^{d+2-2p}, \tag{1}$$

where N_1 and N_2 are fixed constants ≥ 0. Then

$$\int_{\Pi^{3\rho}} f(z)\,dz \le 2^{d+1}\int_\Pi h(z)\,dz + NN_1 + NN_2\rho^{-2p}, \tag{2}$$

where the constants N depend only on d and $|\Pi|$.

Proof. Fix $t_0 \in [0, T-4\rho^2)$ and denote

$$F(x) = \int_{t_0}^{t_0+\rho^2} f(t,x)\,dt, \quad H(x) = \int_{t_0}^{t_0+4\rho^2} H(t,x)\,dt.$$

Then for any $x_0 \in \Omega^{2\rho}$

$$\int_{B_\rho(x_0)} F(x)\,dz \le \int_{B_{2\rho}(x_0)} H(x)\,dz + N_1\rho^{d+2} + N_2\rho^{d+2-2p}.$$

It follows by Lemma 10.4.4 that for any $t_0 \in [0, T-4\rho^2)$

$$\int_{t_0}^{t_0+\rho^2}\int_{\Omega^{3\rho}} f(t,x)\,dxdt \le 2^d \int_{t_0}^{t_0+4\rho^2}\int_\Omega h(t,x)\,dxdt + NN_1\rho^2 + NN_2\rho^{2-2p}. \tag{3}$$

We integrate this estimate with respect to $t_0 \in (0, T-4\rho^2)$ and get

$$\int_{\rho^2}^{T-4\rho^2}\int_{\Omega^{3\rho}} f(t,x)\,dxdt \le 2^d\int_0^T\int_\Omega h(t,x)\,dxdt + NN_1 + NN_2\rho^{-2p}.$$

By adding this and (3) with $t_0 = 0$ and using the fact that $\rho \le 1$ we obtain (2). The lemma is proved. □

Proof of Theorem 1.7. We take the constants $\hat\theta, \theta \in (0,1]$ from Lemma 4.3. By Lemma 4.3 for $\rho \in (0,1]$ and $z \in \Pi^{2\rho}$ we have

$$\|D^2u\|^p_{L_p(C_\rho(z))} \le N\tau_0^p\rho^{d+2} + N\rho^{d+2-2p}\sup_{C_{2\rho}(z)}|u|^p + N\|\bar G\|^p_{L_p(C_{2\rho}(z))},$$

and by Lemma 1, if $\Pi^{3\rho} \neq \emptyset$ and $\rho \in (0,1]$,

$$\int_{\Pi^{3\rho}} |D^2u|^p\,dz \le N\int_{\Pi} |\bar G|^p\,dz + N\tau_0^p + N\rho^{-2p}\sup_{\Pi}|u|^p. \tag{4}$$

On the other hand, if $\Pi^{3\rho}\ne\emptyset$ and $\rho>1$, then $\Pi^{3\rho}\subset\Pi^3$ but $\rho\le T^{1/2}\wedge \operatorname{diam}(\Omega)=:\hat\rho$ and (4) follows because $3^{-2p}\le 3^{-2p}\rho^{-2p}\bar\rho^{2p}$.

Using interpolation inequalities also allows us to estimate the $L_p(\Pi^{3\rho})$-norm of Du and $\partial_t u$ is estimated by using (1.7). The theorem is proved. □

Proof of Theorem 1.10. It is based on Theorem 8.3.3 and, as there, we take P from (6.2.4) and set $P[u](x)=P\big(D^2u(x)\big)$.

Step 1. First we assume that $g\in W^{1,2}_\infty(\mathbb{R}^{d+1})$ and there exist a constant N_0 and a *constant* $\bar H$ such that, for all t,x,u',

$$\big|H(\mathsf{u}',0,t,x)\big|\le N_0|\mathsf{u}'|+\bar H. \tag{5}$$

Then by Theorem 8.3.3 for any $K>0$ there exists a solution $v_K\in W^{1,2}_{\infty,\mathrm{loc}}(\Pi)\cap C(\bar\Pi)$ of the equation

$$\partial_t v_K+\max\big(H[v_K],P[v_K]-K\big)=0 \tag{6}$$

in Π (a.e.) and $v_K=g$ on $\partial'\Pi$. In addition, by Lemma 1.9 applied to (6)

$$\sup_{\Pi}|v_K|\le Ne^{K_0T}\big(\|\bar G\|_{L_{d+1}(\Pi)}+\sup_{\partial'\Pi}|g|\big), \tag{7}$$

where N depends only on d,δ, K_0, and $\operatorname{diam}(\Omega)$.

Next, with the goal of applying Theorem 1.7 set

$$H_K(\mathsf{u},t,x)=\max\big(H(\mathsf{u},x),P(\mathsf{u}'')-K\big),$$

$$F_K\big(\mathsf{u}_0',\mathsf{u}'',t,x\big)=\max\big(F(\mathsf{u}_0',\mathsf{u}'',t,x),P(\mathsf{u}'')-K\big),\quad G_K=H_K-F_K.$$

By closely following our argument below (10.5.4) we convince ourselves that the assumptions of Theorem 1.7 are satisfied for H_K, F_K with the same ingredients as the original ones and

$$\bar F_{K,z,r,\mathsf{u}_0'}(\mathsf{u}'')=\max\big(\bar F_{z,r,\mathsf{u}_0'}(\mathsf{u}''),P(\mathsf{u}'')-K\big)$$

apart from the fact that δ and K_F should be modifies appropriately but based only on the original d,δ, and K_F.

Now estimate (7) and Theorem 9.8.1 guarantee the uniform continuity of the family $\{v_K,K>0\}$ and, in particular, that the functions $\omega_{F,v_K,\Pi}$ are majorated by a continuous function independent of K and vanishing at zero.

As in the elliptic case, as an important observation tacitly used later, we point out that in the above argument guaranteeing the precompactness of $\{v_K:K>0\}$ in $C(\bar\Pi)$, in what concerns g, only the supremum of $|g|$ and the modulus of continuity of g on $\partial'\Pi$ were involved. This is important when we will need to approximate our general $g\in C(\overline{\partial'\Pi})$ with the ones of class $W^{1,2}_\infty(\mathbb{R}^{d+1})$.

Then by Theorem 1.7 we obtain that, provided that an appropriate choice of $\hat\theta$ and θ was made,

$$\sup_{K\ge1}\|v_K\|_{W^{1,2}_p(\Pi^\rho)}<\infty \tag{8}$$

for any $\rho>0$ as long as $\Pi^\rho\ne\emptyset$.

Next, we let $K \to \infty$. In light of the above mentioned precompactness and estimate (8), there is a sequence $K_n \to \infty$ as $n \to \infty$ and $v \in W^{1,2}_{p,\mathrm{loc}}(\Pi)$ such that $v_{K_n} \to v$ weakly in $W^{1,2}_p(\Pi^\rho)$ for any ρ and $v_{K_n} \to v$ uniformly in $\bar{\Pi}$. Then, of course, $v \in C(\bar{\Pi})$, $v = g$ on $\partial'\Pi$, and $v \in W^{1,2}_p(\Pi^\rho)$ for any $\rho > 0$. The said weak convergence implies pointwise convergence $Dv_n \to Dv$ in Π in light of the compactness of the embedding $W^{1,2}_p \subset C^{0,1}$ $(p > d+2)$.

Next, for $m = 1, 2, \ldots$ define

$$H^m(\mathsf{u}'', t, x) = \sup_{n \geq m} \max \big[H\big(v_{K_n}(t,x), Dv_{K_n}(t,x), \mathsf{u}'', t, x\big), P(\mathsf{u}'') - K_n\big].$$

Observe that $H^m(\mathsf{u}'', t, x)$ are Lipschitz continuous in u'' and at all points of differentiability satisfy $D_{\mathsf{u}''}H^m \in \mathbb{S}_{\bar{\delta}}$. Also

$$\big|H^m(0,t,x)\big| \leq K_0 \max_{n \geq m} \Big(\big|v_{K_n}(t,x)\big| + \big|Dv_{K_n}(t,x)\big|\Big) + \bar{G}(t,x),$$

which is in $L_{p,\mathrm{loc}}(\Pi)$. Therefore, the operators $H^m[u]$ fit into the scheme of Section 4.2. Furthermore, for $n \geq m$ obviously

$$\partial_t v_{K_n} + H^m(v_{K_n}, t, x) \geq 0$$

(a.e.) in Π. By Theorem 4.2.6 we conclude that for any m

$$\partial_t v + \sup_{n \geq m} \max \big[H(v_{K_n}, Dv_{K_n}, D^2 v, t, x), P(D^2 v) - K_n\big] \geq 0 \tag{9}$$

in Π (a.e.). We fix (t,x) at which (9) holds for all m (that is, we fix almost any (t,x)) and since $H(\mathsf{u}', \mathsf{u}'', t, x)$ is continuous in u', we have that

$$\big|H\big(v_{K_n}(t,x), Dv_{K_n}(t,x), D^2 v(t,x), t, x\big)$$

$$-H\big(v(t,x), Dv(t,x), D^2 v(t,x), t, x\big)\big| \to 0$$

as $n \to \infty$. Then, in light of (9),

$$\partial_t v(t,x) + \max \big[H\big(v(t,x), Dv(t,x), D^2 v(t,x), t, x\big), P\big(D^2 v(t,x)\big) - K_m\big] \geq o(1),$$

which for $m \to \infty$ yields

$$\partial_t v(t,x) + H\big(v(t,x), Dv(t,x), D^2 v(t,x), t, x\big) = \partial_t v(t,x) + H[v](t,x) \geq 0.$$

The inequality $\partial_t v + H[v] \leq 0$ is proved similarly starting from the function

$$\inf_{n \geq m} \max \big[H\big(v_{K_n}(t,x), Dv_{K_n}(t,x), \mathsf{u}'', t, x\big), P(\mathsf{u}'') - K_n\big].$$

This proves the theorem if condition (5) is satisfied and $g \in W^{1,2}_\infty(\mathbb{R}^{d+1})$.

Step 2. Assume that $g \in W^{1,2}_\infty(\mathbb{R}^{d+1})$ and abandon (5). Let $\eta(t) = t$ for $|t| \leq 1$ and $\eta(t) = \operatorname{sign} t$ for $|t| \geq 1$. For $n = 1, 2, \ldots$ define $\eta_n(t) = n\eta(t/n)$ and

$$\hat{H}^n(\mathsf{u}, t, x) = H(\mathsf{u}, t, x) - H(\mathsf{u}', 0, t, x) + \eta_n\big(H(\mathsf{u}', 0, t, x)\big),$$

$$\hat{G}^n(u, t, x) = \hat{H}^n(u, t, x) - F(\mathsf{u}'_0, \mathsf{u}'', t, x).$$

Then

$$|\hat{G}^n(u,t,x)| = \big|G(u,t,x) + \eta_n\big(H(\mathsf{u}',0,t,x)\big) - H(\mathsf{u}',0,t,x)\big|$$

$$\leq \hat{\theta}|\mathsf{u}''| + 2K_0|\mathsf{u}'| + 2\bar{G}(t,x),$$

so that Assumption 1.3 is satisfied for $\hat{H}^n$ with $2K_0$ and $2\bar{G}$ in place of K_0 and $\bar{G}$. Assumptions 1.4 and 1.8 are also valid for $\hat{H}^n$ with the same parameters.

Furthermore

$$\big|\hat{H}^n(\mathsf{u}',0,t,x)\big| = \big|\eta_n\big(H(\mathsf{u}',0,t,x)\big)\big|,$$

which is bounded.

Hence with the same $\hat{\theta}$ and θ as in Step 1, for any n, there exists $v^n \in W^{1,2}_{p,\mathrm{loc}}(\Pi) \cap C(\bar{\Pi})$ satisfying

$$\partial_t v^n + \hat{H}^n[v^n] = 0$$

in Π (a.e.) and such that $v^n = g$ on $\partial'\Pi$. Estimate (7), applicable to v^n by Lemma 1.9, and Theorem 1.7 and Theorem 9.8.1 again guarantee that the $W^{1,2}_p(\Pi^\rho)$-norms of v^n are bounded for any $\rho > 0$, and v^n are equicontinuous in $\bar{\Pi}$. This enables us to find a subsequence $v^{n'}$ and a function $v \in W^{1,2}_{p,\mathrm{loc}}(\Pi)$ such that $v^{n'} \to v$ weakly in $W^{1,2}_p(\Pi^\rho)$ for any $\rho > 0$ and $v^{n'} \to v$ uniformly in $\bar{\Pi}$. Then, of course, $v \in C(\bar{\Pi})$, $v = g$ on $\partial'\Pi$, and $v \in W^{1,2}_p(\Pi^\rho)$ for any $\rho > 0$.

After that we repeat the rest of Step 1 ignoring, of course $P(\mathsf{u}'') - K$ and taking

$$\sup_{n' \geq m} \Big[H(v^{n'}, Dv^{n'}, \mathsf{u}'', t, x) - H(v^{n'}, Dv^{n'}, 0, t, x) + \eta_{n'}\big(H(v^{n'}, Dv^{n'}, 0, t, x)\big)\Big]$$

in place of $H^m(\mathsf{u}'', t, x)$. One thing which makes the argument here easier is that for any $(t,x) \in \Pi$

$$-H(v^n, Dv^n, 0, t, x) + \eta_{n'}\big(H(v^n, Dv^n, 0, t, x)\big) = 0$$

if n is large enough.

In this way we finish Step 2, and, as in the case of the proof of Theorem 10.1.14, the passage from $g \in W^{1,2}_\infty(\mathbb{R}^{d+1})$ to $g \in C\big(\overline{\partial'\Pi}\big)$ is left to the reader as a routine exercise. The theorem is proved. □

CHAPTER 13

Elements of the $C^{2+\alpha}$-theory of fully nonlinear elliptic and parabolic equations

1. Introduction

We recall some notation from (9.6.14) and introduce some new ones. Define

$$\mathbb{R}^d_+ = \{x = (x^1, x') : x^1 > 0, x' \in \mathbb{R}^{d-1}\}$$

if $d \geq 2$, and set $\mathbb{R}^1_+ = \mathbb{R}_+ = (0, \infty)$. Also for $R \in (0, \infty)$ and $x \in \mathbb{R}^d_+$ define

$$B^+_R(x) = \mathbb{R}^d_+ \cap B_R(x), \quad B^+_R = B^+_R(0),$$

$$\partial_{x^1} B^+_R(x) = \bar{B}^+_R(x) \cap \{x : x^1 = 0\},$$

$$\mathbb{B}^+_R(x) = \bar{\mathbb{R}}^d_+ \cap B_R(x), \quad \mathbb{B}^+_R = \mathbb{B}^+_R(0). \tag{1}$$

Let $\alpha \in (0, 1)$. Observe that, since $\mathbb{B}^+_R$ contains part of the hyperplane $\{x^1 = 0\}$ and, by definition, $C^{2+\alpha}_{\text{loc}}(\mathbb{B}^+_R)$ is the set of functions belonging to $C^{2+\alpha}(\Omega)$ for any open set Ω such that $\bar{\Omega} \subset \mathbb{B}^+_R$, we have

$$C^{2+\alpha}_{\text{loc}}(\mathbb{B}^+_R) = \bigcap_{\rho<r} C^{2+\alpha}(B^+_\rho) = \bigcap_{\rho<r} C^{2+\alpha}(\bar{B}^+_\rho),$$

where $C^{2+\alpha}(\bar{B}^+_\rho)$ is the set of functions u continuous in $\bar{B}^+_\rho$ along with their derivatives Du, D^2u having finite norm $\|u\|_{C^{2+\alpha}(B^+_\rho)}$ and the second equality is due to the fact that any $u \in C^{2+\alpha}(B^+_\rho)$ admits a unique continuation to $\bar{B}^+_\rho$ to become an element of $C^{2+\alpha}(\bar{B}^+_\rho)$.

1. Assumption. We are given a function $F(\mathsf{u}'')$ on $\mathbb{S}$ which is Lipschitz continuous, *convex*, and such that at all points of its differentiability we have

$$D_{\mathsf{u}''} F(\mathsf{u}'') \in \mathbb{S}_\delta,$$

where $\delta \in (0, 1]$ is a fixed number.

In Chapter 14 the following result will play a crucial role similar to the role of Theorem 10.3.1 in Chapter 10. This result is found in [**103**] and can also be extracted from [**61**].

2. Theorem. *Under Assumption 1 there exists $\alpha_0 = \alpha_0(\delta, d) \in (0, 1)$ such that for any $g \in C(\partial B^+_2)$ vanishing on $\partial_{x^1} B^+_2$ there exists a unique u belonging to $C(\bar{B}^+_2) \cap C^{2+\alpha}_{\text{loc}}(\mathbb{B}^+_2)$ for any $\alpha \in (0, \alpha_0)$ and satisfying*

$$F(D^2u) = 0 \quad \textit{in} \quad B^+_2, \quad u = g \quad \textit{on} \quad \partial B^+_2. \tag{2}$$

Furthermore, for any $\alpha \in (0, \alpha_0)$,

$$\sup_{B_2^+} |u| \le \sup_{\partial B_2^+} |g| + N\big|F(0)\big|,$$

$$\big|D^2u(x) - D^2u(y)\big| \le N|x-y|^\alpha \sup_{B_2^+} |u|$$

if $x, y \in B_1^+$, *where the constants* N *depend only on* δ, α, *and* d.

It may look like a surprise that this innocently looking boundary counterpart of Theorem 10.3.1 requires for its proof not only a boundary counterpart of Theorem 10.2.1 but, actually, an almost fully blown solvability theory of fully nonlinear elliptic equations in the Banach spaces $\hat{C}^{2+\alpha}(B_R)$ and $\hat{C}^{2+\alpha}(B_R^+)$.

These spaces are introduced as follows. We define the space $\hat{C}^\alpha(B_R)$ as the set of bounded continuous functions f in B_R which have finite norm

$$\|f\|_{\hat{C}^\alpha(B_R)} := \|f\|_{C(B_R)} + [f]_{\hat{C}^\alpha(B_R)},$$

where

$$[f]_{\hat{C}^\alpha(B_R)} := \sup_{r<R} (R-r)^{2+\alpha}[f]_{C^\alpha(B_r)}.$$

Similarly, we define the space $\hat{C}^\alpha\big(\mathbb{B}_R^+\big)$ as the set of bounded continuous functions f in $\mathbb{B}_R^+$ which have finite norm

$$\|f\|_{\hat{C}^\alpha(\mathbb{B}_R^+)} := \|f\|_{C(B_R^+)} + [f]_{\hat{C}^\alpha(\mathbb{B}_R^+)},$$

where

$$[f]_{\hat{C}^\alpha(\mathbb{B}_R^+)} := \sup_{r<R} (R-r)^{2+\alpha}[f]_{C^\alpha(B_r^+)}.$$

We define the space $\hat{C}^{2+\alpha}(B_R)$ as the set of continuous functions u in $\bar{B}_R$ (notice not in B_R) which are twice continuously differentiable in B_R and have finite norm

$$\|u\|_{\hat{C}^{2+\alpha}(B_R)} := \|u\|_{C(B_R)} + [D^2u]_{\hat{C}^\alpha(B_R)}.$$

Similarly we define the space $\hat{C}^{2+\alpha}\big(\mathbb{B}_R^+\big)$ as the set of continuous functions u in $\bar{B}_R^+$ (notice not only in $\mathbb{B}_R^+$) which are twice continuously differentiable in $\mathbb{B}_R^+$ and have finite norm

$$\|u\|_{\hat{C}^{2+\alpha}(\mathbb{B}_R^+)} := \|u\|_{C(B_R^+)} + [D^2u]_{\hat{C}^\alpha(\mathbb{B}_R^+)}.$$

It is well known that the above introduced spaces are (nonseparable) Banach spaces.

The situation in bounded domains is even more complicated than alluded to above. Theorem 2 will enable us to obtain only a priori estimates in $W_p^2(\Omega)$, but to prove the solvability in $W_p^2(\Omega)$ we need a deeper result which follows.

3. ASSUMPTION. We are given a function $F(\mathsf{u}'', x)$ on $\mathbb{S} \times \mathbb{R}^d$, which is locally Lipschitz continuous in (u'', x), convex with respect to u'', and such that at all point of its differentiability

$$D_{\mathsf{u}''}F(\mathsf{u}'', x) \in \mathbb{S}_\delta, \quad \big|D_xF(\mathsf{u}'', x)\big| \le K_0 + K_1|\mathsf{u}''|,$$

where K_0 and K_1 are fixed constants.

4. THEOREM. *Suppose that Assumption 3 is satisfied. Then there exists* $\alpha_0 = \alpha_0(\delta, d) \in (0,1)$ *such that, for any* $R \in (0,\infty)$ *and* $g \in C(\partial B_R^+)$ *vanishing on* $\partial_{x^1} B_R^+$, *there exists a unique* u, *belonging to* $\hat{C}^{2+\alpha}(\mathbb{B}_R^+)$ *for any* $\alpha \in (0,\alpha_0)$, *such that*

$$F(D^2u, x) = 0 \tag{3}$$

in B_R^+ *and* $u = g$ *on* ∂B_R^+. *Furthermore,*

$$\sup_{B_R^+} |u| \leq \sup_{\partial B_R^+} |g| + \delta^{-1} R^2 \sup_{B_R^+} \big|F(0,x)\big|, \tag{4}$$

and for any $\rho \leq R$ *and* $\alpha \in (0,\alpha_0)$

$$[D^2u]_{\hat{C}^\alpha(\mathbb{B}_\rho^+)} \leq N \sup_{B_\rho^+} |u| + N\big(K_0\rho^3 + K_1\rho\big), \tag{5}$$

where N *depends only on* K_0, K_1, δ, α, *and* d.

An important version of this theorem is presented as Theorem 11.1 where we treat equations of type $F(D^2u) = f(x)$ with f in C^α.

That Theorem 2 follows from Theorem 4 is seen from the fact that

$$[D^2u]_{C^\alpha(B_1^+)} \leq [D^2u]_{\hat{C}^\alpha(\mathbb{B}_2^+)}.$$

5. REMARK. One word of caution is in order. We are going to see, in particular, that, for any $f \in \hat{C}^\alpha(\mathbb{B}_R^+)$ (that is, in particular, bounded) and $g \in C(\partial B_R^+)$ vanishing on $\partial_{x^1} B_R^+$ there exists a unique $u \in \hat{C}^{2+\alpha}(\mathbb{B}_R^+)$ such that $\Delta u = f$ in B_R and $u = g$ on ∂B_R. This result has an unusual feature because the operator Δ is *not* a bounded operator from $\hat{C}^{2+\alpha}(\mathbb{B}_R^+)$ to $\hat{C}^\alpha(\mathbb{B}_R^+)$. Indeed, for $d = 1$ the function $(1-x)^{1-\alpha} - 1$ is in $\hat{C}^{2+\alpha}([0,1))$, but its second derivative is unbounded.

Theorem 4 is proved in Section 10 after we develop necessary a priori estimates for equations like (3) with $f \in C^\alpha$ on the right-hand side in place of 0. These a priori estimates, actually, lead right away to the solvability of such equations. We still keep (3) as is because only in this form we need the result.

On the other hand, the above mentioned a priori estimates are also used in the proof of a parabolic counterpart of Theorem 2 which we need in Chapter 15. To state it we introduce some notation. Recall notation (9.6.15):

$$C_{\tau,r}^+ = [0,\tau) \times B_r^+, \quad \partial_{x^1} C_{\tau,r}^+ = [0,\tau) \times \partial_{x^1} B_r^+,$$

and set

$$\mathbb{C}_{\tau,r}^+ = [0,\tau) \times \mathbb{B}_r^+, \quad \mathbb{C}_r^+ = \mathbb{C}_{r^2,r}^+. \tag{6}$$

Also for $\alpha \in (0,1)$ introduce $C^{2+\alpha}(C_{\tau,r}^+)$ as the set of functions u on $C_{\tau,r}^+$, which are bounded and continuous in $C_{\tau,r}^+$ along with their derivatives $\partial_t u$, Du, D^2u and have finite norm

$$\|u\|_{C^{2+\alpha}(C_{\tau,r}^+)} = \|u, \partial_t u, Du, D^2u\|_{C(C_{\tau,r}^+)} + \big[\partial_t u, D^2u\big]_{C^\alpha(C_{\tau,r}^+)},$$

where

$$[f]_{C^\alpha(C_{\tau,r}^+)} = \sup_{z_1,z_2 \in C_{\tau,r}^+} \frac{\big|f(z_1) - f(z_2)\big|}{\rho^\alpha(z_1,z_2)}.$$

Each $u \in C^{2+\alpha}(C_{\tau,r}^+)$ admits an extension by continuity to $\bar{C}_{\tau,r}^+$ with the derivatives of orders ∂_t, D, D^2 of the extension coinciding with the extensions of $\partial_t u$, Du, D^2u.

Therefore speaking about the values of $u, \partial_t u$, Du, D^2u on the boundary of $C^+_{\tau,r}$ we will always mean the values of the corresponding extensions.

Also as in the case of $C^{2+\alpha}_{\text{loc}}(\mathbb{B}^+_r)$ we explicitly specify that

$$C^{2+\alpha}_{\text{loc}}(\mathbb{C}^+_r) = \bigcap_{\rho<r} C^{2+\alpha}(C^+_\rho).$$

Here is an analog of Theorem 2 for parabolic equations. We consider F independent of t and x.

6. THEOREM. *Under Assumptions* 1 *there exists* $\alpha_0 = \alpha_0(\delta, d) \in (0,1)$ *such that for any* $g \in C(\overline{\partial' C^+_2})$ *vanishing on* $\partial_{x^1} C^+_2$ *there exists a unique* u *belonging to* $C(\bar{C}^+_2) \cap C^{2+\alpha}_{\text{loc}}(\mathbb{C}^+_2)$ *for any* $\alpha \in (0, \alpha_0)$ *and satisfying*

$$\partial_t u(t,x) + F\big(D^2 u(t,x)\big) = 0 \quad in \quad C^+_2, \quad u = g \quad on \quad \partial' C^+_2. \tag{7}$$

Furthermore,

$$\sup_{C^+_2} |u| \le \sup_{\partial' C^+_2} |g| + 4\big|F(0)\big|, \tag{8}$$

and, for any $0 < \rho < 2$ *and* $\alpha \in (0, \alpha_0)$,

$$[\partial_t u]_{C^\alpha(C^+_\rho)} \le N(2-\rho)^{-2-\alpha} \sup_{C^+_R} |u|, \tag{9}$$

$$[D^2 u]_{C^\alpha(C^+_\rho)} \le N(2-\rho)^{-2-\alpha} \sup_{C^+_R} |u|, \tag{10}$$

where N *depends only on* d, δ, *and* α.

The proof of this result is given in Section 12.

7. REMARK. The results of this chapter can be derived in a straightforward manner from the corresponding results in [**45**], [**89**], [**103**], [**104**], or any other source treating general theory of fully nonlinear elliptic and parabolic equations in weighted Hölder spaces under minimal assumptions. Here, mainly following the ideas and techniques from [**103**] and [**104**], we present a short version (about forty pages long) of the theory adapted to our needs and consider only equations in half balls or half cylinders. In particular, it was quite sufficient for us not to introduce the usual weighted Hölder spaces but stick to our spaces like $\hat{C}^{2+\alpha}(\mathbb{B}^+_R)$. One of peculiar implication of that is noted in Remark 5.

2. Characterizations of the spaces $C^{s+\alpha}(B^+_R)$

In this section we give a characterization of $C^{s+\alpha}(B^+_R)$ in the spirit of [**103**] and [**104**] and Section 11.2. Fix

$$\alpha \in (0,1), \quad s \in \{1,2\}, \quad K \in [0,\infty),$$

$$\varepsilon \in [0,1), \quad R \in (0,\infty) \quad \nu_0, \nu_1 \in (1,\infty).$$

Recall that generic points in $\mathbb{R}^d_+$ are denoted by $x = (x^1, x')$, where $x^1 > 0$ and $x' \in \mathbb{R}^{d-1}$.

1. THEOREM. *Let $\mathcal{Q}$ be a linear subset of $\mathcal{P}_s$ and let u be a function on B_R^+ such that*

(*i*) *for any ball $B_{\nu_0 r}(x) \subset B_R^+$ there exists $p \in \mathcal{P}_s$ for which*

$$\sup_{B_r(x)} |u-p| \leq K r^{s+\alpha} + \varepsilon \nu_0^{-s-\alpha} \sup_{B_{\nu_0 r}(x)} |u-q|,$$

where q is any element of $\mathcal{P}_s$;

(*ii*) *for any half ball $B^+_{\nu_1 r}(0,x') \subset B_R^+$ there exists $p \in \mathcal{Q}$ such that the inequality*

$$\sup_{B_r^+(0,x')} |u-p| \leq K r^{s+\alpha} + \varepsilon \nu_1^{-s-\alpha} \sup_{B^+_{\nu_1 r}(0,x')} |u-q| \tag{1}$$

holds for any $q \in \mathcal{Q}$.

Then, for any $\rho \in (0,R)$, we have $u \in C^{s+\alpha}(B_\rho^+)$ and

$$[D^s u]_{C^\alpha(B_\rho^+)} \leq NK + N(R-\rho)^{-s-\alpha} \sup_{B_R^+} |u-q| \quad \forall q \in \mathcal{Q}, \tag{2}$$

where the constants N depend only on $d, \nu_0, \nu_1, \varepsilon$, and α.

Proof. As usual, we only deal with $q \equiv 0$ in (2). If $B_r(x) \subset B_R^+$ introduce

$$\omega_r(x) = r^{-s-\alpha} \inf_{p \in \mathcal{P}_s} \sup_{B_r(x)} |u-p|.$$

In case $B_r^+(0,x') \subset B_R^+$ introduce

$$\omega_r(x') = r^{-s-\alpha} \inf_{p \in \mathcal{Q}} \sup_{B_r^+(0,x')} |u-p|.$$

Observe that the estimate

$$\omega_r(x') \leq NK + N\varepsilon \big(R - |x'|\big)^{-s-\alpha} \sup_{B_R^+} |u| \quad \text{if} \quad |x'| + r < R \tag{3}$$

is obtained quite similarly to (11.2.16). Next, take $x \in B_R^+$ and r such that $B_r(x) \subset B_R^+$. Note that $|x| < R$ and there are four cases:

(a) there exists $n \in \{1, 2, ...\}$ such that $B_{\nu_0^n r}(x) \subset B_R^+$;

(b) $B_{\nu_0 r}(x) \not\subset B_R^+$ and $x^1 \geq \nu_0 r$;

(c) $B_{\nu_0 r}(x) \not\subset B_R^+$, $x^1 < \nu_0 r$, and $B^+_{2\nu_0 r}(0,x') \setminus B_R^+ \neq \emptyset$;

(d) $B_{\nu_0 r}(x) \not\subset B_R^+$, $x^1 < \nu_0 r$, and $B^+_{2\nu_0 r}(0,x') \subset B_R^+$.

In case (b) we have $B_{\nu_0 r}(x) \subset \mathbb{R}^d_+$ and $|x| + \nu_0 r > R$, so that $r \geq \big(R - |x|\big)/\nu_0$ and

$$\omega_r(x) \leq N\big(R - |x|\big)^{-s-\alpha} \sup_{B_R^+} |u|. \tag{4}$$

In case (c) we have $|x'| + 2\nu_0 r > R$, $|x| + 2\nu_0 r > R$, $r \geq \big(R - |x|\big)/(2\nu_0)$, and (4) holds again.

In case (d), owing to the fact that $B^+_{2\nu_0 r}(0,x') \supset B_r(x)$, in light of (3) we have

$$\begin{aligned} \omega_r(x) &\leq (2\nu_0)^{s+\alpha} \omega_{2\nu_0 r}(x') \\ &\leq NK + N\big(R - |x'|\big)^{-s-\alpha} \sup_{B_R^+} |u| \\ &\leq NK + N\big(R - |x|\big)^{-s-\alpha} \sup_{B_R^+} |u|. \end{aligned}$$

In case (a) by assumption

$$\omega_r(x) \le K + \varepsilon\omega_{\nu_0 r}(x),$$

and by iterations, as in the proof of Theorem 11.2.7,

$$\omega_r(x) \le NK + \varepsilon\omega_{\nu_0^n r}(x),$$

where n is such that $B_{\nu_0^n r}(x) \subset B_R^+$ but $B_{\nu_0^{n+1} r}(x) \not\subset B_R^+$. Then one of the three cases (b), (c), or (d) occurs where r is replaced with $\nu_0^n r$. It follows that

$$\omega_{\nu_0^n r}(x) \le NK + N\big(R - |x|\big)^{-s-\alpha} \sup_{B_R^+} |u|,$$

which leads to the conclusion that (4) holds in all cases (a)-(d), that is for any $x \in B_R^+$ and r such that $B_r(x) \subset B_R^+$.

Now, take $\rho \in (0, R)$ and take r, x such that $B_r(x) \subset B_\rho^+$. Then (4) holds and $R - |x| \ge R - \rho$, so that

$$\omega_r(x) \le NK + N(R - \rho)^{-s-\alpha} \sup_{B_R^+} |u|.$$

After that it only remains to refer to Theorem 11.2.4. The theorem is proved. □

3. Solvability of model linear equations in $\hat{C}^{2+\alpha}(\mathbb{B}_R)$

In this section $R \in (0, \infty)$, $\delta, \alpha \in (0, 1)$, and $a = (a^{ij})$ is a constant matrix from $\mathbb{S}_\delta$. Set

$$\mathcal{L}u = a^{ij} D_{ij} u.$$

1. LEMMA. *Let f and g be polynomials in $\bar{B}_R$. Then there exists a unique polynomial u in $\bar{B}_R$ such that $\mathcal{L}u = f$ in B_R and $u = g$ on ∂B_R.*

Proof. By considering $u - g$ we reduce the problem to the one with $g \equiv 0$. Let $f \in \mathcal{P}_n$, where recall that $\mathcal{P}_n$ is the set of polynomials of degree (at most) n. Observe that the operator $T : v \to Tu = \mathcal{L}\big[\big(R^2 - |x|^2\big)v\big]$ maps $\mathcal{P}_n$ into $\mathcal{P}_n$. Furthermore, by the maximum principle, if $v \in \mathcal{P}_n$ and $Tv = 0$, then $\big(R^2 - |x|^2\big)v \equiv 0$, meaning that T is a one-to-one mapping. Since $\mathcal{P}_n$ is a finite-dimensional linear space, the equation $Tv = f$ has a unique solution in $\mathcal{P}_n$, which proves the lemma in light of the fact that $R^2 - |x|^2$ is a polynomial. □

2. REMARK. If $f \equiv 0$ and g is a polynomial of degree n, then u is also a polynomial of degree n.

Indeed, if $n = 0, 1$, then $u = g$, and, if $n \ge 2$, then $\mathcal{L}(u - g) = h$, where h is a polynomial of degree $n - 2$ and it follows from the above proof that $u - g = \big(R^2 - |x|^2\big)v$, where v is a polynomial of degree $n - 2$.

The functions u such that $\mathcal{L}u = 0$ are called $\mathcal{L}$-harmonic.

Any d-tuple $\beta = (\beta_1, ..., \beta_d)$ consisting of $\beta_i \in \{0, 1, ...\}$ is called a multi-index. For any multi-index β we set

$$|\beta| = \sum_{i=1}^d \beta_i, \quad D^\beta = D_1^{\beta_1} \cdot ... \cdot D_d^{\beta_d}.$$

3. LEMMA. *Let f, g, u be as in Lemma 1 and let $f \equiv 0$, that is, let u be $\mathcal{L}$-harmonic in B_R. Then there exists a constant $N = N(d, \delta)$ such that in B_R for any multi-index $\beta \neq 0$*

$$\left|D^\beta u(x)\right| \leq \left(\frac{N|\beta|}{R - |x|}\right)^{|\beta|} \sup_{\partial B_R} |g - p|, \tag{1}$$

where p is an arbitrary polynomial of degree $|\beta|-1$. Furthermore, the same assertion holds for $|\beta| \geq 3$ if f is constant.

Proof. The second assertion follows from the first one since $u(x) - N(R^2 - |x|^2)f$ is $\mathcal{L}$-harmonic if the constant N is chosen appropriately.

To prove the first assertion, find v as a polynomial of degree $|\beta| - 1$ such that $v = p$ on ∂B_R and $\mathcal{L}v = 0$ in B_R. Then the left-hand side of (2) is rewritten as $\left|D^\beta(u - v)(x)\right|$. This shows that without loss of generality we may take $p \equiv 0$, that is, we need only prove that

$$\left|D^\beta u(x)\right| \leq \left(\frac{N|\beta|}{R - |x|}\right)^{|\beta|} \sup_{\partial B_R} |g|. \tag{2}$$

We employ Bernstein's method which uses only the maximum principle. First we note that scalings show that it suffices to concentrate on $R = 1$. In that case take any $\zeta \in C_0^\infty(\mathbb{R}^d)$ with support in B_1, assume that $\zeta(0) = 1$, and consider the function

$$w := \zeta^2 |Du|^2 + \lambda |u|^2,$$

where $\lambda > 0$ is a constant to be chosen later. We have $\mathcal{L}u = 0$, $\mathcal{L}D_i u = 0$, and

$$\begin{aligned}
\mathcal{L}w &= |Du|^2 \mathcal{L}(\zeta^2) + \zeta^2 \left[2D_i u \mathcal{L} D_i u + 2a^{ij} D_{ik} u D_{jk} u\right] \\
&\quad + 8\zeta D_i \zeta D_k u \, a^{ij} D_{jk} u + 2\lambda a^{ij} D_i u D_j u + 2\lambda u \mathcal{L} u \\
&= \left[2\lambda a^{ij} D_i u D_j u + |Du|^2 L(\zeta^2)\right] + 2\zeta^2 a^{ij} D_{ik} u D_{jk} u \\
&\quad + 8a^{ij} [D_i \zeta D_k u] [\zeta D_{jk} u] \\
&\geq |Du|^2 \left[2\lambda\delta + \mathcal{L}(\zeta^2) - 8a^{ij} D_i \zeta D_j \zeta\right]
\end{aligned}$$

(we have used $2a^2 + 8ab \geq -8b^2$ in the matrix form). We see how to take $\lambda = \lambda(d, \delta)$ so that $\mathcal{L}w \geq 0$. Fix such a λ. Then by the maximum principle

$$\left|Du(0)\right|^2 \leq \max_{\bar{B}_1} w \leq \max_{\partial B_1} w = \lambda \max_{\partial B_1} |g|^2.$$

This yields (2) for $|\beta| = 1$, $R = 1$, and $x = 0$. Then (2) holds for $|\beta| = 1$ and any $R \in (0, \infty)$ if $x = 0$. Moving the origin and observing that $\mathcal{L}u = 0$ in $B_{R-|x_0|}(x_0)$ for any $x_0 \in B_R$ we get that (2) holds for $|\beta| = 1$ and any $x \in B_R$.

For higher values of $|\beta|$, one obtains (2) by splitting $B_R \setminus B_{|x|}$ into $|\beta|$ rings of width $(R - |x|)/|\beta|$ and estimating the derivatives $D_{j_1} \cdot \ldots \cdot D_{j_k} u$ inside the k-th ring by using the above result obtained for $|\beta| = 1$ and the fact that $\mathcal{L} D_{j_1} \cdot \ldots \cdot D_{j_{k-1}} u = 0$ in B_R. The lemma is proved. □

4. COROLLARY. *For any $g \in C(\bar{B}_R)$ and constant f there exists a unique $u \in C^2_{\text{loc}}(B_R) \cap C(\bar{B}_R)$, that satisfies $\mathcal{L}u = f$ in B_R and is equal to g on ∂B_R. Furthermore, any such function is in $C^\infty_{\text{loc}}(B_R)$ and the assertions of Lemma 3 hold true.*

Indeed, the case of general f is reduced to the one when $f = 0$ as in the proof of the lemma. Uniqueness is a consequence of the maximum principle, which also assures that, if the polynomials g_n converge uniformly to g on ∂B_R and u_n are the corresponding polynomials from Lemma 3 with $f \equiv 0$, then

$$\sup_{\bar{B}_R} |u_n - u_m| \leq \sup_{\partial B_R} |g_n - g_m| \to 0$$

as $n, m \to \infty$. Hence, u_n converge in $\bar{B}_R$ uniformly to, say u, which is continuous and equal g on ∂B_R. Applying estimates (1) to $u_n - u_m$, we see that also all derivatives of u_n converge uniformly on any compact subset of B_R. Of course, this implies that $u \in C^\infty_{\text{loc}}(B_R)$ and u is the desired solution. After that estimates (1) for u and the last assertion of Lemma 3 are immediate. □

5. COROLLARY. *If $u \in C^2_{\text{loc}}(B_R)$ is $\mathcal{L}$-harmonic in B_R, then it is real analytic.*

This follows from (1) (applied in a smaller ball) and the fact that, for any constant N and $|x| \leq (eN)^{-1}$,

$$\frac{N^k |x|^k k^k}{k!} \to 0$$

as $k \to \infty$.

6. LEMMA. *Let $u \in C^2_{\text{loc}}(B_R) \cap C(\bar{B}_R)$ be such that $f := \mathcal{L}u$ is constant in B_R. Then for any $\varepsilon \in (0,1)$ there exists a constant $\nu \in (1, \infty)$, depending only on d, α, ε, and δ, and there exists $p \in \mathcal{P}_2$ for which*

$$\sup_{B_{R/\nu}} |u - p| \leq \varepsilon \nu^{-(2+\alpha)} \|u - q\|_{C(\partial B_R)} \tag{3}$$

for any $q \in \mathcal{P}_2$.

Proof. Introduce $T(x)$ as the Taylor polynomial of order 2 of u at the origin and observe that by Lemma 3 we have

$$\|u - T\|_{C(B_{R/\nu})} \leq N(R/\nu)^3 \|D^3 u\|_{C(B_{R/\nu})}$$

$$\leq N(R/\nu)^3 \frac{1}{(R - R/\nu)^3} \sup_{\partial B_R} |u - q| = N \frac{1}{(\nu - 1)^3} \sup_{\partial B_R} |u - q|,$$

where q is any quadratic polynomial. Now our assertion follows. The lemma is proved. □

Next, comes a few more a priori estimates.

7. THEOREM. *Let $u \in C^2_{\text{loc}}(B_R) \cap C(\bar{B}_R)$ and $f \in C^\alpha(B_R)$ be such that*

$$\mathcal{L}u = f$$

in B_R. Then

(*i*) *we have*

$$\|u\|_{C(B_R)} \leq \|u\|_{C(\partial B_R)} + R^2 \delta^{-1} \|f\|_{C(B_R)}; \tag{4}$$

(*ii*) *for any* $\varepsilon \in (0,1)$ *there exist constants* $\nu, N \in (1,\infty)$, *depending only on* d, α, ε, *and* δ, *such that, if* $B_{\nu r}(x) \subset B_R$, *then there exists* $p \in \mathcal{P}_2$ *for which*

$$\sup_{B_r(x)} |u-p| \leq N[f]_{C^\alpha(B_{\nu r}(x))} r^{2+\alpha} + \varepsilon \nu^{-(2+\alpha)} \sup_{\partial B_{\nu r}(x)} |u-q| \tag{5}$$

for any $q \in \mathcal{P}_2$;

(*iii*) *for any* $\rho \in (0,R)$, *we have* $u \in C^{2+\alpha}(B_\rho)$ *and there is a constant* $N = N(d,\delta,\alpha)$ *such that for any* $q \in \mathcal{P}_2$

$$[D^2 u]_{C^\alpha(B_\rho)} \leq N[f]_{C^\alpha(B_{(R+\rho)/2})} + N(R-\rho)^{-2-\alpha} \sup_{B_R} |u-q|. \tag{6}$$

Proof. Obviously, $w := u - \big(R^2 - |x|^2\big)\delta^{-1}\|f\|_{C(B_R)}$ satisfies $\mathcal{L}w \geq 0$ and by the maximum principle

$$u - \big(R^2 - |x|^2\big)\delta^{-1}\|f\|_{C(B_R)} \leq \|g\|_{C(\partial B_R)}.$$

By considering $u + \big(R^2 - |x|^2\big)\delta^{-1}\|f\|_{C(B_R)}$, we get a complimentary estimate and conclude that (4) holds indeed.

To prove (ii) take ε and ν from Lemma 6 and take $B_{\nu r}(x_0) \subset B_R$. Then, relying on Corollary 4 find a function $h \in C^2_{\text{loc}}\big(B_{\nu r}(x_0)\big) \cap C\big(\bar{B}_{\nu r}(x_0)\big)$ such that $\mathcal{L}h = f(x_0)$ in $B_{\nu r}(x_0)$ and $h = u$ on $\partial B_{\nu r}(x_0)$. By Lemma 6 there is $p \in \mathcal{P}_2$ such that

$$\sup_{B_r(x_0)} |u-p| \leq \sup_{B_r(x_0)} |u-h| + \varepsilon \nu^{-(2+\alpha)} \sup_{\partial B_{\nu r}(x_0)} |u-q|$$

for any $q \in \mathcal{P}_2$. Since $\mathcal{L}(u-h) = f - f(x_0)$ in $B_{\nu r}(x_0)$ and $u - h = 0$ on $\partial B_{\nu r}(x_0)$, by assertion (i) we get

$$\sup_{B_r(x_0)} |u-h| \leq (\nu r)^2 \delta^{-1} \big\|f - f(x_0)\big\|_{C(B_{\nu r}(x_0))} \leq N r^{2+\alpha}[f]_{C^\alpha(B_{\nu r}(x_0))}.$$

This and the previous estimate of $u - p$ prove assertion (ii).

To prove (iii), take $\varepsilon = 1/2$, find an appropriate ν from assertion (ii), take $\rho \in (0,R)$ and observe that, if $B_{\nu r}(x) \subset B_{(R+\rho)/2}$, then by (5), for an appropriate $p \in \mathcal{P}_2$ and any $q \in \mathcal{P}_2$,

$$\sup_{B_r(x)} |u-p| \leq N[f]_{C^\alpha((R+\rho)/2)} r^{2+\alpha} + \varepsilon \nu^{-(2+\alpha)} \sup_{\partial B_{\nu r}(x)} |u-q|$$

We now get our assertion from Theorem 11.2.7 with $(R-\rho)/2$ in place of ρ and $\Omega = B_{(R+\rho)/2}$. The theorem is proved. □

8. REMARK. If f is constant in $B_{\nu r}(x)$, one can take p in (5) to be the Taylor polynomial of order 2 of u at point x.

9. REMARK. The estimate in assertion (iii) of Theorem 7 is stronger than estimate (11.1.4) in the respect that α in Theorem 7 is an arbitrary number in $(0,1)$ and in (11.1.4) it is restricted to $(0,\alpha_0)$.

Theorem 7 has a nonempty field of applications which is seen if, for instance, f, g, u are polynomials as in Lemma 1. With some extra effort involving the approximations by using Bernstein's polynomials one could extend Theorem 7 to the case of general $f \in C^\alpha(B_R)$. But, actually, Theorem 11.1.6 implies that for any sufficiently small $\beta \in (0,1)$, for any $f \in C^\beta(B_R)$ and $g \in C(\partial B_R)$ there exists a unique $u \in C^2_{\text{loc}}(B_R) \cap C(\bar{B}_R)$ satisfying $\mathcal{L}u = f$ in B_R and equal to g on ∂B_R. This allows us to prove the main result of this section.

10. THEOREM. *Let $f \in \hat{C}^\alpha(B_R)$ and $g \in C(\partial B_R)$. Then there exists a unique $u \in \hat{C}^{2+\alpha}(B_R)$ such that $\mathcal{L}u = f$ in B_R and $u = g$ on ∂B_R. Furthermore, there exists a constant N, depending only on d, α, and δ, such that*

$$[D^2u]_{\hat{C}^\alpha(B_R)} \le N\big([f]_{\hat{C}^\alpha(B_R)} + \|u - p\|_{C(B_R)}\big), \tag{7}$$

where p is any quadratic polynomial.

Proof. Uniqueness is a direct consequences of Theorem 7 (i). To prove the existence part of the theorem we first suppose that $f(x) = 0$ for $|x| \ge \rho_0$, where $\rho_0 \in (0, R)$ is a fixed number. Then $f \in C^\alpha(B_R)$ and by what is said before the theorem we have a unique solution u of class $C^2_{\text{loc}}(B_R) \cap C(\bar{B}_R)$. By Theorem 7 we have $u \in C^{2+\alpha}_{\text{loc}}(B_R)$. We multiply both parts of (6) by $(R-\rho)^{2+\alpha}$ and take supremums with respect to $\rho \in (0, R)$. We also note that

$$\frac{R-\rho}{R-(R+\rho)/2} = 2$$

and then we obviously arrive at (7).

In the case of general $f \in \hat{C}^\alpha(B_R)$ we take a sequence of functions $\zeta_n(t)$ such that $\zeta(t) = 1$ for $t \in [0, R-2/n]$, $\zeta(t)$ is linear on $[R-2/n, R-1/n]$, and $\zeta(t) = 0$ for $t \ge R - 1/n$. Then set $f_n(x) = f(x)\zeta_n\big(|x|\big)$. Let u_n correspond to f_n and g. For $m \ge n$ we have $u_n - u_m = 0$ on ∂B_R and

$$\big|\mathcal{L}(u_n - u_m)\big| \le 2\sup|f| I_{B_R \setminus B_{R-2/n}}$$

in B_R. By the Aleksandrov-Bakel'man estimate $|u_n - u_m| \le N\xi(n)$, where $\xi(n) \to 0$ as $n \to \infty$. Of course, using this powerful tool is an overkill. One can easily construct functions $\phi_n = \phi_n\big(|x|\big)$ such that $L\phi_n \le -I_{B_R \setminus B_{R-2/n}}$ in B_R, $\phi = 0$ on ∂B_R and $\sup \phi_n \to 0$ as $n \to \infty$. Then by the maximum principle

$$|u_n - u_m| \le 2\sup|f| \sup \phi_n \to 0$$

uniformly in $\bar{B}_R$. By Cauchy's criterion, there is a $u \in C(\bar{B}_R)$ to which u_n converge uniformly. Of course, $u = g$ on ∂B_R.

Also, observe that, for $r \in (0, R)$,

$$\begin{aligned}
[f_n]_{C^\alpha(B_r)} &\le [f]_{C^\alpha(B_r)} + \sup_{B_R}|f| [\zeta_n]_{C^\alpha(B_r)} \\
&\le [f]_{C^\alpha(B_r)} + NnI_{r > R-2/n} r^{1-\alpha} \sup_{B_R}|f| \\
&\le [f]_{C^\alpha(B_r)} + Nn^{-1-\alpha}(R-r)^{-2-\alpha} r^{1-\alpha} \sup_{B_R}|f|.
\end{aligned}$$

This provides a uniform estimate of the $\hat{C}^{2+\alpha}(B_R)$-norms of u_n and shows that $u \in \hat{C}^{2+\alpha}(B_R)$ and (7) holds. Finally, passing to the limit in the equations $\mathcal{L}u_n = f_n$ presents no difficulty and the theorem is proved. □

4. Solvability of the Laplace equation in $\hat{C}^{2+\alpha}(\mathbb{B}_R^+)$

In this section, as in Section 3, $R \in (0, \infty)$ and $\alpha \in (0, 1)$. Recall that $\mathbb{B}_R^+$ contains $B_R \cap \{x^1 = 0\}$ and recall that $u \in C^{2+\alpha}_{\text{loc}}(\mathbb{B}_R^+)$ means that $u \in C^{2+\alpha}(\bar{B}_r^+)$ for any $r < R$.

Here is a result, that will serve as a starting point in proving Theorem 1.4.

Let $\mathcal{P}_2^0$ be the subset of $\mathcal{P}_2$ consisting of quadratic polynomials vanishing for $x^1 = 0$.

1. THEOREM. *Let $f \in \hat{C}^\alpha(B_R^+)$ and $g \in C(\partial B_R^+)$ and $g = 0$ on $\partial_{x^1} B_R^+$. Then there exists a unique $u \in \hat{C}^{2+\alpha}(\mathbb{B}_R^+)$ such that $\Delta u = f$ in B_R^+ and $u = g$ on ∂B_R^+. Furthermore,*

$$\|u\|_{C(B_R^+)} \leq \|g\|_{C(\partial B_R^+)} + R^2 \|f\|_{C(B_R^+)}, \tag{1}$$

$$[D^2 u]_{\hat{C}^\alpha(\mathbb{B}_R^+)} \leq N\big([f]_{\hat{C}^\alpha(\mathbb{B}_R^+)} + \|u - p\|_{C(B_R^+)}\big), \tag{2}$$

where N depends only on d and α, and $p \in \mathcal{P}_2^0$ is arbitrary.

To prove the theorem we need some preparations.

2. LEMMA. *Let f be constant and $g \in C(\partial B_R^+)$ be such that $g = 0$ on $\partial_{x^1} B_R^+$. Then there exists a unique*

$$u \in C^2_{\text{loc}}(B_R^+) \cap C(\bar{B}_R^+) \tag{3}$$

such that $\Delta u = f$ in B_R^+ and $u = g$ on ∂B_R^+. Furthermore,

$$u \in C^\infty_{\text{loc}}(\mathbb{B}_R^+) := \bigcap_{\rho < R} C^\infty(\bar{B}_\rho^+).$$

Proof. Uniqueness follows from the maximum principle. To prove the rest, replace u with $u - (1/2)(x^1)^2 f$ to see that without loss of generality we may assume that $f = 0$. In that case extend g to the rest of ∂B_R so that it becomes and odd function with respect to x^1. The extended function, for which we keep the same notation, will be continuous, since $g(x) = 0$ for $x^1 = 0$. Then on the basis of Corollary 3.4 find an infinitely differentiable in B_R harmonic function u in B_R with boundary values g. By antisymmetry of g and uniqueness $u(x^1, x') = -u(-x^1, x')$ and hence u vanishes for $x^1 = 0$. Now u, restricted to B_R^+, is the solution with described properties. The lemma is proved. □

By using the reflection as in the proof of Lemma 2 and applying the estimates in Corollary 3.4, we come to the following (never used in the future).

3. COROLLARY. *For any $g \in C(\partial B_R^+)$, which equals zero on $\partial_{x^1} B_R^+$, there exists a unique $u \in C^2_{\text{loc}}(\mathbb{B}_R^+) \cap C(\bar{B}_R^+)$, which is harmonic in B_R^+ and equal g on ∂B_R^+. Furthermore, there exists a constant $N = N(d)$ such that in B_R^+ for any multi-index $\beta \neq 0$*

$$|D^\beta u(x)| \leq \Big(\frac{N|\beta|}{R - |x|}\Big)^{|\beta|} \sup_{\partial B_R^+} |g - p|, \tag{4}$$

where p is an arbitrary polynomial of degree $|\beta| - 1$ which is odd with respect to x^1.

We draw the reader's attention to the fact that the structure of the right-hand sides of (3.1) and (5) below are different.

4. LEMMA. *Let f, g, u be as in Lemma 2. Then, for any multi-index $\beta \neq 0$, in B_R^+ we have*

$$|D^\beta u(x)| \leq \Big(\frac{N|\beta|}{R-|x|}\Big)^{|\beta|} \|u-q\|_{C(B_R^+)}, \tag{5}$$

for any $q \in \mathcal{P}_2^0$, where $N = N(d)$.

Proof. By replacing u with $u-q$, which will only add a constant to f, we see that it suffices to concentrate on $q \equiv 0$.

Next, we follow some arguments in the proof of Lemma 3.3, but first, by using Lemma 2, we find a function h which is harmonic in B_R^+ and is equal to u on its boundary. Then take $\rho \in (0, R)$ and find a function $\zeta \in C_0^\infty(B_R)$ such that $\zeta = 1$ on B_ρ. By Lemma 5.1.4 one can find such a ζ with the additional properties that

$$\zeta = 1 \quad \text{on} \quad B_\rho, \quad \zeta \leq 1, \quad |D\zeta| \leq N/(R-\rho), \quad |D^2\zeta| \leq N/(R-\rho)^2,$$

where $N = N(d)$.

Then, take $i \in \{2, 3, ..., d\}$ and consider the function

$$w := \zeta^2|D_i u|^2 + \lambda|u|^2 + 2\lambda M(u-h)\,\text{sign}\, f,$$

where $\lambda > 0$ is a constant to be chosen later and

$$M = \sup_{B_R^+} |u|.$$

We have $\Delta u = f$, $\Delta D_i u = 0$,

$$M\Delta(u-h)\,\text{sign}\, f = \sup_{B_R^+} |u||f| \geq -uf.$$

It follows that (no summation with respect to i)

$$\begin{aligned}\Delta w =& 2\zeta^2 D_{ik}uD_{ik}u + 8\big[\zeta D_{ik}u\big]\big[D_k\zeta D_i u\big] + |D_i u|^2\Delta\zeta^2 \\ &+ 2\lambda|Du|^2 + 2\lambda uf + 2\lambda M\Delta(u-h)\,\text{sign}\, f \\ \geq& 2\lambda|Du|^2 + |D_i u|^2\big(\Delta\zeta^2 - 8|D\zeta|^2\big) \\ \geq& |Du|^2\big(2\lambda - N_1(R-\rho)^{-2}\big).\end{aligned}$$

Hence, $\Delta w \geq 0$ if $2\lambda = N_1(R-\rho)^{-2}$. In that case, by the maximum principle, the maximum of w is attained on ∂B_R^+, where w coincides with $\lambda|u|^2$. Furthermore, by the maximum principle $|h| \leq |M|$. Thus, for $|x| \leq \rho$

$$|D_i u(x)|^2 \leq w(x) + 4\lambda M^2 \leq 5\lambda M^2 =: N_2^2(R-\rho)^{-2}M^2. \tag{6}$$

In light of the arbitrariness of ρ and x we obtain (5) when $D^\beta = D_i$.

Now we apply Lemma 3.3 to $D_i u$ upon observing that since it is a harmonic function in B_R^+ which equals zero for $x^1 = 0$, as in the proof of Lemma 2, its odd reflection about $x^1 = 0$ is harmonic in B_R and by Lemma 3.3 for $|\beta| \geq 1$

$$|D^\beta D_i u(x)| \leq \Big(\frac{N_3|\beta|}{\rho-|x|}\Big)^{|\beta|} \|D_i u\|_{C(\partial B_\rho^+)}$$

whenever $|x| < \rho \leq R$. We are going to estimate the last factor by using (6).

Fix $x \in B_R^+$ and set

$$\rho = \big(|\beta|R + |x|\big)/(|\beta| + 1).$$

Since $|x| < R$, $|x| < \rho$, and as is easy to see

$$\frac{\rho - |x|}{|\beta|} = \frac{R - |x|}{|\beta| + 1} = R - \rho,$$

$$\Big(\frac{N_3|\beta|}{\rho - |x|}\Big)^{|\beta|} N_2(R-\rho)^{-1} = N_2 N_3^{|\beta|}\Big(\frac{|\beta|+1}{R-|x|}\Big)^{|\beta|+1}.$$

This proves (5) with $N = \max(N_2, N_3)$ if D^β contains at least one derivative with respect to x'.

In the remaining case $D^\beta = D_1^{|\beta|}$ and if $|\beta| \geq 3$ it suffices to observe that

$$D_1^{|\beta|}u = D_1^{|\beta|-2}D_1^2 u = -D_1^{|\beta|-2}\sum_{i\geq 2} D_i^2 u.$$

To go to lower values of $|\beta|$ (which we will never use in the future) we use interpolation (cf. Remark 11.2.3). Observe that, if $\rho \in (0, R)$ and $x \in B_\rho^+$ and $0 < 2h < R - \rho$, then

$$\big|D_1^2 u(x)\big| \leq \big|D_1^2 u(x) - \Delta_{h,e_1} u(x + he_1)\big| + \big|\Delta_{h,e_1} u(x + he_1)\big|$$

$$\leq 2h \sup_{B_{\rho+2h}^+} \big|D_1^3 u\big| + 4h^{-2}M \leq hN(R-\rho-2h)^{-3}M + 4h^{-2}M. \tag{7}$$

For $h = (R-\rho)/4$ this yields

$$\big|D_1^2 u(x)\big| \leq N(R-\rho)^{-2}M,$$

and again the arbitrariness of ρ and x leads to (5) in this case. Finally, $D_1 u$ is estimated through $D_1^2 u$ and M in a similar fashion. The lemma is proved. □

5. COROLLARY. *For any $\varepsilon \in (0,1)$ there exists a constant $\nu \in (1,\infty)$, depending only on d, ε, and α, and there exists $p \in \mathcal{P}_2^0$ such that*

$$\sup_{B_{R/\nu}^+} |u - p| \leq \varepsilon\nu^{-2-\alpha} \sup_{B_R^+} |u - q| \tag{8}$$

for any $q \in \mathcal{P}_2^0$.

The proof goes the same way as that of Lemma 3.6.

As in the case of Corollary 3.5 we have

6. COROLLARY. *If u is as in (3), is harmonic in B_R^+, and vanishes on $\partial_{x^1} B_R^+$, then it is real analytic in $\mathbb{B}_R^+$.*

Next, comes a few more a priori estimates modeled after Theorem 3.7.

7. THEOREM. *Let u be as in (3) and let u vanish on $\partial_{x^1} B_R^+$. Let $f \in C^\alpha(B_R^+)$ be such that*

$$\Delta u = f$$

in B_R^+. Then

(i) estimate (1) holds true;

(ii) for any $\varepsilon \in (0,1)$ there exist constants $\nu, N \in (1,\infty)$, depending only on d, α, and ε, such that, if $B_{\nu r}^+(0,x') \subset B_R^+$, then there exists $p \in \mathcal{P}_2^0$ for which

$$\sup_{B_r^+(0,x')} |u-p| \le N[f]_{C^\alpha(B_{\nu r}^+(0,x'))} r^{2+\alpha} + \varepsilon \nu^{-2-\alpha} \sup_{B_{\nu r}^+(0,x')} |u-q| \tag{9}$$

for any $q \in \mathcal{P}_2^0$;

(iii) for any $\rho \in (0,R)$, $u \in C^{2+\alpha}(\mathbb{B}_\rho^+)$ and there is a constant $N = N(d,\delta,\alpha)$ such that for any $q \in \mathcal{P}_2^0$

$$[D^2 u]_{C^\alpha(B_\rho^+)} \le N[f]_{C^\alpha(B_{(R+\rho)/2}^+)} + N(R-\rho)^{-2-\alpha} \sup_{B_R^+} |u-q|. \tag{10}$$

Proof. Estimate (1) is proved in the same way as estimate (3.4) by using the same barrier function (or the one with $\delta = 1$).

To prove (ii) take ε and ν from Corollary 5 and take $B_{\nu r}^+(0,x_0') \subset B_R^+$. Then, relying on Lemma 2 find a function $h \in C_{\text{loc}}^2(\mathbb{B}_{\nu r}^+(0,x_0')) \cap C(\bar{B}_{\nu r}^+(0,x_0'))$ such that $\mathcal{L}h = f(0,x_0')$ in $B_{\nu r}^+(0,x_0')$ and $h = u$ on $\partial B_{\nu r}^+(0,x_0')$. By Corollary 5 there is $p \in \mathcal{P}_2^0$ such that

$$\sup_{B_r(0,x_0')} |u-p| \le \sup_{B_r(0,x_0')} |u-h| + \varepsilon\nu^{-(2+\alpha)} \sup_{B_{\nu r}(0,x_0')} |h-q|$$

$$\le 2 \sup_{B_r(0,x_0')} |u-h| + \varepsilon\nu^{-(2+\alpha)} \sup_{B_{\nu r}(0,x_0')} |u-q|$$

for any $q \in \mathcal{P}_2^0$. Since $\Delta(u-h) = f - f(0,x_0')$ in $B_{\nu r}(0,x_0')$ and $u-h=0$ on $\partial B_{\nu r}(0,x_0')$, by assertion (i) we get

$$\sup_{B_r(x_0)} |u-h| \le (\nu r)^2 \delta^{-1} \|f - f(0,x_0')\|_{C(B_{\nu r}(x_0))} \le N r^{2+\alpha} [f]_{C^\alpha(B_{\nu r}(x_0))}.$$

This and the previous estimate of $u-p$ prove assertion (ii).

To prove (iii), take $\varepsilon = 1/2$, find the corresponding $\nu > 1$ from Theorem 3.7 (ii), take $\rho \in (0,R)$, and observe that if $B_{\nu r}(x) \subset B_{(R+\rho)/2}^+$, then by (3.5), for an appropriate $p \in \mathcal{P}_2$ and any $q \in \mathcal{P}_2$,

$$\sup_{B_r(x)} |u-p| \le N[f]_{C^\alpha(B_{(R+\rho)/2}^+)} r^{2+\alpha} + \varepsilon\nu^{-2-\alpha} \sup_{\partial B_{\nu r}(x)} |u-q|.$$

Upon combining this with assertion (ii) we arrive at assertion (iii) by Theorem 2.1 with $(R-\rho)/2$ in place of ρ. The theorem is proved. □

8. REMARK. Assertion (ii) of Theorem 7 is *local* in the sense that if $\Delta u = f$ only in $B_{\nu r}^+(0,x')$, then (9) holds for an appropriate $p \in \mathcal{P}_2^0$ and any $q \in \mathcal{P}_2^0$.

We now give two proofs of Theorem 1, the first one based on a very general Theorem 11.1.4 about the solvability in $C^{2+\alpha}_{\text{loc}}$ of fully nonlinear equations, and the second one based on more elementary facts.

First proof of Theorem 1. Estimate (1), implying uniqueness, is proved in Theorem 7. Thus, we only need to prove the existence of solution $u \in \hat{C}^{2+\alpha}(\mathbb{B}_R^+)$ and estimate (2).

By Theorem 11.1.4 the equation $\Delta u = f$ in B_R^+ with boundary condition $u = g$ on ∂B_R^+ has a solution $u \in C^2_{\text{loc}}(B_R^+) \cap C(\bar{B}_R^+)$. By Theorem 7 for any $\rho \in (0, R)$, $u \in C^{2+\alpha}(\mathbb{B}_\rho^+)$ and (10) holds. This estimate is similar to (3.6) and, as in the proof of Theorem 3.10, we arrive at (2) upon multiplying it by $(R-\rho)^{2+\alpha}$, taking supremums with respect to $\rho \in (0, R)$, and observing that

$$\frac{R-\rho}{R-(R+\rho)/2} = 2.$$

□

Second proof of Theorem 1. As in the first proof we only need to prove the existence of solution $u \in \hat{C}^{2+\alpha}(B_R^+)$ and estimate (2) which as usual reduces to the case that $p \equiv 0$.

First assume that $f = 0$ if $x^1 = 0$. Then we extend $f(x)$ for $x^1 \le 0$ so that it becomes an odd function of x^1. Keep the same notation f for the extended function. We do the same with function g. Obviously, the $\hat{C}^\alpha(B_R)$-norm of f coincides with its $\hat{C}^\alpha(B_R^+)$-norm, and g is a continuous function on ∂B_R.

By Theorem 3.10 there exists a unique solution $u \in \hat{C}^{2+\alpha}(B_R)$ of the equation $\Delta u = f$ in B_R with boundary data $u = g$. Since the equation and the data are odd functions of x^1, by uniqueness, so is u, which implies that $u(x) = 0$ if $x^1 = 0$. Theorem 3.10 also provides an estimate of the $\hat{C}^\alpha(B_R)$-norm of D^2u and thus of its $\hat{C}^\alpha(B_R^+)$-norm. This proves the theorem if $f = 0$ for $x^1 = 0$.

In the general case, introduce $\xi_n(x) = (nx^1) \wedge 1$, $x^1 \ge 0$, $n = 1, 2, \dots$, and let $f_n = f\xi_n$. Denote by u_n the unique $\hat{C}^{2+\alpha}(B_R^+)$-solutions of $\Delta u_n = f_n$ in B_R^+ with boundary data g.

It is not hard to construct smooth nonnegative functions $\phi_n(x^1)$ on $[0, R]$ such that $\phi_n(0) = \phi_n(R) = 0$, $\phi_n'' \le -I_{[0,1/n]}$ and $\phi_n \to 0$ as $n \to \infty$ uniformly on $[0, R]$. Then by the maximum principle for $m \ge n$ we have

$$|u_n - u_m| \le \sup|f| \sup \phi_n \to 0$$

uniformly on $\bar{B}_R^+$ as $n \to \infty$. According to Cauchy, there is $u \in C(\bar{B}_R^+)$, to which u_n converge uniformly.

Estimate (3.5) shows that, if $B_{\nu r}(x) \subset B_R^+$ and $x^1 > \nu r$, so that the ball $\bar{B}_{\nu r}(x)$ is in the interior of B_R^+, then for all large n (such that $f_n = f$ on $\bar{B}_{\nu r}(x)$) there exist $p_n \in \mathcal{P}_2$ such that, for any $q \in \mathcal{P}_2$,

$$\sup_{B_r(x)} |u_n - p_n| \le N[f]_{C^\alpha(B_{\nu r}(x))} r^{2+\alpha} + \varepsilon \nu^{-2-\alpha} \|u_n - q\|_{C(B_{\nu r}(x))}. \tag{11}$$

It follows by Theorem 11.2.7 with $\Omega = B_\gamma \cap \{x^1 > \theta\}$ that for any $\gamma \in (0, R)$ and $\theta \in (0, \gamma)$

$$\varlimsup_{n\to\infty} [D^2 u_n]_{C^\alpha(B_\gamma \cap \{x^1 > \theta\})} < \infty.$$

This yields (cf. the proof of Lemma 11.2.1) that $D^2u_n \to D^2u$ uniformly on compact subsets of B_R^+. Then, of course, $u \in C^2_{\text{loc}}(B_R^+)$, $\Delta u = f$ in B_R^+, and $u = g$ on ∂B_R^+.

To finish the proof of the theorem now it suffices, as in the first proof, to apply Theorem 7 (iii) and draw the same conclusions as there. The theorem is proved. □

5. A priori estimates in $\hat{C}^{2+\alpha}(B_R)$ and $\hat{C}^{2+\alpha}(\mathbb{B}_R^+)$ for linear equations with variable coefficients

Let $\alpha, \delta \in (0,1)$, $R \in (0,\infty)$. Let $a = a(x)$ be an $\mathbb{S}_\delta$-valued function on B_R with

$$M_{\alpha,R}(a) := \sup_{\rho<R}(R-\rho)^\alpha [a]_{C^\alpha(B_\rho)} < \infty. \tag{1}$$

Introduce

$$\mathcal{L}u(x) := a^{ij}(x)D_{ij}u(x).$$

Here is an a priori estimate in $\hat{C}^{2+\alpha}(B_R)$, basically, showing that, if an inverse to $\mathcal{L}$ exists, then it is bounded.

1. Theorem. *Let u be a function of class $\hat{C}^{2+\alpha}(B_R)$. Set $\mathcal{L}u = f$ and assume that $f \in \hat{C}^\alpha(B_R)$. Then*

$$\|u\|_{C(B_R)} \le \|u\|_{C(\partial B_R)} + R^2\delta^{-1}\|f\|_{C(B_R)}, \tag{2}$$

and there exists a constant N, depending only on d, $R^{-2}M_{\alpha,R}(a)$, α, and δ, such that

$$[D^2u]_{\hat{C}^\alpha(B_R)} \le N\big([f]_{\hat{C}^\alpha(B_R)} + \sup_{B_R}|u|\big). \tag{3}$$

Proof. Estimate (2) is obtained in the same way as (3.4). While proving (3) it is convenient to assume that $R = 2$. The general case reduces to this one by using scalings. Then define

$$\rho_1 = 1, \quad \rho_{n+1} = \rho_n + 2^{-n}, \quad n = 1, 2, \dots.$$

Finally, take a $\gamma \in (0, 1/2)$ to be specified later and also introduce

$$V(x_1,x_2) = \frac{|D^2u(x_1) - D^2u(x_2)|}{|x_1 - x_2|^\alpha}, \qquad \Delta\rho_n = \rho_{n+1} - \rho_n.$$

For $n \ge 1$ take $x_1, x_2 \in B_{\rho_n}$, and consider two cases:

(a) $|x_1 - x_2| \ge \gamma\Delta\rho_n$;

(b) $|x_1 - x_2| < \gamma\Delta\rho_n$.

In case (a) we have $V \le (\gamma\Delta\rho_n)^{-\alpha}\big(|D^2u(x_1)| + |D^2u(x_2)|\big)$. We proceed with estimating the right-hand side by using Remark 11.2.3 about interpolation inequalities with $\Omega = B_{\rho_n}$ and $\varepsilon = \gamma^2\Delta\rho_n$. We have

$$\begin{aligned} V &\le N\varepsilon^\alpha(\gamma\Delta\rho_n)^{-\alpha}[D^2u]_{C^\alpha(B_{\rho_n})} + N\varepsilon^{-2}(\gamma\Delta\rho_n)^{-\alpha}\sup_{B_2}|u| \\ &\le N\gamma^\alpha[D^2u]_{C^\alpha(B_{\rho_{n+1}})} + N\gamma^{-2}(\gamma\Delta\rho_n)^{-2-\alpha}\sup_{B_2}|u|. \end{aligned} \tag{4}$$

This holds in case (a) even without using that $\mathcal{L}u = f$.

In case (b) observe that $B_{2\gamma\Delta\rho_n}(x_1) \subset B_2$, write

$$a^{ij}(x_1)D_{ij}u(x) = \big[a^{ij}(x_1) - a^{ij}(x)\big]D_{ij}u(x) + f =: \hat{f},$$

and use (3.6) with $B_\rho(x_1)$ and $B_{2\rho}(x_1)$ in place of B_ρ and B_R, where

$$\rho = \gamma \Delta \rho_n.$$

Then we obtain

$$V \leq N[\hat{f}]_{C^\alpha(B_{2\rho}(x_1))} + N\rho^{-2-\alpha} \sup_{B_2} |u|.$$

Also observe that simple arithmetics show that

$$[\hat{f}]_{C^\alpha(B_{2\rho}(x_1))} \leq [f]_{C^\alpha(B_{2\rho}(x_1))} + N\rho^\alpha [a]_{C^\alpha(B_{2\rho}(x_1))}[D^2u]_{C^\alpha(B_{2\rho}(x_1))}$$
$$+N[a]_{C^\alpha(B_{2\rho}(x_1))}\|D^2u\|_{C(B_{2\rho}(x_1))}.$$

Since $B_{2\rho}(x_1) \subset B_{\rho_{n+1}}$,

$$[\hat{f}]_{C^\alpha(B_{2\rho}(x_1))} \leq [f]_{C^\alpha(B_{\rho_{n+1}})} + N\rho^\alpha (2-\rho_{n+1})^{-\alpha}[D^2u]_{C^\alpha(B_{\rho_{n+1}})}$$
$$+N(2-\rho_{n+1})^{-\alpha}\|D^2u\|_{C(B_{\rho_{n+1}})}.$$

Here $\rho(2-\rho_{n+1})^{-1} = \gamma$ and $2-\rho_{n+1} = \Delta\rho_n$, so that similarly to the derivation of (4) we get

$$[\hat{f}]_{C^\alpha(B_{2\rho}(x_1))} \leq [f]_{C^\alpha(B_{\rho_{n+1}})} + N\gamma^\alpha [D^2u]_{C^\alpha(B_{\rho_{n+1}})}$$
$$+ N\gamma^\alpha(\gamma\Delta\rho_n)^{-2-\alpha}\|u\|_{C(B_2)}.$$

As a result we conclude that in both cases (a) and (b)

$$V \leq N\gamma^\alpha[D^2u]_{C^\alpha(B_{\rho_{n+1}})} + N[f]_{C^\alpha(B_{\rho_{n+1}})} + N\gamma^{-2}(\gamma\Delta\rho_n)^{-2-\alpha}\sup_{B_2}|u|.$$

In light of the arbitrariness of $x_1, x_2 \in B_{\rho_n}$,

$$[D^2u]_{C^\alpha(B_{\rho_n})} \leq N\gamma^\alpha[D^2u]_{C^\alpha(B_{\rho_{n+1}})}$$
$$+ N[f]_{C^\alpha(B_{\rho_{n+1}})} + N\gamma^{-2}(\gamma\Delta\rho_n)^{-2-\alpha}\sup_{B_2}|u|. \tag{5}$$

We multiply both parts of (5) by

$$(2-\rho_n)^{2+\alpha} = 2^{2+\alpha}(\Delta\rho_n)^{2+\alpha} = 2^{2+\alpha}(2-\rho_{n+1})^{2+\alpha}$$

and take supremums with respect to n. This yields

$$\sup_n [D^2u]_{C^\alpha(B_{\rho_n})}(2-\rho_n)^{2+\alpha} \leq N_1\gamma^\alpha \sup_n [D^2u]_{C^\alpha(B_{\rho_n})}(2-\rho_n)^{2+\alpha}$$
$$+ N\big([f]_{\hat{C}^\alpha(B_2)} + \gamma^{-4-\alpha}\sup_{B_2}|u|\big).$$

After choosing γ so that $N_1\gamma^\alpha \leq 1/2$, we conclude

$$\sup_n [D^2u]_{C^\alpha(B_{\rho_n})}(2-\rho_n)^{2+\alpha} \leq N\big([f]_{\hat{C}^\alpha(B_2)} + \sup_{B_2}|u|\big).$$

Now (3) with $R = 2$ is derived immediately by a simple comparison argument. The theorem is proved. □

In order to obtain a similar result for B_R^+ we need the following.

2. LEMMA. *Let u be a real-valued function of class $C^{2+\alpha}_{\text{loc}}(\mathbb{B}^+_R)\cap C(\bar{B}^+_R)$ such that $u(x)=0$ if $x^1=0$. Let $\varepsilon\in(0,1)$, $a_0\in\mathbb{S}_\delta$ (a_0 is a constant matrix), and set*

$$\mathcal{L}_0u=a_0^{ij}D_{ij}u.$$

Then there are constants $\nu\in(1,\infty)$, and $N\in(0,\infty)$, depending only on δ,d,ε, and α, such that, if $B^+_{\nu r}(0,x')\subset B^+_R$, then one can find $p\in\mathcal{P}^0_2$ for which

$$\sup_{B^+_r(0,x')}|u-p|\le N[\mathcal{L}_0u]_{C^\alpha(B^+_{\nu r}(0,x'))}r^{2+\alpha}+\varepsilon\nu^{-(2+\alpha)}\sup_{B^+_{\nu r}(0,x')}|u-q|,\tag{6}$$

where $q\in\mathcal{P}^0_2$ is arbitrary. Furthermore, for any $\rho\in(0,R)$ and any $q\in\mathcal{P}^0_2$

$$[D^2u]_{C^\alpha(B^+_\rho)}\le N[\mathcal{L}_0u]_{C^\alpha(B^+_{(R+\rho)/2})}+N(R-\rho)^{-2-\alpha}\sup_{B^+_R}|u-q|.\tag{7}$$

Proof. As always we take $q\equiv0$. Then first assume that a_0 is diagonal with diagonal elements $\lambda_1,...,\lambda_d$. Define the mapping $x\to Tx$ by $(Tx)^i=\lambda_i^{1/2}x^i$, $i=1,...,d$ (with no summation with respect to i) and set $v(x)=u(Tx)$, $f:=\mathcal{L}_0u$, $h(x)=f(Tx)$. These functions v and h are defined in $T^{-1}\mathbb{B}^+_R$ and satisfy there

$$\Delta v=h.$$

Then take $\tau\in(0,1)$ to be specified later and apply Theorem 4.7 (ii) with $\varepsilon\tau$ in place of ε and Remark 4.8, to see that for any $\varepsilon\in(0,1)$ there exist constants ν_0, N as in the statement of the lemma and also depending on τ such that, if $B^+_{\nu_0r}(T^{-1}(0,x'))\subset T^{-1}B^+_R$, then, for an appropriate $p\in\mathcal{P}^0_2$,

$$\sup_{B^+_r\left(T^{-1}(0,x')\right)}|v-p|\le N[h]_{C^\alpha\left(B^+_{\nu_0r}\left(T^{-1}(0,x')\right)\right)}r^{2+\alpha}$$
$$+\varepsilon\tau\nu_0^{-(2+\alpha)}\sup_{B^+_{\nu_0r}\left(T^{-1}(0,x')\right)}|v|.\tag{8}$$

Furthermore, it is not hard to see that

$$B^+_{\delta^{1/2}r}(T^{-1}(0,x'))\subset T^{-1}B^+_r(0,x')\subset B^+_{\delta^{-1/2}r}(T^{-1}(0,x')).$$

Therefore (8) implies that, if $B^+_{\nu_1r}(0,x')\subset B^+_R$, where $\nu_1=\nu_0\delta^{-1/2}$, then

$$\sup_{T^{-1}B^+_{\delta^{1/2}r}(0,x')}|v-p|\le N[h]_{C^\alpha(T^{-1}B^+_{\nu_1r}(0,x'))}r^{2+\alpha}$$
$$+\varepsilon\tau\nu_0^{-(2+\alpha)}\sup_{T^{-1}B^+_{\nu_1r}(0,x')}|v|.$$

For $r=\rho\delta^{-1/2}$ and $\nu=\nu_1\delta^{-1/2}=\nu_0\delta^{-1}$ this yields, that if $B^+_{\nu\rho}(0,x')\subset B^+_R$

$$\sup_{B^+_\rho(0,x')}|u-p(T\cdot)|\le N[f]_{C^\alpha(B^+_{\nu\rho}(0,x'))}\rho^{2+\alpha}$$
$$+\varepsilon\tau\nu_0^{-(2+\alpha)}\sup_{B^+_{\nu\rho}(0,x')}|u|.$$

After that it only remains to observe that $\varepsilon\tau\nu_0^{-(2+\alpha)}=\varepsilon\nu^{-(2+\alpha)}$ for $\tau=\delta^{2+\alpha}$.

This finishes the proof of (6) if a_0 is diagonal. In the general case one knows that there exists an orthogonal transformation which makes a_0 diagonal. This orthogonal transformation may send $\mathbb{R}^d_+$ to a half plane different from $\mathbb{R}^d_+$, but clearly Theorem 4.7 and the above arguments are applicable to any half space and not only to $\mathbb{R}^d_+$.

Finally, (7) is obtained by repeating the end of the proof of Theorem 4.7. The lemma is proved. □

Here is an a priori estimate in $\hat{C}^{2+\alpha}(\mathbb{B}_R^+)$.

3. THEOREM. *Let u be a function of class $\hat{C}^{2+\alpha}(\mathbb{B}_R^+)$ such that $u(x)=0$ if $x^1=0$. Set $f := \mathcal{L}u(x)$ and assume that $f \in \hat{C}^{\alpha}(\mathbb{B}_R^+)$. Then*

$$\|u\|_{C(B_R^+)} \leq \|u\|_{C(\partial B_R^+)} + R^2\delta^{-1}\|f\|_{C(B_R^+)}, \tag{9}$$

and there exists a constant N, depending only on d, $R^{-2}M_{\alpha,R}(a)$, α, and δ, such that

$$[D^2u]_{\hat{C}^{\alpha}(\mathbb{B}_R^+)} \leq N\big([f]_{\hat{C}^{\alpha}(\mathbb{B}_R^+)} + \sup_{B_R^+}|u|\big). \tag{10}$$

Proof. Estimate (9) is obtained in the same way as (3.4). While proving (10), as in the proof of Theorem 1 we assume that $R=2$ and introduce ρ_n, $\Delta\rho_n$, γ, x_1, x_2, V after replacing there B_{ρ_n} with $B^+_{\rho_n}$, only this time we restrict γ to $(0,1/8)$.

There are three cases:

(a) $|x_1-x_2| \geq \gamma\Delta\rho_n$;

(b) $|x_1-x_2| < \gamma\Delta\rho_n$ and $B_{2\gamma\Delta\rho_n}(x_1) \subset B_2^+$;

(c) $|x_1-x_2| < \gamma\Delta\rho_n$ and $B_{2\gamma\Delta\rho_n}(x_1) \not\subset B_2^+$.

In cases (a) and (b), just by replacing B_{ρ_n} and B_2 in the proof of Theorem 1 with $B^+_{\rho_n}$ and B_2^+, respectively, and using (7) in place of (3.6), we obtain

$$V \leq N\gamma^{\alpha}[D^2u]_{C^{\alpha}(B^+_{\rho_{n+1}})} + N[f]_{C^{\alpha}(B^+_{\rho_{n+1}})} + N\gamma^{-2}(\gamma\Delta\rho_n)^{-2-\alpha}\sup_{B_2^+}|u|. \tag{11}$$

In case (c), since $\gamma < 1/2$ and $x_1 \in B^+_{\rho_n}$, we have $x_1^1 < 2\gamma\Delta\rho_n$. Then set

$$\rho = 4\gamma\Delta\rho_n$$

and observe that, since $8\gamma < 1$,

$$B^+_{2\rho}(0,x_1') \subset B^+_{\rho_{n+1}}.$$

Also $x_1 \in B^+_\rho(0,x_1')$ since $x_1^1 < (1/2)\rho$, and $x_2 \in B^+_\rho(0,x_1')$ since $|(0,x_1')-x_2| \leq x_1^1 + \gamma\Delta\rho_n < \rho$.

We are now in the position to apply (7) to $B^+_\rho(0,x_1')$ and $B^+_{2\rho}(0,x_1')$ in place of B^+_ρ and B^+_R. Then in exactly the same way as we arrived at (11), we conclude that (11) holds in case (c) as well.

The rest of the proof is identical to that of Theorem 1 up to the same replacements of B_{ρ_n} and B_2 with $B^+_{\rho_n}$ and B_2^+, respectively. The theorem is proved. □

6. Solvability in $\hat{C}^{2+\alpha}(\mathbb{B}_R^+)$ of linear equations with variable coefficients

Let α, δ, R, $a=a(x)$, and $\mathcal{L}$ be as in Section 5.

1. THEOREM. *Suppose that there exists $\rho \in (0,R)$ such that $a^{ij}(x) = \delta^{ij}$ for $|x| \geq \rho$. Take $f \in \hat{C}^{\alpha}(\mathbb{B}_R^+)$ and $g \in C(\bar{B}_R^+)$ such that $g(x)=0$ if $x^1=0$. Then there exists a unique $u \in \hat{C}^{2+\alpha}(\mathbb{B}_R^+)$ such that $\mathcal{L}u = f$ in B_R^+ and $u=g$ on ∂B_R^+.*

First, we prove the following.

2. LEMMA. *Let $u \in \hat{C}^{2+\alpha}(\mathbb{B}_R^+)$ and b be a real-valued function of class $C^\alpha(B_R^+)$ such that $b = 0$ outside B_ρ^+. Then*

$$[bD^2u]_{\hat{C}^\alpha(\mathbb{B}_R^+)} \le N[D^2u]_{\hat{C}^\alpha(\mathbb{B}_R^+)} + N\|u\|_{C(B_R^+)},$$

where N depends only on d, α, R, ρ, and the $C^\alpha(B_R^+)$-norm of b.

Proof. For simplicity we assume that $R = 1$. Take $u \in \hat{C}^{2+\alpha}(\mathbb{B}_1^+)$, $\rho \in (0,1)$, and observe that

$$[bD^2u]_{C^\alpha(B_\rho^+)} \le N[D^2u]_{C^\alpha(B_\rho^+)} + N[b]_{C^\alpha(B_\rho^+)} \sup_{B_\rho^+} |D^2u|.$$

We estimate the last supremum by using Remark 11.2.3 about interpolation inequalities and for $\rho \ge 1/2$ and $\varepsilon = 1-\rho$ we get

$$\sup_{B_\rho^+} |D^2u| \le N(1-\rho)^\alpha [D^2u]_{C^\alpha(B_\rho^+)} + N(1-\rho)^{-2}\|u\|_{C(B_R^+)},$$

which leads to ($1-\rho \le 1$, $(1-\rho)^{-2} \le (1-\rho)^{-2-\alpha}$)

$$[bD^2u]_{C^\alpha(B_\rho^+)} \le N[D^2u]_{C^\alpha(B_\rho^+)} + N(1-\rho)^{-2-\alpha}\|u\|_{C(B_R^+)},$$

$$\sup_{\rho\in(1/2,1)} (1-\rho)^{2+\alpha}[bD^2u]_{C^\alpha(B_\rho^+)} \le N[D^2u]_{\hat{C}^\alpha(\mathbb{B}_R^+)} + N\|u\|_{C(B_R^+)}.$$

The last estimate yields the desired result. The lemma is proved. □

Proof of Theorem 1. We are going to use S.N. Bernstein's method of continuity. Denote by I the set of $t \in [0,1]$ such that for any $f \in \hat{C}^\alpha(\mathbb{B}_R^+)$ the equation

$$(1-t)\Delta u + t\mathcal{L}u = f \tag{1}$$

has a unique solution $u^t \in \hat{C}^{2+\alpha}(\mathbb{B}_R^+)$, which equals g on ∂B_R^+. By Theorem 4.1, $0 \in I$. The assertion of the theorem is that $1 \in I$ and to prove this assertion it suffices to prove that I is closed and open in the relative topology of $[0,1]$.

The fact that I is closed trivially follows from Theorem 5.3 and standard arguments (we say more about this in the proof of Theorem 9.1). To prove that I is open, take $t_0 \in I$ and observe that (1) is equivalent to

$$(1-t_0)\Delta u + t_0\mathcal{L}u = f + (t-t_0)(\Delta u - \mathcal{L}u). \tag{2}$$

We are going to prove the solvability of this equation by the method of iterations. Set $u^0 = u^{t_0}$ and for $n = 1, 2, ...$ define u^n as unique solutions of class $\hat{C}^{2+\alpha}(B_R^+)$ of equations

$$(1-t_0)\Delta u^n + t_0\mathcal{L}u^n = f + (t-t_0)(\Delta u^{n-1} - \mathcal{L}u^{n-1}) \tag{3}$$

with boundary condition g. Assume that u^k, $k = 0, ..., n-1$ are already defined. Observe that $\Delta u^{n-1} - \mathcal{L}u^{n-1} = 0$ outside B_ρ^+ and this makes it possible to have that $\Delta u^{n-1} - \mathcal{L}u^{n-1} \in \hat{C}^\alpha(\mathbb{B}_R^+)$. Then, by the choice of t_0, u^n is well defined. Thus u^n are well defined for all $n = 0, 1, ...$

Since

$$\big[(1-t_0)\Delta + t_0\mathcal{L}\big](u^{n+1} - u^n) = (t-t_0)[\Delta - \mathcal{L}](u^n - u^{n-1}),$$

by Theorem 5.3

$$\|u^{n+1}-u^n\|_{C(B_R^+)} \leq |t-t_0|N\big\|D^2(u^n-u^{n-1})\big\|_{C(B_\rho^+)},$$

$$\big[D^2(u^{n+1}-u^n)\big]_{\hat{C}^\alpha(\mathbb{B}_R^+)} \leq |t-t_0|N\big[(\Delta-\mathcal{L})(u^n-u^{n-1})\big]_{\hat{C}^\alpha(\mathbb{B}_R^+)},$$

where and below by N we denote various constants independent of t, t_0, and n.

By interpolation inequalities and by Lemma 2

$$\big\|D^2(u^n-u^{n-1})\big\|_{C(B_\rho^+)} \leq N\|u^n-u^{n-1}\|_{C(B_R^+)} + N\big[D^2(u^n-u^{n-1})\big]_{\hat{C}^\alpha(\mathbb{B}_R^+)}$$

$$\big[(\Delta-\mathcal{L})(u^n-u^{n-1})\big]_{\hat{C}^\alpha(\mathbb{B}_R^+)} \leq N\|u^n-u^{n-1}\|_{C(B_R^+)} + N\big[D^2(u^n-u^{n-1})\big]_{\hat{C}^\alpha(\mathbb{B}_R^+)}.$$

It follows that

$$\|u^{n+1}-u^n\|_{\hat{C}^\alpha(\mathbb{B}_R^+)} \leq |t-t_0|N_1\|u^n-u^{n-1}\|_{\hat{C}^\alpha(\mathbb{B}_R^+)},$$

which allows us to conclude that, for $|t-t_0|N_1 \leq 1/2$, there is a $u \in \hat{C}^\alpha(\mathbb{B}_R^+)$, to which u_n converge in the $\hat{C}^\alpha(\mathbb{B}_R^+)$-norm. By passing to the limit in (3) we get that u is a solution of (1). Since uniqueness follows easily from the maximum principle, $t \in I$, I is open, and the theorem is proved. □

7. About the oscillation of v/x^1 near the flat boundary of B_R^+

Let $\delta, \alpha \in (0,1)$, $R \in (0,\infty)$, and let $a(x)$ be an $\mathbb{S}_\delta$-valued measurable function defined on B_R^+. Introduce

$$\mathcal{L}v(x) = a^{ij}(x)D_{ij}v(x).$$

Denote by M the operator defined by $(Mf)(x) = x^1 f(x)$ and observe that, for any $r > 0$,

$$\mathcal{L}Mu(x) = x^1\mathcal{L}u(x) + 2a^{1i}D_iu(x)$$

is a uniformly nondegenerate elliptic operator with bounded coefficients in the domain $\{x^1 > r\} \cap B_R$. Note that, if a function v satisfies $\mathcal{L}v = 0$ in B_R^+ and $v = 0$ on $\partial_{x^1}B_R^+$ and we are interested in the behavior of D_1v on $\partial_{x^1}B_R^+$, then naturally the function $u = v/x^1$ comes into play satisfying $\mathcal{L}Mu = 0$ in B_R^+.

1. THEOREM. *Assume that we are given two functions*

$$u, v \in W^2_{d,\mathrm{loc}}(B_R^+) \cap B(B_R^+)$$

such that $v \in C^\alpha(B_R^+)$ *and*

$$|\mathcal{L}Mu| \leq \mathcal{L}Mv$$

in B_R^+ *(a.e.). Then there exist constants* $\alpha_0 = \alpha_0(\delta, d) \in (0,1)$ *and* $N = N(\alpha, d, \delta)$ *such that for any* $\beta \in (0, \alpha_0 \wedge \alpha]$ *and any* $r \in (0, R]$ *and* $x, y \in B_r^+$ *we have*

$$\big|u(x)-u(y)\big| \leq Nr^\beta R^{-\beta}\Big(R^\alpha[v]_{C^\alpha(B_R^+)} + \sup_{B_R^+}\big(|u|+|v|\big)\Big). \tag{1}$$

To prove the theorem we need two lemmas, the first of which, actually, shows that for the operator $\mathcal{L}M$, considered on functions given in B_5^+, the set $\partial_{x^1} B_5^+$ should not be treated as part of the boundary: one cannot prescribe, say zero value, on this part of the boundary and find a continuous function w in $\bar{B}_5^+$ which satisfies the equation $\mathcal{L}Mw = 0$ in B_5^+ and equals, say x^1, on the remaining part of the boundary of B_5^+. From the point of view of potential theory this is quite natural, since the function $1/x^1$ is $\mathcal{L}M$-harmonic and the set where it becomes infinite is a polar set, where one cannot prescribe arbitrary values for $\mathcal{L}M$-harmonic bounded continuous functions.

2. LEMMA. *Let $w \in W^2_{d,\text{loc}}(B_5^+) \cap B(B_5^+)$ and let $w \geq 0, \mathcal{L}(Mw) \leq 0$ in B_5^+ (a.e.). Suppose that $w(x) \geq 1$ if $x \in B_3^+ \cap \{x^1 = 1\}$. Then in B_1^+*

$$w \geq \varepsilon = \varepsilon(d, \delta) \in (0, 1).$$

Proof. Take a point $(0, x_0')$ with $|x_0'| \leq 1$ and consider two concentric balls $B_2(2, x_0')$ and $B_1(2, x_0')$. Also take a $\gamma = \gamma(d, \delta) > 1$ such that for

$$\phi(x) := \psi(x - (2, x_0')), \quad \psi(x) = |x|^{-\gamma}$$

it holds that

$$\mathcal{L}\phi \geq 0$$

in $B_2(2, x_0') \setminus \{(2, x_0')\}$. The existence of such a γ is established by the following computation, in which $\lambda = x/|x|$,

$$\begin{aligned}\mathcal{L}\psi &= \gamma(1+\gamma)|x|^{-\gamma-2}a^{ij}\lambda^i\lambda^j - \gamma|x|^{-\gamma-2}[\text{tr}\, a - a^{ij}\lambda^i\lambda^j] \\ &\geq \gamma|x|^{-\gamma-2}[(1+\gamma)\delta - \delta^{-1}d].\end{aligned}$$

After that set $\xi(x) = [\phi(x) - 2^{-\gamma}][1 - 2^{-\gamma}]^{-1}$ and observe that

$$B_2(2, x_0') \subset B_5^+, \quad B_2(2, x_0') \cap \{x^1 = 1\} \subset B_3^+ \cap \{x^1 = 1\},$$

$$\xi(x) \leq 1 \leq x^1 w(x) \quad \text{on} \quad B_2(2, x_0') \cap \{x^1 = 1\},$$

$$\xi(x) = 0 \leq x^1 w(x) \quad \text{on} \quad \{x^1 \leq 1\} \cap \partial B_2(2, x_0').$$

By the maximum principle $\xi(x) \leq x^1 w(x)$ in $\{x^1 \leq 1\} \cap \bar{B}_2(2, x_0')$ and $\xi(x^1, x_0') \leq x^1 w(x^1, x_0')$ for $x^1 \in [0, 1]$, and, since

$$\xi(x^1, x_0') = (|2 - x^1|^{-\gamma} - 2^{-\gamma})(1 - 2^{-\gamma})^{-1} \geq \varepsilon x^1$$

for $x^1 \in [0, 1]$ with $\varepsilon = \varepsilon(\gamma) > 0$, we have that $x^1 w(x^1, x_0') \geq \varepsilon x^1$ for $x^1 \in [0, 1]$. It only remains to note that the set $\{0 \leq x^1 \leq 1\} \times \{x' \in \mathbb{R}^{d-1} : |x'| \leq 1\}$ contains B_1^+. The lemma is proved. □

3. LEMMA. *Let $\xi_r = \eta_r + \zeta_r = \eta_r' + \zeta_r'$ be a nondecreasing nonnegative function on $[0, 5]$. Assume that for $r \in [0, 1]$ we have*

$$\eta_r' \leq \gamma \eta_{5r}, \quad \zeta_r' \leq \zeta_{5r}, \quad \zeta_r' \leq \eta_{5r} + K r^\alpha,$$

where $K \in [0, \infty)$ and $\gamma \in (0, 1)$ are some constants. Then there exist constants $\beta_0 = \beta_0(\gamma) \in (0, 1)$ and $N = N(\gamma, \alpha) \in (0, \infty)$ such that for $\beta = \beta_0 \wedge \alpha$ and $r \in [0, 5]$ we have

$$\xi_r \leq N r^\beta \xi_5 + NK r^\beta.$$

Proof. First, we choose a small $\kappa = \kappa(\gamma) \in (0,1)$ so that

$$\xi^{-1} := 1 - \kappa - \gamma - \kappa\gamma > 0,$$

set $\theta = (1-\gamma)\xi$, and claim that for $r \in (0,1]$ one of the following inequalities holds:

$$(1+\kappa)\xi_r \leq \xi_{5r}, \tag{2}$$

$$\xi_r \leq K\theta r^\alpha. \tag{3}$$

To prove the claim, suppose that (2) is false for an $r \in (0,1]$. Then

$$\eta_{5r} + \zeta_{5r} \leq (1+\kappa)\eta'_r + (1+\kappa)\zeta'_r \leq (1+\kappa)\gamma\eta_{5r} + \zeta_{5r} + \kappa\zeta'_r,$$

$$\eta_{5r} \leq (1+\kappa)\gamma\eta_{5r} + \kappa\zeta'_r \leq (\kappa+\gamma+\kappa\gamma)\eta_{5r} + K\kappa r^\alpha.$$

It follows that

$$\eta_{5r} \leq \xi\kappa K r^\alpha, \quad \eta'_r \leq \xi\kappa\gamma K r^\alpha, \quad \zeta'_r \leq \xi\kappa K r^\alpha + K r^\alpha,$$

$$\xi_r = \eta'_r + \zeta'_r \leq \big(1 + \xi\kappa(1+\gamma)\big) K r^\alpha$$

and the last inequality is just (3).

After we have proved our claim introduce $\beta_0 = \log_5(1+\kappa)$, $\beta = \beta_0 \wedge \alpha$, and

$$\psi_r = \max\big(\xi_r, K\theta r^\alpha\big).$$

It turns out that for $r \in (0,1]$

$$\psi_r \leq 5^{-\beta}\psi_{5r}. \tag{4}$$

Indeed, if (3) is true, then

$$\psi_r = 5^{-\alpha} K\theta (5r)^\alpha \leq 5^{-\alpha}\psi_{5r} \leq 5^{-\beta}\psi_{5r}.$$

However, if (3) is wrong, then (2) holds, and

$$\psi_r = \xi_r \leq (1+\kappa)^{-1}\xi_{5r} \leq (1+\kappa)^{-1}\psi_{5r} = 5^{-\beta_0}\psi_{5r} \leq 5^{-\beta}\psi_{5r}.$$

Inequality (4) allows us to repeat our argument after (9.6.8) and obtain that for $r \in (0,5]$

$$\xi_r \leq \psi_r \leq 5^\beta r^\beta \psi_5 \leq N r^\beta \xi_5 + NK r^\beta.$$

The lemma is proved. □

Proof of Theorem 1. Since $r \leq R$ it suffices to prove (1) with $\beta = \alpha_0 \wedge \alpha$ and appropriate $\alpha_0 = \alpha_0(\delta, d) \in (0,1)$. Using scalings allows us to assume that $R = 5$. Define

$$w^{(\pm)} = v \pm u, \quad m^{(\pm)}(r) = \inf_{B_r^+} w^{(\pm)}, \quad M^{(\pm)}(r) = \sup_{B_r^+} w^{(\pm)},$$

$$\omega^{(\pm)}(r) = M^{(\pm)}(r) - m^{(\pm)}(r), \quad \theta^{(\pm)}(r) = (3/4)M^{(\pm)}(r) + (1/4)m^{(\pm)}(r).$$

In the arguments below we concentrate on $r \in (0,1]$. Clearly, for $x, y \in B_r^+$,

$$\big[w^{(+)}(x) - w^{(+)}(y)\big] + \big[w^{(-)}(x) - w^{(-)}(y)\big] = 2\big[v(x) - v(y)\big] \leq N r^\alpha [v]_{C^\alpha(B_5^+)}. \tag{5}$$

Also,

$$\mathcal{L}Mw^{(\pm)} \geq 0. \tag{6}$$

Next, for $i = +, -$, write $i \in A(r)$ if

$$\big|B_r^+ \cap \{w^{(i)} \le \theta^{(i)}(r)\}\big| \ge (1/2)\big|B_r^+\big|. \tag{7}$$

Observe that, if $A(r) = \emptyset$, then there exists at least one point $x_0 \in B_r^+$ at which

$$w^{(\pm)}(x_0) \ge (3/4)M^{(\pm)}(r) + (1/4)m^{(\pm)}(r). \tag{8}$$

By taking $x = x_0$ in (5) and taking as y a point in B_r^+, at which $w^{(+)}$ almost attains its infimum over B_r^+, and noting that, of course, $w^{(-)}(y) \le M^{(-)}(r)$, we get that

$$(3/4)\omega^{(+)}(r) - (1/4)\omega^{(-)}(r) \le Nr^\alpha [v]_{C^\alpha(B_5^+)}. \tag{9}$$

Here $+$ and $-$ can be swapped yielding another estimate which after being added to (9) leads us to

$$\omega^{(+)}(r) + \omega^{(-)}(r) \le Nr^\alpha [v]_{C^\alpha(B_5^+)}.$$

If $A(r) \ne \emptyset$, $+ \notin A(r)$ and $- \in A(r)$, one can find a point $x_0 \in B_r^+$, for which (8) holds with $+$ in place of $\pm$. Then (9) holds again. Similar thing happens if $- \notin A(r)$ and $+ \in A(r)$ and we summarize the so far obtained result in a form, which is convenient for the future as follows:

$$(3/4)\sum_{i \notin A(r)} \omega^{(i)}(r) \le (1/4)\sum_{i \in A(r)} \omega^{(i)}(5r) + Nr^\alpha [v]_{C^\alpha(B_5^+)},$$

$$\sum_{i \notin A(r)} \omega^{(i)}(r) \le \sum_{i \in A(r)} \omega^{(i)}(5r) + N_1 r^\alpha [v]_{C^\alpha(B_5^+)} \tag{10}$$

understanding the sums over empty sets as zeros.

In the remaining case $\pm \in A(r)$. Recall that $r \in [0, 1]$, define $\hat{w}^{(i)}(x) = w^{(i)}(rx)$ and observe that, in light of (6),

$$\hat{a}^{kn} D_{kn}\hat{w}^{(i)} + 2(x^1)^{-1}\hat{a}^{1k} D_k \hat{w}^{(i)} \ge 0 \tag{11}$$

in B_5^+, where $\hat{a}(x) = a(rx)$. Also (7) implies that

$$\big|B_1^+ \cap \{x^1 > 1/10\} \cap \{\hat{w}^{(i)} \le \theta^{(i)}(r)\}\big| \ge q > 0,$$

where q is an absolute constant. It follows from Theorem 9.5.5 applied to $\Omega = B_5^+ \cap \{x^1 > 1/10\}$ that, if $x^1 = 1$ and $|x'| \le 4$, then

$$\hat{w}^{(i)}(x) - \theta^{(i)}(r) \le \gamma\big(\sup_{B_5^+} \hat{w}^{(i)} - \theta^{(i)}(r)\big) = \gamma\big(M^{(i)}(5r) - \theta^{(i)}(r)\big),$$

where $\gamma \in (0, 1)$ depends only on d and δ. In particular,

$$\check{w}^{(i)} := M^{(i)}(5r) - \hat{w}^{(i)},$$

which is nonnegative in B_5^+, satisfies there

$$\hat{a}^{kn} D_{kn}\check{w}^{(i)} + 2(x^1)^{-1}\hat{a}^{1k} D_k \check{w}^{(i)} \le 0,$$

and, if $x^1 = 1$ and $|x'| \le 4$, then

$$\check{w}^{(i)} \ge M^{(i)}(5r) - \theta^{(i)}(r) - \gamma M^{(i)}(5r) + \gamma\theta^{(i)}(r) = (1-\gamma)\big(M^{(i)}(5r) - \theta^{(i)}(r)\big).$$

By Lemma 2 we have that in B_1^+

$$\check{w}^{(i)} \geq \varepsilon(1-\gamma)\big(M^{(i)}(5r) - \theta^{(i)}(r)\big),$$

which in terms of $w^{(i)}$ means that with $\gamma_1 := \varepsilon(1-\gamma) \in (0,1)$ we have

$$M^{(i)}(5r) - M^{(i)}(r) \geq \gamma_1\big(M^{(i)}(5r) - \theta^{(i)}(r)\big),$$

$$(1-\gamma_1)M^{(i)}(5r) \geq (1 - 3\gamma_1/4)M^{(i)}(r) - \gamma_1 m^{(i)}(r)/4.$$

Upon adding to the last inequality the obvious inequality $(\gamma_1 - 1)m^{(i)}(5r) \geq (\gamma_1 - 1)m^{(i)}(r)$, we obtain

$$(1-\gamma_1)\omega^{(i)}(5r) \geq (1 - 3\gamma_1/4)\omega^{(i)}(r), \quad \omega^{(i)}(r) \leq \gamma_0\omega^{(i)}(5r), \tag{12}$$

where $\gamma_0 = \gamma_0(d,\delta) \in (0,1)$.

Owing to (10) and (12), the assumptions of Lemma 3 are satisfied with

$$\eta_r' = \sum_{i \in A(r)} \omega^{(i)}(r), \quad \eta_r = \sum_{i \in A(r/5)} \omega^{(i)}(r),$$

$$\zeta_r' = \sum_{i \notin A(r)} \omega^{(i)}(r), \quad \zeta_r = \sum_{i \notin A(r/5)} \omega^{(i)}(r).$$

Hence for $\alpha_0 = \beta_0(\gamma_0)$ and $\beta = \alpha_0 \wedge \alpha$ for $r \in (0,5]$ we have

$$\sum_i w^{(i)}(r) \leq \psi(r) \leq Nr^\beta [v]_{C^\alpha(B_5^+)} + Nr^\beta\big[w^{(+)}(5) + w^{(-)}(5)\big],$$

where $N = N(\alpha, d)$. After this it only remains to observe that $2u = w^{(+)} - w^{(-)}$ and

$$2\big|u(x) - u(y)\big| \leq \big|w^{(+)}(x) - w^{(+)}(y)\big| + \big|w^{(-)}(x) - w^{(-)}(y)\big|.$$

The theorem is proved. □

4. REMARK. Theorem 1 is a particular case of Theorem 4.6.1 of [**61**] and although the proofs of both are practically the same, we decided to give a simpler statement which is easier to grasp.

5. COROLLARY. *Assume that we are given a function $u \in C^2_{\text{loc}}\big(\mathbb{B}_R^+\big) \cap C\big(\bar{B}_R^+\big)$ such that $u = 0$ on $\partial_{x^1} B_R^+$ and $|\mathcal{L}u| \leq K$ in B_R^+, where K is a constant. Then there exists a constant $N = N(d,\delta)$ and, for any $r \leq R$, there exists a linear function $l(x) = cx^1$, such that*

$$\sup_{B_r^+} |u - l| \leq Nr^{1+\alpha_0}R^{-\alpha_0}\big(KR + \sup_{B_R^+}|D_1 u|\big).$$

Indeed, define $w(x) = u(x)/x^1$ if $x^1 \neq 0$, $w(x) = D_1 w(x)$ if $x^1 = 0$, and $v(x) = \delta^{-1}Kx^1$. Then $|\mathcal{L}Mw| \leq \mathcal{L}Mv$ and by Theorem 1, in B_r^+

$$|w - w(0)| \leq Nr^{\alpha_0}R^{-\alpha_0}\big(KR + \sup_{B_R^+}|w|\big).$$

After that it only remains to observe that

$$\big|u(x) - w(0)x^1\big| \leq r\big|w(x) - w(0)\big|, \quad \big|w(x)\big| \leq \sup_{B_R^+}|D_1 u|,$$

the latter being true by the mean-value theorem.

By using shifts we extend Corollary 5 in the following way.

6. COROLLARY. *Suppose that the assumptions of Corollary 5 are satisfied. Then for any half ball $B_r^+(0,x_0') \subset B_R^+$, there exists a linear function $l(x) = cx^1$, for which*

$$\sup_{B_r^+(0,x_0')} |u - l| \le N r^{1+\alpha_0}\big(R - |x_0'|\big)^{-\alpha_0}\big(KR + \sup_{B_R^+}|D_1 u|\big), \tag{13}$$

where α_0 and N are the same as in Corollary 5. Furthermore, since $r \le R - |x_0'|$ one can replace α_0 in (13) with any $\alpha \in (0,\alpha_0]$ without modifying N.

8. A priori estimates in $\hat{C}^{2+\alpha}(\mathbb{B}_1^+)$ for equations $F(D^2u, x) = 0$

Let $F(\mathsf{u}'', x)$ be a continuously differentiable function defined for $\mathsf{u}'' \in \mathbb{S}, x \in \mathbb{R}^d$, which is *convex* with respect to u''. Assume that for all values of the arguments

$$D_{\mathsf{u}''}F(\mathsf{u}'', x) \in \mathbb{S}_\delta,$$
$$\big|D_x F(\mathsf{u}'', x)\big| \le K_0\big(1 + |\mathsf{u}''|\big), \tag{1}$$

where $\delta \in (0,1]$ and $K_0 \in [0,\infty)$ are fixed numbers. Take

$$\alpha \in (0,\alpha_0),$$

where $\alpha_0 = \alpha_0(\delta, d)$ is the smallest of the constants called α_0 in Theorems 10.2.1 and 7.1. We are going to deal with a function $u \in C^2_{\text{loc}}(\mathbb{B}_1^+) \cap C(\bar{B}_1^+)$ such that $u = 0$ on $\partial_{x^1} B_1^+$ and

$$F(D^2 u, x) = 0 \quad \text{in} \quad B_1^+. \tag{2}$$

1. THEOREM. *Assume that $u \in \hat{C}^{2+\alpha}(\mathbb{B}_1^+)$. Then there exists a constant $N = N(d,\delta,K_0,\alpha)$ such that*

$$[D^2 u]_{\hat{C}^\alpha(\mathbb{B}_1^+)} \le N \sup_{B_1^+}|u| + NK_0. \tag{3}$$

Furthermore,

$$\sup_{B_1^+}|u| \le \sup_{\partial B_1^+}|u| + \delta^{-1}\sup_{B_1^+}\big|F(0,x)\big|. \tag{4}$$

2. REMARK. In (3) the term NK_0 with N depending on K_0 is written to show that it disappears if $K_0 = 0$.

To prove the theorem, we need two lemmas.

3. LEMMA. *Let $0 < \rho_0 < \rho_1 < 1$, $\beta \in (0,\alpha_0]$. Then for any half ball $B_r^+(0,x_0') \subset B_{\rho_0}^+$, there exists $p \in \mathcal{P}_2$, for which*

$$\sup_{B_r^+(0,x_0')} |u - p| \le N r^{2+\beta} M(\varepsilon,\rho_0,\rho_1), \tag{5}$$

where $N = N(d,\delta,\beta)$, ε is any number in $(0,1]$, and

$$M(\varepsilon,\rho_0,\rho_1) = \varepsilon^\beta [D^2u]_{C^\alpha(B_{\rho_1}^+)} + \varepsilon^{-1}(\rho_1-\rho_0)^{-\beta}\big(K_0 + (K_0+1)\sup_{B_{\rho_1}^+}|D^2u|\big).$$

Proof. Observe that for $x,y\in B^+_{\rho_1}$

$$0=F\big(D^2u(x),x\big)-F\big(D^2u(y),y\big)=a_{ij}\big[D_{ij}u(x)-D_{ij}u(y)\big]+f(x,y), \quad (6)$$

where (Hadamard's formula)

$$a_{ij}=\int_0^1 D_{\mathsf{u}''_{ij}}F\big(tD^2u(x)+(1-t)D^2u(y),tx+(1-t)y\big)\,dt,$$

$$f(x,y)=(x^i-y^i)\int_0^1 D_{x^i}F\big[tD^2u(x)+(1-t)D^2u(y),tx+(1-t)y\big]\,dt$$

and note that $(a_{ij})\in\mathbb{S}_\delta$ and

$$|f(x,y)|\le K_0\big(1+\sup_{B^+_{\rho_1}}|D^2u|\big)|x-y|.$$

Next, fix $k\in\{2,...,d\}$ and apply (6) to $y=x+he_k$. Then we will see that for all sufficiently small $h>0$ the functions $\delta_{h,e_k}u$ satisfy equations

$$a^{ij}D_{ij}\delta_{h,e_k}u=g$$

in $B^+_{\rho_1}$ with the coefficients and g depending on h, but always

$$(a^{ij})\in\mathbb{S}_\delta,\quad |g|\le K_0\big(1+\sup_{B^+_{\rho_1}}|D^2u|\big).$$

Also $\delta_{h,e_k}u\in C^2(\mathbb{B}^+_{\rho_1})$, which by Corollary 7.6 allows us to conclude that, for any half ball $B^+_r(0,x_0')\subset B^+_{\rho_0}$, there exists a linear function $l(x)=cx^1$ (depending on h), for which

$$\sup_{B^+_r(0,x_0')}|\delta_{h,e_k}u-l|\le Nr^{1+\beta}(\rho_1-\rho_0)^{-\beta}\big(K_0+(K_0+1)\sup_{B^+_{\rho_1}}|D^2u|\big).$$

By sending $h\downarrow 0$ and noting that $\delta_{h,e_k}u\to D_ku$ we get that for any $B^+_r(0,x_0')\subset B^+_{\rho_0}$, there exists a linear function $l(x)=cx^1$, for which

$$\sup_{B^+_r(0,x_0')}|D_ku-l|\le Nr^{1+\beta}(\rho_1-\rho_0)^{-\beta}\big(K_0+(K_0+1)\sup_{B^+_{\rho_1}}|D^2u|\big).$$

By Remark 11.2.6 this yields that for any $\varepsilon\in(0,1]$ and $i=1,...,d$

$$\operatorname*{osc}_{B^+_r(0,x_0')}D_{ik}u\le Nr^\beta M(\varepsilon,\rho_0,\rho_1). \quad (7)$$

Here on the left we have oscillations of all second order derivatives of u apart from $D_{11}u$. However, by looking at (6), we see that

$$\operatorname*{osc}_{B^+_r(0,x_0')}D_{11}u\le N\sum_{i=1,k=2}^{d}\operatorname*{osc}_{B^+_r(0,x_0')}D_{ik}u+K_0 2r^\beta\big(1+\sup_{B^+_{\rho_0}}|D^2u|\big).$$

This and (7) imply that

$$\operatorname*{osc}_{B^+_r(0,x_0')}D^2u\le Nr^\beta M(\varepsilon,\rho_0,\rho_1).$$

Now to get (5), it only remains to take as p the second-order Taylor polynomial of u at the origin. The lemma is proved. □

4. LEMMA. *Let $R \in (0,1]$, $v \in C^2(B_R)$, and let $F(D^2v, x) = 0$ in B_R. Then, for any ball $B_r \subset B_R$ there exists $p \in \mathcal{P}_2$ such that*

$$\sup_{B_r} |v - p| \le NK_0 R^{2+\alpha_0}\big(1 + \sup_{B_R} |D^2 v|\big) + N_1 \frac{r^{2+\alpha_0}}{(R-r)^{2+\alpha_0}} \sup_{B_R} |v - l|,$$

where $l = l(x)$ is an arbitrary quadratic function, $N = N(\delta, d)$, $N_1 = N_1(\delta, d)$.

Proof. By using Theorem 10.3.1 find a unique function $w \in C(\bar{B}_R) \cap C^{2+\alpha_0}_{\text{loc}}(B_R)$ such that

$$F(D^2 w, 0) = 0 \quad \text{in} \quad B_R, \quad w = v \quad \text{on} \quad \partial B_R.$$

By Theorem 10.2.1

$$[D^2 w]_{C^{\alpha_0}(B_r)} \le N \frac{1}{(R-r)^{2+\alpha_0}} \sup_{\partial B_R} |v - l|.$$

Next, by taking p as the second-order Taylor polynomial at the origin of w, we obtain

$$\sup_{B_r} |w - p| \le r^{2+\alpha_0} [D^2 w]_{C^{\alpha_0}(B_r)} \le N \frac{r^{2+\alpha_0}}{(R-r)^{2+\alpha_0}} \sup_{\partial B_R} |v - l|. \tag{8}$$

Now, observe that $v - w$ equals zero on ∂B_R and satisfies

$$0 = F\big(D^2 v(x), x\big) - F\big(D^2 w(x), 0\big) + f(x) = a^{ij} D_{ij}(v - w) + f$$

where (a^{ij}) is $\mathbb{S}_\delta$-valued and

$$f(x) = F\big(D^2 v(x), x\big) - F\big(D^2 v(x), 0\big), \quad |f| \le K_0\big(1 + \sup_{B_R} |D^2 v|\big) R^{\alpha_0}.$$

By the maximum principle

$$|v - w| \le N R^{2+\alpha_0} K_0 \big(1 + \sup_{B_R} |D^2 v|\big)$$

even in B_R and this along with (8) yields the result. The lemma is proved. □

Proof of Theorem 1. First observe that

$$0 = \big[F(D^2 u, x) - F(0, x)\big] + F(0, x),$$

which after transforming $F(D^2u, x) - F(0, x)$ as in (6) shows that

$$\big|a^{ij} D_{ij} u\big| \le \sup_{B_1^+} \big|F(0, x)\big|$$

in B_1^+, where (a^{ij}) is an $\mathbb{S}_\delta$-valued function. After that (4) is obtained by a standard application of the maximum principle.

To prove (3), define $\rho_0 = 1/2$, $\rho_{n+1} = \rho_n + 2^{-n-2}$, $n = 0, 1, \ldots$, and

$$M_n = \sup_{B^+_{\rho_n}} |D^2 u|.$$

Next, take $\nu > 1$ and apply Lemma 4 with $R = \nu r$ to the shifts of B_r. Then we see that, whenever $B_{\nu r}(x_0) \subset B^+_{\rho_{n+1}}$, there exists $p \in \mathcal{P}_2$, such that

$$\sup_{B_r(x_0)} |u - p| \le NK_0 (\nu r)^{2+\alpha_0}(1 + M_{n+1}) + \varepsilon_1 \sup_{B_{\nu r}(x_0)} |u - l|$$

for any $l \in \mathcal{P}_2$, where

$$\varepsilon_1 = N_1(\nu-1)^{-2-\alpha_0}.$$

Now using the fact that $\alpha < \alpha_0$ we choose $\nu = \nu(d,\delta,\alpha) > 1$ so that

$$\varepsilon_1 \nu^{2+\alpha} = 1/2.$$

By combining this with Lemma 3, where we take ρ_n and ρ_{n+1} in place of ρ_0 and ρ_1, respectively, by Theorem 2.1 with $R = \rho_{n+1}$ we get that for any $\varepsilon \in (0,1]$ we have

$$\begin{aligned}
[D^2u]_{C^\alpha(B^+_{\rho_n})} \le& NK_0(1+M_{n+1}) + N\varepsilon^\alpha [D^2u]_{C^\alpha(B^+_{\rho_{n+1}})} \\
&+ N\varepsilon^{-1}2^{n\alpha}\big(K_0 + \sup_{B^+_{\rho_{n+1}}} |D^2u|\big) + N2^{(2+\alpha)n}\sup_{B_1^+}|u| \\
\le& N\varepsilon^\alpha [D^2u]_{C^\alpha(B^+_{\rho_{n+1}})} + N\varepsilon^{-1}2^{n\alpha}(K_0 + M_{n+1}) \\
&+ N2^{(2+\alpha)n}\sup_{B_1^+}|u|,
\end{aligned}$$

where the constants N depend only on d, δ, K_0, and α. Here by interpolation inequalities (see Remark 11.2.3)

$$M_{n+1} \le N\varepsilon^{1+\alpha}2^{-n\alpha}[D^2u]_{C^\alpha(B^+_{\rho_{n+1}})} + N(\varepsilon)2^{2n}\sup_{B_1^+}|u|,$$

where and below by $N(\varepsilon)$ we denote various constants depending only on ε, d, δ, K_0, and α. Once $\varepsilon = \varepsilon(d,\delta,K_0,\alpha)$ is chosen, we drop it from the arguments. We now have

$$[D^2u]_{C^\alpha(B^+_{\rho_n})} \le N_2\varepsilon^\alpha[D^2u]_{C^\alpha(B^+_{\rho_{n+1}})} + N(\varepsilon)2^{(2+\alpha)n}\sup_{B_1^+}|u| + NK_0\varepsilon^{-1}2^{n\alpha}.$$

We multiply both parts of this inequality by

$$[1-\rho_n]^{2+\alpha} = 2^{2+\alpha}[1-\rho_{n+1}]^{2+\alpha} = 2^{-(n+1)(2+\alpha)}$$

and take supremums with respect to $n = 0, 1 \ldots$. Then we find

$$I := \sup_{n\ge 0}[D^2u]_{C^\alpha(B^+_{\rho_n})}[1-\rho_n]^{2+\alpha} \le N_2 2^{2+\alpha}\varepsilon^\alpha I + N(\varepsilon)\sup_{B_1^+}|u| + NK_0\varepsilon^{-1}.$$

We finally choose $\varepsilon = \varepsilon(d,\delta,K_0,\alpha) \in (0,1)$ so that $N_2 2^{2+\alpha}\varepsilon^\alpha \le 1/2$ and then conclude that

$$\sup_{n\ge 0}[D^2u]_{C^\alpha(B^+_{\rho_n})}[1-\rho_n]^{2+\alpha} \le N\sup_{B_1^+}|u| + NK_0.$$

This is equivalent to (3) and the theorem is proved. □

9. The method of continuity

Let $F(\mathsf{u}'',x)$ be as in Section 8, and let it be *three times* continuously differentiable with respect to (u'',x). Also suppose that there exists $\rho\in(0,1)$ such that

$$F(\mathsf{u}'',x)=\sum_{i=1}^{d}\mathsf{u}''_{ii}\quad\text{for}\quad |x|>\rho. \tag{1}$$

Take α from Section 8 and let $g\in C(\bar B_1^+)$ be such that $g(x)=0$ on $\partial_{x^1}B_1^+$.

1. THEOREM. *Under the above assumptions the equation*

$$F[u]=0 \tag{2}$$

in B_1^+ with boundary condition $u=g$ on ∂B_1^+ has a unique solution $u\in\hat C^{2+\alpha}(\mathbb{B}_1^+)$.

Proof. Uniqueness is a simple corollary of the maximum principle.

In the proof of existence we are going to use again S.N. Bernstein's method of continuity. Let I be the set of all $t\in[0,1]$ such that the equation

$$(1-t)\Delta u+tF[u]=0 \tag{3}$$

in B_1^+ with boundary condition g has a solution $u^t\in\hat C^{2+\alpha}(B_R^+)$. Zero belongs to I in light of Theorem 4.1. We will prove that $I=[0,1]$ (and then the theorem will be proved), if we can show that I is closed and open in the relative topology of $[0,1]$.

Theorem 8.1 implies that I is closed. Indeed, by this theorem, for any $t\in I$ we have

$$\|u^t\|_{\hat C^{2+\alpha}(\mathbb{B}_1^+)}\le K, \tag{4}$$

where $K<\infty$ is independent of t. Therefore if $t_n\in I$ and $t_n\to t$ as $n\to\infty$, there exists a subsequence $\{s_n\}\subset\{t_n\}$ converging to t such that $u^{s_n},Du^{s_n},D^2u^{s_n}$ uniformly converge on any compact subset of B_1^+ to u, Du, D^2u, where $u\in C^{2+\alpha}_{\text{loc}}(B_1^+)$. Obviously, u is a solution of (3). Moreover, owing to (4),

$$[D^2u]_{\hat C^{\alpha}(\mathbb{B}_1^+)}\le K.$$

Next,

$$0=(1-s_n)\Delta(u^{s_n}-u^{s_m})+s_n\big(F[u^{s_n}]-F[u^{s_m}]\big)+(s_n-s_m)\big(F[u^{s_m}]-\Delta u^{s_m}\big),$$

which in light of Hadamard's formula, (4), and (1) shows that

$$\big|a^{ij}D_{ij}(u^{s_n}-u^{s_m})\big|\le N|s_n-s_m|,$$

where N is independent of n,m and $(a^{ij})\in\mathbb{S}_\delta$. By the maximum principle $|u^{s_n}-u^{s_m}|\to0$ as $n,m\to\infty$ uniformly on $\bar B_1^+$. In particular, $|u^{s_n}-u|\to0$ as $n\to\infty$ uniformly on $\bar B_1^+$, $u\in C(\bar B_1^+)$ and $u=g$ on ∂B_1^+. Hence, $t\in I$.

In order to prove that I is open in $[0,1]$, we follow the arguments in the proof of Theorem 12.6 of [**5**]. Take $t_0\in I$ and denote $u_0=u^{t_0}$. Take $w\in\hat C^{2+\alpha}(\mathbb{B}_1^+)$ and consider the following problem about the unknown function v:

$$(1-t)\Delta v+ta^{ij}D_{ij}v+t\big[F(D^2w,x)-a^{ij}D_{ij}w\big]=0\quad\text{in } B_1^+, \tag{5}$$

with boundary condition

$$v=g\text{ on }\partial B_1^+, \tag{6}$$

where

$$a^{ij} = D_{\mathsf{u}''_{ij}} F(D^2 u_0, x). \tag{7}$$

By assumption $u_0 \in \hat{C}^{2+\alpha}(\mathbb{B}_1^+)$ and, owing to the differentiability properties of F and owing to (1), we have $F(D^2 w, x) - a^{ij} D_{ij} w \in \hat{C}^{\alpha}(\mathbb{B}_1^+)$. By Theorem 6.1 the problem (5)–(6) has a unique solution $v \in \hat{C}^{2+\alpha}(\mathbb{B}_1^+)$. Thus, we have a mapping

$$\Psi_t : w \in \hat{C}^{2+\alpha}(\mathbb{B}_1^+) \to v \in \hat{C}^{2+\alpha}(\mathbb{B}_1^+).$$

Then, for $\varepsilon \in (0,1)$ introduce

$$A(\varepsilon) = \{ w \in \hat{C}^{2+\alpha}(\mathbb{B}_1^+) : \|w - u_0\|_{\hat{C}^{2+\alpha}(\mathbb{B}_1^+)} \le \varepsilon, \quad w = g \quad \text{on} \quad \partial B_1^+ \}.$$

Our next goal is to prove that Ψ_t maps $A(\varepsilon)$ into itself and is a contraction if ε is small enough and t is close enough to t_0.

Simple manipulations based on the fact that u_0 satisfies (3) with $t = t_0$ show that

$$\begin{aligned} &(1-t)\Delta(v - u_0) + t a^{ij} D_{ij}(v - u_0) \\ &= t_0 f + (t - t_0)\big(\Delta - a^{ij} D_{ij}\big) u_0 + (t_0 - t)\big[F(D^2 w, x) - a^{ij} D_{ij} w\big], \end{aligned} \tag{8}$$

where

$$f(x) := F(D^2 u_0, x) - F(D^2 w, x) - a^{ij} D_{ij}(u_0 - w)$$

$$= b^{ij}(x) D_{ij}(u_0 - w)(x),$$

$$b^{ij}(x) := \int_0^1 c^{ij}(t, x)\, dt,$$

$$c^{ij}(t,x) := D_{\mathsf{u}''_{ij}} F\big(tD^2 u_0(x) + (1-t) D^2 w(x), x\big) - D_{\mathsf{u}''_{ij}} F\big(D^2 u_0(x), x\big)$$

$$= d^{ijkr}(t,x) D_{kr}(u_0 - w)(x),$$

$$d^{ijkr}(t,x) := \int_0^1 e^{ijkr}(s,t,x)\, ds,$$

$$e^{ijkr}(s,t,x) := D_{\mathsf{u}''_{ij}} D_{\mathsf{u}''_{kr}} F\big((st + 1 - s) D^2 u_0(x) + s(1-t) D^2 w(x), x\big).$$

Define

$$M = \|u_0\|_{\hat{C}^{2+\alpha}(\mathbb{B}_1^+)},$$

and observe that owing to interpolation inequalities and the fact that $w \in A(\varepsilon)$,

$$\sup_{B_\rho^+} \big(|D^2 u_0| + |D^2 w|\big) \le N(M+1), \tag{9}$$

where N is independent of w and ε. Below we denote by N various constants independent of w, ε, and s, t.

Our assumption about the derivatives of F and (9) imply that

$$\|e^{ijkr}(s,t,\cdot)\|_{C^\alpha(B_\rho^+)} \le N, \quad \|d^{ijkr}(t,\cdot)\|_{C^\alpha(B_\rho^+)} \le N.$$

Then we use an almost obvious inequality

$$\|\phi\psi\|_{C^\alpha(B_\rho^+)} \le N\|\phi\|_{C^\alpha(B_\rho^+)}\|\psi\|_{C^\alpha(B_\rho^+)}$$

to get that

$$\|c^{ij}(t,\cdot)\|_{C^\alpha(B_\rho^+)} \le N\big\|D^2(u_0 - w)\big\|_{C^\alpha(B_\rho^+)} \le N\varepsilon, \quad \|b^{ij}\|_{C^\alpha(B_\rho^+)} \le N\varepsilon,$$

$$\|f\|_{C^\alpha(B_\rho^+)} \le N\varepsilon\|u_0 - w\|_{\hat{C}^\alpha(\mathbb{B}_1^+)}.$$

By noting that $f = 0$ outside B_ρ^+ in light of (1), we conclude that

$$\|f\|_{\hat{C}^\alpha(\mathbb{B}_1^+)} \le N\varepsilon\|u_0 - w\|_{\hat{C}^{2+\alpha}(\mathbb{B}_1^+)}.$$

After that by Theorem 5.3 applied to (8) we find

$$\|u_0 - v\|_{\hat{C}^\alpha(\mathbb{B}_1^+)} \le N_1\big(\varepsilon^2 + |t - t_0|\big). \tag{10}$$

From now on we concentrate on

$$|t - t_0| \le \varepsilon^2, \quad \varepsilon \le \varepsilon_0, \tag{11}$$

where $\varepsilon_0 \in (0,1)$ is chosen so that $2N_1\varepsilon_0 \le 1$, which, due to (10), shows that for such parameters, indeed, Ψ_t is a mapping of $A(\varepsilon_0)$ into itself.

We further show that Ψ_t is a contraction of $A(\varepsilon_0)$ in the space $\hat{C}^{2+\alpha}\big(\mathbb{B}_1^+\big)$ for any t if ε is sufficiently small. Let $w^i \in A(\varepsilon)$ and $v^i = \Psi_t w^i$, $i = 1, 2$. We have

$$(1-t)\Delta(v^1 - v^2) + ta^{ij}D_{ij}(v^1 - v^2) = -tf,$$

where

$$\begin{aligned} f &:= \int_0^1 b^{ij}(t,x)\,dt\,D_{ij}(w^1 - w^2)(x), \\ b^{ij}(t,x) &:= D_{\mathsf{u}''_{ij}}F\big(tD^2w^1(x) + (1-t)D^2w^2(x), x\big) - D_{\mathsf{u}''_{ij}}F\big(D^2u_0(x), x\big) \\ &= \int_0^1 c^{ijrk}(s,t,x)\,ds\,\big[tD_{rk}(w^1 - u_0)(x) \\ &\quad + (1-t)D_{rk}(w^2 - u_0)(x)\big], \\ c^{ijrk}(s,t,x) &= D_{\mathsf{u}''_{ij}}D_{\mathsf{u}''_{rk}}F\Big[s\big(tD^2w^1(x) + (1-t)D^2w^2(x)\big) \\ &\quad + (1-s)D^2u_0(x), x\Big]. \end{aligned}$$

As above

$$\|c^{ijrk}(s,t,\cdot)\|_{C^\alpha(B_\rho^+)} \le N, \quad \|b^{ij}(bt,\cdot)\|_{C^\alpha(B_\rho^+)} \le N\varepsilon,$$

$$\|f\|_{C^\alpha(B_\rho^+)} \le N\varepsilon\|w^1 - w^2\|_{\hat{C}^\alpha(\mathbb{B}_1^+)},$$

$$\|f\|_{\hat{C}^\alpha(\mathbb{B}_1^+)} \le N\varepsilon\|w^1 - w^2\|_{\hat{C}^{2+\alpha}(\mathbb{B}_1^+)},$$

and by Theorem 5.3

$$\|v^1 - v^2\|_{\hat{C}^{2+\alpha}(\mathbb{B}_1^+)} \le N_2\varepsilon\|w^1 - w^2\|_{\hat{C}^{2+\alpha}(\mathbb{B}_1^+)}.$$

This shows that, for ε and t such that $N_2\varepsilon < 1$ and the conditions (11) are satisfied, the mapping Ψ_t is a contraction of $A(\varepsilon)$. By the Banach fixed point theorem there exists $w \in A(\varepsilon)$ such that $\Psi_t w = w$, which means that w is a solution of (3). Again uniqueness is a simple corollary of the maximum principle, $t \in I$, I is open, and the theorem is proved. □

10. Proof of Theorem 1.4

Take $\alpha_0 = \alpha_0(\delta, d)$ from the beginning of Section 8 and take $\alpha \in (0, \alpha_0)$. We will prove Theorem 1.4 after some preparations and we will consider only the case that

$$R = 1.$$

This is an innocent replacement because, if u is a solution of (1.3) in B_R^+, then $v(x) = u(Rx)$ is a solution of $G[v] = 0$ in B_1^+, where $G(\mathsf{u}'', x) = R^2 F(\mathsf{u}''/R^2, Rx)$ possesses the same properties as F. Estimates (1.4) and (1.5) are also scale invariant and, because of that, it suffices to prove (1.5) for $\rho \in (1/2, 1)$ with

$$\bar{K}_0 := K_0 + K_1$$

in place of $K_0\rho^3 + K_1\rho$.

1. LEMMA. *Assume that F is three times continuously differentiable. Take $\rho \in (1/2, 1)$ and take $\zeta \in C_0^\infty(B_1)$ such that $\zeta = 1$ in B_ρ and $0 \leq \zeta \leq 1$. Then the equation*

$$\zeta(x)F(D^2u, x) + \big(1 - \zeta(x)\big)\Delta u = 0 \tag{1}$$

in B_1^+ with boundary condition $u = g$ on ∂B_1^+ has a unique solution $u \in \hat{C}^{2+\alpha}\big(\mathbb{B}_1^+\big)$. Furthermore,

$$[D^2u]_{\hat{C}^\alpha(\mathbb{B}_\rho^+)} \leq N \sup_{B_\rho^+} |u| + N\bar{K}_0, \tag{2}$$

where N depends only on $\bar{K}_0$, d, δ, and α.

Proof. The existence of solution directly follows from Theorem 9.1. One obtains estimate (2) from Theorem 8.1 after observing that u satisfies (1.3) in B_ρ^+. Of course, one should also add a rescaling argument, like the one before the lemma, stretching B_ρ^+ to B_1^+. The lemma is proved. □

2. LEMMA. *The assertions of Theorem 1.4 hold true if, additionally, F is three times continuously differentiable and $g \in C^2(\bar{B}_1)$.*

Proof. Take a sequence $\rho_1 < \rho_2 < ...$ such that $\rho_1 = 1/2$ and $\rho_n \to 1$ as $n \to \infty$ and construct functions $\zeta_n \in C_0^\infty(B_1)$ such that $\zeta_n = 1$ in B_{ρ_n} and $0 \leq \zeta_n \leq 1$. Then let u_n be solutions of (1) with ζ_n in place of ζ and with boundary condition $u_n = g$ on ∂B_1^+.

By Theorem 8.1 functions u_n are uniformly bounded in $\bar{B}_1^+$:

$$\sup_{B_1^+} |u_n| \leq \sup_{\partial B_1^+} |g| + \delta^{-1} \sup_{B_1^+} \big|F(0, x)\big|$$

and owing to (2) their $\hat{C}^\alpha(\mathbb{B}_{\rho_n}^+)$-norms are bounded as well. Hence D^2u_n are uniformly bounded and uniformly continuous on any compact subset of $\mathbb{B}_1^+$. By interpolation inequalities the same is true for u_n and Du_n. Then from Calculus we know that there exist a function $u \in C^{2+\alpha}_{\text{loc}}(\mathbb{B}_1^+)$ and a subsequence $n(k)$ such that

$u_{n(k)}$, $Du_{n(k)}$, $D^2u_{n(k)}$ tend to u, Du, D^2u uniformly on any compact subset of $\mathbb{B}_1^+$. Of course, in light of (2), for any $\rho < 1$

$$[D^2u]_{\hat{C}^\alpha(\mathbb{B}_\rho^+)} \leq \lim_{k\to\infty} \big[D^2u_{n(k)}\big]_{\hat{C}^\alpha(\mathbb{B}_\rho^+)}$$

$$\leq N \lim_{k\to\infty} \sup_{B_\rho^+} |u_{n(k)}| + N\bar{K}_0$$

$$= N \sup_{B_\rho^+} |u| + N\bar{K}_0.$$

We have proved (1.5) for $\rho < 1$. By letting $\rho \uparrow 1$ in (1.5) we get that (1.5) holds with $\rho = 1$ as well.

Obviously, u satisfies (1.3) in B_1^+. So far u is defined only in $\mathbb{B}_1^+$. We set $u = g$ everywhere on ∂B_1^+ and then, as is easy to see, the only thing which remains to be checked is that u is continuous in $\bar{B}_1^+$. Here we use that g is smooth.

Note that, for each n, there exists an $\mathbb{S}_\delta$-valued function a such that

$$\big|a^{ij}(x)D_{ij}u_n(x)\big| \leq \sup_{B_1^+} \big|F(0,x)\big|,$$

$$\big|a^{ij}(x)D_{ij}(u_n - g)(x)\big| \leq N(d,\delta)\big(\sup_{B_1^+} \big|F(0,x)\big| + \|D^2g\|_{C(B_1^+)}\big)$$

in B_1^+. It follows by the maximum principle that

$$|u_n(x) - g(x)| \leq N\big(1 - |x|^2\big), \tag{3}$$

where N is independent of n and x. The same inequality holds for u in place of u_n, and this proves the lemma. □

Proof of Theorem 1.4. First we assume that g is smooth. To get rid of the extra smoothness assumption on F in Lemma 2, we mollify it and find a sequence $F^n(\mathsf{u}'', x)$ of smooth functions, convex in u'', and such that $F^n_{\mathsf{u}''}(\mathsf{u}'',) \in \mathbb{S}_\delta$ for all n and u'', x, and $F^n \to F$ uniformly on any compact subset of $\mathbb{S} \times \mathbb{R}^d$. Then let u_n be the solutions corresponding to F^n. Uniform estimates of u_n in $\hat{C}^{2+\alpha}\big(\mathbb{B}_1^+\big)$ allow us to repeat the argument in the proof of Lemma 2 and show that there exists $u \in C^{2+\alpha}_{\text{loc}}\big(\mathbb{B}_1^+\big)$ and a subsequence $n(k)$ such that $u_{n(k)}$, $Du_{n(k)}$, $D^2u_{n(k)}$ tend to u, Du, D^2 uniformly on any compact subset of $\mathbb{B}_1^+$.

If we extend u to $\bar{B}_1^+$ by setting $u = g$ everywhere on ∂B_1^+, then the fact that u is continuous in $\bar{B}_1^+$ is established as in the the proof of Lemma 2. The estimates of u_n and D^2u_n lead to similar estimates of u and D^2u, and it only remains to observe that u satisfies (1.3) in B_1^+ because at each point $x \in B_1^+$ the set $\big\{D^2u_n(x), n = 1, 2, ...\big\}$ is bounded, $F_n \to F$ uniformly on this set and, hence,

$$F(D^2u(x), x) = \lim_{k\to\infty} F\big(D^2u_{n(k)}(x), x\big) = \lim_{k\to\infty} F^{n(k)}\big(D^2u_{n(k)}(x), x\big) = 0.$$

We have proved the theorem under the assumption that g is smooth. To relax it to the continuity assumption, continue g inside B_1^+ so that it becomes a continuous function in $\bar{B}_1^+$, then extend it for negative x^1, so that it becomes and odd function of x^1 in $\bar{B}_1$. After that we extend it to the whole of $\mathbb{R}^d$ preserving its property to be a bounded continuous function odd with respect to x^1. To do so one can take any bounded continuous extension $\hat{g}$ and take $(1/2)\big[\hat{g}(x^1, x') - \hat{g}(-x^1, x')\big]$. Now we mollify g by using radially symmetric kernels and then we find a sequence of smooth g_n which vanish on $\partial_{x^1}B_1^+$ and uniformly converge to g on ∂B_1^+. If u_n are the corresponding solutions, then the formula

$$0 = F\big(D^2u_n, x\big) - F\big(D^2u_m, x\big) = a^{ij}D_{ij}(u_n - u_m),$$

where (a^{ij}) is an $\mathbb{S}_\delta$-valued function, and the maximum principle show that u_n is a Cauchy sequence in $C\big(\bar{B}_1^+\big)$. By denoting its limit by u and taking into account the estimates of u_n we easily see that u is the function we are after. The theorem is proved. □

11. Solvability in $\hat{C}^{2+\alpha}\big(\mathbb{B}_1^+\big)$ for $F(D^2u) = f(x)$

Take $\alpha_0 = \alpha_0(\delta, d)$ from the beginning of Section 8, that is the same as in Theorem 1.4, and take $\alpha \in (0, \alpha_0)$.

1. THEOREM. *Under Assumption* 1.1 *one can find constants N, depending only on δ, α, and d, such that for any $R \in (0,\infty)$, $g \in C(\partial B_R^+)$ vanishing on $\partial_{x^1}B_R^+$, and $f \in \hat{C}^\alpha\big(\mathbb{B}_R^+\big)$ there exists a unique $u \in \hat{C}^{2+\alpha}\big(\mathbb{B}_R^+\big)$ satisfying*

$$F(D^2u) = f(x) \quad in \quad B_R^+, \quad u = g \quad on \quad \partial B_R^+. \tag{1}$$

Furthermore,

$$\sup_{B_R^+}|u| \le \sup_{\partial B_R^+}|g| + NR^2\big(|F(0)| + \sup_{B_R^+}|f|\big), \tag{2}$$

$$[D^2u]_{C^\alpha(B_\rho^+)} \le N[f]_{C^\alpha(B_r^+)} + N(r-\rho)^{-2-\alpha}\sup_{B_r^+}|u| \quad \forall \rho < r \le R, \tag{3}$$

$$[D^2u]_{\hat{C}^\alpha(\mathbb{B}_R^+)} \le N\big(\sup_{B_R^+}|u| + [f]_{\hat{C}^\alpha(\mathbb{B}_R^+)}\big). \tag{4}$$

In particular,

$$\big|D^2u(x) - D^2u(y)\big| \le N|x-y|^\alpha\big(\sup_{B_R^+}|u| + [f]_{\hat{C}^\alpha(\mathbb{B}_R^+)}\big)$$

if $x, y \in B_{R/2}^+$.

To prove the theorem, we need an a priori estimate.

2. LEMMA. *For any $r > 0$ and $\varepsilon \in (0,1)$ there exist constants $\nu \ge 2$ and N, depending only on ε, α, d, and δ, such that, if the equation $F(D^2u) = f(x)$ in $B_{\nu r}^+$ has a solution $u \in C^{2+\alpha}(B_{\nu r}^+)$ vanishing on $\partial_{x^1}B_{\nu r}^+$, then there exists $p \in \mathcal{P}_2^0$, for which*

$$\sup_{B_r^+}|u-p| \le Nr^{2+\alpha}[f]_{C^\alpha(B_{\nu r}^+)} + \varepsilon\nu^{-(2+\alpha)}\sup_{B_{\nu r}^+}|u-q| \tag{5}$$

for any $q \in \mathcal{P}_2^0$.

Proof. Fix $q \in \mathcal{P}_2^0$ and introduce

$$G(\mathsf{u}'') = F(\mathsf{u}'' + D^2q),$$

so that $D_{\mathsf{u}''}G \in \mathbb{S}_\delta$. Set

$$\alpha' = (1/2)(\alpha_0 + \alpha) \quad (\in (\alpha, \alpha_0))$$

and observe that by Theorem 1.4 there is a function $v \in \hat{C}^{2+\alpha'}(\mathbb{B}^+_{\nu r})$ vanishing on $\partial_{x^1} B^+_{\nu r}$ such that $G(D^2 v) = f(0)$ in $B^+_{\nu r}$ and $v = u - q$ on $\partial B^+_{\nu r}$. Furthermore, for p being the second order Taylor polynomial of v at the origin

$$r^{-(2+\alpha)} \sup_{B^+_r} |v - p| \le N r^{\alpha'-\alpha} [D^2 v]_{C^{\alpha'}(B^+_r)} \le N_1 r^{\alpha'-\alpha} (\nu r - r)^{-(2+\alpha')} \sup_{B^+_{\nu r}} |v|, \quad (6)$$

where $N_1 = N_1(d, \delta, \alpha)$.

Next, note that

$$f(0) - f(x) = G(D^2 v) - G\big(D^2(u - q)\big) = a^{ij} D_{ij}\big(v - (u - q)\big)$$

with an $\mathbb{S}_\delta$-valued (a^{ij}). Since $v = u - q$ on $\partial B^+_{\nu r}$, by the maximum principle

$$\big|v - (u - q)\big| \le \delta^{-1} (\nu r)^{2+\alpha} [f]_{C^\alpha(B^+_{\nu r})}.$$

For $p' = q + p$ now it follows from (6) that (note the replacement of the last supremum in 6)

$$r^{-(2+\alpha)} \sup_{B^+_r} |u - p'| \le N_2 [f]_{C^\alpha(B^+_{\nu r})} + N_3 (\nu r)^{-(2+\alpha)} \sup_{B^+_{\nu r}} |u - q|,$$

where

$$N_2 = N\nu^{2+\alpha} + N r^{\alpha'-\alpha} (\nu r - r)^{-(2+\alpha')} (\nu r)^{2+\alpha} = N\nu^{2+\alpha}\big(1 + (\nu - 1)^{-(2+\alpha')}\big),$$

$$N_3 = N_1 r^{\alpha'-\alpha} (\nu r - r)^{-(2+\alpha')} (\nu r)^{2+\alpha} = N_1 \nu^{2+\alpha} (\nu - 1)^{-(2+\alpha')}.$$

It only remains to chose $\nu = \nu(d, \delta, \alpha, \varepsilon)$ so large that $N_3 \le \varepsilon$, which is possible since $\alpha < \alpha'$. The lemma is proved. □

Proof of Theorem 1. Scalings allow us to concentrate on the case that $R = 1$. Then first assume that $f \in C^1(\mathbb{R}^d)$. Then by Theorem 1.4 there exists a unique u for which the first assertion of Theorem 1 is valid. Estimate (2) is a usual consequence of the maximum principle. Upon combining Remark 11.1.7 and Lemma 2, by Theorem 2.1 we conclude that for any $\rho < r \le 1$

$$[D^2 u]_{C^\alpha(B^+_\rho)} \le N [f]_{C^\alpha(B^+_r)} + N (r - \rho)^{-2-\alpha} \sup_{B^+_r} |u|,$$

where the constants N depend only on d, δ, and α. We have proved (3). Next, take $\rho < 1$, $r = (1/2)(1 + \rho)$ and observe that $1 - \rho = 2(r - \rho)$, $1 - r = (1/2)(1 - \rho)$. Then we get

$$(1 - \rho)^{2+\alpha} [D^s u]_{C^\alpha(B^+_\rho)} \le N (1 - r)^{2+\alpha} [f]_{C^\alpha(B^+_r)} + N \sup_{B^+_1} |u|,$$

which yields (4). This proves the theorem if f is smooth.

In the general case continue f as even function across $\partial_{x^1} B^+_1$, take f_n with compact support in $\mathbb{B}^+_1$ as in the proof of Theorem 3.10 and denote by f^n mollified f_n with kernels supported in $B_{1/(10n)}$. Then by the above we obtain a sequence u^n of solutions of $F(D^2 u^n) = f^n$ in B^+_1 with boundary condition $u^n = g$ on ∂B^+_1 whose $\hat{C}^{2+\alpha}(\mathbb{B}^+_1)$-norms are bounded by a constant independent of n, the latter being true by the same reasons as in the proof of Theorem 3.10. By the Aleksandrov-Bakel'man estimate u^n is a Cauchy sequence in $C(\bar{B}^+_1)$. Then a standard argument based on Arzelà-Ascoli theorem allows us to find a function $u \in \hat{C}^{2+\alpha}(\mathbb{B}^+_1)$, for which the first assertion of Theorem 1 is true. Since this was the only information we used in the first part of the proof, estimates (2), (3), and (4) follow and the theorem is proved. □

12. Proof of Theorem 1.6

We derive Theorem 1.6 from Theorem 11.1 and some other results of the previous chapters. In this section $\alpha_0 = \alpha_0(\delta, d)$ is the smallest of the constants called α_0 in Theorems 10.2.1 and 7.1 and Lemma 9.6.2, and $\alpha \in (0, \alpha_0)$. First we prove some a priori estimates.

1. LEMMA. *Let $u \in C(\bar{C}_2^+) \cap W^{1,2}_{\infty,\text{loc}}(C_2^+)$, $\beta \in (0, \alpha_0]$, and let u satisfy the equation*

$$\partial_t u + F[u] = 0 \tag{1}$$

(a.e.) in C_2^+. Then $\partial_t u \in C^{\alpha_0}_{\text{loc}}(\mathbb{C}_2^+)$ and there exists a constant N, depending only on β, d, δ, such that, if $0 < r < R \leq 2$, then

$$[\partial_t u]_{C^\beta(C_r^+)} \leq N \frac{1}{(R-r)^{2+\beta}} \sup_{C_R^+} |u|. \tag{2}$$

Proof. Take $h \in (0, R^2/10)$ and define

$$u_h(t,x) = \frac{u(t+h,x) - u(t,x)}{h}.$$

Then (recall that F is independent of t) one sees that u_h satisfies

$$\partial_t u_h + a^{ij} D_{ij} u_h = 0$$

in $C^+_{R^2-h,2}$, where (a^{ij}) is a certain $\mathbb{S}_\delta$-valued function on $C^+_{R^2-h,2}$.

Take $0 < \rho_0 < \rho_1 < R$ and concentrate on h so small that u_h is well defined in C_{ρ_1} (that is, such that $C^+_{\rho_1} \subset C^+_{R^2-h,R}$). Then take $z_1 = (t_1, x_1), z_2 = (t_2, x_2) \in C^+_{\rho_0}$, and consider two cases

(a) $8\rho(z_1, z_2) \geq \rho_1 - \rho_0$,

(b) $8\rho(z_1, z_2) < \rho_1 - \rho_0$.

In case (a)

$$\frac{|u_h(z_1) - u_h(z_2)|}{\rho^\beta(z_1, z_2)} \leq \frac{2 \cdot 8^\beta}{|\rho_1 - \rho_0|^\beta} \sup\big(|u_h|, C^+_{\rho_1}\big).$$

We assume that $t_1 \leq t_2$ and split case (b) into two subcases

(c) $4x_1^1 > \rho_1 - \rho_0$,

(d) $4x_1^1 \leq \rho_1 - \rho_0$.

In case (c) define $r = (\rho_1 - \rho_0)/8$ and note that, as is easy to see,

$$z_1, z_2 \in C_r(t_1, x_1) \subset C_{2r}(t_1, x_1) \subset C^+_{\rho_1}.$$

It follows by Lemma 9.6.2 that for all h small enough

$$\frac{|u_h(z_1) - u_h(z_2)|}{\rho^\beta(z_1, z_2)} \leq \frac{N}{|\rho_1 - \rho_0|^\beta} \sup\big(|u_h|, C^+_{\rho_1}\big), \tag{3}$$

where and below by N we denote various constants depending only on d and δ.

In case (d) define $r = (\rho_1 - \rho_0)/2$ and note that

$$z_1, z_2 \in C_r^+(t_1, 0, x_1') \subset C_{2r}^+(t_1, 0, x_1') \subset C^+_{\rho_1}.$$

It follows by Lemma 9.6.7 that (3) holds again.

Thus, for $0<\rho_0<\rho_1<R$ and all sufficiently small h we have

$$[u_h]_{C^\beta(C^+_{\rho_0})} \le \frac{N}{|\rho_1-\rho_0|^\beta} \sup\big(|u_h|, C^+_{\rho_1}\big). \tag{4}$$

Observe that u_h are bounded in $C^+_{\rho_1}$ uniformly with respect to h for small h since $\partial_t u$ is bounded in C_2. Then, as $h\downarrow 0$, (4) shows that $\partial_t u\in C^{\alpha_0}\big(C^+_{\rho_0}\big)$ and the arbitrariness of ρ_0 implies that $\partial_t u\in C^{\alpha_0}_{\text{loc}}\big(\mathbb{C}^+_2\big)$ and

$$[\partial_t u]_{C^\beta(C^+_{\rho_0})} \le \frac{N}{|\rho_1-\rho_0|^\beta} \sup\big(|\partial_t u|, C^+_{\rho_1}\big). \tag{5}$$

Now the only thing remaining to be proved is (2). We will derive it from (5) by using interpolation inequalities very similarly to what is done after (10.2.14). Observe that, for $h\in(0,\rho_1)$ and $z=(t,x)$, $z'=(t',x)\in C^+_{\rho_0}$, such $|t-t'|=h^2$, we have

$$|\partial_t u(z)| \le \Big|\partial_t u(z) - \frac{u(z)-u(z')}{t-t'}\Big| + \frac{2}{h^2}\sup\big(u, C^+_{\rho_1}\big)$$

$$\le h^\alpha[\partial_t u]_{C^\alpha(C^+_{\rho_1})} + \frac{2}{h^2}\sup\big(u, C^+_{\rho_1}\big).$$

By using the notation

$$M(\rho) := [\partial_t u]_{C^\beta(C^+_\rho)},$$

we infer that, if $\rho_n<\rho_{n+1}<R$ and $h\in(0,\rho_{n+1})$, then

$$M(\rho_n) \le \frac{N}{|\rho_{n+1}-\rho_n|^\beta}\sup\big(|\partial_t u|, C^+_{\rho_{n+1}}\big)$$

$$\le N_0^\beta \frac{h^\beta}{(\rho_{n+1}-\rho_n)^\beta} M(\rho_{n+1}) + N\frac{1}{(\rho_{n+1}-\rho_n)^\beta h^2}\sup\big(u, C^+_R\big).$$

By repeating further the arguments following (10.2.14), we arrive at (2). The lemma is proved. □

2. LEMMA. *Under the assumptions of Lemma* 1 *we have that* $u\in C^{2+\alpha}_{\text{loc}}(\mathbb{C}^+_2)$ *and estimate* (1.8) *holds and, for any* $\rho\in(0,2)$,

$$[D^2u]_{C^\alpha(C^+_\rho)} \le N(2-\rho)^{-2-\alpha}\sup_{C_2}|u|, \tag{6}$$

where N *depends only on* d,δ, *and* α.

Proof. Estimate (1.8) is obtained by a standard application of the maximum principle which we saw many times.

To prove that $u\in C^{2+\alpha}_{\text{loc}}(\mathbb{C}^+_2)$, observe that for almost any $t\in(0,4)$ we have $u(t,\cdot)\in W^2_{\infty,\text{loc}}\big(B^+_2\big)$ and $v_t:=u(t,\cdot)$ satisfies

$$F[v_t] = -\partial_t u(t,\cdot) \tag{7}$$

in B^+_2 (a.e.) with boundary condition $v_t=u(t,\cdot)$ on ∂B^+_2. Furthermore, $v_t\in C\big(\bar B^+_\rho\big)$. Owing to Lemma 1, $\partial_t u(t,\cdot)\in \hat C^\alpha\big(\mathbb{B}^+_2\big)$. By Theorem 11.1 equation (7) with boundary condition $u(t,\cdot)$ has a unique solution in $\hat C^{2+\alpha}\big(\mathbb{B}^+_2\big)$. By uniqueness (use Theorem 3.0.2), this new solution coincides with $u(t,\cdot)$. Thus, $u(t,\cdot)\in C^{2+\alpha}_{\text{loc}}\big(\mathbb{B}^+_2\big)$ for almost any $t\in(0,4)$.

Moreover, (11.3) implies that, for $\rho < 2$ and almost all $t \le \rho^2$,

$$\big[D^2u(t,\cdot)\big]_{C^\alpha(B_\rho)} \le N \sup_{s\le\rho^2} \big[\partial_t u(s,\cdot)\big]_{C^\alpha(B_{(\rho+2)/2})} + N(2-\rho)^{-2-\alpha}\sup_{C_2}|u|.$$

Actually, this estimate holds for all $t \le \rho^2$ since $u(t,x)$ is continuous in t. In particular, $u(t,\cdot) \in C^{2+\alpha}_{\text{loc}}(B_2)$ for all $t<4$.

Also by (2)

$$\sup_{t\le\rho^2}\big[\partial_t u(t,\cdot)\big]_{C^\alpha(B_{(\rho+2)/2})}(2-\rho)^{2+\alpha}$$
$$\le 2^{2+\alpha}\sup_{\rho<2}\big[\partial_t u\big]_{C^\alpha(C_\rho)}(2-\rho)^{2+\alpha} \le N\sup_{C_2}|u|, \tag{8}$$

so that

$$\sup_{t\le\rho^2}\big[D^2u(t,\cdot)\big]_{C^\alpha(B_\rho)} \le N(2-\rho)^{-2-\alpha}\sup_{C_2}|u|. \tag{9}$$

Estimates (8) and (9) yield (6) by Lemma 12.2.4. The lemma is proved. □

Proof of Theorem 1.6. As usual uniqueness follows from the maximum principle. To prove the existence, first assume that $g \in W^{1,2}_{\infty,\text{loc}}(\mathbb{R}^{d+1})$. In that case introduce $P[u] = P(D^2u)$, where P is defined by formula (6.2.4), and for $K>0$ consider the equation

$$\partial_t u + \max\big(F[u], P[u]-K\big) = 0$$

in C_2^+ (a.e.) with boundary condition $u=g$ on $\partial' C_2^+$. By Theorem 8.3.3, this problem has a unique solution $u_K \in C(\bar C_2^+) \cap W^{1,2}_{\infty,\text{loc}}(C_2^+)$ and by the maximum principle (cf. (12.2.6))

$$|u_K| \le N, \tag{10}$$

where and below by N we denote various constants independent of K.

Then recall that $\bar\delta$ is introduced in Remark 6.2.2 take $\beta \in \big(0, \alpha_0(\bar\delta, d)\big)$ and use Lemmas 1 and 2 to conclude that $u_K \in C^{2+\beta}_{\text{loc}}(\mathbb{C}_2^+)$ and, for any $r\in(0,2)$,

$$\big[D^2u_K\big]_{C^\beta(C_r)} \le N\frac{1}{(2-r)^{2+\beta}}. \tag{11}$$

Also by Theorem 9.8.1 the family $\{u_K, K>0\}$ is equicontinuous in $\bar C_2^+$. By the Arzelà-Ascoli and interpolation theorems, (10), and (11), we conclude that the set $\{u_K : K\ge 0\}$ is precompact in $C(\bar C_2^+)$ and in $C^2(C_r^+)$ with any $r<2$. Therefore, there is a sequence $K_n\to\infty$ such that u_{K_n}, Du_{K_n}, and $D^2u_{K_n}$ converge uniformly in $\bar C_2^+$, $C(C_r^+)$, and $C(C_r^+)$ for any $r<2$, respectively, to a function v, Dv, and D^2v, respectively. Obviously, $v \in C(\bar C_2^+)\cap C^{2+\beta}_{\text{loc}}(\mathbb{C}_2^+)$ and v satisfies (1.7) and (1.8). But then by Lemmas 1 and 2 the estimates (1.9) and (1.10) hold true. This proves the theorem if $g\in W^{1,2}_{\infty,\text{loc}}(\mathbb{R}^{d+1})$.

As in the elliptic case, we observe that in the above argument concerning the precompactness of $\{u_K : K\ge 0\}$ in $C(\bar C_2^+)$, in what concerns g, only the supremum of $|g|$ and the modulus of continuity of g on $\partial' C_2^+$ were involved.

In the case of general $g\in C\big(\overline{\partial' C_2^+}\big)$ we approximate it in $C\big(\overline{\partial' C_2^+}\big)$-norm by $g^n \in W^{1,2}_{\infty,\text{loc}}(\mathbb{R}^{d+1})$. Then we obtain the functions u^n corresponding to g^n. By the maximum principle

$$\sup_{\bar C_2^+}|u^n-u^m| \le \sup_{\partial' C_2^+}|g^n-g^m|.$$

Hence, there is $u \in C(\bar{C}_2^+)$ such that $u^n \to u$ uniformly in $\bar{C}_2$. Of course, $u = g$ on $\partial' C_2$. Furthermore, the estimates (1.8), (1.9), and (1.10) with u^n and g^n in place of u and g yield similar estimates for u, which shows that $u \in C^{2+\alpha}_{\mathrm{loc}}(\mathbb{C}_2^+)$ and leads to the desired result in an obvious way. □

CHAPTER 14

Nonlinear elliptic equations in $W^2_p(\Omega)$

1. Main results

In this chapter, we complement the results of Chapter 10 by considering elliptic equations in global Sobolev spaces and extending some results in [**79**]. Recall the main setting. We consider elliptic equations

$$H[v](x) := H\big(v(x), Dv(x), D^2v(x), x\big) = 0 \tag{1}$$

in subdomains of $\mathbb{R}^d$. Recall that $\mathbb{S}$ is the set of symmetric $d \times d$ matrices and, for $\delta \in (0, 1]$,

$$\mathbb{S}_\delta = \big\{a \in \mathbb{S} : \delta|\xi|^2 \le a^{ij}\xi^i\xi^j \le \delta^{-1}|\xi|^2, \quad \forall \xi \in \mathbb{R}^d\big\}.$$

In this chapter Ω is a bounded domain with $C^{1,1}$-boundary. We will say more about what we mean by that and what are $\rho_0(\Omega)$ and $M_2(\Omega)$ after we state our main results. Fix

$$p > d \quad \text{and a measurable function} \quad \bar{G} \ge 0 \quad \text{on} \quad \mathbb{R}^d.$$

Also fix some constants $K_0, K_F \in [0, \infty)$, and $\delta \in (0, 1]$.

We are going to work under the same assumptions as in Section 10.1 containing the parameters $\hat{\theta}$ and θ to be specified later, with one exception because we need an analog of Assumption 10.1.6 near the boundary. We mean the following.

1. Assumption. (i) The function F is Lipschitz continuous with respect to u'' with Lipschitz constant K_F and is measurable with respect to x.

There exist $R_0 \in (0, 1]$ and $\tau_0 \in [0, \infty)$ such that for any $\mathsf{u}'_0 \in \mathbb{R}$, $r \in (0, R_0]$, and $z \in \Omega$ one can find a *convex* function $\bar{F}(\mathsf{u}'') = \bar{F}_{z,r,\mathsf{u}'_0}(\mathsf{u}'')$ (independent of x) such that

(ii) We have $\bar{F}(0) = 0$ and $D_{\mathsf{u}''}\bar{F} \in \mathbb{S}_\delta$ at all points of differentiability of $\bar{F}$;

(iii) For any $\mathsf{u}'' \in \mathbb{S}$ with $|\mathsf{u}''| = 1$, we have

$$\int_{B_r(z)\cap\Omega} \sup_{\tau > \tau_0} \tau^{-1}\big|F\big(\mathsf{u}'_0, \tau\mathsf{u}'', x\big) - \bar{F}(\tau\mathsf{u}'')\big|\,dx \le \theta\big|B_r(z) \cap \Omega\big|; \tag{2}$$

(iv) There exists a continuous increasing function $\omega_F(\tau)$, $\tau \ge 0$, such that $\omega_F(0) = 0$ and for any $\mathsf{u}'_0, \mathsf{v}'_0 \in \mathbb{R}$, $x \in \Omega$, and $\mathsf{u}'' \in \mathbb{S}$ we have

$$\big|F\big(\mathsf{u}'_0, \mathsf{u}'', x\big) - F\big(\mathsf{v}'_0, \mathsf{u}'', x\big)\big| \le \omega_F\big(|\mathsf{u}'_0 - \mathsf{v}'_0|\big)|\mathsf{u}''|.$$

2. Assumption. We are given a function $g \in W^2_p(\Omega)$.

We remind the reader that according to Assumption 10.1.4 and Remark 10.1.7 we have

$$F\big(\mathsf{u}'_0, 0, x\big) = 0$$

for any $\mathsf{u}'_0 \in \mathbb{R}$ and $x \in \Omega$.

Recall that $W^2_p(\Omega)$ denotes the set of functions v defined on Ω such that v, Dv, and D^2v are in $L_p(\Omega)$. One knows that, since Ω is smooth enough and $p > d$, for every $u \in W^2_p(\Omega)$, there is a modification $v \in C^1(\bar\Omega)$ such that $u = v$ (a.e.) in Ω (and, of course, $v \in W^2_p(\Omega)$). In the future, speaking about elements of $W^2_p(\Omega)$ we will always mean such modifications.

We now come to the main a priori estimate in Sobolev spaces. Observe that in Theorem 3 we do not even assume that equation (1) is elliptic. Recall that the function $\omega_{F,u,\Omega}$ used in this theorem is introduced in Definition 10.1.9.

3. THEOREM. *Suppose that Assumption* 2 *is satisfied. Then there exist constants* $\hat\theta, \theta \in (0,1]$, *depending only on* d, p, δ, K_F, *and* $M_2(\Omega)$, *such that, if Assumptions* 10.1.4 *and* 1 *are satisfied with these* $\hat\theta$ *and* θ, *respectively, then for any* $u \in W^2_p(\Omega)$, *that satisfies* (1) *in* Ω *(a.e.) and equals* g *on* $\partial\Omega$, *we have*

$$\|u\|_{W^2_p(\Omega)} \leq N\big(\|\bar G\|_{L_p(\Omega)} + \|g\|_{W^2_p(\Omega)} + \|u\|_{C(\Omega)}\big) + N\tau_0, \tag{3}$$

where the constants N *depend only on* K_0, K_F, d, p, δ, R_0, $\rho_0(\Omega)$, $\mathrm{diam}(\Omega)$, $M_2(\Omega)$ *and the functions* $\omega_{F,u,\Omega}$ *and* $\omega_{F,g,\Omega}$.

4. REMARK. Estimate (3) may look ridiculous because the constants depend on u and g. However, note that if F is independent of u'_0, then $\omega_{F,u,\Omega} \equiv \omega_{F,g,\Omega} \equiv 0$ and the constants are *independent* of u and g.

Here is the main result of the chapter concerning the solvability of (1) in global Sobolev spaces.

5. THEOREM. *Suppose that Assumptions* 10.1.11, 10.1.12, *and* 2 *are satisfied and* $\bar G \in L_p(\Omega)$. *Then there exist constants* $\hat\theta, \theta \in (0,1]$, *depending only on* d, p, δ, K_F, *and* $M_2(\Omega)$, *such that, if Assumptions* 10.1.4 *and* 1 *are satisfied with these* $\hat\theta$ *and* θ, *respectively, then there exists* $u \in W^2_p(\Omega)$ *satisfying* (1) *in* Ω *(a.e.) and such that* $u = g$ *on* $\partial\Omega$.

In the same way as Theorem 10.1.16 is derived from Theorem 10.1.14, Theorem 5 yields the following.

6. THEOREM. *Suppose that Assumptions* 10.1.3, 10.1.11, 10.1.12, *and* 2 *are satisfied,* $\bar G \in L_p(\Omega)$, *and* (10.1.5) *holds true. Let* $P(\mathsf{u}'')$ *be a convex function on* $\mathbb{S}$ *such that at each point of its differentiability* $D_{\mathsf{u}''}P \in \mathbb{S}_{\delta'}$, *where* $\delta' \in (0,\delta]$. *Also assume that for any* $a \in \mathbb{S}_\delta$ *and* $\mathsf{u}'' \in \mathbb{S}$ *we have*

$$a^{ij}\mathsf{u}''_{ij} \leq P(\mathsf{u}'') + K,$$

where K *is a constant. Then the equation*

$$\max\big(H[u], P[u]\big) = 0$$

in Ω *(a.e.) with boundary condition* $u = g$ *on* $\partial\Omega$ *has a solution* $u \in W^2_p(\Omega)$.

7. REMARK. We know from Lemma 6.6.10 that under the assumptions of Theorem 5 the term $\|u\|_{C(\Omega)}$ in (3) can be replaced with $\sup_{\partial\Omega}|g|$.

One more solvability result, in which there is no monotonicity assumption on the dependence of H on u'_0, is the following.

8. THEOREM. *There exist constants* $\hat{\theta}, \theta \in (0, 1]$, *depending only on* d, p, δ, *and* K_F, *such that, if Assumptions* 1 *and* 10.1.4 *are satisfied with these* θ *and* $\hat{\theta}$, *respectively, and Assumptions* 10.1.11 *and* 6.6.4 *are also satisfied with* $M_0 \geq 0, \varepsilon > 0$, *and* $|g| \leq M_0$, *and* $\bar{G} \in L_p(\Omega)$, *then there exists* $u \in W_p^2(\Omega)$ *satisfying* (1) *in* Ω *(a.e.) and such that* $u = g$ *on* $\partial\Omega$.

9. REMARK. The reader is sent to comments about the above results and examples of applications to Section 10.1; they can be repeated literally. We only mention the paper by N. Winter [**116**] where a theorem similar to our Theorem 5 is proved by using the theory of L_p-viscosity solutions. Obtaining boundary W_p^2 estimates by using this theory turned out to be extremely challenging: the paper [**116**] appeared only in 2009, twenty years after the work of Caffarelli [**12**].

The most notable distinction between our Theorem 5 and the corresponding result in [**116**] is that H is assumed to be convex in [**116**] and in the case of linear equations the result of [**116**] is only applicable if $D_{u''}H$ is uniformly sufficiently close to uniformly continuous $\mathbb{S}_\delta$-valued functions. By the way, in the case of linear equations Theorem 5 contains Theorem 4.4 of [**21**] (apart from the restriction on p) about equations with VMO leading coefficients. It, actually, extends this result to quasilinear equations (cf. Examples 10.1.24 and 10.1.25).

10. DEFINITION. We say that Ω is a $C^{1,1}$-domain if there exists $\rho_0 = \rho_0(\Omega) \in (0, 1]$ for which at any point $z_0 \in \partial\Omega$ there is an orthonormal system of coordinates $\Psi(z_0)$ with the origin at z_0 such that in the new coordinates $\tilde{x} = (\tilde{x}^1, \tilde{x}')$, $\tilde{x}^1 \in \mathbb{R}$, $\tilde{x}' \in \mathbb{R}^{d-1}$, there exists a function

$$\psi \in C^{1,1}\big(\{\tilde{x}' \in \mathbb{R}^{d-1} : |\tilde{x}'| \leq 8\rho_0\}\big)$$

with the $C^{1,1}(B_{8\rho_0})$-norm majorated by a constant $M_2(\Omega)$ independent of z_0 and such that

$$\psi(0) = 0, \quad D_{\tilde{x}'}\psi(0) = 0, \quad \big|D_{\tilde{x}'}\psi(\tilde{x}')\big| \leq 1 \quad \text{for} \quad |\tilde{x}'| \leq 8\rho_0,$$

$$\big\{\tilde{x} = (\tilde{x}^1, \tilde{x}') : |\tilde{x}'| \leq 8\rho_0,\ \psi(\tilde{x}') - 8\rho_0 \leq \tilde{x}^1 \leq \psi(\tilde{x}') + 8\rho_0\big\} \cap \Omega$$
$$= \{\tilde{x} : |\tilde{x}'| \leq 8\rho_0,\ \psi(\tilde{x}') < \tilde{x}^1 \leq \psi(\tilde{x}') + 8\rho_0\}.$$

Below in this section we assume that

$$0 \in \partial\Omega$$

and that the original system of coordinates in $\mathbb{R}^d$ coincides with the one described above for $z_0 = 0$.

11. LEMMA. *Introduce*

$$\Gamma := \big\{x : |x'| \leq 8\rho_0(\Omega), \psi(x') < x^1 \leq \psi(x') + 8\rho_0(\Omega)\big\} \quad (\subset \Omega),$$

$$\hat{\Gamma} := \big\{y : |y'| \leq 8\rho_0(\Omega), 0 < y^1 \leq 8\rho_0(\Omega)\big\}.$$

Also introduce a mapping $x \to y(x)$ *of* Γ *onto* $\hat{\Gamma}$ *by*

$$x^1 \to y^1 = y^1(x) = x^1 - \psi(x'), \quad x' \to y' = y'(x) = x'. \tag{4}$$

Then this mapping has an inverse $y \to x(y)$. *Furthermore, the Jacobians of both mappings are equal to one.*

This lemma is obvious.

It is convenient to extend $\psi(x')$ for $|x'| \geq 8\rho_0(\Omega)$, so that the extension is smooth and has the magnitude of the gradient bounded by one and define $y(x)$ by the same formula (4) for all $x \in \mathbb{R}^d$. Of course, by $x(y)$ we mean the inverse of $y(x)$. Obviously, the assertions of Lemma 11 hold true for such extensions.

12. Remark. For $r \in (0, \infty)$ and $z \in \mathbb{R}^d$ define

$$\check{B}_r^+ = x\big(B_r^+\big), \quad \check{B}_r^+(z) = x\Big(B_r^+\big(y(z)\big)\Big).$$

Then, as is easy to see

(i) $\check{B}_r^+ \subset \Gamma \subset \Omega$ if $r \leq 8\rho_0(\Omega)$;

(ii) $\check{B}_r^+(z) \subset \check{B}^+_{8\rho_0(\Omega)}$ if $r, \rho > 0$, $\rho + r \leq 8\rho_0(\Omega)$, and $z \in \check{B}_\rho^+$.

13. Lemma. *(i) If $z \in \check{B}^+_{2\rho_0(\Omega)}$ and $r \leq 2\rho_0(\Omega)$, then*

$$\check{B}_r^+(z) \subset B_{2r}(z) \cap \Omega, \quad B_r(z) \cap \Omega \subset \check{B}_{2r}^+(z) \quad (\subset \Gamma); \tag{5}$$

(ii) If $z \in \check{B}_\rho^+$ and $\nu \geq 1$ and $\rho + \nu r \leq 8\rho_0(\Omega)$, then

$$\big|\check{B}_{\nu r}^+(z)\big| \leq \nu^d \big|\check{B}_r^+(z)\big|. \tag{6}$$

Proof. (i). First notice that $\check{B}_r^+(z) \subset \Gamma$. Then, since $|D_{x'}\psi| \leq 1$, for any $x_1, x_2 \in \Gamma$ we have $\big|y(x_1) - y(x_2)\big| \leq 2|x_1 - x_2|$ and $\big|x(y_1) - x(y_2)\big| \leq 2|y_1 - y_2|$ if $y_1, y_2 \in \hat{\Gamma}$. In particular, if $\big|y - y(z)\big| \leq r$, then $\big|x(y) - z\big| \leq 2r$, so that $\check{B}_r^+(z) \subset B_{2r}(z)$ and

$$\check{B}_r^+(z) \subset B_{2r}(z) \cap \Gamma \subset B_{2r}(z) \cap \Omega,$$

which proves the first inclusion in (5). In particular, $\check{B}^+_{2\rho_0(\Omega)} = \check{B}^+_{2\rho_0(\Omega)}(0) \subset B_{4\rho_0(\Omega)}$.

Furthermore, if $|x - z| \leq r \leq 2\rho_0(\Omega)$ and $x \in \Omega$, then, since $z \in B_{4\rho_0(\Omega)}$,

$$x \in B_{6\rho_0(\Omega)} \cap \Omega \subset \Gamma \quad \text{and} \quad y^1(x) > 0.$$

Then $\big|y(x) - y(z)\big| \leq 2r$ and $y(x) \in \hat{\Gamma}$, that is, $y(x) \in B_{2r}^+\big(y(z)\big)$ so that $x \in x\Big(B_{2r}^+\big(y(z)\big)\Big)$, which yields the second inclusion in (5).

To prove (ii), it suffices to note that

$$\big|\check{B}_{\nu r}^+(z)\big| = \big|B_{\nu r}^+\big(y(z)\big)\big| \leq \nu^d \big|B_r^+\big(y(z)\big)\big| = \nu^d \big|\check{B}_r^+(z)\big|.$$

The lemma is proved. □

14. Corollary. *If $z \in \check{B}^+_{2\rho_0(\Omega)}$ and $r \leq \rho_0(\Omega)$, then for any measurable function g*

$$\fint_{\check{B}_r^+(z)} \big|g(x)\big|\,dx \leq N(d) \fint_{B_{2r}(z)\cap\Omega} \big|g(x)\big|\,dx. \tag{7}$$

Indeed, the domain of integration on the right is wider than the one the left owing to (5), and

$$N(d)\big|\check{B}_r^+(z)\big| \geq \big|\check{B}_{4r}^+(z)\big| \geq \big|B_{2r}(z)\cap\Omega\big|$$

in light of (6) and (5).

Another notable property of bounded domains of class $C^{1,1}$, which however we are not going to use, is the following.

15. THEOREM. *There exist $\gamma_0 \in (0,1]$ and N depending only on δ, K_0, d, and Ω such that for any $\mathcal{L} \in \mathbb{L}_{\delta,K_0}$, $\gamma \in (0,\gamma_0]$, and $u \in W^2_{d,\text{loc}}(\Omega)\cap C(\bar{\Omega})$ we have*

$$\int_\Omega \big(|D^2u|^\gamma + |Du|^\gamma\big)\,dx \leq N\|\mathcal{L}u\|^\gamma_{L_d(\Omega)} + N\sup_{\partial\Omega}|u|^\gamma. \tag{8}$$

Proof. One can represent $\bar{\Omega}$ as a finite union of the closures of $C^{1,1}$-domains each of which admits a one-to-one $C^{1,1}$ mapping on B_1 with $C^{1,1}$ inverse. Then after changing coordinates in each domain one can use Theorems 9.4.6 and 9.4.12 applied to appropriately changed operator $\mathcal{L}$. For the transformed operator the constants δ and K_0 may change but still will only depend on δ, K_0, d, and Ω. Then after combining the results of application of Theorems 9.4.6 and 9.4.12 one obtains (8) with $\bar{\Omega}$ in place of $\partial\Omega$. However, Aleksandrov's estimate shows that this replacement can be avoided on the account of, perhaps, increasing the first N on the right in (8). The theorem is proved. □

2. Boundary a priori estimates in the simplest case

Let $F(\mathsf{u}'')$ be a *convex* function of u'' (independent of x) such that

(i) $F(0) = 0$,

(ii) $D_{\mathsf{u}''}F \in \mathbb{S}_\delta$ at all points of differentiability of F, where $\delta \in (0,1]$ is a fixed number.

Recall that

$$B_r^+ = \{x \in \mathbb{R}^d : |x| < r, x^1 > 0\}, \quad B_r^+(z) = \{x \in \mathbb{R}^d : |x - z| < r, x^1 > 0\}.$$

1. LEMMA. *There exist $\bar{\gamma} = \bar{\gamma}(d,\delta) \in (0,1)$ and $\alpha_0 = \alpha_0(\delta,d) \in (0,1)$ such that for any $\gamma \in (0,\bar{\gamma}]$, $\alpha \in (0,\alpha_0)$, $r > 0$, $x_0 \in \bar{\mathbb{R}}^d_+$, $\nu \geq 12$, and*

$$u \in \bigcap_{\rho<\nu r} W^2_d(B_\rho^+(x_0)),$$

such that $u(x)$ vanishes on $\partial_{x^1}B^+_{\nu r}(x_0)$ if this set in nonempty, we have

$$\Big(\fint_{B_r^+(x_0)}\fint_{B_r^+(x_0)}\big|D^2u(x_1) - D^2u(x_2)\big|^\gamma\,dx_1dx_2\Big)^{1/\gamma}$$

$$\leq N\nu^{d/\gamma}\Big(\fint_{B^+_{\nu r}(x_0)}\big|F[u]\big|^d\,dx\Big)^{1/d} + N\nu^{-\alpha}\Big(\fint_{B^+_{\nu r}(x_0)}|D^2u|^d\,dx\Big)^{1/d}, \tag{1}$$

where the constants N depend only on d, α, and δ.

Proof. We take $\bar\gamma$ from (10.3.4) and α_0 from Theorem 13.1.2. Scalings show that it suffices to prove the lemma only for $\nu r = 3$. Furthermore, without loss of generality we may assume that $x_0 = \big(|x_0|, 0, ..., 0\big)$. Then we consider two cases.

Case 1: $|x_0| > 1/2$. In this case, we have

$$B^+_r(x_0) = B^+_{3/\nu}(x_0) = B_r(x_0) \subset B_{\nu' r}(x_0) \subset \mathbb{R}^d_+,$$

where $\nu' = \nu/6 \ (\geq 2)$. Therefore, inequality (1) is an immediate consequence of Lemma 10.3.3.

Case 2: $|x_0| \leq 1/2$. Since $r = 3/\nu \leq 1/2$, we have

$$B^+_r(x_0) \subset B^+_1 \subset B^+_2 \subset B^+_3(x_0) = B^+_{\nu r}(x_0).$$

Define $\hat u(x) := u(x) - x^1 (D_1 u)_{B^+_2}$. Let v be the classical solution of $F[v] = 0$ in B^+_2 with boundary condition $v = \hat u$ on ∂B^+_2. Such a solution exists due to Theorem 13.1.2, which also provides an estimate on $D^2 v$, that combined with Poincaré's inequality yields

$$\fint_{B^+_r(x_0)} \fint_{B^-_r(x_0)} \big|D^2 v(x_1) - D^2 v(x_2)\big| \, dx_1 dx_2 \leq N r^\alpha [D^2 v]_{C^\alpha(B^+_1)}$$

$$\leq N r^\alpha \sup_{\partial B^+_2} |v| = N r^\alpha \sup_{\partial B^+_2} |\hat u| \leq N r^\alpha \sup_{B^+_2} |\hat u| \leq N \nu^{-\alpha} \Big(\fint_{B^+_2} |D^2 u|^d \, dx \Big)^{1/d}.$$

Recall that $\gamma \in (0, 1]$. By Hölder's inequality, we get

$$\fint_{B^+_r(x_0)} \fint_{B^+_r(x_0)} \big|D^2 v(x_1) - D^2 v(x_2)\big|^\gamma \, dx_1 dx_2$$

$$\leq N \nu^{-\gamma\alpha} \Big(\fint_{B^+_{\nu r}(x_0)} |D^2 u|^d \, dx \Big)^{\gamma/d}. \tag{2}$$

Next, use again that

$$f := F(D^2 u) = F(D^2 \hat u) - F(D^2 v) = a^{ij} D_{ij} w$$

in B^+_2 (a.e.) and $w = 0$ on ∂B^+_2, where (a^{ij}) is a measurable $\mathbb{S}_\delta$-valued function and $w = \hat u - v$. By a scaled version of Corollary 9.4.13 we get (recall that $\nu r = 3$)

$$\fint_{B^+_r(x_0)} |D^2 w|^\gamma \, dx \leq N r^{-d} \int_{B^+_2} |D^2 w|^\gamma \, dx \leq N \nu^d \Big(\fint_{B^+_{\nu r}(x_0)} |f|^d \, dx \Big)^{\gamma/d}$$

and

$$\fint_{B^+_r(x_0)} \fint_{B^+_r(x_0)} \big|D^2 w(x_1) - D^2 w(x_2)\big|^\gamma \, dx_1 dx_2 \leq N \nu^d \Big(\fint_{B^+_{\nu r}(x_0)} |f|^d \, dx \Big)^{\gamma/d}.$$

Combining this with (2) and observing that $D^2 u = D^2 v + D^2 w$ yield (1) in Case 2 as well. The lemma is proved. □

As is pointed out on page 229, partitions of unity, commonly used in the theory of linear and nonlinear equations (see, for instance, [**70**] and [**30**]), are not applicable in our situation and we use a different localizing techniques.

Recall that $\rho_0(\Omega)$ is introduced in Definition 1.10. In the rest of the section we assume that

$$0 \in \partial\Omega$$

and that the original system of coordinates in $\mathbb{R}^d$ coincides with the one described in Definition 1.10 with $z_0 = 0$. We will be using the mappings $y(x)$ and $x(y)$ and the sets $\check{B}_r^+$ and $\check{B}_r^+(z)$ introduced in Remark 1.12.

2. Lemma. *There exist $\bar{\gamma} = \bar{\gamma}(d,\delta) \in (0,1)$ and $\alpha_0 = \alpha_0(\delta, d) \in (0,1)$ such that for any $\gamma \in (0,\bar{\gamma}]$ and $\alpha \in (0,\alpha_0)$, whenever*
(i) $r, \rho > 0$, $\nu \geq 12$, $\rho + \nu r \leq 8\rho_0(\Omega)$, $x_0 \in \check{B}_\rho^+$,
(ii) $u \in W_p^2(\check{B}_{\rho+\nu r}^+)$ and $u = 0$ on $\partial\check{B}_{\rho+\nu r}^+ \cap \partial\Omega$,
we have

$$I_r(x_0) := \Big(\fint_{\check{B}_r^+(x_0)} \fint_{\check{B}_r^+(x_0)} \big|D^2u(x_1) - D^2u(x_2)\big|^\gamma \, dx_1 dx_2 \Big)^{1/\gamma}$$

$$\leq N\nu^{d/\gamma} \Big(\fint_{\check{B}_{\nu r}^+(x_0)} \big(\big|F(D^2u)\big|^d + |Du|^d\big)\, dx \Big)^{1/d}$$

$$+ N(\nu^{1+d/\gamma} r + \nu^{-\alpha}) \Big(\fint_{\check{B}_{\nu r}^+(x_0)} |D^2u|^d \, dx \Big)^{1/d}, \tag{3}$$

where the constants N depend only on $M_2(\Omega), d, \gamma$, α, and δ.

Proof. By the change of variables formula recalling that the Jacobian of $x(y)$ is equal to one, we see that

$$I_r(x_0) = \Big(\fint_{B_r^+(y(x_0))} \fint_{B_r^+(y(x_0))} \big|(D^2u)\big(x(y_1)\big) - (D^2u)\big(x(y_2)\big)\big|^\gamma \, dy_1 dy_2 \Big)^{1/\gamma}.$$

In this connection, denote by $A(y) := \partial x(y)/\partial y$ the matrix whose the ij'th entry is $\partial x^i(y)/\partial y^j$, observe that $A(y)$ and its inverse are bounded by a constant depending only on d (recall that $|D_{x'}\psi| \leq 1$), and define

$$A_0 = A\big(y(x_0)\big), \quad \hat{F}(\mathsf{u}'') = F\big((A_0^{-1})^* \mathsf{u}'' A_0^{-1}\big).$$

Notice that

$$D_{\mathsf{u}''_{ij}} \hat{F}(\mathsf{u}'') = (D_{\mathsf{u}''_{rk}} F)\big((A_0^{-1})^* \mathsf{u}'' A_0^{-1}\big)(A_0^{-1})^{ir}(A_0^{-1})^{jk},$$

$$D_{\mathsf{u}''} \hat{F}(\mathsf{u}'') = A_0^{-1}(D_{\mathsf{u}''} F)\big((A_0^{-1})^* \mathsf{u}'' A_0^{-1}\big)(A_0^{-1})^*,$$

which, in particular, implies that $D_{\mathsf{u}''}\hat{F} \in \mathbb{S}_{\hat{\delta}}$, where $\hat{\delta} = \hat{\delta}(d,\delta) \in (0,1]$.

Next, introduce the function

$$\hat{u}(y) = u\big(x(y)\big),$$

which belongs to $W_p^2(B_{\rho+\nu r}^+)$; since $|y(x_0)| < \rho$, it also belongs to $W_p^2\big(B_{\nu r}^+(y(x_0))\big)$ and vanishes on $\partial_{x^1} B_{\nu r}^+\big(y(x_0)\big)$ if this set in nonempty. By Lemma 1, since $\nu \geq 12$, we have

$$\Big(\fint_{B_r^+(y(x_0))} \fint_{B_r^+(y(x_0))} \big|D^2\hat{u}(y_1) - D^2\hat{u}(y_2)\big|^\gamma \, dy_1 dy_2 \Big)^{1/\gamma}$$

$$\le N\nu^{d/\gamma} \Big(\fint_{B_{\nu r}^+(y(x_0))} \big|\hat{F}(D^2\hat{u})\big|^d \, dy \Big)^{1/d} + N\nu^{-\alpha} \Big(\fint_{B_{\nu r}^+(y(x_0))} |D^2\hat{u}|^d \, dy \Big)^{1/d}. \quad (4)$$

Observe also that for $y = y(x)$ and $x = x(y)$

$$D\hat{u}(y) = (Du)(x)A(y),$$

where the D's are row vectors, and

$$D^2\hat{u}(y) = A^*(y)\big[D^2u(x)\big]A(y) + \big[D_k u(x)\big]D^2x^k(y). \quad (5)$$

Since

$$\big|A_0 - A(y)\big| \le N\big|y - y(x_0)\big|,$$

where N depends only on d and the bound on $|D^2\psi|$, for $y_1, y_2 \in B_r^+\big(y(x_0)\big)$, we have

$$\big|D^2\hat{u}(y_1) - D^2\hat{u}(y_2)\big| \ge (1/N)\big|D^2u(x_1) - D^2u(x_2)\big|$$

$$-Nr\big(\big|D^2u(x_1)\big| + \big|D^2u(x_2)\big|\big) - N\big(\big|Du(x_1)\big| + \big|Du(x_2)\big|\big),$$

where $x_i = x(y_i)$ and N depends only on $M_2(\Omega)$ and d. Hence, the left-hand side of (4) is greater than or equal to

$$(1/N)I_r(x_0) - N\Big(\fint_{\check{B}_r^+(x_0)} \big(r|D^2u| + |Du|\big)^\gamma \, dx \Big)^{1/\gamma}$$

$$\ge (1/N)I_r(x_0) - Nr\Big(\fint_{\check{B}_r^+(x_0)} |D^2u|^d \, dx \Big)^{1/d} - N\Big(\fint_{\check{B}_r^+(x_0)} |Du|^d \, dx \Big)^{1/d}$$

$$\ge (1/N)I_r(x_0) - N\nu r\Big(\fint_{\check{B}_{\nu r}^+(x_0)} |D^2u|^d \, dx \Big)^{1/d} - N\nu\Big(\fint_{\check{B}_{\nu r}^+(x_0)} |Du|^d \, dx \Big)^{1/d} \quad (6)$$

where the first inequality follows by Hölder's inequality and the second one is true owing to (1.6).

In what concerns the first term on the right-hand side of (4), observe that, owing to the Lipschitz continuity of F, the fact that $\big|A(y)A_0^{-1} - (\delta^{ij})\big| \le N\big|y - y(x_0)\big|$, and (5), we have (with $x = x(y)$)

$$\big|\hat{F}\big(D^2\hat{u}(y)\big) - F\big(D^2u(x)\big)\big| \le \big|F\big((A_0^{-1})^*A^*(y)\big[D^2u(x)\big]A(y)A_0^{-1}\big)$$

$$-F\big(D^2u(x)\big)\big| + N\big|Du(x)\big| \le N\big|y - y(x_0)\big|\big|D^2u(x)\big| + N\big|Du(x)\big|.$$

This and an easy estimate of the last term in (4) shows that its right-hand side is less than

$$N\nu^{d/\gamma}\Big(\fint_{\check{B}^+_{\nu r}(x_0)}|F(D^2u)|^d\,dx\Big)^{1/d}+N(\nu^{1+d/\gamma}r+\nu^{-\alpha})\Big(\fint_{\check{B}^+_{\nu r}(x_0)}|D^2u|^d\,dx\Big)^{1/d}$$

$$+N(\nu^{d/\gamma}+\nu^{-\alpha})\Big(\fint_{\check{B}^+_{\nu r}(x_0)}|Du|^d\,dx\Big)^{1/d}.$$

Upon combining this result with what was said about (6) we come to (3), thus proving the lemma. □

We see that, basically, Lemma 2 is derived from Lemma 1 by using a change of variables. In the following remark we also use this change of variables.

3. Remark. Suppose that condition (10.1.5) and Assumptions 10.1.3 and 10.1.11 are satisfied. Let $r\le 8\rho_0(\Omega)$ and $u\in W^2_p(\check{B}^+_r)$ (recall that $p\ge d$ and, actually, we fixed $p>d$) be such that $u=0$ on $\partial\Omega\cap\partial\check{B}^+_r$. Then

$$\int_{\check{B}^+_r}(|D^2u|^\gamma+|Du|^\gamma)\,dx\le Nr^{d-\gamma d/p}\|H[u]\|^\gamma_{L_p(\check{B}^+_r)}$$

$$+Nr^{d-\gamma d/p}\|\bar{G}\|^\gamma_{L_p(\check{B}^+_r)}+Nr^{d-2\gamma}\sup_{\check{B}^+_r}|u|^\gamma, \tag{7}$$

where N depend only on δ, K_0, d, p, and $M_2(\Omega)$, and $\gamma\in(0,\gamma_0]$, where $\gamma_0=\gamma_0(d,\delta,K_0,M_2(\Omega))\in(0,1)$ is a certain constant specified below.

Indeed, by using the notation from the above proof and using equation (9.7.2) in Lemma 9.7.2 introduce the operators

$$\mathcal{L}u(x)=a^{ij}(x)D_{ij}u(x)+b^i(x)D_iu(x),\quad \hat{\mathcal{L}}\hat{u}(y)=[\mathcal{L}u]\big(x(y)\big).$$

The operator $\hat{\mathcal{L}}$ can be written as a differential operator with derivatives with respect to y. Clearly, its matrix of second-order derivatives will belong to $\mathbb{S}_{\hat{\delta}}$ for a $\hat{\delta}=\hat{\delta}\big(\delta,M_2(\Omega),d\big)\in(0,1)$ and the drift term by magnitude will be dominated by $N=N\big(K_0,d,M_2(\Omega)\big)$. Since

$$\big|\hat{\mathcal{L}}\hat{u}(y)\big|\le\big|H[u]\big(x(y)\big)\big|+\bar{G}\big(x(y)\big)+K_0\big|u\big(x(y)\big)\big|$$

in $B^+_{8\rho_0(\Omega)}$, by Corollaries 9.4.7 and 9.4.13, for a $\gamma_0=\gamma_0\big(d,\delta,K_0,M_2(\Omega)\big)\in(0,1)$ and $\gamma\in(0,\gamma_0]$, after using scalings (where we use that $\rho_0(\Omega)\le 1$ to take care of the first-order derivatives) and Hölder's inequality (to replace d with p) we get

$$\int_{B^+_r}(|D^2_y\hat{u}|^\gamma+|D_y\hat{u}|^\gamma)\,dy\le Nr^{d-\gamma d/p}\Big(\int_{B^+_r}\big|H[u]\big(x(y)\big)\big|^p\,dy\Big)^{\gamma/p}$$

$$+Nr^{d-\gamma d/p}\Big(\int_{B^+_r}\big|\bar{G}\big(x(y)\big)\big|^p\,dy\Big)^{\gamma/p}+Nr^{d-2\gamma}\sup_{\check{B}^+_r}|u|^\gamma.$$

Now (7) follows after changing variables.

3. A priori estimates in W^2_p near the boundary and the proof of Theorem 1.3

We assume that

$$0 \in \partial\Omega,$$

take $\rho_0 = \rho_0(\Omega)$ from Definition 1.10, and suppose that the original system of coordinates in $\mathbb{R}^d$ coincides with the one described in Definition 1.10 for $z_0 = 0$. We also take $\check{B}^+_r$, $\check{B}^+_r(z)$ from Remark 1.12 and suppose that the assumptions of Theorem 1.3 are satisfied with $\hat{\theta}$ and θ which are yet to be specified.

First we note the following.

1. LEMMA. *For any $q \in [1,\infty)$ and $\mu > 0$ there exists $\theta = \theta(d,\delta,K_F,\mu,q) > 0$ such that, if Assumption 1.1 is satisfied with this θ, then for any $\mathsf{u}'_0 \in \mathbb{R}$, $z \in \check{B}^+_{2\rho_0(\Omega)}$ and $2r \le \rho_0(\Omega) \wedge R_0$, we have*

$$\fint_{\check{B}^+_r(z)} \sup_{\substack{\mathsf{u}''\in\mathbb{S},\\ |\mathsf{u}''|>\tau_0}} \frac{\big|F(\mathsf{u}'_0,\mathsf{u}'',x) - \bar{F}(\mathsf{u}'')\big|^q}{|\mathsf{u}''|^q}\,dx \le \mu^q,$$

where $\bar{F} = \bar{F}_{z,r,\mathsf{u}'_0}$ is taken from Assumption 1.1.

For the proof of this lemma note that, in light of Corollary 1.14 and Assumption 1.1, for any $\mathsf{u}'_0 \in \mathbb{R}$ and $\mathsf{u}'' \in \mathbb{S}$ with $|\mathsf{u}''| = 1$ we have

$$\fint_{\check{B}^+_r(z)} \sup_{\tau>\tau_0} \tau^{-1}\big|F(\mathsf{u}'_0,\tau\mathsf{u}'',x) - \bar{F}(\tau\mathsf{u}'')\big|\,dx \le N(d)\theta$$

if $r \le \rho_0(\Omega)$ and $2r \le R_0$. After that the assertion of the lemma is obtained by repeating word for word the proof of Lemma 10.4.1.

Recall that $\omega_{F,u,\Omega}(\rho)$ is introduced in Definition 10.1.9.

2. LEMMA. *Let $\rho, r > 0$ and $\nu \ge 12$ satisfy $2\nu r \le \rho_0(\Omega) \wedge R_0$ and $\rho \le 2\rho_0(\Omega)$. Take*

$$\mu \in (0,\infty), \quad \beta \in (1,\infty),$$

and suppose that Assumption 1.1 is satisfied with $\theta = \theta(d,\delta,K_F,\mu,\beta d)$ (see Lemma 1). Assume that we are given a function u on Ω such that $u \in W^2_p(\check{B}^+_{\rho+\nu r})$ and $u = 0$ on $\bar{\check{B}}^+_{\rho+\nu r} \cap \partial\Omega$. Use $I_r(x_0)$ introduced in (2.3).

Then, for γ and α from Lemma 2.2 and $x_0 \in \check{B}^+_\rho$ we have

$$I_r(x_0) \le N\nu^{d/\gamma}\Big(\fint_{\check{B}^+_{\nu r}(x_0)} \big(|F[u]|^d + |Du|^d\big)\,dx\Big)^{1/d} + N\tau_0\nu^{d/\gamma}$$

$$+N\Big[\big(\mu + \nu r + \omega_{F,u,\check{B}^+_{\rho+\nu r}}(\nu r)\big)\nu^{d/\gamma} + \nu^{-\alpha}\Big]\Big(\fint_{\check{B}^+_{\nu r}(x_0)} |D^2u|^{\beta' d}\,dx\Big)^{1/(\beta' d)},$$

where $\beta' = \beta/(\beta-1)$ and N depends only on d, K_F, γ, $M_2(\Omega)$, α, and δ.

The proof of this lemma is based on Lemma 2.2 and, in light of Lemma 1, is practically identical to that of Lemma 10.4.2.

We now come to the main a priori estimate near the boundary for nonlinear elliptic equations with "almost VMO coefficients". Recall that $p > d$.

3. THEOREM. *Let $R > 0$ satisfy*

$$4R \leq \rho_0(\Omega) \wedge R_0,$$

and let $u \in W^2_p(\check{B}^+_{2R})$ be such that $u = 0$ on $\partial \check{B}^+_{2R} \cap \partial\Omega$. Then there exist constants $\hat{\theta}, \theta \in (0,1]$, depending only on d, p, δ, $M_2(\Omega)$, and K_F, such that if Assumptions 10.1.4 and 1.1 are satisfied with these $\hat{\theta}$ and θ, respectively, then there exist a constant N, depending only on $R_0, \gamma, d, p, K_0, K_F, \delta$, $M_2(\Omega)$, and the function $\omega_{F,u,\check{B}^+_{2R}}$, such that

$$\|D^2 u\|_{L_p(\check{B}^+_R)} \leq N\big\|H[u]\big\|_{L_p(\check{B}^+_{2R})} + N\|\bar{G}\|_{L_p(\check{B}^+_{2R})} + N\tau_0 R^{d/p}$$
$$+ N\|u\|_{L_p(\check{B}^+_{2R})} + NR^{-\chi}\big\|\,|D^2u|^\gamma\big\|^{1/\gamma}_{L_1(\check{B}^+_{2R})}, \tag{1}$$

$$\|D^2 u\|_{L_p(\check{B}^+_R)} \leq N\big\|H[u]\big\|_{L_p(\check{B}^+_{2R})} + N\|\bar{G}\|_{L_p(\check{B}^+_{2R})}$$
$$+ N\tau_0 R^{d/p} + NR^{d/p-2}\sup_{\check{B}^+_{2R}}|u|, \tag{2}$$

where $\chi = (1/\gamma - 1/p)d$ and γ is the same as in Lemma 2.

Proof. Whenever the following object makes sense, denote

$$h^{\#}_{\Omega,\gamma,\rho}(x) = \sup\big\{I_r(h,x_0) : x_0 \in \Omega, r \in (0,\rho], \check{B}^+_r(x_0) \ni x\big\},$$

where

$$I_r(h,x_0) = \Big(\fint_{\check{B}^+_r(x_0)}\fint_{\check{B}^+_r(x_0)} |h(x_1) - h(x_2)|^\gamma\, dx_1 dx_2\Big)^{1/\gamma}.$$

Then take $\varepsilon \in (0,1]$ to be specified later, take $R_1 < R_2 \leq 2R$ such that

$$R_2 - R_1 \leq \varepsilon R_0, \quad R_2 \leq 2R_1, \tag{3}$$

take $\nu \geq 12$, and set

$$r_0 = (R_2 - R_1)/(\nu + 1).$$

Take x, x_0, r such that

$$r \leq r_0, \quad x \in \check{B}^+_{R_1}, \quad x_0 \in \Omega, \quad x \in \check{B}^+_r(x_0).$$

Observe for the future that $|y(x)| < R_1$, $y(x) \in B^+_r\big(y(x_0)\big)$, $|y(x_0)| < r_0 + R_1$, $|y(x_0)| + \nu r < R_2$,

$$B^+_{\nu r}\big(y(x_0)\big) \subset B^+_{R_2}, \quad \check{B}^+_{\nu r}(x_0) \subset \check{B}^+_{R_2}. \tag{4}$$

Now we claim that for any measurable h,

$$\fint_{\check{B}^+_{\nu r}(x_0)} |h|\,dy \leq N(d)\fint_{B_{2\nu r}(x_0)\cap\Omega} |h| I_{\check{B}^+_{R_2}}\,dy$$
$$\leq N(d, M_2(\Omega))\fint_{B_{2\nu r}(x_0)} |h| I_{\check{B}^+_{R_2}}\,dy \leq N\mathbb{M}(hI_{\check{B}^+_{R_2}})(x), \tag{5}$$

where $\mathbb{M}f$ is introduced in (10.4.5) as the classical maximal function of f.

Indeed, the first inequality follows from (4) and Corollary 1.14 since $x_0 \in \check{B}^+_{2\rho_0(\Omega)}$ and $\nu r \le \rho_0(\Omega)$ because $R_1 \le 2R \le \rho_0(\Omega)$ and $\nu r_0 \le R_1$. The second inequality follows easily from the smoothness of Ω and the fact that $2\nu r_0 \le \rho_0(\Omega)$, and the third inequality follows by the definition of $\mathbb{M}$ since $x \in \check{B}^+_r(x_0) \subset B_{2\nu r}(x_0)$.

This and Lemma 2 show (note that $2\nu r \le 2R_2 \le 4R \le \rho_0(\Omega) \wedge R_0$) that

$$I_r(x_0) \le N\nu^{d/\gamma}\mathbb{M}^{1/d}\big(\big|F[u]\big|^d I_{\check{B}^+_{R_2}}\big)(x) + N\tau_0\nu^{d/\gamma}$$

$$+N\Big[\big(\mu + \nu r + \omega_{F,u,\check{B}^+_{2R}}(\nu r_0)\big)\nu^{d/\gamma} + \nu^{-\alpha}\Big]\mathbb{M}^{1/(\beta' d)}\big(|D^2u|^{\beta' d} I_{\check{B}^+_{R_2}}\big)(x)$$

$$+N\nu^{d/\gamma}\mathbb{M}^{1/d}\big(|Du|^d I_{\check{B}^+_{R_2}}\big)(x).$$

By definition and the fact that $\nu r_0 \le \varepsilon R_0$ we obtain that on $\check{B}^+_{R_1}$

$$(D^2u)^{\#}_{\Omega,\gamma,r_0} \le N\nu^{d/\gamma}\mathbb{M}^{1/d}\big(\big|F[u]\big|^d I_{\check{B}^+_{R_2}}\big) + N\tau_0\nu^{d/\gamma} + N\nu^{d/\gamma}\mathbb{M}^{1/d}\big(|Du|^d I_{\check{B}^+_{R_2}}\big)$$

$$+N\Big[\big(\mu + \nu r_0 + \omega_{F,u,\check{B}^+_{2R}}(\varepsilon R_0)\big)\nu^{d/\gamma} + \nu^{-\alpha}\Big]\mathbb{M}^{1/(\beta' d)}\big(|D^2u|^{\beta' d} I_{\check{B}^+_{R_2}}\big).$$

Next, we are going to use Theorem C.3.3 with $\kappa = r_0/R_1 \le 1/3$ according to which, if $h \in L_p(\check{B}^+_{R_2})$, then

$$\|h\|_{L_p(\check{B}^+_{R_1})} \le N\big\|h^{\#}_{\Omega,\gamma,r_0}\big\|_{L_p(\check{B}^+_{R_1})} + N\nu^{\chi_1}(R_2 - R_1)^{-\chi_1}R_1^{\chi_1-\chi_2}\big\|\,|h|^\gamma\big\|^{1/\gamma}_{L_1(\check{B}^+_{R_1})}, \quad (6)$$

where the constants N depend only on d, γ, and p.

Thanks to (6) and the Hardy-Littlewood maximal function theorem, by taking β so that $p > \beta' d$, we obtain

$$\|D^2u\|_{L_p(\check{B}^+_{R_1})} \le N\nu^{d/\gamma}\big\|F[u]\big\|_{L_p(\check{B}^+_{R_2})} + N\tau_0\nu^{d/\gamma}R^{d/p} + N\nu^{d/\gamma}\|Du\|_{L_p(\check{B}^+_{R_2})}$$

$$+\Big[N\big(\mu + \nu r_0 + \omega_{F,u,\check{B}^+_{2R}}(\varepsilon R_0)\big)\nu^{d/\gamma} + N_0\nu^{-\alpha}\Big]\|D^2u\|_{L_p(\check{B}^+_{R_2})}$$

$$+N\nu^{\chi_1}(R_2 - R_1)^{-\chi_1}R_1^{\chi_1-\chi_2}\big\|\,|D^2u|^\gamma\big\|^{1/\gamma}_{L_1(\check{B}^+_{2R})},$$

where the constants N, N_0 depend only on d, p, K_F, $M_2(\Omega)$, γ, and δ.

This estimate looks almost like (10.4.7). Then we repeat the argument after (10.4.7) and choose and fix ε, ν, $\hat{\theta}$, μ, and θ, recall what r_0 is, and conclude that

$$\|D^2u\|_{L_p(\check{B}^+_{R_1})} \le N\big\|H[u]\big\|_{L_p(\check{B}^+_{R_2})} + N\tau_0\nu^{d/\gamma}R^{d/p}$$

$$+(5/8 + N_1(R_2 - R_1))\|D^2u\|_{L_p(\check{B}^+_{R_2})} + N\|u\|_{L_p(\check{B}^+_{2R})} + N\|\bar{G}\|_{L_p(\check{B}^+_{2R})}$$

$$+N(R_2 - R_1)^{-\chi_1}R_1^{\chi_1-\chi_2}\big\|\,|D^2u\|^\gamma\big\|^{1/\gamma}_{L_1(\check{B}^+_{2R})}. \quad (7)$$

Now we are going to iterate this estimate by defining $R_1 = R$ and for $k \ge 1$

$$R_{k+1} = R_k + cR(n_0 + k)^{-2}, \quad k \ge 1.$$

where we choose large constants c and n_0 in a special way. Namely, $c = O(n_0)$ is chosen so that $R_k \uparrow 2R$ as $k \to \infty$, that is

$$c\sum_{k=1}^{\infty}(n_0+k)^{-2} = 1$$

and n_0 (independent of R since $R \le 1$) is chosen so that, for $k \ge 1$, $N_1(R_{k+1}-R_k) \le 1/8$ and

$$R_{k+1} - R_k = cR(n_0+k)^{-2} \le 2cRn_0^{-2} \le R \le R_k,$$

which is satisfied if n_0 is just an absolute constant, and

$$R_{k+1} - R_k = cR(n_0+k)^{-2} \le cn_0^{-2} \le \varepsilon R_0$$

(this time we need $n_0^{-1} = o(\varepsilon R_0)$ as $\varepsilon R_0 \to 0$). Also recall that $R \le R_k \le 2R$, so that

$$(R_{k+1} - R_k)^{-\chi_1} R_k^{\chi_1-\chi_2} \le N(n_0+k)^{2\chi} R^{-\chi_2}.$$

Then for $k \ge 1$ we get

$$\|D^2u\|_{L_p(\check{B}^+_{R_k})} \le N\big\|H[u]\big\|_{L_p(\check{B}^+_{2R})} + N\tau_0 R^{d/p} + (3/4)\|D^2u\|_{L_p(\check{B}^+_{R_{k+1}})}$$

$$+N(n_0+k)^{2\chi_1}R^{-\chi_2}\big\|\,|D^2u|^{\gamma}\big\|^{1/\gamma}_{L_1(\check{B}^+_{2R})} + N\|u\|_{L_p(\check{B}^+_{2R})} + N\|\bar{G}\|_{L_p(\check{B}^+_{2R})}.$$

We multiply both parts of this inequality by $(3/4)^k$ and sum up the results over $k = 1, 2, \ldots$. Then we cancel the like terms

$$\sum_{k=2}^{\infty}(3/4)^k\|D^2u\|_{L_p(\check{B}^+_{R_k})},$$

which are finite since $u \in W^2_p(\check{B}^+_{2R})$, and finally take into account that

$$\sum_{k=2}^{\infty}(3/4)^k(n_0+k)^{2\chi_1} \le Nn_0^{2\chi_1}\sum_{k=2}^{\infty}(3/4)^k + N\sum_{k=2}^{\infty}(3/4)^k k^{2\chi_1} \le N.$$

Then we come to (1). By using Remark 2.3 we estimate the last term in (1) and then finish the proof of the theorem in the same way as in the case of Lemma 10.4.3. The theorem is proved. □

Proof of Theorem 1.3. To start, assume that $g \equiv 0$. Take $\hat{\theta}$ and θ which suit both Theorems 10.1.10 and 3 and take R such that $4R = \rho_0(\Omega) \wedge R_0$. Theorem 3 allows us to estimate the W^2_p-norm of u in the domain $\check{B}^+_R = \check{B}^+_R(0)$ associated with the origin, that is assumed to belong to $\partial\Omega$. Of course, one can take any point $z_0 \in \partial\Omega$ as the origin and apply Theorem 3 to $\check{B}^+_R(z_0)$ and $\check{B}^+_{2R}(z_0)$ in place of $\check{B}^+_R$ and $\check{B}^+_{2R}$, respectively, where by $\check{B}^+_\rho(z_0)$ we, naturally, mean the sets $\check{B}^+_\rho$ constructed in Lemma 1.13 but with z_0 in place of 0 and relative to the coordinate system $\Psi(z_0)$ associated with z_0 as described in Definition 1.10. According to that, we find finitely many $z_i \in \partial\Omega$ such that

$$\Omega' := \Omega \setminus \bigcup_i \check{B}^+_R(z_i) \subset \Omega^{R/2}.$$

The number of those points, obviously, depends only on $\rho_0(\Omega)$, R_0, $M_2(\Omega)$, and $\operatorname{diam}(\Omega)$.

By Theorem 3, for any i

$$\|D^2u\|^p_{L_p(\check{B}^+_R(z_i))} \leq N\|\bar{G}\|^p_{L_p(\Omega)} + N\tau_0^p + N\sup_\Omega |u|.$$

By Theorem 10.1.10

$$\|D^2u\|^p_{L_p(\Omega')} \leq N\|\bar{G}\|^p_{L_p(\Omega)} + N\tau_0^p + N\sup_\Omega |u|.$$

We sum up these inequalities and come to (1.3). This proves the theorem if $u \in \overset{0}{W}{}^2_p(\Omega)$.

In the general case introduce $w(x) = u(x) - g(x)$,

$$\mathsf{g} = (g, Dg, D^2g)I_\Omega,$$

$$\hat{H}(\mathsf{v}, x) = H\big(\mathsf{v} + \mathsf{g}(x), x\big), \quad \hat{F}(\mathsf{v}_0', \mathsf{v}'', x) = F(\mathsf{v}_0' + \mathsf{g}_0'(x), \mathsf{v}'', x).$$

Observe that $\hat{H}[w] = 0$ in Ω (a.e.) and $w \in \overset{0}{W}{}^2_p(\Omega)$. Furthermore, for

$$\hat{G}(\mathsf{v}, x) := \hat{H}(\mathsf{v}, x) - \hat{F}(\mathsf{v}_0', \mathsf{v}'', x)$$

we have

$$\begin{aligned}\big|\hat{G}(\mathsf{v}, x)\big| &= \big|F\big(\mathsf{v}_0' + \mathsf{g}_0'(x), \mathsf{v}'' + \mathsf{g}''(x), x\big) \\ &\quad - F(\mathsf{v}_0' + \mathsf{g}_0'(x), \mathsf{v}'', x) + G(\mathsf{v} + \mathsf{g}(x), x)\big| \\ &\leq N\big|I_\Omega D^2g(x)\big| + \hat{\theta}|\mathsf{v}''| + \bar{G} + K_0\Big(\big|\mathsf{v}_0' + I_\Omega g(x)\big|^2 \\ &\quad + \sum_{i=1}^d \big|\mathsf{v}_i' + I_\Omega D_i g(x)\big|^2\Big)^{1/2} \leq \hat{\theta}|\mathsf{v}''| + \bar{\hat{G}} + K_0|\mathsf{v}'|,\end{aligned}$$

where

$$\bar{\hat{G}} = N|D^2g|I_\Omega + \bar{G} + K_0\big(g^2 + |Dg|^2\big)^{1/2} I_\Omega$$

and N depends only on K_F and d.

Next, we check that Assumption 1.1 is satisfied for $\hat{F}$. Obviously, we need only deal with its requirements (ii) and (iii) and, accordingly, for $\mathsf{v}_0' \in \mathbb{R}$, $r \in (0, R_0]$, and $z \in \Omega$ we set

$$\bar{\hat{F}}(\mathsf{v}'') = \bar{\hat{F}}_{z,r,\mathsf{v}_0'}(\mathsf{v}'') = \bar{F}_{z,r,\mathsf{v}_0'+\mathsf{g}_0'(z)}(\mathsf{v}'').$$

Then Assumption 1.1 (ii) is satisfied and the left-hand side of (1.2) with $\hat{F}$ and $\bar{\hat{F}}$ in place of F and $\bar{F}$, respectively, is less than $\big|B_r(z) \cap \Omega\big|$ times

$$\theta + \omega_{F,g,\Omega}(r) \leq \theta + \omega_{F,g,\Omega}(R_0).$$

Then with $\theta_0 = \theta\big(d, p, \delta, K_F, M_2(\Omega)\big)$ defined in the first part of the proof where $g \equiv 0$ we find $\hat{R}_0 \leq R_0$ such that $\omega_{F,g,\Omega}(\hat{R}_0) \leq \theta_0/2$ and then require the original Assumption 1.1 (iii) to be satisfied with $\hat{R}_0$ and $\theta_0/2$ in place of R_0 and θ, respectively. Then Assumption 1.1 (iii) will be satisfied with $\hat{R}_0$, $\hat{F}$, $\bar{\hat{F}}$, and θ_0 in place of R_0, F, $\bar{F}$, and θ. This makes the above result applicable to w, and along with the embedding inequality: $|g| \leq N\|g\|_{W^2_p(\Omega)}$, leads to (1.3) in the general case. The theorem is proved. □

4. Investigating when $W^2_{p,\mathrm{loc}}(\check{B}^+_R)$-solutions belong to $W^2_p(\check{B}^+_\rho)$

We assume that

$$0 \in \partial\Omega,$$

take $\rho_0 = \rho_0(\Omega)$ from Definition 1.10, and suppose that the original system of coordinates in $\mathbb{R}^d$ coincides with the one described in Definition 1.10 for $z_0 = 0$. We also take $\check{B}^+_r$, $\check{B}^+_r(z)$ from Remark 1.12.

Let $P(\mathsf{u}'')$ be a convex function on $\mathbb{S}$ such that $D_{\mathsf{u}''}P \in \mathbb{S}_\delta$ at all points of its differentiability, where $\delta \in (0,1]$. Also assume that P is *positive homogeneous* of degree 1. Take $p > d$.

1. THEOREM. *Let $R \leq 8\rho_0(\Omega)$ and let $u \in W^2_{p,\mathrm{loc}}(\check{B}^+_R) \cap C^{0,1}(\Omega) \cap C(\overline{\check{B}^+_R})$ be such that $P[u] = f$ in $\check{B}^+_R$ (a.e.), where f is a* bounded *function, and $u = 0$ on $\partial\check{B}^+_R \cap \partial\Omega$. Then $u \in W^2_p(\check{B}^+_r)$ for any $r \in (0, R)$.*

Proof. Let $y(x)$ and $x(y)$ be the mappings from Lemma 1.11. Denote

$$\hat{u}(y) = u\big(x(y)\big), \quad A(x) = \partial y(x)/\partial x,$$

so that the ijth entry of $A(x)$ is $\partial y^i(x)/\partial x^j$. Observe that for $y = y(x)$ and $x = x(y)$ we have

$$Du(x) = (D\hat{u})(y)\hat{A}(y),$$

$$D^2u(x) = \hat{A}^*(y)\big[D^2\hat{u}(y)\big]\hat{A}(y) + \hat{W}(y).$$

where

$$\hat{A}(y) = A(x), \quad W(x) = \big[Du(x)\big]A^{-1}(x)D^2y^k(x), \quad \hat{W}(y) = W(x).$$

In this connection for $\mathsf{u}'' \in \mathbb{S}$ and $y \in \hat{\Gamma}$ set

$$\hat{f}(y) := f\big(x(y)\big), \quad H(\mathsf{u}'', y) = P\big(\hat{A}^*(y)\mathsf{u}''\hat{A}(y) + \hat{W}(y)\big) - \hat{f}(y).$$

The functions $\hat{A}(y)$, $\hat{W}(y)$, and $\hat{f}(y)$ are bounded in B^+_r and $\hat{A}(y)$ is Lipschitz continuous and has a bounded inverse. We extend them arbitrarily beyond B^+_r preserving these properties along with the associated bounds, take $\zeta \in C^\infty_0(\mathbb{R}^d)$ with unit integral and for $n = 1, 2, ...$ define $\zeta_n(y) = n^d\zeta(ny)$,

$$\hat{W}^n = \hat{W} * \zeta_n, \quad \hat{f}^n = \hat{f} * \zeta_n,$$

$$H^n(\mathsf{u}'', y) = P\big(\hat{A}^*(y)\mathsf{u}''\hat{A}(y) + \hat{W}^n(y)\big) - \hat{f}^n(y).$$

We want to apply the results of Section 3 to solutions of the equation

$$H^n[\hat{u}^n] = 0 \tag{1}$$

by taking there as Ω any smooth domain in $\mathbb{R}^d_+$, having B^+_R as its part, for which $\rho_0(\Omega)$ is preserved (in order not to deal with extra notation). Introduce

$$F(\mathsf{u}'', y) = P\big(\hat{A}^*(y)\mathsf{u}''\hat{A}(y)\big), \quad G^n(\mathsf{u}'', y) = H^n(\mathsf{u}'', y) - F(\mathsf{u}'', y).$$

Observe that

$$\big|G^n(\mathsf{u}'', y)\big| \leq N\big|\hat{W}^n(y)\big| + \big|\hat{f}^n(y)\big| \leq N,$$

where the constants N are independent of u'', y, and n. Hence, Assumption 10.1.4 is satisfied.

Next, Assumption 1.1 (i) is obviously satisfied. Furthermore, if we set $\bar F_{z,r}(\mathsf{u}'') = P\big(\hat A^*(z)\mathsf{u}''(y)\hat A(z)\big)$, then $\bar F_{z,r}(\mathsf{u}'')$ is convex, satisfies Assumption 1.1 (ii) with a $\delta > 0$, and, owing to the Lipschitz continuity of $\hat A$, for $|y-z| \le r$ we have

$$\big|F(\mathsf{u}'', y) - \bar F_{z,r}(\mathsf{u}'')\big| \le Nr|\mathsf{u}''|,$$

where N is independent of u'', y, z, r. It follows that, for any $\theta > 0$, there exists $R_0 \in (0,1]$ such that, for any $r \le R_0$, Assumption 1.1 (iii) is satisfied with $\tau_0 = 0$. We take θ so small that the conclusions of Theorem 3.3 are applicable for the given p and appropriate R_0.

By Theorem 13.1.4, for each n, equation (1) in B^+_R with boundary condition $\hat u$ on ∂B^+_R has a unique solution $\hat u^n \in \hat C^{2+\alpha}\big(\mathbb{B}^+_R\big)$ for some $\alpha \in (0,1)$. Furthermore, in light of the boundary estimates from Theorems 3.3 and interior estimates from Theorem 10.1.10 and the Aleksandrov estimates, the $W^2_p\big(B^+_r\big)$-norms of $\hat u^n$ are bounded for any $r < R$. Therefore, there is a subsequence of $\hat u^n$ which converges weakly in $W^2_p\big(B^+_r\big)$, $r < R$, to a function, say $\hat w$. Of course, $\hat w \in W^2_p\big(B^+_r\big)$ for any $r < R$. Clearly, if we can show that $\hat w = \hat u$, the theorem will be proved.

First, observe that, for any $n \ge m$,

$$\sup_{r \ge m} H^r[\hat u^n] \ge 0$$

in B^+_R (a.e.). By Theorems 4.2.15, 4.2.6 (c), and Lemma 4.2.4 (d) we have

$$\sup_{r \ge m} H^r[\hat w] \ge 0 \tag{2}$$

in B^+_R (a.e.). Since, $\hat f^n \to \hat f$ and $\hat W^n \to \hat W$ in B^+_R (a.e.) as $n \to \infty$, there is a set of full measure in B^+_R such that the left-hand side of (2) tends to $H[\hat w]$. Hence $H[\hat w] \ge 0$ in B^+_R (a.e.), and, since the inequality $H[\hat w] \le 0$ in B^+_R (a.e.) is obtained similarly, $H[\hat w] = 0$ in B^+_R (a.e.).

We conclude that $\hat w$ and $\hat u$ satisfy the same equation in B^+_R and, owing to the uniqueness of $W^2_{p,\mathrm{loc}}\big(B^+_R\big) \cap C\big(\bar B^+_R\big)$-solutions, to prove the theorem, it suffices to prove that $\hat w \in C\big(\bar B^+_R\big)$ and $\hat w = \hat u$ on ∂B^+_R.

Note that

$$0 = H^n[\hat u^n] - H^m[\hat u^m] = \operatorname{tr} \hat A^* a \hat A^* D^2_y[\hat u^n - \hat u^m] + \operatorname{tr} a[\hat W^n - \hat W^m] - \hat f^n + \hat f^m$$

in B^+_R (a.e.), where a is an $\mathbb{S}_\delta$-valued measurable function. By the Aleksandrov estimates, the fact that $\hat u^n = u^n = \hat u$ on ∂B^+_R and the fact that the sequences $\hat W^n$ and $\hat f^n$ are Cauchy in $L_d\big(B^+_R\big)$, we conclude that the sequence $\hat u^n$ is Cauchy in $C\big(\bar B^+_R\big)$. This proves the theorem. □

5. Proof of Theorem 1.5

The proof of Theorem 1.5 is based on Theorem 10.1.14 and is modeled in a big part after the proof of Theorem 10.1.17 in Section 10.5.

Step 1. Assume that $g \in C^{1,1}(\mathbb{R}^d)$ and there exists *constants* $N_0, \bar H$ such that, for all x, u',

$$\big|H(\mathsf{u}', 0, x)\big| \le N_0|\mathsf{u}'| + \bar H. \tag{1}$$

We need this assumption in conjunction with (10.1.5) in order to be able to use Theorem 6.6.9. Once Theorem 4.1 is proved, the argument in Remark 6.6.13 is justified and we can enjoy using all assertions of Theorem 6.6.3 reaffirmed in Theorem 6.6.9. By that theorem for any $K > 0$ there exists a function v_K that is in $W^2_p(\Omega)$ for any $p > 1$, such that $v_K = g$ on $\partial\Omega$ and

$$H_K[v_K] = 0 \quad \text{in} \quad \Omega \text{ (a.e.)}, \tag{2}$$

where

$$H_K(\mathsf{u}, x) = \max\big(H(\mathsf{u}, x), P(\mathsf{u}'') - K\big).$$

Next, we use Theorem 1.3 about a priori estimates in $W^2_p(\Omega)$. Set

$$F_K\big(\mathsf{u}'_0, \mathsf{u}'', x\big) = \max\big(F\big(\mathsf{u}'_0, \mathsf{u}'', x\big), P(\mathsf{u}'') - K\big),$$

$$G_K(u, x) = H_K(u, x) - F_K\big(\mathsf{u}'_0, \mathsf{u}'', x\big).$$

It is shown in the proof of Theorem 10.1.17 in Section 10.5 that the assumptions of Theorem 1.3 are satisfied for H_K, F_K, G_K in place of H, F, G, respectively, with the same $\theta, \hat{\theta}$ and perhaps, new δ and K_F but depending only on the original δ, K_F, and d. It is also noted there (see also the proof of Theorem 10.1.14 in Section 10.5) that the family of the functions v_K is uniformly bounded and is equicontinuous in $\bar{\Omega}$, which, in particular, implies that the functions $\omega_{F,v_K,\Omega}$ are majorated by a continuous function independent of K and vanishing at zero (see more details about the latter in Section 10.5).

Then, provided that an appropriate choice of θ and $\hat{\theta}$ was made, by Theorem 1.3 we obtain that

$$\sup_{K\geq 1} \|v_K\|_{W^2_p(\Omega)} < \infty. \tag{3}$$

This estimate is stronger than (10.5.5). Hence, as in the proof of Theorem 10.1.17, by using an appropriate sequence v_{K_n}, $K_n \to \infty$, we find that its limit is a function $v \in W^2_{p,\text{loc}}(\Omega) \cap C(\bar{\Omega})$ satisfying (1.1) in Ω (a.e.) and equal g on $\partial\Omega$. Owing to (3), of course, $v \in W^2_p(\Omega)$.

This proves the theorem if condition (1) is satisfied and $g \in C^{1,1}(\mathbb{R}^d)$.

Step 2. Assume that $g \in C^{1,1}(\mathbb{R}^d)$ and abandon (1). By repeating word for word what is said in Step 2 of the proof of Theorem 10.1.17, instead of (10.5.9) we find a stronger estimate

$$\sup_{n\geq 1} \|v^n\|_{W^2_p(\Omega)} < \infty, \tag{4}$$

which allows us, by using subsequences of v^n to find a function $v \in W^2_{p,\text{loc}}(\Omega) \cap C(\bar{\Omega})$ satisfying (1.1) in Ω (a.e.) and is equal to g on $\partial\Omega$. Owing to (4), of course, $v \in W^2_p(\Omega)$.

This yields the desired result if $g \in C^{1,1}(\mathbb{R}^d)$. The final step where one approximates $g \in W^2_p(\Omega)$ by polynomials and passes to the limit is much simpler than the above arguments and is left to the reader as a routine exercise. The theorem is proved. □

Proof of Theorem 1.8. It suffices to literally repeat the above proof using Theorem 6.6.5 in place of Theorem 6.6.9. □

CHAPTER 15

Nonlinear parabolic equations in $W^{1,2}_p$

1. Main results

In this chapter as in Chapter 12 we consider parabolic equations

$$\partial_t v(t,x) + H[v](t,x) = 0, \tag{1}$$

where

$$H[v](t,x) = H\big(v(t,x), Dv(t,x), D^2v(t,x), t, x\big),$$

in subdomains of

$$\mathbb{R}^{d+1} = \big\{(t,x) : t \in \mathbb{R}, x \in \mathbb{R}^d\big\}.$$

Let $\Omega \in C^{1,1}$ be an open bounded subset of $\mathbb{R}^d$. Fix $T \in (0,\infty)$ and set

$$\Pi = [0,T) \times \Omega.$$

Fix (later on we will have a stronger restrictions on p)

$$p > d+1 \quad \text{and a measurable function} \quad \bar{G} \geq 0 \quad \text{on} \quad \mathbb{R}^{d+1}.$$

Also fix some constants $K_0, K_F \in [0,\infty)$, $\delta \in (0,1]$.

We are going to work under the same assumptions as in Section 12.1 containing the parameters $\hat{\theta}$ and θ to be specified later, with one exception because we need an analog of Assumption 12.1.4 near the boundary. We mean the following.

1. ASSUMPTION. (i) The function F is Lipschitz continuous with respect to u'' with Lipschitz constant K_F and is measurable with respect to (t,x).

Moreover there exist $R_0 \in (0,1]$ and $\tau_0 \in [0,\infty)$ such that, for any $\mathsf{u}'_0 \in \mathbb{R}$, $z_0 = (t_0,x_0) \in \Pi$ and $r \in (0,R_0]$, one can find a *convex* function $\bar{F}(\mathsf{u}'') = \bar{F}_{z_0,r,\mathsf{u}'_0}(\mathsf{u}'')$ (depending only on u'') for which

(ii) We have $\bar{F}(0) = 0$ and $D_{\mathsf{u}''}\bar{F} \in \mathbb{S}_\delta$ at all points of differentiability of $\bar{F}$;

(iii) For any $\mathsf{u}'' \in \mathbb{S}$ with $|\mathsf{u}''| = 1$, we have

$$\int_{\hat{C}_r(z_0)} \sup_{\tau > \tau_0} \tau^{-1} \big|F\big(\mathsf{u}'_0, \tau\mathsf{u}'', z\big) - \bar{F}(\tau\mathsf{u}'')\big|\, dz \leq \theta |\hat{C}_r(z_0)|, \tag{2}$$

where

$$\hat{C}_r(z_0) = (t_0, t_0 + r^2) \times \big(\Omega \cap B_r(x_0)\big);$$

(iv) There exists a continuous increasing function $\omega_F(\tau)$, $\tau \geq 0$, such that $\omega_F(0) = 0$ and for any $\mathsf{u}'_0, \mathsf{v}'_0 \in \mathbb{R}$, $(t,x) \in \Pi$, and $\mathsf{u}'' \in \mathbb{S}$ we have

$$\big|F(\mathsf{u}'_0, \mathsf{u}'', t, x) - F(\mathsf{v}'_0, \mathsf{u}'', t, x)\big| \leq \omega_F\big(|\mathsf{u}'_0 - \mathsf{v}'_0|\big)|\mathsf{u}''|.$$

2. ASSUMPTION. We are given a function $g \in W^{1,2}_p(\Pi)$.

Recall that the function $\omega_{F,u,\Pi}$ is introduced in Definition 12.1.6. In the following theorem there is no ellipticity assumption on H.

3. THEOREM. *Suppose that Assumption* 2 *is satisfied. Then there exist constants* $\theta, \hat{\theta} \in (0,1]$, *depending only on* d, p, δ, K_F, *and* $M_2(\Omega)$, *such that, if Assumptions* 1 *and* 12.1.3 *are satisfied with these* θ *and* $\hat{\theta}$, *respectively, then for any* $u \in W^{1,2}_p(\Pi)$ *that satisfies* (1) *in* Π (*a.e.*) *and equals* g *on* $\partial'\Pi$ *we have*

$$\|u\|_{W^{1,2}_p(\Pi)} \leq N\|\bar{G}\|_{L_p(\Pi)} + N\|g\|_{W^{1,2}_p(\Pi)} + N\tau_0 + N\sup_{\Pi}|u|, \tag{3}$$

where the constants N *depend only on* T, K_0, K_F, d, p, δ, $\rho_0(\Omega)$, R_0, $M_2(\Omega)$, $\mathrm{diam}(\Omega)$, *and the functions* $\omega_{F,u,\Pi}$ *and* $\omega_{F,g,\Pi}$.

To have the solvability we need ellipticity and more regularity of H.

4. THEOREM. *Let* $p > d+2$ *and suppose that Assumptions* 12.1.8 *and* 2 *are satisfied and* $\bar{G} \in L_p(\Pi)$. *Then there exist constants* $\theta, \hat{\theta} \in (0,1]$, *depending only on* d, p, δ, K_F, *and* $M_2(\Omega)$, *such that, if Assumptions* 1 *and* 12.1.3 *are satisfied with these* θ *and* $\hat{\theta}$, *respectively, then there exists* $u \in W^{1,2}_p(\Pi)$ *satisfying* (1) *in* Π (*a.e.*) *and such that* $u = g$ *on* $\partial'\Pi$.

5. REMARK. Uniqueness of solutions can be investigated by using the results of Subsection 4.1:1.

6. REMARK. In the case of linear equations Theorem 4 contains (apart from the restrictions on p) the corresponding result of [**10**] proved for equations with VMO main coefficients. Note that we can also treat some quasilinear equations (cf. Example 12.1.20 when A and B are singletons). Generally, in what concerns the discussion of our assumptions and results we refer the reader to Section 12.1. Everything there applies here with, of course, rather crucial addition that now we have the solvability in global Sobolev spaces, which first was obtained in [**30**] under much more restrictive assumptions. Exposition here follows [**80**].

In Theorem 5.9 of Wang [**114**] one can find an a priori estimate for *any* viscosity solution in case H is independent of u' and $\Pi = C_1$.

By the way, it will be seen from our proofs that, if H is independent of $[\mathsf{u}']$, we can take $p > d+1$ in Theorem 4.

In the same way as Theorem 10.1.16 is derived from Theorem 10.1.14 we obtain the following.

7. THEOREM. *Let* $p > d+2$ *and suppose that Assumptions* 12.1.2, 12.1.8, *and* 2 *are satisfied,* $\bar{G} \in L_p(\Pi)$, *and* (12.1.4) *holds true. Let* $P(\mathsf{u}'')$ *be a convex function on* $\mathbb{S}$ *such that* $D_{\mathsf{u}''}P \in \mathbb{S}_{\delta'}$ *at each point of its differentiability, where* $\delta' \in (0,\delta]$. *Also assume that for any* $a \in \mathbb{S}_\delta$ *and* $\mathsf{u}'' \in \mathbb{S}$ *we have*

$$a^{ij}\mathsf{u}''_{ij} \leq P(\mathsf{u}'') + K,$$

where K *is a constant. Then the equation*

$$\partial_t u + \max\big(H[u], P[u]\big) = 0$$

in Π (*a.e.*) *with boundary condition* $u = g$ *on* $\partial'\Pi$ *has a solution* $u \in W^{1,2}_p(\Pi)$.

Recall notation (13.1.1) and extend notation (9.6.15) and (13.1.6) by setting

$$C^+_{\tau,r}(t_0,x_0) = [t_0, t_0+\tau) \times B^+_r(x_0),$$

$$\partial_{x^1}C^+_{\tau,r}(t_0,x_0) = \bar{C}^+_{\tau,r}(t_0,x_0) \cap \{x^1 = 0\},$$

$$\mathbb{C}^+_{\tau,r}(t_0,x_0)=[t_0,t_0+\tau)\times \mathbb{B}^+_r(x_0),$$

where $\tau,r\geq 0$, $t_0\in\mathbb{R}$, $x_0\in\bar{\mathbb{R}}^d_+$. If $t_0=0,x_0=0$, we drop (t_0,x_0) in the arguments above. Also, if $\tau=r^2$ we write r in place of τ,r in the subscripts, for instance,

$$C^+_r(t_0,x_0):=C^+_{r^2,r}(t_0,x_0).$$

2. Boundary a priori estimates in the simplest case

Let $F(\mathsf{u}'')$ be a convex function of $\mathsf{u}''\in\mathbb{S}$ (independent of (t,x)) such that

(i) $F(0)=0$,

(ii) $D_{\mathsf{u}''}F\in\mathbb{S}_\delta$ at all points of differentiability of F, where $\delta\in(0,1]$ is a fixed number.

Introduce

$$\mathbb{R}^{d+1}_+=\{(t,x):t\in\mathbb{R},x=(x^1,...,x^d),x^1>0\}.$$

Take $\bar{\gamma}=\bar{\gamma}(d,\delta)$ from (10.3.4), fix $\gamma\in(0,\bar{\gamma}]$, define $\alpha_0=\alpha_0(\delta,d)$ as the least of the constants called α_0 introduced before Theorem 12.2.1 (interior smoothness) and in Theorem 13.1.6 (boundary smoothness), and take $\alpha\in(0,\alpha_0)$.

1. LEMMA. *If $r>0$, $z_0\in\bar{\mathbb{R}}^{d+1}_+$, $\nu\geq 12$,*

$$u\in\bigcap_{\rho<\nu r}W^{1,2}_{d+1}(C^+_\rho(z_0)),$$

and u vanishes on $\partial_{x^1}C^+_{\nu r}(z_0)$ if this set is nonempty, then we have

$$\Big(\fint_{C^+_r(z_0)}\fint_{C^+_r(z_0)}\big|D^2u(z_1)-D^2u(z_2)\big|^\gamma\,dz_1dz_2\Big)^{1/\gamma}$$

$$\leq N\nu^{(d+2)/\gamma}\Big(\fint_{C^+_{\nu r}(z_0)}\big|\partial_t u+F[u]\big|^{d+1}\,dz\Big)^{1/(d+1)}$$

$$+N\nu^{-\alpha}\Big(\fint_{C^+_{\nu r}(z_0)}|D^2u|^{d+1}\,dz\Big)^{1/(d+1)},\tag{1}$$

where N depends only on d and δ.

Proof. Scalings show that it suffices to prove the lemma only for $\nu r=3$. Furthermore, without loss of generality we may assume that $z_0=(0,x_0)$ and $x_0=(|x_0|,0,...,0)\in\mathbb{R}^d$. Then we consider two cases.

Case 1: $|x_0|>1/2$. In this case, we have

$$B^+_r(x_0)=B^+_{3/\nu}(x_0)=B_r(x_0)\subset B_{\nu' r}(x_0)\subset\mathbb{R}^d_+,\quad C_{\nu' r}(z_0)\subset\mathbb{R}^{d+1}_+,$$

where $\nu'=\nu/6$ (≥ 2). Therefore, inequality (1) is an immediate consequence of Lemma 12.3.3.

Case 2: $|x_0| \le 1/2$. Since $r = 3/\nu \le 1/2$, we have

$$B_r^+(x_0) \subset B_1^+ \subset B_2^+ \subset B_3^+(x_0) = B_{\nu r}^+(x_0).$$

Let v be the classical solution of $\partial_t v + F[v] = 0$ in C_2^+ with boundary condition $v = u$ on $\partial' C_2^+$. Such a solution exists due to Theorem 13.1.6, which also provides an estimate on $D^2 v$, so that

$$I := \fint_{C_r^+(z_0)} \fint_{C_r^+(z_0)} |D^2 v(z_1) - D^2 v(z_2)|\, dz_1 dz_2 \le N r^\alpha [D^2 v]_{C^\alpha(C_1^+)}$$

$$\le N r^\alpha \sup_{C_2^+} |v| = N r^\alpha \sup_{\partial' C_2^+} |u|,$$

where the last equality is a consequence of the maximum principle and the fact that $F(0) = 0$. By employing Poincaré's inequality ($u = 0$ on $\partial_{x^1} C_2^+$), we see that

$$I \le N r^\alpha \Big(\fint_{C_2^+} \big(|\partial_t u|^{d+1} + |D^2 u|^{d+1}\big)\, dz \Big)^{1/(d+1)}.$$

Here $r^\alpha = N\nu^{-\alpha}$ and

$$|\partial_t u| \le \big|\partial_t u + F[u]\big| + \big|F[u]\big| \le \big|\partial_t u + F[u]\big| + N|D^2 u|.$$

Therefore,

$$I \le N\nu^{-\alpha} \Big(\fint_{C_2^+} |D^2 u|^{d+1}\, dz \Big)^{1/(d+1)} + N \Big(\fint_{C_2^+} \big|\partial_t u + F[u]\big|^{d+1}\, dz \Big)^{1/(d+1)}.$$

Next, recall that $\gamma \in (0, 1]$. By Hölder's inequality, we get

$$\begin{aligned} &\fint_{C_r^+(z_0)} \fint_{C_r^+(z_0)} \big|D^2 v(z_1) - D^2 v(z_2)\big|^\gamma\, dz_1 dz_2 \\ &\quad \le N\nu^{-\gamma\alpha} \Big(\fint_{C_{\nu r}^+(z_0)} |D^2 u|^{d+1}\, dz \Big)^{\gamma/(d+1)} \\ &\quad\quad + N \Big(\fint_{C_{\nu r}^+(z_0)} \big|\partial_t u + F[u]\big|^{d+1}\, dz \Big)^{\gamma/(d+1)}. \end{aligned} \tag{2}$$

Next, use again that

$$f := \partial u + F(D^2 u) = \partial(u - v) + F(D^2 u) - F(D^2 v) = \partial w + a^{ij} D_{ij}^2 w$$

in C_2^+ (a.e.) and $w = 0$ on $\partial' C_2^+$, where (a^{ij}) is an $\mathbb{S}_\delta$-valued measurable function and $w = u - v$. By the scaled version of Corollary 9.4.11 we get (recall that $\nu r = 3$)

$$\fint_{C_r^+(z_0)} |D^2 w|^\gamma\, dz \le N r^{-d-2} \int_{C_2^+} |D^2 w|^\gamma\, dz \le N\nu^{d+2} \Big(\fint_{C_{\nu r}^+(z_0)} |f|^{d+1}\, dz \Big)^{\gamma/(d+1)}$$

and

$$\fint_{C_r^+(z_0)} \fint_{C_r^+(z_0)} |D^2 w(z_1) - D^2 w(z_2)|^\gamma\, dz_1 dz_2 \le N\nu^{d+2} \Big(\fint_{C_{\nu r}^+(z_0)} |f|^{d+1}\, dz \Big)^{\gamma/(d+1)}.$$

Combining this with (2) and observing that $D^2 u = D^2 v + D^2 w$ yield (1) in Case 2 as well. The lemma is proved. □

Recall that $\rho_0(\Omega)$ is introduced in Definition 14.1.10. In the rest of the section we assume

$$0 \in \partial\Omega$$

and that the original system of coordinates in $\mathbb{R}^d$ coincides with the one described in Definition 14.1.10 with $z_0 = 0$. With the help of the mappings $x(y)$ and $y(x)$, introduced in Lemma 14.1.11, in Remark 14.1.12 we constructed $\check{B}_r^+ = x(B_r^+)$, $r > 0$, and $\check{B}_r^+(z) = x\Big(B_r^+\big(y(z)\big)\Big)$, $z \in \mathbb{R}^d$. By that remark we have $\check{B}_r^+ \subset \Omega$ for $r \leq 8\rho_0(\Omega)$ and $\check{B}_r^+(z) \subset \check{B}_{8\rho_0(\Omega)}^+$ if $\rho > 0$, $\rho + r \leq 8\rho_0(\Omega)$, and $z \in \check{B}_\rho^+$. Generally, these are objects in $\mathbb{R}^d$.

Then set

$$\check{C}_R^+ = [0, R^2) \times \check{B}_R^+$$

and for $\rho + r \leq 8\rho_0(\Omega)$ and $z = (t, x)$ such that $x \in \check{B}_\rho^+$ and $t \in \mathbb{R}$ define

$$\check{C}_r^+(z) = [t, t + r^2) \times \check{B}_r^+(x).$$

Finally, whenever it makes sense, for $\rho \leq \rho_0(\Omega)$, and $z \in \check{C}_{2\rho_0(\Omega)}^+$ introduce

$$h^{\#}_{\Omega,\gamma,\rho}(z) = \sup\big\{I_r(h, z_0) : z_0 \in \mathbb{R}_+ \times \Omega, r \in (0, \rho], \check{C}_r^+(z_0) \ni z\big\}, \tag{3}$$

where

$$I_r(h, z_0) = \Big(\fint_{\check{C}_r^+(z_0)} \fint_{\check{C}_r^+(z_0)} \big|h(z_1) - h(z_2)\big|^\gamma \, dz_1 dz_2 \Big)^{1/\gamma}.$$

The reader should pay attention to the above curved sharp symbol reminding of curved boundaries.

Observe that, if $r \leq \rho \leq \rho_0(\Omega)$ and $z \in \check{C}_{2\rho_0(\Omega)}^+ \cap \check{C}_r^+(z_0)$, then $\check{C}_r^+(z_0) \subset \check{C}_{4\rho_0(\Omega)}^+$, so that $h^{\#}_{\Omega,\gamma,\rho}(z)$ is well defined on $\check{C}_{2\rho_0(\Omega)}^+$ even if h is given only on $\check{C}_{4\rho_0(\Omega)}^+$ $(\subset [0, 16\rho_0(\Omega)) \times \Omega)$.

2. LEMMA. *Let*

(*i*) $r, \rho > 0$, $\nu \geq 12$, $\rho + \nu r \leq 8\rho_0(\Omega)$, $z_0 = (t_0, x_0) \in \check{C}_\rho^+$,

(*ii*) $u \in W_p^{1,2}(\check{C}_{\rho+\nu r}^+)$ *and* $u(t, x) = 0$ *if* $x \in \partial\Omega$.

Then we have

$$I_r(z_0) := \Big(\fint_{\check{C}_r^+(z_0)} \fint_{\check{C}_r^+(z_0)} \big|D^2u(z_1) - D^2u(z_2)\big|^\gamma \, dz_1 dz_2 \Big)^{1/\gamma}$$

$$\leq N\nu^{(d+2)/\gamma} \Big(\fint_{\check{C}_{\nu r}^+(z_0)} \big(\big|\partial_t u + F(D^2u)\big|^{d+1} + |Du|^{d+1}\big) \, dz \Big)^{1/(d+1)}$$

$$+ N(\nu^{1+(d+2)/\gamma} r + \nu^{-\alpha}) \Big(\fint_{\check{C}_{\nu r}^+(z_0)} |D^2u|^{d+1} \, dz \Big)^{1/(d+1)}, \tag{4}$$

where the constants N depend only on d, α, γ, $M_2(\Omega)$, and δ.

Proof. We follow the proof of Lemma 14.2.2. By the change of variables formula we see that $I_r(z_0)$ equals

$$\Big(\fint_{C_r^+(t_0,y(x_0))} \fint_{C_r^+(t_0,y(x_0))} \big|(D^2u)\big(t_1,x(y_1)\big) - (D^2u)\big(t_2,x(y_2)\big)\big|^\gamma \,dz_1dz_2\Big)^{1/\gamma},$$

where $z_i = (t_i, y_i)$.

Then with $A(y) := \partial x(y)/\partial y$ we define

$$A_0 = A\big(y(x_0)\big), \quad \hat F(\mathsf{u}'') = F\big((A_0^{-1})^*\mathsf{u}''A_0^{-1}\big).$$

As in the proof of Lemma 14.2.2, $D_{\mathsf{u}''}\hat F \in \mathbb{S}_{\hat\delta}$, where $\hat\delta = \hat\delta(d,\delta) \in (0,1]$.

Next, introduce the function

$$\hat u(t,y) = u\big(t,x(y)\big),$$

which belongs to $W^{1,2}_p(C^+_{\rho+\nu r})$, and, since $|y(x_0)| < \rho$, it also belongs to the space $W^{1,2}_p\Big(C^+_{\nu r}\big(t_0,y(x_0)\big)\Big)$ and vanishes on $\partial_{x^1}C^+_{\nu r}\big(t_0,y(x_0)\big)$ if this set in nonempty. By Lemma 1, since $\nu \geq 12$, we have

$$\Big(\fint_{C_r^+(t_0,y(x_0))} \fint_{C_r^+(t_0,y(x_0))} \big|D^2\hat u(z_1) - D^2\hat u(z_2)\big|^\gamma \,dz_1dz_2\Big)^{1/\gamma}$$

$$\leq N\nu^{(d+2)/\gamma}\Big(\fint_{C_{\nu r}^+(t_0,y(x_0))} \big|\partial_t\hat u + \hat F(D^2\hat u)\big|^{d+1}\,dz\Big)^{1/(d+1)}$$

$$+N\nu^{-\alpha}\Big(\fint_{C_{\nu r}^+(t_0,y(x_0))} |D^2\hat u|^{d+1}\,dz\Big)^{1/(d+1)}.$$

After that, by doing the same manipulations as in the proof of Lemma 14.2.2 we arrive at (4). The lemma is proved. □

Change of variables helps derive Lemma 14.2.2 and Lemma 2 from their "flat" counterparts. We also allude to it in the following remark, in which $\gamma \in (0,\gamma_0]$, where $\gamma_0 \in (0,1)$ is a constant similar to the one in Remark 14.2.3.

3. REMARK. Suppose that Assumptions 12.1.2, and 12.1.8 and condition (12.1.4) are satisfied. Let $r \leq 8\rho_0(\Omega)$, and $u \in W^{1,2}_p(\check C^+_r)$ (recall that $p \geq d+1$ and, actually, we fixed $p > d+1$) be such that $u(t,x) = 0$ if $x \in \partial\Omega$. Then

$$\int_{\check C_r^+} \big(|D^2u|^\gamma + |Du|^\gamma\big)\,dz \leq N r^{d+2-\gamma(d+2)/p}\big\|\partial_t u + H[u]\big\|^\gamma_{L_p(\check C^+_r)}$$

$$+N r^{d+2-\gamma(d+2)/p}\|\bar G\|^\gamma_{L_p(\check C^+_r)} + N r^{d+2-2\gamma}\sup_{\partial'\check C^+_r}|u|^\gamma,$$

where the constants N depend only on δ, K_0, d, p, γ, and $M_2(\Omega)$.

To prove this, it suffices to repeat the short argument in Remark 14.2.3 using Corollary 9.4.11 instead of Corollaries 9.4.7 and 9.4.13.

3. A priori estimates in $W^{1,2}_p$ near the boundary and the proof of Theorem 1.3

Recall that $p > d+1$, assume that

$$0 \in \partial\Omega,$$

take $\rho_0 = \rho_0(\Omega)$ from Definition 14.1.10, and suppose that the original system of coordinates in $\mathbb{R}^d$ coincides with the one described in Definition 14.1.10 with $z_0 = 0$. Also take $\check{C}^+_r$, $\check{C}^+_r(z)$ from Section 2 and suppose that the assumptions of Theorem 1.3 are satisfied with $\hat{\theta}$ and θ which are yet to be specified.

First we note the following.

1. LEMMA. *For any $q \in [1,\infty)$ and $\mu > 0$ there exists $\theta = \theta(d,\delta,K_F,\mu,q) > 0$ such that, if Assumption 1.1 is satisfied with this θ, then for any $\mathsf{u}_0' \in \mathbb{R}$, $z_0 \in \check{C}^+_{2\rho_0(\Omega)}$ and $2r \le \rho_0(\Omega) \wedge R_0$, we have*

$$\fint_{\check{C}^+_r(z_0)} \sup_{\substack{\mathsf{u}'' \in \mathbb{S},\\ |\mathsf{u}''| > \tau_0}} \frac{|F(\mathsf{u}_0', \mathsf{u}'', z) - \bar{F}(\mathsf{u}'')|^q}{|\mathsf{u}''|^q} \, dz \le \mu^q,$$

where $\bar{F} = \bar{F}_{z,r,\mathsf{u}_0'}$ is taken from Assumption 1.1.

For the proof of this lemma note that, in light of Corollary 14.1.14 and Assumption 1.1, for any $\mathsf{u}'' \in \mathbb{S}$ with $|\mathsf{u}''| = 1$ we have

$$\fint_{\check{C}^+_r(z_0)} \sup_{\tau > \tau_0} \tau^{-1} \big|F(\mathsf{u}_0', \tau\mathsf{u}'', z) - \bar{F}(\tau\mathsf{u}'')\big| \, dz \le N(d)\theta$$

if $2r \le \rho_0(\Omega) \wedge R_0$. After that the assertion of the lemma is obtained by repeating word for word the proof of Lemma 10.4.1.

Recall that $\omega_u(\Pi,\rho)$ and $\omega_{F,u,\Pi}(\rho)$ are introduced in Definition 12.1.6.

2. LEMMA. *Let $r,\rho \in (0,\infty)$ and $\nu \ge 12$ satisfy $2\nu r \le \rho_0(\Omega) \wedge R_0$ and $\rho \le 2\rho_0(\Omega)$. Take*

$$\mu \in (0,\infty), \quad \beta \in (1,\infty),$$

and suppose that Assumption 1.1 is satisfied with $\theta = \theta(d,\delta,K_F,\mu,\beta d + \beta)$ (see Lemma 1). Assume that we are given a function u such that $u \in W^{1,2}_p(\check{C}^+_{\rho+\nu r})$ and $u(t,x) = 0$ if $x \in \partial\Omega$. Use $I_r(z_0)$ introduced in (2.4).

Then, for γ and α from Lemma 2.2, for $z_0 \in \check{C}^+_\rho$, we have

$$I_r(z_0) \le N\eta\Big(\fint_{\check{C}^+_{\nu r}(z_0)} |D^2u|^{\beta'(d+1)} \, dz\Big)^{1/(\beta'(d+1))}$$

$$+N\nu^{(d+2)/\gamma}\Big(\fint_{\check{C}^+_{\nu r}(z_0)} \big(\big|\partial_t u + F[u]\big|^{d+1} + |Du|^{d+1}\big) \, dz\Big)^{1/(d+1)} + N\tau_0\nu^{(d+2)/\gamma},$$

where $\beta' = \beta/(\beta-1)$,

$$\eta = \big(\mu + \nu r + \omega_{F,u,\check{C}^+_{\rho+\nu r}}(\nu r)\big)\nu^{(d+2)/\gamma} + \nu^{-\alpha},$$

and the constants N depend only on d, K_F, α, δ, and $M_2(\Omega)$.

The proof of this lemma is based on Lemma 1 and, in light of Lemma 2.2, is practically identical to that of Lemma 10.4.2.

We now come to the main a priori estimate near the boundary for nonlinear parabolic equations with "almost VMO coefficients".

3. THEOREM. *Take $R > 0$ such that*

$$4R \leq \rho_0(\Omega) \wedge R_0,$$

and let $u \in W^{1,2}_p(\check{C}^+_{2R})$ be such that $u(t,x) = 0$ if $x \in \partial\Omega$. Then there exist constants $\hat{\theta}, \theta \in (0,1]$, depending only on d, p, δ, $M_2(\Omega)$, and K_F, such that if Assumptions 12.1.3 and 1.1 are satisfied with these $\hat{\theta}$ and θ, respectively, then there exist constants N, depending only on $R_0, \gamma, d, p, K_0, K_F, \delta$, $M_2(\Omega)$, and the function $\omega_{F,u,\check{C}^+_{2R}}$, such that

$$\begin{aligned}\|D^2u\|_{L_p(\check{C}^+_R)} \leq{}& N\big\|\partial_t u + H[u]\big\|_{L_p(\check{C}^+_{2R})} + N\|\bar{G}\|_{L_p(\check{C}^+_{2R})} + N\tau_0 R^{(d+2)/p} \\ &+ N\|u\|_{L_p(\check{C}^+_{2R})} + NR^{-\chi}\big\|\,|D^2u|^\gamma\big\|^{1/\gamma}_{L_1(\check{C}^+_{2R})},\end{aligned} \tag{1}$$

$$\begin{aligned}\|D^2u\|_{L_p(\check{C}^+_R)} \leq{}& N\big\|\partial_t u + H[u]\big\|_{L_p(\check{C}^+_{2R})} + N\|\bar{G}\|_{L_p(\check{C}^+_{2R})} \\ &+ N\tau_0 R^{(d+2)/p} + NR^{(d+2)/p-2}\sup_{\check{C}^+_{2R}}|u|,\end{aligned} \tag{2}$$

where $\chi = (d+2)(1/\gamma - 1/p)$ and γ is the same as in Lemma 2.

Proof. Whenever it makes sense, for $\rho \leq \rho_0(\Omega)$, and $z \in \check{C}^+_{2\rho_0(\Omega)}$ introduce

$$h^{\#}_{\Omega,\gamma,\rho}(z) = \sup\big\{I_r(h,z_0) : z_0 \in \mathbb{R}_+ \times \Omega, r \in (0,\rho], \check{C}^+_r(z_0) \ni z\big\},$$

where

$$I_r(h,z_0) = \Big(\fint_{\check{C}^+_r(z_0)}\fint_{\check{C}^+_r(z_0)} \big|h(z_1) - h(z_2)\big|^\gamma\,dz_1dz_2\Big)^{1/\gamma}.$$

Then take $\varepsilon \in (0,1]$ to be specified later, take $R_1 < R_2 \leq 2R$, such that

$$R_2 \leq 2R_1, \quad R_2 - R_1 \leq \varepsilon R_0,$$

take $\nu \geq 12$, and set

$$r_0 = (R_2 - R_1)/(\nu + 1).$$

Take $z = (t,x), z_0 = (t_0,x_0), r$ such that

$$r \leq r_0, \quad z \in \check{C}^+_{R_1}, \quad z_0 \in \Omega, \quad z \in \check{C}^+_r(z_0).$$

Observe that $|y(x)| < R_1$, $y(x) \in B^+_r(y(x_0))$, $|y(x_0)| < r_0 + R_1$, $|y(x_0)| + \nu r < R_2$,

$$B^+_{\nu r}(y(x_0)) \subset B^+_{R_2}, \quad \check{B}^+_{\nu r}(x_0) \subset \check{B}^+_{R_2}, \quad \check{C}^+_{\nu r}(x_0) \subset \check{C}^+_{R_2}. \tag{3}$$

Therefore, (14.3.5) is available. By plugging there $h = h(s,y)$, integrating through the first two inequalities with respect to $s \in (t_0, t_0 + \nu^2 r^2)$, and using the classical parabolic maximal function $\mathbb{M}$ introduced in (12.4.6), we find that, for any measurable h,

$$\fint_{\check{C}^+_{\nu r}(z_0)} |h|\,dyds \le N\mathbb{M}(hI_{\check{C}^+_{R_2}})(z).$$

It follows from Lemma 2 that (just in case, note that $2\nu r \le 2R_2 \le 4R \le \rho_0(\Omega)\wedge R_0$)

$$I_r(D^2u, z_0) \le N\nu^{(d+2)/\gamma}\mathbb{M}^{1/(d+1)}\big(\big|\partial_t u + F[u]\big|^{d+1} I_{\check{C}^+_{R_2}}\big)(z) + N\tau_0\nu^{(d+2)/\gamma}$$

$$+N\eta\mathbb{M}^{1/(\beta'(d+1))}\big(|D^2u|^{\beta'(d+1)} I_{\check{C}^+_{R_2}}\big)(z)$$

$$+N\nu^{(d+2)/\gamma}\mathbb{M}^{1/(d+1)}\big(|Du|^{d+1} I_{\check{C}^+_{R_2}}\big)(z),$$

where (observe that $\nu r \le \varepsilon R_0$)

$$\eta = \big(\mu + \nu r_0 + \omega_{F,u,\check{C}^+_{R_2}}(\varepsilon R_0)\big)\nu^{(d+2)/\gamma} + \nu^{-\alpha}.$$

The arbitrariness of z_0 and r now implies that on $\check{C}^+_{R_1}$

$$(D^2u)^{\#}_{\Omega,\gamma,r_0} \le N\nu^{(d+2)/\gamma}\mathbb{M}^{1/(d+1)}\big(\big|\partial_t u + F[u]\big|^{d+1} I_{\check{C}^+_{R_2}}\big) + N\tau_0\nu^{(d+2)/\gamma}$$

$$+N\eta\mathbb{M}^{1/(\beta'(d+1))}\big(|D^2u|^{\beta'(d+1)} I_{\check{C}^+_{R_2}}\big)$$

$$+N\nu^{(d+2)/\gamma}\mathbb{M}^{1/(d+1)}\big(|Du|^{d+1} I_{\check{C}^+_{R_2}}\big).$$

Next, we are going to use Theorem C.3.2 with $\kappa = r_0/R_1 \le 1/3$, according to which, if $h \in L_p(\check{C}^+_{R_2})$, then

$$\|h\|_{L_p(\check{C}^+_{R_1})} \le N\big\|h^{\#}_{\Omega,\gamma,r_0}\big\|_{L_p(\check{C}^+_{R_1})} + N\nu^{\chi_1}(R_2 - R_1)^{-\chi_1}R_1^{\chi_1-\chi_2}\big\|\,|h|^\gamma\big\|^{1/\gamma}_{L_1(\check{C}^+_{R_1})}, \quad (4)$$

where $\chi_1 = (d+2)/\gamma$, $\chi_2 = (d+2)(1/\gamma - 1/p)$, and the constants N depend only on d, γ, and p.

Thanks to (4) and the Hardy-Littlewood maximal function theorem, by taking β so that $p > \beta'(d+1)$, we get

$$\|D^2u\|_{L_p(\check{C}^+_{R_1})} \le N\nu^{(d+2)/\gamma}\big\|\partial_t u + F[u]\big\|_{L_p(\check{C}^+_{R_2})} + N\tau_0\nu^{(d+2)/\gamma}$$

$$+\Big[N\big(\mu + \nu r_0 + \omega_{F,u,\check{C}^+_{R_2}}(\varepsilon R_0)\big)\nu^{(d+2)/\gamma} + N_0\nu^{-\alpha}\Big]\|D^2u\|_{L_p(\check{C}^+_{R_2})}$$

$$+N\nu^{(d+2)/\gamma}\|Du\|_{L_p(\check{C}^+_{R_2})} + N\nu^{\chi_1}(R_2 - R_1)^{-\chi_1}R_1^{\chi_1-\chi_2}\big\|\,|D^2u|^\gamma\big\|^{1/\gamma}_{L_1(\check{C}^+_{2R})},$$

where the constants N, N_0 depend only on d, p, K_F, γ, $M_2(\Omega)$, and δ.

This estimate looks almost like (10.4.7). Then we repeat the argument after (10.4.7) and choose and fix ε, ν, $\hat{\theta}$, and μ, recall what r_0 is, and conclude that

$$\|D^2u\|_{L_p(\check{C}^+_{R_1})} \leq N\big\|\partial_t u + H[u]\big\|_{L_p(\check{C}^+_{R_2})} + N\tau_0\nu^{d/\gamma}$$

$$+\big(5/8 + N_1(R_2 - R_1)\big)\|D^2u\|_{L_p(\check{C}^+_{R_2})} + N\|u\|_{L_p(\check{C}^+_{2R})} + N\|\bar{G}\|_{L_p(\check{C}^+_{2R})}$$

$$+N(R_2 - R_1)^{-\chi_1}R_1^{\chi_1-\chi_2}\|\,|D^2u|^\gamma\|^{1/\gamma}_{L_1(\check{C}^+_{2R})}.$$

After that, to prove (1), it suffices to repeat literally what follows (14.3.7) (only replacing $\check{B}$ with $\check{C}$ and adding $\partial_t u$ to $H[u]$). By using Remark 2.3 we estimate the last term in (1) and then finish the proof of the theorem in the same way as in the case of Lemma 10.4.3. The theorem is proved. □

Proof of Theorem 1.3. To start, assume that $g \equiv 0$. Observe that in that case we may assume that $u(t, x)$ is defined for $t \geq T$, $x \in \bar{\Omega}$, as zero and still satisfies there (1.1). It suffices for the latter that $H(0, t, x) = 0$ if $t \geq T$, which is easy to accommodate without altering our assumptions just by replacing $G(\mathsf{u}, t, x)$ and $\bar{G}$ with $G(\mathsf{u}, t, x)I_{t<T}$ and $\bar{G}I_{t<T}$, respectively. After such extension $u \in W^{1,2}_p(\mathbb{R}_+ \times \Omega)$.

Take $\hat{\theta} = \hat{\theta}_0 = \hat{\theta}_0\big(d, p, \delta, K_F, M_2(\Omega)\big)$ and $\theta = \theta_0 = \theta_0\big(d, p, \delta, K_F, M_2(\Omega)\big)$ that suit both Theorems 12.1.7 and 3, and take R such that

$$4R = \rho_0(\Omega) \wedge R_0.$$

Theorem 3 allows us to estimate the $W^{1,2}_p$-norm of u in the domain $\check{C}^+_R = \check{C}^+_R(0)$ associated with the origin, that is assumed to belong to $[0, \infty) \times \partial\Omega$. Of course, one can take any point $z_0 = (t_0, x_0) \in [0, \infty) \times \partial\Omega$ as the origin and apply Theorem 3 to $\check{C}^+_{R,z_0}$ and $\check{C}^+_{2R,z_0}$ in place of $\check{C}^+_R$ and $\check{C}^+_{2R}$, respectively, where by $\check{C}^+_{\rho,z_0}$ we, naturally, mean the sets

$$[t_0, t_0 + \rho^2) \times \check{B}^+_{\rho,x_0},$$

where $\check{B}^+_{\rho,x_0}$ is constructed in Lemma 14.1.13 but with x_0 in place of 0 and relative to the coordinate system $\Psi(x_0)$ associated with x_0 as described in Definition 14.1.10. According to that, we find finitely many

$$z_i \in [0, T + R^2] \times \partial\Omega$$

(their number depending only on $\rho_0(\Omega)$, R_0, $\operatorname{diam}(\Omega)$, $M_2(\Omega)$, and T) such that

$$\bigcup_i \check{C}^+_{R,z_i} \cup \Big([0, T + R^2/2) \times \Omega^{R/2}\Big) \supset \Pi.$$

By Theorem 3, for any i (recall that $\bar{G}(t, x) = 0$ for $t \geq T$)

$$\|D^2u\|^p_{L_p(\check{C}^+_{R,z_i})} \leq N\|\bar{G}\|^p_{L_p(\Pi)} + N\tau_0^p + N\sup_\Pi |u|.$$

By Theorem 12.1.7

$$\|D^2u\|^p_{L_p((0,T+R^2/r)\times\Omega^{R/2})} \leq N\|\bar{G}\|^p_{L_p(\Pi)} + N\tau_0^p + N\sup_\Pi |u|.$$

We sum up these estimates and come to (1.3). This proves the theorem if $g \equiv 0$.

In the general case introduce $w(z) = u(z) - g(z)$,

$$\mathsf{g} = (g, Dg, D^2 g) I_\Pi,$$

$$\hat{H}(\mathsf{v}, z) = H\big(\mathsf{v} + \mathsf{g}(z), z\big) + \partial_t g(z) I_\Pi(z), \quad \hat{F}(\mathsf{v}_0', \mathsf{v}'', z) = F(\mathsf{v}_0' + \mathsf{g}_0'(z), \mathsf{v}'', z).$$

Observe that $\partial_t w + \hat{H}[w] = 0$ in Π (a.e.), $w \in W^{1,2}_p(\Pi)$, and $w = 0$ on $\partial'\Pi$.

Also for $\mathsf{v}_0' \in \mathbb{R}$, $r \in (0, R_0]$, and $z \in \Pi$ we set

$$\bar{\hat{F}}(\mathsf{v}'') = \bar{\hat{F}}_{z,r,\mathsf{v}_0'}(\mathsf{v}'') = \bar{F}_{z,r,\mathsf{v}_0' + \mathsf{g}_0'(z)}(\mathsf{v}''),$$

take $\theta_0 = \theta_0\big(d, p, \delta, K_F, M_2(\Omega)\big)$ defined above in the first part of the proof where $g \equiv 0$, find $\hat{R}_0 \le R_0$ such that $\omega_{F,g,\Pi}(\hat{R}_0) \le \theta_0/2$ and then require the original Assumption 1.1 (iii) to be satisfied with $\hat{R}_0$ and $\theta_0/2$ in place of R_0 and θ, respectively.

Then, as in the proof of Theorem 14.1.3 on page 324, it turns out that the assumptions of Theorem 1.3 are satisfied with $\hat{\theta}$ and θ from the beginning of the proof and only one adjustment that, for $\hat{G} = \hat{H} - \hat{F}$,

$$\bar{\hat{G}} = |\partial_t g| I_\Pi + N|D^2 g| I_\Pi + \bar{G} + K_0\big(g^2 + |Dg|^2\big)^{1/2} I_\Pi,$$

where N depends only on K_F and d.

It follows that the above result is applicable to w, and along with the embedding inequality: $|g| \le N\|g\|_{W^{1,2}_p(\Pi)}$, leads to (1.3) in the general case. The theorem is proved. □

4. Proof of Theorem 1.4

The proof of Theorem 1.4 is based on Theorem 8.3.3 and is modeled after the proof of Theorem 14.1.5 in Section 14.5.

Step 1. We take $P(\mathsf{u}'')$ defined in (6.2.4), set $P[u] = P(D^2 u)$, and first we assume that $g \in W^{1,2}_\infty(\Pi)$ and there exists *constants* $N_0, \bar{H}$ such that, for all t, x, u',

$$\big|H(\mathsf{u}', 0, t, x)\big| \le N_0|\mathsf{u}'| + \bar{H}. \tag{1}$$

By Theorem 8.3.3 for any $K > 0$ there exists a function v_K which is in $W^{1,2}_p(\Pi)$ for any $p > 1$, such that $v_K = g$ on $\partial'\Pi$, and

$$\partial v_K + H_K[v_K] = 0 \quad \text{in} \quad \Pi \text{ (a.e.)}, \tag{2}$$

where

$$H_K(\mathsf{u}, t, x) = \max\big(H(\mathsf{u}, t, x), P(\mathsf{u}'') - K\big).$$

Next we use Theorem 1.3 about a priori estimates in $W^{1,2}_p(\Pi)$. Set

$$F_K\big(\mathsf{u}_0', \mathsf{u}'', t, x\big) = \max\big(F\big(\mathsf{u}_0', \mathsf{u}'', t, x\big), P(\mathsf{u}'') - K\big),$$

$$G_K(u, t, x) = H_K(u, x) - F_K\big(\mathsf{u}_0', \mathsf{u}'', t, x\big).$$

By closely following our argument below (10.5.4) we convince ourselves that the assumptions of Theorem 1.3 are satisfied for H_K, F_K with the same ingredients as the original ones and

$$\bar{F}_{K,z,r,\mathsf{u}_0'}(\mathsf{u}'') = \max\big(\bar{F}_{z,r,\mathsf{u}_0'}(\mathsf{u}''), P(\mathsf{u}'') - K\big)$$

apart from the fact that δ and K_F should be modifies appropriately but based only on the original d, δ, and K_F.

It is shown in the proof of Theorem 12.1.10 on page 267 that the family $\{v_K : K > 0\}$ is precompact in $C(\bar{\Pi})$ and, in particular, that the functions $\omega_{F,v_K,\Pi}$ are majorated by a continuous function independent of K and vanishing at zero. Also it is noted that in showing the precompactness, in what concerns g, only the supremum of $|g|$ and the modulus of continuity of g on $\partial'\Pi$ were involved.

Then, provided that an appropriate choice of θ and $\hat{\theta}$ was made, by Theorem 1.3 we obtain that

$$\sup_{K\geq 1} \|v_K\|_{W^{1,2}_p(\Pi)} < \infty. \tag{3}$$

In this way we completed a crucial step consisting of obtaining a uniform control of the $W^{1,2}_p(\Pi)$-norms of v_K.

Estimate (3) is stronger than (12.5.8). In particular, it implies by the proof of Theorem 12.1.10 that there is an appropriate sequence v_{K_n}, $K_n \to \infty$, such that its limit is a function $v \in W^{1,2}_p(\Pi)$ satisfying (1.1) in Π (a.e.) and equal g on $\partial'\Pi$. Owing to (3), of course, $v \in W^{1,2}_p(\Pi)$.

This proves the theorem if condition (1) is satisfied and $g \in W^{1,2}_\infty(\Pi)$.

Step 2. Still assume that $g \in W^{1,2}_\infty(\Pi)$ and abandon (1). The argument in Step 2 of the proof of Theorem 12.1.10 in Section 12.5 is, obviously, applicable in our current setting and, owing to the fact that the right-hand side of (1.3) is independent of N_0 and $\bar{H}$, immediately yields the desired result.

Finally, to treat the general $g \in W^{1,2}_p(\Pi)$ it suffices to use approximations and very simple arguments about passing to the limit, which we have seen a number of times before. This step is left to the reader. The theorem is proved. □

CHAPTER 16

$C^{1+\alpha}$-regularity of viscosity solutions of general parabolic equations

1. Generalities

For a Borel measurable function $H(\mathsf{u}, t, x)$,

$$\mathsf{u} = (\mathsf{u}', \mathsf{u}''), \quad \mathsf{u}' = \big(\mathsf{u}_0', \mathsf{u}_1', ..., \mathsf{u}_d'\big) \in \mathbb{R}^{d+1}, \quad \mathsf{u}'' \in \mathbb{S}, \quad (t, x) \in \mathbb{R}^{d+1},$$

where $\mathbb{S}$ is the set of symmetric $d \times d$ matrices, and sufficiently regular functions $v(t, x)$ we set

$$H[v](t, x) = H\big(v(t, x), Dv(t, x), D^2 v(t, x), t, x\big),$$

and we will be dealing with the parabolic equations

$$\partial_t v(t, x) + H[v](t, x) = 0 \tag{1}$$

in subsets of $[0, T) \times \mathbb{R}^d$, where $T \in (0, \infty)$ is fixed. Recall that

$$\mathbb{R}^d = \big\{x = (x^1, ..., x^d) : x^1, ..., x^d \in \mathbb{R}\big\},$$

$$\partial_t = \frac{\partial}{\partial t}, \quad D^2 u = (D_{ij} u), \quad Du = (D_i u), \quad D_i = \frac{\partial}{\partial x^i}, \quad D_{ij} = D_i D_j,$$

and if $R \in (0, \infty)$ and $(t, x) \in \mathbb{R}^{d+1}$, then

$$B_R = \big\{x \in \mathbb{R}^d : |x| < R\big\}, \quad B_R(x) = x + B_R,$$

$$C_R = [0, R^2) \times B_R, \quad C_R(t, x) = (t, x) + C_R.$$

We also take a bounded domain $\Omega \subset \mathbb{R}^d$ satisfying the exterior ball condition and set

$$\Pi = [0, T) \times \Omega, \quad \partial' \Pi = \bar{\Pi} \setminus \big(\{0\} \times \bar{\Omega}\big)$$

($[0, T) \times \Omega$ looks like a pie if the t-axis is directed upward).

We will be dealing with viscosity solutions of (1) in Π. The following definition is taken from [**25**] and has the same spirit as in [**23**].

1. Definition. For each choice of "regularity" class $\mathcal{R} = C$ or $\mathcal{R} = L_p$, $p \in (1, \infty)$, we say that u is an $\mathcal{R}$-viscosity subsolution of (1) in Π provided that u is continuous in Π and, for any $\bar{C}_r(t_0, x_0) \subset \Pi$ and any function ϕ, that is continuous in $C_r(t_0, x_0)$ and whose generalized derivatives satisfy $\partial_t \phi, D\phi, D^2\phi \in \mathcal{R}\big(C_r(t_0, x_0)\big)$, and is such that $u - \phi$ attains its maximum over $C_r(t_0, x_0)$ at (t_0, x_0), we have

$$\lim_{\rho \downarrow 0} \operatorname*{ess\,sup}_{C_\rho(t_0, x_0)} \big[\partial_t \phi(t, x) + H\big(u(t, x), D\phi(t, x), D^2\phi(t, x), t, x\big)\big] \geq 0. \tag{2}$$

In a natural way one defines $\mathcal{R}$-viscosity supersolutions and calls a function an $\mathcal{R}$-viscosity solution if it is an $\mathcal{R}$-viscosity supersolution and an $\mathcal{R}$-viscosity subsolution.

Note that $C_r(t_0,x_0)$ contains $\{(t,x): t=t_0, |x-x_0|<r\}$, which is part of its boundary. Therefore, the conditions like $D^2\phi \in C\big(C_r(t_0,x_0)\big)$ mean that the second-order derivatives of ϕ are continuous up to this part of the boundary.

2. REMARK. If $H(\mathsf{u},t,x)$ is a continuous function of (u,t,x) and we are talking about the C-viscosity subsolutions, then (2) becomes, of course, just

$$\partial_t\phi(t_0,x_0)+H\big(u(t_0,x_0),D\phi(t_0,x_0),D^2\phi(t_0,x_0),t_0,x_0\big)\geq 0.$$

Also note that L_p-viscosity solutions are automatically C-viscosity solutions.

The reader is referred to [**25**], [**52**], and the references therein for numerous properties of L_p-viscosity solutions and to [**23**], [**51**], [**52**], and the references therein for those of C-viscosity solutions.

The notion of L_p-viscosity solution generalizes that of $W^{1,2}_p$-solution, which is seen from the following well-known fact (see [**25**]).

3. THEOREM. *Suppose that H is a nonincreasing function of u'_0, Lipschitz continuous with respect to $([\mathsf{u}'],\mathsf{u}'')$ with constant independent of u'_0 and (t,x), and $D_{\mathsf{u}''}H\in\mathbb{S}_\delta$ at all points of its differentiability with respect to u'', where the constant $\delta\in(0,1]$. Suppose that v is a continuous in $\bar\Pi$, L_p-viscosity solution of* (1) *and $w\in W^{1,2}_{p,\mathrm{loc}}(\Pi)\cap C(\bar\Pi)$ is a function satisfying* (1) *(a.e.) in Π. Finally, let $v=w$ on $\partial'\Pi$. Then $v=w$ in Π.*

Proof. Denote by K_0 the Lipschitz constant of H with respect to $[\mathsf{u}']$ and take the global barrier ϕ from the proof of Theorem 18.1.2 and for $\varepsilon>0$ denote $w_\varepsilon=w+\varepsilon\psi$. First we want to prove that $v\leq w_\varepsilon$ in $\bar\Pi$. If this is not true then there is a point $(t_0,x_0)\in\Pi$ at which $v-w_\varepsilon$ achieves its strictly positive maximum in $\bar\Pi$. Then by definition

$$\lim_{\rho\downarrow 0}\operatorname*{ess\,sup}_{C_\rho(t_0,x_0)}\big[\partial_t w(t,x)+H\big(v(t,x),Dw_\varepsilon(t,x),D^2w_\varepsilon(t,x),t,x\big)\big]\geq 0.$$

Since v and w are continuous and $v(t_0,x_0)>w_\varepsilon(t_0,x_0)$, this inequality is preserved on a neighborhood of (t_0,x_0). Therefore,

$$\lim_{\rho\downarrow 0}\operatorname*{ess\,sup}_{C_\rho(t_0,x_0)}\big[\partial_t w(t,x)+H\big(w(t,x),Dw_\varepsilon(t,x),D^2w_\varepsilon(t,x),t,x\big)\big]\geq 0. \tag{3}$$

Then observe that

$$H\big(w(t,x),Dw_\varepsilon(t,x),D^2w_\varepsilon(t,x),t,x\big)-H[w](t,x)=\varepsilon a^{ij}D_{ij}\psi+\varepsilon b^iD_i\psi\leq-\varepsilon,$$

where (a^{ij}) is $\mathbb{S}_\delta$-valued and $b=(b^i)$ satisfies $|b|\leq K_0$. This and the fact that (1) holds (a.e.) with w in place of v leads to a contradiction with (3), proves that $v\leq w+\varepsilon\psi$ for any $\varepsilon>0$, and hence $v\leq w$. The inequality $v\geq w$ is proved similarly. □

2. Main results

The main result of this chapter says that under rather general assumptions there exist C-viscosity and L_p-viscosity solutions which are in $C^{1+\alpha}_{\text{loc}}$. We allow H to be just measurable in t and for its principal part to have sufficiently small discontinuities as a function of x. The exposition here is close to the one in [**82**].

For $\kappa \in (0,1]$ and functions $\phi(t,x)$ on Π set

$$[\phi]_{C^\kappa(\Pi)} = \sup_{(t,x),(s,y)\in\Pi} \frac{|\phi(t,x)-\phi(s,y)|}{|t-s|^{\kappa/2}+|x-y|^\kappa}, \quad \|\phi\|_{C(\Pi)} = \sup_{\Pi} |\phi|,$$

$$\|\phi\|_{C^\kappa(\Pi)} = \|\phi\|_{C(\Pi)} + [\phi]_{C^\kappa(\Pi)}.$$

For $\kappa \in (1,2]$ and sufficiently regular ϕ set

$$[\phi]_{C^\kappa(\Pi)} = \sup_{t,s\in[0,T),x\in\Omega} \frac{|\phi(t,x)-\phi(s,x)|}{|t-s|^{\kappa/2}}$$

$$+ \sup_{x,y\in\Omega,t\in[0,T)} \frac{|D\phi(t,x)-D\phi(t,y)|}{|x-y|^{\kappa-1}}, \quad \|\phi\|_{C^\kappa(\Pi)} = \|\phi\|_{C^1(\Pi)} + [\phi]_{C^\kappa(\Pi)}.$$

The set of functions with finite norm $\|\cdot\|_{C^\kappa(\Pi)}$ is denoted by $C^\kappa(\Pi)$. Observe that any $u \in C^\kappa(\Pi)$ admits a unique extension to $\bar{\Pi}$ and we will always consider this extension while dealing with $u(T,x)$. As usual we write $u \in C^\kappa_{\text{loc}}(\Pi)$ if $u \in C^\kappa\big(C_R(t,x)\big)$ for any $C_R(t,x)$ such that $\bar{C}_R(t,x) \subset \Pi$.

1. REMARK. According to the above notation $C^2(\Pi)$ is not what is usually meant. Therefore, we are going to use the symbol $W^{1,2}_\infty(\Pi)$ instead for the space provided with norm $\|\cdot\|_{C^2(\Pi)} = \|\cdot\|_{W^{1,2}_\infty(\Pi)}$. One should keep this in mind when we consider all $\kappa \in (0,2]$ at once. Probably something of the same kind should be said also about $C^1(\Pi)$, but we will never use this space.

For $\kappa \in (1,2)$ we are also going to use the spaces $C^{1+\kappa}(\Pi)$ of functions $u \in W^{1,2}_\infty(\Pi)$ with finite norm

$$\|u\|_{C^{1+\kappa}(\Pi)} = \|v\|_{C^2(\Pi)} + [u]_{C^{1+\kappa}(\Pi)},$$

where

$$[u]_{C^{1+\kappa}(\Pi)} = [\partial_t u]_{C^{\kappa-1}(\Pi)} + [D^2 u]_{C^{\kappa-1}(\Pi)}.$$

2. REMARK. Sometimes it is useful to know that if $\kappa \in (1,2]$ and $\phi \in C^\kappa(C_R)$, then

$$\big|D\phi(t,x) - D\phi(s,x)\big| \le N(d)|t-s|^{(\kappa-1)/2}[\phi]_{C^\kappa(C_R)} \tag{1}$$

whenever $(t,x),(s,x) \in C_R$.

To prove (1) first note that it is scale invariant and, therefore, we may concentrate on $R=1$. In that case take $x \in B_1$, $0 \le s < t \le 1$, and observe that moving if necessary x inside B_1 to the center, so that its distance to the boundary becomes not smaller than $\sqrt{t-s}$ (≤ 1) will change the left-hand side of (1) by not more than $2(t-s)^{(\kappa-1)/2}[\phi]_{C^\kappa(C_1)}$. It follows that we may assume that $1-|x| \ge \sqrt{t-s} =: h$.

Now take $i \in \{1, ..., d\}$, set $y = x + he_i$ ($\in \bar{B}_1$), and notice that

$$\begin{aligned}\big|D_i\phi(t,x) - D_i\phi(s,x)\big| \le & \big|D_i\phi(t,x) - h^{-1}\big[\phi(t,y) - \phi(t,x)\big]\big| \\ & + \big|D_i\phi(s,x) - h^{-1}\big[\phi(s,y) - \phi(s,x)\big]\big| \\ & + h^{-1}\big|\phi(t,y) - \phi(s,y)\big| + h^{-1}\big|\phi(t,x) - \phi(s,x)\big|.\end{aligned}$$

Here the first two terms on the right we estimate by using the mean value theorem. They turn out to be smaller than

$$2|x-y|^{\kappa-1}[\phi]_{C^\kappa(C_1)} = 2|t-s|^{(\kappa-1)/2}[\phi]_{C^\kappa(C_1)}.$$

As far as the remaining terms are concerned we use the fact that by definition

$$h^{-1}|\phi(t,y) - \phi(s,y)| \le h^{-1}|t-s|^{\kappa/2}[\phi]_{C^\kappa(C_1)} = |t-s|^{(\kappa-1)/2}[\phi]_{C^\kappa(C_1)}.$$

This yields (1).

3. Assumption. (i) The function $H(\mathsf{u}, t, x)$ is Lipschitz continuous in u'' for every $\mathsf{u}', (t,x) \in \mathbb{R}^{d+1}$ and $D_{\mathsf{u}''}H \in \mathbb{S}_\delta$ at all points of differentiability of $H(\mathsf{u}, t, x)$ with respect to u'', where δ is a fixed constant in $(0, 1]$.

(ii) Either

(a) for any $\mathsf{u}_0 \in \mathbb{R}^{d+1} \times \mathbb{S}$ and $(t_0, x_0) \in \Pi$ the function $H(\mathsf{u}, t, x)$ is continuous at u_0 as a function of u uniformly with respect to (t,x) belonging to a neighborhood of (t_0, x_0);

or

(b) For any $M \in (0, \infty)$ the function $H(\mathsf{u}, t, x)$ is continuous with respect to $[\mathsf{u}'] = (\mathsf{u}'_1, ..., \mathsf{u}'_d)$ uniformly with respect to $|\mathsf{u}'_0| \le M$, $u'' \in \mathbb{S}$, $(t,x) \in \Pi$.

We fix some constants

$$p \in (d+2, \infty), \quad K_0, K_1, \in [0, \infty), \quad R_0 \in (0, 1], \quad \hat{R}_0 \in [0, \infty],$$

a nonnegative function

$$\bar{G} \in L_p(\Pi),$$

and in the following assumption also use $\theta_0 = \theta_0(d, \delta) \in (0, 1]$, whose value is specified in the proof of Lemma 6.3.

4. Assumption. We have a representation

$$H(\mathsf{u}, t, x) = F\big(\mathsf{u}'_0, \mathsf{u}'', t, x\big) + G(\mathsf{u}, t, x).$$

(i) The functions F and G are Borel measurable functions of their arguments, *continuous* with respect to u for any (t,x).

(ii) For all values of the arguments

$$\big|G(\mathsf{u}, t, x)\big| \le K_0|\mathsf{u}'| + \bar{G}(t,x).$$

(iii) The function F is Lipschitz continuous with respect to u'' and at all points of differentiability of F with respect to u'' and any $\varepsilon \in \mathbb{S}$ such that $|\varepsilon| = 1$ we have

$$D_{\mathsf{u}''}F + \theta_0\varepsilon \in \mathbb{S}_{\delta/2}.$$

(iv) For any $\mathsf{u}'_0, \mathsf{v}'_0 \in \mathbb{R}$, $(t, x_i) \in \mathbb{R}^{d+1}$, $i = 1, 2$, and $\mathsf{u}'' \in \mathbb{S}$ we have

$$\left|F(\mathsf{u}_0', \mathsf{u}'', t, x_1) - F(\mathsf{v}_0', \mathsf{u}'', t, x_2)\right| \le K_1 + \theta_0 |\mathsf{u}''|$$

as long as $|x_1 - x_2| \le R_0$ and $|\mathsf{u}_0' - \mathsf{v}_0'| < \hat{R}_0$.

(v) We have $F(\mathsf{u}_0', 0, t, x) = 0$ for all $\mathsf{u}_0' \in \mathbb{R}$, $(t,x) \in \mathbb{R}^{d+1}$.

5. REMARK. The value ∞ is allowed for $\hat{R}_0$ with the purpose to cover the cases in which F is almost independent of u_0' when conditions like (2) below are automatically satisfied.

6. REMARK. Observe that Assumption 4 (iv) is automatically satisfied if F is a function of only u'' and t.

7. ASSUMPTION. We are given $g \in C(\bar{\Pi})$.

8. THEOREM. *There is a constant $\theta_0 = \theta_0(d, \delta) \in (0, 1]$ such that, if the above assumptions are satisfied with this θ_0, then there exist a $\kappa = \kappa(d, \delta, p) \in (1, 2)$ and a function $v \in C^{\kappa}_{\mathrm{loc}}(\Pi) \cap C(\bar{\Pi})$ that is a C-viscosity or an L_p-viscosity solution of equation (1.1) in Π with boundary condition $v = g$ on $\partial' \Pi$ according as requirements (a) or (b) in Assumption 3 (ii) are satisfied.*

Furthermore, for any $r, R \in (0, R_0]$ satisfying $r < R$ and $(t,x) \in \Pi$ such that $C_R(t,x) \subset \Pi$ and

$$\operatorname*{osc}_{C_R(t,x)} v < \hat{R}_0 \tag{2}$$

we have

$$[v]_{C^{\kappa}(C_r(t,x))} \le N(R - r)^{-\kappa} \sup_{C_R(t,x)} |v| + N\big(K_1 + \|\bar{G}\|_{L_p(C_R(t,x))}\big), \tag{3}$$

where the constants N depend only on d, δ, and K_0.

This theorem is proved in Section 6.

9. REMARK. Wang in Theorem 1.3 of [**115**] assumes a structure condition on H which in the case of linear equations implies that the coefficients of Du are independent of (t,x). On the other hand, he proves the result for *any* C-viscosity solution.

Theorem 8 contains Proposition 5.4 of [**25**] proved there for equations $\partial_t u + H(D^2 u) = 0$ apart from the fact that Proposition 5.4 of [**25**] is valid for *any* C-viscosity solution. Theorem 8 is also close to Theorem 7.3 of [**25**], that is proved for *any* L_p-viscosity solution and not only for the C-viscosity or L_p-viscosity solution we construct. The most significant difference in the assumptions is that H in [**25**] is supposed to be continuous in u_0' uniformly with respect to other variables, be Lipschitz in $[\mathsf{u}']$, and for any u be uniformly in (t,x) close to a function that is uniformly continuous in (t,x) (and not only in x).

In [**26**] the $C^{1+\alpha}$-regularity is investigated when $\bar{G}$ is summable to different powers in t and x. Again H is Lipschitz in u and satisfies the continuity condition like (10.1.11) in t, x.

In what concerns the fully nonlinear *elliptic* equations, Caffarelli [**12**] and Trudinger [**111**], [**112**] were the first authors who proved $C^{1+\alpha}$ regularity for C-viscosity solutions of equations

$$H[u] = f \tag{4}$$

without convexity assumptions on H. The assumptions in these papers are different. In [**12**] the function $H(\mathsf{u}, x)$ is independent of u' and, for each u'', is uniformly sufficiently close to a function which is continuous with respect to x. In [**111**] and [**112**] the function H depends on all arguments but is Hölder continuous in x. If we ignore the difference between C-viscosity and L_p-viscosity solutions, the next step in what concerns $C^{1+\alpha}$-estimates was done by Święch [**110**], who considered general H and imposed the same condition as in [**12**] on the x-dependence, which is much weaker than in [**111**] and [**112**] (but also imposed the Lipschitz condition on the dependence of H on u', whereas in [**111**] and [**112**] only the continuity with respect to u' is assumed). It is worth emphasizing that these results are about *any* continuous viscosity solution and not only about the ones we construct. The same bears on the results in [**107**], where the *boundary* $C^{1+\alpha}$ regularity is obtained.

The best value of α is largely unknown. However, it is proved in [**108**] in case $H = H(\mathsf{u}'')$ that the solutions of $H[u] = 0$ are almost $C^{1+\alpha}$ regular if the solutions of $\hat{H}[u] = 0$ are $C^{1+\alpha}$ regular, where $\hat{H}$ is introduced as $\hat{F}$ in Remark 10.1.23 dropping there u_0'.

By the way, it follows from Theorem 10.1.14 that, if $\hat{H}$ is convex, then the solutions of $H[u] = 0$ are in $W^2_{p,\text{loc}}$ for any $p > 1$, and thus in $C^{1+\alpha}$ for any $\alpha < 1$. This covers Corollary 1.2 of [**108**].

10. EXAMPLE. Let A and B be countable sets and suppose that for any $\alpha \in A$ and $\beta \in B$ we are given an $\mathbb{S}_\delta$-valued functions $a^{\alpha\beta}(\mathsf{u}_0', t, x)$ and a real-valued function $b^{\alpha\beta}(\mathsf{u}', t, x)$ defined for $\mathsf{u}', (t, x) \in \mathbb{R}^{d+1}$. Assume that these functions are Borel measurable as functions of (u', t, x), continuous with respect to u' uniformly with respect to $(\alpha, \beta, (t, x)) \in A \times B \times \Pi$, and $a^{\alpha\beta}(\mathsf{u}_0', t, x)$ is continuous with respect to x uniformly with respect to $(\alpha, \beta, \mathsf{u}_0', t) \in A \times B \times \mathbb{R} \times \mathbb{R}$. Finally, suppose that for all values of indices and arguments

$$\big|b^{\alpha\beta}(\mathsf{u}', t, x)\big| \le K_0|\mathsf{u}'| + \bar{G}(t, x).$$

Then the following equation

$$\partial_t v + \inf_{\alpha\in A}\sup_{\beta\in B}\Big[a^{\alpha\beta}_{ij}(v, t, x)D_{ij}v + b^{\alpha\beta}(v, Dv, t, x)\Big] = 0$$

in Π with boundary condition $v = g$ on $\partial'\Pi$ has an L_p-viscosity solution which belongs to $C^{\kappa}_{\text{loc}}(\Pi) \cap C(\bar{\Pi})$.

This follows immediately from Theorem 8 if one sets

$$F\big(\mathsf{u}_0', \mathsf{u}'', t, x\big) = \inf_{\alpha\in A}\sup_{\beta\in B} a^{\alpha\beta}_{ij}\big(\mathsf{u}_0', t, x\big)\mathsf{u}''_{ij}$$

and observes that, for any θ_0, in particular, for the one from Theorem 8, one can find R_0 and $\hat{R}_0$, for which Assumption 4 (iv) is satisfied with an appropriate K_1.

One can also see that the continuity of $a^{\alpha\beta}(\mathsf{u}_0', t, x)$ with respect to x can be relaxed allowing sufficiently small discontinuities. It is also worth noting that in [**25**] in the case of the Isaacs equations a is independent of u_0' and b is an affine function of u'.

We explain the main rough ideas in the proof of Theorem 8 in case F is independent of u_0' and $\bar{G}$ is bounded. It consists of establishing a priori estimates of the type

$$\sup_{C_r(t_0,x_0)} |v - l| \le N r^{\kappa} \tag{5}$$

for any $C_r(t_0, x_0)$ which is strictly inside Π and any small $r > 0$ as long as an affine function $l = l(x)$ is chosen appropriately. This turns out to be enough to get an estimate of the C^κ-norm of v in small cylinders which are strictly inside Π (see Lemma 6.2). Then, to obtain (5) we represent v as $h+w$, where h is a solution of

$$\partial_t h + F[h] = 0 \tag{6}$$

in $C_r(t_0,x_0)$ with boundary data v and $w = v-h$ is found from

$$0 = \partial_t v + H[v] - \partial_t h - F[h] = \partial_t w + a^{ij}D_{ij}w + G[v],$$

where (a^{ij}) is a certain $\mathbb{S}_\delta$-valued function. Since $w=0$ on $\partial' C_r(t_0,x_0)$ by the maximum principle $|w| \le Nr^2\big(1+\sup(|Dv|, C_r(t_0,x_0))\big)$. The latter supremum is irrelevant because we are going to estimate the C^κ-norm of v and $\kappa > 1$.

Then we see that, to get (5), it suffices to prove that (5) holds with h in place of v. To do this step we want to replace F with the one independent of (t,x). Freezing the coefficients does not help because there is no hope to control the second-order derivatives of solutions of such equations. Therefore, we just replace F with

$$F_0^{(\pm)} = F(\mathsf{u}'', t, x_0) \pm \big(K_1 + \theta_0|\mathsf{u}''|\big)$$

and introduce $v^{(\pm)}$ as solutions of

$$\partial_t v^{(+)} + F_0^{(+)}[v^{(+)}] = 0, \tag{7}$$

$$\partial_t v^{(-)} + F_0^{(-)}[v^{(-)}] = 0 \tag{8}$$

in $C_r(t_0,x_0)$ with boundary condition v. Since $F_0^{(-)} \le F \le F_0^{(+)}$ in $C_r(t_0,x_0)$ if r is small enough, we have $v^{(-)} \le h \le v^{(+)}$ by the maximum principle. Furthermore, since $F_0^{(\pm)}$ is independent of x, one can differentiate the equations (7) and (8) with respect to x and get estimates of the Hölder constants of $Dv^{(\pm)}$ by the Krylov-Safonov theorem. These estimates guarantee that $v^{(\pm)}$ can be approximated by affine functions of x as in (5).

Then the only thing which remains is to estimate $|v^{(+)} - v^{(-)}|$. To this end, for $\theta \in [-\theta_0, \theta_0]$ we introduce

$$F_0(\mathsf{u}'', t, \theta) = F_0(\mathsf{u}'', t) + \theta|\mathsf{u}''| + K_1\theta/\theta_0$$

and define v^θ from the equation

$$\partial_t v^\theta + F_0(D_x^2 v^\theta, t, \theta) = 0 \tag{9}$$

in $C_r(t_0,x_0)$ with boundary condition v. Since $v^{\pm\theta_0} = v^{(\pm)}$ to estimate $|v^{(+)} - v^{(-)}|$, it suffices to estimate $D_\theta v^\theta$.

Now comes an idea originated in the theory of diffusion processes. We look at (9) as an equation in variables (t,x,θ). There is no derivatives with respect to θ, so that it is a degenerate equation, but this was never a problem in such matters in that theory, that suggests that the function

$$V(t,\theta,x,\tau,\xi) := \tau D_\theta v^\theta(t,x) + \xi^i D_{x^i} v^\theta(t,x)$$

satisfies a parabolic equation with respect to the variables (t,θ,x,τ,ξ). Indeed

$$\partial_t V + a^{ij}D_{x^ix^j}V + \tau\varepsilon^{ij}D_{x^i\xi^j}V + N\tau^2\delta^{ij}D_{\xi^i\xi^j}V + K_1\tau/\theta_0 = 0,$$

where (a^{ij}) is an $\mathbb{S}_{\delta/2}$-valued function (see Assumption 4 (iii)),

$$\varepsilon^{ij} = D_{x^ix^j}v^\theta / |D_x^2 v^\theta|,$$

and N is any constant. Since there is no derivatives with respect to τ, it is just a parameter and for $W(t,\theta,x,\xi) := V(t,\theta,x,1,\xi)$ we obtain the equation

$$\partial_t W + a^{ij} D_{x^i x^j} W + \varepsilon^{ij} D_{x^i \xi^j} W + N\delta^{ij} D_{\xi^i \xi^j} W + K_1/\theta_0 = 0,$$

which is parabolic if N is sufficiently large. We consider this equation in $(t,x,\xi) \in C_r(t_0,x_0) \times \mathbb{R}^d$ and see that to estimate W, it suffices to have a good control of it on the parabolic boundary of this set, where $D_\theta v^\theta(t,x) = 0$ by construction. Thus we see that we need to estimate $|D_x v^\theta|$ in $C_r(t_0,x_0)$. This will be done by differentiating (9) with respect to x and using the maximum principle, which reduces the matter to estimating $|D_x v^\theta|$ on $\partial' C_r(t_0,x_0)$.

As a general comment we point out that, since we have to have sufficiently smooth solutions in the above argument, we use cut-off equations and use finite-differences to avoid using third-order derivatives.

3. Auxiliary results about linear equations

We will be using here and in the future a common way of approximating functions in $C^\kappa(C_1)$ by infinitely differentiable ones.

1. LEMMA. *Let $\kappa \in (0,2)$, $r \in (0,\infty)$, $g,h \in C^\kappa(C_r)$, and $h \in C^{1+\kappa}(C_r)$. Then for any $\varepsilon > 0$ there exist infinitely differentiable functions g^ε and h^ε on $\mathbb{R}^{d+1}$ such that in C_r with $N = N(d)$*

$$|g - g^\varepsilon| \le N[g]_{C^\kappa(C_r)}(r\varepsilon)^\kappa, \quad |Dg - Dg^\varepsilon| \le N[g]_{C^\kappa(C_r)}(r\varepsilon)^{\kappa-1},$$

$$\big|\partial_t g^\varepsilon\big| + |D^2 g^\varepsilon| + r\varepsilon|D^3 g^\varepsilon| + r\varepsilon\big|D\partial_t g^\varepsilon\big| \le N[g]_{C^\kappa(C_r)}(r\varepsilon)^{\kappa-2}, \tag{1}$$

$$[h^\varepsilon]_{C^{1+\kappa}(C_r)} \le N[h]_{C^\kappa(C_r)}(r\varepsilon)^{-1}.$$

Proof. Parabolic scalings reduce the general situation to the one in which $r = 1$. Then this well-known result is obtained by first continuing $g(t,x)$, $h(t,x)$ as functions of t to $\mathbb{R}$ to become even, 2-periodic functions, then continuing thus obtained functions across $|x| = 1$ almost preserving $[g]_{C^\kappa(C_1)}$, $[h]_{C^\kappa(C_1)}$ in the whole space and then taking convolutions with δ-like kernels. The lemma is proved. □

2. THEOREM. *Let $\kappa \in (1,2)$ and let $g \in C^\kappa(C_1)$. Then there exists a unique $u \in C^\infty_{\text{loc}}(C_1) \cap C^\kappa(C_1)$ which satisfies the heat equation*

$$\partial_t u + \Delta u = 0 \tag{2}$$

in C_1 and equals g on $\partial' C_1$. Furthermore, there exists a constant $N = N(d,\kappa)$ such that

$$[u]_{C^\kappa(C_1)} \le N[g]_{C^\kappa(C_1)}, \tag{3}$$

$$\big|D^2 u(t,x)\big| \le N\Big[\big(1 - |x|\big) \wedge \sqrt{1-t}\Big]^{\kappa-2} |g|_{C^\kappa(C_1)}. \tag{4}$$

Proof. One can subtract an affine function of x from g and reduce the general situation to the one where

$$g(1,0)=0, \quad Dg(1,0)=0. \tag{5}$$

In that case take g^ε from Lemma 1. Then by a classical result (see, for instance, Theorem 5.14 in [**89**] or Theorems 10.3.3 and 10.2.2 in [**65**]) there exists a unique $u^\varepsilon \in C^{1+\kappa}(C_1)$ satisfying (2) in C_1 and equal to g on $\partial' C_1$. In addition,

$$\|u^\varepsilon\|_{C^{1+\kappa}(C_1)} \le N\|g^\varepsilon\|_{C^{1+\kappa}(C_1)}. \tag{6}$$

Furthermore, by classical results (see, for instance, Theorem 8.12.1 in [**65**]) the functions u^ε are infinitely differentiable with respect to x in $\bar C_r$ for any $r<1$ and by Theorem 8.4.4 in [**65**] and the maximum principle any derivative of u^ε of any order with respect to x is bounded in $\bar C_r$ for any $r<1$ by a constant independent of ε. As it follows from equation (2) itself, the same holds for derivatives with respect to t of any derivative of any order with respect to x (cf. Exercise 8.12.4 in [**65**]).

In addition, by the maximum principle and (1),

$$|u^{\varepsilon_1}-u^{\varepsilon_2}| \le |g^{\varepsilon_1}-g^{\varepsilon_2}| \le N[g]_{C^\kappa(C_1)}(\varepsilon_1^\kappa+\varepsilon_2^\kappa). \tag{7}$$

It follows that, as $\varepsilon \downarrow 0$, u^ε converges uniformly on $\bar C_1$ to a continuous function, which is equal to g on $\partial' C_1$, is infinitely differentiable in C_1 and satisfies (2). In light of (7) we have

$$|u-u^\varepsilon| \le N[g]_{C^\kappa(C_1)}\varepsilon^\kappa. \tag{8}$$

Also (6) and (5) along with (1) imply that

$$[u^\varepsilon]_{C^{1+\kappa}(C_1)} \le N[g^\varepsilon]_{C^{1+\kappa}(C_1)} \le N[g]_{C^\kappa(C_1)}\varepsilon^{-1}.$$

Now take $x_0 \in \mathbb{R}^d$, unit $l \in \mathbb{R}^d$ and $0<h_1\le h$ such that

$$x_0, x_0+h_1l_1, x_0+2hl, x_0+2hl+h_1l_1 \in \bar B_1.$$

Observe that for any $t \in [0,1]$

$$u(t,x) = \big[u(t,x)-u^\varepsilon(t,x)\big] + u^\varepsilon(t,x_0) + (x^i-x_0^i)D_iu^\varepsilon(t,x_0)$$

$$+(1/2)(x^i-x_0^i)(x^j-x_0^j)D_{ij}u^\varepsilon(t,x_0)+v(t,x),$$

where

$$|v(t,x)| \le N|x-x_0|^{1+\kappa}[D^2u^\varepsilon]_{C^{\kappa-1}(C_1)} \le N|x-x_0|^{1+\kappa}[g]_{C^\kappa(C_1)}\varepsilon^{-1}.$$

Since the third-order finite difference of any quadratic polynomial is zero and

$$\big|u(t,x)-u^\varepsilon(t,x)\big| \le N[g]_{C^\kappa(C_1)}\varepsilon^\kappa,$$

we have

$$\big|(T_{h,l}-1)^2(T_{h_1,l_1}-1)u(t,x_0)\big| \le N[g]_{C^\kappa(C_1)}\varepsilon^\kappa + Nh^{1+\kappa}[g]_{C^\kappa(C_1)}\varepsilon^{-1}.$$

By taking $\varepsilon = h$ we arrive at

$$\big|(T_{h,l}-1)^2(T_{h_1,l_1}-1)u(t,x_0)\big| \leq N[g]_{C^\kappa(C_1)}h^\kappa.$$

Now by Theorem 11.2.10 and by the maximum principle for any $t \in [0,1]$

$$\big[u(t,\cdot)\big]_{C^\kappa(B_1)} \leq N\big([g]_{C^\kappa(C_1)} + \operatorname*{osc}_{C_1} u\big) \leq N\big([g]_{C^\kappa(C_1)} + \operatorname*{osc}_{\partial' C_1} g\big),$$

where $\operatorname{osc}_{\partial' C_1} g$ can be replaced with $[g]_{C^\kappa(C_1)}$ in light of (5).

Next, fix $x \in B_1$ and take $t_0 \in (0,1)$, $h > 0$, such that $t_0 + 2h^2 \in (0,1)$. Observe that

$$u(t,x) = \big[u(t,x) - u^\varepsilon(t,x)\big] + u^\varepsilon(t_0,x) + (t-t_0)\partial_t u^\varepsilon(t_0,x) + w(t,x),$$

where

$$\big|w(t,x)\big| \leq |t-t_0|^{(\kappa+1)/2}[\partial_t u^\varepsilon]_{C^{\kappa-1}(C_1)} \leq N|t-t_0|^{(\kappa+1)/2}[g]_{C^\kappa(C_1)}\varepsilon^{-1}.$$

Since the second-order differences of linear function are equal to zero,

$$\big|u(t_0+2h^2,x) - 2u(t_0+h^2,x) + u(t_0,x)\big|$$

$$\leq N[g]_{C^\kappa(C_1)}\varepsilon^\kappa + Nh^{1+\kappa}[g]_{C^\kappa(C_1)}\varepsilon^{-1}.$$

Here, for $\varepsilon = h$, the right-hand side becomes

$$N[g]_{C^\kappa(C_1)}h^\kappa$$

which by Theorem 11.2.10, applied to $(0,1)$ in place of B_R, shows that

$$\big|u(t,x) - u(s,x)\big| \leq |t-s|^{\kappa/2}N\big([g]_{C^\kappa(C_1)} + \operatorname*{osc}_{(0,1)} u(\cdot,x)\big)$$

if $s,t \in [0,1]$. Again the last oscillation can be replaced by $[g]_{C^\kappa(C_1)}$, and this proves (3).

By Theorem 8.4.4 in [**65**] for any $(t_0,x_0) \in C_1$

$$\big|\partial_t u(t_0,x_0)\big| + \big|D^2 u(t_0,x_0)\big| \leq NR^{-2} \sup_{C_R(t_0,x_0)} |u|,$$

where $R = (1-|x_0|) \wedge \sqrt{1-t_0}$. This holds for any sufficiently regular solution of (2) and not only for the one constructed above. Therefore u on the right can be replaced with $u - \hat{u}$, where $\hat{u}$ is any affine function of x. By taking $\hat{u}$ as the first-order Taylor polynomial of $u(t_0,x)$ with respect to x at x_0 and using (3) we come to (4). The theorem is proved. □

3. LEMMA. *For $\kappa \in (1,2)$ there is a function $\Phi \in C^{1,2}_{\mathrm{loc}}(C_1) \cap C(\bar{C}_1)$ and a constant $N = N(\kappa,\delta)$ such that*

$$\partial_t \Phi + a^{ij}D_{ij}\Phi \leq -\Big[\big(1-|x|\big) \wedge \sqrt{1-t}\Big]^{\kappa-2} \tag{9}$$

in C_1 for any $(a^{ij}) \in \mathbb{S}_\delta$ and

$$0 \leq \Phi(t,x) \leq N\big(1-|x|\big).$$

Proof. Set $\rho(x) = 1 - |x|^2$, $\beta = (\kappa+1)/2$, and, for a constant N_0 to be determined later, define

$$\Phi(t,x) = N_0\big[\rho(x) - (1/\kappa)\rho^\kappa(x)\big] + \beta^{-1}(1-t)^\beta\rho(x).$$

We have

$$D_i\rho = -2x^i, \quad D_{ij}\rho = -2\delta^{ij},$$

$$D_{ij}\Phi = -2N_0[1-\rho^{\kappa-1}]\delta^{ij} - 4N_0(\kappa-1)\rho^{\kappa-2}x^ix^j - 2\beta^{-1}(1-t)^\beta\delta^{ij}.$$

Hence,

$$\partial_t\Phi + a^{ij}D_{ij}\Phi \le -\delta I\rho^{\kappa-2} - (1-t)^{\beta-1}\rho,$$

where

$$I := 2N_0[\rho^{2-\kappa} - \rho] + 4N_0(\kappa-1)|x|^2.$$

Obviously, there exists $N_0 = N_0(\delta,\kappa)$ such that in C_1 we have $I \ge 2/\delta$, in which case

$$-\delta I\rho^{\kappa-2} \le -2\rho^{\kappa-2} = -2(1+|x|)^{\kappa-2}(1-|x|)^{\kappa-2} \le -(1-|x|)^{\kappa-2}$$

and (9) holds if $1-|x| \le \sqrt{1-t}$. In case $1-|x| \ge \sqrt{1-t}$ we have $\rho \ge \sqrt{1-t}$ and

$$-(1-t)^{\beta-1}\rho \le -(1-t)^{\beta-1/2} = -(1-t)^{\kappa/2-1},$$

so that (9) holds again. The lemma is proved. □

4. Estimating the difference of solutions of two different equations

Let $F_0(\mathsf{u}'',t)$ be a function satisfying Assumption 2.4 (iii), measurable in t, and such that $F_0(0,t)=0$. Fix a constant $K_1 \ge 0$ and set

$$F^{(\pm)}(\mathsf{u}'',t) = F_0(\mathsf{u}'',t) \pm \big(K_1 + \theta_0|\mathsf{u}''|\big),$$

take $P(\mathsf{u}'')$ from (6.2.4) with $\delta/2$ in place of δ, set $P[v] = P(D^2v)$, and for fixed $R \in (0,\infty)$, $K \ge 0$ consider the equations

$$\partial_t v + \max\big(F^{(\pm)}[v], P[v] - K\big) = 0$$

in C_R (a.e.) with boundary data $v = g$ on $\partial' C_R$, where $g \in W^{1,2}_\infty(C_R)$ is a given function. By $v^{(\pm)}$ we denote their solutions that exist by Theorem 8.3.3 and belong to $W^{1,2}_p(C_R)$ for any $p \in [1,\infty)$.

1. THEOREM. *For any $\kappa \in (1,2)$ there exists a constant $N = N(\kappa,\delta,d)$ such that in $\bar C_R$ we have*

$$\big|v^{(+)} - v^{(-)}\big| \le NR^2K_1 + NR^\kappa\theta_0[g]_{C^\kappa(C_R)}.$$

The proof of this theorem is based on the following two auxiliary results.

2. LEMMA. *Let a be a measurable $\mathbb{S}_\delta$-valued function, $p > d+2$, and let $v \in W^{1,2}_p(C_1)$ be a solution of*

$$\partial_t v + a^{ij} D_{ij} v + f = 0$$

in C_1 (a.e.) with boundary condition $v = g$ on $\partial' C_1$, where $|f| \le \bar{f}$ for a constant $\bar{f} \in [0, \infty)$. Assume that (3.5) holds. Then, for any $\kappa \in (1,2)$, there exists $N = N(\delta, d, \kappa)$ such that

$$\sup_{(0,1)\times\partial B_1} |Dv| \le N[Dg]_{C^\kappa(C_1)} + N\bar{f}. \tag{1}$$

Proof. By the embedding Lemma 2.3.3 of [**87**], the function Dv is continuous in $\bar{C}_1$. Then take the function u from Theorem 3.2 and set $w(t,x) = v(t,x) - u(t,x)$. We have

$$\partial_t w + a^{ij} D_{ij} w + h = 0,$$

where $h := f + \partial_t u + a^{ij} D_{ij} u$. By using Lemma 3.3 we conclude

$$\begin{aligned}\big|\partial_t w + a^{ij} D_{ij} w\big| &\le \bar{f} + N\Big[\big(1 - |x|\big) \wedge \sqrt{1-t}\Big]^{\kappa-2} |g|_{C^\kappa(C_1)} \\ &\le -\partial_t \Psi - a^{ij} D_{ij} \Psi,\end{aligned}$$

where $\Psi = N|g|_{C^\kappa(C_1)}\Phi + \bar{f}\delta^{-1}(1 - |x|^2)$. By the maximum principle

$$\big|w(t,x)\big| \le \Psi(t,x) \le N_0\big[|g|_{C^\kappa(C_1)} + \bar{f}\,\big]\big(1 - |x|\big).$$

Consequently, the normal derivative of $v(t,x)$ at a point $x_0 \in \partial B_1$ by magnitude is less than the absolute value of the normal derivative of $u(t,x)$ plus $N_0\big(|g|_{C^\kappa(C_1)} + \bar{f}\,\big)$. By interpolation inequalities $|Du|$ is estimated by $[u]_{C^\kappa(C_1)}$ and $\operatorname{osc}_{C_1} u \le \operatorname{osc}_{C_1} g$, where the former is estimated in (3.3) by $|g|_{C^\kappa(C_1)}$ and the latter is estimated by the same quantity due to (3.5). Thus, the normal derivative of $v(t,x)$ admits the estimate we are after. By noting that the tangential derivatives of $v(t,x)$ coincide with those of $g(t,x)$, we finally come to (1). The lemma is proved. □

3. LEMMA. *Let $R, \chi \in (0,\infty)$, and $\kappa \in (1,2)$ be constants and let $F(\theta, \mathsf{u}'', t)$ be a Borel measurable function on $[-\theta_0, \theta_0] \times \mathbb{S} \times \mathbb{R}$, which is Lipschitz continuous in (θ, u'') for any t and such that at all points of its differentiability*

$$|D_\theta F| \le \chi + |\mathsf{u}''|, \quad D_{\mathsf{u}''} F \in \mathbb{S}_\delta.$$

Also suppose that $F(\theta, 0, t)$ is bounded. Take $g \in W^{1,2}_\infty(C_R)$, and assume that for any $\theta \in [-\theta_0, \theta_0]$ the equation

$$\partial_t v + F(\theta, D^2 v, t) = 0 \tag{2}$$

in C_R (a.e.) with boundary condition $v = g$ on $\partial' C_R$ has a solution $v = v(\theta, \cdot) \in W^{1,2}_p(C_R)$, where $p > d+2$, $p \ge 2d+1$. Then for any $(t,x) \in C_R$ the function $v(\theta, t, x)$ is Lipschitz continuous on $[-\theta_0, \theta_0]$ and at all points of its differentiability with respect to θ

$$|D_\theta v| \le R^2\big(\chi + N \sup_t |F(\theta, 0, t)|\big) + NR^\kappa [g]_{C^\kappa(C_R)}, \tag{3}$$

where the constant N depends only on δ, d, κ.

Proof. The idea of the proof comes from the theory of diffusion processes and is explained at the end of Section 2. The parabolic equation for the directional derivative of $v(\theta, t, x)$ with respect to (θ, x) in appropriate directions will be what we are interested in. Since in our situation there is no guarantee that v is smooth enough, we follow Trudinger's method (see [**111**]) based on finite-differences.

As usual, parabolic scalings reduce the general situation to the one in which $R = 1$. Also by subtracting from g and v the same affine function of x we may assume that (3.5) holds.

In that case fix $\theta^0 \in (-\theta_0, \theta_0)$ and for sufficiently small h introduce

$$w(t, x, \xi) = v(\theta^0 + h, t, x + \xi) - v(\theta^0, t, x),$$

where $t, x, \xi \in \bar{Q}$ with

$$Q := [0, 1) \times \big\{(x, \xi) : x, x + \xi \in B_1\big\}.$$

Note for the future that, by embedding theorems $D_x v(\theta, t, x)$ is a continuous function in $\bar{C}_1$ for any θ.

Next, observe that

$$\begin{aligned} &F\big(\theta^0 + h, D_x^2 v(\theta^0 + h, t, x + \xi), t\big) - F\big(\theta^0, D_x^2 v(\theta^0, t, x), t\big) \\ &= \big[F\big(\theta^0, D_x^2 v(\theta^0 + h, t, x + \xi), t\big) - F\big(\theta^0, D_x^2 v(\theta^0, t, x), t\big)\big] + I \\ &= a^{ij} D_{x^i x^j} w + I, \end{aligned}$$

in Q (a.e.) where (a^{ij}) is a measurable $\mathbb{S}_\delta$-valued function and

$$I = F\big(\theta^0 + h, D_x^2 v(\theta^0 + h, t, x + \xi), t\big) - F\big(\theta^0, D_x^2 v(\theta^0 + h, t, x + \xi), t\big).$$

Since $|D_\theta F| \leq \chi + |\mathsf{u}''|$ by assumption, we have

$$I = \tau h + h\varepsilon^{ij} D_{x^i x^j} v(\theta^0 + h, t, x + \xi) = \tau h + h\varepsilon^{ij} D_{x^i \xi^j} w(t, x, \xi),$$

where $|\tau| \leq \chi$ and (ε^{ij}) is an $\mathbb{S}$-valued function with norm majorated by one. Hence, in Q (a.e.) we have

$$\partial_t w + \mathcal{L} w + \tau h = 0,$$

where

$$\mathcal{L} = \big[a^{ij} - N_0 h^2 \delta^{ij}\big] D^2_{x^i x^j} + h\varepsilon^{ij} D_{x^i \xi^j} + N_0 h^2 \delta^{ij} D_{\xi^i \xi^j}$$

is a uniformly elliptic operator for an appropriate $N_0 = N_0(\delta, d)$ and all sufficiently small $h \neq 0$.

Notice that on $\partial' Q$ either $t = 1$, and then

$$|w| \leq \sup_{|x| \leq 1} |Dg|\,|\xi| = \sup_{|x| \leq 1} \big|Dg - Dg(1, 0)\big|\,|\xi|,$$

or $|x| \leq 1$ and $|x + \xi| = 1$, in which case $v(\theta^0 + h, t, x + \xi) = v(\theta^0, t, x + \xi)$ and

$$\big|w(t, x, \xi)\big| \leq \sup_{C_1} \big|D_x v(\theta^0, \cdot)\big|\,|\xi|,$$

or else $|x + \xi| \leq 1$ and $|x| = 1$, in which case $v(\theta^0, t, x) = v(\theta^0 + h, t, x)$ and

$$\big|w(t, x, \xi)\big| \leq \sup_{C_1} \big|D_x v(\theta^0 + h, \cdot)\big|\,|\xi|.$$

In all cases on $\partial' Q$ we have

$$\big|w(t,x,\xi)\big| \le N_1|\xi| \le N_1 h + N_1 h^{-1}|\xi|^2,$$

where

$$N_1 = [Dg]_{C^\kappa(C_1)} + \max_{\theta=\theta^0,\theta^0+h} \sup_{C_1} \big|D_x v(\theta,\cdot)\big|.$$

As is easy to see, there is a constant $N_2 = N_2(d,\delta)$ such that, for the function

$$\phi(t,x,\xi) = h\chi(1-t) + N_1 h + N_1 h^{-1}\big[|\xi|^2 + N_2 h^2\big(1-|x|^2\big)\big]$$

we have

$$\partial_t \phi + \mathcal{L}\phi + \tau h \le 0$$

in Q and, of course, $|w| \le \phi$ on $\partial' Q$. By the maximum principle $|w| \le \phi$ in $\bar{Q}$, in particular, (take $\xi = 0$) in $\bar{C}_1$

$$\big|v(\theta^0+h,t,x) - v(\theta^0,t,x)\big| \le h\chi + NN_1 h,$$

where $N = N(\delta,d)$.

It follows that to prove the lemma, it suffices to show that for any $\theta \in [-\theta_0,\theta_0]$, with a constant $N = N(\delta,d,\kappa)$, we have

$$\sup_{C_1} \big|D_x v(\theta,\cdot)\big| \le N[Dg]_{C^\kappa(C_1)} + N\sup_t \big|F(\theta,0,t)\big|. \tag{4}$$

By applying finite-difference operators with respect to x to (2), we see that, for small h and unit $l \in \mathbb{R}^d$, the function $\delta_{h,l} v(\theta,t,x)$ satisfies a parabolic equation with zero free term in a domain slightly smaller than C_1. Hence, its sup over the domain is achieved on the parabolic boundary. By letting $h \to 0$ we conclude that

$$\sup_{C_1} \big|D_x v(\theta,\cdot)\big| = \sup_{\partial' C_1} \big|D_x v(\theta,\cdot)\big|,$$

and since $v(\theta,1,x) = g(1,x)$ for $|x| \le 1$, to prove (4), it suffices to prove that

$$\sup_{(0,1)\times\partial B_1} \big|D_x v(\theta,\cdot)\big| \le N[Dg]_{C^\kappa(C_1)} + N\sup_t \big|F(\theta,0,t)\big|. \tag{5}$$

We fix θ and observe that

$$\begin{aligned} 0 &= \partial_t v(\theta,t,x) + \big[F\big(\theta, D^2 v(\theta,t,x), t\big) - F(\theta,0,t)\big] + f \\ &= \partial_t v(\theta,t,x) + a^{ij} D_{ij} v(\theta,t,x) + f, \end{aligned}$$

(a.e.), where (a^{ij}) is a measurable $\mathbb{S}_\delta$-valued function and $f = F(\theta,0,t)$. After that (5) immediately follows from Lemma 2. The lemma is proved. □

Proof of Theorem 1. For $\theta \in [-\theta_0,\theta_0]$ introduce

$$F(\theta,\mathsf{u}'',t) = \max\Big(F_0(\mathsf{u}'',t) + \theta\big(K_1/\theta_0 + |\mathsf{u}''|\big), P(\mathsf{u}'') - K\Big).$$

By Theorem 8.3.3 equation (2) in C_R with boundary condition $v = g$ on $\partial' C_R$ admits a solution $v = v(\theta,\cdot) \in W^{1,2}_p(C_R)$ for any $p \ge 1$. By the maximum principle the solution is unique. Obviously, $\big|D_\theta F(\theta,\mathsf{u}'')\big| \le K_1/\theta_0 + |\mathsf{u}''|$ whenever the left-hand side is well defined. Also $\big|F(\theta,0,t)\big| \le K_1$. After that, it only remains to observe that, in light of Lemma 3,

$$\begin{aligned}|v^{(+)}-v^{(-)}| &\le \int_{-\theta_0}^{\theta_0}\Big(R^2\big(K_1/\theta_0+NK_1\big)+NR^\kappa[g]_{C^\kappa(C_r)}\Big)\,d\theta\\ &\le NR^2K_1+NR^\kappa\theta_0[g]_{C^\kappa(C_r)}.\end{aligned}$$

The theorem is proved. □

Later on we will need one more piece of information about $v^{(\pm)}$, before which we prove the following two auxiliary facts.

4. LEMMA. *Take $k\in\{1,...,d\}$, $\rho\in\mathbb{R}$, $\lambda\in(0,\infty)$, and introduce*

$$V(t,x)=\exp\big(-\lambda|x^k-\rho|^2/(r^2-t)\big).$$

Let $(a^{ij})\in\mathbb{S}$ be such that $0\le a^{kk}\le 1/(4\lambda)$. Then for $t<r^2$ it holds that

$$\partial_tV(t,x)+a^{ij}D_{ij}V(t,x)\le 0. \tag{6}$$

The proof is achieved by a straightforward computation showing that the left-hand side of (6) equals

$$V(t,x)\Big[-\lambda\frac{|x^k-\rho|^2}{(r^2-t)^2}+4\lambda^2a^{kk}\frac{|x^k-\rho|^2}{(r^2-t)^2}-2\lambda a^{kk}\frac{1}{r^2-t}\Big]\le 0.$$

5. LEMMA. *Let $r\in(0,\infty)$, $v\in W^{1,2}_{d+1,\mathrm{loc}}(C_r)\cap C(\bar C_r)$, and assume that, for some constants $\gamma\in(0,2]$, $M\ge 0$, we have*

$$\big|v(r^2,x)\big|\le M|x|^\gamma\quad\text{if}\quad |x|\le r.$$

Also assume that $\big|\partial_tv+a^{ij}D_{ij}v\big|\le\theta$ in C_r (a.e.) for an $\mathbb{S}_\delta$-valued function $a=(a^{ij})$ and a constant $\theta\in[0,\infty)$. Then there exists a constant $N=N(d,\delta,\gamma)$ such that for $t\in[0,r^2]$

$$\big|v(t,0)\big|\le N\big(M+r^{-\gamma}\sup_{C_r}|v|\big)(r^2-t)^{\gamma/2}+\theta(r^2-t). \tag{7}$$

Proof. Take $\varepsilon>0$, set $\rho=r/\sqrt{d}$, $\lambda=\delta/4$, and consider the function

$$\begin{aligned}V(t,x)&=M\varepsilon+M\varepsilon^{1-2/\gamma}\big(|x|^2+2(r^2-t)d/\delta\big)+\theta(r^2-t)\\ &\quad+\sup_{C_r}|v|\sum_{k=1}^d\big(V^k_{(+)}(t,x)+V^k_{(-)}(t,x)\big),\end{aligned}$$

where

$$V^k_{(\pm)}(t,x)=\exp\big(-\lambda|x^k\pm\rho|^2/(r^2-t)\big).$$

Observe that, by Lemma 4, we have

$$\partial_tV(t,x)+a^{ij}D_{ij}V\le-\theta$$

in C_r, in particular, in the cylinder $(0,r^2)\times(-\rho,\rho)^d$. On the parabolic boundary of this cylinder either $t=r^2$ and $V\ge M|x|^\gamma\ge|v|$ by Young's inequality ($\gamma\le 2$), or one of x^i is equal to either ρ or $-\rho$, when the corresponding V^i equals 1, and again $V \geq |v|$. By the maximum principle, for $t \in [0, r^2]$ we have

$$v(t,0) \leq V(t,0) \leq NM\big(\varepsilon + \varepsilon^{1-2/\gamma}(r^2 - t)\big)$$

$$+\theta(r^2 - t) + N \sup_{C_r} |v| \exp\big(-\lambda\rho^2/(r^2 - t)\big).$$

After that, to estimate $v(t,0)$ by the right-hand side of (7), it only remains to take the inf with respect to $\varepsilon > 0$ and observe that

$$\exp\big(-\lambda\rho^2/(r^2 - t)\big) \leq N\big(r^2 - t\big)^{\gamma/2}/r^\gamma.$$

Similarly $-v(t,0)$ is estimated from above. The lemma is proved. □

Recall that $v^{(\pm)}$ are introduced before Theorem 1.

6. THEOREM. *There exist constants $\kappa_0 = \kappa_0(d,\delta) \in (1,2)$ and $N \in (0,\infty)$ depending only on d and δ such that for any $r \in (0,R)$*

$$\big[v^{(\pm)}\big]_{C^{\kappa_0}(C_r)} \leq N(R-r)^{-\kappa_0}\big[\operatorname*{osc}_{\partial' C_R}(g - \hat{g}) + K_1 R^2\big],$$

where $\hat{g} = \hat{g}(x)$ is any affine function of x.

Proof. As usual we may take $\hat{g} = 0$. Recall that $Dv^{(\pm)}$ is bounded and even Hölder continuous in C_R since $v^{(\pm)} \in W^{1,2}_p(C_R)$ for any $p \geq 1$. Then observe that for any $\gamma \in (0,1)$ and function $f(x)$ of one variable $x \in [0,\varepsilon]$, $\varepsilon > 0$, we have

$$\big|f'(0)\big| \leq \big|f'(0) - \big(f(\varepsilon) - f(0)\big)/\varepsilon\big| + \varepsilon^{-1}\operatorname*{osc}_{[0,\varepsilon]} f \leq \varepsilon^\gamma [f']_{C^\gamma[0,\varepsilon]} + \varepsilon^{-1}\operatorname*{osc}_{[0,\varepsilon]} f.$$

By applying this fact to functions $v(x)$ given in B_R we obtain that for any $r_{n+1} < r_{n+2} \leq R$ and any $\varepsilon \in (0,1)$ and smooth $v = v(x)$

$$|Dv| \leq \varepsilon^\gamma (r_{n+2} - r_{n+1})^\gamma [Dv]_{C^\gamma(B_{r_{n+2}})} + \varepsilon^{-1}(r_{n+2} - r_{n+1})^{-1}\operatorname*{osc}_{B_R} v \tag{8}$$

in $\bar{B}_{r_{n+1}}$.

Next, for unit $l \in \mathbb{R}^d$ and $h > 0$ define

$$\delta_{h,l}u(t,x) = h^{-1}\big[u(t,x+hl) - u(t,x)\big]$$

and note that for any $r_1 \in (0,R)$ the function $\delta_{h,l}v^{(+)}$ satisfies an equation of the type

$$\partial_t \delta_{h,l}v^{(+)} + a^{ij}D_{ij}\delta_{h,l}v^{(+)} = 0$$

in C_{r_1} (a.e.) with some measurable (a^{ij}) taking values in $\mathbb{S}_{\bar{\delta}}$ if h is sufficiently small. By Corollary 9.6.4 for $r_0 \in (0, r_1)$ and perhaps even smaller h we have

$$\big[\delta_{h,l}v^{(+)}\big]_{C^\gamma(C_{r_0})} \leq N(r_1 - r_0)^{-\gamma} \sup_{C_{r_1}} \big|\delta_{h,l}v^{(+)}\big|,$$

where $\gamma \in (0,1)$ and N depend only on δ and d. By letting $h \to 0$ we conclude

$$[Dv^{(+)}]_{C^\gamma(C_{r_0})} \leq N(r_1 - r_0)^{-\gamma} \sup_{C_{r_1}} |Dv^{(+)}|. \tag{9}$$

By using (8) and (9) and setting

$$r_0 = r, \quad r_n = r + (R-r)\sum_{k=1}^{n} 2^{-k}, \quad n \geq 1,$$

we obtain

$$A_n := \sup_{[0,r_n^2]} \big[Dv^{(+)}(t,\cdot)\big]_{C^\gamma(B_{r_n})} \leq N(r_{n+1}-r_n)^{-\gamma} \sup_{C_{r_{n+1}}} |Dv^{(+)}|$$

$$\leq N_1\varepsilon^\gamma A_{n+2} + N_2(R-r)^{-(1+\gamma)}\varepsilon^{-1}2^{(1+\gamma)n} \operatorname*{osc}_{C_R} v^{(+)}, \tag{10}$$

where the constants N_i are different from the one in (9) but still depend only on δ and d. Without losing generality we may assume that $N_1 \geq 1$ and we first take ε so that

$$N_1\varepsilon^\gamma = 2^{-5},$$

then take $n = 2k$, $k = 0, 1, \ldots$, multiply both parts of (10) by 2^{-5k} and sum up with respect to k. Then upon observing that $(1+\gamma)2k \leq 4k$ we get

$$\sum_{k=0}^{\infty} A_{2k}2^{-5k} \leq \sum_{k=1}^{\infty} A_{2k}2^{-5k} + N(R-r)^{-(1+\gamma)}\sum_{k=0}^{\infty} 2^{-k} \operatorname*{osc}_{C_R} v^{(+)}.$$

By canceling (finite) like terms we find

$$\sup_{[0,r^2]} \big[Dv^{(+)}(t,\cdot)\big]_{C^\gamma(B_r)} \leq N(R-r)^{-(1+\gamma)} \operatorname*{osc}_{C_R} v^{(+)}. \tag{11}$$

Note for the future that (11) and the second inequality in (10) also imply that

$$\sup_{C_r} |Dv^{(+)}| \leq N(R-r)^{-1} \operatorname*{osc}_{C_R} v^{(+)}. \tag{12}$$

Next, we use the fact that $v^{(+)}$ itself satisfies the equation

$$0 = \partial_t v^{(+)} + \max\big(F^{(+)}[v^{(+)}], P[v^{(+)}] - K\big) - \max(K_1, -K) + K_1 \tag{13}$$

$$= \partial_t v^{(+)} + a_{ij}D_{ij}v^{(+)} + K_1$$

in C_R (a.e.) with some measurable (a_{ij}) taking values in $\mathbb{S}_{\bar\delta}$. Furthermore, for any $(t_0, x_0) \in C_R$ the function

$$v(t,x) := v^{(+)}(t,x) - v^{(+)}(t_0,x_0) - (x^i - x_0^i)D_i v^{(+)}(t_0,x_0)$$

satisfies the same equation and, owing to (11),

$$|v(t_0,x)| \leq \big[Dv^{(+)}(t_0,\cdot)\big]_{C^\gamma(B_r)}|x-x_0|^{1+\gamma} \leq N(R-r)^{-(1+\gamma)}|x-x_0|^{1+\gamma} \operatorname*{osc}_{C_R} v^{(+)}$$

if $(t_0, x_0) \in C_r$ and $|x - x_0| \leq \rho := (R-r)/2$.

Also, for $t_1 = 0 \vee (t_0 - \rho^2)$, due to (12), we have that

$$\sup_{C_{t_0-t_1,\rho}(t_1,x_0)} |v| \leq \operatorname*{osc}_{C_{t_0-t_1,\rho}(t_1,x_0)} v^{(+)} + N\rho\big|Dv^{(+)}(t_0,x_0)\big| \leq N \operatorname*{osc}_{C_R} v^{(+)}.$$

Then we apply Lemma 5 with $C_{t_0-t_1,\rho}(t_1,x_0)$ in place of C_r and take into account that $\rho^{-(1+\gamma)} \le N(R-r)^{-(1+\gamma)}$. Then for $t \in [t_1,t_0]$ we obtain

$$\frac{|v^{(+)}(t_0,x_0)-v^{(+)}(t,x_0)|}{(t_0-t)^{(1+\gamma)/2}} \le \frac{N}{(R-r)^{1+\gamma}} \operatorname*{osc}_{C_R} v^{(+)} + K_1(t_0-t)^{(1-\gamma)/2}.$$

Here $t_0 - t \le \rho^2$ and $(t_0-t)^{(1-\gamma)/2} \le R^2(R-r)^{-(1+\gamma)}$. Therefore

$$\frac{|v^{(+)}(t_0,x_0)-v^{(+)}(t,x_0)|}{(t_0-t)^{(1+\gamma)/2}} \le \frac{N}{(R-r)^{1+\gamma}}\big[\operatorname*{osc}_{C_R} v^{(+)} + K_1R^2\big]. \tag{14}$$

If x_0, t_0 are as above but $0 \le t < t_1$, then $t_0 - t \ge \rho^2 = (R-r)^2/4$ and

$$|v^{(+)}(t_0,x_0)-v^{(+)}(t,x_0)| \le \operatorname*{osc}_{C_R} v^{(+)} = (R-r)^{-(1+\gamma)}\big(4\rho^2\big)^{(1+\gamma)/2} \operatorname*{osc}_{C_R} v^{(+)},$$

so that (14) holds again.

This provides the necessary estimate of the oscillation of $v^{(+)}$ in the time variable and along with (11) shows that

$$[v^{(+)}]_{C^{1+\gamma}(C_r)} \le N(R-r)^{-(1+\gamma)}\big[\operatorname*{osc}_{C_R} v^{(+)} + K_1R^2\big].$$

Now the assertion of the theorem about $v^{(+)}$ with $\kappa_0 = 1+\gamma$ follows from the fact that

$$\operatorname*{osc}_{C_R} v^{(+)} \le \operatorname*{osc}_{\partial' C_R} g + NK_1R^2.$$

The function $v^{(-)}$ is considered similarly with the only difference that it satisfies an equation similar to (13) with $-\max(-K_1,-K)+\max(-K_1,-K)$ in place of $-\max(K_1,-K)+K_1$. Since $|\max(-K_1,-K)| \le K_1$ this difference is irrelevant. The theorem is proved. □

5. Main estimate for solutions of an auxiliary cut-off equation

Let $F(\mathsf{u}_0', \mathsf{u}'', t, x)$ satisfy Assumptions 2.4 (i), (iii), (iv), and (v). Take $K \in (0,\infty)$, $R \in (0,R_0]$, and $g \in W^{1,2}_\infty(C_R)$ and take $P[u]$ as in the beginning of Section 4.

By Theorem 8.3.3 there exists $u \in W^{1,2}_p(C_R)$ for all $p \ge 1$, such that $u = g$ on $\partial' C_R$ and the equation

$$\partial_t u + \max\big(F[u], P[u]-K\big) = 0. \tag{1}$$

holds (a.e.) in C_R.

Here is the result of this section.

1. THEOREM. *Suppose that*

$$\operatorname*{osc}_{\partial' C_R} g < \hat{R}_0.$$

Then there exist a constant $\kappa_0 = \kappa_0(d,\delta) \in (1,2]$ *such that, if* $r < R \le R_0$, *one can find an affine function* $\hat{u} = \hat{u}(x)$ *for which*

$$|u - \hat{u}| \le NK_1R^2 + N\theta_0[g]_{C^\kappa(C_R)}R^\kappa$$

$$+ N r^{\kappa_0} (R-r)^{-\kappa_0} \big[\operatorname*{osc}_{\partial' C_R} (g - \hat{g}) + K_1 R^2 \big]$$

in $\bar{C}_r$ *for any* $\kappa \in (0,2)$, *where* $\hat{g} = \hat{g}(x)$ *is any affine function of* x *and* $N = N(d, \delta, \kappa)$.

Proof. Set

$$F_0(\mathsf{u}'', t) = F\big(g(R^2, 0), \mathsf{u}'', t, 0\big),$$

and take $v^{(\pm)}$ from Section 4.

Observe that since $\max\big(F\big(\mathsf{u}_0', 0, t, x\big), P(0) - K\big) = 0$, as we have seen many times in the past,

$$\partial_t u + a^{ij} D_{ij} u = 0$$

(a.e.) in C_R for an appropriate measurable $\mathbb{S}_{\bar{\delta}}$-valued (a^{ij}). It follows that

$$\operatorname*{osc}_{C_R} u = \operatorname*{osc}_{\partial' C_R} g < \hat{R}_0,$$

and in C_R by Assumption 2.4 (iv)

$$F\big(u(t,x), D^2 u(t,x), t, x\big) \le F_0\big(D^2 u(t,x), t\big) + K_1 + \theta_0 \big|D^2 u(t,x)\big|,$$

$$F_0\big(D^2 u(t,x), t\big) - K_1 - \theta_0 \big|D^2 u(t,x)\big| \le F\big(u(t,x), D^2 u(t,x), t, x\big).$$

Since the operators $F^{(\pm)}$ satisfy the maximum principle, the above inequalities imply that $v^{(-)} \le u \le v^{(+)}$.

Next, set $p^{(\pm)}(x) = v^{(\pm)}(0) + x^i D_i v^{(\pm)}(0)$ and observe that by Theorem 4.6 in $\bar{C}_r$ we have

$$|p^{(+)} - v^{(+)}| + |p^{(-)} - v^{(-)}| \le N r^{\kappa_0} (R-r)^{-\kappa_0} \big[\operatorname*{osc}_{\partial' C_R} (g - \hat{g}) + K_1 R^2 \big],$$

where κ_0 is taken from Theorem 4.6. Furthermore by Theorem 4.1 in $\bar{C}_R$

$$|p^{(+)} - p^{(-)}| \le |p^{(+)} - v^{(+)}| + |p^{(-)} - v^{(-)}|$$

$$+ N R^2 K_1 + N R^{\kappa} \theta_0 [g]_{C^{\kappa}(C_R)}.$$

After that it only remains to note that, since $v^{(-)} \le u \le v^{(+)}$, we have

$$p^{(+)} - v^{(+)} \le p^{(+)} - u \le [p^{(+)} - p^{(-)}] + [p^{(-)} - v^{(-)}] + v^{(-)} - u$$

$$\le [p^{(+)} - p^{(-)}] + [p^{(-)} - v^{(-)}].$$

The theorem is proved. □

6. Proof of Theorem 2.8

We first take a constant $K > 0$, assume that $g \in W^{1,2}_{\infty}(\mathbb{R}^{d+1})$, $\bar{G}$ is bounded, and investigate solutions v_K of the cut-off equation (8.3.1). Take $\kappa_0 = \kappa_0(d, \delta) \in (1, 2]$ from Theorem 5.1. Naturally, we suppose that the assumptions of Theorem 2.8: Assumptions 2.3, 2.4, and 2.7, are satisfied with θ_0 yet to be specified.

1. LEMMA. *Let $R \in (0, R_0]$ and let $v \in W^{1,2}_{\infty}(C_R)$ be a solution of (8.3.1) in $\bar{C}_R$ (a.e.) such that*

$$\underset{\partial' C_R}{\operatorname{osc}}\, v < \hat{R}_0. \tag{1}$$

Then for each $r \in (0, R)$ one can find an affine function $\hat{v}(x)$ such that in C_r for any $\kappa \in (1, 2)$

$$|v - \hat{v}| \leq N\theta_0 [v]_{C^{\kappa}(C_R)} R^{\kappa} + N r^{\kappa_0}(R - r)^{-\kappa_0}\big[R^{\kappa}[v]_{C^{\kappa}(C_R)} + K_1 R^2\big]$$

$$+ N K_0 R^2 \sup_{C_R}\big(|v| + |Dv|\big) + N K_1 R^2 + N R^{2-(d+2)/p}\|\bar{G}\|_{L_p(C_R)},$$

where the constants N depend only on d, δ, and κ.

Proof. Observe that

$$\max\big(H[v], P[v] - K\big) = \max\big(F[v], P[v] - K\big) + h$$

where h, defined by the above equality, satisfies

$$|h| \leq \big|H[v] - F[v]\big| \leq K_0\big(|v| + |Dv|\big) + \bar{G}.$$

Next define $u \in W^{1,2}_{d+1}(C_R)$ as a solution of

$$\partial_t u + \max\big(F[u], P[u] - K\big) = 0$$

(a.e.) with boundary data $u = v$ on $\partial' C_R$, which exists by Theorem 8.3.3. Then, in light of the fact that

$$\max\big(F(u(t,x), 0, t, x), P(0) - K\big) = 0,$$

there exists an $\mathbb{S}_{\bar{\delta}}$-valued measurable function a such that in C_R (a.e.) we have

$$\partial_t(v - u) + a^{ij} D_{ij}(v - u) + h = 0.$$

By the parabolic Aleksandrov estimate (see Theorem 3.0.1)

$$|v - u| \leq N R^{d/d+1}\|h\|_{L_{d+1}(C_R)} = N R^2\Big(\fint_{C_R} |h|^{d+1}\, dxdt\Big)^{1/(d+1)}$$

$$\leq N K_0 R^2 \sup_{C_R}\big(|v| + |Dv|\big) + N R^{2-(d+2)/p}\|\bar{G}\|_{L_p(C_R)}.$$

After that our assertion follows from Theorem 5.1 and the lemma is proved. □

We need one more characterization of $C^{1+\alpha}$-functions.

2. LEMMA. *Let $r_0 \in (0,\infty)$, $\kappa \in (1,2)$, and let ϕ be a function on C_{r_0}. Assume that there is a constant N_0 such that for any $(t,x) \in C_{r_0}$ and $r \in (0, 2r_0]$ there exists an affine function $\hat{\phi} = \hat{\phi}(x)$ such that*

$$\sup_{C_r(t,x)\cap C_{r_0}} |\phi - \hat{\phi}| \le N_0 r^\kappa .$$

Then $\phi \in C^\kappa(C_{r_0})$ and

$$[\phi]_{C^\kappa(C_{r_0})} \le NN_0,$$

where N depends only on d and κ.

Proof. The fact that, for any $t \in (0, r^2)$, we have

$$\big[D\phi(t,\cdot)\big]_{C^{\kappa-1}(B_{r_0})} \le NN_0$$

follows from Theorem 11.2.4. To estimate $|\phi(t,x) - \phi(s,x)|$ we may assume that $t > s$, so that $(t,x),(s,x) \in \bar{C}_r(s,x)$, where $r = \sqrt{t-s}$. Then, for an appropriate $\hat{\phi}(x)$

$$\big|\phi(t,x) - \phi(s,x)\big| \le \big|\phi(t,x) - \hat{\phi}(x)\big| + \big|\phi(t,x) - \hat{\phi}(x)\big| \le 2N_0 r^\kappa = 2N_0(t-s)^{\kappa/2}.$$

The lemma is proved. □

3. LEMMA. *Take $r_1 \in (0, R_0]$, $r_0 \in (0, r_1)$, and define*

$$\kappa = \kappa(d,\delta,p) = \frac{1+\kappa_0}{2} \wedge \Big(2 - \frac{d+2}{p}\Big).$$

Let $v \in W^{1,2}_\infty(C_{r_1})$ be a solution of (8.3.1) in C_{r_1} (a.e.). Then there exists $\theta_0 = \theta_0(d,\delta) \in (0,1]$ such that, if Assumption 2.4 (iv) is satisfied with this θ_0 and

$$\operatorname*{osc}_{C_{r_1}} v < \hat{R}_0, \tag{2}$$

then

$$[v]_{C^\kappa(C_{r_0})} \le (1/2)[v]_{C^\kappa(C_{r_1})} + N(K_0+1)(r_1-r_0)^{-\kappa} \sup_{C_{r_1}} |v|$$

$$+ N(K_0+1)(r_1-r_0)^{-(\kappa-1)} \sup_{C_{r_1}} |Dv| + N(K_1 + \|G\|_{L_p(C_{r_1})}), \tag{3}$$

where $N = N(d,\delta)$.

Proof. Take $(t_0,x_0) \in C_{r_0}$, $\varepsilon \in (0,1)$ to be specified later, define

$$r_0' = \frac{\varepsilon}{3}(r_1 - r_0),$$

and notice that for any $(t,x) \in C_{r_0'}(t_0,x_0)$, $r \in (0, 2r_0']$, and $R = \varepsilon^{-1} r$, we have $R \le r_1 \le R_0 \le 1$ and

$$C_R(t,x) \subset C_{r_1}.$$

Therefore, by Lemma 1 we can find an affine function $\hat{v}(x)$ such that

$$\sup_{C_r(t,x)\cap C_{r_0'}(t_0,x_0)} |v-\hat{v}| \le \sup_{C_r(t,x)} |v-\hat{v}| \le N\theta_0[v]_{C^\kappa(C_R(t,x))}\varepsilon^{-\kappa}r^\kappa$$

$$+N\varepsilon^{\kappa_0-\kappa}(1-\varepsilon)^{-\kappa_0}r^\kappa[v]_{C^\kappa(C_R(t,x))}+N\varepsilon^{\kappa_0-2}(1-\varepsilon)^{-\kappa_0}K_1r^2$$

$$+NK_0\varepsilon^{-2}r^2\sup_{C_R(t,x)}\big(|v|+|Dv|\big)+NK_1\varepsilon^{-2}r^2$$

$$+N\varepsilon^{-\kappa}r^\kappa\|\bar{G}\|_{L_p(C_{r_1})}\le Nr^\kappa I(\theta_0,\varepsilon,r_1),$$

where the constants N depend only on d and δ and

$$I(\theta_0,\varepsilon,r_1):=\big(\theta_0\varepsilon^{-\kappa}+\varepsilon^{\kappa_0-\kappa}(1-\varepsilon)^{-\kappa_0}\big)[v]_{C^\kappa(C_{r_1})}+\varepsilon^{\kappa_0-2}(1-\varepsilon)^{-\kappa_0}K_1$$

$$+\varepsilon^{-2}K_0\sup_{C_{r_1}}\big(|v|+|Dv|\big)+K_1\varepsilon^{-2}+\varepsilon^{-\kappa}\|\bar{G}\|_{L_p(C_{r_1})}.$$

It follows by Lemma 2 that

$$[v]_{C^\kappa(C_{r_0'}(t_0,x_0))}\le N_1I(\theta_0,\varepsilon,r_1),$$

where N_1 depends only on d and δ. We can now specify θ_0 and ε. First we chose $\varepsilon\in(0,1)$ so that

$$N_1\varepsilon^{\kappa_0-\kappa}(1-\varepsilon)^{-\kappa_0}=1/4.$$

Since $\kappa_0-\kappa\ge(\kappa_0-1)/2>0$ and κ_0 depends only on d and δ and N_1 depends only on d and δ, ε also depends only on d and δ. After that we take $\theta_0=\theta_0(d,\delta)\in(0,1]$ so that $N_1\theta_0\varepsilon^{-\kappa}\le 1/4$.

Then

$$[v]_{C^\kappa(C_{r_0'}(t_0,x_0))}\le(1/2)[v]_{C^\kappa(C_{r_1})}+NJ, \tag{4}$$

where $N=N(d,\delta)$ and

$$J=K_0\sup_{C_{r_1}}\big(|v|+|Dv|\big)+K_1+\|\bar{G}\|_{L_p(C_{r_1})}.$$

Now observe that if $(t,x),(s,x)\in C_{r_0}$ and $t>s$, then either $|t-s|<\big(r_0'\big)^2$, in which case $(t,x)\in C_{r_0'}(s,x)$ and

$$(t-s)^{-\kappa/2}\big|v(t,x)-v(s,x)\big|\le(1/2)[v]_{C^\kappa(C_{r_1})}+NJ$$

owing to (4), or $|t-s|\ge\big(r_0'\big)^2$ when

$$\big|v(t,x)-v(s,x)\big|\le 2(t-s)^{\kappa/2}(r_0')^{-\kappa}\sup_{C_{r_1}}|v|$$

$$\le N(t-s)^{\kappa/2}(r_1-r_0)^{-\kappa}\sup_{C_{r_1}}|v|.$$

Next if $(t,x),(t,y)\in C_{r_0}$ and $x\ne y$, then either $|x-y|<r_0'$, in which case

$(t, y) \in C_{r_0'}(t, x)$ and

$$|x - y|^{-(\kappa-1)}\big|Dv(t, x) - Dv(t, y)\big| \leq (1/2)[v]_{C^\kappa(C_{r_1})} + NJ,$$

or $|x - y| \geq r_0'$ and

$$|Dv(t, x) - Dv(t, y)| \leq 2|x - y|^{\kappa-1}\big(r_0'\big)^{-(\kappa-1)} \sup_{C_{r_1}} |Dv|$$

$$\leq N|x - y|^{\kappa-1}(r_1 - r_0)^{-(\kappa-1)} \sup_{C_{r_1}} |Dv|.$$

This proves (3) and the lemma. □

4. THEOREM. *Take $0 < r < R \leq R_0$ and take κ and θ_0 from Lemma 3 and suppose that Assumption 2.4 (iv) is satisfied with this θ_0. Let $v \in W^{1,2}_\infty(C_R)$ be a solution of (8.3.1) in C_R (a.e.) such that*

$$\operatorname*{osc}_{C_R} v < \hat{R}_0.$$

Then

$$[v]_{C^\kappa(C_r)} \leq N(R - r)^{-\kappa} \sup_{C_R} |v| + N\big(K_1 + \|\bar{G}\|_{L_p(C_R)}\big), \tag{5}$$

where N depends only on d, δ, and K_0.

Proof. Fix a number $c \in (0, 1)$ such that $c^4 > 3/4$ and introduce

$$r_0 = r, \quad r_n = r + c_0(R - r)\sum_{k=1}^{n} c^k, \quad n \geq 1,$$

where c_0 is chosen in such a way that $r_n \to R$ as $n \to \infty$. Then Lemma 3 and (4.8) allow us to find constants N_1 and N depending only on d, δ, and K_0, such that for all n and $\varepsilon \in (0, 1)$

$$A_n := [v]_{C^\kappa(C_{r_n})} \leq \big(2^{-1} + N_1\varepsilon^{\kappa-1}\big)A_{n+2} + N(K_1 + \|\bar{G}\|_{L_p(C_R)})$$

$$+ N(R - r)^{-\kappa}c^{-n\kappa}(1 + \varepsilon^{-1}) \sup_{C_R} |v|.$$

We choose $\varepsilon < 1$ so that $2^{-1} + N_1\varepsilon^{\kappa-1} \leq 3/4$ and then recalling that $\kappa \leq 2$ conclude that

$$\sum_{m=0}^{\infty} (3/4)^m A_{2m} \leq \sum_{k=1}^{\infty} (3/4)^m A_{2m} + N(K_1 + \|\bar{G}\|_{L_p(C_R)})$$

$$+N(R - r)^{-\kappa} \sup_{C_R} |v| \sum_{m=0}^{\infty} (3/4)^m c^{-4m},$$

where the last series converges since $3c^{-4}/4 < 1$. By canceling like terms we come to (5) and the theorem is proved. □

Proof of Theorem 2.8. We keep assuming that $g \in W^{1,2}_\infty(\mathbb{R}^{d+1})$ and $\bar{G}$ is bounded. By Lemma 12.1.9

$$|v_K| \leq N(\|g\|_{C(\Pi)} + \|\bar{G}\|_{L_{d+1}(\Pi)}), \tag{6}$$

where N depends only on d, δ, K_0, T, and the diameter of Ω.

Also by Theorem 9.8.1 the family $\{v_K; K \geq 1\}$ is equicontinuous on $\bar{\Pi}$. More precisely there exists a function $\bar{\omega}(\varepsilon)$, depending only on ε, $\|\bar{G}\|_{L_{d+1}(\Pi)}$, δ, d, K_0, $\rho_{\text{ext}}(\Omega)$, $\|g\|_{C(\Pi)}$, and the modulus ω of continuity of g on $\partial'\Pi$, such that $\bar{\omega}(\varepsilon) \to 0$ as $\varepsilon \downarrow 0$ and

$$\big|v_K(t,x) - v_K(s,y)\big| \leq \bar{\omega}\Big(\rho\big((t,x),(s,y)\big)\Big) \tag{7}$$

for any $K \geq 1$, $x, y \in \bar{\Omega}$ and $s, t \in [0,T]$.

It follows that there is a sequence $K_n \to \infty$ and a function v such that $v^n := v_{K_n} \to v$ uniformly in $\bar{\Pi}$. Of course, (2.3) holds, owing to Theorem 4, which also implies that $Dv^n \to Dv$ locally uniformly in Π.

The following lemma, in which the boundedness of $\bar{G}$ is not used, will allow us to prove that v is a viscosity solution. Introduce

$$H_0(\mathsf{u}'', t, x) = H\big(v(t,x), Dv(t,x), \mathsf{u}'', t, x\big).$$

5. LEMMA. *There is a constant N, depending only on d and δ, such that for any $C_r(t,x)$ satisfying $\bar{C}_r(t,x) \subset \Pi$ and $\phi \in W^{1,2}_{d+1}(C_r(t,x))$ we have on $C_r(t,x)$ that*

$$v \leq \phi + Nr^{d/(d+1)}\big\|\big(\partial_t\phi + H_0[\phi]\big)_+\big\|_{L_{d+1}(C_r(t,x))} + \max_{\partial' C_r(t,x)}(v-\phi)_+, \tag{8}$$

$$v \geq \phi - Nr^{d/(d+1)}\big\|\big(\partial_t\phi + H_0[\phi]\big)_-\big\|_{L_{d+1}(C_r(t,x))} - \max_{\partial' C_r(t,x)}(v-\phi)_-. \tag{9}$$

Proof. Introduce

$$H_m(\mathsf{u}'', t, x) = \sup_{n \geq m} H\big(v^n(t,x), Dv^n(t,x), \mathsf{u}'', t, x\big)$$

and observe that for $n \geq m$

$$\partial_t v_n + \max\big(H_m[v^n], P[v^n] - K_n\big) \geq 0,$$

implying that

$$\begin{aligned}-\partial_t\phi - \max\big(H_m[\phi], P[\phi] - K_n\big) &\leq -\partial_t\phi - \max\big(H_m[\phi], P[\phi] - K_n\big)\\ &\quad + \partial_t v^n + \max\big(H_m[v^n], P[v^n] - K_n\big)\\ &= \partial_t(v^n - \phi) + a^{ij}D_{ij}(v^n - \phi)\end{aligned}$$

(a.e.) in $C_r(t,x)$, where $a = (a^{ij})$ is an $\mathbb{S}_{\bar{\delta}}$-valued function.

It follows by Theorem 3.1.6 that

$$\begin{aligned}v^n &\leq \phi + \max_{\partial' C_r(t,x)}(v^n - \phi)_+\\ &\quad + Nr^{d/(d+1)}\big\|\big\{\partial_t\phi + \max\big(H_m[\phi], P[\phi] - K_n\big)\big\}_+\big\|_{L_{d+1}(C_r(t,x))},\end{aligned}$$

where $N = N(d,\delta)$. By sending $n \to \infty$ and using the dominated convergence theorem, we obviously get

$$v \leq \phi + \max_{\partial' C_r(t,x)}(v-\phi)_+ + Nr^{d/(d+1)}\big\|\big(\partial_t\phi + H_m[\phi]\big)_+\big\|_{L_{d+1}(C_r(t,x))}. \tag{10}$$

Since $v^n \to v$ and $Dv^n \to Dv$ uniformly in $C_r(t,x))$ we have that $\big|H_m[\phi]\big| \leq \big|H_m[0]\big| + N|D^2\phi|$, where $\big|H_m[0]\big|$ is dominated by an L_{d+1}-function independent of m. Furthermore, the continuity of $H(\mathsf{u},t,x)$ with respect to u implies that $H_m[\phi] \to H_0[\phi]$ as $m \to \infty$ at any point in $C_r(t,x)$. By using the dominated convergence theorem once more and sending $m \to \infty$ in (10) we come to (8).

Similarly (9) is established. The lemma is proved. □

Now we prove that v is a C-viscosity solution of (1.1) if part (a) of Assumption 2.3 (ii) is satisfied. Let $(t_0,x_0) \in \Pi$, $r > 0$, and $\phi \in C^{1,2}\big(C_r(t_0,x_0)\big)$ be such that $\bar{C}_r(t_0,x_0) \subset \Pi$ and $v - \phi$ attains a local maximum at (t_0,x_0). Then for $\varepsilon > 0$ and all small $r > 0$ for

$$\phi_{\varepsilon,r}(t,x) = \phi(t,x) - \phi(t_0,x_0) + v(t_0,x_0) + \varepsilon(|x-x_0|^2 + t - t_0 - r^2)$$

we have that

$$\max_{\partial' C_r(t_0,x_0)} (v - \phi_{\varepsilon,r})_+ = 0.$$

Hence, by Lemma 5

$$\varepsilon r^2 = (v - \phi_{\varepsilon,r})(t_0,x_0) \le N r^{d/(d+1)} \big\| \big(\partial_t \phi_{\varepsilon,r} + H_0[\phi_{\varepsilon,r}]\big)_+ \big\|_{L_{d+1}(C_r(t_0,x_0))}$$

$$= N r^{d/(d+1)} \big\| \big(\partial_t \phi_{\varepsilon} + H_0[\phi_{\varepsilon}]\big)_+ \big\|_{L_{d+1}(C_r(t_0,x_0))},$$

where $\phi_\varepsilon = \phi + \varepsilon\big(|x|^2 + t\big)$. It follows that

$$N r^{-(d+2)} \big\| \big(\partial_t \phi_{\varepsilon} + H_0[\phi_{\varepsilon}]\big)_+ \big\|^{d+1}_{L_{d+1}(C_r(t_0,x_0))} \ge \varepsilon^{d+1},$$

$$N \operatorname*{ess\,sup}_{C_r(t_0,x_0)} \big(\partial_t \phi_\varepsilon + H_0[\phi_\varepsilon]\big) \ge \varepsilon,$$

$$N \lim_{r \downarrow 0} \operatorname*{ess\,sup}_{C_r(t_0,x_0)} \big(\partial_t \phi_\varepsilon + H_0[\phi_\varepsilon]\big) \ge \varepsilon.$$

By letting $\varepsilon \downarrow 0$ we conclude that

$$\lim_{r \downarrow 0} \operatorname*{ess\,sup}_{C_r(t_0,x_0)} \big[\partial_t \phi(t,x) + H\big(v(t,x), Dv(t,x), D^2\phi(t,x), t, x\big)\big] \ge 0. \tag{11}$$

Now, note that, as $(t,x) \to (t_0,x_0)$, we have $v(t,x) \to v(t_0,x_0)$ and (see Remark 2.2) $Dv(t,x) \to Dv(t_0,x_0)$. Also $D\phi(t,x) \to D\phi(t_0,x_0) = Dv(t_0,x_0)$ and $D^2\phi(t,x) \to D^2\phi(t_0,x_0)$. It follows by Assumption 2.3 (ii) (a) that one can replace $Dv(t,x)$ in (11) with $D\phi(t,x)$. Then, so modified (11) implies (1.2) meaning that v is a C-viscosity subsolution.

The fact that it is also a C-viscosity supersolution is proved similarly on the basis of (9).

In case Assumption 2.3 (ii) (b) is satisfied, we still come to (11) and can replace $Dv(t,x)$ with $D\phi(t,x)$ just because of the continuity of $H(\mathsf{u},t,x)$ in $[u']$ uniform with respect to u_0', u'', and (t,x).

This proves Theorem 2.8 in our particular case that $g \in W^{1,2}_\infty(\mathbb{R}^{d+1})$ and $\bar{G}$ is bounded. In the same case we also have estimates (6), (7) with v in place of v_K and the same N and $\bar{\omega}$. Also (2.3) holds by the above.

In the case of general $\bar{G}$, for $n = 1, 2, \ldots$, we replace $H(\mathsf{u}, t, x)$ in (1.1) with

$$H(\mathsf{u}, t, x) I_{\bar{G}(t,x) \leq n} + F(\mathsf{u}'_0, \mathsf{u}'', t, x) I_{\bar{G}(t,x) > n}$$

$$= F(\mathsf{u}'_0, \mathsf{u}'', t, x) + G(\mathsf{u}, t, x) I_{\bar{G}(t,x) \leq n}$$

and apply the already proved version of Theorem 2.8 to introduce u_n as the C-viscosity or L_p-viscosity solutions in Π of so obtained equations with the same boundary condition $u_n = g$ on $\partial'\Pi$.

From the above we see that the estimates (6) and (7) with u_n in place of v_K and the estimates of $[u_n]_{C^\kappa(C_r(t,x))}$ are uniform with respect to n. This and the fact that the boundedness of $\bar{G}$ is not used in Lemma 5 allow us to repeat what was said about v^n with obvious changes and proves the theorem for general H but still assuming that $g \in W^{1,2}_\infty(\mathbb{R}^{d+1})$.

One drops this assumption by using uniform approximations of g by smooth ones preserving the modulus of continuity on $\partial'\Pi$. This guarantees that for the approximating solutions the estimates like (6) and (7) will hold and referring to the argument in the previous paragraph brings the proof of Theorem 2.8 to an end. $\square$

CHAPTER 17

$C^{1+\alpha}$-regularity of L_p-viscosity solutions of the Isaacs parabolic equations with almost VMO coefficients

We closely follow [**77**] and adopt the framework of Section 16.1.

1. Main results

The main goal of this chapter is to establish the existence of L_p-viscosity solutions of boundary value problems associated with (16.1.1), solutions, that are in $C^{1+\alpha}$ for an $\alpha \in (0,1)$. In contrast with Chapter 16 here we allow the main part of $H(\mathsf{u},t,x)$ to be not close to functions that are continuous with respect to x but be almost in VMO, in the class of functions which can be wildly discontinuous. However, allowing less regular H in x forces us to restrict ourselves, basically, to the Isaacs equations only.

Fix a constant $\delta \in (0,1]$, and recall that

$$\mathbb{S}_\delta = \{a \in \mathbb{S} : \delta|\xi|^2 \leq a^{ij}\xi^i\xi^j \leq \delta^{-1}|\xi|^2, \quad \forall \xi \in \mathbb{R}^d\},$$

where as everywhere in the book the summation convention is enforced. Recall that

$$[\mathsf{u}'] = \big(\mathsf{u}'_1, ..., \mathsf{u}'_d\big).$$

1. Assumption. (i) The function $H(\mathsf{u},t,x)$ is measurable with respect to $(t,x) \in \mathbb{R}^{d+1}$ for any $\mathsf{u} = (\mathsf{u}', \mathsf{u}''), \mathsf{u}' = \big(\mathsf{u}'_0, \mathsf{u}'_1, ..., \mathsf{u}'_d\big) \in \mathbb{R}^{d+1}, \mathsf{u}'' \in \mathbb{S}$, *continuous* with respect to u for every $(t,x) \in \mathbb{R}^{d+1}$, and Lipschitz continuous in u'' for every $\mathsf{u}', (t,x) \in \mathbb{R}^{d+1}$.

(ii) For any $\mathsf{u}', (t,x) \in \mathbb{R}^{d+1}$, $D_{\mathsf{u}''}H \in \mathbb{S}_\delta$ at all points of differentiability of $H(\mathsf{u},t,x)$ with respect to u''.

(iii) There is a function $\bar{H}(t,x)$ and a constant $K_0 \geq 0$ such that

$$\big|H(\mathsf{u}',0,t,x)\big| \leq K_0|\mathsf{u}'| + \bar{H}(t,x)$$

for every $\mathsf{u}', (t,x) \in \mathbb{R}^{d+1}$.

(iv) For any $M \in (0,\infty)$ the function $H(\mathsf{u},t,x)$ is continuous with respect to $[\mathsf{u}'] = \big(\mathsf{u}'_1, ..., \mathsf{u}'_d\big)$ uniformly with respect to $|\mathsf{u}'_0| \leq M$, $\mathsf{u}'' \in \mathbb{S}$, $(t,x) \in \Pi$.

For measurable sets Γ in $\mathbb{R}^{d+1}$ by $|\Gamma|$ we denote the Lebesgue measure of Γ. Also for a function f on Γ we set

$$\fint_\Gamma f(t,x)\,dxdt = \frac{1}{|\Gamma|}\int_\Gamma f(t,x)\,dxdt$$

in case Γ has a nonzero Lebesgue measure in $\mathbb{R}^{d+1}$. Similar notation is used in the case of functions $f(x)$ on $\mathbb{R}^d$.

We fix a constant $R_0 \in (0,1]$ and for $\kappa \in (0,2]$ and measurable $f(t,x)$ introduce

$$f_\kappa = \sup_{R \le R_0, t, x} R^{2-\kappa}\Big(\fint_{C_R(t,x)} \big|f(s,y)\big|^{d+1}\, dyds\Big)^{1/(d+1)}. \tag{1}$$

2. Remark. By Hölder's inequality for $p \ge d+1$

$$\Big(\fint_{C_R(t,x)} \big|f(s,y)\big|^{d+1}\, dyds\Big)^{1/(d+1)} \le NR^{-(d+2)/p}\Big(\int_{\mathbb{R}^{d+1}} \big|f(s,y)\big|^{p}\, dyds\Big)^{1/p},$$

which shows that $f_\kappa < \infty$ if $f \in L_p(\mathbb{R}^{d+1})$ and $\kappa \le 2-(d+2)/p$. It is useful to observe that one can take $\kappa > 1$ if $f \in L_p(\mathbb{R}^{d+1})$ for $p > d+2$.

In the following assumption there are three fixed objects

$$\kappa_1 = \kappa_1(d,\delta) \in (1,2), \quad \kappa \in (1,\kappa_1], \quad \theta = \theta(\kappa,d,\delta) \in (0,1].$$

The values of κ_1 and θ are specified later in the proof of Lemma 4.2.

3. Assumption. We have a representation

$$H(\mathsf{u},t,x) = F(\mathsf{u}'',t,x) + G(\mathsf{u},t,x).$$

(i) The functions F and G are Borel measurable functions of their arguments.
(ii) For all values of the arguments

$$\big|G(\mathsf{u},t,x)\big| \le K_0|\mathsf{u}'| + \bar{H}(t,x) \quad \text{and} \quad \bar{H}_\kappa < \infty.$$

(iii) The function F is positive homogeneous of degree one with respect to u'', is Lipschitz continuous with respect to u'' and $D_{\mathsf{u}''}F \in \mathbb{S}_\delta$ at all points of differentiability of F with respect to u''.

(iv) For any $R \in (0,R_0]$, $(t,x) \in \mathbb{R}^{d+1}$, and $\mathsf{u}'' \in \mathbb{S}$ with $|\mathsf{u}''| = 1$ ($|\mathsf{u}''| := (\operatorname{tr} \mathsf{u}''\mathsf{u}'')^{1/2}$), we have

$$\theta_{R,t,x} := \fint_{C_R(t,x)} \big|F(\mathsf{u}'',s,y) - \bar{F}_{B_R(x)}(\mathsf{u}'',s)\big|\, dsdy \le \theta,$$

where

$$\bar{F}_B(\mathsf{u}'',s) = \fint_B F(\mathsf{u}'',s,y)\, dy.$$

4. Remark. Assumption 3 (ii) is stronger than Assumption 1 (iii) and is not needed in order to apply Theorem 9.8.1. It is worth emphasizing that G depends on u and not only on u', and this will be important in an application to the Isaacs equations.

Also observe that one can take $\theta = 0$ in Assumption 3 (iv) if F is independent of x.

For sufficiently regular functions $\phi(t,x)$ we set

$$H[\phi](t,x) = H\big(\phi(t,x), D\phi(t,x), D^2\phi(t,x), t, x\big). \tag{2}$$

Similarly we introduce $F[\phi]$ and other operators if we are given functions of u,t,x.

Let Ω be a bounded domain in $\mathbb{R}^d$ satisfying the exterior ball condition, $T \in (0,\infty)$, and $\Pi = [0,T) \times \Omega$.

The following is the first main result of this chapter. Recall that $\kappa \in (1,\kappa_1]$.

5. THEOREM. *Let $g \in C(\bar{\Pi})$. Then there is a function $v \in C^{\kappa}_{\text{loc}}(\Pi) \cap C(\bar{\Pi})$ which, for any $p > d+2$, is an L_p-viscosity solution of equation* (16.1.1) *in Π with boundary condition $v = g$ on $\partial'\Pi$.*

Furthermore, for any $r, R \in (0, R_0]$ satisfying $r < R$ and $(t,x) \in \Pi$ such that $C_R(t,x) \subset \Pi$ we have

$$[v]_{C^{\kappa}(C_r(t,x))} \le N(R-r)^{-\kappa} \sup_{C_R(t,x)} |v| + N\bar{H}_{\kappa}, \tag{3}$$

where the constants N depend only on d, δ, K_0, and κ.

6. EXAMPLE. A typical example of applications of Theorem 5 arises in connection with the theory of stochastic differential games where the so-called Isaacs equations play a major role. To describe a particular case of these equations, assume that we are given countable sets A and B and, for each $\alpha \in A$ and $\beta \in B$, we have an $\mathbb{S}_\delta$-valued function $a^{\alpha\beta}(t,x)$ defined on $\mathbb{R}^{d+1}$ and a real-valued function $G^{\alpha\beta}(\mathsf{u}', t, x)$ defined for $\mathsf{u}', (t,x) \in \mathbb{R}^{d+1}$. Suppose that these functions are Borel measurable and Assumption 3 (ii) is satisfied with $G^{\alpha\beta}$ in place of G for any $\alpha \in A$ and $\beta \in B$ (and $\bar{H}$ independent of $\alpha \in A$ and $\beta \in B$). Assume that $G^{\alpha\beta}$ are continuous with respect to u' uniformly with respect to α, β for any (t,x). Also suppose that for any $M \in (0,\infty)$ the function $G^{\alpha\beta}(\mathsf{u}', t, x)$, as a function of $[\mathsf{u}']$, is continuous in $[\mathsf{u}']$ uniformly with respect to $|\mathsf{u}'_0| \le M$, $(t,x) \in \Pi$, $\alpha \in A$, and $\beta \in B$. Finally, suppose that for any $R \in (0, R_0]$ and $(t,x) \in \mathbb{R}^{d+1}$

$$\fint_{C_R(t,x)} \sup_{\alpha\in A} \sup_{\beta \in B} \left| a^{\alpha\beta}(s,y) - \bar{a}^{\alpha\beta}_{B_R(x)}(s) \right| \, dsdy \le \theta,$$

where

$$\bar{a}^{\alpha\beta}_{B}(s) = \fint_B a^{\alpha\beta}(s,y)\, dy.$$

Introduce

$$F(\mathsf{u}'', t, x) = \sup_{\alpha\in A} \inf_{\beta\in B} a^{\alpha\beta}_{ij}(t,x)\mathsf{u}''_{ij}$$

and observe that

$$\theta_{R,t,x} \le \fint_{(t,t+R^2)} \fint_{B_R(x)} \fint_{B_R(x)} |F(\mathsf{u}'', s, y_1) - F(\mathsf{u}'', s, y_2)| \, dy_1 dy_2$$

$$\le \fint_{(t,t+R^2)} \fint_{B_R(x)} \fint_{B_R(x)} \sup_{\alpha\in A} \sup_{\beta\in B} \left| a^{\alpha\beta}(s,y_1) - a^{\alpha\beta}(s,y_2) \right| dy_1 dy_2$$

$$\le 2 \fint_{C_R(t,x)} \sup_{\alpha\in A} \sup_{\beta\in B} \left| a^{\alpha\beta}(s,y) - \bar{a}^{\alpha\beta}_{B_R(x)}(s) \right| dsdy,$$

where the last inequality is obtained by using

$$\left| a^{\alpha\beta}(s,y_1) - a^{\alpha\beta}(s,y_2) \right| \le \left| a^{\alpha\beta}(s,y_1) - \bar{a}^{\alpha\beta}_{B_R(x)}(s) \right| + \left| a^{\alpha\beta}(s,y_2) - \bar{a}^{\alpha\beta}_{B_R(x)}(s) \right|.$$

Also set

$$G(\mathsf{u}, t, x) = \sup_{\alpha\in A} \inf_{\beta\in B} \left[a^{\alpha\beta}_{ij}(t,x)\mathsf{u}''_{ij} + G^{\alpha\beta}(\mathsf{u}', t, x) \right] - F(\mathsf{u}'', t, x).$$

Then one easily sees that Theorem 5 is applicable to the equation

$$\partial_t v + \sup_{\alpha\in A}\inf_{\beta\in B}\big[a^{\alpha\beta}_{ij}(t,x)D_{ij}v + G^{\alpha\beta}(v,Dv,t,x)\big] = 0.$$

7. REMARK. This example is close to Theorem 9.2 of [**25**] and is more general, because $G^{\alpha\beta}$ are not assumed to be linear or even Lipschitz continuous in u'. Also $a^{\alpha\beta}$ are not assumed to be continuous in (t,x). On the other hand, $p \le d+2$ is allowed in [**25**] and the result applies to *any* L_p-viscosity solution and not only to the one we construct.

8. REMARK. Assumption 3 (iii), (iv) can be replaced with the following, which turns out to be basically weaker (cf. (2.2)): There exist countable sets A and B and functions $a^{\alpha\beta}(t,x)$ satisfying the conditions of Example 6 and there are numbers $f^{\alpha\beta}$ (independent of (t,x)) such that

$$F(\mathsf{u}'',t,x) = \sup_{\alpha\in A}\inf_{\beta\in B}\big[a^{\alpha\beta}_{ij}(t,x)\mathsf{u}''_{ij} + f^{\alpha\beta}\big] \quad\text{and}\quad F(0,t,x)\equiv 0.$$

9. REMARK. As in the case of Theorem 16.2.8 the proof of Theorem 5 is based on approximations by affine functions and on a very simple idea from the theory of diffusion processes, which is, however, very hard to explain to nonprobabilists. This idea is behind the proof of Lemma 3.3.

2. Auxiliary facts

We are going to rely on our favorite way of approximating nonlinear parabolic equations based on Theorem 8.3.3 and we start with the equation

$$\partial_t u + \max\big(F[u], P[u]-K\big) = 0, \tag{1}$$

where, as in Theorem 8.3.3, by $P[v]$ we mean $P(D^2u(t,x))$ with P defined in (6.2.4), and $K\ge 0$ is a constant.

We want to rewrite (1) in a different form.

Notice that, in light of Assumption 1.3 (iii), we have $F(\cdot,t,x)\in\mathfrak{H}_{\mathcal{L}_1}$ for any t,x in the sense of Definition 6.3.1, where $\mathcal{L}_1=\{0\}\times\mathbb{S}_\delta$. Then observe that, if in Section 6.3 we take a convex closed set $\mathcal{L}_0'$ in $\mathbb{S}$ such that $\mathbb{S}_{\delta/2}\subset\mathcal{L}_0'\subset\mathbb{S}_{\delta/4}$ and define $\mathcal{L}_0=\{0\}\times\mathcal{L}_0'$, then by Theorem 6.3.10 we have

$$F(\mathsf{u}'',t,x) = \inf_{\alpha\in\mathbb{S}}\sup_{\beta\in\mathcal{L}_1} a^{\alpha\beta}_{ij}(t,x)\mathsf{u}''_{ij}, \tag{2}$$

where, for $\alpha\in\mathbb{S}$ and $\beta=(0,\beta')\in\mathcal{L}_1$,

$$a^{\alpha\beta}(t,x) = \lambda^{\alpha\beta}(t,x)\beta' + \big(1-\lambda^{\alpha\beta}(t,x)\big)\partial G(\alpha),$$

$$G(\mathsf{u}'') = \inf_{\beta'\in\mathcal{L}_0'}\beta'_{ij}\mathsf{u}''_{ij},$$

$$\lambda^{\alpha\beta}(t,x) = 1\wedge\frac{F(\alpha,t,x)-G(\alpha)}{\beta'_{ij}\alpha_{ij}-G(\alpha)} \qquad (0/0:=1).$$

Also (6.3.5) in our situation implies that

$$\beta'_{ij}\alpha_{ij}-G(\alpha)\ge\gamma(\alpha)\ge\varepsilon|\alpha| \quad \forall\alpha\in\mathbb{S},\ \beta'\in\mathbb{S}_\delta, \tag{3}$$

where $\varepsilon=\varepsilon(d,\delta)>0$ is a constant.

Next, since $P(\mathsf{u}'')$ is positive homogeneous, convex, and $D_{\mathsf{u}''}P\in\mathbb{S}_{\bar\delta}$, there exists a closed set $\mathcal{L}_2\subset\mathbb{S}_{\bar\delta}$ such that

$$P(\mathsf{u}'') = \sup_{\beta \in \mathcal{L}_2} \beta_{ij} \mathsf{u}''_{ij}.$$

For uniformity of notation introduce $\hat{\mathcal{L}}$ as a disjoint union of $\mathcal{L}_1$ and $\mathcal{L}_2$ and for $\beta \in \mathcal{L}_2$ and $\alpha \in \mathbb{S}$ set

$$a^{\alpha\beta}(t,x) = \beta.$$

Note the following simple fact.

1. LEMMA. *For all* $\mathsf{u}'' \in \mathbb{S}$ *and* $(t,x) \in \mathbb{R}^{d+1}$ *we have*

$$\max\big(F(\mathsf{u}'',t,x), P(\mathsf{u}'') - K\big) = \inf_{\alpha \in \mathbb{S}} \sup_{\beta \in \hat{\mathcal{L}}} \big[a^{\alpha\beta}_{ij}(t,x)\mathsf{u}_{ij} + f^{\alpha\beta}_K(t,x)\big],$$

where $f^{\alpha\beta}_K = -K I_{\beta \in \mathcal{L}_2}$.

We also have the following which is an adaptation of Lemma 12.4.1 to our case of $\tau_0 = 0$ and positive homogeneous functions satisfying Assumptions 1.3 (iii), (iv).

2. LEMMA. *There is a function* $\theta = \theta(\mu) = \theta(\mu, d, \delta) > 0$ *defined for* $\mu > 0$ *such that Assumptions* 1.3 *(i), (iii), (iv) being satisfied with this* $\theta(\mu)$ *implies that for any* $R \in (0, R_0]$ *and* $(t,x) \in \mathbb{R}^{d+1}$

$$\fint_{C_R(t,x)} \sup_{\mathsf{u}'' \neq 0} \frac{\big|F(\mathsf{u}'',s,y) - \bar{F}_{B_R(x)}(\mathsf{u}'',s)\big|}{|\mathsf{u}''|}\, dsdy \leq \mu.$$

Next, observe that in light of (3), for any $(t,x) \in \mathbb{R}^{d+1}$, $R>0$, $(s,y),(s,z) \in C_R(t,x)$, $\alpha \in \mathbb{S}$, and $\beta \in \hat{\mathcal{L}}$ we have

$$\big|a^{\alpha\beta}(s,y) - a^{\alpha\beta}(s,z)\big| \leq N \frac{\big|F(\alpha,s,y) - F(\alpha,s,z)\big|}{|\alpha|}$$

$$\leq N \frac{\big|F(\alpha,s,y) - \bar{F}_{B_R(x)}(\alpha,s)\big|}{|\alpha|} + N \frac{\big|F(\alpha,s,z) - \bar{F}_{B_R(x)}(\alpha,s)\big|}{|\alpha|},$$

where the constant N depends only on d and δ. It follows that

$$\fint_{(t,t+R^2)} \fint_{B_R(x)} \fint_{B_R(x)} \sup_{\alpha \in \mathbb{S}} \sup_{\beta \in \hat{\mathcal{L}}} \big|a^{\alpha\beta}(s,y) - a^{\alpha\beta}(s,z)\big|\, dydzds$$

$$\leq N \fint_{C_R(t,x)} \sup_{\mathsf{u}'' \neq 0} \frac{\big|F(\mathsf{u}'',s,y) - \bar{F}_{B_R(x)}(\mathsf{u}'',s)\big|}{|\mathsf{u}''|}\, dsdy.$$

By taking into account that

$$\sup_{\alpha \in \mathbb{S}} \sup_{\beta \in \hat{\mathcal{L}}} \big|a^{\alpha\beta}(s,y) - \bar{a}^{\alpha\beta}_{B_R(x)}(s)\big|$$

$$\leq \fint_{B_R(x)} \sup_{\alpha \in \mathbb{S}} \sup_{\beta \in \hat{\mathcal{L}}} \big|a^{\alpha\beta}(s,y) - a^{\alpha\beta}(s,z)\big|\, dz$$

and using Lemma 2 we arrive at

3. LEMMA. *There is a function $\theta = \theta(\mu) = \theta(\mu, d, \delta) > 0$ defined for $\mu > 0$ such that Assumptions 1.3 (i), (iii), (iv) being satisfied with this $\theta(\mu)$ implies that for any $R \in (0, R_0]$ and $(t, x) \in \mathbb{R}^{d+1}$*

$$\mu_R(t,x) := \fint_{C_R(t,x)} \sup_{\alpha \in \mathbb{S}} \sup_{\beta \in \hat{\mathcal{L}}} \left| a^{\alpha\beta}(s,y) - a^{\alpha\beta}_{B_R(x)}(s) \right| dsdy \leq \mu. \tag{4}$$

3. Main estimate for solutions of (2.1)

Take $R, K \in (0, \infty)$ and $g \in W^{1,2}_\infty(C_R)$. Take P from (6.2.4), define $P[u] = P(D^2u)$ and recall that by Theorem 8.3.3 there exists $u \in W^{1,2}_{\infty,\mathrm{loc}}(C_R) \cap C(\bar{C}_R)$ such that $u = g$ on $\partial' C_R$ and equation (2.1) holds (a.e.) in C_R. By the maximum principle such u is unique. Naturally, F in (2.1) is the same as in Section 1.

Here is the main result of this section.

1. THEOREM. *There exist constants $\kappa_0 \in (1, 2]$ and $N \in (0, \infty)$ depending only on d and δ such that for each $r \in (0, R)$ one can find an affine function $\hat{u} = \hat{u}(x)$ for which*

$$|u - \hat{u}| \leq N\big(\mu_R^{\kappa/(6d+6)} \vee \mu_R^{1/(6d+6)}\big)[g]_{C^\kappa(\bar{C}_R)} R^\kappa + N r^{\kappa_0}(R - r)^{-\kappa_0} \operatorname*{osc}_{\partial' C_R}(g - \hat{g})$$

in $\bar{C}_r$ for any $\kappa \in (0, 2]$, where

$$\mu_R = \mu_R(0,0)$$

(see (2.4)) and $\hat{g} = \hat{g}(x)$ is any affine function of x.

By using parabolic scalings one easily sees that in the proof of Theorem 1 one may take $R = 1$. In that case we need some preparations. Introduce $\bar{u}$ as a unique solution of (2.1) (in C_1) with $\bar{F} := \bar{F}_{B_1}$ in place of F and the same boundary condition g on $\partial' C_1$.

Below by N with occasional subscripts we denote various constants depending only on d and δ. Also use the functions g^ε introduced in Lemma 16.3.1.

2. LEMMA. *Let $\kappa \in (0, 2]$ and*

$$[g]_{C^\kappa(C_1)} = 1. \tag{1}$$

Then for $w = u, \bar{u}$ in $\bar{C}_1$ we have

$$\big|w(t,x) - g^\varepsilon(t,x)\big| \leq N\varepsilon^{\kappa-2}\big(1 - |x|^2\big) + N\varepsilon^\kappa. \tag{2}$$

Proof. Observe that since $K \geq 0$, for $w^\varepsilon = u - g^\varepsilon$ we have in C_1 (a.e.)

$$\partial_t w^\varepsilon + \partial_t g^\varepsilon + \max\big[F(D^2 w^\varepsilon + D^2 g^\varepsilon), P(D^2 w^\varepsilon + D^2 g^\varepsilon)\big] \geq 0,$$

which in light of (16.3.1) implies that

$$\partial_t w^\varepsilon + \max\big(F[w^\varepsilon], P[w^\varepsilon]\big) \geq -N_1 \varepsilon^{\kappa-2}.$$

Next, it is easily seen that there is a constant N $(= N(d, \delta))$ such that for $\phi^\varepsilon(t,x) = NN_1\varepsilon^{\kappa-2}(1 - |x|^2)$ we have

$$\partial_t \phi^\varepsilon + \max\big(F[\phi^\varepsilon], P[\phi^\varepsilon]\big) \leq -N_1\varepsilon^{\kappa-2}$$

in C_1. It follows by the parabolic Aleksandrov maximum principle that in C_1

$$w^\varepsilon \le \phi^\varepsilon + \sup_{\partial' C_1}(w^\varepsilon - \phi^\varepsilon), \quad u \le g^\varepsilon + N\varepsilon^{\kappa-2}(1-|x|^2) + N\varepsilon^\kappa. \tag{3}$$

On the other hand, in C_1 (a.e.)

$$\partial_t w^\varepsilon + \partial_t g^\varepsilon + F(D^2 w^\varepsilon + D^2 g^\varepsilon) \le 0,$$

$$\partial_t w^\varepsilon + F[w^\varepsilon] \le N\varepsilon^{\kappa-2},$$

and with perhaps a different constant in the formula for ϕ^ε

$$w^\varepsilon \ge -\phi^\varepsilon + \inf_{\partial' C_1}(\phi^\varepsilon + w^\varepsilon), \quad u \ge g^\varepsilon - N\varepsilon^{\kappa-2}(1-|x|^2) - N\varepsilon^\kappa,$$

which along with (3) yields (2) for $w = u$. The proof of (2) for $w = \bar{u}$ is identical and the lemma is proved. □

3. Lemma. *For any $\kappa \in (0,2]$ in $\bar{C}_1$ we have*

$$|u - \bar{u}| \le N\big(\mu_1^{\kappa/(6d+6)} \vee \mu_1^{1/(d+1)}\big)[g]_{C^\kappa(C_1)}. \tag{4}$$

Proof. To simplify some formulas observe that if $[g]_{C^\kappa(C_1)} = 0$, then g is an affine function of x independent of t, so that $u = \bar{u} = g$ and we have nothing to prove. However, if $[g]_{C^\kappa(C_1)} > 0$, we can divide equation (2.1) and $u, \bar{u}$ by this quantity, and, since our assertion means, in particular, that N in (4) is independent of K, we can reduce the general situation to the one in which (1) holds. Therefore, below we assume (1).

For $\alpha \in \mathbb{S}$ and $\beta \in \hat{\mathcal{L}}$ introduce $\bar{a}^{\alpha\beta} = \bar{a}^{\alpha\beta}_{B_1}$,

$$\sigma^{\alpha\beta}(t,x) = \big[a^{\alpha\beta}(t,x)\big]^{1/2}, \quad \bar{\sigma}^{\alpha\beta}(t) = \big[\bar{a}^{\alpha\beta}(t)\big]^{1/2},$$

$$\mathcal{L}^{\alpha\beta} v(t,x) = a^{\alpha\beta}_{ij}(t,x) D_{ij} v(t,x), \quad \bar{\mathcal{L}}^{\alpha\beta} v(t,x) = \bar{a}^{\alpha\beta}_{ij}(t) D_{ij} v(t,x).$$

Also on sufficiently regular functions $u(t,x,\bar{x})$, $t \in \mathbb{R}$, $x, \bar{x} \in \mathbb{R}^d$, introduce

$$\Phi[u](t,x,\bar{x}) = \inf_{\alpha\in\mathbb{S}} \sup_{\beta\in\hat{\mathcal{L}}} \big[\mathfrak{L}^{\alpha\beta} u(t,x,\bar{x}) + f^{\alpha\beta}_K\big],$$

where

$$\mathfrak{L}^{\alpha\beta} u(t,x,\bar{x}) = a^{\alpha\beta}_{ij}(t,x) D^x_{ij} u(t,x,\bar{x}) + \hat{a}^{\alpha\beta}_{ij}(t,x,\bar{x}) D^{x\bar{x}}_{ij} u(t,x,\bar{x})$$

$$+ \check{a}^{\alpha\beta}_{ij}(t,x,\bar{x}) D^{\bar{x}x}_{ij} u(t,x,\bar{x}) + \bar{a}^{\alpha\beta}_{ij}(t) D^{\bar{x}}_{ij} u(t,x,\bar{x}),$$

$$D^x_{ij} = \frac{\partial^2}{\partial x^i \partial x^j}, \quad D^{x\bar{x}}_{ij} = \frac{\partial^2}{\partial x^i \partial \bar{x}^j}, \quad D^{\bar{x}x}_{ij} = \frac{\partial^2}{\partial \bar{x}^i \partial x^j}, \quad D^{\bar{x}}_{ij} = \frac{\partial^2}{\partial \bar{x}^i \partial \bar{x}^j},$$

$$\hat{a}^{\alpha\beta}(t,x,\bar{x}) = \sigma^{\alpha\beta}(t,x)\bar{\sigma}^{\alpha\beta}(t), \quad \check{a}^{\alpha\beta}(t,x,\bar{x}) = \bar{\sigma}^{\alpha\beta}(t)\sigma^{\alpha\beta}(t,x).$$

Observe that for $\lambda, \bar{\lambda} \in \mathbb{R}^d$ we have

$$a^{\alpha\beta}_{ij}\lambda^i\lambda^j + \hat{a}^{\alpha\beta}_{ij}\lambda^i\bar{\lambda}^j + \check{a}^{\alpha\beta}_{ij}\bar{\lambda}^i\lambda^j + \bar{a}^{\alpha\beta}_{ij}\bar{\lambda}^i\bar{\lambda}^j = |\sigma^{\alpha\beta}\lambda + \bar{\sigma}^{\alpha\beta}\bar{\lambda}|^2 \ge 0,$$

so that Φ is a (degenerate) elliptic operator.

Next let $w \in W^{1,2}_{d+1}(C_1)$ be a solution of the equation

$$\partial_t w + \sup_{\alpha\in\mathbb{S}} \sup_{\beta\in\hat{\mathcal{L}}} \mathcal{L}^{\alpha\beta} w = -\sup_{\alpha\in\mathbb{S}} \sup_{\beta\in\hat{\mathcal{L}}} |a^{\alpha\beta} - \bar{a}^{\alpha\beta}| =: -h$$

in C_1 (a.e.) with zero boundary condition on $\partial' C_1$. Such solution exists by Theorem 15.1.4 and, by the parabolic Aleksandrov estimate and definition (2.4), we have in C_1 that

$$0 \le w \le N\mu_1^{1/(d+1)}. \tag{5}$$

One of the reasons we need the function w for is that, as is easy to see, there is a $\lambda > 0$ depending only on d and δ such that for all α, β on $(0,1)\times B_1^2$ (a.e.) we have

$$\partial_t\big(\lambda w(t,x) + |x-\bar{x}|^2\big) + \mathfrak{L}^{\alpha\beta}(t,x,\bar{x})\big(\lambda w(t,x) + |x-\bar{x}|^2\big)$$

$$= \lambda\big(\partial_t w + \mathcal{L}^{\alpha\beta} w\big)(t,x) + 2\big|\sigma^{\alpha\beta}(t,x) - \bar{\sigma}^{\alpha\beta}(t)\big|^2 \le -h,$$

where the inequality follows from the fact that $a^{1/2}$ is a Lipschitz continuous function on $\mathbb{S}_{\delta/4}$, so that $|\sigma^{\alpha\beta} - \bar{\sigma}^{\alpha\beta}|^2 \le N|a^{\alpha\beta} - \bar{a}^{\alpha\beta}|^2 \le N|a^{\alpha\beta} - \bar{a}^{\alpha\beta}|$.

After that we proceed in two steps.

Step 1. Estimate of $u - \bar{u}$ from above. According to Lemma 2 for $|\bar{x}| = 1$ and $(t,x) \in \bar{C}_1$ we have

$$u(t,x) \le g^{\varepsilon}(t,x) + N\varepsilon^{\kappa-2}\big(1 - |x|^2\big) + N\varepsilon^{\kappa}$$

$$\le g^{\varepsilon}(t,x) + N\varepsilon^{\kappa-2}|x-\bar{x}| + N\varepsilon^{\kappa},$$

where

$$\varepsilon^{\kappa-2}|x-\bar{x}| \le \varepsilon^{\kappa-4}|x-\bar{x}|^2 + \varepsilon^{\kappa},$$

so that

$$u(t,x) \le g^{\varepsilon}(t,x) + N\varepsilon^{\kappa-4}|x-\bar{x}|^2 + N\varepsilon^{\kappa}.$$

This inequality also obviously holds if $|x| = 1$ and $|\bar{x}| \le 1$ or if $t = 1$ and $|x|, |\bar{x}| \le 1$. This shows that for $\varepsilon \in (0,1)$

$$u^{\varepsilon}(t,x,\bar{x}) := u(t,x) - \big[g^{\varepsilon}(t,x) - g^{\varepsilon}(t,\bar{x}) + N\varepsilon^{\kappa-6}e^{1-t}|x-\bar{x}|^2 + N\varepsilon^{\kappa}\big] \le g^{\varepsilon}(t,\bar{x})$$

on $\partial'[(0,1)\times B_1^2]$. Actually, above we could replace $\varepsilon^{\kappa-6}$ with $\varepsilon^{\kappa-4}$ but later on we will need to deal with terms of order $\varepsilon^{\kappa-6}|x-\bar{x}|^2$ anyway. Also observe that for $\varepsilon \in (0,1)$,

$$I^{\varepsilon}(t,x,\bar{x}) := \partial_t u^{\varepsilon}(t,x,\bar{x}) + \Phi[u^{\varepsilon}](t,x,\bar{x}) = \partial_t u(t,x) + \partial_t g^{\varepsilon}(t,\bar{x}) - \partial_t g^{\varepsilon}(t,x)$$

$$+N\varepsilon^{\kappa-6}e^{1-t}|x-\bar{x}|^2 + \inf_{\alpha\in\mathbb{S}} \sup_{\beta\in\hat{\mathcal{L}}} \Big[a^{\alpha\beta}_{ij} D_{ij}u(t,x) + \bar{a}^{\alpha\beta}_{ij} D_{ij} g^{\varepsilon}(t,\bar{x}) - a^{\alpha\beta}_{ij} D_{ij} g^{\varepsilon}(t,x)$$

$$-N\varepsilon^{\kappa-6}e^{1-t}\big|\sigma^{\alpha\beta}(t,x) - \bar{\sigma}^{\alpha\beta}(t)\big|^2 + f^{\alpha\beta}_K\Big],$$

where

$$\begin{aligned}\big|\bar a^{\alpha\beta}_{ij}D_{ij}g^\varepsilon(t,\bar x)-a^{\alpha\beta}_{ij}D_{ij}g^\varepsilon(t,x)\big| &\le \big|\bar a^{\alpha\beta}(t)-a^{\alpha\beta}(t,x)\big|\,\big|D^2g^\varepsilon(t,x)\big|\\ &\quad+N\big|D^2g^\varepsilon(t,x)-D^2g^\varepsilon(t,\bar x)\big|\\ &\le N\varepsilon^{\kappa-2}h(t,x)+N\varepsilon^{\kappa-3}|x-\bar x|,\end{aligned}$$

$$\big|\partial_t g^\varepsilon(t,\bar x)-\partial_t g^\varepsilon(t,x)\big|\le N\varepsilon^{\kappa-3}|x-\bar x|,$$

$$-N\varepsilon^{\kappa-3}|x-\bar x|+N\varepsilon^{\kappa-6}|x-\bar x|^2\ge -N\varepsilon^\kappa.$$

By also taking into account Lemma 2.1 we conclude that, for $\varepsilon\in(0,1)$,

$$\begin{aligned}I^\varepsilon(t,x,\bar x)&\ge \partial_t u(t,x)+\inf_{\alpha\in\mathbb{S}}\sup_{\beta\in\hat{\mathcal{L}}}\big[\mathcal{L}^{\alpha\beta}u(t,x)+f^{\alpha\beta}_K\big]\\ &\quad-N_1\varepsilon^{\kappa-6}h-N_1\varepsilon^\kappa=-N_1\varepsilon^{\kappa-6}h-N_1\varepsilon^\kappa\end{aligned}$$

in $(0,1)\times B_1^2$ (a.e.).

On the other hand,

$$\bar u(t,\bar x)\ge g^\varepsilon(t,\bar x)-N\varepsilon^{\kappa-2}(1-|\bar x|^2)-N\varepsilon^\kappa,$$

which implies that

$$\begin{aligned}\bar u^\varepsilon(t,x,\bar x)&:=\bar u(t,\bar x)+N_2\varepsilon^{\kappa-6}\big((1+\lambda)w(t,x)+|x-\bar x|^2\big)\\ &\quad+N_2(2-t)\varepsilon^\kappa\ge g^\varepsilon(t,\bar x)\end{aligned}$$

on $\partial'[(0,1)\times B_1^2]$. It is also easily seen that by increasing N_2 if needed (which does not violate the above inequality) we may assume that in $(0,1)\times B_1^2$ (a.e.)

$$\partial_t\bar u^\varepsilon(t,x,\bar x)+\Phi[\bar u^\varepsilon](t,x,\bar x)\le -N_1\varepsilon^{\kappa-6}h-N_1\varepsilon^\kappa.$$

Hence, by the maximum principle (see, for instance, Theorem 4.1.11 in combination with Corollary 4.1.4) in $[0,1]\times\bar B_1^2$ we have

$$\begin{aligned}\bar u(t,\bar x)+N\varepsilon^{\kappa-6}(w(t,x)+|x-\bar x|^2)+N\varepsilon^\kappa&\ge u(t,x)\\ &\quad-\big[g^\varepsilon(t,x)-g^\varepsilon(t,\bar x)+N\varepsilon^{\kappa-6}|x-\bar x|^2+N\varepsilon^\kappa\big],\end{aligned}$$

which for $x=\bar x$ in light of (5) yields

$$u(t,x)-\bar u(t,\bar x)\le N\big(\varepsilon^\kappa+\varepsilon^{\kappa-6}\mu_1^{1/(d+1)}\big).$$

If $\mu_1\le 1$, then for $\varepsilon=\mu_1^{1/(6d+6)}$ (≤ 1) we get $u-\bar u\le N\mu_1^{\kappa/(6d+6)}$ and if $\mu_1\ge 1$, then for $\varepsilon=1$ we obtain $u-\bar u\le N\mu_1^{1/(d+1)}$, so that generally

$$u-\bar u\le N\big(\mu_1^{\kappa/(6d+6)}\vee\mu_1^{1/(d+1)}\big).$$

Step 2. Estimate of $u-\bar u$ from below. Observe that (cf. the beginning of Step 1)

$$\bar v^\varepsilon(t,x,\bar x):=\bar u(t,\bar x)-N\varepsilon^{\kappa-4}\big(\lambda w(t,x)+|x-\bar x|^2\big)-N\varepsilon^\kappa\le g^\varepsilon(t,\bar x)$$

on $\partial'[(0,1)\times B_1^2]$. It is also easily seen that in $(0,1)\times B_1^2$ (a.e.)

$$\partial_t \bar{v}^\varepsilon(t,x,\bar{x}) + \Phi[\bar{v}^\varepsilon](t,x,\bar{x}) \geq 0.$$

On the other hand,

$$v^\varepsilon(t,x,\bar{x}) := u(t,x) - \big[g^\varepsilon(t,x) - g^\varepsilon(t,\bar{x})\big]$$

$$+N\varepsilon^{\kappa-6}e^{1-t}\big(\lambda w(t,x) + |x-\bar{x}|^2\big) + N(2-t)\varepsilon^\kappa \geq g^\varepsilon(t,\bar{x})$$

on $\partial'[(0,1)\times B_1^2]$ and the above computations show that (for sufficiently large N)

$$\partial_t v^\varepsilon(t,x,\bar{x}) + \Phi[v^\varepsilon](t,x,\bar{x}) \leq 0$$

in $(0,1)\times B_1^2$ (a.e.). By the maximum principle $\bar{v}^\varepsilon \leq v^\varepsilon$, which leads to the desired estimate of $u-\bar{u}$ from below and the lemma is proved. □

4. LEMMA. *There exist constants $\kappa_0 \in (1,2]$ and $N \in (0,\infty)$ depending only on d and δ such that for any $r \in (0,1)$*

$$[\bar{u}]_{C^{\kappa_0}(C_r)} \leq N(1-r)^{-\kappa_0} \operatorname*{osc}_{\partial' C_1}(g-\hat{g}), \tag{6}$$

where $\hat{g}=\hat{g}(x)$ is any affine function of x.

Proof. By Theorem 16.6.4 applied to $\bar{u}-\hat{g}$ and $\bar{F}$ in place of v and H, respectively, estimate (6) holds if we replace the oscillation on its right-hand side with $\sup\{|\bar{u}-\hat{g}|;C_1\}$. Since $\bar{u}-\hat{g}$ satisfies a homogeneous parabolic equation without lower order terms, the last supremum can be evaluated only over $\partial' C_1$, where $\bar{u}=g$ and this immediately yields the result. The lemma is proved. □

Proof of Theorem 1. Take $\hat{u}(t,x) = \bar{u}(0,0) + x^i D_i \bar{u}(0,0)$ and observe that in C_r by Lemma 3

$$|u-\hat{u}| \leq |u-\bar{u}| + |\bar{u}-\hat{u}| \leq N\big(\mu_1^{\kappa/(6d+6)} \vee \mu_1^{1/(6d+6)}\big)[g]_{C^\kappa(C_1)} + I,$$

where by Lemma 4

$$I = |\bar{u}-\hat{u}| \leq 2r^{\kappa_0}[\bar{u}]_{C^{\kappa_0}(C_r)} \leq Nr^{\kappa_0}(1-r)^{-\kappa_0} \operatorname*{osc}_{\partial' C_1}(g-\hat{g})$$

so that the theorem is proved. □

4. Estimating the C^κ-norm of solutions of (8.3.1)

In this section we assume that $\bar{H}$ is bounded, take a constant $K \geq 0$, and investigate solutions of the cut-off equation (8.3.1). We take $\kappa_0 \in (1,2]$ from Theorem 3.1, take a $\mu \in (0,1]$, and suppose that Assumption 1.3 (iv) is satisfied with $\theta = \theta(\mu)$ so that (2.4) holds for any $R \in (0,R_0]$ and $(t,x) \in \mathbb{R}^{d+1}$.

1. LEMMA. *Let $R \in (0,R_0]$ and let $v \in W^{1,2}_\infty(C_R)$ be a solution of (8.3.1) in C_R. Then for each $r \in (0,R)$ one can find an affine function $\hat{v}(x)$ such that in C_r for any $\kappa \in [1,2]$*

$$|v-\hat{v}| \leq N\mu^{1/(6d+6)}[v]_{C^\kappa(C_R)}R^\kappa + Nr^{\kappa_0}(R-r)^{-\kappa_0}R^\kappa[v]_{C^\kappa(C_R)}$$

$$+NK_0R^2\sup_{C_R}\big(|v|+|Dv|\big) + NR^\kappa \bar{H}_\kappa,$$

where the constants N depend only on d and δ.

The proof of this lemma coincides with that of Lemma 16.6.1, only instead of using Theorem 16.5.1 one should use Theorem 3.1 and take $p=d+1$.

2. LEMMA. *Take* $r_1 \in (0, R_0]$, $r_0 \in (0, r_1)$, *and define*

$$\kappa_1 = \frac{1 + \kappa_0}{2}.$$

Let $v \in W^{1,2}_{\infty}(C_{r_1})$ *be a solution of* (8.3.1) *in* C_{r_1} *and let* $\kappa \in (1, \kappa_1]$. *Then there exists* $\theta = \theta(\kappa, d, \delta) \in (0, 1]$ *such that, if Assumption* 1.3 *(iv) is satisfied with this* θ, *then*

$$[v]_{C^{\kappa}(C_{r_0})} \leq (1/2)[v]_{C^{\kappa}(C_{r_1})} + N(K_0 + 1)(r_1 - r_0)^{-\kappa} \sup_{C_{r_1}} |v|$$
$$+ N(K_0 + 1)(r_1 - r_0)^{-(\kappa - 1)} \sup_{C_{r_1}} |Dv| + N\bar{H}_{\kappa}, \tag{1}$$

where $N = N(d, \delta, \kappa)$.

Proof. We closely follow the proof of Lemma 16.6.3. To specify θ we first take a $\mu \in (0, 1]$ and suppose that Assumption 1.3 (iv) is satisfied with $\theta = \theta(\mu)$ so that (2.4) holds for any $R \in (0, R_0]$ and $(t, x) \in \mathbb{R}^{d+1}$.

Then take $(t_0, x_0) \in C_{r_0}$, $\varepsilon \in (0, 1)$, define

$$r_0' = \frac{\varepsilon}{3}(r_1 - r_0),$$

and notice that for any $(t, x) \in C_{r_0'}(t_0, x_0)$, $r \in \big(0, 2r_0'\big]$, and $R = \varepsilon^{-1} r$, we have $R \leq r_1 \leq R_0 \leq 1$ and

$$C_R(t, x) \subset C_{r_1}.$$

Therefore, by Lemma 1 we can find an affine function $\hat{v}(x)$ such that

$$\sup_{C_r(t,x) \cap C_{r_0'}(t_0,x_0)} |v - \hat{v}| \leq \sup_{C_r(t,x)} |v - \hat{v}|$$

$$\leq N\mu^{1/(6d+6)} [v]_{C^{\kappa}(C_R(t,x))} \varepsilon^{-\kappa} r^{\kappa} + N\varepsilon^{\kappa_0 - \kappa}(1 - \varepsilon)^{-\kappa_0} r^{\kappa} [v]_{C^{\kappa}(C_R(t,x))}$$

$$+ NK_0 \varepsilon^{-2} r^2 \sup_{C_R(t,x)} \big(|v| + |Dv|\big) + N\varepsilon^{-\kappa} r^{\kappa} \bar{H}_{\kappa} \leq N r^{\kappa} I(\varepsilon, r_1),$$

where the constants N depend only on d and δ and

$$I(\varepsilon, r_1) := \big(\mu^{1/(6d+6)} \varepsilon^{-\kappa} + \varepsilon^{\kappa_0 - \kappa}(1 - \varepsilon)^{-\kappa_0}\big)[v]_{C^{\kappa}(C_{r_1})}$$

$$+ \varepsilon^{-2} K_0 \sup_{C_{r_1}} \big(|v| + |Dv|\big) + \varepsilon^{-\kappa} \bar{H}_{\kappa}.$$

It follows by Lemma 16.6.2 that

$$[v]_{C^{\kappa}(C_{r_0'}(t_0,x_0))} \leq N_1 I(\varepsilon, r_1),$$

where N_1 depends only on d, κ, and δ. We can now specify θ and ε. First we chose $\varepsilon \in (0, 1)$ so that

$$N_1 \varepsilon^{\kappa_0 - \kappa}(1 - \varepsilon)^{-\kappa_0} = 1/4.$$

Since $\kappa_0 - \kappa \geq (\kappa_0 - 1)/2 > 0$ and κ_0 depends only on d and δ and N_1 depends only on d, κ, and δ, ε also depends only on d, κ, and δ. After that we take

$\mu = \mu(d, \kappa, \delta) \in (0, 1]$ so that

$$N_1 \mu^{1/(6d+6)} \varepsilon^{-2} \le 1/4$$

and set $\theta = \theta(\mu(d, \kappa, \delta))$. Then

$$[v]_{C^\kappa(C_{r_0'}(t_0,x_0))} \le (1/2)[v]_{C^\kappa(C_{r_1})} + NJ, \tag{2}$$

where $N = N(d, \delta, \kappa)$ and

$$J = K_0 \sup_{C_{r_1}} \big(|v| + |Dv|\big) + \bar{H}_\kappa.$$

Now observe that if $(t, x), (s, x) \in C_{r_0}$ and $t > s$, then either $|t - s| \le (r_0')^2$, in which case $(t, x) \in C_{r_0'}(s, x)$ and

$$(t-s)^{-\kappa/2}\big|v(t,x) - v(s,x)\big| \le (1/2)[v]_{C^\kappa(C_{r_1})} + NJ$$

owing to (2), or $|t - s| \ge (r_0')^2$ when

$$\begin{aligned}\big|v(t,x) - v(s,x)\big| &\le 2(t-s)^{\kappa/2}(r_0')^{-\kappa} \sup_{C_{r_1}} |v| \\ &\le N(t-s)^{\kappa/2}(r_1 - r_0)^{-\kappa} \sup_{C_{r_1}} |v|.\end{aligned}$$

Next if $(t, x), (t, y) \in C_{r_0}$ and $x \ne y$, then either $|x - y| < r_0'$, in which case $(t, y) \in C_{r_0'}(t, x)$ and

$$|x-y|^{-(\kappa-1)}\big|Dv(t,x) - Dv(t,y)\big| \le (1/2)[v]_{C^\kappa(C_{r_1})} + NJ,$$

or else $|x - y| \ge r_0'$ and

$$\begin{aligned}|Dv(t,x) - Dv(t,y)| &\le 2|x-y|^{\kappa-1}\big(r_0'\big)^{-(\kappa-1)} \sup_{C_{r_1}} |Dv| \\ &\le N|x-y|^{\kappa-1}(r_1 - r_0)^{-(\kappa-1)} \sup_{C_{r_1}} |Dv|.\end{aligned}$$

This proves (1) and the lemma. □

3. THEOREM. *Take $0 < r < R \le R_0$ and take κ_1, $\kappa \in (1, \kappa_1]$, and θ from Lemma 2 and suppose that Assumption 1.3 (iv) is satisfied with this θ. Let $v \in W^{1,2}_\infty(C_R)$ be a solution of (8.3.1) in C_R. Then*

$$[v]_{C^\kappa(C_r)} \le N(R-r)^{-\kappa} \sup_{C_R} |v| + N\bar{H}_\kappa, \tag{3}$$

where N depends only on d, δ, K_0, and κ.

The proof of this theorem is identical to that of Theorem 16.6.4 with the only change that one should use Lemma 2 instead of Lemma 16.6.3.

5. Proof of Theorem 1.5

First assume that $\bar{H}$ is bounded and $g \in W^{1,2}_{\infty}(\mathbb{R}^{d+1})$. For $K > 0$ denote by v_K a solution of (8.3.1) with boundary condition $v = g$ on $\partial'\Pi$. By Theorem 8.3.3 such a solution exists is continuous in $\bar{\Pi}$ and has locally bounded derivatives.

1. REMARK. Theorem 4.3 says that, for any $R \in (0, R_0]$, such that $C_R \subset \Pi$, and any $r \in (0, R)$ we have

$$[v_K]_{C^{\kappa}(C_r)} \leq N(R-r)^{-\kappa} \sup_{C_R} |v_K| + N\bar{H}_{\kappa}, \tag{1}$$

where N depends only on d, δ, κ, and K_0. Of course, a similar statement holds for shifted C_R's like (1.3). Furthermore, by Lemma 12.1.9

$$|v_K| \leq N\big(\|g\|_{C(\Pi)} + \|\bar{H}\|_{L_{d+1}(\Pi)}\big), \tag{2}$$

where N depends only on d, δ, K_0, T, and the diameter of Ω. This provides interior estimates of the C^{κ}-norms of v_K uniform with respect to K.

Also by Theorem 9.8.1 the family $\{v_K; K \geq 1\}$ is equicontinuous on $\bar{\Pi}$. More precisely there exists a function $\bar{\omega}(\varepsilon)$, depending only on ε, $\|\bar{H}\|_{L_{d+1}(\Pi)}$, δ, d, K_0, $\rho_{\text{ext}}(\Omega)$, $\|g\|_{C(\Pi)}$, and the modulus ω of continuity of g on $\partial'\Pi$, such that $\bar{\omega}(\varepsilon) \to 0$ as $\varepsilon \downarrow 0$ and

$$\big|v_K(t,x) - v_K(s,y)\big| \leq \bar{\omega}\big[\rho\big((t,x),(s,y)\big)\big] \tag{3}$$

for any $x, y \in \bar{\Omega}$ and $s, t \in [0, T]$.

It follows that there is a sequence $K_n \to \infty$ and a function v such that $v^n := v_{K_n} \to v$ uniformly in $\bar{\Pi}$. Of course, (1.3) holds, owing to Theorem 4.3, which also implies that $Dv^n \to Dv$ locally uniformly in Π.

We keep repeating the proof of Theorem 16.2.8 on page 364, observing that, obviously, Lemma 16.6.5 is valid in our situation, and for any $(t_0, x_0) \in \Pi$, $r > 0$, such that $\bar{C}_r(t_0, x_0) \subset \Pi$, and $\phi \in W^{1,2}_p\big(C_r(t_0, x_0)\big)$ (recall that $p > d+2$) such that $v - \phi$ attains a local maximum at (t_0, x_0), we come to (16.6.11):

$$\lim_{r \downarrow 0} \operatorname*{ess\,sup}_{C_r(t_0,x_0)} \big[\partial_t \phi(t,x) + H\big(v(t,x), Dv(t,x), D^2\phi(t,x), t, x\big)\big] \geq 0. \tag{4}$$

Observe that, as $(t,x) \to (t_0, x_0)$, $Dv(t,x) \to Dv(t_0, x_0)$. Also the inclusion $\phi \in W^{1,2}_p\big(C_r(t_0, x_0)\big)$ with $p > d+2$ by embedding theorems implies that $\phi \in C^{1+\alpha}\big(\bar{C}_r(t_0, x_0)\big)$, where $\alpha \in (0,1)$, and hence $D\phi(t,x) \to D\phi(t_0, x_0)$. Finally, the fact that $v - \phi$ attains a local maximum at (t_0, x_0) yields $Dv(t_0, x_0) = D\phi(t_0, x_0)$. Now by Assumption 1.1 (iv) we can replace $Dv(t,x)$ in (4) by $D\phi(t,x)$ and then, by definition, v is an L_p-viscosity subsolution.

The rest of the proof of Theorem 1.5 is identical to that of the proof of Theorem 16.2.8 with some obvious changes in notation and references. □

CHAPTER 18

Uniqueness and existence of extremal viscosity solutions for parabolic equations

We are going to work in the framework of Section 16.1. Sections 1 and 2 closely follow [**78**] and [**81**], Sections 3 and 4 closely follow [**74**] and [**81**].

1. Uniqueness of C-viscosity solutions of parabolic equations

Fix some constants $\delta \in (0, 1]$, $K_0 \in (0, \infty)$, and recall that we are given $T \in (0, \infty)$, a bounded domain $\Omega \subset \mathbb{R}^d$ satisfying the exterior ball condition, and

$$\Pi = [0, T) \times \Omega.$$

In the following assumption there is $\gamma = \gamma(d, \delta) \in (1/4, 1/2)$ which we specify in the following way. In Lemma 16.6.3

$$\kappa(d, \delta, p) \in (1, 2)$$

is defined. We take $\bar{\delta} = \bar{\delta}(d, \delta)$ from Remark 6.2.2 and set

$$\kappa = \kappa(d, \bar{\delta}, d+3), \quad \gamma = \frac{7 - 3\kappa}{12 - 4\kappa} \quad \big(\in (1/4, 1/2) \big).$$

1. Assumption. (i) The function $H(\mathsf{u}, t, x)$ is a *continuous* function of (u, t, x) and is Lipschitz continuous with respect to u with Lipschitz constant K_0.

(ii) $D_{\mathsf{u}''} H \in \mathbb{S}_\delta$ at all points of differentiability of H with respect to u''.

(iii) For all values of the arguments we have

$$\big| H(\mathsf{u}, t, x) - H(\mathsf{u}, t, y) \big| \le K_0 |x - y|^\gamma |\mathsf{u}''| + \big(|\mathsf{u}'| + 1 \big) \omega\big(|x - y| \big), \tag{1}$$

where $\omega(\tau)$, $\tau \ge 0$, is a continuous functions vanishing at the origin.

(iv) $\sup \big\{ |H(0, t, x)| : (t, x) \in \mathbb{R}^{d+1} \big\} =: \bar{H} < \infty$.

(iv) We are given a $g \in C(\overline{\partial' \Pi})$.

2. Theorem. *Under the above assumptions there exists a unique $v \in C(\bar{\Pi})$ which is a C-viscosity solution of* (16.1.1) *in Π with boundary condition $v = g$ on $\partial' \Pi$. Furthermore, there exists a constant $N \in (0, \infty)$ such that for any $\rho > 0$*

$$\|v\|_{C^\kappa(\Pi^\rho)} \le N \rho^{-\kappa}. \tag{2}$$

3. Remark. The assumptions of Theorem 2 are almost identical to the assumptions made in the elliptic case in [**111**], that, to the best of our knowledge, provides the most general result to date concerning the uniqueness of C-viscosity solutions for the uniformly elliptic case (see Remark 3.1 there). Our Theorem 2 is a parabolic counterpart of Trudinger's result from [**111**].

In the parabolic case the uniqueness of L_p-viscosity solutions is proved in Lemma 6.2 of [**25**] when H is independent of (t,x). In the case of the Isaacs equations, under the assumptions on the coefficients guaranteeing that our assumptions are satisfied as well, the statement about the uniqueness of C-viscosity solutions is found in Theorem 9.3 of [**25**]. However, this statement is not provided with a proof with the excuse that its proof is similar to the one known in the elliptic case.

One of the features of our proof is that it also allows one to establish an algebraic rate of convergence of numerical approximations (see [**78**]).

4. EXAMPLE. This example is close to Example 10.1.2. For $\varepsilon > 0$ introduce

$$H(\mathsf{u},t,x) = \Big(1 + \varepsilon \cos \sqrt{|\ln|\mathsf{u}''||}\Big) a^{ij}(t,x)\mathsf{u}''_{ij},$$

where $a(t,x) = (a^{ij}(t,x))$ is a continuous $\mathbb{S}_{(\delta+1)/2}$-valued function satisfying

$$|a(t,x) - a(t,y)| \le |x-y|^\gamma$$

and ε is taken so small that $D_{\mathsf{u}''}H \in \mathbb{S}_\delta$ wherever $D_{\mathsf{u}''}H$ exists. Then obviously the above assumptions are satisfied and the corresponding equation has a unique C-viscosity solution.

At the same time it is not hard to check that H is not a boundedly inhomogeneous function of u'' (see Remark 6.3.2) and therefore does not admit the representation as

$$\inf_{\alpha\in A} \sup_{\beta\in B} \big[a^{\alpha\beta}_{ij}(t,x)\mathsf{u}''_{ij} + f^{\alpha\beta}(t,x)\big]$$

with uniformly bounded $a^{\alpha\beta}_{ij}$ and $f^{\alpha\beta}$. Therefore, Theorem 9.3 of [**25**] is not applicable in this case.

The proof of Theorem 2 is based on a few auxiliary results.

5. REMARK. It is easy to see that if $v(t,x)$ is a C-viscosity subsolution of (16.1.1) in Π, then, for any constant c, the functions $w(t,x) := e^{ct}v(t,x)$ is a C-viscosity subsolution of

$$\partial_t w + H^c[w] = 0$$

in Π, where

$$H^c(\mathsf{u},t,x) := e^{ct}H(e^{-ct}\mathsf{u},t,x) - c\mathsf{u}'_0.$$

The function $H^c(u,t,x)$ has the same Lipschitz constant with respect to $\big([\mathsf{u}'],\mathsf{u}''\big)$ and its derivative with respect to u'_0, wherever it exists, is

$$D_{\mathsf{u}'_0}H(e^{-ct}\mathsf{u},t,x) - c \le K_0 - c.$$

If we take $c = K_0 + 1$ and redefine H^c for $t \notin [0,T]$ as its value at the closest end point of $[0,T]$, then H^c will satisfy all assumptions of Theorem 2 with $K_0 + 1$ in place of K_0 and additionally satisfy $D_{\mathsf{u}'_0}H_c \le -1$.

That is why without loss of generality we suppose that not only Assumption 1 is satisfied but also for all values of arguments

$$D_{\mathsf{u}'_0}H \le -1 \tag{3}$$

wherever the left-hand side exists.

Below we suppose that the above assumptions are satisfied, take $P(\mathsf{u}'')$ from (6.2.4), and set $P[u](t,x) = P\big(D^2u(t,x)\big)$. Recall that κ is introduced before Assumption 1.

6. LEMMA. *(i) For any $K \geq 1$ each of the equations*

$$\partial_t u_K + \max\big(H[u_K], P[u_K] - K\big) = 0, \tag{4}$$

$$\partial_t u_{-K} + \min\big(H[u_{-K}], -P[-u_{-K}] + K\big) = 0 \tag{5}$$

in Π *(a.e.) with boundary condition* $u_{\pm K} = g$ *on* $\partial'\Pi$ *has a unique solution* $u_{\pm K} \in W^{1,2}_{p,\text{loc}}(\Pi^\rho) \cap C(\bar{\Pi})$ *for any* $p \geq 1$ *and* $\rho > 0$.

(ii) We have $u_{-K} \leq u_K$ *and, as* K *increases,* u_{-K} *increase and* u_K *decrease.*

(iii) The family $\{u_{-K}, u_K : K \geq 1\}$ *is equicontinuous in* $\bar{\Pi}$.

(iv) There exists a constant $N \in (0, \infty)$ *such that for any* $\rho > 0$ *and* $K \geq 1$

$$\|u_K, u_{-K}\|_{C^\kappa(\Pi^\rho)} \leq N\rho^{-\kappa}. \tag{6}$$

Proof. The existence part in assertion (i) for the sign + follows from Theorem 12.1.13. For the sign − it suffices to replace $H(\mathsf{u}, t, x)$ with $-H(-\mathsf{u}, t, x)$. Uniqueness and assertion (ii) are direct consequences of the maximum principle.

Assertion (iii) follows from Theorem 9.8.1 and the observation that

$$\big|\max\big(H(\mathsf{u}', 0, t, x), -K\big)\big| \leq \big|H(\mathsf{u}', 0, t, x)\big| \leq K_0|\mathsf{u}'| + \bar{H}.$$

To prove assertion (iv), in what concerns u_K, we introduce

$$H_K(\mathsf{u}, t, x) = \max\big(H(\mathsf{u}, t, x), P(\mathsf{u}'') - K\big),$$

$$F_K(\mathsf{u}'', t, x) = H_K(0, \mathsf{u}'', t, x) - H_K(0, t, x)$$

and check the assumptions of Theorem 16.6.4 (that is of Theorem 16.2.8).

Assumption 16.2.3 [(i), (ii) a)] is obviously satisfied for H_K with $\bar{\delta}$ in place of δ owing to the choice of $\bar{\delta}$ and the fact that H is Lipschitz continuous in u with constant K_0.

Assumption 16.2.4 (i) is satisfied because H and H_K are continuous, and (ii) is satisfied because

$$|H_K(\mathsf{u}, t, x) - F_K(\mathsf{u}'', t, x)| \leq |H(\mathsf{u}, t, x) - H(0, \mathsf{u}'', t, x)| + |H(0, t, x)|$$

$$\leq K_0|u'| + \bar{H}.$$

Assumption 16.2.4 (iii) is satisfied with $\bar{\delta}$ in place of δ if $|\theta_0| \leq \bar{\delta}/2$.

To check Assumption 16.2.4 (iv) observe that

$$\big|F_K\big(\mathsf{u}'', t, x_1\big) - F_K\big(\mathsf{u}'', t, x_2\big)\big| \leq \big|H(0, \mathsf{u}'', t, x_1) - H(0, \mathsf{u}'', t, x_2)\big|$$

$$+ \big|H(0, t, x_1) - H(0, t, x_2)\big|$$

$$\leq K_0|x_1 - x_2|^\gamma|\mathsf{u}''| + 2\omega\big(|x_1 - x_2|\big).$$

This shows that for any $\theta_0 > 0$ (in particular, for $\theta_0 = \theta_0(d, \delta)$ mentioned before Assumption 16.2.4) there exists $R_0 \in (0, 1]$ ($K_0 R_0^\gamma \leq \theta_0$) such that Assumption 16.2.4 (iv) is satisfied for F_K with $K_1 = 2\omega(R_0)$ and $\hat{R}_0 = \infty$.

Assumption 16.2.4 (v) is also obviously satisfied.

It follows that Theorem 16.6.4 is applicable to H_K, which immediately (along with assertion (iii)) yields estimate (6) for u_K.

For u_{-K} it is obtained again by replacing $H(\mathsf{u},t,x)$ with $-H(-\mathsf{u},t,x)$. The lemma is proved. □

7. LEMMA. *Under the assumptions of this section, for $K\to\infty$ we have $|u_K - u_{-K}|\to 0$ uniformly in $\bar\Pi$.*

This lemma is proved in Section 2.

Proof of Theorem 2. First we prove uniqueness. Introduce $\psi\in C^2(\mathbb{R}^d)$ as a global barrier for Ω, that is, in Ω we have $\psi\ge 1$ and

$$a_{ij}D_{ij}\psi + b_iD_i\psi \le -1$$

for any $(a_{ij})\in\mathbb{S}_{\bar\delta}$, $|(b_i)|\le K_0$. Such a ψ can be found in the form $\cosh\mu R - \cosh\mu|x|$ for sufficiently large μ and R (see Lemma 5.2.6).

Then we take and fix a radially symmetric with respect to x function $\zeta=\zeta(t,x)$ of class $C_0^\infty(\mathbb{R}^{d+1})$ with support in $(-1,0)\times B_1$ and unit integral. For $\varepsilon>0$ we define $\zeta_\varepsilon(t,x)=\varepsilon^{-d-2}\zeta(\varepsilon^{-2}t,\varepsilon^{-1}x)$ and for locally summable $u(t,x)$ introduce

$$u^{(\varepsilon)}(t,x)=u(t,x)*\zeta_\varepsilon(t,x). \tag{7}$$

Fix a sufficiently small $\varepsilon_0>0$ such that $\Pi^{\varepsilon_0}\ne\emptyset$ and for $\varepsilon\in(0,\varepsilon_0]$ define

$$\xi_{\varepsilon,K}=\partial_t u_K^{(\varepsilon)}+\max\big(H\big[u_K^{(\varepsilon)}\big],P\big[u_K^{(\varepsilon)}\big]-K\big)$$

in Π^{ε_0}. Since the second-order derivatives with respect to x and the first derivative with respect to t of u_K are in $L_p(\Pi^\rho)$ for any p and ρ, we have $\xi_{\varepsilon,K}\to 0$ as $\varepsilon\downarrow 0$ in any $L_p(\Pi^{\varepsilon_0})$ for any K and any $p>1$. Furthermore, $\xi_{\varepsilon,K}$ are continuous *because $H(\mathsf{u},t,x)$ is continuous*. Therefore, there exist smooth functions $\zeta_{\varepsilon,K}$ on $\bar\Pi^{\varepsilon_0}$ such that

$$-\varepsilon\le\xi_{\varepsilon,K}+\zeta_{\varepsilon,K}\le 0$$

in Π^{ε_0}.

It is not hard to see that Ω^{ε_0} satisfies the exterior ball condition (with the same ρ_0). Therefore, by Theorem 12.1.10 there exists

$$w_{\varepsilon,K}\in\bigcap_{\rho>\varepsilon_0,p>1}W_p^{1,2}(\Pi^\rho)\cap C(\bar\Pi^{\varepsilon_0})$$

satisfying

$$\partial_t w_{\varepsilon,K}+\sup_{\substack{a\in\mathbb{S}_{\bar\delta},|b|\le K_0\\0\le c\le K_0}}\big[a_{ij}D_{ij}w_{\varepsilon,K}+b_iD_iw_{\varepsilon,K}-cw_{\varepsilon,K}\big]=\zeta_{\varepsilon,K}$$

in Π^{ε_0} (a.e.) and such that $w_{\varepsilon,K}=0$ on $\partial'\Pi^{\varepsilon_0}$. By the maximum principle such $w_{\varepsilon,K}$ is unique. Then owing to the continuity of $\zeta_{\varepsilon,K}$, for any ε and K, for all sufficiently small $\beta>0$, we have

$$\partial_t w_{\varepsilon,K}^{(\beta)}+\sup_{\substack{a\in\mathbb{S}_{\bar\delta},|b|\le K_0\\0\le c\le K_0}}\big[a_{ij}D_{ij}w_{\varepsilon,K}^{(\beta)}+b_iD_iw_{\varepsilon,K}^{(\beta)}-cw_{\varepsilon,K}^{(\beta)}\big]\le\zeta_{\varepsilon,K}^{(\beta)}\le\zeta_{\varepsilon,K}+\varepsilon \tag{8}$$

in $\Pi^{2\varepsilon_0}$.

For $w^{\beta}_{\varepsilon,K}(t,x) := w^{(\beta)}_{\varepsilon,K} + \varepsilon(T-t)$ now, obviously,

$$\partial_t\big(u^{(\varepsilon)}_K + w^{\beta}_{\varepsilon,K}\big) + \max\big(H\big[u^{(\varepsilon)}_K + w^{\beta}_{\varepsilon,K}\big], P\big[u^{(\varepsilon)}_K + w^{\beta}_{\varepsilon,K}\big] - K\big)$$

$$\le \partial_t u^{(\varepsilon)}_K + \max\big(H\big[u^{(\varepsilon)}_K\big], P\big[u^{(\varepsilon)}_K\big] - K\big) + \partial_t w^{\beta}_{\varepsilon,K}$$

$$+ \sup_{\substack{a\in\mathbb{S}_{\bar\delta},|b|\le K_0\\ 0\le c\le K_0}} \big[a_{ij}D_{ij}w^{(\beta)}_{\varepsilon,K} + b_iD_iw^{(\beta)}_{\varepsilon,K} - cw^{(\beta)}_{\varepsilon,K}\big] \le \xi_{\varepsilon,K} + \zeta_{\varepsilon,K} \le 0 \qquad (9)$$

in $\Pi^{2\varepsilon_0}$. Hence, in $\Pi^{2\varepsilon_0}$

$$\partial_t\big(u^{(\varepsilon)}_K + w^{\beta}_{\varepsilon,K}\big) + H\big[u^{(\varepsilon)}_K + w^{\beta}_{\varepsilon,K}\big] \le 0.$$

Since $H[v] - H[v+\beta\psi] = \beta(a^{ij}D_{ij}\psi + b^iD_i\psi - c\psi)$ for some $\mathbb{S}_\delta$-valued a, $\mathbb{R}^d$-valued b, such that $|b| \le K_0$, and $c \ge 1$, we have in $\Pi^{2\varepsilon_0}$ that

$$\partial_t\big(u^{(\varepsilon)}_K + w^{\beta}_{\varepsilon,K}\big) + H\big[u^{(\varepsilon)}_K + w^{\beta}_{\varepsilon,K} + \beta\psi\big] \le -\beta < 0.$$

This and the definition of C-viscosity solutions imply that, if v is a continuous in $\bar\Pi$, C-viscosity solution of $\partial_t v + H[v] = 0$ with boundary data g, then the minimum of $u^{(\varepsilon)}_K + w^{\beta}_{\varepsilon,K} + \beta\psi - v$ in $\bar\Pi^{2\varepsilon_0}$ is either nonnegative or is attained on the parabolic boundary of $\Pi^{2\varepsilon_0}$. The same conclusion holds after letting first $\beta \downarrow 0$ and then $\varepsilon \downarrow 0$. Combining this with the Aleksandrov estimates showing that $w_{\varepsilon,K} \to 0$ as $\varepsilon \downarrow 0$ uniformly on $\bar\Pi^{\varepsilon_0}$, we get that in Π

$$u_K - v \ge -\sup_{\Pi\setminus\Pi^{2\varepsilon_0}} |u_K - v|,$$

which after letting $\varepsilon_0 \downarrow 0$ and then $K \to \infty$ yields $v \le u$, where u is the common limit of u_K, u_{-K}, which exists by Lemma 7. By comparing v with u_{-K}, we get $v \ge u$, and hence uniqueness. After that estimate (2) follows immediately from (6), and the theorem is proved. □

2. Proof of Lemma 1.7

To prove Lemma 1.7 we need an auxiliary result. In the following theorem Ω can be just any bounded domain. Below by $C^{1,2}_{\text{loc}}(\Pi)$ we mean, as usual, the space of functions $u = u(t,x)$ which are continuous in Π along with their derivatives $\partial_t u$, $D_{ij}u$, D_iu. We recall that κ is introduced before Assumption 1.1 and fix a constant

$$\tau \in (0,1).$$

1. THEOREM. *Let $u, v \in C^{1,2}_{\text{loc}}(\Pi) \cap C(\bar\Pi)$ be such that for a constant $K \ge 1$*

$$\partial_t u + \max\big(H[u], P[u] - K\big) \ge 0 \ge \partial_t v + \min\big(H[v], -P[-v] + K\big) \qquad (1)$$

in Π and $v \ge u$ on $\partial'\Pi$. Also assume that, for a constant $M \in [1,\infty)$,

$$\|u, v\|_{C^{\kappa}(\Pi)} \le M. \qquad (2)$$

Then there exist a constant $N \in (0,\infty)$, depending only on τ, the diameter of Ω, d, K_0, $\bar H$, and δ, and a constant $\eta > 0$, depending only on τ, d, and δ, such that, if $K \ge NM^{\eta}$ and

$$K \ge T^{-1}, \quad \Pi^{2/\sqrt{K}} \ne \emptyset, \qquad (3)$$

then in Π

$$u(t,x) - v(t,x) \leq NK^{-(\kappa-1)/4} + NM\omega(M^{-1/\tau}K^{-1}). \tag{4}$$

2. REMARK. The purpose of introducing τ is that for $\omega = t^\tau$ estimate (4) becomes $u - v \leq NK^{-(\kappa-1)/4} + NK^{-\tau}$, which was used in [**78**] to estimate the rate of convergence of finite-difference approximations for (16.1.1).

To prove this theorem, we are going to adapt to our situation an argument from Section 5.A of [**23**]. For that we need a construction and two lemmas. From the start throughout the section we will only concentrate on K satisfying (3).

We take and fix a function $\zeta = \zeta(t,x)$ as before (1.7) and use the notation $u^{(\varepsilon)}$ introduced in (1.7). Recall some standard properties of parabolic mollifiers in which no regularity properties of Ω are required: If $u \in C^\kappa(\Pi)$, then in Π^ε

$$\varepsilon^{-\kappa}|u - u^{(\varepsilon)}| + \varepsilon^{-(\kappa-1)}|Du - Du^{(\varepsilon)}| \leq N\|u\|_{C^\kappa(\Pi)},$$

$$|u^{(\varepsilon)}| + |Du^{(\varepsilon)}| + \varepsilon^{2-\kappa}|D^2u^{(\varepsilon)}| + \varepsilon^{2-\kappa}|\partial_t u^{(\varepsilon)}|$$

$$+\varepsilon^{3-\kappa}|D^3u^{(\varepsilon)}| + \varepsilon^{3-\kappa}|D\partial_t u^{(\varepsilon)}| + \varepsilon^{4-\kappa}|\partial_t D^2 u^{(\varepsilon)}| \leq N\|u\|_{C^\kappa(\Pi)}, \tag{5}$$

where the constants N depend only on d and κ.

Next, take the constants $\nu, \varepsilon_0 \in (0,1)$, specified below in Lemma 4 and (31), respectively, depending only on d, K_0, $\bar{H}$, δ, and the diameter of Ω, introduce

$$\varepsilon = \varepsilon_0 M^{-1/(\kappa-1)}K^{-(1-\gamma)/(2\gamma)},$$

and consider the function

$$W(t,x,y) = u(t,x) - u^{(\varepsilon)}(t,x) - \big[v(t,y) - u^{(\varepsilon)}(t,y)\big] - \nu K|x-y|^2$$

for $(t,x),(t,y) \in \bar{\Pi}^\varepsilon$. Note that $\Pi^\varepsilon \neq \emptyset$ and even $\Pi^{2\varepsilon} \neq \emptyset$ owing to (3) and the fact that $1-\gamma > \gamma$ and $K, M \geq 1$.

We will need the following simple observation.

3. LEMMA. *For any* $\chi > 0$ *there exists* $N = N(\chi, d, \delta)$ *such that, if* $K \geq N$, *then*

$$\varepsilon M \leq \chi K^{-(\kappa-1)/4}. \tag{6}$$

Proof. Since $\varepsilon_0 < 1$, $\kappa \leq 2$, and $M \geq 1$, the left-hand side of (6) is less than

$$K^{-(1-\gamma)/(2\gamma)} = K^{-(5-\kappa)/(14-6\kappa)}.$$

One easily checks that $(5-\kappa)/(14-6\kappa) > 1/2 > (\kappa-1)/4$ for $\kappa \in (1,2)$ and this proves the lemma. $\square$

Denote by $(\bar t,\bar x,\bar y)$ a maximum point of W in $[0,T-\varepsilon^2]\times(\bar\Omega^\varepsilon)^2$. Observe that, obviously,

$$u(\bar t,\bar x)-u^{(\varepsilon)}(\bar t,\bar x)-\big[v(\bar t,\bar y)-u^{(\varepsilon)}(\bar t,\bar y)\big]-\nu K|\bar x-\bar y|^2$$
$$\geq u(\bar t,\bar x)-u^{(\varepsilon)}(\bar t,\bar x)-\big[v(\bar t,\bar x)-u^{(\varepsilon)}(\bar t,\bar x)\big],$$

which implies that

$$|\bar x-\bar y|\leq NM/(\nu K). \tag{7}$$

where and below by N with indices or without them we denote various constants depending only on d, K_0, $\bar H$, δ, τ, and the diameter of Ω, unless specifically stated otherwise. By the way, recall that κ and, hence, γ depend only on d and δ.

4. Lemma. *There exist a constant $\nu\in(0,1)$, depending only on d, K_0, $\bar H$, δ, and the diameter of Ω, and a constant $N\in[1,\infty)$ such that, if*

$$K\geq N\nu^{-\eta_1}\varepsilon_0^{(\kappa-2)\eta_1}M^{\eta_1(2-\kappa)/(\kappa-1)}, \tag{8}$$

where $\eta_1^{-1}=1-(2-\kappa)(1-\gamma)(2\gamma)^{-1}$ (>0), and $\bar x,\bar y\in\Omega^\varepsilon$ and $\bar t<T-\varepsilon^2$, then

(i) we have

$$2\nu|\bar x-\bar y|\leq NMK^{-1}\varepsilon^{\kappa-1}=N\varepsilon_0^{\kappa-1}K^{-(3+\kappa)/(8\gamma)},\quad |\bar x-\bar y|\leq\varepsilon/2; \tag{9}$$

(ii) for any $\xi,\eta\in\mathbb{R}^d$

$$D_{ij}[u-u^{(\varepsilon)}](\bar t,\bar x)\xi^i\xi^j-D_{ij}[v-u^{(\varepsilon)}](\bar t,\bar y)\eta^i\eta^j\leq 2\nu K|\xi-\eta|^2, \tag{10}$$

$$\partial_t[u-u^{(\varepsilon)}](\bar t,\bar x)\leq\partial_t[v-u^{(\varepsilon)}](\bar t,\bar y); \tag{11}$$

(iii) we have

$$\partial_t u(\bar t,\bar x)+H[u](\bar t,\bar x)\geq 0, \tag{12}$$

$$\partial_t v(\bar t,\bar y)+H[v](\bar t,\bar y)\leq 0. \tag{13}$$

Proof. The first inequality in (9) follows from (5) and the fact that the first derivatives of W with respect to x vanish at $\bar x$, that is,

$$D[\bar u-\bar u^{(\varepsilon)}](\bar t,\bar x)=2\nu K(\bar x-\bar y).$$

Also the matrix of second-order derivatives of W with respect to (x,y) is nonpositive at $(\bar t,\bar x,\bar y)$ as well as its (at least one sided if $\bar t=0$) derivative with respect to t, which yields (ii).

By taking $\eta=0$ in (10) and using the fact that $|D^2u^{(\varepsilon)}|\leq NM\varepsilon^{\kappa-2}$ we see that

$$D^2u(\bar t,\bar x)\leq 2\nu K+NM\varepsilon^{\kappa-2}.$$

Similarly,

$$D^2v(\bar t,\bar x)\geq -N(\nu K+M\varepsilon^{\kappa-2}),$$

which yields

$$P[u](\bar{t},\bar{x}) \le N_1(\nu K + M\varepsilon^{\kappa-2}), \quad -P[-v](\bar{t},\bar{y}) \ge -N_1(\nu K + M\varepsilon^{\kappa-2}).$$

Since $H(\mathsf{u},t,x) = a^{ij}\mathsf{u}''_{ij} + H(\mathsf{u}',0,t,x)$ where $(a^{ij}) \in \mathbb{S}_\delta$ ((a^{ij}) depends on u,t,x) and $|H(\mathsf{u}',0,t,x)| \le K_0|\mathsf{u}'| + \bar{H}$ and $M \ge 1$ and $\varepsilon < 1$, we also have (by increasing the above N_1 if necessary)

$$H[u](\bar{t},\bar{x}) \le N_1(\nu K + M\varepsilon^{\kappa-2}), \quad H[v](\bar{t},\bar{y}) \ge -N_1(\nu K + M\varepsilon^{\kappa-2}). \tag{14}$$

Also it follows from (11) and (5) that

$$\partial_t u(\bar{t},\bar{x}) \le \partial_t v(\bar{t},\bar{y}) + N_2 M\varepsilon^{\kappa-2}. \tag{15}$$

Now, if $H[u](\bar{t},\bar{x}) \le P[u](\bar{t},\bar{x}) - K$, then at $(\bar{t},\bar{x})$

$$0 \le \partial_t u + \max\big(H[u], P[u] - K\big) \le \partial_t u + N_1(\nu K + M\varepsilon^{\kappa-2}) - K,$$

$$\partial_t u \ge K - N_1(\nu K + M\varepsilon^{\kappa-2})$$

and at $(\bar{t},\bar{y})$

$$0 \ge \partial_t v + \min\big(H[v], -P[-v] + K\big) \ge K - 2N_1(\nu K + M\varepsilon^{\kappa-2}) - N_2 M\varepsilon^{\kappa-2}.$$

Here $M\varepsilon^{\kappa-2} \le \nu K$, which is equivalent to (8) with $N = 1$. Hence

$$K \le 4N_1\nu K + N_2\nu K,$$

which is impossible if we choose and fix ν such that

$$(4N_1 + N_2)\nu \le 1/2.$$

It follows that $H[u](\bar{t},\bar{x}) \ge P[u](\bar{t},\bar{x}) - K$,

$$\partial_t u(\bar{t},\bar{x}) + H[u](\bar{t},\bar{x}) \ge 0,$$

and this proves (12).

Similarly, if $-P[-v](\bar{t},\bar{y}) + K \le H[v](\bar{t},\bar{y})$, then at $(\bar{t},\bar{y})$

$$0 \ge \partial_t v - N_1(\nu K + M\varepsilon^{\kappa-2}) + K,$$

and at $(\bar{t},\bar{x})$ we have $\partial_t u \le N_1(\nu K + M\varepsilon^{\kappa-2}) - K + N_2 M\varepsilon^{\kappa-2}$ and

$$0 \le \partial_t u + \max\big(H[u], P[u] - K\big) \le -K + 2N_1(\nu K + M\varepsilon^{\kappa-2}) + N_2 M\varepsilon^{\kappa-2},$$

which again is impossible with the above choice of ν for K satisfying (8). This yields (13).

Moreover, not only $M\varepsilon^{\kappa-2} \le \nu K$ for K satisfying (8), but we also have $NM\varepsilon^{\kappa-2} \le \nu K$, where N is taken from (9), if we increase N in (8). This yields the second inequality in (9).

The lemma is proved. □

Everywhere below in this section ν is the constant from Lemma 4.

5. LEMMA. *There exist a constant N such that, for any $\mu \geq 0$, if $K \geq NM^{\eta_2}$, where $\eta_2 = 8/(5-\kappa)$, and*

$$W(\bar{t},\bar{x},\bar{y}) \geq 2K^{-(\kappa-1)/4} + \mu M\omega(M^{-1/\tau}K^{-1}), \tag{16}$$

then

$$u(\bar{t},\bar{x}) - v(\bar{t},\bar{y}) - \nu K|\bar{x}-\bar{y}|^2 \geq K^{-(\kappa-1)/4} + \mu M\omega(M^{-1/\tau}K^{-1}). \tag{17}$$

Furthermore, $\bar{x},\bar{y} \in \Omega^{2\varepsilon}$ and $\bar{t} < T - \varepsilon^2$.

Proof. It follows from (7) that (recall that ν is already fixed)

$$\left|u^{(\varepsilon)}(\bar{t},\bar{x}) - u^{(\varepsilon)}(\bar{t},\bar{y})\right| \leq M|\bar{x}-\bar{y}| \leq N_0 M^2/K. \tag{18}$$

Hence we have from (16) that

$$u(\bar{t},\bar{x}) - v(\bar{t},\bar{y}) - \nu K|\bar{x}-\bar{y}|^2 \geq 2K^{-(\kappa-1)/4} - N_0M^2/K + \mu M\omega(M^{-1/\tau}K^{-1}),$$

and (17) follows provided that

$$N_0M^2/K \leq (1/4)K^{-(\kappa-1)/4}, \tag{19}$$

which indeed holds if

$$K \geq NM^{\eta_2}. \tag{20}$$

Now, if $\bar{x}$ or $\bar{y}$ is in $\bar{\Omega}^{\varepsilon}\setminus\Omega^{2\varepsilon}$, then for appropriate $\hat{x}\in\partial\Omega$ and $\hat{y}\in\partial\Omega$ either

$$u(\bar{t},\bar{x}) - v(\bar{t},\bar{y}) \leq u(\bar{t},\bar{x}) - u(\bar{t},\hat{x}) + v(\bar{t},\hat{x}) - v(\bar{t},\bar{y}) \leq M\big(4\varepsilon + |\bar{x}-\bar{y}|\big)$$

or

$$u(\bar{t},\bar{x}) - v(\bar{t},\bar{y}) \leq \bar{u}(\bar{t},\bar{x}) - u(\bar{t},\hat{y}) + v(\bar{t},\hat{y}) - v(\bar{t},\bar{y}) \leq M\big(4\varepsilon + |\bar{x}-\bar{y}|\big).$$

In any case in light of (17), (7), and (19)

$$4\varepsilon M + N_0M^2/K - \nu K|\bar{x}-\bar{y}|^2 \geq K^{-(\kappa-1)/4},$$

$$4\varepsilon M \geq (3/4)K^{-(\kappa-1)/4}. \tag{21}$$

Owing to Lemma 3, inequality (21) is impossible if $K \geq N$ or, upon adjusting N in (20) appropriately, if (20) holds (recall that $M \geq 1$). Below we assume that (20) holds after the modification, so that we have $\bar{x},\bar{y}\in\Omega^{\varepsilon}$.

Furthermore, if $\bar{t} = T - \varepsilon^2$, then (recall (18) and that $u \leq v$ on $\partial'\Pi$)

$$W(\bar{t},\bar{x},\bar{y}) \leq NM\varepsilon^{\kappa} + v(\bar{t},\bar{x}) - v(\bar{t},\bar{y}) + NM^2/K$$

$$\leq NM\varepsilon + M|\bar{x}-\bar{y}| + NM^2/K \leq NM^2/K + NM\varepsilon,$$

which is less than $(1/2)K^{-(\kappa-1)/4}$ in light of (19) and Lemma 3, again after perhaps further adjusting the constant in (20). This is impossible due to (16). Hence, $\bar{t} < T - \varepsilon^2$ and this finishes the proof of the lemma. □

We also need a simple result based on solving quadratic inequalities.

6. LEMMA. *Let $\beta \in [0, \delta/2]$. Then for any $\alpha \geq -4\nu$ we have*

$$\beta\alpha - \delta\frac{\alpha^2}{\alpha + 4\nu} \leq 2(\nu/\delta)\beta^2.$$

Proof of Theorem 1. Fix a (large) constant $\mu > 0$ to be specified later as a constant, depending only on d, K_0, $\bar{H}$, δ, τ, and the diameter of Ω, recall that ν is found in Lemma 4 and first assume that

$$W(t,x,y) \leq 2K^{-(\kappa-1)/4} + \mu M\omega(M^{-1/\tau}K^{-1})$$

for $(t,x), (t,y) \in \bar{\Pi}^\varepsilon$. Observe that for any point $(t,x) \in \bar{\Pi}$ one can find a point $(s,y) \in \bar{\Pi}^\varepsilon$ with $|x-y| \leq \varepsilon$ and $|t-s| \leq \varepsilon^2$ and then

$$u(t,x) - v(t,x) \leq u(s,y) - v(s,y) + NM\varepsilon$$

$$\leq W(s,y,y) + NM\varepsilon \leq 2K^{-(\kappa-1)/4} + NM\varepsilon + \mu M\omega(M^{-1/\tau}K^{-1}).$$

In that case, as follows from Lemma 3, (4) holds for K satisfying (20) with any $N \geq 1$ in (20).

It is clear now that, to prove the theorem, it suffices to find μ such that the inequality (16) is impossible if $K \geq NM^\eta$ with N and η as in the statement of the theorem and at least not smaller than those in (20). Of course, we will argue by contradiction and suppose that (16) holds, despite (20) is valid and (8) is satisfied with ν fixed in Lemma 4 and $\varepsilon_0 \in (0,1)$, which is yet to be specified.

Then (17) holds, and, in particular,

$$u(\bar{t},\bar{x}) \geq v(\bar{t},\bar{y}). \tag{22}$$

Also by Lemma 5 the points $\bar{x}, \bar{y}$ are in Ω^ε (even in $\Omega^{2\varepsilon}$) and $\bar{t} < T - \varepsilon^2$, so that we can use the conclusions of Lemma 4.

Denote

$$A = D^2[u - u^{(\varepsilon)}](\bar{t},\bar{x}), \quad B = D^2[v - u^{(\varepsilon)}](\bar{t},\bar{y})$$

and interpret matrices as linear operators in a usual way and constants as operators of multiplications by these constants. Observe that (10) implies that the operator $B + 2\nu K$ is nonnegative and, hence, $B + 4\nu K$ is strictly positive. Then for

$$\eta = 4\nu K(B + 4\nu K)^{-1}\xi, \quad \eta - \xi = -B(B + 4\nu K)^{-1}\xi$$

inequality (10) yields

$$\langle A\xi, \xi\rangle \leq \big\langle B4\nu K(B + 4\nu K)^{-1}\xi, 4\nu K(B + 4\nu K)^{-1}\xi\big\rangle$$

$$+ 4\nu K\big|B(B + 4\nu K)^{-1}\xi\big|^2$$

$$= \big\langle 4\nu KB(B + 4\nu K)^{-1}\xi, \xi\big\rangle.$$

Hence $A \le 4\nu KB(B+4\nu K)^{-1}$ and

$$D^2u(\bar t,\bar x) \le D^2u^{(\varepsilon)}(\bar t,\bar x) + 4\nu KB(B+4\nu K)^{-1}$$

$$= D^2v(\bar t,\bar y) - B + D^2u^{(\varepsilon)}(\bar t,\bar x) - D^2u^{(\varepsilon)}(\bar t,\bar y) + 4\nu KB(B+4\nu K)^{-1}$$

$$= D^2v(\bar t,\bar y) - B^2(B+4\nu K)^{-1} + D^2u^{(\varepsilon)}(\bar t,\bar x) - D^2u^{(\varepsilon)}(\bar t,\bar y).$$

We now use (9) to get that $|\bar x - \bar y| \le NMK^{-1}\varepsilon^{\kappa-1}$, and in light of (5) that

$$\big|D^2u^{(\varepsilon)}(\bar t,\bar x) - D^2u^{(\varepsilon)}(\bar t,\bar y)\big| \le NM\varepsilon^{\kappa-3}|\bar x-\bar y| \le NC,$$

where

$$C = M^2K^{-1}\varepsilon^{2\kappa-4}.$$

Also (11) reads

$$\partial_t u(\bar t,\bar x) \le \partial_t v(\bar t,\bar y) + \partial_t u^{(\varepsilon)}(\bar t,\bar x) - \partial_t u^{(\varepsilon)}(\bar t,\bar y)$$

and as is easy to see

$$\big|\partial_t u^{(\varepsilon)}(\bar t,\bar x) - \partial_t u^{(\varepsilon)}(\bar t,\bar y)\big| \le NC.$$

Thus far, we have

$$D^2u(\bar t,\bar x) \le D^2v(\bar t,\bar y) - B^2(B+4\nu K)^{-1} + NC,$$

$$\partial_t u(\bar t,\bar x) \le \partial_t v(\bar t,\bar y) + NC. \tag{23}$$

Next,

$$D_iu(\bar t,\bar x) = 2\nu K(\bar x^i - \bar y^i) + D_iu^{(\varepsilon)}(\bar t,\bar x),$$

$$D_iv(\bar t,\bar y) = 2\nu K(\bar x^i - \bar y^i) + D_iu^{(\varepsilon)}(\bar t,\bar y),$$

$$D_iu(\bar t,\bar x) - D_i\bar v(\bar t,\bar y) = D_iu^{(\varepsilon)}(\bar t,\bar x) - D_iu^{(\varepsilon)}(\bar t,\bar y),$$

where

$$\big|Du^{(\varepsilon)}(\bar t,\bar x) - D\ u^{(\varepsilon)}(\bar t,\bar y)\big| \le NM\varepsilon^{\kappa-2}|\bar x-\bar y| \le NC$$

and, therefore,

$$\big|Du(\bar t,\bar x) - Dv(\bar t,\bar y)\big| \le NC. \tag{24}$$

It follows from (23), (24), (12), and (22) by Assumption 1.1 and the fact that $D_{\mathsf{u}_0'}H \le -1$ (see (1.3)) that

$$0 \le \partial_t u(\bar t,\bar x) + H\big(u(\bar t,\bar x), Du(\bar t,\bar x), D^2v(\bar t,\bar y) - B^2(B+4\nu K)^{-1} + NC, \bar t, \bar x\big)$$

$$\le \partial_t v(\bar t,\bar y) + NC + H\big(v(\bar t,\bar y), Dv(\bar t,\bar y), D^2v(\bar t,\bar y) + NC, \bar t, \bar x\big)$$

$$- \big[u(\bar t,\bar x) - v(\bar t,\bar y)\big] - \delta \operatorname{tr} B^2(B+4\nu K)^{-1}$$

$$\le \partial_t v(\bar t,\bar y) + NC + H\big(v(\bar t,\bar y), Dv(\bar t,\bar y), D^2v(\bar t,\bar y), \bar t, \bar x\big)$$

$$- \big[u(\bar t,\bar x) - v(\bar t,\bar y)\big] - \delta \operatorname{tr} B^2(B+4\nu K)^{-1} + NC.$$

This along with (13) and Assumption 1.1 (iii) yields (recall that $M \ge 1$)

$$u(\bar{t},\bar{x})-v(\bar{t},\bar{y})\le K_0|\bar{x}-\bar{y}|^{\gamma}\big|D^2v(\bar{t},\bar{y})\big|+NM\omega\big(|\bar{x}-\bar{y}|\big)$$

$$-\,\delta\operatorname{tr}B^2(B+4\nu K)^{-1}+NC.$$

Upon combining this with (17) we arrive at

$$2K^{-(\kappa-1)/4}+\mu M\omega\big(M^{-1/\tau}K^{-1}\big)\le K_0|\bar{x}-\bar{y}|^{\gamma}\big|D^2v(\bar{t},\bar{y})\big|+NC$$

$$+N_1M\omega\big(|\bar{x}-\bar{y}|\big)-\delta\operatorname{tr}B^2(B+4\nu K)^{-1}. \tag{25}$$

Now we choose $\mu=N_1$ and observe that (see (9))

$$|\bar{x}-\bar{y}|\le N_2K^{-(3+\kappa)/(8\gamma)}\le M^{-1/\tau}K^{-1}$$

for

$$K^{\theta_1}\ge N_2M^{1/\tau}, \tag{26}$$

where $\theta_1=(\kappa-1)(5-\kappa)(14-6\kappa)^{-1}$. Then for such K we infer from (25) that

$$2K^{-(\kappa-1)/4}\le K_0|\bar{x}-\bar{y}|^{\gamma}\big|D^2v(\bar{t},\bar{y})\big|-\delta\operatorname{tr}B^2(B+4\nu K)^{-1}+N_3C. \tag{27}$$

Here, owing to (5) and (9),

$$|\bar{x}-\bar{y}|^{\gamma}\big|D^2v(\bar{t},\bar{y})\big|\le N\varepsilon_0^{(\kappa-1)\gamma}K^{-(3+\kappa)/8}\big(|B|+\big|D^2u^{(\varepsilon)}(\bar{t},\bar{y})\big|\big)$$

$$\le N\varepsilon_0^{(\kappa-1)\gamma}K^{-(3+\kappa)/8}|B|+N_4M^{1/(\kappa-1)}\varepsilon_0^{(\kappa-1)\gamma+\kappa-2}K^{\theta_2(\kappa-1)/4},$$

where $\theta_2=(7\kappa-19)/(14-6\kappa)$, and as is easy to see $\theta_2\le-3/2$ for $\kappa\in[1,2]$, so that

$$N_4M^{1/(\kappa-1)}\varepsilon_0^{(\kappa-1)\gamma+\kappa-2}K^{\theta_2(\kappa-1)/4}\le(1/4)K^{-(\kappa-1)/4}$$

if, for instance,

$$K\ge\big(4N_4M^{1/(\kappa-1)}\varepsilon_0^{(\kappa-1)\gamma+\kappa-2}\big)^{8/(\kappa-1)}. \tag{28}$$

In what concerns the last term in (27), note that

$$N_3C=N_3M^{2/(\kappa-1)}\varepsilon_0^{2\kappa-4}K^{\theta_3(\kappa-1)/4},$$

where $\theta_3=(4\kappa-12)/(7-3\kappa)\le-2$ for $\kappa\in[1,2]$, so that

$$N_3C\le(1/4)K^{-(\kappa-1)/4}$$

if

$$K\ge\big(4N_3M^{2/(\kappa-1)}\varepsilon_0^{2\kappa-4}\big)^{4/(\kappa-1)}. \tag{29}$$

We conclude that, for K satisfying (28) and (29), relation (27) yields

$$K^{-(\kappa-1)/4}\le N\varepsilon_0^{(\kappa-1)\gamma}K^{-(3+\kappa)/8}|B|-\delta\operatorname{tr}B^2(B+4\nu K)^{-1}. \tag{30}$$

Next, observe that, by Lemma 6 applied after we diagonalize B and set $B = K\alpha$ implies that the right-hand side of (30) is

$$K\Big[N\varepsilon_0^{(\kappa-1)\gamma}K^{-(3+\kappa)/8}|\alpha| - \delta\operatorname{tr}\alpha^2(\alpha+4\nu)^{-1}\Big]$$
$$\le N_5\varepsilon_0^{2(\kappa-1)\gamma}K^{-(\kappa-1)/4},$$

where the inequality holds owing to Cauchy's inequality. We can certainly assume that $N_5 \ge 1$ and then we can choose $\varepsilon_0 \in (0,1)$ so that

$$N_5\varepsilon_0^{2(\kappa-1)\gamma} = 1/2, \tag{31}$$

which along with (30) leads to the desired contradiction:

$$K^{-(\kappa-1)/4} \le (1/2)K^{-(\kappa-1)/4}.$$

With so specified ε_0 we rewrite condition (8) (with ν fixed in Lemma 4), conditions (20), (26), (28), and (29) as

$$K \ge NM^{\eta_1(2-\kappa)/(\kappa-1)}, \quad K \ge NM^{\eta_2}, \quad K \ge NM^{1/(\theta_1\tau)},$$
$$K \ge NM^{8/(\kappa-1)^2}.$$

Since $M \ge 1$, for η' defined as the sum of the above powers of M and N' defined as the sum of the above N's, the inequality (16) is impossible for $K \ge N'M^{\eta'}$ and this brings the proof of the theorem to an end. □

Proof of Lemma 1.7. Fix a sufficiently small $\varepsilon_0 > 0$ such that $\Pi^{2\varepsilon_0} \ne \emptyset$ and for $\varepsilon \in (0, \varepsilon_0)$ define

$$\xi_{\varepsilon,K} = \partial_t u_K^{(\varepsilon)} + \max\big(H\big[u_K^{(\varepsilon)}\big], P\big[u_K^{(\varepsilon)}\big] - K\big),$$

$$\xi_{\varepsilon,-K} = \partial_t u_{-K}^{(\varepsilon)} + \min\big(H\big[u_{-K}^{(\varepsilon)}\big], -P\big[-u_{-K}^{(\varepsilon)}\big] + K\big)$$

in Π^{ε_0}. Since the second-order derivatives with respect to x and the first derivative with respect to t of $u_{\pm K}$ are in $L_p(\Pi^\rho)$ for any p and ρ, we have $\xi_{\varepsilon,\pm K} \to 0$ as $\varepsilon \downarrow 0$ in any $L_p(\Pi^{\varepsilon_0})$ for any K and any $p > 1$. Furthermore, $\xi_{\varepsilon,\pm K}$ are continuous. Therefore, there exist smooth functions $\zeta_{\varepsilon,K}$ on $\bar\Pi^{\varepsilon_0}$ such that

$$-\varepsilon \le \zeta_{\varepsilon,K} - \min\big(\xi_{\varepsilon,K}, -\xi_{\varepsilon,-K}\big) \le 0$$

in Π^{ε_0}.

By Theorem 12.1.10 there exists

$$w_{\varepsilon,K} \in \bigcap_{\rho>\varepsilon_0, p>1} W_p^{1,2}(\Pi^\rho) \cap C(\bar\Pi^{\varepsilon_0})$$

satisfying

$$\partial_t w_{\varepsilon,K} + \sup_{\substack{a\in\mathbb{S}_{\bar\delta}, |b|\le K_0\\ 0\le c\le K_0}}\big[a_{ij}D_{ij}w_{\varepsilon,K} + b_iD_iw_{\varepsilon,K} - cw_{\varepsilon,K}\big] = \zeta_{\varepsilon,K}$$

in Π^{ε_0} (a.e.) and such that $w_{\varepsilon,K} = 0$ on $\partial'\Pi^{\varepsilon_0}$. Then owing to the continuity of

$\zeta_{\varepsilon,K}$, for any ε and K for all sufficiently small $\beta > 0$, we have

$$\partial_t w^{(\beta)}_{\varepsilon,K} + \sup_{\substack{a\in\mathbb{S}_{\bar\delta},|b|\le K_0\\0\le c\le K_0}} \big[a_{ij}D_{ij}w^{(\beta)}_{\varepsilon,K} + b_iD_iw^{(\beta)}_{\varepsilon,K} - cw^{(\beta)}_{\varepsilon,K}\big] \le \zeta^{(\beta)}_{\varepsilon,K} \le \zeta_{\varepsilon,K} + \varepsilon$$

in $\Pi^{2\varepsilon_0}$.

For $w^{\beta}_{\varepsilon,K}(t,x) := w^{(\beta)}_{\varepsilon,K} + \varepsilon(T-t)$ now similarly to (1.9)

$$\partial_t\big(u^{(\varepsilon)}_K - w^{\beta}_{\varepsilon,K}\big) + \max\big(H\big[u^{(\varepsilon)}_K - w^{\beta}_{\varepsilon,K}\big], P\big[u^{(\varepsilon)}_K - w^{\beta}_{\varepsilon,K}\big] - K\big)$$

$$\ge \partial_t u^{(\varepsilon)}_K + \max\big(H\big[u^{(\varepsilon)}_K\big], P\big[u^{(\varepsilon)}_K\big] - K\big) + \varepsilon$$

$$- \sup_{\substack{a\in\mathbb{S}_{\bar\delta},|b|\le K_0\\0\le c\le K_0}} \big[a_{ij}D_{ij}w^{(\beta)}_{\varepsilon,K} + b_iD_iw^{(\beta)}_{\varepsilon,K} - cw^{(\beta)}_{\varepsilon,K}\big] \ge \xi_{\varepsilon,K} - \zeta_{\varepsilon,K} \ge 0,$$

$$\partial_t\big(u^{(\varepsilon)}_{-K} + w^{\beta}_{\varepsilon,K}\big) + \min\big(H\big[u^{(\varepsilon)}_{-K} + w^{\beta}_{\varepsilon,K}\big], -P\big[-u^{(\varepsilon)}_{-K} - w^{\beta}_{\varepsilon,K}\big] + K\big)$$

$$\le \partial_t u^{(\varepsilon)}_{-K} + \min\big(H\big[u^{(\varepsilon)}_{-K}\big], -P\big[-u^{(\varepsilon)}_{-K}\big] + K\big) + \partial_t w^{\beta}_{\varepsilon,K}$$

$$+ \sup_{\substack{a\in\mathbb{S}_{\bar\delta},|b|\le K_0\\0\le c\le K_0}} \big[a_{ij}D_{ij}w^{(\beta)}_{\varepsilon,K} + b_iD_iw^{(\beta)}_{\varepsilon,K} - cw^{(\beta)}_{\varepsilon,K}\big] \le \xi_{\varepsilon,-K} + \zeta_{\varepsilon,K} \le 0$$

in $\Pi^{2\varepsilon_0}$.

After setting

$$\mu^{\beta}_{\varepsilon,K} = \sup_{\partial'\Pi^{2\varepsilon_0}} \big(u^{(\varepsilon)}_K - u^{(\varepsilon)}_{-K} - 2w^{\beta}_{\varepsilon,K}\big)_+$$

we conclude by Theorem 1 applied to $u^{(\varepsilon)}_K - w^{\beta}_{\varepsilon,K}$ and $u^{(\varepsilon)}_{-K} + w^{\beta}_{\varepsilon,K} + \mu^{\beta}_{\varepsilon,K}$ in place of u and v, respectively, that there exist a constant $N \in (0,\infty)$, depending only on τ, the diameter of Ω, d, K_0, $\bar H$, and δ, and a constant $\eta > 0$, depending only on τ, d, and δ, such that, if $K \ge N\big(M^{\beta}_{\varepsilon,K}\big)^{\eta}$, then

$$u^{(\varepsilon)}_K - u^{(\varepsilon)}_{-K} \le \mu^{\beta}_{\varepsilon,K} + 2w^{\beta}_{\varepsilon,K} + NK^{-(\kappa-1)/4} + NM^{\beta}_{\varepsilon,K}\omega\Big(\big(M^{\beta}_{\varepsilon,K}\big)^{-1/\tau}K^{-1}\Big)$$

in $\Pi^{2\varepsilon_0}$, where $M^{\beta}_{\varepsilon,K} \ge 1$ is any number satisfying

$$M^{\beta}_{\varepsilon,K} \ge \big\|u^{(\varepsilon)}_K - w^{\beta}_{\varepsilon,K}, u^{(\varepsilon)}_{-K} + w^{\beta}_{\varepsilon,K} + \mu^{\beta}_{\varepsilon,K}\big\|_{C^{\kappa}(\Pi^{2\varepsilon_0})}.$$

By letting $\beta \downarrow 0$ we obviously obtain that, if $K \ge N\big(M_{\varepsilon,K}\big)^{\eta}$, then

$$u^{(\varepsilon)}_K - u^{(\varepsilon)}_{-K} \le \mu_{\varepsilon,K} + 2w_{\varepsilon,K} + NK^{-(\kappa-1)/4} + NM_{\varepsilon,K}\omega\big(M^{-1/\tau}_{\varepsilon,K}K^{-1}\big)$$

in $\Pi^{2\varepsilon_0}$, where

$$\mu_{\varepsilon,K} = \sup_{\partial'\Pi^{2\varepsilon_0}} \big(u^{(\varepsilon)}_K - u^{(\varepsilon)}_{-K} - 2w_{\varepsilon,K}\big)_+,$$

and $M_{\varepsilon,K} \ge 1$ is any number satisfying

$$M_{\varepsilon,K} \ge \big\|u^{(\varepsilon)}_K - w_{\varepsilon,K}, u^{(\varepsilon)}_{-K} + w_{\varepsilon,K} + \mu_{\varepsilon,K}\big\|_{C^{\kappa}(\Pi^{2\varepsilon_0})}.$$

First we discuss what is happening as $\varepsilon \downarrow 0$. By Theorem 12.1.7 and Lemma 12.1.9 we obtain $w_{\varepsilon,K} \to 0$ in $W^{1,2}_p(\Pi^{3\varepsilon_0/2})$ for any $p > 1$, which by embedding theorems implies that $w_{\varepsilon,K} \to 0$ in $C^\kappa(\Pi^{2\varepsilon_0})$. Obviously, the constants $\mu_{\varepsilon,K}$ converge in $C^\kappa(\Pi^{2\varepsilon_0})$ to

$$\sup_{\partial'\Pi^{2\varepsilon_0}} (u_K - u_{-K})_+.$$

Now Lemma 1.6 (iv) implies that for sufficiently small ε one can take $N\varepsilon_0^{-\kappa}$ as $M_{\varepsilon,K}$, where N is independent of K and ε_0. Thus, for sufficiently small ε, if $K \geq N\varepsilon_0^{-\kappa\eta}$, then

$$u^{(\varepsilon)}_K - u^{(\varepsilon)}_{-K} \leq \mu_{\varepsilon,K} + 2w_{\varepsilon,K} + NK^{-(\kappa-1)/4} + N\varepsilon_0^{-\kappa}\omega\big(N\varepsilon_0^{\kappa/\tau}K^{-1}\big)$$

in $\Pi^{2\varepsilon_0}$, which after letting $\varepsilon \downarrow 0$ yields

$$u_K - u_{-K} \leq NK^{-(\kappa-1)/4} + N\varepsilon_0^{-\kappa}\omega\big(N\varepsilon_0^{\kappa/\tau}K^{-1}\big) + \sup_{\partial'\Pi^{2\varepsilon_0}} (u_K - u_{-K})_+$$

in $\Pi^{2\varepsilon_0}$. Hence in Π

$$u_K - u_{-K} \leq NK^{-(\kappa-1)/4} + N\varepsilon_0^{-\kappa}\omega\big(N\varepsilon_0^{\kappa/\tau}K^{-1}\big) + \sup_{\Pi\setminus\Pi^{2\varepsilon_0}} (u_K - u_{-K})_+$$

$$\leq NK^{-(\kappa-1)/4} + N\varepsilon_0^{-\kappa}\omega\big(N\varepsilon_0^{\kappa/\tau}K^{-1}\big) + \xi(\varepsilon_0)$$

where $\xi(\varepsilon_0) \to 0$ as $\varepsilon_0 \downarrow 0$ by Lemma 1.6 (iii). This obviously proves the lemma because, as is noted in Lemma 1.6 (ii), we have $u_{-K} \leq u_K$. □

3. Existence of maximal and minimal L_p-viscosity solutions

Fix constants, $K_0, T \in (0,\infty)$, $p > d+2$, $\delta \in (0,1]$, and fix a nonnegative

$$\bar{H} \in L_p(\mathbb{R}^{d+1}).$$

Also according to the setting in Section 16.1 we take a bounded domain $\Omega \subset \mathbb{R}^d$ satisfying the exterior ball condition and set $\Pi = [0,T) \times \Omega$.

1. ASSUMPTION. (i) The function H is a nonincreasing function of u_0', is continuous with respect to u_0', uniformly with respect to other variables $[\mathsf{u}'], (t,x) \in \mathbb{R}^{d+1}$, $\mathsf{u}'' \in \mathbb{S}$, is measurable with respect to (t,x) for any u, and is Lipschitz continuous in $[\mathsf{u}']$ with Lipschitz constant independent of $\mathsf{u}_0', \mathsf{u}'', (t,x)$.

(ii) For any $\mathsf{u}', (t,x) \in \mathbb{R}^{d+1}$, the function $H(\mathsf{u},t,x)$ is Lipschitz continuous with respect to u'' and $D_{\mathsf{u}''}H \in \mathbb{S}_\delta$ at all points of differentiability of $H(\mathsf{u},t,x)$ with respect to u''.

(iii) for all $\mathsf{u}', (t,x) \in \mathbb{R}^{d+1}$

$$\big|H(\mathsf{u}',0,t,x)\big| \leq K_0|\mathsf{u}'| + \bar{H}(t,x).$$

2. ASSUMPTION. We are given a function $g \in C(\overline{\partial'\Pi})$.

Introduce

$$P(\mathsf{u}'') = \sup_{a\in\mathbb{S}_\delta} \operatorname{tr} a\mathsf{u}'', \tag{1}$$

define $P[u] = P(D^2u)$, take a constant $K > 0$, and recall that by Theorem 12.1.13 there exists $v \in W^{1,2}_{p,\mathrm{loc}}(\Pi) \cap C(\bar{\Pi})$ such that $u = g$ on $\partial'\Pi$ and the equation

$$\partial_t v + \max\big(H[v], P[v] - K\big) = 0 \tag{2}$$

holds in Π (a.e.). By the maximum principle the solutions $v = v_K$ are unique and decrease as $K \to \infty$.

3. THEOREM. *Under the above assumptions, as $K \to \infty$, v_K converges uniformly on $\bar{\Pi}$ to a continuous function v which is an L_{d+1}-viscosity solutions of* (16.1.1) *with boundary condition $v = g$ on $\partial'\Pi$. Furthermore, for any $q \in [1, \infty)$, v is greater than any L_q-viscosity subsolution of* (16.1.1) *of class $C(\bar{\Pi})$ with given boundary condition. In particular, v is the maximal L_{d+1}-viscosity solution of* (16.1.1) *of class $C(\bar{\Pi})$ with given boundary condition.*

4. REMARK. To obtain an L_{d+1}-viscosity solution which is smaller than L_q-viscosity supersolutions, it suffices to consider

$$\partial_t v + \min\big(H[v], -P[-v] + K\big) = 0, \tag{3}$$

which reduces to (2) if we replace v with $-v$ and $H(\mathsf{u}, t, x)$ with $-H(-\mathsf{u}, t, x)$.

This yields the following result.

5. THEOREM. *Let $v_{-K} \in W^{1,2}_{p,\mathrm{loc}}(\Pi) \cap C(\bar{\Pi})$ denote a unique solution of* (3) *(a.e.) in Π with boundary data $v_{-K} = g$ on $\partial'\Pi$. Then, as $K \to \infty$, v_{-K} converges uniformly on $\bar{\Pi}$ to a continuous function w which is an L_{d+1}-viscosity solutions of* (16.1.1) *with boundary condition $w = g$ on $\partial'\Pi$. Furthermore, for any $q \in [1, \infty)$, w is smaller than any L_q-viscosity supersolution of* (16.1.1) *of class $C(\bar{\Pi})$ with given boundary condition. In particular, w is the minimal L_{d+1}-viscosity solution of* (16.1.1) *of class $C(\bar{\Pi})$ with given boundary condition.*

6. REMARK. The existence of extremal C-viscosity solutions is proved in the elliptic and parabolic cases in [**24**] when H is a continuous function. Our function $H(\mathsf{u}, t, x)$ is just measurable in (t, x) and we are dealing with L_{d+1}-viscosity solutions.

Also note that the existence of the extremal L_p-viscosity solution for the elliptic case was proved in [**50**] with no continuity assumption on H with respect to x. We provide a method which in principle allows one to find it.

Here is a stability result for the extremal L_{d+1}-viscosity solutions. Take $\kappa_1 = \kappa_1(d, \delta) \in (1, 2)$ and $\theta = \theta(\kappa, d, \delta) \in (0, 1]$ that are introduced before Assumption 17.1.3. Then fix a κ satisfying

$$1 < \kappa < [2 - (d+2)/p] \wedge \kappa_1.$$

Observe that, in notation 17.1.1 we have $\bar{H}_\kappa < \infty$ by Remark 17.1.2.

7. THEOREM. *Let H_n, $n = 0, 1, ...$, satisfy Assumptions* 1 *and* 17.1.3 *with $R_0 \in (0, 1]$ independent of n, with K_0 and $\bar{H}$ from the beginning of the section, $\theta/2$ in place of θ (specified above) and have, perhaps, different Lipschitz constants with respect to $[\mathsf{u}']$ for different n. Suppose that H_0 is Lipschitz continuous in u with a constant independent of (t, x). Let v_n, $n = 0, 1, ...$, be the maximal L_{d+1}-viscosity solutions of class $C(\bar{\Pi})$ of $\partial_t v_n + H_n[v_n] = 0$ in Π with boundary condition $v_n = g_n$ on $\partial'\Pi$, where $g_n \in C(\bar{\Pi})$ and $g_n \to g_0$ in $C(\bar{\Pi})$ as $n \to \infty$.*

Assume that for any $M > 0$

$$\Delta_{n,M}(t, x) := \sup_{|\mathsf{u}| \le M} \big|H_n(\mathsf{u}, t, x) - H_0(\mathsf{u}, t, x)\big| \to 0 \tag{4}$$

in $L_{d+1}(\Pi)$ *as* $n \to \infty$. *Also assume that for all values of the arguments and* n

$$\big|H_n(\mathsf{u},t,x) - H_0(\mathsf{u},t,x)\big| \le \bar{H}(t,x)\big(1+|\mathsf{u}'|\big). \tag{5}$$

Then $v_n \to v_0$ *in* $C(\bar{\Pi})$ *as* $n \to \infty$. *The same holds true if* v_n *are minimal* L_{d+1}*-viscosity solutions of class* $C(\bar{\Pi})$.

Proof. According to Theorem 3, it suffices to show that

$$\sup_{K\ge 1}\big(v_{n,K} - v_{0,K}\big) \to 0, \tag{6}$$

in $C(\bar{\Pi})$, where $v_{n,K}$ are the solutions of

$$\partial_t v_{n,K} + \max\big(H_n[v_{n,K}], P[v_{n,K}] - K\big) = 0$$

in Π with boundary condition $v_{n,K} = g_n$ on $\partial'\Pi$.

By Remark 6.6.16, obviously applicable in the parabolic case,

$$\partial_t v_{n,K} + a^{ij}D_{ij}v_{n,K} + b^iD_iv_{n,K} + cv_{n,K} + \tau\bar{H} = 0 \tag{7}$$

(a.e.) in Π for certain measurable $\mathbb{S}_\delta$-valued (a^{ij}), measurable $\mathbb{R}^d$-valued b, and measurable real-valued τ and c, such that $|b|,|c| \le K_0$, and $|\tau| \le 1$. By the parabolic Aleksandrov estimates (7) implies that $|v_{n,K}|$ are uniformly bounded in $\bar{\Pi}$ and by by Theorem 9.8.1 we conclude that the family $\{v_{n,K} : K \ge 1, n \ge 0\}$ is precompact in $C(\bar{\Pi})$.

Next, fix a $\rho > 0$ such that $\Pi^\rho \ne \emptyset$ and observe that by a scaled version of Theorem 9.4.9 there is a number $\gamma = \gamma(d,\delta,K_0) \in (0,1)$ and a constant N, that depends only on ρ, d, δ, and K_0, such that for any cylinder $C_\rho(t_0,x_0) \subset \Pi$ we have due to (7) that

$$\int_{C_\rho(t_0,x_0)} \big|D^2v_{n,K}\big|^\gamma\,dxdt \le N\sup_{\Pi}|v_{n,K}| + N\Big(\int_{\Pi}|cv_{n,K} + \tau\bar{H}|^{d+1}\,dxdt\Big)^{\gamma/(d+1)}$$

for all n,K. Here the right-hand side is dominated by a constant independent of n,K, and it follows, by Chebyshev's inequality that there is a constant N (perhaps depending on ρ) for which

$$\big|\Pi^\rho \cap \big\{|D^2v_{n,K}| \ge M\big\}\big| \le NM^{-\gamma} \tag{8}$$

for all $n \ge 0, K \ge 1, M > 0$.

To finish with preparations, set $H_{n,K} = \max(H_n, P-K)$, $F_{n,K} = \max(F_n, P)$, $G_{n,K} = H_{n,K} - F_{n,K}$, where F_n is taken from Assumption 17.1.3 written for H_n. Then $F_{n,K}$ and $G_{n,K}$ satisfy Assumption 17.1.3 (i) and (ii) with the same K_0 and $\bar{H}$. Assumption 17.1.3 (iii) also is satisfied with the same δ. Finally, easy manipulations, using the fact that in the assumptions of the theorem we suppose that Assumption 17.1.3 (iv) is satisfied for F_n with $\theta(\kappa,d,\delta)/2$ in place of θ, show that Assumption 17.1.3 (iv) is satisfied for $F_{n,K}$ with $\theta = \theta(\kappa,d,\delta)$.

Thus, all the assumptions of Theorem 17.1.5 are satisfied and hence there exists a continuous in $\bar{\Pi}$, L_p-viscosity solutions of (2). By Theorem 16.1.3 this viscosity solution coincides with $v_{n,K}$. Therefore, by Theorem 17.1.5 and interpolation theorems we get

$$\sup_{\Pi^\rho}|Dv_{n,K}| \le N, \tag{9}$$

where and below by N we denote various constants independent of K and n, perhaps depending on ρ.

Now set $w_{n,K} = v_{0,K} - v_{n,K}$, and observe that in Π (a.e.)

$$0 = \partial_t w_{n,K} + I_1 + I_2 + I_3,$$

where

$$I_1 = \max\big(H_0[v_{0,K}], P[v_{0,K}] - K\big)$$

$$- \max\Big(H_0\big(v_{0,K}, Dv_{0,K}, D^2 v_{n,K}\big), P[v_{n,K}] - K\Big) = a^{ij} D_{ij} w_n,$$

$$I_2 = \max\Big(H_0\big(v_{0,K}, Dv_{0,K}, D^2 v_{n,K}\big), P[v_{n,K}] - K\Big)$$

$$- \max\Big(H_0\big(v_{n,K}, Dv_{n,K}, D^2 v_{n,K}\big), P[v_{n,K}] - K\Big),$$

$$I_3 = \max\big(H_0[v_{n,K}], P[v_{n,K}] - K\big)$$

$$- \max\big(H_n[v_{n,K}], P[v_{n,K}] - K\big),$$

and (a^{ij}) is an $\mathbb{S}_\delta$-valued function. By assumption

$$|I_2| \le N\big(|w_{n,K}| + |Dw_{n,K}|\big).$$

Therefore, in Π (a.e.)

$$0 = \partial_t w_{n,K} + a^{ij} D_{ij} w_{n,K} + b^i D_i w_{n,K} + c w_{n,K} + I_3,$$

where b and c are bounded uniformly with respect to n, K.

Upon observing that by assumption and (9) in Π^ρ for any $M > 0$ we have

$$|I_3| \le \bar{H} N I_{|D^2 v_{n,K}| \ge M} + \Delta_{n,M+N} I_{|D^2 v_{n,K}| \le M}$$

and using the parabolic Aleksandrov estimates in Π^ρ we conclude that there exists a constant N such that for all n, K, M in Π

$$|v_{0,K} - v_{n,K}| \le N \|\bar{H} I_{|D^2 v_{n,K}| \ge M}\|_{L_{d+1}(\Pi^\rho)}$$

$$+ N\|\Delta_{n,M+N}\|_{L_{d+1}(\Pi^\rho)} + \sup_{\Pi \setminus \Pi^\rho} |v_{0,K} - v_{n,K}|. \tag{10}$$

Here

$$\sup_{n,K} \|\bar{H} I_{|D^2 v_{n,K}| \ge M}\|_{L_{d+1}(\Pi^\rho)} \to 0$$

as $M \to \infty$, since $\bar{H} \in L_{d+1}(\Pi)$ and (8) holds. Therefore, by first taking the sup's with respect to $K \ge 1$ in (10), then sending $n \to \infty$, using assumption (4), and then sending $M \to \infty$, we infer from (10) that, for any small $\rho > 0$

$$\varlimsup_{n \to \infty} \sup_{\Pi} |v_{0,K} - v_{n,K}| \le \varlimsup_{n \to \infty} \sup_{\Pi \setminus \Pi^\rho} |v_{0,K} - v_{n,K}|.$$

After that it only remains to set $\rho \downarrow 0$ and use the equicontinuity of $v_{n,K}$ and the fact that $g_n \to g_0$ uniformly in $\overline{\partial' \Pi}$. The theorem is proved. □

8. REMARK. It follows from the above proof that $\bar{H}$ in (5) can be replaced with $\bar{H}_n$, provided that the family $|\bar{H}_n|^{d+1}$ is uniformly integrable over Π.

An obvious consequence of this theorem is the stability of uniqueness.

9. COROLLARY. *Suppose that for any $n = 1, 2, ...$ there is only one L_{d+1}-viscosity solutions of class $C(\bar\Pi)$ of $\partial_t v_n + H_n[v_n] = 0$ in Π with boundary condition $v_n = g_n$ on $\partial'\Pi$. Then the same holds for $n = 0$.*

Coming back to Theorem 3, observe that, as we have mentioned above, by the maximum principle v_K decreases as K increases. Theorem 9.8.1 and Lemma 12.1.9 (cf. (12.5.7)) guarantee that v_K converges uniformly on $\bar\Pi$ as $K \to \infty$ to a function $v \in C(\bar\Pi)$. To prove that v is an L_{d+1}-viscosity solution we need the following, which is similar to Lemma 16.6.5.

10. LEMMA. *There is a constant N depending only on d, δ, and the Lipschitz constant of H with respect to $[\mathsf{u}'] = (\mathsf{u}'_1, ..., \mathsf{u}'_d)$ such that for any $r \in (0, 1]$ and $C_r(t, x)$ satisfying $C_r(t, x) \subset \Pi$ and $\phi \in W^{1,2}_{d+1}(C_r(t, x))$ we have on $C_r(t, x)$ that*

$$v \le \phi + N r^{d/(d+1)} \|(\partial_t \phi + H[\phi])^+\|_{L_{d+1}(C_r(t,x))} + \max_{\partial' C_r(t,x)} (v - \phi)^+, \tag{11}$$

$$v \ge \phi - N r^{d/(d+1)} \|(\partial_t \phi + H[\phi])^-\|_{L_{d+1}(C_r(t,x))} - \max_{\partial' C_r(t,x)} (v - \phi)^-. \tag{12}$$

Proof. Observe that in $C_r(t, x)$ (a.e.)

$$-\partial_t \phi - \max\big(H[\phi], P[\phi] - K\big) = -\partial_t \phi - \max\big(H[\phi], P[\phi] - K\big)$$

$$+\partial_t v_K + \max\big(H[v_K], P[v_K] - K\big)$$

$$= \partial_t(v_K - \phi) + a_{ij} D_{ij}(v_K - \phi) + b_i D_i(v_K - \phi) - c(v_K - \phi),$$

where measurable $a = (a_{ij})$ is an $\mathbb{S}_\delta$-valued function, $b = (b_i)$ is bounded by the Lipschitz constant of H with respect to $[\mathsf{u}']$, and $c \ge 0$. It follows by Theorem 3.0.1 that

$$v_K \le \phi + \max_{\partial' C_r(t,x)} (v_K - \phi)^+$$

$$+ N r^{d/(d+1)} \big\| \big(\partial_t \phi + \max\big(H[\phi], P[\phi] - K\big)\big)^+ \big\|_{L_{d+1}(C_r(t,x))}, \tag{13}$$

where the constant N is of the type described in the statement of the present lemma. We obtain (11) from (13) by letting $K \to \infty$. In the same way (12) is established. The lemma is proved. □

Proof of Theorem 3. First we prove that v is an L_{d+1}-viscosity solution. Let $(t_0, x_0) \in \Pi$ and $\phi \in W^{1,2}_{d+1,\mathrm{loc}}(\Pi)$ be such that $v - \phi$ attains a local maximum at (t_0, x_0). Then for $\varepsilon > 0$ and all small $r > 0$ for

$$\phi_{\varepsilon,r}(t, x) = \phi(t, x) - \phi(t_0, x_0) + v(t_0, x_0) + \varepsilon\big(|x - x_0|^2 + t - t_0 - r^2\big)$$

we have that

$$\max_{\partial' C_r(t_0,x_0)} (v - \phi_{\varepsilon,r})^+ = 0.$$

Hence, by Lemma 10

$$\varepsilon r^2 = (v - \phi_{\varepsilon,r})(t_0, x_0) \le N r^{d/(d+1)} \big\| \big(\partial_t \phi_{\varepsilon,r} + H[\phi_{\varepsilon,r}]\big)^+ \big\|_{L_{d+1}(C_r(t_0,x_0))},$$

$$\varepsilon^{d+1} \le N r^{-(d+2)} \big\| \big(\partial_t \phi_{\varepsilon,r} + H[\phi_{\varepsilon,r}]\big)^+ \big\|^{d+1}_{L_{d+1}(C_r(t_0,x_0))}.$$

By letting $r \downarrow 0$ and using the continuity of $H(\mathsf{u}, t, x)$ in u_0', which is assumed to be uniform with respect to other variables and also using the continuity of ϕ (embedding theorems) and v, we obtain

$$N \lim_{r\downarrow 0} \operatorname*{ess\,sup}_{C_r(t_0,x_0)} \Big(\partial_t\phi_\varepsilon(t,x) + H\big(v(t,x), D\phi_\varepsilon(t,x), D^2\phi_\varepsilon(t,x)\big)\Big) \geq \varepsilon. \tag{14}$$

where $\phi_\varepsilon = \phi + \varepsilon\big(|x - x_0|^2 + t - t_0\big)$. Finally, observe that $H(\mathsf{u}, t, x)$ is Lipschitz continuous with respect to $\big([\mathsf{u}'], \mathsf{u}''\big)$ with Lipschitz constant independent of u_0', (t, x) by assumption. Then letting $\varepsilon \downarrow 0$ in (14) proves that v is an L_{d+1}-viscosity subsolution. The fact that it is also an L_{d+1}-viscosity supersolution is proved similarly on the basis of (12).

Finally, we prove the maximality of v. Let w be an L_q-viscosity subsolution of (16.1.1) of class $C(\bar\Pi)$ with boundary data g. To prove that $v \geq w$, it suffices to show that for any $\varepsilon > 0$ and $K > 1$ we have $u_K + \varepsilon(T - t) \geq w$ in $\bar\Pi$.

Assume the contrary and observe that, since $u_K + \varepsilon(T - t) - w \geq 0$ on $\partial'\Pi$, there is a point $(t_0, x_0) \in \Pi$ such that

$$\gamma := u_K(t_0, x_0) + \varepsilon(T - t_0) - w(t_0, x_0) < 0 \tag{15}$$

and $u_K + \varepsilon(T - t) - w \geq \gamma$ in $C_\rho(t_0, x_0)$ for sufficiently small $\rho > 0$, so that by definition

$$-\varepsilon + \lim_{r\downarrow 0} \operatorname*{ess\,sup}_{C_r(t_0,x_0)} \Big(\partial_t u_K(t,x) + H\big(w(t,x), Du_K(t,x), D^2u_K(t,x), t, x\big)\Big) \geq 0.$$

Since H is a decreasing function of u_0', in light of (15),

$$-\varepsilon + \lim_{r\downarrow 0} \operatorname*{ess\,sup}_{C_r(t_0,x_0)} \Big(\partial_t u_K(t,x) + H\big(u_K(t,x), Du_K(t,x), D^2u_K(t,x), t, x\big)\Big) \geq 0.$$

This is however impossible since $\partial_t u_K + H[u_K] \leq 0$ in Π (a.e.). This contradiction finishes proving the theorem. □

4. Uniqueness of L_p-viscosity solutions for parabolic Isaacs equations

Fix some constants $\delta \in (0, 1]$, $K_0, T \in (0, \infty)$, $p > d + 2$. Assume that we are given countable sets A and B, and, for each $\alpha \in A$ and $\beta \in B$, we are given an $\mathbb{S}_\delta$-valued function $a^{\alpha\beta}$ on $\mathbb{R}^{d+1} = \big\{(t, x) : t \in \mathbb{R}, x \in \mathbb{R}^d\big\}$, a real-valued function $b^{\alpha\beta}(\mathsf{u}', t, x)$ on $\mathbb{R}^{d+1} \times \mathbb{R}^{d+1}$, and a real-valued $\bar H(t, x) \geq 0$ on $\mathbb{R}^{d+1}$.

1. Assumption. (i) The above introduced functions are measurable.

(ii) The function $a^{\alpha\beta}(t, x)$ is uniformly continuous with respect to (t, x) uniformly with respect to α, β and, with γ introduced before Assumption 1.1, for all values of indices and arguments

$$\big|a^{\alpha\beta}(t, x) - a^{\alpha\beta}(t, y)\big| \leq K_0|x - y|^\gamma.$$

(iii) The function $b^{\alpha\beta}(\mathsf{u}', t, x)$ is nonincreasing with respect to u_0', is Lipschitz continuous with respect to u' with Lipschitz constant K_0, and for all values of indices and arguments

$$\big|b^{\alpha\beta}(0, t, x)\big| \leq \bar H(t, x).$$

(iv) We have $\bar{H} \in L_p(\mathbb{R}^{d+1})$.

For

$$\mathsf{u} = (\mathsf{u}', \mathsf{u}''), \quad \mathsf{u}' = (\mathsf{u}'_0, \mathsf{u}'_1, ..., \mathsf{u}'_d) \in \mathbb{R}^{d+1}, \quad \mathsf{u}'' \in \mathbb{S},$$

and $(t,x) \in \mathbb{R}^{d+1}$ introduce

$$H(\mathsf{u}, t, x) = \sup_{\alpha \in A} \inf_{\beta \in B} \left[a^{\alpha\beta}_{ij}(t,x) \mathsf{u}''_{ij} + b^{\alpha\beta}(\mathsf{u}', t, x) \right]$$

(where as everywhere the summation convention is enforced and the summations are done inside the brackets).

For sufficiently smooth functions $u = u(t,x)$ define

$$H[u](t,x) = H\big(u(t,x), Du(t,x), D^2u(t,x), t, x\big). \tag{1}$$

As is usual in this chapter we take an open bounded subset Ω of $\mathbb{R}^d$ satisfying the exterior ball condition and set

$$\Pi = [0,T) \times \Omega.$$

Here is the main results of this section.

2. THEOREM. *Under the above assumption for any $g \in C(\overline{\partial' \Pi})$ there exists a unique continuous in $\bar{\Pi}$, L_{d+1}-viscosity solution u of the Isaacs equation*

$$\partial_t u + H[u] = 0 \tag{2}$$

in Π with boundary condition $u = g$ on $\partial' \Pi$.

3. REMARK. Under Assumption 1 (i), (ii), in case

$$b^{\alpha\beta}(\mathsf{u}', t, x) = \sum_{i=1}^{d} b^{\alpha\beta}_i(t,x) \mathsf{u}'_i - c^{\alpha\beta}(t,x) \mathsf{u}'_0 + f^{\alpha\beta}(t,x)$$

with uniformly continuous (in (t,x) uniformly in α, β) and uniformly bounded coefficients and the free terms, the uniqueness of C-viscosity solutions is stated without proof in Theorem 9.3 in [**25**]. This case is covered by Theorem 1.2.

For general H, not necessarily related to Isaacs equations, uniqueness is claimed for $\partial_t u + H(D^2 u) = 0$ in Lemma 4.7 of [**115**]. It is proved for L_p-viscosity solutions in Lemma 6.2 of [**25**] in case H is independent of (t,x) with no reference to Wang's Lemma 4.7 of [**115**].

In the elliptic case Jensen and Święch [**50**] proved the uniqueness of continuous L_p-viscosity solutions for Isaacs equations, assuming that $b^{\alpha\beta}_i$, $c^{\alpha\beta}$ are bounded, $\sup_{\alpha,\beta} |f^{\alpha\beta}| \in L_p$ and Assumptions 1 (i), (ii) are satisfied. Their proof uses a remarkable Corollary 1.6 of Święch [**110**] of which the parabolic counterpart is given in [**25**].

An important difference with [**50**] here is that we consider lower-order terms in a more general form, but in [**50**] the summability assumption is weaker (some $p < d$ are allowed).

We prove Theorem 2 by using Corollary 3.9 after some preparations.

Fix a nonnegative $\zeta \in C_0^\infty(\mathbb{R}^{d+1})$ which integrates to one and for $n = 1, 2, ...,$ introduce $\zeta_n(t,x) = n^{d+1} \zeta(nt, nx)$. Also, for real-valued $\xi = \xi^{\alpha\beta}$ given on $A \times B$ and $\mathsf{u}', (t,x) \in \mathbb{R}^{d+1}$ define

$$\mathcal{H}_0(\xi, \mathsf{u}', t, x) = \sup_{\alpha \in A} \inf_{\beta \in B} \big[\xi^{\alpha\beta} + b^{\alpha\beta}(\mathsf{u}', t, x)\big],$$

$$\mathcal{H}_n(\xi, \mathsf{u}', t, x) = \mathcal{H}_0(\xi, \mathsf{u}', t, x) * \zeta_n(t, x), \quad n = 1, 2, ...,$$

where the convolution is performed with respect to (t, x). We will also use the notation

$$|\xi| = \sup_{\alpha \in A} \sup_{\beta \in B} |\xi^{\alpha\beta}|.$$

Here are a few properties of $\mathcal{H}_n$.

4. LEMMA. *(i) For each $n \geq 1$, the function $\mathcal{H}_n(\xi, \mathsf{u}', t, x)$ is continuous, is Lipschitz continuous in u' with Lipschitz constant K_0, and the function $|\mathcal{H}_n(0, 0, t, x)|$ is bounded.*

(ii) For each $n \geq 1$, the function $\mathcal{H}_n(\xi, \mathsf{u}', t, x)$ is infinitely differentiable with respect to x (and t), and there exists a constant N (depending on n) such that $|D_x \mathcal{H}_n| \leq N(1 + |\mathsf{u}'|)$ for all values of the arguments.

(iii) For all values of the arguments

$$\big|\mathcal{H}_n(\xi, \mathsf{u}', t, x) - \mathcal{H}_0(\xi, \mathsf{u}', t, x)\big| \leq 2K_0|\mathsf{u}'| + \bar{H} + \hat{H},$$

where

$$\hat{H} = \sup_n \hat{H} * \zeta_n.$$

(iv) For each $M > 0$

$$\delta_{n,M}(t, x) := \sup_{|\mathsf{u}'| \leq M} \sup_{|\xi| \leq M} \big|\mathcal{H}_n(\xi, \mathsf{u}', t, x) - \mathcal{H}_0(\xi, \mathsf{u}', t, x)\big| \to 0 \tag{3}$$

in $L_{d+1}(\Pi)$ as $n \to \infty$.

Proof. Assertions (i)-(iii) are quite elementary and their proofs are left to the reader. To prove (iv) set

$$\delta_n(\xi, \mathsf{u}', t, x) = \mathcal{H}_n(\xi, \mathsf{u}', t, x) - \mathcal{H}_0(\xi, \mathsf{u}', t, x)$$

and observe that for any (ξ, u')

$$\lim_{n \to \infty} \big\|\delta_n(\xi, \mathsf{u}', \cdot)\big\|_{L_{d+1}(\Pi)} \to 0$$

by the L_{d+1}-continuity of L_{d+1}-functions. Furthermore, by the Lipschitz continuity of $\mathcal{H}_n$ with respect to (ξ, u') uniform with respect to n, t, x, for any $\varepsilon > 0$, one can find m and (ξ_k, u'_k), $k = 1, ..., m$, such that $|\xi_k|, |\mathsf{u}'_k| \leq M$ and any (ξ, u') satisfying $|\xi|, |\mathsf{u}'| \leq M$ has a neighbor (ξ_k, u'_k) such that

$$\big|\mathcal{H}_n(\xi, \mathsf{u}', t, x) - \mathcal{H}_n(\xi_k, \mathsf{u}'_k, t, x)\big| \leq \varepsilon$$

in Π for any $n \geq 0$. It follows that

$$\delta_{n,M}(t, x) \leq \max_{k=1,...,m} \big|\delta_n(\xi_k, \mathsf{u}'_k, t, x)\big| + \varepsilon.$$

Then

$$\varlimsup_{n \to \infty} \|\delta_{n,M}\|_{L_{d+1}(\Pi)} \leq \varlimsup_{n \to \infty} \sum_{k=1}^{m} \big\|\delta_n(\xi_k, \mathsf{u}'_k, \cdot)\big\|_{L_{d+1}(\Pi)} + N\varepsilon = N\varepsilon,$$

where N is independent of ε. This certainly proves (iv) and the lemma. □

Proof of Theorem 2. First we check that the assumptions of Theorem 3.7 are satisfied for

$$H_n(\mathsf{u}, t, x) = \mathcal{H}_n(\xi, \mathsf{u}', t, x), \quad \xi^{\alpha\beta} = a^{\alpha\beta}_{ij} \mathsf{u}''_{ij}.$$

Assumptions (3.4), (3.5) are taken care of in Lemma 4. Assumption 3.1 is easily seen to be satisfied with $\bar{H} + \hat{H}$ in place of $\bar{H}$.

In what concerns Assumption 17.1.3 we set

$$F_n(\mathsf{u}', t, x) = F_0(\mathsf{u}', t, x) = \sup_{\alpha \in A} \inf_{\beta \in B} a^{\alpha\beta}_{ij}(t, x) \mathsf{u}''_{ij}.$$

Then Assumption 17.1.3 (iii) is obviously satisfied, and its part (ii) is satisfied due to $\bar{H} \in L_p(\Pi)$, $p > d + 2$, Remark 17.1.2, and the choice of κ before Theorem 3.7. The remaining assumptions of Theorem 3.7 is Assumption 17.1.3 (iv), which is satisfied for any $\theta > 0$ if R_0 is chosen appropriately in light of Assumption 1 (ii).

Thus Theorem 3.7 is applicable. Furthermore, for each $n \geq 1$ the assumptions of Theorem 1.2 are satisfied in light of Lemma 4 (ii) and Assumption 1 (ii). Also uniqueness of C-viscosity solutions implies that of L_p-viscosity solutions.

Now, the combination of Theorem 1.2 and Corollary 3.9 (proved under the assumptions of Theorem 3.7) immediately yields the desired result. □

Appendix A. Proof of Theorem 6.2.1

We follow parts of [**67**] and [**71**].

5. LEMMA. *Let P^o be an open bounded polyhedron in a Euclidean space E of points y with vertices, say $y_1, ..., y_n$. Then there exist real-analytic functions $p_k(y) > 0$, $k = 1, ..., n$, on P^0 such that for any $y \in P^o$*

$$y = \sum_k p_k(y) y_k, \quad \sum_k p_k(y) = 1. \tag{1}$$

Proof. We use the induction on the dimension of P^o. First, without losing generality one may assume that the volume of P^o in E is strictly greater than zero. Then, assume that the lemma is true for any face of P^o and then solve Laplace's equation $\Delta p_k = 0$ in P^o with boundary condition $p_k(y) = \bar{p}_k(y)$ on ∂P^o, where $\bar{p}_k(y)$ is the weight of the vertex y_k in the representation of $y \in \partial P^o$, which is supposed to exist by the induction hypothesis. Of course, if $y \in \partial P^o$ and y_k do not belong to the same face, we set $\bar{p}_k(y) = 0$. Then by the well-known properties of harmonic functions $p_k > 0$ in P^o, they are real analytic in P^o and since

$$\Delta \sum_k p_k(y) y_k = 0, \quad \text{in} \quad P^o \quad \text{and} \quad \sum_k p_k(y) y_k = y \quad \text{on} \quad \partial P^o$$

and $\Delta y = 0$, by uniqueness we have the first relation in (1). The second one is obtained similarly from the fact that it holds on ∂P^o and $\Delta 1 = 0$. Since our assertions are trivial if the dimension of P^o is one, the induction argument goes through and the lemma is proved. □

As is easy to see to prove Theorem 6.2.1 it suffices to prove the following.

6. THEOREM. *There exists an integer $n = n(\delta, d)$ and vectors $l_1, ..., l_n \in \bar{B}_1 \subset \mathbb{R}^d$ with rational coordinated such that for any extension $\{l_1, ..., l_m\}$, $m \geq n$, of the set $\{l_1, ..., l_n\}$ there exist real-analytic functions $\lambda_1(a), ..., \lambda_m(a)$ on $\mathbb{S}_\delta$ possessing there the following properties:*

$$a \equiv \sum_{k=1}^{m} \lambda_k(a) l_k l_k^*, \quad \lambda_k(a) \geq \delta_1, \quad \forall k, \tag{2}$$

where the constant $\delta_1 > 0$.

Proof. We will be looking for $l_1, ..., l_n$ with rational coordinates on the unit sphere in $\mathbb{R}^d$. We rely on a well-known fact that points with rational coordinates are dense on the unit sphere (a nice one-line proof of this can be found in [**106**] and consists of just saying that the inverse of the stereographic projection maps rational points into rational points).

First we deal with the case that $m = n$. Introduce $\mathbb{S}^1_{\delta'}$, $\delta' = \delta^2/d$, as a subset of $\mathbb{S}_{\delta'}$ consisting of matrices with unit trace. Take a convex closed polyhedron P belonging to the convex set $\mathbb{S}^1_{\delta'/2}$ and such that its relative interior contains the closed set $\mathbb{S}^1_{\delta'}$. Let $a_1, ..., a_r$ be the vertices of P and ℓ_{ki}, $i = 1, ..., d$, be the unit eigenvectors of a_k corresponding to eigenvalues $\nu_{ki} > 0$, $i = 1, ..., d$. Then for each $k = 1, ..., r$

$$a_k = \sum_{i=1}^{d} \nu_{ki} \ell_{ki} \ell_{ki}^* \tag{3}$$

(no summation in k) and

$$\sum_{i=1}^{d} \nu_{ki} = 1,$$

so that a_k is a convex combination of $\ell_{ki}\ell_{ki}^*$ (no summation in k, i). By slightly moving the ℓ_{ki}'s along the unit sphere, if necessary, so that they will become all distinct and have rational coordinates, and defining the new a_k's according to (3) with the new ℓ_{ki}'s, we can still have that the new $a_1, ..., a_r$ are the vertices of a convex closed polyhedron belonging to $\mathbb{S}^1_{\delta'/2}$ and such that its relative interior contains the closed set $\mathbb{S}^1_{\delta'}$. In order to simplify the notation we will assume that the moving and the replacements are not necessary.

Now we set $\{l_1, ..., l_n\} = \{\ell_{ki} : k = 1, ..., r, i = 1, ..., d\}$. Observe that none of the matrices $l_k l_k^*$ can be expressed as a convex combination of the remaining ones in $l_1 l_1^*, ..., l_n l_n^*$. Indeed, for instance, the first relation in

$$l_n l_n^* = \sum_{k=1}^{n-1} \tau_k l_k l_k^*, \quad \tau_k \geq 0, \quad \sum_{k=1}^{n-1} \tau_k = 1$$

after been multiplied from the left by l_n^* and from the right by l_n leads to

$$1 = \sum_{k=1}^{n-1} \tau_k \langle l_k, l_n \rangle^2,$$

which is impossible since $\langle l_k, l_n \rangle < 1$ for $k \neq n$. It follows that the matrices $l_k l_k^*$ are the vertices of the smallest convex polyhedron P' containing $l_1 l_1^*, ..., l_n l_n^*$.

By the above, any a_k is expressed as a convex combination of $l_1 l_1^*, ..., l_n l_n^*$. Since $\mathbb{S}^1_{\delta'}$ is contained in the interior of P it is also contained in the interior of P'. Then by Lemma 5 there exist real-analytic functions $\lambda_1(a), ..., \lambda_n(a)$ on $\mathbb{S}^1_{\delta'}$ such that (2) holds on $\mathbb{S}^1_{\delta'}$ with $m = n$.

To finish the proof in case $m = n$ it only remains to observe that, for any $a \in \mathbb{S}_\delta$, we have $a/\operatorname{tr} a \in \mathbb{S}^1_{\delta'}$ and

$$a = \sum_{k=1}^{n} (\operatorname{tr} a) \lambda_k (a/\operatorname{tr} a) l_k l_k^*.$$

In the general case, if for some $k > n$, we have $l_k = 0$, then one can take $\lambda_k \equiv 1$ without affecting (2). It follows that it suffices to concentrate on nonzero l_k's. Then normalizing them, if necessary, and accordingly changing λ_k we reduce the general case to the one in which all l_k, $k > n$, are also unit vectors. The normalization may make some of l_m's to coincide. One can always spread $\lambda_k(a)$ evenly between coinciding vectors and preserve both relations in (2). This convinces us that it suffices to consider the case where all $l_1, \ldots, l_m$ are distinct unit vectors.

In that case it suffices to repeat the above argument after the point where we introduced P' and take it as the smallest convex polyhedron with vertices $l_1 l_1^*, \ldots, l_m l_m^*$. The theorem is proved. □

Appendix B. Proof of Lemma 9.2.6

Recall that in Lemma 9.2.6 $\Gamma \subset C_1$ is a measurable set,

$$q, \eta, \zeta \in (0,1),$$

$\mathcal{B}$ is the family of open cylinders Q of type $C^o_\rho(t_0, x_0)$ which lie in C_1 and satisfy $|Q \cap \Gamma| \geq q|Q|$. If $Q = C^o_\rho(t_0, x_0) \in \mathcal{B}$, we set

$$Q' = (t_0, x_0) - C^o_{\eta^{-1}\rho^2, \rho}, \quad Q'' = (t_0 - \eta^{-1}\rho^2, x_0) + C^o_{\eta^{-1}\rho^2\zeta^2, \rho\zeta}.$$

Finally, denote

$$\hat{\Gamma} = \bigcup_{Q \in \mathcal{B}} Q, \quad \Gamma' = \bigcup_{Q \in \mathcal{B}} Q', \quad \Gamma'' = \bigcup_{Q \in \mathcal{B}} Q''.$$

We present the proof of Lemma 9.2.6 as a sequence of lemmas that are proved as in Krylov and Safonov [**83**, **84**] and Safonov [**100**].

1. Lemma. *We have* $|\Gamma \setminus \hat{\Gamma}| = 0$ *and if* $|\Gamma| \leq q|C_1|$, *then*

$$|\hat{\Gamma}| \geq \left(1 - \frac{1-q}{3^{d+1}}\right)^{-1} |\Gamma|.$$

Proof. Since almost every point of a measurable set is its Lebesgue point (with respect to parabolic cylinders, see Guzmán [**48**]) and $q < 1$, we have $|\Gamma \setminus \hat{\Gamma}| = 0$.

To prove the second assertion, introduce

$$\mathcal{B}^0 = \left\{Q \in \mathcal{B} : |Q \cap \Gamma| = q|Q|\right\}.$$

For $Q \in \mathcal{B}$ and $\theta \in [0,1]$ define $Q(\theta)$ as continuously expanding cylinders of standard form in such a way that $Q(0) = Q$, $Q(1) = C_1$. Then $\varphi(\theta) := \big|Q(\theta) \cap \Gamma\big| \, \big|Q(\theta)\big|^{-1}$ is a continuous function on $[0,1]$, $\varphi(0) \geq q$, $\varphi(1) \leq q$, and there exists θ such that $\varphi(\theta) = q$ and $Q \subset Q(\theta) \in \mathcal{B}^0$. Since $Q \in \mathcal{B}$ is arbitrary, we conclude that

$$\hat{\Gamma} = \bigcup_{Q \in \mathcal{B}^0} Q. \tag{4}$$

Then we choose a (countable or finite) family of cylinders Q_i as follows. If $Q = C^o_\rho(t_0, x_0) \in \mathcal{B}^0$, set $\rho(Q) = \rho$ and introduce

$$\rho_1 = \sup\left\{\rho(Q) : Q \in \mathcal{B}^0\right\}.$$

The dominated convergence theorem easily implies that there exists $Q_1 \in \mathcal{B}^0$ with $\rho(Q_1) = \rho_1$. If, for a $k \geq 1$, Q_i with $i = 1, ..., k$ have already been found, set

$$\mathcal{B}^k = \left\{Q \in \mathcal{B}^0 : Q \cap Q_i = \emptyset, i = 1, ..., k\right\}.$$

In case $\mathcal{B}^k \neq \emptyset$, introduce

$$\rho_{k+1} = \sup\left\{\rho(Q) : Q \in \mathcal{B}^k\right\}.$$

As above, there exists $Q_{k+1} \in \mathcal{B}^k$ such that $\rho(Q_{k+1}) = \rho_{k+1}$. If $\mathcal{B}^i \neq \emptyset$ for all $i = 1, 2, ...,$ we obtain a countable family Q_i, $i = 1, ...$ In the case that $\mathcal{B}^i \neq \emptyset$ for $i = 1, 2, ..., k$, while $\mathcal{B}^{k+1} = \emptyset$, we only have Q_i, $i = 1, 2, ..., k+1$, and define $\rho_{k+2} = ... = 0$. Note that the Q_i's are disjoint and $\rho_i \downarrow 0$ as $i \to \infty$.

Next dilate Q_i three times with center of dilation being its geometric center and denote by $\tilde{Q}_i$ the result of dilation. We claim that

$$\hat{\Gamma} \subset \bigcup_{i \geq 1} \tilde{Q}_i. \tag{5}$$

If the contrary is true then (see (4)) there exists $Q \in \mathcal{B}^0$, such that $Q \not\subset \tilde{Q}_i$, $i = 1, 2, ...$ For $\rho_i \geq \rho(Q)$, this obviously implies $Q \cap Q_i = \emptyset$. On the other hand, there exists k such that $\rho_1 \geq ... \geq \rho_k \geq \rho(Q) > \rho_{k+1}$. We have that $Q \cap Q_i = \emptyset$ for $i = 1, ..., k$, $Q \in \mathcal{B}^k$ and then, by the definition of ρ_{k+1}, we get that $\rho(Q) \leq \rho_{k+1}$ in contradiction with $\rho(Q) > \rho_{k+1}$ thus proving (5).

Now (5) implies that

$$|\hat{\Gamma}| \leq 3^{d+1} \sum_i |Q_i|.$$

Next, since the cylinders Q_i are disjoint and

$$\left|(\hat{\Gamma} \setminus \Gamma) \cap Q_i\right| = |Q_i| - |Q_i \cap \Gamma| = (1-q)|Q_i|,$$

we conclude that

$$|\hat{\Gamma} \setminus \Gamma| \geq \sum_i \left|(\hat{\Gamma} \setminus \Gamma) \cap Q_i\right| = (1-q) \sum_i |Q_i|.$$

Therefore,

$$|\hat{\Gamma}| = |\Gamma| + |\hat{\Gamma} \setminus \Gamma| \geq |\Gamma| + (1-q)\, 3^{-d-1} |\hat{\Gamma}|.$$

The lemma is proved. □

2. Lemma. *Let $\kappa \geq 1$, let $-\infty \leq t_1 < t_2 \leq \infty$, let Σ be the set of open subintervals of (t_1, t_2), let $\sigma \subset \Sigma$, and let a function $g : \Sigma \to \Sigma$ be such that*

$$\left|g(I)\right| \leq \kappa\, |I|$$

for $I \in \Sigma$, where $\left|g(I)\right|$ and $|I|$ are the (one-dimensional) Lebesgue measures of $g(I)$ and I. Also, let $g(I_1) \subset g(I_2)$ if $I_1 \subset I_2$. Then

$$\Big|\bigcup_{I \in \sigma} g(I)\Big| \leq \kappa \Big|\bigcup_{I \in \sigma} I\Big|.$$

Proof. The open set $\cup\{I : I \in \sigma\}$ is the union of disjoint intervals denoted by I_n. Hence,

$$\Big|\bigcup_{I \in \sigma} g(I)\Big| \leq \Big|\bigcup_n \bigcup_{I \in \sigma, I \subset I_n} g(I)\Big| \leq \Big|\bigcup_n g(I_n)\Big| \leq \sum_n k\, |I_n| = \kappa \Big|\bigcup_{I \in \sigma} I\Big|$$

and the lemma is proved. □

3. Lemma. $|\hat{\Gamma}| \leq (1+\eta)|\Gamma'|$.

Proof. It obviously suffices to show that

$$|\hat{\Gamma} \cup \Gamma'| \le (1+\eta)|\Gamma'|.$$

Bearing in mind Fubini's theorem, we convince ourselves that it suffices to show that for any x, the one-dimensional Lebesgue measure of the set

$$\{t : (t,x) \in \hat{\Gamma} \cup \Gamma'\}$$

is less than $(1+\eta)\left|\{t : (t,x) \in \Gamma'\}\right|$. To this end take $\kappa = 1+\eta$, $t_1 = -t_2 = -\infty$,

$$\sigma = \Big\{\{t : (t,x) \in Q'\} \, : \, Q \in \mathcal{B}\Big\}$$

in Lemma 2 and introduce $g\,(I)$ as the κ-fold dilation of I with the center of dilation being the lower end point of I. Then the one-dimensional result follows from Lemma 2 and the fact that

$$g\big(\{t : (t,x) \in Q'\}\big) = \{t \, : \, (t,x) \in Q \cup Q'\},$$

$$\bigcup_{I \in \sigma} I = \{t \, : \, (t,x) \in \Gamma'\}.$$

The lemma is proved. □

4. LEMMA. $|\Gamma'| \le \zeta^{-d-2}|\Gamma''|$.

Proof. The set Γ'' is obtained by $d+1$ consecutive contractions of Γ' first by a factor of ζ^{-2} along the t-axis and then by a factor of ζ^{-1} along each of the remaining x-axes.

One shows as in the proof of Lemma 3 that the first contraction reduces the volume of Γ' no more than ζ^{-2} times. Each of the remaining d contractions reduces the volume no more than ζ^{-1} times. This is shown as in the proof of Lemma 3. This proves the lemma. □

Appendix C. Some tools from Real Analysis

Various parts of exposition here are taken from [**68**], [**70**], and [**75**].

1. Partitions, maximal and sharp functions

In this section $(\Omega, \mathcal{F}, \mu)$ is a complete measure space with a σ-finite measure μ, such that

$$\mu(\Omega) = \infty.$$

Generic points in Ω will be denoted by $x, y, z, ...$ and generic subsets of Ω we denote by ω, ω_i,.... Let $\mathcal{F}^0$ be the subset of $\mathcal{F}$ consisting of all sets $\omega \subset \Omega$ such that $\mu(\omega) < \infty$. By $\mathbb{L}$ we denote a fixed dense subset of $L_1(\Omega) = L_1(\Omega, \mathcal{F}, \mu)$. For any $\omega \in \mathcal{F}$ we set

$$|\omega| = \mu(\omega).$$

For $\omega \in \mathcal{F}^0$ and functions f summable on ω we use the notation

$$f_\omega = \fint_\omega f \, \mu(dx) := \frac{1}{|\omega|} \int_\omega f(x) \, \mu(dx) \quad (0/0 := 0)$$

for the average value of f over ω. We write $f \in L_{1,\mathrm{loc}}(\Omega)$ if $f I_\omega \in L_1(\Omega)$ for any $\omega \in \mathcal{F}^0$.

1. Remark. As is easy to see $L_1(\Omega) \cap L_p(\Omega)$ is dense in $L_p(\Omega)$ for any $p \in (1, \infty)$.

1:1. Partitions and stopping times.

2. Definition. Let $\mathbb{Z} = \{n : n = 0, \pm 1, \pm 2, ...\}$ and let $(\mathcal{F}_n, n \in \mathbb{Z})$ be a sequence of partitions of Ω each consisting of countably many disjoint sets and such that $\mathcal{F}_n \subset \mathcal{F}^0$ for each n. For each $x \in \Omega$ and $n \in \mathbb{Z}$ there exists (a unique) $\omega \in \mathcal{F}_n$ such that $x \in \omega$. We denote this ω by $\omega_n(x)$.

We call the sequence $(\mathcal{F}_n, n \in \mathbb{Z})$ *a filtration of partitions* if the following conditions are satisfied.

(i) The elements of partitions are "large" for big negative n's and "small" for big positive n's:

$$\inf_{\omega \in \mathcal{F}_n} |\omega| \to \infty \quad \text{as} \quad n \to -\infty, \qquad \lim_{n \to \infty} f_{\omega_n(x)} = f(x) \quad \text{(a.e.)} \quad \forall f \in \mathbb{L}.$$

(ii) The partitions are nested: for each n and $\omega \in \mathcal{F}_n$ there is a (unique) $\omega' \in \mathcal{F}_{n-1}$ such that $\omega \subset \omega'$.

(iii) The following regularity property holds: for any n, ω, and ω' as in (ii) we have

$$|\omega'| \le N_0 |\omega|,$$

where N_0 is a constant independent of n, ω, ω'.

Observe that since the elements of partition $\mathcal{F}_n$ become large as $n \to -\infty$, we have $N_0 > 1$. One obvious but useful property of filtrations is that if $n \le m$ and $\omega_n \in \mathcal{F}_n$ and $\omega_m \in \mathcal{F}_m$, then $\omega_n \cap \omega_m$ is either ω_m or empty.

3. EXAMPLE. The simplest example of a filtration of partitions in the case $\Omega = \mathbb{R}^d$ with Lebesgue measure μ is given by *dyadic cubes*, that is, by

$$\mathcal{F}_n = \{\omega_n(i_1, ..., i_d), i_1, ..., i_d \in \mathbb{Z}\},$$

where

$$\omega_n(i_1, ..., i_d) = [i_1 2^{-n}, (i_1 + 1)2^{-n}) \times ... \times [i_d 2^{-n}, (i_d + 1)2^{-n}).$$

In this case, to satisfy requirement (i) in Definition 2, one can take $\mathbb{L}$ as the set of continuous functions with compact support.

4. EXAMPLE. Another example, frequently used in the theory of parabolic equations, is given by *parabolic dyadic cubes* in

$$\Omega = \mathbb{R}^{d+1} = \{(t, x) : t \in \mathbb{R}, x \in \mathbb{R}^d\},$$

namely,

$$\mathcal{F}_n = \{[i_0 4^{-n}, (i_0 + 1)4^{-n}) \times \omega_n(i_1, ..., i_d), i_0, i_1, ..., i_d \in \mathbb{Z}\}.$$

Here again μ is Lebesgue measure but in $\mathbb{R}^{d+1}$.

5. DEFINITION. Let $\mathcal{F}_n$, $n \in \mathbb{Z}$, be a filtration of partitions of Ω.

(i) Let $\tau = \tau(x)$ be a function on Ω with values in $\{\infty, 0, \pm 1, \pm 2, ...\}$. We call τ *a stopping time* (relative to the filtration) if, for each $n = 0, \pm 1, \pm 2, ...$, the set

$$\{x : \tau(x) = n\}$$

is either empty or else is the union of some elements of $\mathcal{F}_n$.

(ii) For a function $f \in L_{1,\text{loc}}(\Omega)$ and $n \in \mathbb{Z}$, we denote

$$f_{|n}(x) = \fint_{\omega_n(x)} f(y)\, \mu(dy).$$

We read $f_{|n}$ as "f given $\mathcal{F}_n$", continuing to borrow the terminology from probability theory. If we are also given a stopping time τ, we let

$$f_{|\tau}(x) = f_{|\tau(x)}(x)$$

for those x for which $\tau(x) < \infty$ and $f_{|\tau}(x) = f(x)$ otherwise.

The simplest example of a stopping time is given by $\tau(x) \equiv 0$.

To define a stopping time, we usually choose, for each $n \in \mathbb{Z}$, elements $\omega_{n1}, \omega_{n2}, ... \in \mathcal{F}_n$ and define $\tau = n$ on each of them. Of course, we need to have all ω_{ni} disjoint for different n, i. If the union of all ω_{ni} is not Ω, then one lets $\tau = \infty$ on the complement of the union. The most important property of $f_{|n}$ is that, obviously, it is constant on each $\omega \in \mathcal{F}_n$, so that, for any $x \in \omega \in \mathcal{F}_n$, we have

$$\int_\omega f_{|n}(y)\, \mu(dy) = |\omega| f_{|n}(x) = \int_\omega f(y)\, \mu(dy).$$

We are going to use the following simple properties of the objects introduced above. By $I_{\tau<\infty}$ we mean the indicator of the set $\{x : \tau(x) < \infty\}$.

6. LEMMA. *Let $\mathcal{F}_n$, $n \in \mathbb{Z}$, be a filtration of partitions of Ω.*

(i) Let $f \in L_{1,\mathrm{loc}}(\Omega)$, $f \geq 0$, and let τ be a stopping time. Then

$$\int_\Omega f_{|\tau}(x) I_{\tau<\infty}\, \mu(dx) = \int_\Omega f(x) I_{\tau<\infty}\, \mu(dx), \tag{1}$$

$$\int_\Omega f_{|\tau}(x)\, \mu(dx) = \int_\Omega f(x)\, \mu(dx). \tag{2}$$

(ii) Let $g \in L_{1,\mathrm{loc}}(\Omega)$, $g \geq 0$, and suppose that $g_{|n}(x) \to 0$ as $n \to -\infty$ at any $x \in \Omega$. Also let $\lambda > 0$ be a constant. Then

$$\tau(x) := \inf\big\{n : g_{|n}(x) > \lambda\big\} \quad (\inf \emptyset := \infty) \tag{3}$$

is a stopping time. Furthermore, we have

$$0 \leq g_{|\tau}(x) I_{\tau<\infty} \leq N_0 \lambda, \quad \big|\{x : \tau(x) < \infty\}\big| \leq \lambda^{-1} \int_\Omega g(x) I_{\tau<\infty}\, \mu(dx). \tag{4}$$

Proof. (i) Equation (2) follows immediately from (1) since $f_{|\tau} = f$ on the set where $\tau = \infty$. Then notice that, for any $n \in \mathbb{Z}$ and set Γ which is the union of some elements $\omega_i \in \mathcal{F}_n$, obviously

$$\int_\Gamma f_{|n}\, \mu(dx) = \sum_i \int_{\omega_i} f_{|n}\, \mu(dx) = \sum_i \int_{\omega_i} f\, \mu(dx) = \int_\Gamma f\, \mu(dx).$$

Hence,

$$\int_\Omega f_{|\tau} I_{\tau<\infty}\, \mu(dx) = \sum_{n=-\infty}^{\infty} \int_{\{\tau=n\}} f_{|n}\, \mu(dx)$$

$$= \sum_{n=-\infty}^{\infty} \int_{\{\tau=n\}} f\, \mu(dx) = \int_\Omega f I_{\tau<\infty}\, \mu(dx).$$

(ii) First notice that $\tau > -\infty$ since $g_{|n} \to 0$ as $n \to -\infty$. Next, observe that

$$\omega_n(x) \subset \omega_m(x)$$

for all $m \leq n$ since the partitions are nested. It follows that, if $y \in \omega_n(x)$, then

$$\omega_m(y) = \omega_m(x), \quad g_{|m}(y) = g_{|m}(x), \quad \forall m \leq n.$$

By adding that

$$\tau(x) = n \Longleftrightarrow g_{|n}(x) > \lambda, \quad g_{|m}(x) \leq \lambda \quad \forall m < n,$$

we conclude that the set $\{\tau = n\}$ contains $\omega_n(x)$ along with each x. Therefore, $\{\tau = n\}$ is indeed the union of some elements of $\mathcal{F}_n$.

To prove the first relation in (4), it suffices to notice that, if $\tau(x) = n$, then

$$g_{|\tau}(x) = g_{|n}(x) = \frac{1}{|\omega_n(x)|} \int_{\omega_n(x)} g(y)\, \mu(dy) \leq N_0 \frac{1}{|\omega_{n-1}(x)|} \int_{\omega_n(x)} g(y)\, \mu(dy)$$

$$\leq N_0 \frac{1}{|\omega_{n-1}(x)|} \int_{\omega_{n-1}(x)} g(y)\, \mu(dy) = N_0 g_{|n-1}(x) \leq N_0 \lambda.$$

The second inequality in (4) follows from Chebyshev's inequality and (i):

$$\big|\{x : \tau < \infty\}\big| = \big|\{x : g_{|\tau} I_{\tau<\infty} > \lambda\}\big|$$

$$\leq \lambda^{-1} \int_\Omega g_{|\tau} I_{\tau<\infty}\, \mu(dx) = \lambda^{-1} \int_\Omega g I_{\tau<\infty}\, \mu(dx).$$

The lemma is proved. □

It is worth clarifying the structure of $g_{|\tau}$ in Lemma 6. First we take any m so largely negative that

$$g_{|m} \leq \lambda \quad \text{on} \quad \Omega.$$

Let ω_j, $j = 1, 2, \ldots$, be the set of all "cubes" in the family $\mathcal{F}_m$, so that

$$\Omega = \bigcup_j \omega_j.$$

Then we divide each "cube" ω_j into smaller "cubes" $\omega_{jk} \in \mathcal{F}_{m+1}$, so that $\omega_j = \bigcup_k \omega_{jk}$ and we look for those

$$\omega_{jk} \quad \text{on which} \quad g_{|m+1} > \lambda.$$

We set those aside and set $\tau = m + 1$ and accordingly $g_{|\tau} = g_{|m+1}$ on them. With the *remaining* "cubes" on which $g_{|m+1} \leq \lambda$ we proceed in the same way splitting each of them into smaller "cubes" $\omega_{jkp} \in \mathcal{F}_{m+2}$ and defining $\tau = m + 2$ and $g_{|\tau} = g_{|m+2}$ on those ω_{jkp} on which $g_{|m+2} > \lambda$. By continuing in this way, we define τ and $g_{|\tau}$ on a subset of Ω, which may not coincide with Ω (it is just empty if $g \leq \lambda$ everywhere). The remaining set is the set of points x at which $g_{|n}(x) \leq \lambda$ for all n and we let $\tau = \infty$ and $g_{|\tau} = g$ at those points.

1:2. Maximal functions. Define *the ("dyadic") maximal function* of f by

$$\mathcal{M}f(x) = \sup_{n<\infty} |f|_{|n}(x),$$

so that $\mathcal{M}f = \mathcal{M}|f|$.

Notice that Lemma 6 implies the following.

7. Corollary (Maximal inequality). *For $\lambda > 0$ and nonnegative $g \in L_{1,\mathrm{loc}}(\Omega)$ such that $g_{|n} \to 0$ as $n \to -\infty$, the* maximal inequality *holds:*

$$\big|\{x : \mathcal{M}g(x) > \lambda\}\big| \leq \lambda^{-1} \int_\Omega g(x) I_{\mathcal{M}g>\lambda}\, \mu(dx). \tag{5}$$

Indeed, for τ as in (3), we have

$$\{x : \mathcal{M}g(x) > \lambda\} = \{x : \tau(x) < \infty\}.$$

8. Corollary. *For any $f \in L_{1,loc}(\Omega)$ we have $f_{|n} \to f$ on Ω (a.e.) as $n \to \infty$.*

To prove this, first observe that for any $n \geq 0$ and $\omega \in \mathcal{F}_0$ we have $(fI_\omega)_{|n} = f_{|n} I_\omega$ and this easily implies that it suffices to concentrate on the case that $f \in L_1(\Omega)$.

In that case there exists a sequence $f_k \in \mathbb{L}$ such that $f_k \to f$ in $L_1(\Omega)$ and almost everywhere as $k \to \infty$. Owing to (5), $\mathcal{M}(f_k - f) \to 0$ in measures and by choosing a subsequence we may assume that this convergence holds almost everywhere as well. Now (a.e.)

$$\varlimsup_{n\to\infty} f_{|n} \le \varlimsup_{n\to\infty} f_{k|n} + \sup_n |f_{|n} - f_{k|n}| \le f_k + \mathcal{M}(f - f_k),$$

which after letting $k \to \infty$ yields

$$\varlimsup_{n\to\infty} f_{|n} \le f$$

(a.e.). By changing f to $-f$ we obtain

$$\varliminf_{n\to\infty} f_{|n} \ge f$$

(a.e.) and our assertion is proved. □

9. REMARK. Our interest in estimating $\big|\{\mathcal{M}g > \lambda\}\big|$ as in Corollary 7 is based on the following formula valid for any measurable $f \ge 0$ in light of Fubini's theorem:

$$\int_\Omega f(x)\,\mu(dx) = \int_\Omega \Big(\int_0^{f(x)} dt\Big)\,\mu(dx) = \int_\Omega \Big(\int_0^\infty I_{f(x)>t}\,dt\Big)\,\mu(dx)$$

$$= \int_0^\infty \Big(\int_\Omega I_{f(x)>t}\,\mu(dx)\Big)\,dt = \int_0^\infty \big|\{x : f(x) > t\}\big|\,dt. \qquad (6)$$

10. COROLLARY. *Let* $p \in (1,\infty)$, $g \in L_1(\Omega)$, $g \ge 0$. *Then*

$$\|\mathcal{M}g\|_{L_p(\Omega)} \le q\|g\|_{L_p(\Omega)},$$

where $q = p/(p-1)$.

Indeed, from (6), (5), and Fubini's theorem we conclude that, for any finite constant $\nu > 0$,

$$\|\nu \wedge \mathcal{M}g\|^p_{L_p(\Omega)} = \int_0^\infty \big|\{x : \nu \wedge \mathcal{M}g(x) > \lambda^{1/p}\}\big|\,d\lambda$$

$$= \int_0^{\nu^p} \big|\{x : \mathcal{M}g(x) > \lambda^{1/p}\}\big|\,d\lambda \le \int_\Omega g(x)\Big(\int_0^{\nu^p} \lambda^{-1/p} I_{\mathcal{M}g(x)>\lambda^{1/p}}\,d\lambda\Big)\,\mu(dx)$$

$$= \int_\Omega g(x)\Big(\int_0^{(\nu\wedge\mathcal{M}g(x))^p} \lambda^{-1/p}\,d\lambda\Big)\,\mu(dx) = q\int_\Omega (\nu \wedge \mathcal{M}g)^{p-1} g\,\mu(dx).$$

This and $g \in L_1(\Omega)$ imply that $\|\nu \wedge \mathcal{M}g\|_{L_p(\Omega)} < \infty$. Then upon using Hölder's inequality, we get

$$\|\nu \wedge \mathcal{M}g\|^p_{L_p(\Omega)} \le q\|g\|_{L_p(\Omega)}\|\nu \wedge \mathcal{M}g\|^{p-1}_{L_p(\Omega)}, \quad \|\nu \wedge \mathcal{M}g\|_{L_p(\Omega)} \le q\|g\|_{L_p(\Omega)}$$

and it only remains to let $\nu \to \infty$ and use Fatou's theorem.

Next we extend Corollary 10 to $g \in L_p(\Omega)$.

11. THEOREM. *For any $p \in (1,\infty)$ and $g \in L_p(\Omega)$*

$$\|\mathcal{M}g\|_{L_p(\Omega)} \leq q\|g\|_{L_p(\Omega)}.$$

Proof. Since

$$\mathcal{M}g = \mathcal{M}|g| \quad \text{and} \quad \|g\|_{L_p(\Omega)} = \|\,|g|\,\|_{L_p(\Omega)},$$

we may concentrate on $g \geq 0$.

In that case take an increasing sequence of $A_m \in \mathcal{F}^0, m \geq 1$, such that $A_m \uparrow \Omega$ and introduce $g^m(x) = g(x)I_{A_m}$. Then $g^m \in L_1(\Omega)$ and

$$\|\mathcal{M}g^m\|_{L_p(\Omega)} \leq q\|g^m\|_{L_p(\Omega)} \leq q\|g\|_{L_p(\Omega)}$$

by Corollary 10. It only remains to use Fatou's theorem along with the observation that for any x and n, since $|C_n(x)| < \infty$, we have

$$(g^m)_{|n}(x) \to g_{|n}(x) \quad \text{as} \quad m \to \infty,$$

which implies

$$g_{|n}(x) \leq \varliminf_{m\to\infty} \sup_r (g^m)_{|r}(x), \quad \mathcal{M}g \leq \varliminf_{m\to\infty} \mathcal{M}g^m.$$

The theorem is proved. □

1:3. Sharp functions. In the remaining part of this section we consider three measurable functions u, v, g on Ω, such that $g \geq 0$.

The most relevant case of Lemma 12 for the purposes of the book is when $u^\omega = |u|$. In the form it is stated and for $\gamma = 1$ the lemma was used in [**69**] while treating linear elliptic equations with rather rough coefficients.

12. LEMMA. *Let $\gamma \in (0,1]$, $v \in L_{1,\text{loc}}(\Omega)$, and let $v_{|n} \to 0$ as $n \to -\infty$ on Ω. Assume that $|u| \leq v$ and for any $n \in \mathbb{Z}$ and $\omega \in \mathcal{F}_n$ there exists a measurable function u^ω given on ω such that $|u| \leq u^\omega \leq v$ on ω and*

$$⨍_\omega ⨍_\omega |u^\omega(x) - u^\omega(y)|^\gamma \, \mu(dx)\mu(dy) \leq ⨍_\omega g^\gamma(x)\, \mu(dx). \tag{7}$$

Then for any $\lambda > 0$ we have

$$\big|\{x : |u(x)| \geq \lambda\}\big| \leq \nu^{-1}\lambda^{-\gamma} \int_\Omega g^\gamma(x) I_{\mathcal{M}v(x)>\alpha\lambda}\, \mu(dx), \tag{8}$$

where $\alpha = (2N_0)^{-1}$ and $\nu = 1 - 2^{-\gamma}$.

Proof. Obviously we may assume that $u \geq 0$. Fix a $\lambda > 0$ and define

$$\tau(x) = \inf\{n \in \mathbb{Z} : v_{|n}(x) > \alpha\lambda\}.$$

We know that τ is a stopping time and if $\tau(x) < \infty$, then

$$v_{|n}(x) \leq \lambda/2, \quad \forall n \leq \tau(x).$$

We also know that $v_{|n} \to v \geq u$ (a.e.) as $n \to \infty$. It follows that (a.e.)

$$\{x : u(x) \geq \lambda\} = \{x : u(x) \geq \lambda, \tau(x) < \infty\}$$

$$= \{x : u(x) \geq \lambda, v_{|\tau}(x) \leq \lambda/2\} = \bigcup_{n \in \mathbb{Z}} \bigcup_{\omega \in \mathcal{F}_n^\tau} A_n(\omega),$$

where

$$A_n(\omega) := \{x \in \omega : u(x) \geq \lambda, v_{|n}(x) \leq \lambda/2\},$$

and $\mathcal{F}_n^\tau$ is the family of disjoint elements of $\mathcal{F}_n$ such that

$$\{x : \tau(x) = n\} = \bigcup_{\omega \in \mathcal{F}_n^\tau} \omega.$$

Next, for each $n \in \mathbb{Z}$ and $\omega \in \mathcal{F}_n$ on the set $A_n(\omega)$, if it is not empty, we have $v_{|n} = v_\omega$ and on $A_n(\omega)$

$$u^\gamma - (v_\omega)^\gamma \geq \lambda^\gamma(1 - 2^{-\gamma}) = \nu\lambda^\gamma.$$

We use this and the inequality $|a - b|^\gamma \geq |a|^\gamma - |b|^\gamma$ and conclude that for $x \in A_n(\omega)$

$$⨍_\omega |u^\omega(x) - u^\omega(y)|^\gamma \, \mu(dy) \geq (u^\omega(x))^\gamma - ⨍_\omega (u^\omega(y))^\gamma \, \mu(dy)$$

$$\geq u^\gamma(x) - ⨍_\omega v^\gamma(y) \, \mu(dy) \geq u^\gamma(x) - (v_\omega(x))^\gamma \geq \nu\lambda^\gamma,$$

so that by Chebyshev's inequality

$$|A_n(\omega)| \leq \nu^{-1}\lambda^{-\gamma} \int_\omega ⨍_\omega |u^\omega(x) - u^\omega(y)|^\gamma \, \mu(dy)\mu(dx).$$

It follows by assumption (7) that

$$|A_n(\omega)| \leq \nu^{-1}\lambda^{-\gamma} \int_\omega g^\gamma \, \mu(dx),$$

$$|\{x : u(x) \geq \lambda\}| \leq \nu^{-1}\lambda^{-\gamma} \sum_{n \in \mathbb{Z}} \sum_{\omega \in \mathcal{F}_n^\tau} \int_\omega g^\gamma \, \mu(dx)$$

$$= \nu^{-1}\lambda^{-\gamma} \int_\Omega g^\gamma I_{\tau < \infty} \, \mu(dx).$$

It only remains to observe that $\{\tau < \infty\} = \{\mathcal{M}v > \alpha\lambda\}$. The lemma is proved. □

13. COROLLARY. *Under the assumptions of Lemma* 12 *for any* $p > \gamma$ *we have*

$$\int_\Omega |u|^p \, \mu(dx) \leq \beta\Big(\int_\Omega (\mathcal{M}v)^p \, \mu(dx)\Big)^{1-\gamma/p} \Big(\int_\Omega g^p \, \mu(dx)\Big)^{\gamma/p},$$

where $\beta = \nu^{-1}(1 - \gamma/p)^{-1}\alpha^{\gamma - p}$.

Indeed,

$$\int_\Omega |u|^p \, \mu(dx) = \int_0^\infty \big|\{x : |u(x)| \geq \lambda^{1/p}\}\big| \, d\lambda$$

$$\leq \nu^{-1} \int_\Omega g^\gamma \int_0^\infty \lambda^{-\gamma/p} I_{\mathcal{M}v > \alpha\lambda^{1/p}} \, d\lambda \, \mu(dx) = \beta \int_\Omega g^\gamma (\mathcal{M}v)^{p-\gamma} \, \mu(dx)$$

and it only remains to use Hölder's inequality.

The second statement of the following theorem for $\gamma = 1$ is, actually, the Fefferman-Stein theorem in the "dyadic" setting. Before stating it observe that, if $w \in L_p(\Omega)$, then by Hölder's inequality

$$|w|_{|n}(x) \leq |\omega_n(x)|^{-1/p} \|w\|_{L_p(\Omega)} \to 0$$

as $n \to -\infty$.

14. Theorem. *For $\gamma \in (0,1]$ define the sharp function $u^\sharp_\gamma$ of u by the formula*

$$u^\sharp_\gamma(x) = \sup_n \sup_{\omega \in \mathcal{F}_n : \omega \ni x} \Big(\fint_\omega \fint_\omega \big|u(z) - u(y)\big|^\gamma \, \mu(dz)\mu(dy) \Big)^{1/\gamma}. \tag{9}$$

Let $u \in L_{1,\mathrm{loc}}(\Omega)$ and suppose that $|u|_{|n} \to 0$ as $n \to -\infty$. Then for $p > \gamma$

$$\int_\Omega |u|^p \, \mu(dx) \leq \beta \Big(\int_\Omega (\mathcal{M}u)^p \, \mu(dx) \Big)^{1-\gamma/p} \Big(\int_\Omega \big(u^\sharp_\gamma\big)^p \, \mu(dx) \Big)^{\gamma/p}.$$

In particular, by Theorem 11, if $p > 1$ and $w \in L_p(\Omega)$ we have

$$\|w\|_{L_p(\Omega)} \leq N \big\| w^\sharp_\gamma \big\|_{L_p(\Omega)}, \tag{10}$$

where $N = \beta^{1/\gamma} q^{(p-\gamma)/\gamma}$, $q = p/(p-1)$.

This is a simple consequence of Corollary 13 since, obviously, for $n \in \mathbb{Z}$, $\omega \in \mathcal{F}_n$, and $v = u^\omega = |u|$ we have $|u| \leq u^\omega \leq v$, and for any $x \in \omega$

$$\fint_\omega \fint_\omega \big|u^\omega(z) - u^\omega(y)\big|^\gamma \, \mu(dz)\mu(dy)$$

$$\leq \fint_\omega \fint_\omega \big|u(z) - u(y)\big|^\gamma \, \mu(dz)\mu(dy) \leq \big(u^\sharp_\gamma(x)\big)^\gamma,$$

so that

$$\fint_\omega \fint_\omega \big|u^\omega(z) - u^\omega(y)\big|^\gamma \, \mu(dz)\mu(dy) \leq \fint_\omega \big(u^\sharp_\gamma\big)^\gamma \, \mu(dx).$$

15. Corollary. *Let $p > 1$, $\gamma \in (0,1]$, $A \in \mathcal{F}^0$, $|u|^\gamma \in L_1(\Omega)$ and $u = 0$ outside A. Then $u^\sharp_\gamma$ is well defined and*

$$\|u\|_{L_p(\Omega)} \leq N \big\| u^\sharp_\gamma \big\|_{L_p(\Omega)}$$

with N from (10).

Indeed, by Fatou and the fact that the Lipschitz constants of the functions $f_n(t) = -n \vee t \wedge n$ are equal to one we have

$$\|u\|_{L_p(\Omega)} \leq \varliminf_{n\to\infty} \|f_n(u)\|_{L_p(\Omega)} \leq \varliminf_{n\to\infty} N \big\| [f_n(u)]^\sharp_\gamma \big\|_{L_p(\Omega)} \leq N \big\| u^\sharp_\gamma \big\|_{L_p(\Omega)}.$$

16. REMARK. Note a peculiar feature of Corollary 15 that for $u^{\sharp}_{\gamma}$ to be defined u need not be even locally summable to the first power (if $\gamma < 1$). Still its L_p-norm dominates the L_p-norm of u (both could be infinite).

2. Sharp functions in a more classical settings

Fix $\gamma \in (0, 1]$, $r \in (0, \infty)$, and $z \in \mathbb{R}^{d+1}$ define

$$I_r(h, z) = \Big(\fint_{C_r(z)} \fint_{C_r(z)} \big|h(z_1) - h(z_2)\big|^{\gamma} \, dz_1 dz_2 \Big)^{1/\gamma} \tag{1}$$

whenever the right-hand side makes sense.

The maximal and sharp functions introduced in Subsections 1:2 and 1:3 are related to the underlying filtration of partitions. In particular applications the following sharp functions close to the classical one are used:

$$h^{\#}_{\gamma}(z) = \sup \big\{ I_r(h, z_0) : z_0 \in \mathbb{R}^{d+1}, r > 0, C_r(z_0) \ni z \big\}$$

(the supremum is taken over all cylinders $C_r(z_0)$, $r > 0$, $z_0 \in \mathbb{R}^{d+1}$, which contain z). One obtains the corresponding "elliptic" counterparts of these functions by taking h independent of t and one easily sees that

$$g^{\#}_{\gamma}(x) = \sup_{B_r(x_0) \ni x} \Big(\fint_{B_r(x_0)} \fint_{B_r(x_0)} \big|g(x_1) - g(x_2)\big|^{\gamma} \, dx_1 dx_2 \Big)^{1/\gamma}.$$

In the future we are also going to use the sharp functions of functions defined in domains $Q \subset \mathbb{R}^{d+1}$ by the formula

$$h^{\#}_{Q,\gamma}(z) = \sup \big\{ I_r(Q, h, z_0) : z_0 \in Q, r > 0, C_r(z_0) \ni z \big\}, \tag{2}$$

where

$$I_r(Q, h, z) = \Big(\fint_{C_r(z) \cap Q} \fint_{C_r(z) \cap Q} \big|h(z_1) - h(z_2)\big|^{\gamma} \, dz_1 dz_2 \Big)^{1/\gamma} \tag{3}$$

(of course $C_r(z) \cap Q$ can be replaced with $C_r(z)$ if $z \in Q = (0, \infty) \times \mathbb{R}^d$).

Let $g^{\sharp}_{\gamma}$ be the dyadic sharp function of g associated with the filtration $(\mathcal{F}_n, n \in \mathbb{Z})$ of parabolic dyadic cubes from Example 1.4 in case $g = g(t, x)$ and let it be associated with the filtration $(\mathcal{F}_n, n \in \mathbb{Z})$ of dyadic cubes from Example 1.3 in case $g = g(x)$. The reader should note the difference in the shapes of the "sharp" symbol in $g^{\sharp}$ and $g^{\#}$.

1. LEMMA. *Let g be a measurable function of $(t, x) \in \mathbb{R}^{d+1}$ or of $x \in \mathbb{R}^d$. Then there is a constant $N = N(d)$ such that $\big(g^{\sharp}_{\gamma}\big)^{\gamma} \leq N \big(g^{\#}_{\gamma}\big)^{\gamma}$ everywhere.*

Proof. We will only concentrate on the case of $g = g(t, x)$. If $g = g(x)$ one can just let $g(t, x) = g(x)$.

For fixed $(t, x) \in \mathbb{R}^{d+1}$, $n \in \mathbb{Z}$, and $\omega_n \in \mathcal{F}_n$ such that $(t, x) \in \omega_n$ we have

$$\omega_n \subset (t - 4^{-n}, t + 4^{-n}) \times B_{2^{-n} 2d}(x) \subset C_{\alpha 2^{-n}}(t_0, x),$$

where $\alpha = 2d$ and $t_0 = t - 4^{-n}$. Hence,

$$\fint_{\omega_n} \fint_{\omega_n} |g(z_1) - g(z_2)|^\gamma \, dz_1 dz_2$$

$$\le N_1 \fint_{C_{\alpha 2^{-n}}(t_0,x)} \fint_{C_{\alpha 2^{-n}}(t_0,x)} |g(z_1) - g(z_2)|^\gamma \, dz_1 dz_2 \le N_1 \big[g_\gamma^\#(t,x)\big]^\gamma,$$

where

$$N_1 = \frac{|C_{\alpha 2^{-n}}(t_0,x)|^2}{|\omega_n|^2}$$

depends only on d. This obviously yields the desired result and proves the lemma. □

The following theorem is just a corollary of Theorem 1.14 and Lemma 1. In case of $\mathbb{R}^d$ and $\gamma = 1$ we recover the classical Fefferman-Stein theorem.

2. THEOREM. *If $p \in (1,\infty)$, $g \in L_p(\mathbb{R}^{d+1})$, and $f \in L_p(\mathbb{R}^d)$, then*

$$\|g\|_{L_p(\mathbb{R}^{d+1})} \le N \|g_\gamma^\#\|_{L_p(\mathbb{R}^{d+1})}, \quad \|f\|_{L_p(\mathbb{R}^d)} \le N \|f_\gamma^\#\|_{L_p(\mathbb{R}^d)},$$

where N is independent of f and g.

Quite similarly by considering parabolic dyadic cubes in $\mathbb{R}_+ \times \mathbb{R}^d$ one obtains the following.

3. THEOREM. *Denote $Q = \mathbb{R}_+ \times \mathbb{R}^d$ and let $p \in (1,\infty)$ and $g \in L_p(Q)$. Then*

$$\|g\|_{L_p(Q)} \le N \|g_{Q,\gamma}^\#\|_{L_p(Q)},$$

where N depends only on d, p, and γ.

For $\rho > 0$ introduce the restricted sharp function of h by the formula

$$h_{G,\gamma,\rho}^\#(z) = \sup\big\{I_r(G,h,z_0) : z_0 \in G, r \in (0,\rho], C_r(z_0) \ni z\big\} \tag{4}$$

whenever it makes sense. Note that, if $Q = \mathbb{R}_+ \times \mathbb{R}^d$, $h_{Q,\gamma,\rho}^\#$ is well defined in C_R for measurable h even defined only in $C_{R+2\rho}$.

4. THEOREM. *Let $p \in (1,\infty)$, $\kappa \in (0,1]$, $R \in (0,\infty)$, and $h \in L_p(C_{R(1+2\kappa)})$. Let $Q = \mathbb{R}_+ \times \mathbb{R}^d$. Then*

$$\|h\|_{L_p(C_R)} \le N \|h_{Q,\gamma,\kappa R}^\#\|_{L_p(C_R)} + N\kappa^{-\chi_1} R^{-\chi_2} \| |h|^\gamma \|_{L_1(C_R)}^{1/\gamma}, \tag{5}$$

where $\chi_1 = (d+2)/\gamma$, $\chi_2 = (d+2)(1/\gamma - 1/p)$ and the constants N depend only on d, γ, and p.

We prove the theorem after a short discussion.

5. REMARK. Theorem 4 is a "local" version of Theorem 3 and will be derived from it. But actually Theorem 3 also follows from Theorem 4, which is seen by sending $R \to \infty$.

If h is independent of t, then

$$h^{\#}_{Q,\gamma,\rho}(x) = \sup_{\substack{r\in(0,\rho] \\ B_r(x_0)\ni x}} \Big(\fint_{B_r(x_0)} \fint_{B_r(x_0)} \big|h(x_1) - h(x_2)\big|^{\gamma} \, dx_1 dx_2 \Big)^{1/\gamma} =: h^{\#}_{\gamma,\rho}(x)$$

is also independent of t and the following is just a corollary of Theorem 4.

6. THEOREM. *Let $p \in (1,\infty)$, $\kappa \in (0,1]$, $R \in (0,\infty)$, and $h \in L_p(B_{R(1+2\kappa)})$. Then*

$$\|h\|_{L_p(B_R)} \le N \big\|h^{\#}_{\gamma,\kappa R}\big\|_{L_p(B_R)} + N\kappa^{-\chi_1} R^{-\chi_2} \big\| \, |h|^{\gamma} \big\|^{1/\gamma}_{L_1(B_R)},$$

where $\chi_1 = (d+2)/\gamma$, $\chi_2 = (1/\gamma - 1/p)d$ and the constants N depend only on d, γ, and p.

Proof of Theorem 4. First we note that, if we define $f(t,x) = h(R^2 t, Rx)$, then, as is easy to see,

$$f^{\#}_{Q,\gamma,\kappa}(t,x) = h^{\#}_{Q,\gamma,\kappa R}(R^2 t, Rx),$$

$$R^{-(d+2)/p}\|h\|_{L_p(C_R)} = \|f\|_{L_p(C_1)},$$

$$R^{-(d+2)/p}\big\|h^{\#}_{Q,\gamma,\kappa R}\big\|_{L_p(C_R)} = \big\|f^{\#}_{Q,\gamma,\kappa}\big\|_{L_p(C_1)},$$

$$R^{-d-2}\int_{C_R} |h|^{\gamma}\,dz = \int_{C_1} |f|^{\gamma}\,dz.$$

It follows that if (5) holds true for $R = 1$, then it also holds true for all $R > 0$ with the same constants N. Furthermore, because of monotonicity if (5) holds for $\kappa \in (0, 1/10]$, then it also holds with the second N slightly increased for $\kappa \in (0,1]$ as well. Finally, setting $h = 0$ outside $C_{R(1+2\kappa)}$ does not affect any term in (5). Therefore, in the rest of the proof we assume that

$$R = 1, \quad \kappa \in (0, 1/10], \quad h = 0 \quad \text{outside} \quad C_{1+2\kappa}.$$

Introduce a mapping

$$\psi_{(1)} : \big[0, (1+4\kappa)^2\big] \to [0,1]$$

by defining $\psi_{(1)}(t) = t$ for $t \in [0,1]$ and setting $\psi_{(1)}(t) = 2 - t$ for $(1+4\kappa)^2 \ge t \ge 1$. Notice that $\psi_{(1)}$ preserves $[0,1]$ and for $(1+4\kappa)^2 \ge t \ge 1$ it maps t into the point in $[0,1]$ that is symmetric to t with the center of symmetry being the point 1. Here we used that $2 > (1+4\kappa)^2$ to be sure that $0 \le \psi_{(1)} \le 1$. Obviously, $\psi_{(1)}$ is Lipschitz continuous with Lipschitz constant one.

Then, introduce a mapping

$$\psi_{(2)} : \bar{B}_{1+4\kappa} \to \bar{B}_1$$

by defining $\psi_{(2)}(x) = x$ for $x \in \bar{B}_1$ and setting

$$\psi_{(2)}(x) = x/|x| - \big(x - x/|x|\big)$$

for $x \in B_{1+4\kappa} \setminus B_1$. Notice that $\psi_{(2)}$ preserves B_1 and outside of B_1 it maps x into the point inside B_1 on the same radius as x and at the same distance from ∂B_1 as x. Of course $\psi_{(2)}$ is Lipschitz continuous in $\bar{B}_{1+4\kappa}$.

For $z=(t,x)\in C_{1+4\kappa}$ define

$$\psi(t,x)=\big(\psi_{(1)}(t),\psi_{(2)}(x)\big),\quad \check{h}(z)=h\big(\psi(z)\big).$$

Then take $\zeta_{(1)}\in C_0^\infty(\mathbb{R})$ and $\zeta_{(2)}\in C_0^\infty(\mathbb{R}^d)$ such that $\zeta_{(1)}=1$ on $[0,1]$, $\zeta_{(1)}(t)=0$ for $t\geq(1+\kappa)^2$, $\zeta_{(2)}=1$ on B_1, $\zeta_{(2)}=0$ outside $B_{1+\kappa}$, $0\leq\zeta_{(i)}\leq 1$, $|D\zeta_{(i)}|\leq N\kappa^{-1}$, $i=1,2$, for $z=(t,x)\in C_{1+4\kappa}$ define

$$\zeta(z)=\zeta_{(1)}(t)\zeta_{(2)}(x),\quad \hat{h}(z)=\check{h}(z)\zeta(z),$$

and continue $\hat{h}$ as zero outside $C_{1+4\kappa}$, where formally ψ is not defined.

For $z_0=(t_0,x_0)\in Q$ define

$$I_r(z_0):=\Big(\fint_{C_r(z_0)}\fint_{C_r(z_0)}\big|\hat{h}(z_1)-\hat{h}(z_2)\big|^\gamma\,dz_1dz_2\Big)^{1/\gamma}.$$

Note that, generally,

$$I_r(z_0)\leq Nr^{-(d+2)/\gamma}\Big(\int_{C_{1+\kappa}}|\hat{h}|^\gamma\,dz\Big)^{1/\gamma}\leq Nr^{-(d+2)/\gamma}K, \tag{6}$$

where

$$K=\big\|\,|h|^\gamma\big\|_{L_1(C_1)}^{1/\gamma}.$$

Now take $z=(t,x)$, z_0, $r>0$ such that

$$z,z_0\in Q,\quad z\in C_r(z_0).$$

We claim that, for any $\rho\in(0,\kappa]$,

$$I_r(z_0)\leq N\nu^{-(d+2)/\gamma}(z)I_{C^c_{1+2\kappa}}(z)K+N\big[\kappa^{-(d+2)/\gamma}+\rho^{-(d+2)/\gamma}\big]I_{C_{1+2\kappa}}(z)K$$

$$+N\rho\kappa^{-1}\big[\mathbb{M}(\check{h}^\gamma)(z)\big]^{1/\gamma}+Nh^{\#}_{Q,\gamma,\kappa}\big(\psi(z)\big)I_{C_{1+2\kappa}}(z), \tag{7}$$

where $\mathbb{M}$ is taken from (12.4.6),

$$\nu(z)=\big[t-(1+\kappa)^2\big]_+^{1/2}+\big(|x|-1-\kappa\big)_+,$$

and the constants N depend only on d and γ.

To prove the claim, we consider two cases.

Case 1:

$$z\notin C_{1+2\kappa}. \tag{8}$$

Observe that $I_r(z_0)\neq 0$ only if $C_{1+\kappa}\cap C_r(z_0)\neq\emptyset$, that may only happen if the parabolic distance of z to $C_{1+\kappa}$ is less that $2r$. It follows easily that $3r\geq\nu(z)$ in that case and this along with (6) yields (7).

Case 2:

$$z\in C_{1+2\kappa}. \tag{9}$$

Additionally we may assume that $r\leq\rho$ $(\leq\kappa)$, because otherwise (7) holds owing to (6). Note now that, for $z_1=(t_1,x_1),z_2=(t_2,x_2)\in C_r(z_0)$, we have

$$\big|\hat{h}(z_1)-\hat{h}(z_2)\big|\leq\big|\check{h}(z_1)\big|\,\big|\zeta(z_1)-\zeta(z_2)\big|+\big|\check{h}(z_1)-\check{h}(z_2)\big|.$$

Here the first term on the right is less than $Nr\kappa^{-1}|\check{h}(z_1)|$, and we conclude

$$I_r(z_0) \le N\rho\kappa^{-1}\big[\mathbb{M}(\check{h}^\gamma)(z)\big]^{1/\gamma} + Nr^{-(2d+4)/\gamma}J_r(z_0), \tag{10}$$

where

$$J_r^\gamma(z_0) = \int_{C_r(z_0)}\int_{C_r(z_0)} \big|\check{h}(z_1) - \check{h}(z_2)\big|^\gamma\, dz_1dz_2.$$

We represent $J_r^\gamma(z_0)$ as

$$J_r^\gamma(z_0) = \sum_{i,j=1}^4 J^{ij}$$

according to integrating with respect to $z_1 \in \Gamma_i$ and $z_2 \in \Gamma_j$, where

$$\Gamma_1 = C_r(z_0) \cap C_1, \quad \Gamma_2 = C_r(z_0) \cap [1,\infty) \times B_1,$$

$$\Gamma_3 = C_r(z_0) \cap [1,\infty) \times B_1^c, \quad \Gamma_4 = C_r(z_0) \cap [0,1] \times B_1^c.$$

Since $C_r(z_0) \subset C_{1+4\kappa}$, the Γ_i's are cylinders (perhaps not round ones, some of which could be empty), in which the mapping ψ is one-to-one with bounded Jacobian separated from zero. By performing obvious changes of variables, we see that for any i, j

$$\begin{aligned} J^{ij} &\le N\int_{\psi(\Gamma_i)}\int_{\psi(\Gamma_j)} \big|h(z_1) - h(z_2)\big|^\gamma\, dz_1dz_2 \\ &\le N\int_{\psi(C_r(z_0))}\int_{\psi(C_r(z_0))} \big|h(z_1) - h(z_2)\big|^\gamma\, dz_1dz_2. \end{aligned} \tag{11}$$

Furthermore, $\psi(C_r(z_0))$ is contained in a cylinder, say $C_r\big(z_0'\big)$ with $z_0' \in C_1$ (containing $\psi(z)$, recall $z \in C_r(z_0)$), and we increase the right-hand side of (11) by replacing $\psi\big(C_r(z_0)\big)$ with $C_r\big(z_0'\big) \subset C_{1+\kappa}$ (recall $r \le \rho \le \kappa$). Then by definition we conclude from (11) that

$$J^{ij} \le r^{2d+4}\Big(h^\#_{Q,\gamma,\kappa}\big(\psi(z)\big)\Big)^\gamma I_{C_{1+2\kappa}}(z). \tag{12}$$

In combination with (10) this leads to (7) in Case 2 as well and proves our claim.

Now according to our definitions we get that $\hat{h}^\#_{Q,\gamma}(z)$ is less than the right-hand side of (7). By Theorem 3 and by the Hardy-Littlewood maximal function theorem we obtain that

$$\|h\|_{L_p(C_1)} \le \big\|\hat{h}\big\|_{L_p(Q)} \le N\big\|\hat{h}^\#_{Q,\gamma}\big\|_{L_p(Q)} \le NKF_1 + F_2 + N\rho\kappa^{-1}F_3 + NF_4,$$

where

$$F_1 = \|\nu^{-(d+2)/\gamma} I_{C^c_{1+2\kappa}}\|_{L_p(Q)},$$

$$F_2 = NK[\kappa^{-(d+2)/\gamma} + \rho^{-(d+2)/\gamma}],$$

$$F_3 = \|\check{h}\|_{L_p(Q)} \le 3\|h\|_{L_p(C_1)},$$

$$F_4 = \|h^{\#}_{Q,\gamma,\kappa}(\psi(\cdot)) I_{C_{1+2\kappa}}\|_{L_p(Q)} \le 3\|h^{\#}_{Q,\gamma,\kappa}\|_{L_p(C_1)}.$$

Observe that

$$F_1^p = \int_{(1+2\kappa)^2}^{\infty}\int_{|x|\ge 1+2\kappa} \frac{1}{\left(\left[t-(1+\kappa)^2\right]^{1/2} + |x| - 1 - \kappa\right)^{\beta}}\,dtdx$$

$$+\int_{(1+2\kappa)^2}^{\infty}\int_{|x|\le 1+2\kappa} \frac{1}{\left[t-(1+\kappa)^2\right]^{\beta/2}}\,dtdx$$

$$+(1+2\kappa)^2\int_{x|\ge 1+2\kappa} \frac{1}{\left(|x|-1-\kappa\right)^{\beta}}\,dx =: G_1 + G_2 + (1+2\kappa)^2 G_3,$$

where $\beta = (d+2)p/\gamma$.

While dealing with G_1 we substitute $t = s^2 + (1+\kappa)^2$ $(s > 0)$ and note that the inequality $s^2 + (1+\kappa)^2 \ge (1+2\kappa)^2$ implies that $s \ge \sqrt{\kappa}$. We also use polar coordinates and get that

$$G_1 \le N\int_{s\ge\sqrt{\kappa}}\int_{r\ge 1+2\kappa} \frac{sr^{d-1}}{\left(s+r-1-\kappa\right)^{\beta}}\,dsdr$$

$$= N\int_{s\ge\sqrt{\kappa}}\int_{r\ge\kappa} \frac{s(r+1+\kappa)^{d-1}}{\left(s+r\right)^{\beta}}\,dsdr.$$

The numerator in the last integrand we estimate by $Nsr^{d-1} + Ns \le N(s+r)^d + N(s+r)$ and we see that

$$G_1 \le N\int_{s\ge\sqrt{\kappa}}\int_{r\ge\kappa} \left[(s+r)^{d-\beta} + (s+r)^{1-\beta}\right] dsdr$$

$$= \int_{s\ge\sqrt{\kappa}} \left[N(s+\kappa)^{d-\beta+1} + N(s+\kappa)^{2-\beta}\right] ds$$

$$= N(\sqrt{\kappa}+\kappa)^{d-\beta+2} + N(\sqrt{\kappa}+\kappa)^{3-\beta}$$

$$\le N\kappa^{(3-\beta)/2}.$$

The term G_2 is obviously less than

$$N\left[(1+2\kappa)^2 - (1+\kappa)^2\right]^{1-\beta/2} \le N\kappa^{1-\beta/2}.$$

While estimating G_3 we again use polar coordinates and see that it is less than a constant times

$$\int_{1+2\kappa}^{\infty} \frac{r^{d-1}}{(r-1-\kappa)^{\beta}}\,dr \le 2^{d-1}\int_{1+2\kappa}^{2} \frac{1}{(r-1-\kappa)^{\beta}}\,dr + N \le N\kappa^{1-\beta}.$$

Now since $\beta-1 \ge \beta/2-1, (\beta-3)/2$, the above estimates imply that $F_1 \le N\kappa^{(1-\beta)/p}$ and lead to the conclusion that

$$\|h\|_{L_p(C_1)} \le N\big[\kappa^{(1-\beta)/p} + \kappa^{-(d+2)/\gamma} + \rho^{-(d+2)/\gamma}\big]K$$
$$+N_1\rho\kappa^{-1}\|h\|_{L_p(C_1)} + N\big\|h^{\#}_{\gamma,\kappa}\big\|_{L_p(C_1)},$$

where ρ is any number in $(0,\kappa]$. As usual we may assume that $N_1 \ge 1$ and then take $\rho = \varepsilon\kappa$, where $N_1\varepsilon = 1/2$. Then we come to (5) after observing that $(\beta-1)/p < (d+2)/\gamma$ and the theorem is proved. □

7. REMARK. The reader might have noticed that the last expression in (11) increases even if we replace $\psi\big(C_r(z_0)\big)$ $(\subset C_1)$ not with $C_r\big(z_0'\big)$ but with $C_r\big(z_0'\big)\cap C_1$. Since $\big|C_r\big(z_0'\big)\cap C_1\big| \le \big|C_r\big(z_0'\big)\big| = \big|C_r(z_0)\big|$ this shows that in (12) we can replace Q with C_1 and then it shows that (5) holds with C_R in place of Q. This result looks neater, because the values of h outside C_R are not involved anywhere. However, it turns out to be much less useful in our applications than (12).

3. Restricted sharp functions in the presence of lateral boundaries

Fix $\gamma \in (0,1]$ and recall definition (2.4) of $h^{\#}_{G,\gamma,\rho}$. First we concentrate on the case that the lateral boundary is flat.

Recall that $C_r^+(t,x) = [t,t+r^2)\times B_r^+(x)$ and $C_r^+ = C_r^+(0,0)$.

1. THEOREM. *Let* $p \in (1,\infty)$, $\kappa \in (0,1]$, $R \in (0,\infty)$, *and* $g \in L_p\big(C^+_{R(1+2\kappa)}\big)$. *Let* $G = \mathbb{R}_+ \times \mathbb{R}^d_+$. *Then*

$$\|g\|_{L_p(C_R^+)} \le N\big\|g^{\#}_{G,\gamma,\kappa R}\big\|_{L_p(C_R^+)} + N\kappa^{-\chi_1}R^{-\chi_2}\big\|\,|g|^{\gamma}\big\|^{1/\gamma}_{L_1(C_R^+)}, \tag{1}$$

where $\chi_1 = (d+2)/\gamma$, $\chi_2 = (d+2)(1/\gamma - 1/p)$ *and the constants* N *depend only on* $d,\gamma,$ *and* p.

Proof. The argument in the beginning of the proof of Theorem 2.4 is still valid in our case and shows that it suffices to concentrate on $R=1$.

In that case, we use reflection about the hyperplane $\{x^1=0\}$. Define a mapping

$$\eta : (t,x) = (t,x^1,x') \in Q := \mathbb{R}_+ \times \mathbb{R}^d \to \eta(t,x) = \big(t,|x^1|,x'\big)$$

and extend g to $C_{1+2\kappa}$ by the formula $g(z) = g\big(\eta(z)\big)$. We claim that in C_1

$$g^{\#}_{Q,\gamma,\kappa}(z) \le N(\gamma)g^{\#}_{G,\gamma,\kappa}\big(\eta(z)\big). \tag{2}$$

Notice that (2) is unaffected on C_1 if we change arbitrarily g outside $C_{1+2\kappa}$. It is convenient to set $g=0$ there and prove (2) for all $z \in Q$.

Since g is symmetric with respect to x^1, it suffices to prove (2) for all $z \in G$. For $z_0 = (t_0,x_0) \in Q$, $r>0$, set

$$I_r(z_0) = \Big(\fint_{C_r(z_0)} \fint_{C_r(z_0)} \big|g(z_1) - g(z_2)\big|^\gamma \, dz_1 dz_2 \Big)^{1/\gamma}.$$

Observe that, if $z \in C_r(z_0)$, $x_0^1 \ge r$, and $r \le \kappa$, then

$$I_r(z_0) = \Big(\fint_{C_r^+(z_0)} \fint_{C_r^+(z_0)} \big|g(z_1) - g(z_2)\big|^\gamma \, dz_1 dz_2 \Big)^{1/\gamma} \le g^{\#}_{G,\gamma,\kappa}(z).$$

However, if $x_0^1 \le r \le \kappa$, then

$$C_r(z_0) \subset C_r(z_0) \cup C_r\big(t_0, -x_0^1, x_0'\big) = C_r^+(\hat z_0) \cup \operatorname{ref} C_r^+(\hat z_0)$$

(the last equality holds up to a set of measure zero) where $\hat z_0 = \big(t_0, |x_0^1|, x_0'\big)$ and ref is the reflection operator in $\mathbb{R}^{d+1}$ about the hyperplane $x^1 = 0$. Also note that $C_r^+(\hat z_0) \ni z$ and $\big|C_r^+(\hat z_0)\big| \le \big|C_r(z_0)\big|$. It follows from these arguments and from the symmetry of g that in this case

$$I_r^\gamma(z_0) \le 4\big|C_r(z_0)\big|^{-2} \int_{C_r^+(\hat z_0)} \int_{C_r^+(\hat z_0)} \big|g(z_1) - g(z_2)\big|^\gamma \, dz_1 dz_2$$

$$\le 4 \fint_{C_r^+(\hat z_0)} \fint_{C_r^+(\hat z_0)} \big|g(z_1) - g(z_2)\big|^\gamma \, dz_1 dz_2 \le 4\big(g^{\#}_{G,\gamma,\kappa}(z)\big)^\gamma.$$

We see that the inequality between the extreme terms holds for all $z \in G$ and $r \le \kappa$, and this is equivalent to (2). We now get (1) in light of Theorem 2.4. The theorem is proved. □

The remaining results of this section treat smooth cylinders or smooth domains. If $\Omega \in C^{1,1}$ and $0 \in \partial\Omega$, we assume that the original system of coordinates in $\mathbb{R}^d$ coincides with the one described in Definition 14.1.10 and with the help of the mappings $x(y)$ and $y(x)$ introduced in Lemma 14.1.11, for $r > 0$ and $z \in \mathbb{R}^d$, we construct

$$\check B_r^+ = x\big(B_r^+\big), \quad \check B_r^+(z) = x\big(B_r^+(y(z))\big).$$

By Remark 14.1.12 we have $\check B_r^+ \subset \Omega$ for $r \le 8\rho_0(\Omega)$ and $\check B_r^+(z) \subset \check B^+_{8\rho_0(\Omega)}$ if $\rho > 0$, $\rho + r \le 8\rho_0(\Omega)$, and $z \in \check B_\rho^+$. Generally, these are objects in $\mathbb{R}^d$.

Then set

$$\check C_R^+ = [0, R^2) \times \check B_R^+$$

and for $\rho, r \le 2\rho_0(\Omega)$ and $z = (t, x)$ such that $x \in \check B_\rho^+$ and $t \in \mathbb{R}$ define

$$\check C_r^+(z) = [t, t + r^2) \times \check B_r^+(x).$$

Finally, whenever it makes sense, for $\rho \le \rho_0(\Omega)$, and $z \in \check C^+_{2\rho_0(\Omega)}$ introduce

$$h^{\#}_{\Omega,\gamma,\rho}(z) = \sup\big\{I_r(h, z_0) : z_0 \in \mathbb{R}_+ \times \Omega, r \in (0, \rho], \check C_r^+(z_0) \ni z\big\}, \tag{3}$$

where

$$I_r(h, z_0) = \Big(\fint_{\check C_r^+(z_0)} \fint_{\check C_r^+(z_0)} \big|h(z_1) - h(z_2)\big|^\gamma \, dz_1 dz_2 \Big)^{1/\gamma}.$$

The reader should pay attention to the above curved sharp symbol, reminding of curved boundaries.

Observe that, if $r \le \rho \le \rho_0(\Omega)$ and $z \in \check{C}^+_{2\rho_0(\Omega)} \cap \check{C}^+_r(z_0)$, then

$$\check{C}^+_r(z_0) \subset \check{C}^+_{4\rho_0(\Omega)},$$

so that $h^\#_{\Omega,\gamma,\rho}(z)$ is well defined on $\check{C}^+_{2\rho_0(\Omega)}$ even if h is given only on $\check{C}^+_{4\rho_0(\Omega)}$ $(\subset [0, 16\rho_0^2(\Omega)) \times \Omega)$.

2. THEOREM. *If $p \in (1,\infty)$, $\kappa \in (0,1/2]$, $0 < R \le 2\rho_0(\Omega)$, and $h \in L_p(\check{C}^+_{R(1+2\kappa)})$, then*

$$\|h\|_{L_p(\check{C}^+_R)} \le N\|h^\#_{\Omega,\gamma,\kappa R}\|_{L_p(\check{C}^+_R)} + N\kappa^{-\chi_1}R^{-\chi_2}\| \, |h|^\gamma \|^{1/\gamma}_{L_1(\check{C}^+_R)}, \tag{4}$$

where $\chi_1 = (d+2)/\gamma$, $\chi_2 = (d+2)(1/\gamma - 1/p)$ and the constants N depend only on d, γ, and p.

Proof. Take $h \in L_p(\check{C}^+_{R(1+2\kappa)})$ and introduce $g(s,y) = h(s, x(y))$, which is well defined, belongs to $L_p(C^+_{R(1+2\kappa)})$, and

$$\|h\|_{L_p(\check{C}^+_R)} = \|g\|_{L_p(C^+_R)}.$$

Furthermore, as is easy to see, for $(s,y) \in C^+_R$ and $G = \mathbb{R}_+ \times \mathbb{R}^d_+$,

$$g^\#_{G,\gamma,\kappa R}(s,y) = h^\#_{\Omega,\gamma,\kappa R}(s, x(y)),$$

so that also

$$\|h^\#_{\Omega,\gamma,\kappa R}\|_{L_p(\check{C}^+_R))} = \|g^\#_{G,\gamma,\kappa R}\|_{L_p(C^+_R)}.$$

The same relation also holds for the last terms in (4) and (1). Thus, (4) is just another form of (1) and the theorem is proved. □

By taking h independent of t we reduce definition (3) to

$$h^\#_{\Omega,\gamma,\rho}(x) = \sup\{I_r(h,x_0) : x_0 \in \Omega, r \in (0,\rho], \check{B}^+_r(x_0) \ni x\},$$

where

$$I_r(h,x_0) = \Big(\fint_{\check{B}^+_r(x_0)} \fint_{\check{B}^+_r(x_0)} |h(x_1) - h(x_2)|^\gamma \, dx_1 dx_2 \Big)^{1/\gamma},$$

and get from Theorem 2 the following "elliptic" version.

3. THEOREM. *If $p \in (1,\infty)$, $\kappa \in (0,1/2]$, $0 < R \le 2\rho_0(\Omega)$, and $h \in L_p(\check{B}^+_{R(1+2\kappa)})$, then*

$$\|h\|_{L_p(\check{B}^+_R)} \le N\|h^\#_{\Omega,\gamma,\kappa R}\|_{L_p(\check{B}^+_R)} + N\kappa^{-\chi_1}R^{-\chi_2}\| \, |h|^\gamma \|^{1/\gamma}_{L_1(\check{B}^+_R)}, \tag{5}$$

where $\chi_1 = (d+2)/\gamma$, $\chi_2 = (1/\gamma - 1/p)d$ and the constants N depend only on d, γ, and p.

Bibliography

[1] A. D. Aleksandrov, *Dirichlet's problem for the equation* $\operatorname{Det}||z_{ij}|| = \varphi(z_1, \cdots, z_n, z, x_1, \cdots, x_n)$. *I* (Russian, with English summary), Vestnik Leningrad. Univ. Ser. Mat. Meh. Astr. **13** (1958), no. 1, 5–24. MR0096903

[2] A. D. Aleksandrov, *Certain estimates for the Dirichlet problem*, Soviet Math. Dokl. **1** (1961), 1151–1154. English translation in Soviet Math. Dokl., Vol. 1 (1961), 1151–1154. MR0147776

[3] A. D. Aleksandrov, *Uniqueness conditions and bounds for the solution of the Dirichlet problem* (Russian, with English summary), Vestnik Leningrad. Univ. Ser. Mat. Meh. Astronom. **18** (1963), no. 3, 5–29. English translation in Amer. Mat. Soc. Transl., Vol. 68 (1968), no. 2, 89-119. MR0164135

[4] A. D. Aleksandrov, *Majorants of solutions of linear equations of order two* (Russian, with English summary), Vestnik Leningrad. Univ. **21** (1966), no. 1, 5–25. English translation in Amer. Mat. Soc. Transl., Vol. 68 (1968), no. 2, 120–143. MR0199540

[5] S. Agmon, A. Douglis, and L. Nirenberg, *Estimates near the boundary for solutions of elliptic partial differential equations satisfying general boundary conditions. I*, Comm. Pure Appl. Math. **12** (1959), 623–727, DOI 10.1002/cpa.3160120405. MR0125307

[6] I. Ja. Bakel′man, *On the theory of quasilinear elliptic equations* (Russian), Sibirsk. Mat. Ž. **2** (1961), 179–186. MR0126604

[7] I. Ja. Bakel'man, *Geometrical methods for solving elliptic equations*, Moscow, Nauka, 1965 in Russian.

[8] J. M. Ball and L. C. Evans, *Weak convergence theorems for nonlinear partial differential equations of first and second order*, J. London Math. Soc. (2) **25** (1982), no. 2, 332–346, DOI 10.1112/jlms/s2-25.2.332. MR653391

[9] J.-M. Bony, *Principe du maximum dans les espaces de Sobolev* (French), C. R. Acad. Sci. Paris Sér. A-B **265** (1967), A333–A336. MR0223711

[10] M. Bramanti and M. C. Cerutti, $W_p^{1,2}$ *solvability for the Cauchy-Dirichlet problem for parabolic equations with VMO coefficients*, Comm. Partial Differential Equations **18** (1993), no. 9-10, 1735–1763, DOI 10.1080/03605309308820991. MR1239929

[11] H. Brézis and L. C. Evans, *A variational inequality approach to the Bellman-Dirichlet equation for two elliptic operators*, Arch. Rational Mech. Anal. **71** (1979), no. 1, 1–13, DOI 10.1007/BF00250667. MR522704

[12] L. A. Caffarelli, *Interior a priori estimates for solutions of fully nonlinear equations*, Ann. of Math. (2) **130** (1989), no. 1, 189–213, DOI 10.2307/1971480. MR1005611

[13] L. A. Caffarelli, *Interior* $W^{2,p}$ *estimates for solutions of the Monge-Ampère equation*, Ann. of Math. (2) **131** (1990), no. 1, 135–150, DOI 10.2307/1971510. MR1038360

[14] L. A. Caffarelli and X. Cabré, *Fully nonlinear elliptic equations*, American Mathematical Society Colloquium Publications, vol. 43, American Mathematical Society, Providence, RI, 1995. MR1351007

[15] L. Caffarelli, M. G. Crandall, M. Kocan, and A. Swięch, *On viscosity solutions of fully nonlinear equations with measurable ingredients*, Comm. Pure Appl. Math. **49** (1996), no. 4, 365–397, DOI 10.1002/(SICI)1097-0312(199604)49:4⟨365::AID-CPA3⟩3.3.CO;2-V. MR1376656

[16] L. Caffarelli and L. Silvestre, *On the Evans-Krylov theorem*, Proc. Amer. Math. Soc. **138** (2010), no. 1, 263–265, DOI 10.1090/S0002-9939-09-10077-1. MR2550191

[17] L. Caffarelli, J. J. Kohn, L. Nirenberg, and J. Spruck, *The Dirichlet problem for nonlinear second-order elliptic equations. II. Complex Monge-Ampère, and uniformly elliptic, equations*, Comm. Pure Appl. Math. **38** (1985), no. 2, 209–252, DOI 10.1002/cpa.3160380206. MR780073

[18] L. Caffarelli and L. Silvestre, *The Evans-Krylov theorem for nonlocal fully nonlinear equations*, Ann. of Math. (2) **174** (2011), no. 2, 1163–1187, DOI 10.4007/annals.2011.174.2.9. MR2831115

[19] S. Y. Cheng and S. T. Yau, *On the regularity of the Monge-Ampère equation* $\det(\partial^2 u/\partial x_i \partial s x_j) = F(x,u)$, Comm. Pure Appl. Math. **30** (1977), no. 1, 41–68, DOI 10.1002/cpa.3160300104. MR0437805

[20] F. Chiarenza, M. Frasca, and P. Longo, *Interior* $W^{2,p}$ *estimates for nondivergence elliptic equations with discontinuous coefficients*, Ricerche Mat. **40** (1991), no. 1, 149–168. MR1191890

[21] F. Chiarenza, M. Frasca, and P. Longo, $W^{2,p}$*-solvability of the Dirichlet problem for nondivergence elliptic equations with VMO coefficients*, Trans. Amer. Math. Soc. **336** (1993), no. 2, 841–853, DOI 10.2307/2154379. MR1088476

[22] M. G. Crandall, K. Fok, M. Kocan, and A. Święch, *Remarks on nonlinear uniformly parabolic equations*, Indiana Univ. Math. J. **47** (1998), no. 4, 1293–1326, DOI 10.1512/iumj.1998.47.1561. MR1687138

[23] M. G. Crandall, H. Ishii, and P.-L. Lions, *User's guide to viscosity solutions of second order partial differential equations*, Bull. Amer. Math. Soc. (N.S.) **27** (1992), no. 1, 1–67, DOI 10.1090/S0273-0979-1992-00266-5. MR1118699

[24] M. G. Crandall, M. Kocan, P. L. Lions, and A. Święch, *Existence results for boundary problems for uniformly elliptic and parabolic fully nonlinear equations*, Electron. J. Differential Equations (1999), no. 24, 22 pp. MR1696765

[25] M. G. Crandall, M. Kocan, and A. Święch, L^p*-theory for fully nonlinear uniformly parabolic equations*, Comm. Partial Differential Equations **25** (2000), no. 11-12, 1997–2053, DOI 10.1080/03605300008821576. MR1789919

[26] J. V. da Silva and E. V. Teixeira, *Sharp regularity estimates for second order fully nonlinear parabolic equations*, Math. Ann. **369** (2017), no. 3-4, 1623–1648, DOI 10.1007/s00208-016-1506-y. MR3713553

[27] E. De Giorgi, *Sulla differenziabilità e l'analiticità delle estremali degli integrali multipli regolari* (Italian), Mem. Accad. Sci. Torino. Cl. Sci. Fis. Mat. Nat. (3) **3** (1957), 25–43. MR0093649

[28] G. De Philippis, A. Figalli, and O. Savin, *A note on interior* $W^{2,1+\varepsilon}$ *estimates for the Monge-Ampère equation*, Math. Ann. **357** (2013), no. 1, 11–22, DOI 10.1007/s00208-012-0895-9. MR3084340

[29] H. Dong and N. V. Krylov, *On the existence of smooth solutions for fully nonlinear parabolic equations with measurable "coefficients" without convexity assumptions*, Comm. Partial Differential Equations **38** (2013), no. 6, 1038–1068, DOI 10.1080/03605302.2012.756013. MR3053993

[30] H. Dong, N. V. Krylov, and X. Li, *On fully nonlinear elliptic and parabolic equations with VMO coefficients in domains*, Algebra i Analiz **24** (2012), no. 1, 53–94, DOI 10.1090/S1061-0022-2012-01231-8; English transl., St. Petersburg Math. J. **24** (2013), no. 1, 39–69. MR3013294

[31] N. Dunford and J. T. Schwartz, *Linear Operators. I. General Theory*, With the assistance of W. G. Bade and R. G. Bartle. Pure and Applied Mathematics, Vol. 7, Interscience Publishers, Inc., New York; Interscience Publishers, Ltd., London, 1958. MR0117523

[32] L. Escauriaza, $W^{2,n}$ *a priori estimates for solutions to fully nonlinear equations*, Indiana Univ. Math. J. **42** (1993), no. 2, 413–423, DOI 10.1512/iumj.1993.42.42019. MR1237053

[33] L. C. Evans, *A convergence theorem for solutions of nonlinear second-order elliptic equations*, Indiana Univ. Math. J. **27** (1978), no. 5, 875–887, DOI 10.1512/iumj.1978.27.27059. MR503721

[34] L. C. Evans, *Classical solutions of fully nonlinear, convex, second-order elliptic equations*, Comm. Pure Appl. Math. **35** (1982), no. 3, 333–363, DOI 10.1002/cpa.3160350303. MR649348

[35] L. C. Evans, *Some estimates for nondivergence structure, second order elliptic equations*, Trans. Amer. Math. Soc. **287** (1985), no. 2, 701–712, DOI 10.2307/1999671. MR768735

[36] L. C. Evans and A. Friedman, *Optimal stochastic switching and the Dirichlet problem for the Bellman equation*, Trans. Amer. Math. Soc. **253** (1979), 365–389, DOI 10.2307/1998203. MR536953

[37] L. C. Evans and R. F. Gariepy, *Measure theory and fine properties of functions*, Revised edition, Textbooks in Mathematics, CRC Press, Boca Raton, FL, 2015. MR3409135

[38] L. C. Evans and P.-L. Lions, *Résolution des équations de Hamilton-Jacobi-Bellman pour des opérateurs uniformément elliptiques* (French, with English summary), C. R. Acad. Sci. Paris Sér. A-B **290** (1980), no. 22, A1049–A1052. MR582494

[39] E. Fabes, N. Garofalo, and S. Salsa, *A control on the set where a Green's function vanishes*, Colloq. Math. **60/61** (1990), no. 2, 637–647, DOI 10.4064/cm-60-61-2-637-647. MR1096402

[40] E. Fabes and C. Kenig, *Examples of singular parabolic measures and singular transition probability densities*, Duke Math. J. **48** (1981), no. 4, 845–856, DOI 10.1215/S0012-7094-81-04846-8. MR782580

[41] E. Fabes and D. Stroock, *The L^p-integrability of Green's functions and fundamental solutions for elliptic and parabolic equations*, Duke Math. J. **51** (1984), no. 4, 997–1016, DOI 10.1215/S0012-7094-84-05145-7. MR771392

[42] W. Feller, *An Introduction to Probability Theory and Its Applications. Vol. I*, John Wiley & Sons, Inc., New York, N.Y., 1950. MR0038583

[43] W. H. Fleming and R. W. Rishel, *Deterministic and stochastic optimal control*, Springer-Verlag, Berlin-New York, 1975. Applications of Mathematics, no. 1. MR0454768

[44] A. Figalli, Y.-H. Kim, and R. J. McCann, *Hölder continuity and injectivity of optimal maps*, Arch. Ration. Mech. Anal. **209** (2013), no. 3, 747–795, DOI 10.1007/s00205-013-0629-5. MR3067826

[45] D. Gilbarg and N. S. Trudinger, *Elliptic partial differential equations of second order*, Classics in Mathematics, Springer-Verlag, Berlin, 2001. Reprint of the 1998 edition. MR1814364

[46] K. K. Golovkin, *On equivalent normalizations of fractional spaces* (Russian), Trudy Mat. Inst. Steklov. **66** (1962), 364–383. MR0154028

[47] K. K. Golovkin and V. A. Solonnikov, *Estimates for integral operators in translation invariant norms. II* (Russian), Trudy Mat. Inst. Steklov. **92** (1966), 5–30. English translation in Proc. Steklov Inst. Math., 92 (1968), 3–32. MR0205005

[48] M. de Guzmán, *Differentiation of integrals in R^n*, Lecture Notes in Mathematics, Vol. 481, Springer-Verlag, Berlin-New York, 1975. With appendices by Antonio Córdoba, and Robert Fefferman, and two by Roberto Moriyón. MR0457661

[49] A. V. Ivanov, *Second-order quasilinear degenerate and nonuniformly elliptic and parabolic equations* (Russian), Trudy Mat. Inst. Steklov. **160** (1982), 285. English translation by American Mathematical Soc., 1984. MR702172

[50] R. Jensen and A. Święch, *Uniqueness and existence of maximal and minimal solutions of fully nonlinear elliptic PDE*, Commun. Pure Appl. Anal. **4** (2005), no. 1, 199–207. MR2126286

[51] N. Katzourakis, *An introduction to viscosity solutions for fully nonlinear PDE with applications to calculus of variations in L^∞*, SpringerBriefs in Mathematics, Springer, Cham, 2015. MR3289084

[52] S. Koike, A beginner guide to the theory of viscosity solutions, June 28, 2012, 2nd edition, available online at http://www.math.tohoku.ac.jp/~koike/evis2012version.pdf.

[53] N. V. Krylov, *Boundedly inhomogeneous nonlinear elliptic and parabolic equations on the plane* (Russian), Mat. Sb. (N.S.) **82 (124)** (1970), 99–110. English translation in Math. USSR Sbornik, Vol. 11 (1970), no. 1, 89–99. MR0262662

[54] N. V. Krylov, *The uniqueness of the solution of Bellman's equation* (Russian), Izv. Akad. Nauk SSSR Ser. Mat. **35** (1971), 1377–1388. English translation in Math. USSR Izvestija, Vol. 5 (1971), no. 6, 1387–1398. MR0292571

[55] N. V. Krylov, *Certain estimates in the theory of stochastic integrals*, Teoriya Veroyatnostei i ee Primeneniya, Vol. 18 (1973), no. 1, 56–65, Russian; English translation in Theor. Probability Appl., Vol. 18 (1973), no. 1, 54–63.

[56] N. V. Krylov, *Some estimates for the density of the distribution of a stochastic integral* (Russian), Izv. Akad. Nauk SSSR Ser. Mat. **38** (1974), 228–248. English translation in Math. USSR Izvestija, Vol. 8 (1974), no. 1, 233–254. MR0345206

[57] N. V. Krylov, *Controlled diffusion processes*, Applications of Mathematics, vol. 14, Springer-Verlag, New York-Berlin, 1980. Translated from the Russian by A. B. Aries. MR601776

[58] N. V. Krylov, *Some new results in the theory of controlled diffusion processes* (Russian), Mat. Sb. (N.S.) **109(151)** (1979), no. 1, 146–164, 166. English translation in Math. USSR Sbornik, Vol. 37 (1980), no. 1, 133–149. MR538554

[59] N. V. Krylov, *Boundedly inhomogeneous elliptic and parabolic equations* (Russian), Izv. Akad. Nauk SSSR Ser. Mat. **46** (1982), no. 3, 487–523, 670. English translation in Math. USSR Izvestija, Vol. 20 (1983), no. 3, 459–492. MR661144

[60] N. V. Krylov, *G-convergence of elliptic operators in nondivergence form* (Russian), Mat. Zametki **37** (1985), no. 4, 522–527, 600. English translation in Math. Notes, Vol. 37 (1985), no. 3-4, 290–292. MR790976

[61] N. V. Krylov, *Nonlinear elliptic and parabolic equations of the second order*, Mathematics and its Applications (Soviet Series), vol. 7, D. Reidel Publishing Co., Dordrecht, 1987. Translated from the Russian by P. L. Buzytsky [P. L. Buzytskiĭ]. MR901759

[62] N. V. Krylov, *Estimates of the maximum of the solution of a parabolic equation and estimates of the distribution of a semimartingale* (Russian), Mat. Sb. (N.S.) **130(172)** (1986), no. 2, 207–221, 284. English translation in Math. USSR Sbornik, Vol. 58 (1987), no. 1, 207–222. MR854972

[63] N. V. Krylov, *Smoothness of the payoff function for a controllable diffusion process in a domain* (Russian), Izv. Akad. Nauk SSSR Ser. Mat. **53** (1989), no. 1, 66–96, DOI 10.1070/IM1990v034n01ABEH000603; English transl., Math. USSR-Izv. **34** (1990), no. 1, 65–95. MR992979

[64] N. V. Krylov, *A theorem on degenerate elliptic Bellman equations in bounded domains*, Differential Integral Equations **8** (1995), no. 5, 961–980. MR1325541

[65] N. V. Krylov, *Lectures on elliptic and parabolic equations in Hölder spaces*, Graduate Studies in Mathematics, vol. 12, American Mathematical Society, Providence, RI, 1996. Russian translation, "Nauchnaya kniga", Novosibirsk, 1998. MR1406091

[66] N. V. Krylov, *Fully nonlinear second order elliptic equations: recent development*, Ann. Scuola Norm. Sup. Pisa Cl. Sci. (4) **25** (1997), no. 3-4, 569–595 (1998). Dedicated to Ennio De Giorgi. MR1655532

[67] N. V. Krylov, *On factorizations of smooth nonnegative matrix-values functions and on smooth functions with values in polyhedra*, Appl. Math. Optim. **58** (2008), no. 3, 373–392, DOI 10.1007/s00245-008-9040-2. MR2456852

[68] N. V. Krylov, *Lectures on elliptic and parabolic equations in Sobolev spaces*, Graduate Studies in Mathematics, vol. 96, American Mathematical Society, Providence, RI, 2008. MR2435520

[69] N. V. Krylov, *Second-order elliptic equations with variably partially VMO coefficients*, J. Funct. Anal. **257** (2009), no. 6, 1695–1712, DOI 10.1016/j.jfa.2009.06.014. MR2540989

[70] N. V. Krylov, *On Bellman's equations with VMO coefficients*, Methods Appl. Anal. **17** (2010), no. 1, 105–121, DOI 10.4310/MAA.2010.v17.n1.a4. MR2735102

[71] N. V. Krylov, *On a representation of fully nonlinear elliptic operators in terms of pure second order derivatives and its applications*, Problemy Matemat. Analiza, Vol. 59, July 2011, p. 3–24 in Russian; English translation in Journal of Mathematical Sciences, New York, Vol. 177 (2011), no. 1, 1-26.

[72] N. V. Krylov, *On the existence of smooth solutions for fully nonlinear elliptic equations with measurable "coefficients" without convexity assumptions*, Methods Appl. Anal. **19** (2012), no. 2, 119–146, DOI 10.4310/MAA.2012.v19.n2.a2. MR3011597

[73] N. V. Krylov, *Some L_p-estimates for elliptic and parabolic operators with measurable coefficients*, Discrete Contin. Dyn. Syst. Ser. B **17** (2012), no. 6, 2073–2090, DOI 10.3934/dcdsb.2012.17.2073. MR2924453

[74] N. V. Krylov, *An ersatz existence theorem for fully nonlinear parabolic equations without convexity assumptions*, SIAM J. Math. Anal. **45** (2013), no. 6, 3331–3359, DOI 10.1137/120903762. MR3124896

[75] N. V. Krylov, *On the existence of W_p^2 solutions for fully nonlinear elliptic equations under relaxed convexity assumptions*, Comm. Partial Differential Equations **38** (2013), no. 4, 687–710, DOI 10.1080/03605302.2012.741177. MR3040680

[76] N. V. Krylov, *On the rate of convergence of difference approximations for uniformly nondegenerate elliptic Bellman's equations*, Appl. Math. Optim. **69** (2014), no. 3, 431–458, DOI 10.1007/s00245-013-9228-y. MR3197305

[77] N. V. Krylov, *On $C^{1+\alpha}$ regularity of solutions of Isaacs parabolic equations with VMO coefficients*, NoDEA Nonlinear Differential Equations Appl. **21** (2014), no. 1, 63–85, DOI 10.1007/s00030-013-0238-x. MR3158801

[78] N. V. Krylov, *To the theory of viscosity solutions for uniformly parabolic Isaacs equations*, Methods Appl. Anal. **22** (2015), no. 3, 259–280, DOI 10.4310/MAA.2015.v22.n3.a2. MR3404762

[79] N. V. Krylov, *On the existence of W_p^2 solutions for fully nonlinear elliptic equations under either relaxed or no convexity assumptions*, Commun. Contemp. Math. **19** (2017), no. 6, 1750009, 39, DOI 10.1142/S0219199717500092. MR3691503

[80] N. V. Krylov, *On the existence of $W_p^{1,2}$ solutions for fully nonlinear parabolic equations under either relaxed or no convexity assumptions*, to appear in Harvard University, Center of Mathematical Sciences and Applications, Nonlinear Equations Publication, http://arxiv.org/abs/1705.02400

[81] N.V. Krylov, *Uniqueness for L_p-viscosity solutions for uniformly parabolic Isaacs equations with measurable lower order terms*, Comm. on Pure Appl. Analysis **17** (2018), no. 6, 2495–2516. DOI:3934/cpaa.2018119.

[82] N. V. Krylov, *$C^{1+\alpha}$-regularity of viscosity solutions of general nonlinear parabolic equations*, Problemy Matematicheskogo Analiza, no. 93, June 2018, 3–23; English translation in Journal of Mathematical Sciences, New York (Springer), Vol. 232 (2018), no. 4, 403–427.

[83] N. V. Krylov and M. V. Safonov, *An estimate for the probability of a diffusion process hitting a set of positive measure* (Russian), Dokl. Akad. Nauk SSSR **245** (1979), no. 1, 18–20. English translation in Soviet Math. Dokl., Vol. 20 (1979), 253–255. MR525227

[84] N. V. Krylov and M. V. Safonov, *A property of the solutions of parabolic equations with measurable coefficients* (Russian), Izv. Akad. Nauk SSSR Ser. Mat. **44** (1980), no. 1, 161–175, 239. English translation in Math. USSR Izvestija, Vol. 16 (1981), no. 1, 151–164. MR563790

[85] H. J. Kuo and N. S. Trudinger, *Discrete methods for fully nonlinear elliptic equations*, SIAM J. Numer. Anal. **29** (1992), no. 1, 123–135, DOI 10.1137/0729008. MR1149088

[86] O. A. Ladyzhenskaya and N. N. Ural′tseva, *Lineĭnye i kvazilineĭnye uravneniya èllipticheskogo tipa.* (Russian), Izdat. "Nauka", Moscow, 1973. Second edition, revised. MR0509265

[87] O. A. Ladyženskaja, V. A. Solonnikov, and N. N. Ural′ceva, *Linear and quasilinear equations of parabolic type*, Translated from the Russian by S. Smith. Translations of Mathematical Monographs, Vol. 23, American Mathematical Society, Providence, R.I., 1968. MR0241822

[88] E. M. Landis, *Second order equations of elliptic and parabolic type*, Translations of Mathematical Monographs, vol. 171, American Mathematical Society, Providence, RI, 1998. English translation by Amer. Math. Soc., Providence, RI, 1997. MR1487894

[89] G. M. Lieberman, *Second order parabolic differential equations*, World Scientific Publishing Co., Inc., River Edge, NJ, 1996. MR1465184

[90] F.-H. Lin, *Second derivative L^p-estimates for elliptic equations of nondivergent type*, Proc. Amer. Math. Soc. **96** (1986), no. 3, 447–451, DOI 10.2307/2046592. MR822437

[91] P.-L. Lions, *Équations de Hamilton-Jacobi-Bellman dégénérées* (French, with English summary), C. R. Acad. Sci. Paris Sér. A-B **289** (1979), no. 5, A329–A332. MR552287

[92] P.-L. Lions, *Optimal control of diffusion processes and Hamilton-Jacobi-Bellman equations. II. Viscosity solutions and uniqueness* (English, with French summary), Comm. Partial Differential Equations **8** (1983), no. 11, 1229–1276, DOI 10.1080/03605308308820301. MR709162

[93] J. Moser, *A Harnack inequality for parabolic differential equations*, Comm. Pure Appl. Math. **17** (1964), 101–134, DOI 10.1002/cpa.3160170106. MR0159139

[94] A. I. Nazarov, *Interpolation of linear spaces and estimates for the maximum of a solution for parabolic equations* (Russian), Partial differential equations (Russian), Akad. Nauk SSSR Sibirsk. Otdel., Inst. Mat., Novosibirsk, 1987, pp. 50–72, 132. English translation in https://arxiv.org/abs/1507.05232. MR994027

[95] A. I. Nazarov and N. N. Ural′tseva, *Convex-monotone hulls and an estimate of the maximum of the solution of a parabolic equation* (Russian, with English summary), Zap. Nauchn. Sem. Leningrad. Otdel. Mat. Inst. Steklov. (LOMI) **147** (1985), 95–109, 204–205. English translation in Journal of Soviet Mathematics April 1987, Vol. 37, no. 1, 851–859. MR821477

[96] N. Nadirashvili and S. Vlăduţ, *Singular viscosity solutions to fully nonlinear elliptic equations* (English, with English and French summaries), J. Math. Pures Appl. (9) **89** (2008), no. 2, 107–113, DOI 10.1016/j.matpur.2007.10.004. MR2391642

[97] E. A. Pimentel and E. V. Teixeira, *Sharp Hessian integrability estimates for nonlinear elliptic equations: an asymptotic approach*, J. Math. Pures Appl. (9) **106** (2016), no. 4, 744–767, DOI 10.1016/j.matpur.2016.03.010. MR3539473

[98] A. V. Pogorelov, *The Dirichlet problem for the multidimensional analogue of the Monge-Ampère equation* (Russian), Dokl. Akad. Nauk SSSR **201** (1971), 790–793. MR0293228

[99] A. V. Pogorelov, *The Minkowski multidimensional problem*, V. H. Winston & Sons, Washington, D.C.; Halsted Press [John Wiley & Sons], New York-Toronto-London, 1978. Translated from the Russian by Vladimir Oliker; Introduction by Louis Nirenberg; Scripta Series in Mathematics. MR0478079

[100] M. V. Safonov, *Harnack's inequality for elliptic equations and Hölder property of their solutions* (Russian), Zap. Nauchn. Sem. Leningrad. Otdel. Mat. Inst. Steklov. (LOMI) **96** (1980), 272–287, 312. English translation in Journal of Soviet Mathematics, Vol. 21 (March 1983), no. 5, 851–863. MR579490

[101] M. V. Safonov, *An example of the diffusion processes with singular distribution at a fixed time instant*, Abstr. Comm. Third Intern. Vilnius Conf. Probability Theory and Math. Statist., 1981, 133–134.

[102] M. V. Safonov, *The classical solution of the elliptic Bellman equation* (Russian), Dokl. Akad. Nauk SSSR **278** (1984), no. 4, 810–813. English translation in Soviet Math. Dokl., Vol. 30 (1984), no. 2, 482–485. MR765302

[103] M. V. Safonov, *Classical solution of second-order nonlinear elliptic equations* (Russian), Izv. Akad. Nauk SSSR Ser. Mat. **52** (1988), no. 6, 1272–1287, 1328, DOI 10.1070/IM1989v033n03ABEH000858; English transl., Math. USSR-Izv. **33** (1989), no. 3, 597–612. MR984219

[104] M. V. Safonov, *On the boundary value problems for fully nonlinear elliptic equations of second order*, The Australian National University, Mathematics Research Report no. MRR 049-94, Canberra, 1994. http://www-users.math.umn.edu/~safon002/NOTES/notes.html

[105] M. V. Safonov, *Non-divergence elliptic equations of second order with unbounded drift*, Nonlinear partial differential equations and related topics, Amer. Math. Soc. Transl. Ser. 2, vol. 229, Amer. Math. Soc., Providence, RI, 2010, pp. 211–232, DOI 10.1090/trans2/229/13. MR2667641

[106] E. Schmutz, *Rational points on the unit sphere*, Cent. Eur. J. Math. **6** (2008), no. 3, 482–487, DOI 10.2478/s11533-008-0038-4. MR2425007

[107] L. Silvestre and B. Sirakov, *Boundary regularity for viscosity solutions of fully nonlinear elliptic equations*, Comm. Partial Differential Equations **39** (2014), no. 9, 1694–1717, DOI 10.1080/03605302.2013.842249. MR3246039

[108] L. Silvestre and E. V. Teixeira, *Regularity estimates for fully non linear elliptic equations which are asymptotically convex*, Contributions to nonlinear elliptic equations and systems, Progr. Nonlinear Differential Equations Appl., vol. 86, Birkhäuser/Springer, Cham, 2015, pp. 425–438, DOI 10.1007/978-3-319-19902-3_25. MR3494911

[109] J. Streets and M. Warren, *Evans-Krylov estimates for a nonconvex Monge-Ampère equation*, Math. Ann. **365** (2016), no. 1-2, 805–834, DOI 10.1007/s00208-015-1293-x. MR3498927

[110] A. Święch, *$W^{1,p}$-interior estimates for solutions of fully nonlinear, uniformly elliptic equations*, Adv. Differential Equations **2** (1997), no. 6, 1005–1027. MR1606359

[111] N. S. Trudinger, *Comparison principles and pointwise estimates for viscosity solutions of nonlinear elliptic equations*, Rev. Mat. Iberoamericana **4** (1988), no. 3-4, 453–468, DOI 10.4171/RMI/80. MR1048584

[112] N. S. Trudinger, *On regularity and existence of viscosity solutions of nonlinear second order, elliptic equations*, Partial differential equations and the calculus of variations, Vol. II, Progr. Nonlinear Differential Equations Appl., vol. 2, Birkhäuser Boston, Boston, MA, 1989, pp. 939–957. MR1034037

[113] L. Wang, *On the regularity theory of fully nonlinear parabolic equations*, Bull. Amer. Math. Soc. (N.S.) **22** (1990), no. 1, 107–114, DOI 10.1090/S0273-0979-1990-15854-9. MR999619

[114] L. Wang, *On the regularity theory of fully nonlinear parabolic equations. I*, Comm. Pure Appl. Math. **45** (1992), no. 1, 27–76, DOI 10.1002/cpa.3160450103. MR1135923

[115] L. Wang, *On the regularity theory of fully nonlinear parabolic equations. II*, Comm. Pure Appl. Math. **45** (1992), no. 2, 141–178, DOI 10.1002/cpa.3160450202. MR1139064

[116] N. Winter, *$W^{2,p}$ and $W^{1,p}$-estimates at the boundary for solutions of fully nonlinear, uniformly elliptic equations*, Z. Anal. Anwend. **28** (2009), no. 2, 129–164, DOI 10.4171/ZAA/1377. MR2486925

[117] V. V. Zhikov and M. M. Sirazhudinov, *G-compactness of a class of second-order nondivergence elliptic operators* (Russian), Izv. Akad. Nauk SSSR Ser. Mat. **45** (1981), no. 4, 718–733, 927. English translation in Mathematics of the USSR-Izvestiya, Vol. 19 (1982), no. 1, 27–40. MR631435

[118] V. V. Zhikov, S. M. Kozlov, and O. A. Oleĭnik, *G-convergence of parabolic operators* (Russian), Uspekhi Mat. Nauk **36** (1981), no. 1(217), 11–58, 248. English translation in Russian Mathematical Surveys, Vol. 36 (1981), no. 1, 9–60. MR608940

Index

SELECTED PUBLISHED TITLES IN THIS SERIES

233 **N. V. Krylov,** Sobolev and Viscosity Solutions for Fully Nonlinear Elliptic and Parabolic Equations, 2018

230 **Stephen D. Smith,** Applying the Classification of Finite Simple Groups, 2018

229 **Alexander Molev,** Sugawara Operators for Classical Lie Algebras, 2018

228 **Zhenbo Qin,** Hilbert Schemes of Points and Infinite Dimensional Lie Algebras, 2018

227 **Roberto Frigerio,** Bounded Cohomology of Discrete Groups, 2017

226 **Marcelo Aguiar and Swapneel Mahajan,** Topics in Hyperplane Arrangements, 2017

225 **Mario Bonk and Daniel Meyer,** Expanding Thurston Maps, 2017

224 **Ruy Exel,** Partial Dynamical Systems, Fell Bundles and Applications, 2017

223 **Guillaume Aubrun and Stanisław J. Szarek,** Alice and Bob Meet Banach, 2017

222 **Alexandru Buium,** Foundations of Arithmetic Differential Geometry, 2017

221 **Dennis Gaitsgory and Nick Rozenblyum,** A Study in Derived Algebraic Geometry, 2017

220 **A. Shen, V. A. Uspensky, and N. Vereshchagin,** Kolmogorov Complexity and Algorithmic Randomness, 2017

219 **Richard Evan Schwartz,** The Projective Heat Map, 2017

218 **Tushar Das, David Simmons, and Mariusz Urbański,** Geometry and Dynamics in Gromov Hyperbolic Metric Spaces, 2017

217 **Benoit Fresse,** Homotopy of Operads and Grothendieck–Teichmüller Groups, 2017

216 **Frederick W. Gehring, Gaven J. Martin, and Bruce P. Palka,** An Introduction to the Theory of Higher-Dimensional Quasiconformal Mappings, 2017

215 **Robert Bieri and Ralph Strebel,** On Groups of PL-homeomorphisms of the Real Line, 2016

214 **Jared Speck,** Shock Formation in Small-Data Solutions to 3D Quasilinear Wave Equations, 2016

213 **Harold G. Diamond and Wen-Bin Zhang (Cheung Man Ping),** Beurling Generalized Numbers, 2016

212 **Pandelis Dodos and Vassilis Kanellopoulos,** Ramsey Theory for Product Spaces, 2016

211 **Charlotte Hardouin, Jacques Sauloy, and Michael F. Singer,** Galois Theories of Linear Difference Equations: An Introduction, 2016

210 **Jason P. Bell, Dragos Ghioca, and Thomas J. Tucker,** The Dynamical Mordell–Lang Conjecture, 2016

209 **Steve Y. Oudot,** Persistence Theory: From Quiver Representations to Data Analysis, 2015

208 **Peter S. Ozsváth, András I. Stipsicz, and Zoltán Szabó,** Grid Homology for Knots and Links, 2015

207 **Vladimir I. Bogachev, Nicolai V. Krylov, Michael Röckner, and Stanislav V. Shaposhnikov,** Fokker–Planck–Kolmogorov Equations, 2015

206 **Bennett Chow, Sun-Chin Chu, David Glickenstein, Christine Guenther, James Isenberg, Tom Ivey, Dan Knopf, Peng Lu, Feng Luo, and Lei Ni,** The Ricci Flow: Techniques and Applications: Part IV: Long-Time Solutions and Related Topics, 2015

205 **Pavel Etingof, Shlomo Gelaki, Dmitri Nikshych, and Victor Ostrik,** Tensor Categories, 2015

204 **Victor M. Buchstaber and Taras E. Panov,** Toric Topology, 2015

203 **Donald Yau and Mark W. Johnson,** A Foundation for PROPs, Algebras, and Modules, 2015

For a complete list of titles in this series, visit the
AMS Bookstore at **www.ams.org/bookstore/survseries/**.